INSTANTONS IN GAUGE THEORIES

INSTANTONS IN GAUGE THEORIES

Edited by

M. Shifman

Theoretical Physics Institute
University of Minnesota

World Scientific
Singapore • New Jersey • London • Hong Kong

Published by

World Scientific Publishing Europe Ltd.
57 Shelton Street, Covent Garden, London WC2H 9HE
Head office: 5 Toh Tuck Link, Singapore 596224
USA office: 27 Warren Street, Suite 401-402, Hackensack, NJ 07601

We are grateful to the following publishers for their permission to reproduce the articles found in this volume:

American Institute of Physics
  Sov. J. Nucl. Phys.
  Sov. Phys. Usp.
American Physical Society
  Phys. Rev. D
  Phys. Rev. Lett.
Elsevier Science Publishers
  Nucl. Phys. A
  Nucl. Phys. B
  Phys. Lett. A
  Phys. Lett. B
  Phys. Rep.
Plenum Press
Springer-Verlag
  Z. Phys. C

INSTANTONS IN GAUGE THEORIES

ISBN-13 978-981-02-1681-8
ISBN-10 981-02-1681-5
ISBN-13 978-981-02-1826-3 (pbk)
ISBN-10 981-02-1826-5 (pbk)

PREFACE

In this volume you will find a compilation of works which, taken together, give a complete and consistent presentation of instanton calculus in non-Abelian gauge theories, as it exists now. Some of the papers reproduced are instanton classics. Among other things, they show from a historical perspective how the instanton solution has been found, the motivation behind it and how the physical meaning of instantons has been revealed. Other papers are devoted to different aspects of the instanton formalism. Only topics which have a direct relation to the formalism *per se* are included. No attempt is made to discuss numerous applications in non-Abelian gauge theories or other models. The papers are organized into several sections that are linked both logically and historically, accompanied by brief comments.

M. Shifman
November 28, 1993

CONTENTS

SECTION III. MULTI-INSTANTON SOLUTIONS

SECTION IV. GREEN FUNCTIONS IN THE INSTANTON BACKGROUND

SECTION V. FERMIONS

SECTION VI. CONSTRAINED INSTANTONS, I–A INTERACTION AND THE VALLEY METHOD

SECTION VII. SUPERSYMMETRIC INSTANTONS

SECTION VIII. REVIEWS

INTRODUCTION

Since the early seventies, non-Abelian gauge theories have played a major role in the theory of fundamental interactions. With their roots in quantum electrodynamics (QED) they inherited some features of the latter, such as gauge invariance, for instance. However, their similarity to QED is not the reason why they are so cherished.

Non-Abelian theories possess truly unique properties that make them much richer and more interesting than QED. They present the only example of field theories with asymptotic freedom in four dimensions [1]. The effective gauge coupling constant is *anti*screened. As a result, it falls off at short distances and blows up at large distances, in the infrared domain. Moreover, peculiar infrared properties of non-Abelian theories manifest themselves in a complicated structure of the vacuum state, not fully understood even at present. The fact that the vacuum "wave function" is nontrivial and is delocalized in one of the directions in the space of fields warms the hearts of theorists for almost two decades [2] and produces high expectations that one day, the phenomenon of color confinement in quantum chromodynamics (QCD) will be fully understood.

Instantons discovered in 1975 [3] gave the first indication of a nontrivial vacuum structure in non-Abelian theories. After this discovery, the spiral began unwinding very fast and within a year people realized that there is an infinite set of degenerate classical minima of the potential energy labeled by an integer (which goes under the name of the winding number), so that the genuine vacuum "wave function" is a linear superposition of the Bloch type [4, 5]. A hidden parameter, the vacuum angle θ (analogous to the quasimomentum of the Bloch theory), has been introduced [5]. And with it came the understanding that QCD is not automatically CP-invariant as was believed previously!

Technically, instantons are solutions of classical equations of motion in the Euclidean space–time. They describe a tunneling trajectory interpolating between the classical minima with different topology mentioned above. The action of the instanton field configuration is finite; the instanton contribution to physical quantities is proportional to $\exp(-C/g^2)$ where C is a positive number and g is the coupling constant. This contribution is well defined in the quasiclassical limit of small g^2 when the exponential factor $\exp(-C/g^2) << 1$. Instantons are singled out from all other field configurations with the given boundary conditions by the minimal action.

The first surprising finding — the issue of the strong CP violation in QCD — was not the last. Inclusion of massless fermions in the instanton formalism which followed shortly [6, 7] brought new surprises. One of the $U(1)$ symmetries valid for any finite order in perturbation theory turned out to be explicitly broken in the instanton field. The impact of this observation, both on QCD and on the Standard Model of electroweak interactions, was drastic.

In QCD it was known for a long time [8] that a "superfluous" axial $U(1)$ symmetry required the existence of a light $SU(3)_{flavor}$ singlet pseudoscalar meson, η', with $m_{\eta'} \sim m_{\pi}$. Since there is no such meson in nature the puzzle of η' [sometimes also called the $U(1)$ problem] was one of the major challenges, a source of headache for many theorists. The fact that instantons effectively eliminate the superfluous $U(1)$ symmetry solves the problem [6].

In the Standard Model the corresponding $U(1)$ is responsible for the baryon number conservation. The absence of this symmetry automatically means that the baryon charge is not conserved in the nonperturbative sector, an intriguing consequence of the nontrivial vacuum structure. This result, the baryon number nonconservation, was obtained in Refs. [6, 7]. Let me add that QCD and — to a lesser extent — the Standard Model of electroweak interactions were the major targets in the pioneering papers.

After such a promising overture instantons were considered by many, at least for some time, as a prospective candidate for the solution of the confinement problem, the most intriguing aspect of QCD. The excitement was fueled by the fact that in the three-dimensional Georgi–Glashow model they do indeed provide a mechanism ensuring color confinement [9].

Alas, this dream never came true, and we are not much closer now to the final solution of the confinement problem than we were 20 years ago. The idea that nonperturbative vacuum fields are responsible for the peculiar infrared behavior of QCD proved to be fruitful but nobody succeeded in building the confining mechanism based on instantons. The fundamental difficulty is due to the growth of the effective gauge coupling at large distances: when it becomes of order one, the quasiclassical picture becomes inapplicable — instantons lose their special position and get buried in the multitude of other nonperturbative fluctuations.

Still, instantons turned out to be a real breakthrough in many aspects. They were absolutely instrumental in achieving the qualitative understanding of the vacuum structure in the non-Abelian theories that we now have, demonstrating a nontrivial topology of the space of fields. Moreover, they were the first explicit example of nonperturbative fluctuations distinguishing non-Abelian theories from, say, QED - the example one can work with. The novelty and elegance of this theoretical construction is undeniable.

Apart from the qualitative insights mentioned above there is a certain range of phenomena where instantons can be used in the quantitative aspect. True, problems of this type as a rule go beyond the original scope of applications as was formulated in pioneering works (see below.)

Although the instanton solution did not turn out to be the remedy people were expecting in connection with the infrared problem in QCD, instanton studies became a vast field both in pure theory and in applications. Let me mention a few topics (without any attempt to order the list).

Instantons have been found and investigated in many other models, for instance, in two-dimensional σ models [10, 11]. They are used in supersymmetric (SUSY) generalizations of gauge theories where they provide a possible mechanism for spontaneous SUSY breaking [12, 13], and even in the theory of strings [14]. It is worth mentioning that in supersymmetric theories, instanton calculus allows one to find gaugino condensates [15], a calculation which served as a first prototype for analogous calculations in topological field theories [16] widely discussed at present. On the technical side instanton calculus was applied in supersymmetric gluodynamics to obtain the exact Gell–Mann–Low function [17].

Among other more traditional applications I would like to mention

(i) axion physics [18, 19]

— this topic is deeply related to QCD instantons on the one hand and the structure of fundamental interactions at very short distances on the other;

(ii) calculation of divergencies of perturbation theory in high orders [20, 21, 22];

(iii) instanton-based models of the QCD vacuum [23, 24, 25]

— the ongoing efforts aimed at building such models are, for obvious reasons, of special practical importance. The first attempt of this type was undertaken as early as 1978 by Callan, Dashen and Gross [23], whose work became, in a sense, a blueprint for further constructions. Unfortunately, the simplest approximation of the instanton gas suggested there (an ensemble of noninteracting pseudoparticles), seemingly promising at that moment, turned out to be absolutely unrealistic. In the eighties, people went much further, introducing the instanton liquid model [24, 25] which they continue to analyze and improve until now. The details of the model would lead us too far away from the main purpose of this volume, and therefore I will just refer the reader to numerous original publications and reviews. Work in this direction is far from completion, and any conclusion at this stage would be premature.

Last but not least, the discovery of instantons triggered new thorough analyses of the topologically nontrivial structure of non-Abelian theories, and other topologically nontrivial solutions were soon constructed. The most well known of them are torons [26]. The Gribov horizons [27] can also be considered as an offshoot from the same tree.

If, in the beginning, the development of the subject was rapid, in the last decade, the enthusiasm of instanton practitioners had declined steadily and the topic was in a rather dormant state, apart from a few sporadic outbursts, say, in topological field theory. The situation has drastically changed after 1989. Interest has been revived by the observation [28] that instanton-induced cross sections of multiparticle processes in the Standard Model exponentially grow with energy. Therefore, the baryon number violation suppressed by $\exp(-4\pi/\alpha_2) \sim \exp(-356)$ in the GeV range [6] can jump, say, by 100 orders of magnitude at energy of the order of a few TeV [α_2 is the gauge coupling constant corresponding to the $SU(2)$ subgroup, $\alpha_2 = \alpha/\sin^2(\theta_W)$]. The growth is incredibly fast!

At first there were suspicions [29] that the exponential suppression might be lifted at all at $E \sim 10$ to 15 TeV. If true, these extrapolations might lead to revolutionary changes both in theory and phenomenology. It is therefore not surprising that the original observation generated a significant flow of works devoted to electroweak instantons at high energies (for a review see e.g. Ref. [30]). At present it seems to be a well-established fact that the exponential suppression is not all gone; a part of it persists at any energy. At the same time the central question — what is the value of the instanton-induced cross section at the maximum — is still open. The problem is still alive. Active work in the last few years in this direction created a demand for the compilation of classical literature on instanton calculus. As a matter of fact, I felt this need myself, and this was one of the motivations for the publication of this reprint volume.

I shall now explain the selection of topics in this volume. After much hesitation, I decided, with a heavy heart, to limit the selection only to instanton calculus *per se*, leaving applications aside. Also not reflected are all achievements in related directions, such as toron physics, instantons as a mathematical tool in topological field theory, etc.

As it follows from the above remarks, the situation with applied problems is very inhomogeneous, sometimes incomplete, so that any adequate discussion would require a very substantial amount of supplementary literature and extensive commentaries. What is even more important is the fact that here, it is difficult to formulate criteria which would allow one to separate the results which can be considered as well established from those which seem controversial. In contrast, the formalism itself is very clean and clear, and will stay with us forever — there is nothing in it which would require a change.[1]

The only exception to the above statement is the series of elegant and *exact* results stemming from instantons in supersymmetric gauge theories: calculation of the chiral condensates, the exact Gell–Mann–Low function and superpotentials [12, 13, 15, 17, 31, 32]. It was very tempting to discuss this topic in a separate section, but this would violate the principle of leaving all applications out in this volume. Hence, this application was excluded as well.

In selecting the papers to be reproduced I have tried to maintain a balance between the works treating conceptual issues and the more technical ones. For convenience the original papers are supplemented by two reviews [33, 34], with strong emphasis on the pedagogical aspects, explaining the basic ideas as well as numerous technical details. (It is worth noting that a nice introduction to this subject is also given in Ref. [35].) Each section opens with brief comments which are mostly intended to provide a proper perspective and cohesion to the selected papers presented in the section. Following these the reader will find a short list of recommended literature. All papers reproduced in this reprint volume are marked by an asterisk in the references.

---

[1]Occasionally, though, one or two applied questions will slip in, among other issues in the papers reproduced, e.g. the 't Hooft qualitative explanation of the η' problem [6].

I am grateful to Ian Balitsky, and especially, to Arkady Vainshtein, who read my notes and made valuable remarks. I would like to thank my editor, Ms. Lim Feng Nee, for her assistance in the preparation of the manuscript.

References

[1] D. Gross and F. Wilczek, *Phys. Rev. Lett.* **30** (1973) 1343; H. D. Politzer, *Phys. Rev. Lett.* **30** (1973) 1346.
[2] *A. Polyakov, *Phys. Lett.* **59B** (1975) 82.
[3] *A. Belavin, A. Polyakov, A. Schwarz and Yu. Tyupkin, *Phys. Lett.* **59B** (1975) 85.
[4] V. N. Gribov, 1975, unpublished.
[5] *C. Callan, R. Dashen and D. Gross, *Phys. Lett.* **63B** (1976) 334; *R. Jackiw and C. Rebbi, *Phys. Rev. Lett.* **37** (1976) 172.
[6] *G. 't Hooft, *Phys. Rev. Lett.* **37** (1976) 8.
[7] *G. 't Hooft, *Phys. Rev.* **D14** (1976) 3432.
[8] S. Weinberg, *Phys. Rev.* **D11** (1975) 3583.
[9] A. Polyakov, *Nucl. Phys.* **B120** (1977) 429.
[10] A. Belavin and A. Polyakov, *Pisma ZhETF* **22** (1975) 503 [*JETP Lett.* **22** (1975) 245].
[11] For a review see e.g. V. Novikov, M. Shifman, A. Vainshtein and A. Zakharov, *Phys. Rep.* **116** (1984) 103.
[12] D. Amati, G. Rossi and G. Veneziano, *Nucl. Phys.* **B249** (1985) 1.
[13] I. Affleck, M. Dine and N. Seiberg, *Phys. Rev. Lett.* **52** (1984) 1677; *Nucl. Phys.* **B256** (1985) 557.
[14] M. Dine, N. Seiberg, X. Wen and E. Witten, *Nucl. Phys.* **B278** (1986) 769; *Nucl. Phys.* **B289** (1987) 319.
[15] V. Novikov, M. Shifman, A. Vainshtein and A. Zakharov, *Nucl. Phys.* **B229** (1983) 407.
[16] E. Witten, *Comm. Math. Phys.* **117** (1988) 353.
[17] V. Novikov, M. Shifman, A. Vainshtein and A. Zakharov, *Nucl. Phys.* **B229** (1983) 381; *Phys. Lett.* **166B** (1986) 329.
[18] R. Peccei and H. Quinn, *Phys. Rev. Lett.* **38** (1977) 1440; *Phys. Rev.* **D16** (1977) 1791.
[19] S. Weinberg, *Phys. Rev. Lett.* **40** (1978) 223; F. Wilczek, *Phys. Rev. Lett.* **40** (1978) 279.
[20] L. Lipatov, *ZhETF* **72** (1977) 411; *Sov. Phys. JETP* **45** (1977) 216; E. Brezin, G. Parisi and J. Zinn-Justin, *Phys. Rev.* **D16** (1977) 408; A. Bukhvostov, L. Lipatov and E. Malkov, *Pisma ZhETF* **27** (1978) 594 [*JETP Lett.* **27** (1978) 561]; E. Bogomolny, *Phys. Lett.* **91B** (1980) 431; E. Bogomolny and V. Fateev, *Phys. Lett.* **71B** (1977) 93.
[21] "Large Order Behavior of Perturbation Theory," eds. J. Le Guillou and J. Zinn-Justin (North-Holland, Amsterdam, 1990).
[22] I. Balitsky, *Phys. Lett.* **B273** (1991) 282.
[23] *C. Callan, R. Dashen and D. Gross, *Phys. Rev.* **D17** (1978) 2717; *Phys. Rev.* **D19** (1979) 1826.
[24] D. Diakonov and V. Petrov, *ZhETF* **89** (1985) 751 [*Sov. Phys. JETP* **62** (1985) 431]; *Nucl. Phys.* **B272** (1986) 475; E. Shuryak, *Nucl. Phys.* **B214** (1983) 237; *Nucl. Phys.* **B319** (1989) 521; *Nucl. Phys.* **B328** (1989) 85; 102.
[25] E. Shuryak, *The QCD Vacuum, Hadrons and the Superdense Matter* (World Scientific, Singapore, 1988).
[26] G. 't Hooft, *Comm. Math. Phys.* **81** (1981) 267.

[27] V. N. Gribov, *Nucl. Phys.* **B139** (1978) 1.
[28] A. Ringwald, *Nucl. Phys.* **B330** (1990) 1; O. Espinosa, *Nucl. Phys.* **B343** (1990) 310.
[29] L. McLerran, A. Vainshtein and M. Voloshin, *Phys. Rev.* **D42** (1990) 171; 180; *Phys. Lett.* **249B** (1990) 261.
[30] M. Mattis, *Phys. Rep.* **214** (1992) 159.
[31] I. Affleck, M. Dine and N. Seiberg, *Phys. Lett.* **137B** (1984) 187; Y. Meurice and G. Veneziano, *Phys. Lett.* **141B** (1984) 69; A. Vainshtein, V. Zakharov, V. Novikov and M. Shifman, *Pisma ZhETF* **39** (1984) 494 [*JETP Lett.* **39** (1984) 601]; D. Amati, Y. Meurice, G. Rossi and G. Veneziano, *Nucl. Phys.* **B263** (1986) 591.
[32] For a review of the instanton mechanism of the spontaneous breaking of the gauge and/or supersymmetry, see A. Vainshtein, V. Zakharov and M. Shifman, *Usp. Fiz. Nauk* **146** (1985) 683 [*Sov. Phys. Uspekhi* **28** (1985) 709]; D. Amati, K. Konishi, Y. Meurice, G. Rossi and G. Veneziano, *Phys. Rep.* **162** (1988) 169.
[33] *S. Coleman, "The Uses of Instantons," in *Proc. 1977 Int. School of Subnuclear Physics, Erice, 1977* [reproduced in S. Coleman, *Aspects of Symmetry* (Cambridge University Press, 1985)].
[34] *A. Vainshtein, V. Zakharov, V. Novikov and M. Shifman, *Usp. Phys. Nauk* **136** (1982) 553 [*Sov. Phys. Uspekhi* **25** (1982) 195].
[35] R. Rajaraman, *Solitons and Instantons* (North-Holland, Amsterdam, 1982).

I. BPST INSTANTONS: DISCOVERY AND PHYSICAL INTERPRETATION

INTRODUCTION

Instantons have many faces. Historically they were found as topologically non-trivial solutions of the duality equations of the Euclidean Yang–Mills theory with finite action [1].

The Lagrangian of pure gluodynamics (the Yang–Mills theory with no matter fields) in the Euclidean space–time can be written as

$$\mathcal{L} = \frac{1}{4g^2} G^a_{\mu\nu} G^a_{\mu\nu} \ . \tag{1}$$

Here $G^a_{\mu\nu}$ is the gluon field strength tensor, [2]

$$G^a_{\mu\nu} = \partial_\mu A^a_\nu - \partial_\nu A^a_\mu + f^{abc} A^b_\mu A^c_\nu , \tag{2}$$

f^{abc} are the structure constants of the gauge group considered [$SU(2)$ throughout this section], and g is the gauge coupling constant. Note that the normalization of the fields A^a_μ in Eqs. (1), (2) is such that $1/g^2$ appears only in Eq. (1) as an overall factor.

The classical action of the Yang–Mills fields can be identically rewritten as

$$S = \frac{1}{8g^2} \int d^4x \{(G^a_{\mu\nu} \pm \tilde{G}^a_{\mu\nu})^2\} \mp \frac{8\pi^2}{g^2} Q \ , \tag{3}$$

where

$$Q = \frac{1}{32\pi^2} \int d^4x G^a_{\mu\nu} \tilde{G}^a_{\mu\nu} \tag{4}$$

and

$$\tilde{G}^a_{\mu\nu} = \frac{1}{2} \epsilon_{\mu\nu\alpha\beta} G^a_{\alpha\beta} \ .$$

The quantity defined in Eq. (4) is called the topological charge for reasons which will become clear shortly (the Pontryagin index in mathematical literature).

As was explained in Ref. [2] (this work provided an ideological motivation for searches of instantons and opened the field), the action of the field configuration we are interested in must be finite. Since $G^a_{\mu\nu} G^a_{\mu\nu}$ is positive-definite finiteness of the action implies that on large sphere

$$|x| = R \to \infty \ , \tag{5}$$

the gluon field strength tensor $G^a_{\mu\nu}$ must vanish:

$$G^a_{\mu\nu}(|x| = R) \to 0, \ \text{faster than} \ R^{-3} \ . \tag{6}$$

In other words, on large sphere, A^a_μ must be pure gauge:

$$A^a_\mu T^a \to iU(x)\partial_\mu U^{-1}(x), \ \text{at} \ R \to \infty \ , \tag{7}$$

[2] In passing from the Minkowski to the Euclidean space and back, all sign conventions must be carefully adjusted, see e.g. the review paper of Vainshtein *et al.* in Sec. VIII.

where T^a are the generators of the gauge group in the representation considered [if, as is usually done, the fundamental representation is chosen, then $T^a = (1/2)\sigma^a$ for $SU(2)$, σ^a are the Pauli matrices]. Furthermore, $U(x)$ is a matrix, an element of the gauge group,

$$U \in SU(2) .$$

It is worth noting that $iU\partial_\mu U^{-1}$ is automatically an element of the algebra, see Eq. (7).

For what follows it is crucial that the topological charge (4) is representable as an integral over a full derivative,

$$Q = \frac{1}{32\pi^2}\int d^4x \partial_\mu K_\mu = \frac{1}{32\pi^2}\int_{|x|=R} K_\mu dS_\mu \,, \tag{8}$$

where the vector K_μ in the last expression is given by the formula

$$K_\mu = 2\epsilon_{\mu\nu\alpha\beta}(A^a_\nu \partial_\alpha A^a_\beta + \frac{1}{3} f^{abc} A^a_\nu A^b_\alpha A^c_\beta) \tag{9}$$

and dS_μ is the element of the surface on large sphere. The second equation in (8) is due to the Gauss theorem. The vector K_μ is sometimes called the Chern–Simons current.

We see that the topological charge is totally determined by the asymptotic behavior of the field A^a_μ. Moreover, if we limit ourselves to the class of fields satisfying Eq. (6), then Q actually does not depend on the *local* behavior of A_μ i.e. small variations of A_μ do not change the topological charge. This fact is readily proved by calculating the variation of Q under continuous deformations of A^a_μ,

$$\delta Q = \frac{1}{32\pi^2}\int_{|x|=R} dS_\mu \delta K_\mu = \frac{1}{8\pi^2}\int_{|x|=R} dS_\mu \tilde{G}^a_{\mu\nu}\delta A^a_\nu \,. \tag{10}$$

The right hand side vanishes since $\tilde{G}_{\mu\nu} \to 0$ at $|x| \to \infty$ faster than $|x|^{-3}$.

Thus, Q depends only on global properties of the function $A^a_\mu(|x| = R)$. If $A^a_\mu = 0$ on large sphere, then Q is obviously zero. All other functions $U(x)$ on large sphere continuously deformable to $U(x) = \text{const.}$ (or $A^a_\mu = 0$) produce the same vanishing value of the topological charge. Hence, the possibility of having $Q \neq 0$ depends on the existence of classes of functions not deformable (continuously) to $A^a_\mu = 0$. Topological arguments tell us that such classes do exist.

Indeed, as it follows from Eq. (7), on large sphere, $A^a_\mu(x)$ is unambiguously determined in terms of $U(x)$. Moreover, any $U \in SU(2)$ can be written as

$$U = A + i\vec{\sigma}\vec{B}, \;\; A^2 + \vec{B}^2 = 1 \,.$$

In other words, the group space of $SU(2)$ is a S_3, a three-dimensional sphere. This means, in turn, that the three-dimensional sphere in the coordinate space, $|x| = R$, is being mapped onto S_3 in the group space. Intuitively, it is clear that all continuous mappings $S_3 \to S_3$ naturally fall in distinct classes corresponding to different numbers of coverings of the group manifold when the coordinate x sweeps large sphere. The intuitive understanding is backed up by a rigorous topological theorem proving this assertion. In mathematical language the theorem reads $\pi_3(S_3) = \mathbf{Z}$. $\mathbf{Z}$ includes all integer numbers. Since the mappings $S_3 \to S_3$ are orientable the integer numbers can be both positive and negative. Normalization of the topological charge is such that Q takes integer values corresponding to the different numbers of coverings, $Q = 0, 1, 2, \ldots; -1, -2, \ldots$.

As it follows from Eq. (3), for any given value of Q, the minimal action in the given class of functions is achieved provided that

$$G^a_{\mu\nu} = \pm \tilde{G}^a_{\mu\nu} \,. \tag{11}$$

This is the famous self-duality condition derived in Ref. [1]. The plus sign must be chosen for positive Q (instantons), while the minus sign is chosen for negative Q (anti-instantons).

The solution of the duality equation is automatically the solution of the classical equations of motion. Indeed,

$$D_\mu G_{\mu\nu} = \pm D_\mu \tilde{G}_{\mu\nu} = 0 \,, \tag{12}$$

where the last equation in this line is merely the consequence of the Jacobi identity. Note that, generally speaking, the opposite assertion is not true. Not every solution of the classical equations of motion satisfies the duality relation.

The solutions of Eq. (11) with $|Q| = 1$ are called the BPST instantons. The instanton action is obviously equal to $8\pi^2/g^2$. The term "instanton" has been suggested in Ref. [3]; originally, they were called pseudoparticles [1] (perhaps due to the fact that they are localized in the four-dimensional Euclidean space). One can encounter references to "pseudoparticles" in early works devoted to this subject.

The consideration presented above is based on the topological classification of four-dimensional field configurations [1] and is related to the ideology of pseudoparticles [2]. As a matter of fact, there are two different topological aspects associated with instantons in gauge theories. An alternative aspect outlined below reveals, in an absolutely transparent form, the physical meaning of the instantons as the (Euclidean time) tunneling trajectories connecting the degenerate classical minima of the potential energy. The simplest analogy one can keep in mind in this context is the quantum mechanics of a particle living on a vertically oriented circle and subject to a constant gravitational force (Fig. 1).

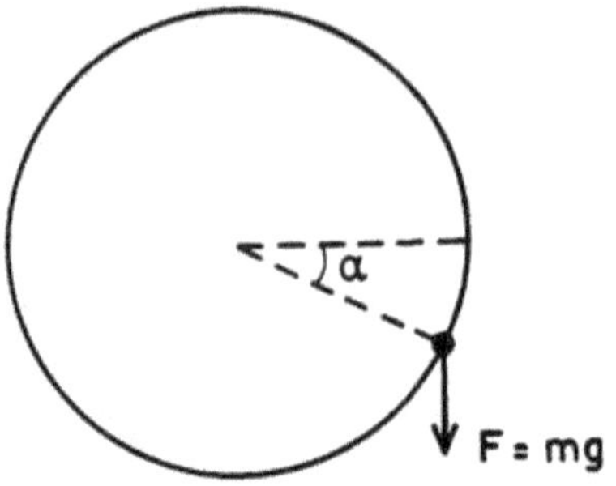

Fig. 1. Quantum mechanics of a particle (•) on a one-dimensional topologically nontrivial manifold, circle.

To see that this is indeed the case, first of all, it is necessary to proceed to the Hamiltonian formulation of the Yang–Mills theory which implies, of course, that the time component of the four-potential has to be gauged away, $A_0 = 0$. Then,

$$\mathcal{H} = \frac{1}{2g^2}\int d^3x\{\vec{E}^a\vec{E}^a + \vec{B}^a\vec{B}^a\}\ , \tag{13}$$

where $\mathcal{H}$ is the Hamiltonian and $E_i^a = \dot{A}_i^a$ are to be treated as canonical momenta. (More exactly, $\vec{E} = g^2\vec{\pi}$.)

Two subtle points are to be mentioned in connection with the Hamiltonian (13). First, the equation div$\vec{E} = \rho$, inherent to the original Yang–Mills theory, does not stem from this Hamiltonian *per se*. This equation must be imposed as a constraint on the states from the Hilbert space by hand. Second, the gauge freedom is not fully eliminated in the system (13). Gauge transformations which depend on $\vec{x}$ but not t are still allowed. This freedom is reflected in the fact that, instead of two transverse degrees of freedom $\vec{A}_\perp$, the Hamiltonian (13) has three (three components of $\vec{A}$). Imposing, say, the Coulomb gauge condition

$$\partial_i A_i = 0\ ,$$

we could get rid of the "superfluous" degree of freedom, a procedure quite standard in perturbation theory (in the Coulomb gauge). Alas! If we want to keep and reveal the topologically nontrivial structure of the space of fields, the Coulomb gauge condition *cannot* be imposed. We have to work, with certain care, with the "undergauged" Hamiltonian (13).

Quasiclassically, the state of the system described by Eq. (13) at any given moment of time is characterized by the field configuration $A_i^a(\vec{x})$. Since we are interested in the zero-energy states, A_i^a must be pure gauge:

$$A_i^a(\vec{x})\ |_{vac} = iU(\vec{x})\partial_i U^{-1}(\vec{x})\ , \tag{14}$$

where U is a matrix belonging to $SU(2)$ and depending on the *spatial* components of the coordinate. [Warning: although one and the same letter is used in Eqs. (14) and (7), the matrices U in these equations do not coincide and depend on different variables.] Moreover, we are interested only in those zero-energy states which might be connected to each other by tunneling transitions (i.e. the corresponding action must be finite). As explained in Ref. [4] reproduced below, the latter requirement results in the boundary condition

$$U(|\vec{x}| \to \infty) = 1$$

or any other constant matrix U_0 independent of the direction in the three-dimensional space along which $\vec{x}$ tends to infinity. This boundary condition compactifies our *three-dimensional* space which becomes topologically equivalent to the three-dimensional sphere.

We already know that all mappings $S_3 \to S_3$ are classified according to the number of coverings [the latter S_3 represents the $SU(2)$ group space]. In other words the matrices $U(\vec{x})$ can be sorted out in distinct classes labeled by an integer number, $U_n(\vec{x})$, $n = 0, \pm 1, \pm 2, \ldots$, which in this case is referred to as the winding number. All matrices belonging to a given class are reducible to each other by a continuous $\vec{x}$-dependent gauge transformation. At the same time no continuous gauge transformation can transform $U_n(\vec{x})$ into $U_{n'}(\vec{x})$ if $n \neq n'$. The unit matrix represents the class $U_0(\vec{x})$. For $n = 1$, one can take, for instance,

$$U_1(\vec{x}) = \exp\left[i\pi \frac{\vec{x}\vec{\sigma}}{(\vec{x}^2 + \rho^2)^{1/2}}\right] . \tag{15}$$

Let us now return to the quantum-mechanical analogy mentioned above. The space of fields is of course infinite-dimensional. In order to reduce the field-theoretic problem to that from quantum mechanics it is necessary to single out one direction in the space of fields which would play the role of the angle variable on Fig. 1.

Consider to this end the charge $\mathcal{K}$ corresponding to the Chern–Simons current (9),

$$\mathcal{K} = \int K_0(\vec{x}) d^3x . \tag{16}$$

It is not difficult to show (see e.g. Coleman's lecture reproduced in Sec. VIII) that $\mathcal{K}$ measures the winding number; for any gauge field $A_i(\vec{x}) = iU_n(\vec{x})\partial_i U_n^{-1}(\vec{x})$, we have $\mathcal{K} = n$. In the "direction of $\mathcal{K}$" the space of fields has the topology of the circle; the points $\mathcal{K}$ and $\mathcal{K}+1$, and $\mathcal{K}-1, \ldots$, are physically one and the same point, so that $\mathcal{K}$ is literally an angle variable, and the Bloch type superselection rule associated with the angle variables is absolute. Any given field configuration generates a gauge *equivalent* configuration, with a shifted value of $\mathcal{K}$,

$$A_i \to A_i^{\Omega} \equiv U^{-1} A_i U + iU^{-1}\partial_i U, \quad \mathcal{K} \to \mathcal{K} \pm 1 ,$$

provided the matrix of the gauge transformation is chosen in the topologically nontrivial form (15).

Now, if one tries to depict the (classical) potential energy of the Yang–Mills system as a function of $\mathcal{K}$, one arrives at the plot of Fig. 2. We did here what is usually done in problems with angle type coordinates. We cut the circle and map it many times onto a straight line. In other words, we pretend that the variable $\mathcal{K}$ lives on the line. To take into account the fact that the original problem is formulated on the circle, we impose the (quasi)periodic Bloch boundary condition on the wave function. Any integer value of $\mathcal{K}$ on this plot corresponds to a pure gauge configuration with zero energy. On the other hand, if $\mathcal{K} \neq n$, the field strength tensor is nonvanishing and the energy of the field configuration is positive. Viewed as a function on the line, the potential energy $V(\mathcal{K})$ is, of course, periodic — with the unit period.

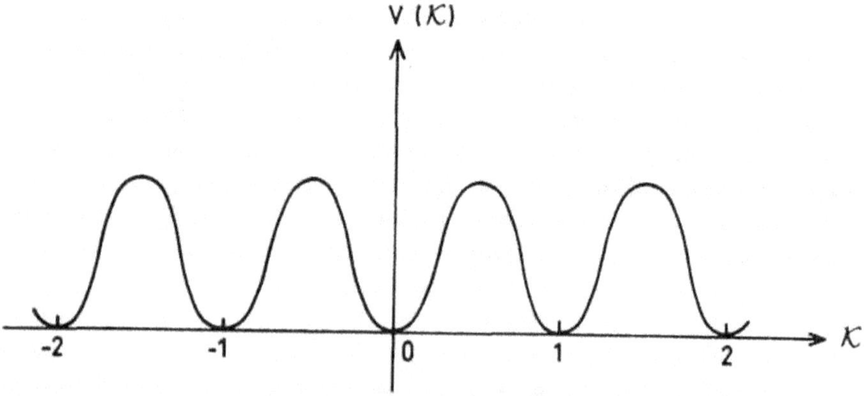

Fig. 2. Potential energy of the classical Yang–Mills field as a function of the Chern–Simons charge.

To conclude we emphasize once again that the field dynamics in the "$\mathcal{K}$ direction" is an exact parallel to quantum mechanics of the particle presented in Fig. 1. The angle variable in Fig. 1 should be identified with $2\pi\mathcal{K}$.

Assume that at $t = -\infty$ and at $t = +\infty$, our system is at the classical minimum (zero-energy state). Assume also that at $t = -\infty$ the winding number $\mathcal{K} = n$ while at $t = +\infty$, $\mathcal{K} = n \pm 1$. Consider now a field configuration $A_\mu(t, \vec{x})$ continuously interpolating between these two states (in the Euclidean time). Then the topological charge Q of this field configuration is

$$Q = \mathcal{K}(t = +\infty) - \mathcal{K}(t = -\infty) . \tag{17}$$

Choosing the interpolating trajectory in an "optimal" way so that the corresponding action is minimal, $(8\pi^2/g^2)$, we arrive at the BPST instanton. In the analog problem of Fig. 1 the counterpart of the instanton would be a path starting at the bottom of the circle, making a full rotation (in the clockwise or anticlockwise direction) and returning to the very same point at the bottom. In both cases we deal with tunneling under the barrier described by a solution of the classical equations of motion in the Euclidean time.

The fact that instantons describe tunnelings in the space of fields which possesses noncontractible paths was realized by V. Gribov shortly after the discovery of the BPST instantons and was popularized by him in private discussions.[3] Gribov's observation remained unpublished. The first published works where the issue was treated and the tunneling picture revealed [6, 7] are reproduced in this section. A very pedagogical and illustrative discussion of the tunneling interpretation given in the *Minkowski* space is presented in Ref. [8] which may be recommended to the reader who is just beginning to study instanton calculus.

The analysis outlined above (the one based on the Hamiltonian formulation) is convenient for establishing the existence of topologically nonequivalent vacuum states and, hence, the existence of the "winding" field configurations corresponding to tunneling. In practice, however, the Hamiltonian gauge $A_0 = 0$ is rarely used in constructing particular solutions of the duality equations. After all, the only thing important in the search of the nontrivial solutions is the value of the topological charge Q. The standard procedure is based on a specific ansatz for $A_\mu(x)$ in which all four Lorentz components of A_μ are nonvanishing. This ansatz entangles the color and Lorentz indices; the field configurations emerging in this way are generically referred to as "hedgehogs," following Polyakov. More concretely, the Euclidean Lorentz group $O(4)$ is represented as $SU(2) \times SU(2)$, where the two $SU(2)$ subgroups correspond to the left-handed and right-handed chiral rotations. Under this decomposition the vector A_μ is actually the representation $\{\frac{1}{2}, \frac{1}{2}\}$. Now, one couples the spinor indices of one of these $SU(2)$ subgroups to those of $SU(2)_{gauge}$. Thus, for instance, for the anti-instanton, we have

$$A^{\eta\xi}_{\alpha\dot\beta} = -i(x^\eta_{\dot\beta}\delta^\xi_\alpha + x^\xi_{\dot\beta}\delta^\eta_\alpha)f(x^2)\ , \tag{18}$$

where the upper indices refer to the color group, $A^a \to A^{\eta\xi} \equiv A^a(\sigma^a/2)^\eta_\delta\epsilon^{\xi\delta}$, while the lower (dotted + undotted) indices describe the Lorentz vector, $A_\mu \to A_{\alpha\dot\beta} = (\sigma_\mu)_{\alpha\dot\beta}A_\mu$. Moreover, $\sigma_\mu = (i, \vec\sigma)$ and $\epsilon^{\alpha\beta}$ is the antisymmetric tensor $(\alpha, \beta, \ldots, = 1, 2)$. The function $f(x^2)$ appearing in Eq. (18) is a function of x^2 only. The spinor notations are extremely convenient in those cases when the fermion degrees of freedom are involved, e.g. in the supersymmetric Yang–Mills theories. They will be explained in more detail in Secs. 5 and 7. In pure Yang–Mills theory, vector notations are more often used due to 't Hooft, who introduced [3] the so-called 't Hooft symbols,

$$A^a_\mu = 2\eta_{a\mu\nu}x_\nu f(x^2)\ , \tag{19}$$

where

$$\begin{aligned} &\eta_{a\mu\nu} = \epsilon_{a\mu\nu}\ \ (a, \mu, \nu = 1, 2, 3)\ , \\ &\eta_{a4\nu} = -\delta_{a\nu},\ \ \eta_{a\nu 4} = \delta_{4\nu}\ \ (\nu = 1, 2, 3)\ , \\ &\eta_{a44} = 0\ . \end{aligned} \tag{20}$$

[3] See e.g. a remark in Ref. [5] where Polyakov acknowledges Gribov's suggestion of the tunneling interpretation.

The details of the instanton formalism will be gradually presented in subsequent sections. The excursion may begin, however, from Ref. [4], the first part of which is reproduced below. In this part, along with the general aspects of instanton calculus, the authors discuss the BPST solution in the nonsingular and singular gauges. The latter is singled out by the fact that in the singular gauge the instanton field A^a_μ is characterized by a fast falloff at infinity, and the nonvanishing topological charge is ensured by a singularity of A^a_μ at the instanton center. A sufficiently fast falloff at infinity is needed in applications. The authors also address, for the first time, the problem of the instanton–anti-instanton (I–A) interaction in a well-separated pair. It is shown that this interaction is of the dipole type and at large separations R behaves as R^{-4}. The issue of the I–A interaction will be considered in more detail in Sec. VI. In the second (omitted) part of Ref. [4], technicalities of the instanton gas model of the QCD vacuum are treated — a model which, unfortunately, turned out to be unsubstantiated and did not survive. (It is worth noting, though, that it gave rise to more sophisticated models of the instanton liquid still under development now; for a review see e.g. Ref. [9].)

Let us return to the schematic plot of the potential energy depicted in Fig. 2. The fact that the potential energy is periodic in $\mathcal{K}$ ($\mathcal{K}$ is the angle type variable) implies that the vacuum "wave function" must be of the Bloch type. In this way one discovers a new fundamental parameter of QCD, the vacuum angle θ[6, 7, 4] appearing through the Bloch (quasi)periodic boundary condition. Although this parameter is absent in the QCD Lagrangian, still, physics depends on it in an essential way! [4]

Another legitimate question which may come to mind is as follows: "If the instanton solution is the tunneling trajectory connecting two zero-energy states under the barrier, what is the height of the barrier?" In the pure Yang–Mills theory at the classical level there are no dimensional parameters — the theory is classically conformally invariant. As we will see in Sec. II (see also Ref. [4]) each particular instanton solution breaks the conformal invariance and is characterized by a scale ρ called the instanton size. The scale invariance is restored only if one considers the whole family of instantons with all possible values of ρ. Roughly speaking, the height of the barrier "seen" by the instanton is $\propto \rho^{-1}$ — different for pseudoparticles of different sizes. That is why the graph of Fig. 2 should not be taken too literally: it misses all complexities of the field-theoretic problem with the infinite number of degrees of freedom. The instanton size ρ is an auxiliary parameter not reducible to any physically observable quantity. The only physical parameter through which the barrier height could be expressed is Λ_{QCD}, the scale parameter defining

[4] In particular, CP-violating effects in strong interactions can be directly expressed in terms of θ [10]. Since the latter are known to be tiny, it is obvious that nature somehow ensures that the vacuum angle θ is sufficiently small. This story, the smallness of θ, goes beyond QCD and even beyond the Standard Model of electroweak interactions, however, and we will not dwell on this topic.

the running coupling constant. At this scale, however, the quasiclassical approximation is definitely inapplicable.

The height of the barrier acquires an unambiguous meaning, inside the domain of validity of the quasiclassical approximation, in the Yang–Mills theory with the spontaneously broken gauge symmetry. In such models the Higgs fields H are introduced to generate the mass for the gauge bosons *via* the Higgs mechanism, the Standard Model of electroweak interactions being the most important example of this type. Then the scale invariance is explicitly broken by the vacuum expectation value of the Higgs field, $< H >= v$, gauge bosons are massive, $M \propto gv$, so that the theory is regularized in the infrared and the gauge coupling constant is frozen at the scale gv. Strictly speaking, if $v \neq 0$, the BPST instantons are no more the exact solutions of the classical equations of motion and they do not provide the minimum of the action. (Certain constraints can be imposed, however, and then the instantons minimize the action under these constraints. For more details, see Ref. [11] reproduced in Sec. VI.) All topological classification outlined above is still valid in the case $v \neq 0$. The picture of the degenerate zero-energy states and tunnelings remains intact.

Due to the cutoff provided by the gauge boson mass the maximal size of the instantons in such theories is of the order v^{-1}. As has been first shown in Ref. [12], if $v \neq 0$, the classical equations of motion have a static (time-independent) solution for the gauge and Higgs fields corresponding to the state of the system right at the top of the periodic curve of Fig. 2, the so-called sphaleron. The sphaleron solution has a finite mass and is unstable, of course, since this field configuration realizes the maximum of the energy, not the minimum. The mass of the sphaleron m_{sph} is the height of the barrier in the quasiclassical tunneling picture associated with the Yang–Mills theory. Parametrically, $m_{sph} \propto v/g$. A more detailed discussion of the sphalerons and their relation to instanton calculus can be found in Ref. [13].

In conclusion I would like to add that some questions which might arise in reading the original publications presented in this section are elucidated in the reviews [14, 15] (see Sec. VIII).

References

[1] *A. Belavin, A. Polyakov, A. Schwarz and Yu. Tyupkin, *Phys. Lett.* **59B** (1975) 85.
[2] *A. Polyakov, *Phys. Lett.* **59B** (1975) 85.
[3] *G. 't Hooft, *Phys. Rev. Lett.* **37** (1976) 8.
[4] *C. Callan, R. Dashen and D. Gross, *Phys. Rev.* **D17** (1978) 2717.
[5] A. Polyakov, *Nucl. Phys.* **B120** (1977) 429.
[6] *R. Jackiw and C. Rebbi, *Phys. Rev. Lett.* **37** (1976) 172.
[7] *C. Callan, R. Dashen and D. Gross, *Phys. Lett.* **63B** (1976) 334.
[8] K. M. Bitar and S.-J. Chang, *Phys. Rev.* **D17** (1978) 486.
[9] E. Shuryak, *The QCD Vacuum, Hadrons and Superdense Matter* (World Scientific, Singapore, 1988).
[10] R. Crewther, P. Di Vecchia, G. Veneziano and E. Witten, *Phys. Lett.* **88B** (1979) 123; M. Shifman, A. Vainshtein and V. Zakharov, *Nucl. Phys.* **B166** (1980) 493.

[11] *I. Affleck, *Nucl. Phys.* **B191** (1981) 429.
[12] F. Klinkhamer and N. Manton, *Phys. Rev.* **D30** (1984) 2212.
[13] L. G. Yaffe, *Phys. Rev.* **D40** (1989) 3463; F. Klinkhamer, "Sphalerons and Energy Barriers in the Weinberg–Salam Model," in *Proc. Int. Conf. on High Energy Physics, Singapore, 1990*, p. 913.
[14] *S. Coleman, "The Uses of Instantons," in S. Coleman, *Aspects of Symmetry* (Cambridge University Press, 1985).
[15] *A. Vainshtein, V. Zakharov, V. Novikov and M. Shifman, *Sov. Phys.—Uspekhi* **25** (1982) 195.

Volume 59B, number 1 PHYSICS LETTERS 13 October 1975

COMPACT GAUGE FIELDS AND THE INFRARED CATASTROPHE

A.M. POLYAKOV
Landau Institute for Theoretical Physics, Moscow, USSR

Received 19 August 1975

It is shown that infrared phenomena in the gauge theories are guided by certain classical solutions of the Yang-Mills equations. The existence of such solutions can lead to a finite correlation length which stops infrared catastrophe. In the present paper we deal only with theories with a compact but abelian gauge group. In this case the problems of correlation length and charge confinement are completely solved.

It was pointed out by different authors [1] several years ago that the infrared phenomena, occurring with a gauge field, might provide a natural explanation for the confinement of quarks. At the same time there exist no methods for analyzing the interaction of gauge fields in the deep infrared region. It is the purpose of the present paper to work out a formalism which permits, at least partly, to take into account the infrared effects in gauge-field interactions. Our main idea is that the system of gauge fields acquires a finite correlation length through the following phenomenon.

Imagine that we are calculating a certain correlation function in the euclidean formulation of the gauge theory. This means averaging over all possible fields A_μ with the weight equal to:

$$\exp\{-S(A)\} = \exp\left\{-\frac{1}{4g^2}\,\mathrm{Sp}\int F_{\mu\nu}^2\,\mathrm{d}^4x\right\} \tag{1}$$

$$F_{\mu\nu} = \partial_\mu A_\nu - \partial_\nu A_\mu + [A_\mu, A_\nu]\,.$$

Assume that the charge $g^2 \ll 1$; then the leading role in the averaging will be played by the fields close to that defined by the equation:

$$\frac{\delta S}{\delta \bar{A}_\mu(x)} = 0; \quad S[\bar{A}] < \infty. \tag{2}$$

Usually one takes into account only the trivial minima of S, i.e. $A_\mu = 0$, and developes the perturbation theory as a small deviation from this. For the correlation function with the distance R the parameter of the perturbation expansion is $g^2 \log R/a$ where a is the inverse cut-off. Hence, for very large R, perturbation theory is not applicable and another $\bar{A}$ might become essential. Indeed, though the weight with which non-trivial minima enters the averaging is small being proportional to

$$\exp\{-S(\bar{A})\} = \exp(-E/g^2) \tag{3}$$

(where E is certain constant) their influence on the correlation is large if the classical field $\bar{A}$ is long ranged. (In fact, the contribution to the correlation will be shown to be proportional to $\exp\{-E/g^2\}R^4$).

Now assume that the fields $\bar{A}_\mu$ are such as if they were produced by certain "particles" in the four-dimensional euclidean space. In other words there exist the "one-particle" minima of S, the "two-particle" and so on. Of course, the "energy" E depends on the number of the above mentioned pseudo-particles. The average density of pseudo-particles in our system is very small, being proportional to $\exp(-E/g^2)$. However, their existence creates long range random fields in our system. Due to these random fields, the correlation length becomes finite. This is precisely the phenomena we are going to investigate.

The above discussion was based on the crucial assumption that there exists pseudo-particle solutions of the gauge field equations. It will be proved in the second paper of this series that such solutions indeed exist for every compact nonabelian gauge group.

In this first paper we confine ourselves to the problem of realizing the above program in the case of compact but abelian gauge fields. The purpose of this consideration is two fold. First, it is a good and simple model for trying our program on. Second, the compactness of quantum electrodynamics seems to be an attractive hypothesis and our results may have physical applications. For example we shall prove the existence of a certain critical charge in QED.

Volume 59B, number 1 PHYSICS LETTERS 13 October 1975

The definition of the theory is as follows. Let us introduce a lattice in the four dimensional space, necessary in the definition of functional integrals. Generally, the action should have the form:

$$S = \sum_{x,\mu,\nu} f(F_{x,\mu\nu}) \tag{3}$$

$$F_{x,\mu\nu} = A_{x,\mu} + A_{x+a_\mu,\nu} - A_{x+a_\nu,\mu} - A_{x,\nu}$$

where a_μ is lattice vector,

$$f(x) \underset{x\to 0}{\approx} \frac{1}{4g^2} x^2.$$

The hypothesis of the compactness of the gauge group means that $A_{x,\mu}$ are the angular variables, and the group is the circle and not the line. This is equivalent to the hypothesis that:

$$f(x + 2\pi) = f(x). \tag{4}$$

Gauge theories on the lattice have been considered earlier by Wilson [2] and the present author (unpublished). See also [3].

The immediate consequence of the periodicity of $f(x)$ is that the nearest neighbours $A_{x+a\lambda,\mu}$ and $A_{x,\mu}$ can be different by $2\pi N$ (where N is integer) without producing large action. Hence, in the continuous limit $F_{\mu\nu}$ may have the following singularities:

$$F_{\mu\nu}(x) = F_{\mu\nu}^{\text{reg}} + 2\pi \sum_i N_{i\mu\nu}\, \delta^{(S_i)}(x) \tag{5}$$

where $\delta^{(S)}(x)$ is the surface δ-function. The second term in (5) will not contribute to the action, due to the periodicity.

It will be convenient for us to analyze first the three dimensional theory. In this case there exists quasiparticle solutions of Maxwell equations which simply coincide with the Dirac monopole solution. If we introduce the field:

$$F_\alpha \equiv \tfrac{1}{2}\epsilon_{\alpha\beta\gamma} F_{\beta\gamma} \tag{6}$$

then the general pseudo-particle solution will be given by:

$$F_\alpha = \sum_a \frac{q_a}{2} \cdot \frac{(x - x_a)_\alpha}{|x - x_a|^3} - 2\pi\delta_{\alpha 3} \sum q_a \theta(x_3 - x_{3a})\delta(x - x_{1a})\delta(x - x_{2a}). \tag{7}$$

If $\{q_a\}$ are integers then the singularities in (7) are just of the permitted type.

The action is given by:

$$S(\bar{A}) = E/g^2$$
$$E = \frac{\pi}{2} \sum_{a\neq b} \frac{q_a q_b}{|x_a - x_b|} + \epsilon \sum q_a^2 \tag{8}$$

(the value of the constant ϵ depends on the lattice type and is not essential for us).

Now let us analyze the correlation function introduced in [2] which is most convenient in the confinement problem:

$$F(C) \equiv \exp\{-W(C)\} = \langle \exp\{\mathrm{i} \oint_C A_\mu\, \mathrm{d}x_\mu\}\rangle \tag{9}$$

(here C is some large contour).

For the evaluation of (9) let us substitute $A_\mu = \bar{A}_\mu + a_\mu$. Since the integral over a_μ is gaussian we get:

$$F(C) = F_0(C) \frac{\Sigma \exp\{-S(\bar{A})\} \exp\{\mathrm{i}\oint \bar{A}_\mu\, \mathrm{d}x_\mu\}}{\Sigma \exp\{-S(\bar{A})\}} \tag{10}$$

(Here F_0 is the contribution of $\bar{A} = 0$).

The sum in (10) goes over all possible configurations of pseudo-particles. Now, let us use the formula:

$$\exp\{\mathrm{i}\oint A_\mu\, \mathrm{d}x_\mu\} = \exp\{\mathrm{i}\int F_\alpha\, \mathrm{d}\sigma_\alpha\} \tag{11}$$

in which, due to the periodicity of the exponent, only the first term from (7) should be substituted.

The problem is reduced now to the calculation of the free energy of the monopoles plasma with the "temperature" g^2 in the external field:

$$\varphi^c(x) = \mathrm{i}\, \frac{\partial}{\partial x_\alpha} \int \frac{\mathrm{d}\sigma_\alpha}{|x - y|}. \tag{12}$$

This problem was solved by using Debye method which is correct for sufficiently small g^2. The result is two fold. First, there exist the Debye correlation length and the corresponding photon mass m equal to:

$$m^2 = \exp\{-\epsilon/g^2\} \tag{13}$$

(in the units of the inverse lattice length).

Secondly:

$$W[C] = \mathrm{const}(g^2 m A) \tag{14}$$

where A is the area of the contour C. Eq. (14) was derived for arbitrary planar contour. According to

Wilson [1] this result means "charge confinement" in the three dimensional QED with the compact gauge group.

In the case of the four dimensional QED it can be shown that the only classical solutions with finite action are closed rings. This follows from the fact that singular points in this case should form lines. To prove this, assume that it is not so, and consider the pseudo particle solution with $x = 0$. Consider the cube K with $x_4 = 0$. Then it should be:

$$\oint F_{\mu\nu}\, d\sigma_{\mu\nu} = 2\pi q \tag{15}$$

But, after small variation of the x_4, our pseudo-particle will be outside the cube, and this contradicts (15).

Since the closed rings produce only dipole forces their influence on the correlation are rather weak. We showed that in this case the correlation length remains infinite and that

$$W[C] = \text{const}\, \exp(-B/g^2)\cdot L \tag{16}$$

where L is the length of the contour C, and B is some constant. This result means the absence of the charge confinement for small g^2. Since it was proved in [2] that for large g^2 the charge confinement exist there are some critical charge g_c^2 at which the phase transition occurs. It is not clear now whether this critical charge is connected with the fine structure constant.

The extension of the above ideas on the nonabelian theory will be presented in the other papers of this series.

References

[1] S. Weinberg, D. Gross and F. Wilezek, Phys. Rev. D8 (1973) 3633; Phys. Rev. Lett. 31 (1973) 494.

[2] K. Wilson, Phys. Rev. D10 (1974) 2445.

[3] R. Balian et al., Phys. Rev. D10 (1974) 3376.

Volume 59B, number 1 PHYSICS LETTERS 13 October 1975

PSEUDOPARTICLE SOLUTIONS OF THE YANG-MILLS EQUATIONS

A.A. BELAVIN, A.M. POLYAKOV, A.S. SCHWARTZ and Yu.S. TYUPKIN
Landau Institute for Theoretical Physics, Academy of Sciences, Moscow, USSR

Received 19 August 1975

We find regular solutions of the four dimensional euclidean Yang-Mills equations. The solutions minimize locally the action integrals which is finite in this case. The topological nature of the solutions is discussed.

In the previous paper by one of the authors [1] the importance of the pseudoparticle solutions of the gauge field equations for the infrared problems was shown. By "pseudoparticle" solutions we mean the long range fields A_μ which minimize locally the Yang-Mills actions S and for which $S(A) < \infty$. The space is euclidean and four-dimensional. In the present paper we shall find such a solution. Let us start from the topological consideration which shows the existence of the desired solutions.

All fields we are interested in satisfy the condition:

$$F_{\mu\nu} = \partial_\mu A_\nu - \partial_\nu A_\mu + [A_\mu, A_\nu] \underset{x\to\infty}{\to} 0 \,. \tag{1}$$

Consider a very large sphere S^3 in our 4-dimensional space. The sphere itself is of course 3-dimensional. From (1) it follows that

$$A_\mu|_{S^3} \approx g^{-1}(x) \left.\frac{\partial g(x)}{\partial x_\mu}\right|_{S^3} \tag{2}$$

where $g(x)$ are matrices of the gauge group. Hence every field $A_\mu(x)$ produce a certain mapping of the sphere S^3 onto the gauge group G. It is clear that if two such mappings belong to different homotopy classes then the corresponding fields $A_\mu^{(1)}$ and $A_\mu^{(2)}$ cannot be continuously deformed one into another. It is well known [2] that there exists an infinite number of different classes of mappings of $S^3 \to G$ if G is a nonabelian simple Lie group. Hence, the phase space of the Yang-Mills fields are divided into an infinite number of components, each of which is characterized by some value of q, where q is a certain integer.

Our idea is to search for the absolute minimum of the given component of the phase space. In order to do this we need the formula expressing the integer q through the field A_μ [‡]. It is easy to check that

$$q = \frac{1}{8\pi^2} \epsilon_{\mu\nu\lambda\gamma} \,\mathrm{Sp} \int F_{\mu\nu} F_{\lambda\gamma} \,\mathrm{d}^4x. \tag{3}$$

To prove this let us use the identity:

$$\epsilon_{\mu\nu\lambda\gamma} \,\mathrm{Sp}\, F_{\mu\nu} F_{\lambda\gamma} = \partial_\alpha J_\alpha$$
$$J_\alpha = \epsilon_{\alpha\beta\gamma\delta} \,\mathrm{Sp}(A_\beta(\partial_\gamma A_\delta + \tfrac{2}{3} A_\gamma A_\delta)). \tag{4}$$

From (4) follows:

$$q = \frac{1}{8\pi^2} \oint_{S^3} J_\alpha \,\mathrm{d}^3\sigma^\alpha$$
$$= \frac{1}{8\pi^2} \tfrac{4}{3} \epsilon_{\alpha\beta\gamma\delta} \oint \mathrm{Sp}(A_\beta A_\gamma A_\delta) \mathrm{d}^3\sigma_\alpha \tag{5}$$

where

$$A_\mu = g^{-1}(x) \partial g / \partial x_\mu \tag{6}$$

Now consider the case $G = \mathrm{SU}(2)$. In this case it is clear that:

$$\mathrm{d}\mu(g) = \mathrm{Sp}(g^{-1}\mathrm{d}g \times g^{-1}\mathrm{d}g \times g^{-1}\mathrm{d}g) \tag{7}$$

is just the invariant measure on this group, since it is the invariant differential form of the appropriate dimension. The meaning of the notation in (7) is as follows. Let $g(\xi_1 \xi_2 \xi_3)$ be some parametrization of SU(2), say, through the Euler angles. Then the invariant measure will be:

‡ Formulas like (3) are known in topology by the name of "Pontryagin class".

$$d\mu = \mathrm{Sp}\left(g^{-1}\frac{\partial g}{\partial \xi_1} g^{-1}\frac{\partial g}{\partial \xi_2} g^{-1}\frac{\partial g}{\partial \xi_3}\right) d\xi_1 \dots d\xi_3 \,. \quad (8)$$

Comparing (8) with (5) we see that the integrand in (5) is precisely the Jacobian of the mapping of S^3 on SU(2). Hence q is the number of times the SU(2) is covered under this mapping. It is just the definition of the mapping degree. In the case of the arbitrary group G one should consider the mapping of S^3 on its SU(2) subgroup and repeat the above. There exists an important inequality which will be extensively used below. Consider the following relation:

$$\mathrm{Sp}\int (F_{\mu\nu} - \widetilde{F}_{\mu\nu})^2 d^4x \geqslant 0 \quad (9)$$

where $\widetilde{F}_{\mu\nu} = \frac{1}{2}\epsilon_{\mu\nu\lambda\gamma}F_{\lambda\gamma}$. From (9) and (3) it follows that:

$$E \geqslant 2\pi^2 |q| \quad (10)$$

where

$$S(A) \equiv E(A)/g^2$$

and g^2 is a coupling constant.

The formula (10) gives the lower bound for the energy of the quasiparticles in each homotopy class. We shall show now that for $q = 1$ this bound can be saturated. In other words one can search the solution of the equation, which replace the usual Yang-Mills one:

$$F_{\alpha\beta} = \pm\tfrac{1}{2}\epsilon_{\alpha\beta\gamma\delta}F_{\gamma\delta}$$
$$F_{\alpha\beta} = \partial_\alpha A_\beta - \partial_\beta A_\alpha + [A_\alpha A_\beta] \,. \quad (11)$$

Again it is sufficient to consider the case $G = \mathrm{SU}(2)$. In this case it is convenient though not necessary to extend this group up to $\mathrm{SU}(2) \times \mathrm{SU}(2) \approx \mathrm{O}(4)$. The gauge fields for O(4) are $A_\mu^{\alpha\beta}$ where A_μ are antisymmetric on $\alpha\beta$. The SU(2) gauge field are connected with $A_\mu^{\alpha\beta}$ by the formulas:

$$\pm A_\mu^i = \tfrac{1}{2}(A_\mu^{oi} \pm \tfrac{1}{2}\epsilon_{ikl}A_\mu^{kl}). \quad (12)$$

Now, two equations:

$$\pm F_{\mu\nu}^i = \pm\tfrac{1}{2}\epsilon_{\mu\nu\lambda\gamma} \pm F_{\lambda\gamma}^i$$

are equivalent to the following one:

$$\epsilon_{\alpha\beta\gamma\delta}F_{\mu\nu}^{\gamma\delta} = \epsilon_{\mu\nu\lambda\gamma}F_{\lambda\gamma}^{\alpha\beta}. \quad (13)$$

Let us search the solution of (13) which is invariant under simultaneous rotations of space and isotopic space. The only possibility is:

$$A_\mu^{\alpha\beta} = f(\tau)(x_\alpha\delta_{\mu\beta} - x_\beta\delta_{\mu\alpha}). \quad (14)$$

It is easy to calculate F:

$$F_{\mu\nu}^{\alpha\beta} = (2f - \tau^2 f^2)(\delta_{\mu\alpha}\delta_{\nu\beta} - \delta_{\mu\beta}\delta_{\nu\alpha})$$
$$+ (f'/\tau + f^2)(x_\alpha x_\mu \delta_{\nu\beta} - x_\alpha x_\nu \delta_{\mu\beta} \quad (15)$$
$$+ x_\beta x_\nu \delta_{\mu\alpha} - x_\beta x_\mu \delta_{\nu\alpha}).$$

It is evident that the first tensor structure (15) satisfies the equation (13) and the second does not. Hence we are to choose:

$$f'/\tau + f^2 = 0, \quad f(\tau) = \frac{1}{\tau^2 + \lambda^2} \quad (16)$$

where τ is an arbitrary scale. The quasi-energy E is given by

$$E = \tfrac{1}{4}\mathrm{Sp}\int_{\pm} F_{\mu\nu}^2 d^4x$$
$$= \tfrac{1}{32}\mathrm{Sp}\int (F_{\mu\nu}^{\alpha\beta})^2 d^4x = 2\pi^2 \quad (17)$$

Comparison of (17) and (10) shows that we find absolute minimum for $q = 1$.

Another representation for the solution (14) is given by the formulas:

$$A_\mu = \frac{\tau^2}{\tau^2 + \lambda^2} g^{-1}(x)\frac{\partial g(x)}{\partial x_\mu}$$

$$g(x) = (x_4 + i\boldsymbol{x}\cdot\boldsymbol{\sigma})(x_4^2 + \boldsymbol{x}^2)^{-1/2}$$

$$g^\dagger g = 1, \quad \tau^2 = x_4^2 + \boldsymbol{x}^2$$

($\boldsymbol{\sigma}$ are Pauli matrixes).

For arbitrary group G one should consider its subgroup SU(2) for which A_μ is given by (18) and all other matrix elements of A_μ let be zero.

Our solution, as is evident from the scale invariance, contains the arbitrary scale λ. Hence these fields are long range and are essential in the infrared problems.

We do not know whether any solutions of (13) exist with $q > 1$. One may consider of course several pseudoparticles with $q = 1$. However, we do not know whether they are attracted to each other and form the pseudoparticle with $q > 1$ or whether there exists repulsion and no stable pseudoparticle.

One of us (A.M.P.) is indebted to S.P. Novikov for explanation of some topological ideas.

References

[1] A.M. Polyakov, Phys. Lett. 59B (1975) 82.

Volume 37, Number 3 PHYSICAL REVIEW LETTERS 19 July 1976

Vacuum Periodicity in a Yang-Mills Quantum Theory*

R. Jackiw and C. Rebbi
Laboratory for Nuclear Science and Physics Department, Massachusetts Institute of Technology, Cambridge, Massachusetts
(Received 1 June 1976)

We propose a description of the vacuum in Yang-Mills theory and arrive at a physical interpretation of the pseudoparticle solution and the attendant violation of symmetries. The existence of topologically inequivalent classical gauge fields gives rise to a family of quantum mechanical vacua, parametrized by a *CP*-nonconserving angle. The requirement of vacuum stability against gauge transformations renders the vacua chirally noninvariant.

A classical pseudoparticle solution to the SU(2) Yang-Mills theory in Euclidean four-dimensional space has been given by Belavin, Polyakov, Schwartz, and Tyupkin,[1] with the suggestion that it be used to dominate the functional integral which describes a quantum field theory continued to Euclidean space. 't Hooft[2] has shown that these nontrivial minima of the action give nonvanishing contributions to amplitudes which would be zero in the ordinary sector. Specifically in a theory of fermions coupled to Yang-Mills fields, with chiral U(1) and *CP* symmetries, symmetry-nonconserving effects are found through the presence of the axial-vector-current anomaly.[3] Thus he provides a possible resolution of the long-standing U(1) problem[4] and an intriguing suggestion for the origin of *CP* nonconservation. The phenomena are $O(\exp(-8\pi^2/g^2))$, where g is the gauge coupling constant; they are nonperturbative.

The fact that the classical field configuration which is responsible for the new results is in Euclidean four-dimensional space, i.e., imaginary time, leads one to suspect that the pseudoparticle is associated with quantum-mechanical tunneling by which field configurations in the ordinary three-dimensional space are joined in the course of the (real-time) evolution through the penetration of an energy barrier.[5] Also the exponentially small magnitude is indicative of tunneling. Here we wish to present a further explanation of this point, which we hope, will clarify the physical interpretation of the pseudoparticle solution and will supplement 't Hooft's more formal computations. Our considerations lead to a description of the quantum mechanical vacuum state of a Yang-Mills theory which is unexpectedly rich.

In the quantum field theory, a state of the system can be represented by a wave functional $\Psi[\vec{A}]$ of the field configuration. Having in mind a Yang-Mills theory, we have taken the potentials $\vec{A}(\vec{x})$ (anti-Hermitian matrices in the space of the infinitesimal group generators) as argument of the functional, excluding the time components $A^0(\vec{x})$, because they are dependent variables. In defining scalar products and matrix elements of observables one must avoid infinities associated with the volume of the gauge group. Without repeating details of the well-kown gauge-fixing procedure, let us only recall that it removes from the functional integral over $\vec{A}$ configurations of the fields which can be joined by a *continuous* gauge transformation to configurations already counted. In particular, one does not integrate over potentials of the form

$$\vec{A}(\vec{x}) = g^{-1}(\vec{x})\nabla g(\vec{x}), \tag{1}$$

where g is the unitary matrix of a gauge transformation that can be joined to the identity through a one-parameter continuous family of transformations $g(\vec{x},\alpha)$:

$$g(\vec{x},1) = g(\vec{x}); \quad g(\vec{x},0) = I. \tag{2}$$

The potentials of Eq. (1) are of course gauge equivalent to $\vec{A}=0$.

But it is important to realize that there are values of $\vec{A}$ that can be obtained from each other by gauge transformations which cannot be continuously joined with the identity transformation. For instance, we may consider

$$g_1(\vec{x}) = \frac{\vec{x}^2 - \lambda^2}{\vec{x}^2 + \lambda^2} - \frac{2i\lambda\vec{\sigma}\cdot\vec{x}}{\vec{x}^2+\lambda^2} \tag{3}$$

which gives origin to

$$\vec{A}(\vec{x}) = g_1^{-1}(\vec{x})\nabla g_1(\vec{x}) = \frac{2i\lambda}{(\vec{x}^2+\lambda^2)^2}[\sigma(\lambda^2 - \vec{x}^2) + 2\vec{x}(\sigma\cdot\vec{x}) + 2\lambda\vec{x}\times\vec{\sigma}] \tag{4}$$

and of course to vanishing field strengths F_{ij}. Values of the potentials like those of Eq. (4), although gauge equivalent to $\vec{A} = 0$, should *not* be removed from the integrations over the field configurations by the gauge fixing procedure, and indeed we shall argue that physical effects are associated with them.

Before proceeding, let us characterize the classes of gauge-equivalent, but not continuously gauge-equivalent, potentials. We study effects which are local in space and therefore, when we consider a gauge transformation g, we require

$$g(\vec{x}) \xrightarrow[|\vec{x}|\to\infty]{} I. \tag{5}$$

Thus g defines a mapping of the three-dimensional space, *with all the directions at* ∞ *identified*, into the group space. From the topological point of view, the Euclidean space E^3 with points at ∞ identified is equivalent (homeomorphic) to a three-dimensional sphere S^3; but the manifold of SU(2) is also homeomorphic to S^3, so that g defines a mapping

$$S^3 \xrightarrow[g]{} S^3. \tag{6}$$

It is known that these mappings fall into homotopy classes (mappings belonging to different classes cannot be continuously distorted into each other) classified by an integer n,

$$g_n(\vec{x}) = [g_1(\vec{x})]^n \tag{7}$$

with g_1 given in Eq. (3) being a representative of the nth class.

We can make contact now with the pseudoparticle solution.[1] Observe that the field configuration of Eq. (4) has zero potential energy, and that there is no energy-conserving evolution of the system which adiabatically connects that configuration with $\vec{A} = 0$. Such an evolution should be a continuous gauge transformation; but this is impossible because g_1 and the identity belong to different homotopy classes. All paths joining the two field configurations in real time must go over an energy barrier. To exemplify this, let us multiply the potentials of Eq. (4) by $\frac{1}{2} - \alpha$ and increase α adiabatically from $-\frac{1}{2}$ to $+\frac{1}{2}$. Now the field strength is nonvanishing, but proportional to $\frac{1}{4} - \alpha^2$. The energy, $-\frac{1}{8}\int d^3x\, \mathrm{Tr}\, F_{ij}F^{ij} \geq 0$, becomes proportional to $(\alpha^2 - \frac{1}{4})^2$ and exhibits a barrier shape as α varies from $-\frac{1}{2}$ to $\frac{1}{2}$.

In the quantum theory, tunneling will occur across this barrier. It is well known that a semiclassical description of tunneling can be given by solving the classical equations of motion with *imaginary* time, thus achieving an evolution which would be classically forbidden for real time.[5] The pseudoparticle solution[1] serves precisely this purpose: It carries zero energy (the Euclidean stress tensor vanishes); it can be arranged to connect $g = g_1$ at $x_4 = -it = -\infty$ with $g = I$ at $x_4 = \infty$. The physical implication of the pseudoparticle solution is that the quantal description of the vacuum state cannot be limited to fluctuations around any definite classical configuration of zero energy.

Let us now describe in greater detail the nature of the vacuum wave functional. Consider any of the field configurations

$$\vec{A}_n(\vec{x}) = g_n^{-1}(\vec{x})\nabla g_n(\vec{x}) \tag{8}$$

with vanishing F^{ij}. Neglecting tunneling effects we might expect the vacuum to be of the form

$$\psi_n[\vec{A}] = \varphi[\vec{A} - \vec{A}_n], \tag{9}$$

where the wave functional φ is peaked about zero and has a spread due to quantum fluctuations and any $\vec{A}_n$ can be chosen as representative of the classical vacuum, i.e., the classical zero-energy configuration.

But the pseudoparticle solution connects $\vec{A}_n$ with $\vec{A}_{n+1}$, giving origin to tunneling between the different ψ_n. The true quantal vacuum state will therefore be a superposition of the form

$$\Psi[\vec{A}] = \sum_n c_n \psi_n[\vec{A}] + O(\exp(-8\pi^2/g^2)). \tag{10}$$

To determine the coefficients c_n in this equation let us observe that the finite gauge transformation g_1 changes ψ_n into ψ_{n+1}. Requiring the vacuum state to be stable against gauge transforma-

VOLUME 37, NUMBER 3 PHYSICAL REVIEW LETTERS 19 JULY 1976

tions determines the coefficients to be

$$c_n = e^{in\theta} \tag{11}$$

Thus we find a family of vacua, parametrized by an angle θ, where under the gauge transformation g_1

$$\Psi_\theta[\vec{A}] \underset{g_1}{\to} e^{-i\theta}\Psi_\theta[\vec{A}]. \tag{12}$$

The occurrence of multiple vacua is intriguing and is reminiscent of the situation encountered in the Schwinger model.[6]

The significance of the phase θ in Eqs. (10) and (11) becomes apparent when massless fermions are coupled to the Yang-Mills fields. One may then introduce the U(1) axial-vector current

$$J_5^{\mu} = i\bar{\psi}\gamma^{\mu}\gamma^5\psi$$

which however is not conserved because of the anomaly.[3] A conserved, but gauge-variant, current is given by

$$\tilde{J}_5^{\mu} = J_5^{\mu} - 4\pi^{-2}\epsilon^{\mu\nu\alpha\beta} \times \mathrm{Tr}(A_\nu \partial_\alpha A_\beta + \tfrac{2}{3} A_\nu A_\alpha A_\beta) \tag{13a}$$

and the conserved axial charge is

$$\tilde{Q}_5 = \int d^3x\, \tilde{J}_5^{0}. \tag{13b}$$

To exhibit the gauge dependence of $\tilde{Q}_5$, we perform a finite gauge transformation g, with $g(\vec{x}) \underset{|\vec{x}|\to\infty}{\longrightarrow} I$, and find

$$\Delta\tilde{Q}_5 = \frac{1}{12\pi^2}\int d^3x\, \mathrm{Tr}\epsilon_{ijk}(g^{-1}\partial_i g)(g^{-1}\partial_j g)(g^{-1}\partial_k g) = \frac{1}{12\pi^2}\int d\mu(g) = 2n, \tag{14}$$

where $d\mu(g)$ is the invariant measure of the group and n is the integer which characterizes the homotopy class of g. (g belongs to the nth homotopy class when it is continuously deformable to g_n.) The fact that $\tilde{Q}_5$ commutes with the Hamiltonian, and that it changes by two units under the gauge transformation g_1, together with Eq. (11), implies that

$$\exp(-\tfrac{1}{2}i\theta'\tilde{Q}_5)\Psi_\theta[\vec{A}] = \Psi_{\theta+\theta'}[\vec{A}] \tag{15}$$

which in turn shows that all the vacua are degenerate in energy and define the same theory. Equation (15) also demonstrates the possibility of symmetry breaking without Goldstone bosons: This may provide a solution to the U(1) problem.[2,4]

An explanation of the nonconservation of the gauge-invariant fermionic axial charge

$$Q_5(t) = \int d^3x\, J_5^{0}(t,\vec{x}) \tag{16}$$

may be given. The tunneling process between adjacent vacuum components $\psi_n[\vec{A}]$ and $\psi_{n+1}[\vec{A}]$ is equivalent to a gauge transformation g_1 which changes $\tilde{Q}_5 - Q_5$ by two units; see Eq. (14). But $\tilde{Q}_5$ is conserved, and therefore $\Delta Q_5 = -2$.

Finally, we remark that, whereas in the massless case conservation of $\tilde{Q}_5$ renders all vacua degenerate, we expect that if the fermions are massive, so that $\tilde{Q}_5$ is no longer conserved, different values of θ define nonequivalent theories as in the Schwinger model.[6] A nonzero value of θ could describe CP nonconservation,[2] but in the theory as developed thus far there is no indication how to compute θ.

We are happy to acknowledge our indebtedness to G. 't Hooft, whose calculations made our observations possible. We also thank S. Coleman and L. Susskind for discussions.

Added note.—After completion of this manuscript, we received a paper by C. Callan, R. Dashen, and D. Gross (to be published) who arrive at conclusions similar to ours.

---

*This work is supported in part through funds provided by the U. S. Energy Research and Development Administration under Contract No. E(11-1)-3069.

[1]A. Belavin, A. Polyakov, A. Schwartz, and Y. Tyupkin, Phys. Lett. 59B, 85 (1975); A. Polyakov, Phys. Lett. 59B, 82 (1975).

[2]G. 't Hooft, Phys. Rev. Lett. 37, 8 (1976), and lectures at various universities.

[3]S. L. Adler, in *Lectures on Elementary Particles and Quantum Field Theory*, edited by S. Deser, M. Grisaru, and H. Pendleton (The MIT Press, Cambridge, Mass., 1970); R. Jackiw, in *Lectures on Current Algebra and Its Applications*, edited by S. Treiman, R. Jackiw, and D. Gross (Princeton Univ. Press, Princeton, N. J., 1972).

[4]H. Pagels, Phys. Rev. D 13, 343 (1976).

[5]Use of imaginary time in discussions of tunneling is well known; see, e.g., K. Freed, J. Chem. Phys. 56, 692 (1972); D. McLaughlin, J. Math. Phys. (N.Y.) 13, 1099 (1972).

[6]J. Schwinger, Phys. Rev. 128, 2425 (1962); J. Lowenstein and A. Swieca, Ann. Phys. (N.Y.) 68, 172 (1971); S. Coleman, R. Jackiw, and L. Susskind, Ann. Phys. (N.Y.) 93, 267 (1975). (This analogy was developed in conversations with Coleman and Susskind.) There is no pseudoparticle solution for *spinor* electrodynamics in two space-time dimensions. However, by adding to the model charged spinless fields, with a Higgs potential and minimal electromagnetic coupling, a pseudoparticle solution exists in Euclidean two-dimension space—it is just the Nielsen-Olesen string

VOLUME 37, NUMBER 3 PHYSICAL REVIEW LETTERS 19 JULY 1976

[Nucl. Phys. B61, 45 (1973)]. The topologically conserved object is $\int d^2x \epsilon_{\mu\nu} F^{\mu\nu}$, and $\epsilon_{\mu\nu} F^{\mu\nu}$ is also proportional to the anomalous divergence of the axial vector current; K. Johnson, Phys. Lett. 5, 253 (1963); R. Jackiw, in *Laws of Hadronic Matter*, edited by A. Zichichi (Academic, New York, 1975).

Volume 63B, number 3 PHYSICS LETTERS 2 August 1976

THE STRUCTURE OF THE GAUGE THEORY VACUUM*

C.G. CALLAN, Jr.
Joseph Henry Laboratories, Princeton University, Prinecton, New Jersey 08540, USA

R.F. DASHEN*
Institute for Advanced Study, Princeton, New Jersey 08540, USA

and

D.J. GROSS
Joseph Henry Laboratories, Princeton University, Princeton, New Jersey 08540, USA

Received 20 May 1976

The finite action Euclidean solutions of gauge theories are shown to indicate the existence of tunneling between topologically distinct vacuum configurations. Diagonalization of the Hamiltonian then leads to a continuum of vacua. The construction and properties of these vacua are analyzed. In non-abelian theories of the strong interactions one finds spontaneous symmetry breaking of axial baryon number without the generation of a Goldstone boson, a mechanism for chiral SU(N) symmetry breaking and a possible source of T violation.

Polyakov [1] has recently pointed out that the Euclidean classical equations of motion of gauge theories have soliton-like solutions and has suggested that when properly included in the Euclidean functional integral they may have a bearing on the dynamics of confinement. The physical interpretation of these solutions has, however, been obscure since they are localized in time as well as space. In this letter we shall show that Euclidean gauge solitons describe events in which topologically distinct realizations of the gauge vacuum *tunnel* into one another and that this process radically changes the nature of the vacuum state. In fact, we find a continuum of vacua, each one of which is a superposition of the vacua with difinite topology and stable under the tunnelling process. The new vacua are the ground states of independent, and in general, inequivalent worlds (most striking, P and T are spontaneously violated in some of them!). When massless fermions are present, the vacuum tunnelling process forces a redefinition of the fermion vacuum as well and leads directly to spontaneous breakdown of chiral invariance without generating a "ninth" Goldstone boson.

* Research supported by the National Science Foundation under Grant Number MPS 75-22514.
* Research sponsored in part by the ERDA under Grant No. E(11-1)-2220.

We have, in effect, shown that the vacuum "seizes" as suggested by Kogut and Susskind [2], and identified the mechanism by which it does so. Our primary aim in this letter will be to give arguments for the existence of the new vacuum structure and to present the correct form of the functional integral appropriate to studying the properties of a particular vacuum. In the spirit of displaying qualitative consequence of the new vacuum structure we shall also briefly summarize results obtained from rather crude approximations to the functional integral.

To explore the structure of the vacuum we study the Euclidean functional integral

$$\langle 0|\exp(-Ht)|0\rangle \underset{t\to\infty}{\longrightarrow} \int [DA_\mu D\psi] \times \exp\{-\int d^dx[\mathcal{L}(A_\mu, \psi_i \ldots) + \mathcal{L}_{gf}]\} \qquad (1)$$

where d is the dimension of space time, $\mathcal{L}$ is the Langrange density of the theory, $\mathcal{L}_{gf}$ is a gauge-fixing term and the integration is to be done over all fields that approach vacuum values ($F_{\mu\nu} = 0$) at infinity. Now since $F_{\mu\nu} = 0$ implies $A_\mu = g^{-1}(x)\partial_\mu g(x)$ takes on values in the gauge group, G, any gauge field included in the functional integration defines a map of the sphere at Euclidean infinity into G. As pointed out by

Volume 63B, number 3 PHYSICS LETTERS 2 August 1976

Belavin et al. [3] these maps fall into homotopy classes corresponding to elements of the homotopy group, $\Pi_{d-1}(G)$. For most non-Abelian groups in four dimensional space time and U(1) in two dimensional spacetime, this homopoty group is Z. In these theories (the only ones we consider) the gauge fields integrated over in eq. (1) fall into discrete classes indexed by an integer ν running from $-\infty$ to $+\infty$ (we shall use the notation $[DA_\mu]_\nu$ to denote functional integration over the νth class). Thus there is actually a discrete infinity of funtional integrals and one must ask which, if any, is the "right" one.

One can clearly see what is going on by working in the gauge $A_0 = 0$ and requiring $F_{\mu\nu}$ to vanish outside a large, but finite, spacetime volume, V (this boundary condition is, of course, gauge invariant). The dynamical variables are now just the space components, A_i, for the vector potential and at large negative and positive times they must take on time independent vacuum values, $A_i(x) = g^{-1}(x)\partial_i g(x)$. The topological quantum number, ν, associated with any particular Euclidean gauge field time history may be written as a gauge invariant volume integral

$$\begin{aligned}\nu &= \frac{1}{8\pi^2}\int d^4x \, \mathrm{tr}\,(F_{\mu\nu}\tilde{F}^{\mu\nu}), \quad d=4\\ &= \frac{1}{4\pi}\int d^2x \, \epsilon_{\mu\nu}F^{\mu\nu} \quad , \quad d=2.\end{aligned} \tag{2}$$

In both cases, the integrand is a total divergence and $A_0 = 0$ gauge ν may be rewritten as a surface integral, $\nu = n(t=+\infty) - n(t=-\infty)$, where

$$\begin{aligned}n &= \frac{1}{6\pi^2}\,\epsilon_{ijk}\int d^3x \, \mathrm{tr}\,(A_i A_j A_k), \; d=4\\ &= \frac{1}{2\pi}\int dx A_1 \; , \qquad d=2.\end{aligned} \tag{3}$$

With no loss of generality (we have the freedom of making time independent gauge transformations) we may choose $n(t=-\infty)$ to be an integer. Then since ν is integral the gauge vacuum configuration at $t=-\infty$ must also have integral winding number $n(+\infty) = n(-\infty) + \nu$.

Therefore we must admit the existence of a discrete infinity of vacuum states, $|n\rangle$, labelled by a winding number taking on integral values from $-\infty$ to $+\infty$. The interpretation of the multiplicity of Euclidean functional integrals corresponding to different ν-classes, is then straightforward:

$$\begin{aligned}\langle n|\exp(-Ht)|m\rangle \underset{t\to\infty}{\longrightarrow} &\int [DA_\mu \cdots]_{(n-m)}\\ &\times \exp\{-\int d^dx[\mathcal{L}(A_\mu) + \mathcal{L}_{\mathrm{gf}}]\}\,.\end{aligned} \tag{4}$$

The functional integral over homotopy class ν describes a vacuum-to-vacuum transition in which the vacuum winding number changes by ν! Now the minimum action for $\nu = 0$ is zero (corresponding to $A_\mu \equiv 0$) so that in the WKB sense the $|n\rangle \to |n\rangle$ amplitude is O(1). In the $\nu \neq 0$ sectors the minimum action is in general non-zero – for $\nu = 1$ in four dimensions it corresponds to the Belavin et al. instanton [3], whose action is $8\pi^2/g^2$. Thus in the same WKB sense the $|n\rangle \to |n+1\rangle$ amplitude is $O(\exp(-8\pi^2/g^2))$. This is a typical "tunnelling" amplitude vanishing exponentially for small coupling and unseen by standard perturbation theory. Indeed, perturbation treatments of gauge theories expand about $A_\mu = 0$ and pretend that the vacuum $|n=0\rangle$ is true vacuum. Because of vacuum tunnelling, this is completely wrong and causes perturbation theory to miss qualitatively significant effects.

What then is the true vacuum? A convenient way of constructing it is to consider the generators of time independent gauge transformations characterized by a gauge function $\lambda^a(x)$:

$$Q_\lambda \; \int d^{d-1}x[F^a_{0i}D_i\lambda^a + g\,J^a_0\lambda^a]$$

where D_i is the covariant derivative and J^a_0 is the gauge source of fields other than the gauge field itself. In order to satisfy Gauss' law, $D_iF^a_{0i} = gJ^a_0$, it is sufficient to restrict the state space by $Q_\lambda|\psi\rangle = 0$ for all gauge functions λ which vanish at infinity. In particular, all our vacuum states $|n\rangle$ are annihilated by such local gauge transformations. There also exist gauge functions which do not vanish at infinity and generate gauge transformations, T, which change the vacuum topology. One can easily construct a unitary T effecting such a non-local gauge transformation: $T = \exp(iG_\infty)$, with $G_\infty \; (2\pi/g)[E(\infty) + E(-\infty)]$ for the two-dimensional abelian theory or $G_\infty = (2\pi/g)\int d^2S_iE^a_i\hat{x}^a$ for the four-dimensional non-Abelian theory, and T satisfies $T|n\rangle = |n+1\rangle$.

Since T is a gauge transformation, the hamiltonian

Volume 63B, number 3 PHYSICS LETTERS 2 August 1976

commutes with it and energy eigenstates must be T eigenstates. Since T is unitary, its eigenvalues are $e^{i\theta}$, $0 \leq \theta \leq 2\pi$, and the eigenstates are $|\theta\rangle = \Sigma e^{in\theta}|n\rangle$. This diagonalization of H is obviously unaffected by including in $\mathcal{L}$ sources coupled to gauge invariant densities. Thus, each $|\theta\rangle$ vacuum is the ground state of an independent and in general physically inequivalent sector within which we may study the propagation of gauge invariant disturbances. Since the different θ-worlds do not communicate with each other, there is no a-priori way of deciding which world is the right one. It is gratifying that this multiplicity of worlds is known to exist in the Schwinger model, corresponding there to different values of background electric field [4].

Finally, we must express the functional integral, eq. (4), in θ basis:

$$\langle\theta'|\exp(-Ht)|\theta\rangle \xrightarrow[t\to\infty]{} \delta(\theta-\theta')I(\theta)$$

$$I(\theta) = \sum_{\nu} \exp(-i\nu\theta)$$

$$\times \int [DA_{\mu}\, ...]_{\nu} \exp\{-\int d^d x [\mathcal{L}(A_{\mu}\, ...) + \mathcal{L}_{gf}]\}$$

$$\int [DA_{\mu} ...] \exp\{-\int d^d x [\mathcal{L}_{gf} + \mathcal{L}_{\theta}]\} \qquad (5)$$

where $\mathcal{L}_{\theta} = (i\theta/8\pi^2)\,\mathrm{tr}\,(F_{\mu\nu}\tilde{F}^{\mu\nu})$ for $d = 4$, $\mathcal{L}_{\theta} = (i\theta/4\pi)\,\epsilon_{\mu\nu}F_{\mu\nu}$ for $d = 2$ and in the second expression for $I(\theta)$ all gauge field topologies are summed over. $I(\theta)$ contains all possible information about physics in the θ-world and requires no further modification. The second form for $I(\theta)$ makes manifest one of the peculiar ways in which the θ-worlds differ from one another. In four dimensional pure Yang-Mills theory, re-expressed in Minkowski coordinates, the effective Lagrangian is $\mathrm{tr}\,[F_{\mu\nu}F^{\mu\nu} + (\theta/8\pi^2 F_{\mu\nu}\tilde{F}^{\mu\nu}]$. This clearly breaks P and T invariance (except for $\theta = 0$) and we must in general expect spontaneous breaking of space-time symmetries in all but a few special θ-worlds!

As a concrete illustration of these general remarks we should like to present an approximate evaluation of $I(\theta)$ in two-dimensional charged scalar electrodynamics. In the sector with $\nu = \pm 1$ the field configuration with minimum action is just the Nielson-Olesen vortex [5] in which there is a localized region of non-zero field of flux $\pm 2\pi/g$, radius μ^{-1} (μ is the heavy photon mass), arbitrary location and total action, S_0, proportional to μ^2/g^2. We shall construct the sectors with topological quantum number ν by superposing n_+ $\nu = +1$ vortices and n_- $\nu = -1$ vortices with $n_+ - n_- = \nu$, neglecting any interactions between vortices (since fields decrease exponentially this is not too bad for low vortex density, which turns out to mean small g). In this "dilute gas" approximation, the functional integral is

$$\langle\theta'|\exp(-Ht)|\theta\rangle \sim \delta(\theta'-\theta) \sum_{n_+,n_-=0}^{\infty} \exp\{-(n_+ + n_-)S_i\}$$

$$\times \frac{\exp\{i\theta(n_+ - n_-)\}}{n_+!\, n_-!} \left(\frac{V}{V_0}\right)^{n_+ + n_-} \qquad (6)$$

where the factors of V come from integrating over vortex locations and V_0 is a normalization factor which can be calculated from the quantum corrections to this basically semiclassical approximation. The sum is trivial and yields $\exp\,[2(V/V_0)e^{-S_0}\cos\theta)$. We have normalized the energy so that the naive perturbation theory vacuum energy is zero. By contrast, the θ vacua have an energy per unit volume equal to $-2V_0^{-1}\cos\theta\; e^{-S_0}$. Because $S_0 \propto 1/g^2$, this energy difference is a non-perturbative effect (a tunnelling effect) but potentially important nonetheless. Although the $\theta = 0$ vacuum has lowest energy (and no parity violation) we can't conclude that it is *the* vacuum since the other θ-vacua, though higher in energy are stable to gauge invariant perturbations. Having constructed a vacuum one can then calculate Green' functions of gauge invariant operators perturbatively. In the path integral this corresponds to performing ordinary perturbation theory about the appropriate classical solution for each topologically distinct sector and summing.

If we try the above sort of approximation on the non-abelian theory in four dimensions, there is a problem. The classical theory is scale invariant and the basic $\nu = 1$ solution (instanton) has an arbitrary scale parameter, λ, as well as an arbitrary position. The integration over λ need not diverge since scale invariance is broken by quantum corrections. Indeed, the renormalization group should tell us whether the integral converges at the short distance end. In the dilute gas approximation one finds for the vacuum energy density

$$\mathcal{E}_\theta = -2\cos\theta \int_0^\infty \lambda^3\, \mathrm{d}\lambda\, \exp\left[-\frac{8\pi^2}{g(\lambda/\mu)^2} + \mathcal{F}(\bar{g}(\lambda/\mu))\right] \tag{7}$$

where $\bar{g}(\lambda/\mu)$ is the usual effective coupling, normalized so that $\bar{g}(\lambda = \mu) = g$, μ is arbitrary, and $\mathcal{F}$ summarizing the effect of loop corrections, can be computed perturbatively. If the theory is asymptotically free and there are not too many quark multiplets, $\bar{g}$ can vanish rapidly enough for the integral to converge in the limit of large λ (small instanton size). This condition is met for any pure SU(N) gauge theory and for SU(3) with no more than *ten* flavors of quark.

On the other hand, in the limit of large instanton size, one is driven to large coupling (unless β has a small infrared fixed point) and the dilute gas approximation breaks down (instanton overlap and have long range interactions). Thus the attempt to construct the vacuum may run into an essential strong coupling problem because the quantum corrections to vacuum tunnelling will be large for large instanton size. In fact, there may not be a sensible way of perturbatively calculating even Green's functions of gauge invariant operators, no matter how small one makes g. This phenomenon is typical of a theory with no inherent mass scale which produces masses dynamically. If one sets the renormalization mass scale, μ, equal to some physical mass (e.g. $4\sqrt{\mathcal{E}_\theta}$), then g is determined (dimensional transmutation) and typically of order 1.

These problems should not, however, affect the standard applications of asymptotic freedom which rely on one's ability to compute operator product expansion coefficients at short distances. Precisely because of asymptotic freedom, vacuum tunnelling is suppressed at arbitrarily small scales and leading short distance behavior will agree with conventional calculations. There will, however, be calculated non-leading terms suppressed by powers of momentum, which reflect the mass scale set non-perturbatively by the tunnelling phenomenon.

The arguments presented above require some modification when massless fermions are present. We again confine non-zero $F_{\mu\nu}$ to a large but finite space-time volume, V, and again encounter a discrete infinity, $\{|n\rangle\}$, of vacuum states characterized by a vacuum gauge field with winding number n and a standard fermi vacuum with all negative energy states filled. In principle we must allow for transitions between vacuum, and evaluate $\langle n|\mathrm{e}^{-Ht}|m\rangle$ for general n and m. In fact, for massless quarks, $\langle n|\mathrm{e}^{-Ht}|m\rangle \propto \delta_{nm}$!

The reason for this is that, because of the anomaly, the *conserved* axial charge

$$Q_5 = \int \mathrm{d}\bar{x} J^5_c,$$

$$J^5_\mu = \sum_{\text{flavor,color}} \bar{\psi}\gamma_\mu\gamma_5\psi - \mathrm{tr}\left\{\frac{g^2N}{32\pi^2}\epsilon_{\mu\nu\lambda\sigma}A_\nu(\partial_\lambda A_\sigma + \tfrac{2}{3}A_\lambda A_\sigma)\right\} \tag{8}$$

while invariant under local gauge transformations, is not invariant under global gauge transformations. In particular, one has $TQ_5T^{-1} = Q_5 - 2N$ where T is the global gauge transformation, introduced earlier, which changes gauge field winding number by one unit and N is the number of flavors. If the vacuum states of different topology are defined by $|n\rangle = T^n|0\rangle$, with $Q_5|0\rangle = 0$ one finds that $Q_5|n\rangle = 2N\cdot n|n\rangle$. However, Q_5 is *conserved*, so that it must be true that $\langle n|\mathrm{e}^{-Ht}|m\rangle \propto \delta_{n,m}$. In general, we must find $\langle n|\mathrm{e}^{-Ht}D_\nu|m\rangle \propto \delta_{n-m,\nu}$ where D_ν is any operator of chirality $2N\nu$ (D_ν may stand for multiple insertions of local operators at different times – all that matters is net chirality). Therefore, we may replace eq. (4) by

$$\langle n|\mathrm{e}^{-Ht}|m\rangle \to \delta_{nm}\int[DA_\mu\ldots]_0 \exp\{-\int \mathrm{d}^dx[\mathcal{L}(A_\mu\ldots) + \mathcal{L}_{\mathrm{gf}}]\}$$

and

$$\langle n|\mathrm{e}^{-Ht}D_\nu|m\rangle \to \delta_{n+\nu,m}\int[DA_\mu\ldots]_\nu D_\nu \times \exp\{-\int \mathrm{d}^dx[\mathcal{L}(A_\mu\ldots) + \mathcal{L}_{\mathrm{gf}}]\} \tag{10}$$

with the same meaning still attached to D_ν. The restriction on the topology of the gauge field histories would actually have emerged directly from a mindless application of eq. (4): Doing the fermion integrations for fixed A_μ yields $[\det(\partial\!\!\!/ - A\!\!\!/)]^{+1}$. This determinant vanishes whenever $(\partial\!\!\!/ - A\!\!\!/)$ has a zero eigenvalue. 't Hooft [6] has noted that if A_μ is taken equal to the $\nu = +1$ or $\nu = -1$ instanton there is a zero eigenvalue, and our argument is just telling us that whenever A_μ belongs to a $\nu \neq 0$ class, $(\partial\!\!\!/ - A\!\!\!/)$ has a zero eigenvalue, eliminating the $\nu \neq 0$ sectors from the integration.

Though vacuum tunnelling is now suppressed, the

$|n\rangle$ vacua are not acceptable because they violate cluster decomposition for operators of non-zero chirality. Consider an operator D of chirality $2N$. The arguments of the preceeding paragraph show that $\langle n|D|n\rangle = 0$, $\langle n+1|D|n\rangle \neq 0$. Then $\langle n|D^+(x)D(y)|n\rangle$ will not vanish for large $|x-y|$ as required by cluster decomposition and the vanishing of the "vacuum" expectation $\langle n|D|n\rangle$: it obviously approached $\langle n|D^+|n+1\rangle\langle n+1|D|n\rangle$. The solution to this problem is obvious (it was solved in the Schwinger model years ago!): The proper vacuum states are the $|\theta\rangle$ vacua, in which basis the functional integrals have the form

$$\langle\theta'|e^{-Ht}|\theta\rangle \rightarrow \delta(\theta'-\theta)\int[DA_\mu\cdots]_{(0)}$$
$$\times \exp\{-\int d^dx[\mathcal{L}(A_\mu\cdots)+\mathcal{L}_{gf}]\}$$
$$\langle\theta'|e^{-Ht}D_\nu|\theta\rangle \rightarrow \delta(\theta'-\theta)\int[DA_\mu\cdots]_{(\nu)}$$
$$\times \exp\{-\int d^dx[\mathcal{L}(A_\mu\cdots)+\mathcal{L}_{gf}]\}D_\nu\,. \qquad (11)$$

The cluster problem is resolved by the non-vanishing vacuum expectation value of D in the true vacuum state. The fact that only one topological class of gauge field history contributes to each functional integral makes physical quantities have a trivial dependence on θ: The vacuum energy density, while non-zero, is independent of θ. The variation of the vacuum energy with respect to θ is just $\langle \mathrm{tr}\, F_{\mu\nu}\tilde{F}^{\mu\nu}\rangle$, the quantity whose non-zero value is the signal for P and T violation. In the massless fermion case, P and T appear not to be spontaneously violated and, indeed, all the $|\theta\rangle$ vacua are physically equivalent. Finally, the axial baryon number invariance of the original Lagrangian is violated by a vacuum expectation value of operators with non-zero chirality. It should be said that at this stage we have in the N-flavor case, only broken axial U(1) and *not* axial SU(N). Actually, since the axial charge rotates θ, a discrete subgroup of order $2N$ of U(1) is left unbroken, consisting of those elements which rotate θ by a multiple of 2π. There is no associated Goldstone boson because the conserved, but gauge-variant, U(1) charge takes one out of a given $|\theta\rangle$ sector ($e^{i\alpha Q_5}|\theta\rangle = |\theta+2N\alpha\rangle$) while $\mathrm{tr}(F\tilde{F})$, the divergence of the gauge invariant axial current, has non-vanishing matrix elements. That Q_5 causes transitions between different vacua is characteristic of the "vacuum seizing" mechanism postulated by Kogut and Susskind [2] while the non-vanishing of $F\tilde{F}$ in instanton solutions as a possible escape from the U(1) problem was noted by G. 't Hooft [7].

Although the presence of zero mass fermions suppresses vacuum tunnelling in the strict asymptotic sense, tunnelling does have a profound effect on the vacuum energy and other physically relevant quantities. When the vacuum tunnels, fermion pairs are produced. Although the pair must ultimately be absorbed by an anti-tunnelling, since the fermions are massless the pair may live for a long time and tunnelling occurs freely in intermediate states. To get some notion of what goes on it is instructive to attempt a crude calculation of the basic functional integrals of eq. (11) in the case of a single flavor.

We shall assume that the integral over A_μ is dominated by configurations of widely separated instantons (n_+ in number) and anti-instantons (n_- in number). To compute the vacuum energy we must set $n_+ = n_-$ (configurations with $\nu = 0$), sum over n_+ and integrate over instanton locations. We will ignore the integration over instanton sizes. For a given gauge field configuration, the integration over fermi fields yields $\det(\not\partial - \not A)$. This determinant must also be approximated.

Now, as 't Hooft has pointed out [7], individual instantons have a zero-energy eigenfunction $\psi_0^{\pm}(x, x_\pm)$ ($x_\pm$ is the instanton location and the $\pm$ label distinguishes instanton from anti-instanton). Since the interesting physical effects arise precisely from these zero energy solutions, we shall compute the determinant of $(\not\partial - \not A)$ in the subspace spanned by the $2n_+$ functions $\psi_0^+(x, x_i^+)$, $\psi_0^-(x, x_i^-)$. In the widely separated instanton approximation, these functions are orthonormal and one has to compute the determinant of the matrix

$$M_{ij} = (\psi_{0,i}^+|(\not\partial - \not A)|\psi_{0,j}^-).$$

In this approximation A_μ differs from a gauge transformation only in the neighborhood of each instanton and M_{ij} can be approximated by

$$M_{ij} \approx (\phi_{0,i}^+|(-\not\partial)^{-1}|\phi_{0,j}^-)$$

where $\phi_0 = \not\partial\psi_0$. The γ_5 structure of the ψ_0 forbids i and j to be both instantons or both anti-instantons. The ψ_0 fall off at large distances exactly like the free fermi propagator, which is why we introduce the ϕ_0's which are localized as well as the instanton itself. The cycle expansion of the determinant then gives a sum of terms with a graphical interpretation in terms of closed fermion loops. For instance, if $n_+ = n_- = 1$ we get

$$\mathrm{Det} \approx (\phi_0^+|(\not{\partial})^{-1}|\phi_0^-)(\phi_0^-|(\not{\partial})^{-1}|\phi_0^+)\,,$$

which has the obvious interpretation of massless quark propagators connecting non-local vertices, $V^\pm = \phi_{0\pm}^*(x)\phi_{0\pm}(x')$, associated with the instantions. The functional integral weights each vertex with a factor proportional to $e^{-S_{\mathrm{cl}}}$, where S_{cl} is the instanton classical action. The γ_5 structure of $\phi_0^\pm$ is such that $V^\pm$ is like $\bar{\psi}(1 \pm \gamma_5)\psi$ in its Dirac matrix structure, which is to say that $V^\pm$ looks like a non-local, or momentum-dependent mass term. Summing over numbers and locations of instantons simply completes the vacuum fermion loop analogy by providing all possible insertions of the pseudo mass terms, $V^\pm$, on massless quark loops. Then calculations of physical quantities proceed in a perfectly conventional way so long as we remember to add the mass term $\sim e^{-S_{\mathrm{cl}}}(V_+ + V_-)$ to the massless quark propagator. Anything which directly depends on the mass term, such as the vacuum expectation of $\bar{\psi}\psi$, will be proportional to $e^{-S_{\mathrm{cl}}}$ and will have only the by now familiar dependence on g characteristic of a tunnelling process. In the "scale invariant" four dimensional theory $\langle\bar{\psi}\psi\rangle \neq 0$ implies spontaneous generation of mass and dimensional transmutation as before.

If the fermion is given a bare mass tunnelling is allowed and one is driven directly to a θ vacuum (whose energy depends on θ). The limit of zero mass is smooth. If the bare mass is small compared to the spontaneously generated mass, it acts as a small perturbation on the $m_0 = 0$ theory.

If there are N flavors, the above discussion is modified in an important way: the effective instanton-quark interaction is no longer billinear, but $2N$-linear. Indeed, the instanton (anti-instanton) vertex has the structure $V_+ = \Pi_{i=1}^N \bar{\psi}_i(1+\gamma_5)\psi_i$ ($V_- = \Pi_{i=1}^N \bar{\psi}_i(1-\gamma_5)\psi_i$). As a result, summing over instantons in the dilute gas approximation will not just produce a mass term in the quark propagator, but does something more complicated. To produce a quark mass term, one must break the global chiral SU(N), while we have argued that the vacuum tunnelling phenomenon is only guaranteed to violate the chiral U(1) symmetry. On the other hand, the effective interactions between quarks of different helicity generated by the instanton provide new ways of identifying sums of graphs which can lead to the desired symmetry breakdown. We have constructed a simple Hartree–Fock type argument for $N = 2$ which has a chance of being correct in a weak coupling theory and which seems, on superificial examination, to generate quark masses. The inevitable Goldstone bosons arise in this case from iterated bubble graphs generated by the four-fermion interactions, $V_\pm$. We do not wish to make too much of these crude arguments other than to suggest that the new interactions generated by vacuum tunnelling are likely to play a key role in generation of quark masses and Goldstone bosons.

In terms of the picture presented here Polyakov's ideas about confinement appear as follows. For an isolated quark located at x the tunnelling amplitude $|n, x\rangle \to |n+1, x\rangle$ will be reduced relative to the vacuum to vacuum amplitude. A quark state will then have more energy than the vacuum, as it should. In the dilute gas approximation the energy difference is proportional to the integral over all instantons which overlap the quark. Integrating over instanton locations and then over the scale size λ^{-1} leads to an integral which tends to diverge at small λ. For the large instantons (small λ), however, the dilute gas approximation is not valid, and one is again confronted with a strong coupling problem.

Obviously, much remains to be done to fully exploit the phenomena we have found. The major difficulty, of course, is that the theory we are really interested in, quantum chromodynamics, is basically a strong coupling theory and reliable calculations are difficult, if not impossible. However, one may hope that a new understanding of the qualitative physics will suggest new methods of calculation. We are especially encouraged by the appearance, already in semiclassical approximations, of a vacuum that breaks chiral symmetry and sets a dynamical mass scale. We are also intrigued by the natural appearance of spontaneous violation of P and T invariance but have so far not seen how to understand why these effects are small in the real world or how to exploit them to explain observed violations of these symmetries. Perhaps super-unified theories will shed some light on these questions.

One of us (D.J. Gross) would like to acknowledge V.N. Gribov for stimulating conversations and in particular for the suggestion that the Euclidean solitons might be relevant to the structure of the vacuum.

References

[1] A.M. Polyakov, Phys. Lett. 59B (1975) 82.
[2] J. Kogut and L. Susskind, Phys. Rev. D11 (1975) 3594.
[3] A.A. Belavin, A.M. Polyakov, A.S. Schwartz and Yu. S. Tyupkin, Phys. Lett. 59B (1975) 85.
[4] J. Lowenstein and A. Swieca, Ann. Phys. 68 (1971) 172. S. Coleman, Harvard preprint (1975).
[5] H. Nielsen and P. Olesen, Nucl. Phys. B61 (1973) 45.
[6] T.S. Bell and R. Jackiw, Nuovo Cimento 60 (1969) 47; S. Adler, Phys. Rev. 177 (1969) 2426.
[7] G. 't Hooft, Harvard preprint (1976).

PHYSICAL REVIEW D VOLUME 17, NUMBER 10 15 MAY 1978

Toward a theory of the strong interactions

Curtis G. Callan, Jr.* and Roger Dashen
Institute for Advanced Study, Princeton, New Jersey 08540

David J. Gross†
Joseph Henry Laboratories, Princeton University, Princeton, New Jersey 08540
(Received 26 August 1977)

A systematic study is made of the relevant degrees of freedom and the dynamics of quantum chromodynamics (QCD). We find that the dynamical properties of QCD are, to a large extent, a consequence of the structure of the vacuum arising from the tunneling between degenerate, classically stable, vacuums, and that the relevant degrees of freedom can be taken to be the Euclidean path histories that can be used to calculate the tunneling in the semiclassical approximation. This nonperturbative vacuum structure appears well suited to the major features of QCD, i.e., the dimensional transmutation that determines the size of the hadrons and the strong-interaction coupling constant, the source of dynamical chiral symmetry breaking, and the mechanism responsible for quark confinement.

I. INTRODUCTION

It is widely believed that the strong interactions are generated by a non-Abelian [SU(3)] gauge theory of quarks and gluons, permanently confined in color-singlet hadronic bound states. This theory is called quantum chromodynamics (QCD). It can be described by the Lagrangian density

$$\mathcal{L} = -\tfrac{1}{4}F^a_{\mu\nu}F_a^{\mu\nu} + \prod_{i=1}^{F} \bar{\psi}_i(i\not{D} - m_i)\psi_i\,, \qquad (1.1)$$

where

$$F^a_{\mu\nu} = \partial_\mu A^a_\nu - \partial_\nu A^a_\mu + g f_{abc} A^b_\mu A^c_\nu\,,$$

A^a_μ is an SU(3) gauge field, and ψ_i are quark fields with the index i labeling the various quark types, or flavors. The theory is thus parametrized by the one coupling constant g and the values of the quark mass parameters m_i. The total number of quark flavors is so far unknown. In addition to the established up, down ($m_u = m_d \sim 10-100$ MeV), strange ($m_s \simeq 100-300$ MeV), and charmed ($m_c = 1.3-1.4$ GeV) quarks, there might very well exist many heavier quarks with new quantum numbers. Fortunately this is of little relevance to the bulk of hadronic physics, although it is of fundamental importance in understanding the structure of the weak interactions. The properties of light hadrons will not be affected by such heavy quarks. Charmed quarks can be neglected to a very good approximation in describing the properties of noncharmed hadrons.

Our knowledge of the nature of the constituents of hadrons and their interactions derives from their symmetries, the success of various phenomenological quark models, and, most importantly, the observed short-distance behavior of hadronic currents. The success of the SU(3)-symmetry scheme and simple nonrelativistic quark models leads to the picture of hadrons as bound states of spin-$\frac{1}{2}$ triplet colored quarks,[1,2] The absence of colored states leads to the hypothesis of confinement–namely that the only physical states are color singlets.

That the strong interactions are mediated by vector mesons coupled to color is strongly indicated if one is to explain confinement. Even proponents of dynamical schemes outside the framework of QCD invoke colored gauge vector mesons to explain why quark-antiquark bound states occur whereas quark-quark states do not. The observation of scaling in the deep-inelastic scattering of leptons off hadrons singled out non-Abelian gauge theories as the only ones capable of possessing the asymptotic freedom necessary to produce such free-field-like behavior at short distances.[3] In addition these experiments, as well as e^+e^- annihilation to hadrons, allow us to observe directly the quantum numbers of quarks and to derive qualitative predictions that test the validity of the theory.[4,5]

Although one cannot claim that the precise predictions of the theory as to the short-distance structure of hadrons have been experimentally confirmed, the qualitative picture is in remarkable agreement with the data. The asymptotic freedom of QCD has the enormous benefit of allowing one to control the short-distance behavior of the theory. Thus all properties of the theory are calculable in terms of an effective coupling which can be made arbitrarily small by going to short enough distances. As one goes to larger distances, the effective coupling increases, leading to the hypothesis that the increasing coupling in the infrared domain leads to quark confinement–infrared slavery.[4,6]

However, we are still far from possessing a quantitative theory of hadrons in which we could calculate their masses, couplings, and scattering amplitudes. We do not even have a qualitative understanding of the dynamical mechanism for quark confinement. In this paper we shall propose a new approach to QCD, based on an improved understanding of the nature of the QCD vacuum, which might ultimately lead to a quantitative theory of strong interactions.

Before describing the nature of our program we would like to focus on a few of the most difficult and important dynamical problems which must be faced in any serious attempt to solve QCD.

A. Dimensional transmutation

QCD possesses few adjustable parameters. Indeed we shall argue that to a good approximation it has no adjustable parameters. As discussed above the free paramters in the QCD Lagrangian are the coupling g and the various masses, m_i, of the quarks. If we restrict attention to noncharmed hadrons we need three flavors of quark. We should obtain an excellent (10 – 20%) approximation to the real world by setting $m_u = m_d = m_s = 0$. This chiral SU(3)×SU(3)-symmetric approximation to the world is quite reasonable as is evident from the small value of the pion mass. Even if we wish to include the effects of the explicit chiral and SU(3)-symmetry breaking generated by nonvanishing quark masses we do not believe that they would have an important effect on the dynamics and could be treated perturbatively.

Thus, to a good approximation, the theory is described by a single dimensionless parameter g, and contains no relevant dimensional parameter to set the scale of masses. In such a scale-invariant theory the parameter that sets the scale of all dimensional quantities is the renormalization scale parameter μ. This is the arbitrary parameter (with dimension of mass) that is introduced to get the length scale at which the normalization of the quantum fields and the coupling g are defined. Thus all Green's functions will depend on g and μ, although physically measurable quantities $P(g, \mu)$ must depend only on a combination of g and μ invariant under the renormalization, since a change in μ merely indicates a change in what one means by g. Thus

$$\left(\mu\frac{\partial}{\partial\mu}+\beta(g)\frac{\partial}{\partial g}\right)P(g,\mu)=0\,. \tag{1.2}$$

In particular, masses of physical particles, or any physical parameter with dimensions of masses is given by

$$m(g,\mu)=\mu\exp\left(-\int^{g}\frac{dx}{\beta(x)}\right), \tag{1.3}$$

and all dimensionless physical parameters must be independent of μ and therefore calculable numbers independent of g.[7]

Consequently, except for the overall mass scale of the theory, there are no adjustable parameters. Given some definite renormalization prescription (or definition of g) QCD will be characterized by an effective coupling $\bar{g}(p)$ for a given range of momenta, where

$$p\frac{d\bar{g}}{d(p)}=+\beta(\bar{g})\,.$$

The value of this coupling for momenta of the order of hadronic masses is a calculable number, $\bar{g}(m_H)$.

This phenomenon of dimensional transmutation has serious implications. It means that ordinary perturbation theory can be of little value in any attempt to calculate the properties of hadrons. This is because Eq. (1.3) implies that hadronic masses behave as

$$m(g,\mu)\sim\mu e^{-b_0/g^2}$$

for small g. Ordinary perturbation theory, or even fancy resummation techniques, will not produce masses that have zero asymptotic expansions in powers of g. Even to construct the vacuum state of QCD, nonperturbative techniques will be required.

One of our major tasks will therefore be to identify nonperturbative mechanisms which could set the mass or length scale of hadronic states, and to attempt to evaluate the magnitude of the hadronic coupling at lengths comparable to the hadronic size.

B. Dynamical symmetry breaking

Owing to the remarkable success of approximate chiral SU(3) symmetry and partial conservation of axial-vector current (PCAC), we believe that the strong interactions possess an almost exact chiral symmetry which is realized in the Goldstone mode. Thus to solve QCD we must not only understand the confinement mechanism but also we must construct the true, chirally asymmetric, ground state. In a theory such as QCD, which does not contain elementary scalar meson fields, the chiral symmetry must be broken dynamically. Thus the fields that acquire nonvanishing expectation values in the true vacuum will be composite (i.e., $\bar{\psi}\psi$), the Goldstone bosons will be quark-antiquark bound states, and the dynamically produced masses will obey Eq. (1.3).

The problem of dynamical symmetry breaking is very difficult, particularly in a gauge theory.

Until now little progress has been made in identifying the mechanism that generates the symmetry breaking, and most attempts to understand confinement have sidestepped the problem by explicitly breaking the chiral symmetry. It is clear that to obtain an accurate description of hadrons it will be necessary to solve this problem. In addition it might very well be the case that the problems of confinement and symmetry breaking are interrelated. In fact, as we shall argue below, the mechanism that we envisage for confinement requires that the quarks acquire a mass dynamically, through a mechanism that we shall identify. The properties of massless quarks might very well differ substantially from those of massive quarks.

C. The U(1) problem

The U(1) problem arises in any quark model which does not contain fundamental scalar fields. It arises due to the fact that in addition to the desirable SU(N) chiral symmetry that obtains for zero quark masses, there also exists a chiral U(1) symmetry generated by the axial baryon number current.[8] If the only explicit breaking of this symmetry is due to the nonvanishing of the quark masses then the dynamical symmetry breaking of chiral U(N) would be expected to generate an isoscalar Goldstone boson with a mass comparable to that of the isovector pion. Weinberg has given plausible arguments which show that in the absence of spontaneous symmetry breaking of SU(3) this would lead to an η particle whose mass would be bounded by $\sqrt{3}\, m_\pi$, in blatant contradiction with experiment.[9]

Consequently any attempt to derive the properties of hadrons and to understand the mechanism of dynamical symmetry breaking in a quark-gluon gauge theory must contain a solution to the U(1) problem or face disaster.

D. Confinement

The central problem in QCD is to understand the mechanism that confines quarks and gluons in color-singlet hadronic bound states. An understanding of this mechanism should then allow one to calculate the properties of the hadronic bound states. At first, however, one wants a simple criterion for confinement. Such a criterion is provided by considering a pure gauge theory (no quarks) and evaluating in such a theory the energy $E(R)$ of a singlet state consisting of two colored external sources separated by a distance R.[10] These sources can be regarded as infinitely massive quarks. A necessary condition for confinement is that $E(R)$ grows with R for large enough separation, so that colored states cannot be produced with a finite expenditure of energy. The advantage of this criterion is that it can be addressed in a gauge-invariant fashion and investigated by means of Euclidean path integrals. Of course, in order to calculate the properties of the hadronic bound states one must contend with real, light, quarks. However, $E(R)$ might be of some physical relevance in treating the low-lying bound states of very massive quarks (e.g. charmonium).

How is one to make progress toward a dynamical understanding of these problems within the framework of QCD? It is clear that a straightforward perturbation theory is useless. Because of infrared slavery, there is no way in which one can ensure a small coupling for low-momentum states. Furthermore, owing to dimensional transmutation and dynamical symmetry breaking, one expects the theory to contain terms which have no asymptotic expansion in powers of g. Such terms, of the form $g^{-p}\exp(-\text{const}/g^2)$, can be large even if g is very small. Of course they will never show up to any order in perturbation theory. It is therefore extremely unlikely that any approach based on perturbation theory, even one that utilizes summation techniques to sum divergent asymptotic series, will be useful in solving QCD.

The most ambitious attempts to calculate within the framework of QCD to date have utilized lattice formulations of the theory.[11] The advantages of this approach are many. By introducing a space-time lattice (or spatial lattice in the Hamiltonian approach) one renders finite the number of degrees of freedom in a finite volume and introduces extra parameters (cutoffs) that can be varied. In addition, lattice QCD has an extremely simple strong-coupling limit which exhibits confinement and many of the qualitative features of hadrons. The systematic approach to lattice QCD envisages the utilization of renormalization-group techniques to proceed from the known dynamics at short distances of the order of the lattice spacing to an effective Hamiltonian appropriate for large (hadronic) distances. At these distances one will presumably be in the strong-coupling regime where other techniques are available (i.e., strong-coupling or high-temperature expansions).

However, the introduction of a space-time lattice has many severe disadvantages. In addition to destroying manifest Lorentz invariance the lattice approach has difficulty in accommodating fermions without explicitly breaking chiral symmetry. Whereas these unphysical features might disappear in the continuum limit, they produce problems at any finite stage of a lattice theory calculation.

In an asymptotically free theory such as QCD

there is no need for an ultraviolet cutoff of the type introduced in lattice approximations since the short-distance structure of the theory is completely calculable and under control. Thus it might prove profitable to construct other approximations to the theory which, similar to the lattice models, reduce the number of degrees of freedom but do not spoil the short-distance behavior of the theory or introduce new parameters. Our approach to QCD is an attempt to systematically explore the relevant degrees of freedom, starting from short distances. What we have found is that as one proceeds from short distances physically significant effects are generated by instantons.

Historically the interest in instantons arose because of the discovery of an exact finite-action solution to the classical Yang-Mills equations in Euclidean space-time,[12] and the realization that the existence of such finite-action field configurations indicates that the structure of the vacuum in QCD is much more complicated than one would have surmised from straightforward perturbation theory.[13-15] Thus the classical ground state of QCD is infinitely degenerate and the true quantum-mechanical vacuum is a coherent superposition of these classically degenerate vacuums. For sufficiently weak coupling the true vacuum can be constructed by semiclassical techniques, where the role of multiple-instanton field configurations is to give the dominant contribution in summing over path histories that travel from one classical vacuum to another.

In a scale-invariant theory, such as QCD, there is no way one can adjust the coupling to be small for all distances. Indeed, as we have remarked above QCD has no relevant adjustable parameters. Thus one cannot use semiclassical or weak-coupling methods to determine the structure of the vacuum, which involves all scale lengths. However, all physically relevant questions involve some external length parameter or momentum which sets the scale of the field configurations. Our approach is to explore physical quantities, such as the "potential" between massive quarks $[E(R)]$, characterized by a scale length R, as a function of R. For small enough R, asymptotic freedom will ensure that the effective coupling will be sufficiently small that one can use semiclassical techniques (saddle-point approximations) to evaluate functional integrals. The net effect in this small-distance region, as we shall explain below, is that the path integrals can be replaced by the partition function of a "gas" of instantons characterized by their position, scale size, and SU(3) orientation. The density of instantons of size ρ will be given by

$$D(\rho)\sim\frac{1}{\bar{g}^{12}(\rho)}\exp\left(-\frac{\text{const}}{\bar{g}^{2}(\rho)}\right),$$

where $\bar{g}^2(\rho)$ is the effective coupling at distance ρ. For small R only instantons of size $\rho\lesssim R$ will matter and since $\bar{g}^2(\rho)\sim\text{const}/\ln(1/\rho)$ as $\rho\to 0$, for small R the analog gas will be sufficiently dilute that one can trust a virial expansion of the partition function in powers of D. As one increases R, the effective size of the relevant instantons increases and the effective coupling increases.

Thus there are two sources of corrections to the free-field asymptotic behavior that occurs for $R\approx 0$. First there are the ordinary perturbation theory corrections to the integration about a given saddle point (including perturbation theory about the ordinary vacuum). These can in principle be calculated by standard perturbation theory (summing Feynman diagrams) and are of order $\bar{g}^2(R)$. In addition, there are the nonperturbative effects due to tunneling, proportional to the density of instantons, $D(R)$. Now one would think that the quantum-mechanical corrections to the ordinary vacuum sector would be much greater than the tunneling effects for small $\bar{g}^2$. This is the case in ordinary quantum mechanics, where tunneling can be neglected for small coupling ($g^2\gg e^{-1/g^2}$), and while for large coupling the tunneling effects may become substantial, they cannot be calculated by semiclassical techniques. In QCD we find, however, that this is not the case. Owing to the large phase space available to instantons, the density of instantons, $D(R)$, becomes large (≈ 1) for distances at which the coupling $\bar{g}^2/8\pi^2$ is still very small ($\approx\frac{1}{10}$). Essentially there exist so many distinct paths to tunnel between the degenerate vacuums, that even though the individual tunneling amplitudes are small, the net amplitude for tunneling can be of order one even for very small coupling. Thus as one proceeds from small distances one first arrives at a region where instanton effects are substantial, yet reliably calculable using the dilute-gas approximation and ordinary perturbative corrections are small.

As one goes to larger distances the density of instantons rises rapidly. One must then take into account the interaction between instantons and anti-instantons. In this highly nonlinear gas the interactions are quite complicated, and at present one can only make semi-quantitative estimates of their effects. Crudely speaking we find that the gas is analogous to a paramagnetic medium of dipolar objects. In this medium there is an effective renormalization of the coupling constant associated with large instantons due to the screening caused by smaller instantons. The result is to cause the coupling $\bar{g}(\rho)$ to grow dramatically with distance, leading to the density rising rapidly to large values at a sharply defined distance, $\rho_c \approx 0.2\mu$.

It is this distance that we associate with confinement. The actual confinement mechanism requires the consideration of field configurations other than instantons. In particular, instantons of scale comparable to ρ_c have a tendency to dissociate into pairs of half-instantons – or "merons." These merons have logarithmic attractive interactions, proportional to $[1/\bar{g}^2(R)]\ln R$, where R is the separation of a meron pair. However, for large enough R the entropy (log of the volume in function space) of a meron pair is proportional to $\ln R$ and can dominate – leading to a phase transition to a plasma of merons. These liberated merons, we shall argue, will confine quarks. To be sure we are still unable to treat the confinement phase very precisely or calculate hadronic masses. However, it seems inevitable that if our mechanism does lead to confinement the size of hadrons will be of order ρ_c and the hadronic mass scale will be of order $1/\rho_c$. Thus the expression of dimensional transmutation in QCD is that the renormalization-group-invariant equation which determines the scale size of hadrons is $D(\bar{g}(\rho_c)) \simeq 1$.

Now one of the consequences of dimensional transmutation is to fix the size of the hadronic coupling constant – i.e., the effective coupling at distances corresponding to hadronic size. This is of course the coupling that is of physical interest. The effective coupling for larger distances is of little interest since as one pulls quarks farther apart than the size of a hadron it will be energetically favorable to produce a quark-antiquark pair from the vacuum, and the stretched hadron will split into two smaller hadrons. Now the phase transition to a meron plasma already occurs for small $\bar{g}^2/8\pi^2$ and it is conceivable (although we certainly are unable to prove this) that confinement will occur on this distance scale and one will never need to go to larger distances where $g^2/8\pi^2$ becomes substantial. We therefore conjecture that *the hadronic coupling is always small.*

Thus we have discovered a small parameter in QCD. This has major consequences. First, the small size of $\bar{g}^2/8\pi^2$ ensures that we can use semiclassical techniques to calculate the properties of hadrons, without having to include quantum corrections beyond one or two orders. This reduces QCD to a semiclassical problem, albeit an extremely difficult one. Second, a small hadronic coupling has many phenomenological consequences, and may shed light on some of the unresolved mysteries of hadronic physics. For example, it is this coupling that determines, in our approach, the probability of producing a quark antiquark pair from the vacuum. The small magnitude of this probability may explain much of the success of the naive quark and parton models – namely that the hadrons fall into SU(6) multiplets as if they were made of valence quarks alone, that the hadrons as seen in deep-inelastic scattering contain very few quark-antiquark pairs (for $x \neq 0$), the success of the free-field theory Melosh transformation, the rapid approach to scaling, etc. We emphasize that the small value of $\bar{g}^2(\rho_c\mu)/8\pi^2$ does not mean that the strong interactions are weak since nonperturbative instanton and meron effects are large.

Until now we have ignored the problem of generating the dynamical symmetry breaking of chiral symmetry and the U(1) problem. These problems cannot be separated from the problem of confinement. As was originally pointed out by 't Hooft[13] the existence of massless fermions has dramatic effects on instantons. Because of the Adler-Bell-Jackiw anomaly,[16] the conserved axial baryon number, Q_5, is not invariant under gauge transformations. This results in the suppression of tunneling between the classically degenerate vacuums, since they now are eigenstates of Q_5 with different eigenvalues. One must still construct the vacuum as a coherent superposition of the degenerate classical vacuums to satisfy cluster decomposition; however, the only path histories that can now contribute have net topological number zero – i.e., contain an equal number of instantons and anti-instantons.

This phenomenon has two important effects. First, the U(1) problem is eliminated. The axial baryon number charge is no longer a symmetry of the theory constructed about the true vacuum.[13-15] Second, if we try to replace path integrals by a partition function for a gas of instantons we will find strong long-range attractive interactions between instantons and anti-instantons (with a logarithmic dependence on their separation). Consequently in the presence of massless or very light fermions, instantons and antiinstantons will be closely bound in pairs. Such field configurations differ little from the ordinary vacuum and do not have much effect on the dynamics. In short, massless fermions confine instantons, and unless the fermions acquire a mass through dynamical chiral-symmetry breaking, tunneling effects are negligible.

Now we have suggested that the structure of the θ vacuum in QCD is such that there is a natural mechanism for dynamical symmetry breaking,[14] namely the effective determinantal interaction for massless fermions in a θ vacuum. This interaction can be regarded as the source of the U(1) symmetry breaking. Such an interaction, which is of the form $(\bar{\psi}\psi)^2$, if there are but two massless fermions, does not break chiral SU(N). However, it can provide a mechanism for the dynamical breaking of chiral symmetry. Indeed the only known mechanism for dynamical chiral-symmetry breaking utilizes such $(\bar{\psi}\psi)^2$ interactions.[17,7] In a two-dimensional model with two massless fermions we have shown that this type of interaction, generated by instantons, leads to the Goldstone realization of chiral symmetry and gives the fermion a mass.[18]

In the case of QCD the problem is of course much more difficult. To explore the possibility that the true ground state breaks chiral symmetry and to construct this state as well as the quark "mass" requires controlling the physics over arbitrarily large scale sizes. However, we have found it possible, in lieu of this, to demonstrate that the chirally symmetric vacuum is unstable under perturbations that would shift the vacuum expectation value of $\sigma = \bar{\psi}\psi$. To see the instability we calculate the propagator of σ for large momentum. For large enough momentum of course, asymptotic freedom determines this propagator perturbatively in powers of $\bar{g}^2(p)$. As the momentum decreases instanton effects (generated by the effective determinantal interaction) come into play. We find that these are very large for a range of momenta where the instanton gas is still dilute and the effective coupling is still small. In fact they are so large as to generate a tachyon pole in the σ propagator. This tachyon indicates the instability of the θ vacuum under shifts in σ. To be sure we are yet unable to construct the true, chirally asymmetric vacuum state. However, unless the theory is total nonsense, such a ground state will exist. Our calculation indicates that dynamical chiral-symmetry breaking does occur via the mechanism that we have suggested, and that it occurs at rather short distances (compared with what we regard as the confinement scale). This mechanism for chiral-symmetry breaking will not suffer from the U(1) problem. The determinantal interaction which is attractive in the π channel and thus will produce, in the true vacuum, a zero mass pion is repulsive in the η' channel.

The occurence of spontaneous symmetry breaking of chiral symmetry at distances short compared to the confinement scale is crucial to the success of our program. It ensures that as the quarks are pulled apart they acquire a mass at some distance where the dynamics is still manageable. Once this occurs instantons are liberated and, as we proceed to larger distances, begin interacting strongly to produce confinement. Thus, as a function of distance, there are two "phase transitions." At very short distances the vacuum can be described by a gas of tightly bound instanton-anti-instanton pairs, which at distance ρ_A (the asymptotically free chirally symmetric phase) undergo a phase transition to a dilute gas of instantons (the chirally asymmetric phase) due to chiral symmetry breaking. At a somewhat larger distance, ρ_c, the instantons themselves dissociate into meron pairs (the confining phase).

This paper is structured as follows. In Sec. II we present a general introduction to instanton physics. We show that four-dimensional gauge theories possess an infinity of classical vacuum states with a finite-energy barrier separating them, and discuss how instantons can be used to construct the amplitude for tunneling between these states in the semiclassical approximation. The analog gas model of instantons is constructed and the methods used to calculate in the dilute-gas approximation are presented. We give a qualitative picture in real space-time of the effect of tunneling on the interaction between quarks and on the structure of the vacuum wave function. The physical effects produced by massless fermions are discussed including the resolution of the U(1) problem. Finally we discuss briefly the problems associated with going beyond the dilute-gas approximation.

In Sec. III we begin to explore the dynamical effects of instantons in QCD. We first analyze the nature of instanton interactions, show that these give rise to a nonperturbative coupling-constant renormalization. We argue that a sharply defined infrared cutoff on instanton scale size is produced which we identify with the confinement radius or hadronic size.

Section IV is devoted to an evaluation of the instanton contribution to the quark-antiquark "potential," which for small separations can reliably be evaluated (in the absence of massless quarks). We find a large effect but one which is unlikely to produce confinement.

Section V deals with the effects of massless quarks and a discussion of dynamical chiral-symmetry breaking. We argue that the chirally symmetric vacuum is unstable, and identify the mode in which the instability occurs and the mechanism responsible for generating a dynamical quark mass.

Section VI is devoted to a discussion of confinement. Here we describe the phase transition from a gas of instantons to a plasmalike configuration of merons (half-instantons) and argue that this produces a linear potential between quarks.

Finally in Sec. VII we discuss the (many) unsolved problems in our approach and suggest various ways of proceeding to solve these problems.

II. INTRODUCTORY INSTANTON PHYSICS

Our approach to the dynamics of Yang-Mills theories amounts to nothing more than an elaboration of the standard perturbation theory treatment in which the vacuum is regarded as associated with a (typically unique) classical field corresponding to minimum potential energy, and quantum mechanics amounts to including the effects of *Gaussian* fluctuations about this classical vacuum configuration. In Yang-Mills theories there is a new kind of vacuum fluctuation which must be considered. Despite initial appearances, there is a countable infinity of classical vacuum states with only a finite barrier separating them. Consequently there is a finite amplitude for tunneling back and forth *between* such states. Such spontaneous vacuum fluctuations are large in amplitude and potentially much more important than the standard Gaussian zero-point fluctuations. Our first task will be to bring this multiple vacuum structure to light and to show that the tunnelings are conveniently described by certain Euclidean classical solutions of the Yang-Mills equations called instantons.[11] In later sections we will discuss in detail the consequences for physics of including this new class of quantum fluctuation.

A. Tunneling

We consider the pure Yang-Mills theory based on a Lie group G [in practice SU(2) or SU(3)] and described by the Lagrangian

$$\mathcal{L}=\tfrac{1}{4}\sum F^a_{\mu\nu}F^a_{\mu\nu},$$
$$F^a_{\mu\nu}=\partial_\mu A^a_\nu-\partial_\nu A^a_\mu+f_{abc}A^b_\mu A^c_\nu. \tag{2.1}$$

For the purposes we have in mind it is simplest to eliminate the gauge freedom by setting $A^a_0=0$. Then the dynamical variables are the space components A^a_i, the canonical momenta are the electric field components $E^a_i=\dot{A}^a_i$, and the Hamiltonian density is

$$\mathcal{H}=\tfrac{1}{2}\left[(E^a_i)^2+(B^a_i)^2\right],$$
$$B^a_i=\tfrac{1}{2}(\nabla_j A^a_k-\nabla_k A^a_j+f_{abc}A^b_j A^c_k)\epsilon_{ijk}. \tag{2.2}$$

The equations of motion derived from the Hamiltonian are the usual Yang-Mills equation *except* for the analog of Gauss's law

$$C^a(A)=\nabla_i\dot{A}^a_i+f_{abc}A^b_i\dot{A}^c_i=0, \tag{2.3}$$

which is conjugate to A_0 in the usual treatment. On the other hand, $[\mathcal{H},C]=0$ so that if the initial configuration satisfies $C(A)=0$, then so does the time evolution of that configuration. Thus, Gauss's law may be regarded as a *constraint* on the initial p's and q's selecting physical configurations out of a larger manifold.

The quantum-mechanical version of this theory has states $|A^a_i(\vec{x})\rangle$ (eigenstates of the field operator) between which matrix elements of the time evolution operator may be computed by the path-integral technique

$$\langle\tilde{A}^a_i(\vec{x})|e^{-iHT}|A^a_i(\vec{x})\rangle=\int_{A^a_i(\vec{x})}^{\tilde{A}^a_i(\vec{x})}[DA^a_i(\vec{x},t)]\exp\left\{\frac{i}{2g^2}\int_0^T dt\,d^3x[E^a_i(\vec{x},t)^2-B^a_i(\vec{x},t)^2]\right\}. \tag{2.4}$$

The path integral is over all time histories connecting the configurations $A^a_i(\vec{x})$ to $\tilde{A}^a_i(\vec{x})$ in time T and g is the coupling constant. The quantum-mechanical version of Gauss's law constraint is that $C^a(A)$, regarded as an operator, should annihilate physical states: $C^a(A)|\psi\rangle=0$. The field eigenstates, $|A^a_i(\vec{x})\rangle$, which are convenient for the path integral do not satisfy this condition and we must characterize the states which do.

At this point it is convenient to write the fields as elements of the Lie algebra of G: $A_i(\vec{x})=A^a_i(\vec{x})T_a$, $[T_a,T_b]=if_{abc}T_c$. Now the theory is invariant under time-independent gauge transformations

$$A_i\to U_\lambda A_i U_\lambda^{-1}+iU_\lambda\nabla_i U_\lambda^{-1}, \tag{2.5}$$

where

$$U_\lambda=e^{i\lambda^a(\vec{x})T_a}$$

is an arbitrary function of $\vec{x}$ taking values in the group G [we have parametrized it by as many functions of position, $\lambda_a(\vec{x})$, as there are generators]. The above transformation on the fields is implemented by the unitary operator e^{iQ_λ} where

$$Q_\lambda=\int d^3x\dot{A}^a_i(\nabla_i\lambda^a+f_{abc}A^b_i\lambda^c). \tag{2.6}$$

If the function $\lambda^a(\vec{x})$ vanishes at spatial infinity we may integrate by parts to find that

$$Q_\lambda=-\int d^3x\,\lambda^a(\vec{x})(\nabla_i\dot{A}^a_i+f_{abc}A^b_i\dot{A}^c_i). \tag{2.7}$$

The term in parentheses is just Gauss's law and annihilates physical states. Thus physical states are also characterized by

$$e^{iQ_\lambda}|\psi\rangle = |\psi\rangle \tag{2.8}$$

if

$$\lambda^a(\vec{x}) \xrightarrow[|\vec{x}|\to\infty]{} 0,$$

which is to say that they are eigenstates with eigenvalue 1 of the subclass of all gauge transformations characterized by $\lambda^a \to 0$ at spatial infinity. Since a field eigenstate transforms under e^{iQ_λ} as

$$\begin{aligned} e^{iQ_\lambda}|A_i(\vec{x})\rangle &= |A_i(\vec{x})_\lambda\rangle, \\ A_i(\vec{x})_\lambda &= U_\lambda A_i U_\lambda^{-1} + iU_\lambda \nabla_i U_\lambda^{-1}, \\ U_\lambda &= e^{i\lambda^a(\vec{x})T_a}, \end{aligned} \tag{2.9}$$

it is apparent that the only way to construct a physical state is to sum over gauge transforms:

$$|A_i(\vec{x})\rangle_{\text{physical}} = \int [D\lambda^a(\vec{x})] e^{iQ_\lambda}|A_i(\vec{x})\rangle,$$

where the functional integral is over all λ's which vanish at $|\vec{x}|\to\infty$ and the integration measure is locally gauge invariant.

The notation here is a bit clumsy since we have not identified the variables which actually specify the physical state. In standard perturbation theory treatments it would be convenient to regard the above state as being parametrized by A_i^{tr}, that A_i, among all the A_i's summed over to form the physical state, which satisfies $\vec{\nabla}\cdot\vec{A}=0$. A_i^{tr} has precisely the correct degrees of freedom to describe the massless gluons which are normally thought to express the physical content of the theory. We could adopt the same strategy here, defining a function $U(\vec{x},t)$ by

$$\begin{aligned} A_i(\vec{x},t) &= UA_i^{\text{tr}}U^{-1} + iU\nabla_i U^{-1}, \\ \nabla_i A_i^{\text{tr}} &= 0, \end{aligned}$$

and writing the $A_0 = 0$ functional integral in terms of A_i^{tr} and U instead of A_i (and regarding the states as functions of A_i^{tr}). Since U turns out to be essentially a cyclic variable (this follows directly from our ability to impose Gauss's law as a constraint, i.e., an equation which when imposed at one time remains true for all time) we could think of eliminating it from the system and writing the path integral entirely in terms of the physical variables, A_i^{tr}. This would amount to casting the theory in the Coulomb gauge. The fact that only restricted classes of gauge transformations U (those which go to a constant at spatial infinity) may be integrated over makes this a rather tricky enterprise and we will not attempt it here.

Although it is not necessary to so limit our attention, we will for convenience now consider physical *vacuum* states $|\omega\rangle$. A vacuum gauge field configuration, corresponding to zero energy density, must have $B_i = 0$ which is possible only if $A_i = ig(\vec{x})\nabla_i g^{-1}(\vec{x})$ [$g(\vec{x})$ takes values in G]. For simplicity of notation let the corresponding state be written $|g(\vec{x})\rangle$. Under a gauge transformation e^{iQ_λ}, $|g(\vec{x})\rangle \to |U_\lambda g(\vec{x})\rangle = |\tilde{g}(\vec{x})\rangle$ and $|g(\vec{x})\rangle_{\text{phys}}$ will be a sum over all $|g(\vec{x})\rangle$ obtainable this way. The equivalence relation $g(\vec{x}) \approx \tilde{g}(\vec{x}) = U_\lambda g(x)$ [$\lambda^a(\vec{x})\to 0$ as $|\vec{x}|\to\infty$] divides the set of all possible elements of G into equivalence classes and possible physical vacuum states are just

$$|\text{vac}\rangle = \sum_{g\in(\text{equivalence class})} |g(\vec{x})\rangle. \tag{2.10}$$

This large class of possible vacuums can be cut down to manageable size, while bringing topological notions into the game, by the following dynamical remarks. Consider a transition amplitude $\langle\overline{\text{vac}}|e^{-iHT}|\text{vac}\rangle$ between two of these possible vacuums. Because the initial and final states may be *simultaneously* gauge transformed without affecting the value of the amplitude, we may without harm choose the initial vacuum to be in the equivalence class of $g(\vec{x}) = 1$. Clearly, all the states in this equivalence class are characterized by $g(\vec{x}) \to g_0$ [or $A_i(\vec{x}) \to 0$] as $|\vec{x}|\to\infty$. We would like to argue that there will be a finite transition amplitude from this equivalence class only to other equivalence classes also characterized by $g(\vec{x}) \to g_0$ [or $A_i(\vec{x}) \to 0$] as $|\vec{x}|\to\infty$. For any other type of final equivalence class $A_i(\vec{x}) \not\to 0$ as $|\vec{x}|\to\infty$ and the histories entering the path integral must have $\dot{A} \neq 0$ over an infinite spatial volume. Such inifinite energy transitions must in fact have zero amplitude and we may limit our attention, as far as dynamics is concerned, to vacuum equivalence classes characterized by $g(\vec{x}) \to 1$ as $|\vec{x}|\to\infty$.

If $g(\vec{x}) \to 1$ as $|\vec{x}|\to\infty$, then the domain of $g(\vec{x})$ may be thought of as three-space with points at infinity identified. This manifold is topologically equivalent to S_3. The vacuum equivalence classes are now seen to be classes of maps from S_3 to G which may be continuously deformed into another: homotopy classes. It is known that the homotopy classes of $G = \text{SU}(n)$ are countable and characterized by a positive and negative integer-valued topological invariant (winding number or Pontryagin class)

$$n = \frac{1}{24\pi^2}\int d^3x\, \epsilon_{ijk}\, \text{tr}[g^{-1}\nabla_i g(\vec{x}) g^{-1}\nabla_j g(\vec{x}) g^{-1}\nabla_k g(\vec{x})]. \tag{2.11}$$

A typical element [for $G = \text{SU}(2)$] of the $n = 0$ class is $g(\vec{x}) = I$ while a typical element for $n = 1$ is

$$g(\vec{x}) = \exp\left[i\pi \frac{\vec{x}\cdot\vec{\tau}}{(\vec{x}^2+a^2)^{1/2}}\right] = g_1 .$$

For reference, a $g(\vec{x})$ which belongs to *no* homotopy class (and is representative of the vacuums we threw out with our dynamical argument) is

$$g(\vec{x}) = \exp\left[i\alpha \frac{\vec{x}\cdot\vec{\tau}}{(x^2+a^2)^{1/2}}\right]$$

with α not an integer multiple of π. For convenience we will now label the surviving vacuum states by the appropriate winding numbers: $|n\rangle$.

It can be shown that multiplying an element of the n_1 homotopy class by an element of the n_2 class gives an element of the n_1+n_2 class. Therefore if we call R the unitary operator which implements a typical gauge transformation of the $n = 1$ class ($g_1 = \exp[i\pi\vec{x}\cdot\vec{\tau}/(x^2+a^2)^{1/2}]$, say) we must have $R|n\rangle = |n+1\rangle$ $(R^{-1}|n\rangle = |n-1\rangle)$. Now $R = U_{\lambda_R}$ with $\vec{\lambda}_R = 2\pi\vec{x}/(\vec{x}^2+a^2)^{1/2}$ if we choose R to correspond to g_1. Although R is a gauge transformation, since $\lambda_R \not\to 0$ as $|\vec{x}| \to \infty$ Gauss's law constraint does not specify how physical states behave under R. The physical requirement of gauge invariance is met so long as the physical states are eigenstates of R: $R|\psi\rangle = e^{i\theta}|\psi\rangle$. No physical principle determines what θ must be, although since $[H,R] = 0$, neither time evolution nor local gauge-invariant perturbations in general will change θ. In short, θ labels superselection sectors of the theory and the Hamiltonian must be block diagonal in θ. One easily constructs θ eigenstates from n eigenstates by the rule

$$|\theta\rangle = \sum_{-\infty}^{\infty} e^{in\theta}|n\rangle , \tag{2.12}$$

and it is apparent that the θ-vacuum states are nondegenerate. What is not altogether obvious at this stage, although true, is that the theories based on the different $|\theta\rangle$ states are physically different from each other, so that this multiplicity of θ worlds is a nontrivial property of the theory and not some gauge artifact. For $\theta \neq 0$, $|\theta\rangle$ is not an eigenstate of parity or time reversal. Therefore for QCD θ must be very small and in all likelihood is equal to zero. The physical principle (if any) which determines θ is unknown.

We have argued that there will be a finite quantum-mechanical transition amplitude between states $|n\rangle$ and $|n'\rangle$ because the time variation of A_i needed to effect the transition can be localized in space and does not require infinite kinetic energy at any point. On the other hand, if we look at the vacuum $A_i = 0$, say, it is apparent that the classical equations of motion will leave the system in the state $A_i = 0$ forever. Thus transitions from $|n\rangle$ to $|n'\rangle \neq |n\rangle$ look like tunneling processes: classically forbidden but quantum mechanically allowed barrier penetration processes.

To bring the tunneling interpretation into clearer view, it is helpful to pass to the imaginary time picture, i.e., to discuss $\langle n'|e^{-HT}|n\rangle$ instead of $\langle n'|e^{-iHT}|n\rangle$. The reason for this is that we know from experience with ordinary quantum mechanics that imaginary time solutions of the classical equations of motion can be used to obtain a WKB (or small $\hbar$) treatment of barrier penetration problems. In real time a classically forbidden process defines no stationary path which dominates the functional integral and there is no simple way to study tunneling.

The new path integral is

$$\langle n'|\exp(-HT)|n\rangle = \int_n^{n'} [DA_i]\exp\left\{-\int_0^T dt\, d^3x[(\dot{A}_i^a)^2+(B_i^a)^2]\frac{1}{2g^2}\right\}, \tag{2.13}$$

with the implied boundary condition of first computing the path integral between a representative pair $(A_i^a, A_i^{a\prime})$ of functions belonging to the homotopy classes (n,n') and then integrating over all representatives in the (n,n') classes. Since the action is positive-definite, the dominant history is one of *minimum* action consistent with the boundary conditions. Such a path satisfies all the Euclidean Yang-Mills equations—except Gauss's law constraint. Upon varying the end points of the A_i path history, one will finally pick out the path which satisfies the constraint as well. This path is guaranteed to have the *absolute* minimum action consistent with the constraint that it describes a transition $n \to n'$ and satisfies the full set of Euclidean Yang-Mills equations.

A rather large class of Euclidean Yang-Mills solutions are known by now,[19] but we need only discuss the original one of Belavin *et al.*[12] out of which, in a sense, all the others are constructed. In Landau the gauge $(\partial\cdot A = 0)$ the particular solution is $[G = \mathrm{SU}(2)]$

$$A_\mu^a = 2\frac{\eta_{a\mu\nu}x_\nu}{x^2+\rho^2}, \quad F_{\mu\nu}^a = \frac{4\eta_{a\mu\nu}\rho^2}{(x^2+\rho^2)^2} , \tag{2.14}$$

where $\eta_{a\mu\nu}$ is a numerical tensor coupling two SU(2)'s to O(4) $(\eta_{a\mu\nu} = \epsilon_{0a\mu\nu} + \frac{1}{2}\epsilon_{abc}\epsilon_{bc\mu\nu})$ and ρ is an arbitrary scale parameter (arising from the scale invariance of the classical theory). The Euclidean action of this solution is $8\pi^2/g^2$ (independent of ρ) so that the magnitude of the path integral dominated by this path is $\exp(-8\pi^2/g^2)$. Remembering that $g^2 \sim \hbar$ one sees that this amplitude vanishes like $\exp(-1/\hbar)$ as $\hbar \to 0$, the sort of essential singularity characteristic of vacuum tunneling in ordinary quantum mechanics. Finally, if we pass to the gauge $A_0 = 0$ in order to make contact with the earlier discussion in this section, we find

$$A_i(x_0 = -\infty) = 0,$$
$$A_i(x_0 = +\infty) = U^{-1}\nabla_i U, \quad U = \exp\left[i\pi \frac{\vec{x}\cdot\vec{\tau}}{(x^2+\rho^2)^{1/2}}\right] \tag{2.15}$$

indicating that this solution describes the vacuum tunneling $|0\rangle \to |1\rangle$. A more convenient and gauge-invariant way of identifying the topological class of a trajectory is

$$\Delta n = \frac{1}{8\pi^2}\int d^4x\, \mathrm{tr}(F_{\mu\nu}\tilde{F}_{\mu\nu}), \tag{2.16}$$

where $F_{\mu\nu} = \partial_\mu A_\nu - \partial_\nu A_\mu + i[A_\mu, A_\nu]$ is the four-dimensional field strength tensor and $\tilde{F}_{\mu\nu} = \frac{1}{2}\epsilon_{\mu\nu\lambda\sigma}F_{\lambda\sigma}$. In general, $\mathrm{tr}(F_{\mu\nu}\tilde{F}_{\mu\nu})$ may be written as a total divergence, and when evaluated in the $A_0 = 0$ gauge the $t = \pm\infty$ surface contributions are seen to be identical to the vacuum winding numbers defined earlier. This solution, called the instanton, exists in a conjugate version, called anti-instanton, describing tunneling from 0 to -1. One simply replaces $\eta_{a\mu\nu}$ by $\bar{\eta}_{a\mu\nu}$ $(\bar{\eta}_{a\mu\nu} = \epsilon_{0a\mu\nu} - \frac{1}{2}\epsilon_{abc}\epsilon_{bc\mu\nu})$. The same solutions describe the basic tunneling event in any theory based on SU(n). In the explicit formulas for A_i, SU(2) generators are simply replaced by an SU(2) subset of the SU(n) generators.[20]

There is a slight conceptual complication that this field-theoretic vacuum tunneling process has over and above its analog in ordinary quantum mechanics, namely, the multiplicity of Euclidean tunneling saddle points. The classical solution may be centered anywhere in space-time and may be of any scale size (small scale size ρ means the tunneling event has large field strengths but happens quickly; vice versa for large scale size).

For weak coupling the detailed tunneling amplitude can be computed by the following device. The matrix element

$$\langle n+1|\exp(-HT)|n\rangle = \langle 1|\exp(-HT)|0\rangle$$

is expressed as a functional integral as in Eq. (2.13). For weak coupling, $g^2 \approx 0$, one then performs the Euclidean functional integral by a saddle-point approximation. This requires a solution to the Euclidean equations of motion satisfying the appropriate boundary conditions—which is just the Belavin-Polyakov-Schwartz-Tyupkin (BPST) one-instanton solution, $A_\mu^a(x - x_1, \rho)$, located at an arbitrary point x_1, with arbitrary scale size ρ. Expanding the quantum field as $A_\mu^a(x - x_I, \rho) + gQ_\mu^a(x)$, one has

$$\langle 1|\exp(-HT)|0\rangle = \exp\left(-\frac{8\pi^2}{g_0^2}\right)\int [DQ_\mu^a] \times \exp\left(-\frac{1}{2}\int d^4x\, \mathcal{L}''[A_\mu^a]Q^2\right), \tag{2.17}$$

where $\mathcal{L}''$ is the second functional derivative of the Lagrangian evaluated at the instanton solution, and higher-order terms, proportional to g_0^2 (g_0 is the bare coupling), are neglected.

The normalization of the tunneling amplitude arises from the Gaussian integral around the saddle point. Evaluation of this integral requires calculating the determinant of the operator $\mathcal{L}''(A_\mu^a)$. By exploiting the conformal invariance of the classical theory, 't Hooft has computed this determinant explicitly.[21] The following points are worth mentioning:

(1) For every symmetry of the original Lagrangian there will exist a zero eigenvalue of $\mathcal{L}''(A_\mu^a)$. These zero energy modes can be dealt with by introducing collective coordinates for the degrees of freedom corresponding to the appropriate symmetries, and yield factors of the volume of the corresponding symmetry groups. In the case of QCD the SU(2) instanton possesses 4 translational, 1 scale, and 3 group degrees of freedom. [In the case of an SU(N) instanton, constructed using an SU(2) subgroup, there are $4N - 5$ group degrees of freedom.] For each degree of freedom a factor of $1/g$ will result from the introduction of collective coordinates. Thus the matrix element will take the form [for an SU(N) gauge group]

$$\langle 1|\exp(-HT)|0\rangle = V_N\left(\frac{8\pi^2}{g_0^2}\right)^{2N}\int d^4x\int\frac{d\rho}{\rho^5}\exp\left(-\frac{8\pi^2}{g_0^2}\right)\int[DQ']\exp\left[-\frac{1}{2}\int d^4x\, \mathcal{L}''(A^a)Q'^2\right], \tag{2.18}$$

where V_N is a numerical constant and Q' refers to the quantum field with the zero-energy modes removed.

(2) Owing to the standard ultraviolet divergences of ordinary perturbation theory, the remaining determinant (which is simply the exponential of the sum of connected vacuum-to-vacuum diagrams in the background instanton field) will require renormalization. This renormalization is standard since the ultraviolet divergences do not depend on the smooth background field. The net effect is to replace the bare action, $8\pi^2/g_0^2$, with the renormalized value, $8\pi^2/\bar{g}^2(1/\rho\mu)$, where μ is the renormalization mass and $\bar{g}^2$ is the ef-

fective coupling constant of the renormalization group, satisfying

$$\frac{d\bar{g}}{d\ln\rho} = -\beta(\bar{g}) = +b_0\bar{g}^3 + O(\bar{g}^5). \tag{2.19}$$

We thus have

$$\langle 1|\exp(-HT)|0\rangle = VT\int\frac{d\rho}{\rho^5}\left(\frac{8\pi^2}{\bar{g}^2(1/\mu\rho)}\right)^{2N}\exp\left(-\frac{8\pi^2}{\bar{g}^2(1/\rho\mu)}\right)C[1+O(\bar{g}^2(1/\rho\mu))], \tag{2.20}$$

where VT is the volume of space-time, and C is a numerical constant. For SU(2), $C_{SU(2)} = 0.26$ (Ref. 21) whereas for SU(3) we find that $C_{SU(3)} = 0.10$.

The integration over scale sizes is rendered convergent for $\rho \to 0$ by asymptotic freedom, since [for a pure SU(N) gauge theory]

$$8\pi^2/\bar{g}^2(1/\rho\mu) \underset{\rho\to 0}{\sim} (\tfrac{11}{3}N)\ln(1/\rho\mu). \tag{2.21}$$

However, the integration also extends to arbitrarily large scale sizes, where $\bar{g}^2(1/\rho\mu)$ increases. Thus in a scale-invariant theory such as QCD one cannot adjust the coupling to be small, and even constructing the vacuum requires an understanding of the infrared, perhaps strong-coupling, behavior of the theory.

For the moment we shall ignore this problem, and consider the contribution to the tunneling amplitude of instantons of sizes ρ to $\rho + d\rho$, replacing Eq. (2.20) with

$$\langle 1|\exp(-HT)|0\rangle = DVT\exp\left(-\frac{8\pi^2}{\bar{g}^2(1/\rho\mu)}\right), \quad D = C_N\frac{d\rho}{\rho^5}\left(\frac{8\pi^2}{\bar{g}^2(\rho)}\right)^{2N} \tag{2.22}$$

This is what one would obtain in a superrenormalizable field theory and will be instructive as to the physical picture of tunneling. In the following section, we shall discuss in great detail the physical consequences that arise from the existence of instantons of all sizes in QCD.

The scale size ρ can be considered as the time which it takes to tunnel and the coefficient of T, $VD\exp(-8\pi^2/\bar{g}^2)$ can be interpreted as the inverse of the mean time between tunnelings. Then for $\rho \ll T \ll (VD)^{-1}\exp(8\pi^2/\bar{g}^2)$ we can expand the matrix element in Eq. (2.22) according to

$$\begin{aligned}\langle 1|\exp(-HT)|0\rangle &\approx \langle 1|1 - HT + \cdots|0\rangle\\ &\cong -\langle 1|H|0\rangle T\\ &\cong TVD\exp\left(-\frac{8\pi^2}{\bar{g}^2}\right).\end{aligned} \tag{2.23}$$

One can then read off the tunneling Hamiltonian which for general $n \to n+1$ is

$$\langle n+1|H|n\rangle = -VD\exp\left(-\frac{8\pi^2}{\bar{g}^2}\right). \tag{2.24}$$

Taking this as the tunneling Hamiltonian it is straightforward to compute the energy of a θ vacuum. If $E(\theta)\delta(\theta-\theta') = \langle\theta'|H|\theta\rangle$ one finds

$$E(\theta) = E_0 - 2\cos\theta\, VD\exp\left(-\frac{8\pi^2}{\bar{g}^2}\right), \tag{2.25}$$

where E_0 is the energy in the absence of tunneling. Note that the energy is decreased and proportional to V as it should be.

It is important to note that there are two steps involved in constructing the θ vacuum. First, the tunneling amplitude $\langle n|H|n+1\rangle$ is determined from a single instanton and then the tunneling Hamiltonian is diagonalized. It is the second step that brings in multi-instanton effects. This is shown schematically in Fig. 1 where the single instanton and anti-instanton (anti-tunneling $n \to n-1$) of Fig. 1(a) are iterated in Fig. 1(b) to produce a θ vacuum. Observe that the θ vacuum looks like a gas of instantons in 4 dimensions. Below we will use this analogy to develop a more powerful method for handling instantons.

B. The analog gas

The above method of constructing a tunneling Hamiltonian gives correct answers for weak coupling but suffers from conceptual difficulties for larger $\bar{g}$ where the mean time between tunnelings is not so small. In particular, it is totally inadequate for QCD.

A systematic approach is based on the long-time, $T \gg (VD)^{-1}\exp(8\pi^2/g^2)$, Euclidean functional integral $\exp(-HT)$,

$$\langle\theta'|\exp(-HT)|\theta\rangle = \sum_{n,n'}\exp[i(n\theta - n'\theta')] \times\langle n'|\exp(-HT)|n\rangle, \tag{2.26}$$

with $\langle n'|\exp(-HT)|n\rangle$ given by Eq. (2.13). A deductive approach to vacuum tunneling would begin with this functional integral. As $T\to\infty$ this be-

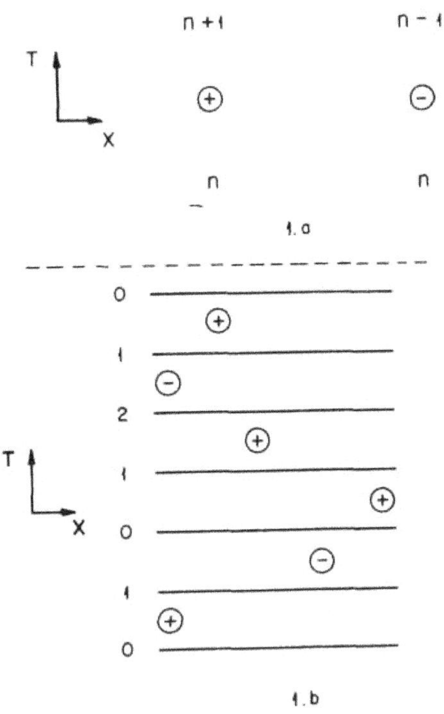

FIG. 1. Construction of a θ vacuum. The single instanton and anti-instanton shown in (a) causes transitions from n to $n+1$ or $n-1$. The θ vacuum is built upon successive transitions back and forth between different n states as in (b). One should imagine integrating over the locations in imaginary time and space of the instantons and anti-instantons.

comes const $\times\,\delta(\theta-\theta')\exp[-E(\theta)T]$. Let us see how this comes about in the weak-coupling limit where the mean time between tunneling is large (but small compared to T). One can then classify the configurations, A^μ, that contribute to $\langle n|\exp(-HT)|0\rangle$ according to the number of well-separated tunnelings (instantons) n_+, and antitunnelings (anti-instantons) n_-, such that $n=n_+-n_-$. We choose the normalization so that the naive vacuum term, $n_+=n_-=0$, is equal to 1. The term with $n_+=1$, $n_-=0$ can be evaluated by the saddle-point method described above yielding

$$VTD\exp\left(-\frac{8\pi^2}{\bar{g}^2}\right)\exp(i\theta)\,.$$

The term with $n_+=0$, $n_-=1$ is the same with θ replaced by $-\theta$. The first nontrivial term has $n_+=2$, $n_-=0$ and could also be computed by a standard saddle-point technique, since there exist exact two-instanton solutions[19] depending on the right number of parameters to describe two independent tunnelings. The action for these solutions is simply twice $8\pi^2/\bar{g}^2$, but the determinant D_2 has not yet been calculated for two instantons. However, when the instantons are far apart, as will be the case for small $\bar{g}^2$, D_2 must reduce to the square of the determinant for $n_+=1$, $n_-=0$, and the dominant contribution to the $n_+=2$, $n_-=0$ sector is

$$\frac{1}{2!}(VTD)^2\exp\left(-\frac{16\pi^2}{g^2}\right)e^{2i\theta}\,,$$

where the 1/2! is necessary to avoid double counting. Since the determinant D_2 presumably does depend on the instanton separation when they are close this form is only approximate. There are corrections to it which are proportional to one power of the volume VT, which for small coupling will yield corrections to $E(\theta)$ which are suppressed by $\exp(-8\pi^2/g^2)$. The $n_+=0$, $n_-=2$ contribution is then obtained in the obvious way ($\theta\to-\theta$).

The $n_+=n_-=1$ term introduces some new physics. This time we cannot evaluate the functional integral by a strict saddle-point method. The minimum action configuration for $n_+-n_-=0$ is simply the naive vacuum $A_\mu^a=0$, corresponding to a tunneling on top of an antitunneling, or no tunneling at all. However, we can certainly imagine configurations corresponding to a well-separated instanton–anti-instanton pair.

Such a configuration is not an exact solution to the Euclidean equations of motion but will nevertheless make a nontrivial contribution equal to $(VTD)^2\exp(-16\pi^2/g^2)$, when

$$VTD\exp\left(-\frac{8\pi^2}{g^2}\right)\gg 1\,.$$

Generalizing to arbitrary n_+ and n_- then yields

$$\langle\theta|\exp(-HT)|\theta\rangle=\sum_{n_+,n_-=0}^{\infty}\frac{1}{n_+!}\frac{1}{n_-!}(TVD)^{n_++n_-}\exp\left[-\frac{8\pi^2}{g^2}(n_++n_-)+i\theta(n_+-n_-)\right]$$
$$=\exp\left[2TVD\cos\theta\exp\left(-\frac{8\pi^2}{g^2}\right)\right]\,, \qquad (2.27)$$

and we recover the previous result

$E(\theta)=(\text{const})-2VD\cos\theta\exp(-8\pi^2/g^2)$.

Although it is valid only for weak coupling, Eq. (2.27) contains some important lessons. Specializing for simplicity to $\theta=0$, we have

$$\langle 0|\exp(-HT)|0\rangle=\sum_{n_+,n_-=0}^{\infty}\frac{1}{n_+!}\frac{1}{n_-!}(TVD)^{n_++n_-}\exp\left[-\frac{8\pi^2}{g^2}(n_++n_-)\right]. \tag{2.28}$$

For large T the dominant term in this series is the one for which

$$n_+=n_-=TVD\exp\left(-\frac{8\pi^2}{g^2}\right), \tag{2.29}$$

and as $T\to\infty$ essentially the entire sum comes from this term alone. Observe that

(i) the dominant term contains both instantons and anti-instantons and cannot be computed by a strict saddle-point method that relies on exact solutions to the Euclidean equations of motion,

(ii) the dominant term is not the one for which the classical action $\exp[-(8\pi^2/g^2)(n_++n_-)]$ is minimal.

One conclusion is that although the remarkable exact multi-instanton solutions[19] may indicate a new structure or symmetry to the theory, they are of essentially no relevance when it comes to constructing the vacuum state. In fact the sum over all terms with either n_+ or n_- equal to zero yields

$$2\exp\left[TVD\exp\left(-\frac{8\pi^2}{g^2}\right)\right],$$

which for large T is exponentially small compared to the complete sum,

$$\exp\left\{2\left[TVD\exp\left(-\frac{8\pi^2}{g^2}\right)\right]\right\}.$$

The nature of the long time functional integral is most easily understood in terms of an analog gas. The sum in Eq. (2.28) is precisely the grand partition function for a classical, four-dimensional perfect gas containing two species of particles with equal fugacities $\exp[-8\pi^2/g^2]$ and volume measured in units of D^{-1}. The energy (action) for a configuration with n_+ and n_- members of each species is $(n_++n_-)8\pi^2/g^2$ while entropy of the configuration is $\ln[(TVD)^{n_++n_-}/n_+!n_-!]$. The dominant term is the one for which the free energy (energy minus entropy) is smallest.

More generally, the entropy of a field configuration can be thought of as the log of the volume in function space occupied by similar configurations. For large coupling the action of a given field configuration decreases like g^{-2} while its entropy is generally less sensitive to g. Thus for moderate or strong coupling the entropy of a field configuration can be a more important consideration than its action. This will become increasingly evident in later sections. It also has its effects at small g: The exact multi-instanton solutions are in a sense uninteresting because they have so little entropy.

When g is small the analog gas is *extremely* dilute and the physics is the same as the tunneling picture discussed above. For larger g when instantons and anti-instantons are closer together the language of tunneling is at best picturesque. However, the idea of an interacting gas where instantons and anti-instantons interfere and the action for an n_+,n_- configuration is not just $(n_++n_-)8\pi^2/g^2$ still makes sense.

C. Wave functions and the functional integral

Consider a gauge where the gauge condition ($A^0=0$, $A^3=0$, or $\vec\nabla\cdot\vec A=0$) does not involve time explicitly. One can then consider the wave function $\Phi_\theta[A(\vec x)]$ of a θ vacuum as a functional of the independent components of A at one fixed time. The modulus of Φ is easily related to the Euclidean functional integral. If the fields at $T/2$ are fixed to be $A(x)$ then

$$\lim_{T\to\infty}\int[dA^\mu]\exp\left[\int_0^T dt\int d^3x\left(-\tfrac{1}{4}\mathrm{tr}F^2+\frac{i\theta}{8\pi^2}\mathrm{tr}\,F\tilde F\right)\right]\Big|_{A(T/2,\vec x)=A(\vec x)}=\text{const}\times\exp[-E(\theta)T]\,|\Phi_\theta[A(\vec x)]|^2. \tag{2.30}$$

Thus by examining a time slice of the Euclidean functional integral one can determine which field configurations are important in the vacuum. For a dilute instanton gas in the gauge $A^0=0$ most time slices will see only a gauge-rotated-vacuum $\vec{A}=i\Omega^{-1}\vec{\nabla}\Omega$ consistent with the wave function $|\theta\rangle=\sum\exp(in\theta)|n\rangle$ obtained from the tunneling picture. In the dense gas that occurs in QCD most time slices will intersect one or more instantons adding a new component to Φ_θ. For strong enough coupling further objects such as merons (Sec. VI) will also be present.

D. New quark interactions

Until now we have discussed tunneling in a pure Yang-Mills theory. Other fields are easily incorporated without significant effect on the tunneling picture (except for massless fermions as we shall see below). For example, in the presence of massive fermions the action will contain a term $\bar{\psi}(i\not{D}-m)\psi$. In constructing the θ vacuum, however, this will only slightly affect the normalization of the tunneling amplitudes. One will still, in the saddle-point approximation, expand about the multiple-instanton gauge fields; however, the one-loop quantum fluctuation will now include those of the fermion fields in the background instanton field. This will modify the functional integral by a factor of $\{\det[i\not{D}(A_\mu^a)-m]\}^{n_++n_-}$ for well-separated instantons.

It is possible to calculate, using the dilute-gas approximation, the Green's functions of any number of gauge-invariant operators. One simply adds to the Lagrangian density a term $\sum_i J_i(x)\theta_i(x)$, where $J_i(x)$ is an external c-number source coupled to the operator in question, $\theta_i(x)$, constructed out of the gauge and matter fields. Euclidean Green's functions of θ_i are then defined as

$$\langle\theta|T(\theta_1(x_1)\cdots\theta_N(x_N))|\rangle=\left(\frac{\delta}{\delta J_1(x_1)}\right)\cdots\left(\frac{\delta}{\delta J_N(x_N)}\right)\times\frac{\int[DA_\mu]\exp[-S(A)+\int\sum J_i(x)\theta_i(x)d^4x]}{\int[DA_\mu]\exp[-S(A)]}\Bigg|_{J_i=0}. \tag{2.31}$$

In performing the saddle-point integrations for the various multiple-instanton sectors in Eq. (2.31), one treats the term $\sum J_i\theta_i$ as a small perturbation and expands the gauge fields in θ_i about the multiple-instanton configurations.

Of particular concern to us is the effect of tunneling on the quark-antiquark interaction. Let us first give a physical picture of how tunneling modifies the interaction of quarks.

A quark-antiquark state constructed in perturbation theory would be built on one of the n states $|n\rangle$ rather than a θ vacuum. Let $|n\vec{r}\rangle$ be such a state where the $q\bar{q}$ pair is separated by $\vec{r}$. There will be tunneling to other states $|m\vec{r}\rangle$ and the true $q\bar{q}$ state built on a θ vacuum is $|\theta\vec{r}\rangle=\sum_m\exp(im\theta)|m\vec{r}\rangle$. Because the tunneling amplitude $\langle m\vec{r}|H|n\vec{r}\rangle$ will differ from the vacuum amplitude $\langle m|H|n\rangle$ and depend on $\vec{r}$, nonperturbative $q-\bar{q}$ interactions will appear. In a weak-coupling approximation the energy $E(\theta,\vec{r})$ of $|\theta\vec{r}\rangle$ relative to the energy of the θ vacuum is

$$E(\vec{r},\theta)=E_0(\vec{r})+2\,\mathrm{Re}[\exp(i\theta)(\langle 1\vec{r}|H|0\vec{r}\rangle-\langle 1|H|0\rangle)], \tag{2.32}$$

where $E_0(\vec{r})$ includes the perturbative terms, e.g. Coulomb interaction, which are diagonal and the same for each n. For heavy "test quarks" the state $|n\vec{r}\rangle$ can be constructed in a gauge-invariant way to be

$$|n\vec{r}\rangle=\bar{q}(\vec{r})U(\vec{r},\vec{0})q(\vec{0})|n\rangle, \tag{2.33}$$

where

$$U(\vec{r},\vec{0})=T\exp\left(i\int_{\vec{0}}^{\vec{r}}\vec{A}\cdot d\vec{x}\right)$$

is the ordered exponential integrated along a straight line from $\vec{0}$ to $\vec{r}$. The tunneling just shifts $U(\vec{r},\vec{0})$ by a gauge and for a given instanton location (and scale size and orientation) $\langle 1\vec{r}|H|0\vec{r}\rangle$ differs from $\langle 1|H|0\rangle$ only by a factor $\mathrm{tr}[U_1^{-1}(\vec{r},\vec{0})U_0(\vec{r},\vec{0})]$ where the subscripts refer to the values of U in the initial and final n states. Because we are in the gauge $A^0=0$, we can write

$$\mathrm{tr}[U_1^{-1}(\vec{r},\vec{0})U_0(\vec{r},\vec{0})]=T\exp\left(i\oint A^\mu_{\rm inst}\,dx^\mu\right), \tag{2.34}$$

where the ordered line integral of the instanton field runs around a rectangular Euclidean loop with corners at $(\vec{0},0)$, $(\vec{r},0)$, $(\vec{r},T)$, $(\vec{0},T)$. This ordered loop integral then has to be integrated over all instanton locations (and scale sizes and orientations) to get $\langle 1\vec{r}|H|0\vec{r}\rangle$ and then from Eq. (2.32) $E(\theta,\vec{r})$ can be obtained.

The tunneling calculation outlined above is, for weak coupling where it is valid, equivalent to the following dilute-instanton-gas calculation. Let $E(\theta,\vec{r})$ be defined by

$$\exp[-E(\theta,\vec{r})T]=\frac{\int[dA^\mu]\operatorname{tr}[T\exp(i\oint A^\mu dx^\mu)]\exp\{\int[-\frac{1}{4}\operatorname{tr}F^2+(i\theta/8\pi^2)\operatorname{tr}F\tilde{F}]d^4x\}}{\int d[A^\mu]\exp\{\int[-\frac{1}{4}\operatorname{tr}F^2+(i\theta/8\pi^2)\operatorname{tr}F\tilde{F}]d^4x\}}, \quad (2.3$$

where the Euclidean time T is assumed to be large and the ordered exponential runs around the same Euclidean loop as before. If this ratio of functional integrals is evaluated in exactly the same (dilute gas) approximation as in Eq. (2.32) except that in the numerator each multiple instanton-anti-instanton configuration is multiplied by the loop integral for that configuration (before integrating over locations, scale sizes, etc.) then the resulting approximation to $E(\theta,\vec{r})$ will be equivalent to the tunneling calculation.

A deductive derivation of the effect of instantons on the $q\bar{q}$ interaction would begin with the functional integral in Eq. (2.34). Wilson has argued that the average Euclidean loop integral does in fact yield the interaction energy of a heavy "test quark" pair[10] (not to be confused with the potential between real light quarks). Unlike the tunneling picture the averaged loop integral makes perfectly good sense when the instanton gas is not dilute but rather dense as in QCD.

For the moment we will refrain from commenting on whether or not this new tunneling interaction has anything to do with quark confinement. It is, however, definitely there and will turn out to be non-negligible as we shall see in Sec. IV.

E. Light quarks

The Wilson loop is not directly relevant to the binding of ordinary light quarks. The problem is not just a kinematic one, since, as 't Hooft has emphasized, instantons do qualitatively new things when light fermions are present.[13] It is simplest to discuss the situation in the limit of massless fermions. As before we will develop the physics in a simple tunneling picture and then pass to the long-time Euclidean functional integral.

In a theory with fermions a "classical" state is specified by giving the boson field configuration and saying which fermion states are occupied. A perturbation-theoretic n state is thus one for which the gauge field belongs to class n and all the negative-energy fermion levels are occupied. One can think of tunneling as being an adiabatic process as far as the fermions are concerned. For each value of Euclidean time t, the time-independent Dirac equation in an instanton field will have eigenvalues $\epsilon_i(t)$ and eigenfunctions $\Psi_i(t,\vec{x})$ where t enters only as a parameter. Since the net effect of tunneling is just a gauge transformation the original eigenvalues of $\epsilon_i(0)$ must be in one-to-one correspondence with the final eigenvalues. *If the correspondence* $\epsilon_i(0)-\epsilon_i(T)$ *is the trivial* one $\epsilon_i(0)=\epsilon_i(T)$ then the tunneling can connect two distinct classical vacuums. For an instanton and massless fermions, however, this mapping of the spectrum of the Dirac equation into itself is nontrivial and in particular the highest right-handed negative-energy state crosses zero and becomes the lowest right-handed positive-energy state while the lowest positive-energy left-handed state becomes the highest negative-energy left-handed state. Thus if the initial configuration had all its negative-energy states occupied and all positive-energy states empty, then in the final configuration there will be one right-handed positive-energy state which is occupied and one left-handed negative-energy state which is empty. The tunneling process is then $n\to m+\bar{q}_L\bar{q}_R$ rather than just $n\to m$. Strictly speaking there are then no vacuum tunnelings but only virtual tunnelings $n\to m+\bar{q}q\to n$.

The above picture is easy to demonstrate explicitly in two-dimensional models (the reader can easily work it out, for the two-dimensional Abelian model discussed in Ref. 18) and is known to follow directly from the topology of the instanton field in four dimensions.[22]

A canonical picture of the suppression of tunneling in the presence of massless fermions can also be given. As is well known[8] in QCD the axial baryon number current contains an anomaly, and is not conserved (α labels color, i labels flavor):

$$J^5_\mu=\sum_{\alpha,i}\bar{\psi}_{\alpha i}\gamma_\mu\gamma_5\psi_{\alpha i},$$
$$\partial^\mu J^5_\mu=\frac{1}{8\pi^2}F^a_{\mu\nu}\tilde{F}^a_{\mu\nu}. \quad (2.36)$$

The above current is the gauge-invariant regulated current (defined, say, by a point-separation technique). One can define a gauge-variant current which is conserved:

$$\tilde{J}^5_\mu=J^5_\mu-\frac{g^2N}{32\pi^2}\epsilon_{\mu\nu\lambda\sigma}\operatorname{Tr}A_\nu(\partial_\lambda A_\sigma+\tfrac{2}{3}A_\lambda A_\sigma),$$
$$Q_5=\int d^3x\tilde{J}^5_0,\quad \frac{d}{dt}Q_5=0 \quad (2.37)$$

Q_5 is now conserved but not gauge invariant. It is easy to see that Q_5 is gauge invariant under gauge transformations that vanish at infinity, but not under gauge transformations that change the topological class of A_μ.

If we consider $A_0 = 0$ gauge we find that the unitary operator R which implements a gauge transformation of the $n=1$ class ($R|n\rangle = |n+1\rangle$) has the following effect on Q_5:

$$R Q_5 R^\dagger = Q_5 - 2N\,, \tag{2.38}$$

where N is the number of flavors. If the vacuum states of different topology $|n\rangle$ are defined by $|n\rangle = R^n|0\rangle$, with $Q_5|0\rangle = 0$, then $Q_5|n\rangle = 2Nn|0\rangle$. However, Q_5 is conserved, $[Q_5, H] = 0$, so that $\langle n|\exp(-HT)|m\rangle \sim \delta_{n,m}$, namely tunneling is suppressed in the absence of non-chiral-invariant sources.

However, virtual tunneling does occur and again the true vacuum will be the coherent superposition of the n vacuums. Indeed it is only in such a state that one recovers cluster decomposition of operators of chirality $= 2N$.[14] However, unlike the case of massive fermions, physical quantities will have no dependence on θ: The vacuum energy will be independent of θ, and $\partial E(\theta)/\partial\theta = \langle\theta|\mathrm{tr}F_{\mu\nu}\tilde{F}_{\mu\nu}|\theta\rangle = 0$. Finally axial baryon number conservation is violated. The operator $\exp(i\alpha Q_5)$ (for $\alpha \neq \pi/N$) is ill defined, and takes one out of the Hilbert space constructed about a θ vacuum. Alternatively the gauge-invariant current, J^5_μ, has a hard divergence, $\mathrm{Tr}F_{\mu\nu}\tilde{F}_{\mu\nu}$, which has nonvanishing matrix elements ($\langle\theta|\bar\psi\psi\,\mathrm{tr}F\tilde{F}|\theta\rangle \neq 0$ in the case of one flavor) thus vitiating Goldstone's theorem.

Let us now proceed to examine the effects of massless fermions on the long-time Euclidean functional integral after making the following observation. In an adiabatic approximation the solutions to the time-dependent Dirac equation in an instanton field will be

$$\psi_i(t,\vec{x}) = \Psi_i^0(t,x)\exp\left[-\int_0^t \epsilon_i(t')dt'\right].$$

Such a solution will be normalizable if ϵ_i crosses zero and ϵ_i is positive for $t\to+\infty$ and negative for $t\to-\infty$. Thus the shift in a right-handed state from negative to positive energy in an instanton field can be expected to produce a normalizable solution to the time-dependent Dirac equation.

The existence of normalizable solutions to the massless Dirac equation in fields with nonzero topological quantum number Q has been demonstrated by a number of authors.[23] This has a dramatic effect on the functional integral. The fermion determinant then vanishes identically for any configuration containing an unequal number of instantons and anti-instantons, in agreement with the above argument that only virtual tunnelings are allowed. In the analog gas picture a virtual tunneling is a closely correlated instanton–anti-instanton pair. We have in fact shown in a previous paper that in the presence of several massless fermions a sufficiently dilute instanton gas is actually a "molecular" gas composed of instantons and anti-instantons permanently bound into "diatomic molecules,"[14] by the exchange of massless fermions. This exchange gives rise to a logarithmic attractive interaction proportional to the number of flavors, of the form $6N\ln R$, where R is the distance between the instanton and the anti-instanton. In QCD it is probable that for strong enough coupling there is a phase transition at which point the "molecules" disassociate, liberating free instantons and anti-instantons. This would lead to a spontaneous breakdown of chiral symmetry and will be discussed in Sec. V. In any case the quasitunneling vac → fermions and antifermions (which crosses to fermions → fermions) leads to qualitatively new nonperturbative interactions among massless fermions.

F. Beyond the dilute-gas approximation

The analog model developed above of a perfect gas of instantons and anti-instantons is only a valid approximation for very small coupling (or small $\hbar$). As the nonlinear coupling g increases *one must improve this approximation.* In order to see the effects which emerge when g increases let us consider the contribution to the functional integral of the $n_+ = n_- = 1$ sector. This sector has net topological quantum number equal to zero, i.e., the same as the naive vacuum sector $n_+ = n_- = 0$, and strictly speaking the only saddle point (solution of the classical field equations) is the naive vacuum configuration. However, a superposition of widely separated instanton and anti-instanton configurations is very close to a saddle point.

One way of including such approximate saddle points in a systematic fashion is to introduce constraints into the functional integral. Thus we write (schematically)

$$\int_{n_+=n_-=1}[DA_\mu]\,e^{-S(A)}$$
$$\int[DA_\mu]\int \pi da_i^+ C_i^+(A_\mu, a_i^+)\pi da_i^- C_i^-(A_\mu, a_i^-)e^{-S(A)}\,, \tag{2.39}$$

where $\int \pi da_i^\pm C_i^\pm(A_\mu, a_i^\pm) = 1$, the a_i^+ (a_i^-) are "collective coordinates" for the individual instanton (anti-instanton) corresponding to the relevant translational, scale, and group degrees of freedom, and the $C_i^\pm$ are functions of the field that, for given a_i, fix these degrees of freedom. Interchanging orders of integration one now expands, for fixed $a_i^\pm$, about a true saddle point of the constrained functional integral. For values of $a_i^\pm$ corresponding to well-separated instanton configurations the saddle-point configuration will be approximately given by the superposition of the instanton and anti-instanton solution, and the action will be simply $-16\pi^2/g^2$, independent of the $a_i^{\pm}$. This is what yields, upon integration over $a_i^{\pm}$, the term proportional to the square of the volume of space-time $(VT)^2$ reproducing, with the correct normalization, the perfect gas approximation.

Moreover, one can now imagine improving on this approximation, taking into account that the action at the saddle point does depend on $a_i^{\pm}$ when the instantons are at a finite separation. We can interpret

$$S(a_i^{\pm})=\frac{16\pi^2}{g^2}+\delta S(a_i^{+},a_i^{-}) \qquad (2.40)$$

as consisting of, in the analog gas model, a term corresponding to the chemical potential ($8\pi^2/g^2$ a piece) for the instanton and anti-instanton and an interaction energy, $\delta S(a_i^{+},a_i^{-})$. A similar procedure can be carried out in the multiple instanton–anti-instanton sectors. The result of performing the saddle-point integrations will be to replace the functional integral by the partition function of a gas of interacting instantons and anti-instantons. Since the Lagrangian is a nonlinear functional of the fields there will in general be multibody interactions.

The nature of these interactions for well separated configurations is easily understood. In general instantons will attract anti-instantons and repel instantons, since in the first case one reduces the action by bringing the configurations together and in the latter the action is a minimum for infinite separation. In special theories, such as QCD, instantons will have no interaction with instantons, but an attractive instanton–anti-instanton interaction will always exist. The interaction energy will also depend on the group orientation and scale size. The interaction will vanish for infinite separation, exponentially in non-scale-invariant superrenormalizable theories, and according to a power law in scale-invariant theories such as QCD.

Since for weak coupling we have shown that the density of instantons is proportional to exp $(-8\pi^2/g^2)$ one can perform a virial expansion of the partition function. To first approximation we have the perfect gas described above, where δS has been neglected. Including the effects of the interactions will yield corrections proportional to the density of instantons. Thus one might hope to set up a systematic virial expansion of the analog gas in powers of $\exp(-8\pi^2/g^2)$. There are, however, severe technical and conceptual problems in attempting to do this. The first problem is that of double counting. A well-separated instanton–anti-instanton pair clearly gives an important contribution to the functional integral. However, it is clearly nonsense to consider both the ordinary vacuum field configuration as well as an instanton anti-instanton pair close together with equal weight. The second problem is how to systematically sum the quantum fluctuations about a given field configuration when tunneling exists. We now know that in the presence of tunneling the ordinary perturbation theory is not Borel-summable.[24] Thus the perturbation theory about the ordinary vacuum will yield an asymptotic power series $\sum_n C_n g^n$ which we do not know precisely how to define when $g\neq 0$.

These problems are interrelated. The lack of Borel summability arises from the existence of real instantons, and the ambiguity in dealing with overlapping instantons is related to whether such configurations have been included as fluctuations in other sectors. At the moment we lack the solution to both of these problems and do not have a systematic formalism for dealing with tunneling for large coupling.

To illustrate the nature of the dilute-gas approximation as well as the double-counting problem, it is instructive to consider for following model, in which all quantum fluctuations have been suppressed. Consider replacing the field variables $A_\mu^a(\vec{x}_i)$ by discrete spins $\sigma_i=\sigma(t_i)$ on a discrete (Euclidean) time lattice. At each discrete time the field will be constrained to be in one of an infinity of possible $|n\rangle$ states, $\sigma_i=-\infty,\ldots,-1,0,+1,2,\ldots$. Thus a Euclidean field configuration is represented by a lattice of spins that take integer values: $(\sigma_1,\sigma_2,\ldots,\sigma_T)$, where T = total time. A static $|n\rangle$ vacuum state is represented by the $\sigma_i=n$, $i=1,\ldots,T$ configuration $(n,n,\ldots,n)$. Obviously an instanton that effects $|n\rangle\to|n+1\rangle$ at time $t=i$, is represented by $(\sigma_1=\sigma_2=\cdots=\sigma_i=n;\ \sigma_{i+1}=\cdots=\sigma_T=n+1)$. The topological quantum number is clearly $\nu=\sigma_T-\sigma_1$. We shall choose the action to be

$$S=\sum_{i=1}^{T}\frac{1}{g^2}|\sigma_i-\sigma_{i+1}|, \qquad (2.41)$$

so that a configuration consisting of n_+ instantons and n_- anti-instantons has action $=(n_++n_-)g^2$. The θ vacuum to θ vacuum amplitude is then given by

$$\langle\theta|e^{-HT}|\theta\rangle=\sum_{(\sigma_i),\,\sigma_1=0}e^{-S(\sigma_i)+i\theta(\sigma_T-\sigma_1)}. \qquad (2.42)$$

This is simply the partition function for a one-dimensional system of infinite-component classical spins with nearest-neighbor interactions. For $\theta=0$ the Hamiltonian is simply $H=\sum_i|\sigma_i-\sigma_{i+1}|$, and the temperature is $kT=1/g^2$.

For this system

$$E(\theta)=\frac{1}{T}\lim_{T\to\infty}(-\ln\langle\theta|e^{-HT}|\theta\rangle)$$

can easily be evaluated [by the change of variable $x_i=(\sigma_i-\sigma_{i+1})$]:

$$E(\theta)=\ln\left(\frac{1-2\cos\theta e^{-1/g^2}+e^{-2/g^2}}{1-e^{-2/g^2}}\right)$$

$$\simeq -2\cos\theta e^{-1/g^2}+2\sin^2\theta e^{-2/g^2}+O(e^{-3/g^2}). \quad (2.43)$$

Alternatively one can replace the spin variables with instanton and anti-instanton variables. Every configuration of spins corresponds to a configuration of instantons and anti-instantons (which are simply domain walls in the one-dimensional lattice). Thus the configuration ($\sigma_1=\sigma_2=\ldots=\sigma_a=0$, $\sigma_{a+1}=\sigma_{a+2}=\ldots=\sigma_b=1$, $\sigma_{b+1}=\ldots=\sigma_c=0$, $\sigma_{c+1}=\ldots=\sigma_d=2$) corresponds to an instanton at $t=a$, an anti-instanton at $t=b$, and two instantons at $t=c$. We can then rewrite Eq. (2.42) as the partition function for a lattice gas of instantons and anti-instantons with a chemical potential equal to $1/g^2$. The instantons and anti-instantons may be placed anywhere without affecting the action (energy) except that we cannot allow an instanton to sit on top of an anti-instanton. Thus, in effect, the only interaction is an infinite repulsive core between instanton and anti-instanton; instantons do not interact among themselves.

The perfect-gas approximation ignores the double-counting interaction, yielding

$$\langle\theta|e^{-HT}|\theta\rangle=\sum_{n_+,n_-=0}^{\infty}\frac{1}{n_+!n_-!}\times T^{n_++n_-}e^{-(n_++n_-)/g^2}e^{i\theta(n_+-n_-)}$$

$$=\exp(2T\cos\theta e^{-1/g^2}). \quad (2.44)$$

The correction to this, which gives the first term in Eq. (2.43) can be calculated by taking into account the double-counting interactions of the gas. Thus in the two-instanton sector, a configuration consisting of two instantons on top of each other contributes an amount $Te^{-2/g^2}e^{2i\theta}$, not $\frac{1}{2}Te^{-2/g^2}e^{2i\theta}$ as given by Eq. (2.44). Also, an instanton-anti-instanton cannot sit on top of each other and one must subtract a term Te^{-2/g^2} from Eq. (2.44). Thus to order e^{-2/g^2} we have, including the above two corrections,

$$\langle\theta|e^{-HT}|\theta\rangle=1+2T\cos\theta e^{-1/g^2}+\tfrac{1}{2}(2T\cos\theta)^2e^{-2/g^2}$$

$$+T\cos2\theta e^{-2/g^2}-Te^{-2/g^2}$$

$$\simeq\exp\{T[2\cos\theta e^{-1/g^2}-\sin^2\theta e^{-2/g^2}+O(e^{-3/g^2})]\}, \quad (2.45)$$

in agreement with Eq. (2.43).

In a continuum theory, such as QCD, there will of course be additional corrections due to long-range interactions between instantons and anti-instantons, as well as to the quantum fluctuations about the saddle points. Fortunately the dominant corrections to the perfect gas in QCD arise from the long-range interactions which can be evaluated in the dilute-gas approximation. Furthermore, as we shall see below, instantons never get too dense nor does the coupling constant get large enough to seriously invalidate the dilute-gas approximation. Thus we shall be able to proceed to include effects of instanton interaction even though we lack a systematic procedure for dealing with the problems of double counting and quantum fluctuations which would arise for large densities and couplings.

Finally let us note that, when the coupling increases, other field configurations in addition to instantons might become important. That instantons determine the vacuum structure for arbitrarily small coupling is due to the fact that although the "energy" necessary to create an instanton in the analog gas is $(1/g^2)S_{cl}$, the "entropy" is even larger [$S\sim\ln V$ (V = volume of space time)] and the "free energy" $F=S_{cl}-g^2\ln V$ is dominated by the entropy term. The "temperature" corresponding to a phase transition from an n vacuum is roughly given by

$$\sim g^2=S_{cl}/\ln V\xrightarrow[V\to\infty]{}0.$$

On the other hand, one might consider other field configurations whose action is in some sense infinite. If such field configurations occupy a volume in function space which is large enough they might be important at some finite g^2. Of particular importance are configurations consisting of pairs of configurations (molecules) whose action depends logarithmically on the separation of the pair. The free energy of such a pair, separated by R, will behave as $F\sim C\ln R-g^2\ln R$. For small g^2 the pair will be close together, while for large g^2 the entropy term will dominate and the pair will separate. We might then expect a phase transition at a finite temperature ($=g^2$) from a gas of tightly bound molecules to a gas of dissociated molecules. Such phase transitions are well known in one- and two-dimensional systems with logarithmic interactions.[25]

In QCD there are two important cases of such molecular field configurations. First, there are instantons themselves in the presence of massless fermions. The fermions effectively bind instantons to anti-instantons with an attractive logarithmic *potential, thus suppressing tunneling. The phase* transition which is responsible for chiral-symmetry breaking, from a gas of instanton–anti-instanton molecules to a dilute gas of instantons, is discussed in Sec. V. The second case concerns merons. As we shall see, instantons themselves can be regarded as tightly bound pairs of merons, which are localized lumps of one-half topological charge, with independent entropy of position and logarithmic interactions. The phase transition in which instantons dissociate into merons and which may be responsible for confinement is discussed in Sec. VI.

III. INSTANTON INTERACTIONS

In this section we will study some simple features of the instanton gas in QCD. As explained in the Introduction, this theory has no free parameters and we cannot vary the relative importance of instantons by varying some convenient coupling constant. However, by asking questions which emphasize different distance scales one may accomplish much the same effect. At short distances the effective coupling becomes small and the density of instantons is very low. Our strategy will be to start at the small scales and work our way outward. Inevitably, we reach a scale where the instantons are close enough that their interactions become significant. This produces new phenomena which we will discuss in this section. We will find that large new effects begin to appear at scales where the density of instantons is reasonably low and quantitative calculations are possible. Still larger scales where confinement presumably becomes manifest will be probed in Sec. VI.

In order to focus on pure QCD effects, we will discuss an unrealistic theory with no light fermions (the properties of light quarks will be studied in Sec. V). We will also treat SU(2) and SU(3) in parallel since they differ in their behavior in instructive ways. Let us begin with the role of the BPST instanton in the SU(2) vacuum. As shown by 't Hooft its contribution to the functional integral is, to one-loop order,

$$\langle n|H|n+1\rangle = \int d^4z \int \frac{d\rho}{\rho^5}(0.26)\left(\frac{8\pi^2}{g_0^2}\right)^4 \times \exp\left(-\frac{8\pi^2}{g_0^2}+\tfrac{22}{3}\ln\mu\rho\right), \tag{3.1}$$

where μ is a renormalization mass introduced by the Pauli-Villars procedure. It of course makes sense to express this in terms of a running coupling constant, $g(\rho)$, but this is ambiguous until some precise definition of g has been adopted. To one-loop order, however, the ambiguity is just an additive constant in $8\pi^2/g^2$:

$$\frac{8\pi^2}{g^2(\rho)} = \frac{8\pi^2}{g_0^2} + \tfrac{22}{3}\ln\frac{1}{\rho\mu} + C. \tag{3.2}$$

For simplicity, we shall adopt the coupling-constant definition which sets $C=0$. This is not the same definition as the dimensional regularization definition of g [which amounts in the SU(2) case to setting $C=-6.9$] but does not appear particularly unnatural—indeed, for the values of g we shall be interested in, it makes two-loop contributions to anomalous dimensions of low-lying twist-two operators rather smaller than does the dimensional regularization definition. This is quite important since we shall claim that interesting strong-interaction effects occur at sufficiently small coupling constant that we may neglect higher-loop effects—but the same physics may correspond to large or small g depending on what definition of g has been adopted. In what follows we shall replace μ and g_0 by $\bar{\mu}$ so that

$$\frac{8\pi^2}{g^2(\rho)} = \tfrac{22}{3}\ln\frac{1}{\bar{\mu}\rho} = \tfrac{22}{3}\ln\frac{1}{\mu\rho} + \frac{8\pi^2}{g_0^2}. \tag{3.3}$$

Eventually we will see that $\bar{\mu}$ can be related to the hadron scale size [we expect this one-loop expression for $g(\rho)$ to be useful as long as $\bar{\mu}\rho$ is small]. To be consistent with the requirements of the renormalization group we must also assume that the determinantal factor of $(8\pi^2/g_0^2)^4$ is actually $[8\pi^2/g^2(\rho)]^4$. To be more precise about this it would be necessary to do a two-loop calculation of the instanton determinant, a worthy exercise which has not yet been carried out. In summary, our expression for the one-instanton contribution to the vacuum functional in the pure SU(2) gauge theory is

$$\langle n|H|n+1\rangle_{SU(2)} = \int d^4z \int \frac{d\rho}{\rho^5}(0.26)\left(\frac{8\pi^2}{g^2(\rho)}\right)^4 \times \exp\left[-\frac{8\pi^2}{g^2(\rho)}\right] \tag{3.4}$$

A modest extension of 't Hooft's calculation of the SU(2) instanton determinant allows us to conclude that, for SU(3),

$$\langle n|H|n+1\rangle_{SU(3)} = \int d^4z \int \frac{d\rho}{\rho^5}(0.10)\left(\frac{8\pi^2}{g^2(\rho)}\right)^6 \times \exp\left[-\frac{8\pi^2}{g^2(\rho)}\right], \tag{3.5}$$

where this time $8\pi^2/g^2(\rho) = 11\ln(1/\rho\bar{\mu})$. The 12 powers of g^{-1} arise from the zero modes: one dilatation, four translations, and seven gauge modes (the λ_8 generator does not induce a change in the instanton field and does not induce a zero mode).

Next we construct the analog gas. To be systematic, we should invent a set of constraints such that solving the Yang-Mills equations under them yields the desired multi-instanton–anti-instanton configurations, evaluate the functional integral about the saddle point, and then integrate out the constraints. Since we do not know a good way to do this we resort to a crude procedure which should be adequate in the limit of low pseudoparticle density. First, we convert the standard instanton to singular gauge by inversion:

$$A_\mu^a = \frac{2}{g}\frac{\rho^2}{x^2+\rho^2}\bar{\eta}_{a\mu\nu}\frac{x_\nu}{x^2}\,. \tag{3.6}$$

Since A_μ now falls as r^{-3} at infinity it makes sense to construct multi-instanton configurations by superposition:

$$A_\mu^a = \sum_i R_{ab}^{(i)} A_{\mu b}^{(i)}(x - z_i, \rho_i)\,. \tag{3.7}$$

In this expression $A_\mu^{(i)}$ is the basic instanton or anti-instanton solution of Eq. (3.6) (for anti-instanton $\bar{\eta} \to \eta$) and R_{ab} is a matrix from the adjoint representation of the group representing the group orientation degree of freedom. This is a solution of the Yang-Mills equations only in the limit of infinite separation, but should be a decent approximation to the dominant configuration so long as the pseudoparticles do not overlap significantly. In this case the functional determinant should just be the product of the individual instanton determinants, and we know what weight to give these configurations when we integrate over $z_i, \rho_i, R_{ab}^{(i)}$. In what follows, we shall in first approximation neglect interactions, taking the action of N pseudoparticles to be $N8\pi^2/g^2$. After exploring the consequences of this assumption, we will turn to a computation and discussion of the effects of instanton–anti-instanton interactions.

The dilute gas arguments of Sec. II then imply that the dominant contribution to the functional integral comes from configurations where the space-time density of pseudoparticles of scale size between ρ and $\rho + d\rho$ is [for SU(2)]

$$\frac{d\rho}{\rho^5} D(\rho) = \frac{d\rho}{\rho^5}(0.26)\left(\frac{8\pi^2}{g^2(\rho)}\right)^4 e^{-8\pi^2/g^2(\rho)} \tag{3.8}$$

To check the consistency of the dilute-gas approximation we may compute the fraction, $f(\rho)$, of space-time occupied by pseudo-particles of scale size less than ρ. If $f(\rho)$ is less than unity we will have a dilute gas at scale size ρ. We take the "volume" of an instanton of scale size ρ to be that of a sphere of radius $\rho[(\pi^2/2)\rho^4]$ and find

$$f(\rho) = \pi^2 \int_0^\rho \frac{d\rho'}{\rho'} D(\rho') \tag{3.9}$$

(this is the sum of equal contributions from instantons and anti-instantons). In the asymptotic freedom regime (small ρ) we may reexpress f as an integral over $x = 8\pi^2/g^2$ by using the [SU(2)] relation

$$\frac{dx}{d\ln 1/\rho\mu} = \tfrac{22}{3}\,. \tag{3.10}$$

The result is

$$f(x) = \frac{3\pi^2}{22}\int_x^\infty dx\, D(x)\,, \tag{3.11}$$

where now $D(x) = 0.26x^4e^{-x}$. Figure 2 displays f as a function of x as well as $(\rho\mu)$. It is clearly a very rapidly varying function: f increases from 0.01 to 1 as ρ increases by only a factor of 2, from $0.15\mu^{-1}$ to $0.35\mu^{-1}$. We shall find that when f is less than one, but not vanishingly small (greater than 0.1, say) instantons cause significant modifications of vacuum properties, in spite of the smallness of the effective coupling in this region ($x \sim 10$). When f is greater than 1, however, the instanton gas picture must break down and some new vacuum physics must take over. Everything that we will find strongly suggests that this transition is associated with confinement, or at least with the physics which sets the scale size of hadrons, and we will provisionally make that identification. Since the rise in f is so rapid as a function of x or ρ one gets a rather sharp definition of the hadronic coupling constant and scale size: $x_c \sim 8$, $\rho_c\mu \sim 0.3$. Thus, through the equation $f(x(\rho_c\mu)) \sim 1$, one realizes dimensional transmutation, eliminates μ in favor of the hadron scale size, and identifies the hadronic coupling constant as a pure number. By most usual measures $x \sim 8$ corresponds to a rather small coupling constant and measurements which

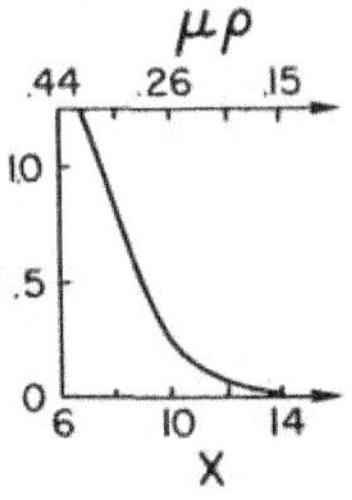

FIG. 2. The fraction f of space-time occupied by instantons smaller than a given scale size ρ in an SU(2) (no quarks) gauge theory. We plot f as a function of ρ and $x(\rho)$.

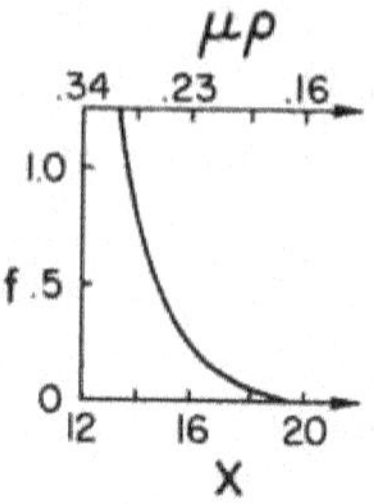

FIG. 3. The fraction f of space-time occupied by instantons smaller than a given scale size ρ in an SU(3) (no quarks) gauge theory. We plot f as a function of ρ and $x(\rho)$.

probe distances smaller than the hadron scale will see even smaller coupling. This picture of what happens in SU(2) requires some modification because it will turn out that at $x \lesssim 11$, the instantons ionize into a new kind of pseudoparticle, dubbed meron, which is very directly related to the dynamics of confinement. This would lead us to revise our estimate of the critical values of ρ and x slightly to $x_c \sim 11$, $\rho_c \sim 0.2\mu^{-1}$. We will amplify this remark in Sec. VI.

The situation for SU(3) is significantly different. Now

$$\frac{dx}{d\ln(1/\rho\mu)} = 11 \tag{3.12}$$

and we have [in contrast to Eq. (3.11)]

$$f(x) = \frac{\pi^2}{11}\int_x^\infty dx\, D_{\mathrm{SU}(3)}(x)\,. \tag{3.13}$$

This function, plotted in Fig. 3, attains the critical value at a much larger value of x and grows much more rapidly with ρ than the SU(2) case. The same kind of argument as before would lead us to believe that the hadron coupling constant and scale size are given by $x_c \sim 16$, $\rho_c \sim 0.25\mu^{-1}$. Again, the analysis of various effects arising from merons or instanton interactions will cause us to revise these numbers slightly, but the basic point remains: The dilute instanton gas picture reveals how and where dimensional transmutation occurs and shows, most importantly, that the coupling constant at the hadron scale size is a *small* number which gets smaller as the group gets larger. The value provisionally associated with SU(3), $\alpha \sim 0.4$, is not far from numbers which have been extracted from optimistic studies of scaling in electroproduction.

We now would like to discuss various effects which arise within the dilute-gas picture when the density, while still small, becomes large enough for the pseudoparticles to influence each other. To study this problem, we imagine imposing on the system from the outside a *weak* slowly varying external field, $F^{\mathrm{ext}}_{\mu\nu}$. In the no-instanton vacuum the action of such a configuration is just

$$\delta S = \frac{1}{4g^2}\int d^4x \sum_a (F^{\mathrm{ext}}_{\mu\nu a})^2\,. \tag{3.14}$$

In fact A^{ext} must be regarded as a perturbation on $[A_\mu]^0$, the multiple-instanton configuration which dominates the dilute gas vacuum. The interaction of the external field with the instantons must be included in δS and, as we shall see, the net effect for a weak external field is just a coupling-constant renormalization.

First, we examine the problem of a single instanton in a weak, slowly varying $F^{\mathrm{ext}}_{\mu\nu}$. Thus, $A_\mu = [A_\mu]^0 + \delta A_\mu$ where $[A_\mu]^0$ is the standard single instanton of scale size ρ and δA_μ approaches the potential of a weak constant $F^{\mathrm{ext}}_{\mu\nu}$ at distances large compared to the instanton scale size. In fact we will divide space into two regions by a sphere of radius R, many times ρ, such that inside, in retion I, $[A_\mu]^0$ is larger than δA_μ and outside, in retion II, δA_μ is larger than $[A_\mu]^0$. For $|x|$ comparable to R we may choose $\delta A_\mu = -\frac{1}{2}F^{\mathrm{ext}}_{\mu\nu}x_\nu$, the Landau gauge potential of a weak constant $F_{\mu\nu}$.

Inside R we could find δA_μ explicitly as the solution of the linearized equations of motion in the instanton background field subject to the condition of regularity at the origin and linear growth at large x. On the other hand, if we define

$$S_{\mathrm{I}} = \tfrac{1}{4}\int_{|x|<R} d^4x (F^a_{\mu\nu})^2\,, \tag{3.15}$$

then

$$S_{\mathrm{I}}([A]^0 + \delta A) - S_{\mathrm{I}}([A]^0)$$
$$\int_{|x|<R} d^4x (D^0_\mu \delta A^a_\nu) F^{a0}_{\mu\nu} + O(\delta A^2)\,. \tag{3.16}$$

Upon integrating by parts and using the equation of motion for the instanton background field, $D^0_\mu F^0_{\mu\nu} = 0$, we can express the interaction energy as a surface term,

$$S_{\mathrm{I}}(A^0 + \delta A) - S_{\mathrm{I}}(A^0) = \int_{|x|=R} d\Omega\, \hat{x}_\mu \delta A^a_\nu (F^a_{\mu\nu})^0 + O(\delta A^2)\,. \tag{3.17}$$

Now A^0_μ is the instanton field in singular gauge so that

$$[F^a_{\mu\nu}]^0 = \frac{4}{g} M_{\mu\mu'} M_{\nu\nu'} \frac{\bar{\eta}_{a\mu'\nu'}\rho^2}{(x^2+\rho^2)^2}\,, \tag{3.18}$$

where $M_{\mu\nu} = g_{\mu\nu} - 2\hat{x}_\mu \hat{x}_\nu$. Also, on the surface $|x| = R$ we may set $\delta A_\mu = -\frac{1}{2}F^{\mathrm{ext}}_{\mu\nu}x^\nu$. The angular averages may then be carried out explicitly to evaluate the $O(\delta A)$ surface term, with the result

$$S_1(A^0+\delta A)-S_1(A^0)=-\frac{\pi^2}{g}\rho^2 F^{a\,\text{ext}}_{\mu\nu}\bar{\eta}_{a\mu\nu} + O(\delta A^2). \tag{3.19}$$

The $O(\delta A^2)$ piece can also be seen to reduce to a surface term and turns out to give the integral of $\frac{1}{4}(F^{\text{ext}}_{\mu\nu})^2$ over the region $|x|<R$. A similar analysis of region II, with A^0 now regarded as the perturbing field and A^{ext} as the background field, yields an identical surface term to Eq. (3.19), assuming only that A^{ext} is a solution by itself of the Yang-Mills equations and that it goes to zero at infinity.

The net result of all this is that an instanton in a weak, slowly varying background field may be assigned an interaction energy

$$S_{\text{int}}=-\frac{2\pi^2}{g}\rho^2\bar{\eta}_{a\mu\nu}F^{\text{ext}}_{a\mu\nu}, \tag{3.20}$$

where F^{ext} is the value of background field at the instanton position. Note that if F^{ext} is self-dual, S_{int} vanishes because $\bar{\eta}_{a\mu\nu}$ is an anti-self-dual tensor. If F^{ext} is itself taken to be an anti-instanton field of scale size $\bar{\rho}$ and centered a distance R from the instanton, one finds

$$S_{\text{int}}=+\frac{32\pi^2}{g^2}\frac{\rho^2\bar{\rho}^2}{R^4}R_{ab}\bar{\eta}_{a\mu\nu}\eta_{b\mu\nu'}\hat{R}_\nu\hat{R}_{\nu'}, \tag{3.21}$$

where $\hat{R}$ is the unit vector pointing from the instanton to the anti-instanton and R_{ab} is the matrix describing the relative group orientation of the two pseudoparticles. The maximum value of $-S_{\text{int}}$ (obtained by varying R_{ab}) is $96\pi^2 g^{-2}\rho^2\bar{\rho}^2R^{-4}$, independent of $\hat{R}$. This agrees with Forster's calculation of the action of a pair constrained only as to location, but not as to orientation.[34] Since, in general, S_{int} depends on instanton group orientation and falls with separation like R^{-4}, instantons look like objects carrying a *color magnetic dipole* moment proportional to $\rho^2\bar{\eta}_{a\mu\nu}$. [Recall that a magnetic dipole moment is really an antisymmetric tensor $\sim\int(j^\mu x^\nu - j^\nu x^\mu)$.] This interpretation leads us to expect that the dilute instanton gas will behave like a dilute gas of spins with its response to an external field described by a susceptibility. This effect will, among other things, lead to a "classical coupling-constant renormalization" of a rather interesting nature which we will now study.

In the low density limit we may use a virial expansion to find the effect of the medium on an external field: Given the interaction, Eq. (3.20), of a single pseudoparticle with the external field, one first computes

$$-S_{\text{eff}}=\langle e^{-S_{\text{int}}}-1\rangle\,,$$

averaging over group orientation of the pseudoparticle, and then weights this effective single-particle action with the appropriate pseudoparticle density and integrates over scale size and location to get the full effective action. For weak external fields, one finds from Eq. (3.20) that

$$\langle e^{-S_{\text{int}}}-1\rangle=\left(\frac{F_{\text{ext}}-\tilde{F}_{\text{ext}}}{2}\right)^2\frac{8\pi^2}{g^2}\frac{\pi^2\rho^4}{\Delta(n)} + O(F_{\text{ext}}^4), \tag{3.22}$$

where n refers to the SU(n) gauge group, $\Delta(n)$ is 3 (8) for $n=2$ (3), and $\tilde{F}_{\mu\nu}=\frac{1}{2}\epsilon_{\mu\nu\lambda\sigma}F_{\lambda\sigma}$. The appearance of $F-\tilde{F}$ simply reflects the fact that the instanton interacts only with the anti-self-dual part of F_{ext}. The analog of Eq. (3.22) for an anti-instanton simply replaces $(F-\tilde{F})^2$ by $(F+\tilde{F})^2$. Since instantons and anti-instantons occur with equal probability, the net effect of the medium is proportional to $[(F+\tilde{F})^2+(F-\tilde{F})^2]/4=F^2$. In other words, the effect of the medium is to renormalize the original external field action density by a multiplicative constant K^{-1}, where

$$K^{-1}=1-\int\frac{d\rho}{\rho^5}D_n(\rho)\frac{8\pi^2}{g^2(\rho)}\frac{4\pi^2\rho^4}{\Delta(n)}\,. \tag{3.23}$$

Since $K>1$, the instantons cause the vacuum to behave like a paramagnetic medium and increase the interaction energy between fixed external sources. In fact, one easily sees from our discussion of the integrated density, $f(\rho)$, that the integral in Eq. (3.20) will be large even in the dilute gas region where f is less than one. If K is large enough, it may even be energetically favorable in the presence of external sources (quarks) to form a flux tube (or bag) in which the flux is expelled from the region of normal vacuum (K large) and confined to a region of abnormal vacuum ($K\simeq 1$) where no instantons are present. Expelling the instantons costs vacuum energy which is made up for by the lowered interaction energy between the quarks. In this picture, the confinement or hadron scale will have directly to do with where K begins to depart significantly from one. This will be different from our earlier criterion based on integrated density, but not dramatically so.

If we focus our attention on the instantons themselves, the above effects can be interpreted as a coupling-constant renormalization. Evidently, the action of an instanton of scale size ρ is decreased by the presence within it of smaller scale instantons. Very crudely

$$\frac{8\pi^2}{\bar{g}^2(\rho)}=\frac{8\pi^2}{g^2(\rho)} - \int_0^\rho\frac{d\rho'}{\rho'}\frac{4\pi^2}{\Delta(n)}\left(\frac{8\pi^2}{g^2(\rho')}\right)^2 D_n(\rho')\,, \tag{3.24}$$

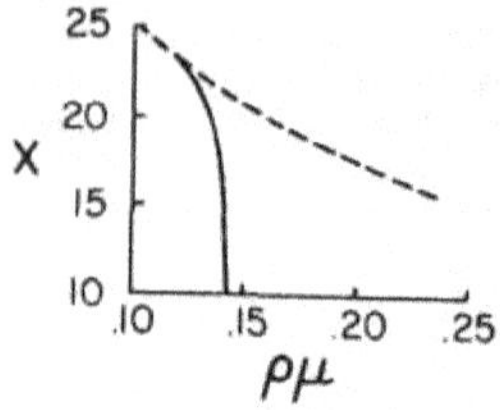

FIG. 4. The dependence of the effective coupling $x = 8\pi^2/g^2$ as a function of distance $\rho\mu$, taking the classical renormalization effects into account for SU(3) (no quarks). The dashed line represents the value of x using asymptotic freedom alone.

where $g(\rho)$ is, at this stage, the standard asymptotic freedom running coupling constant and $\bar{g}(\rho)$ is the running coupling constant including the "classical" renormalization effects of the medium. As expected, the medium amplifies the ordinary perturbative asymptotic freedom effects and causes the effective coupling to increase more rapidly with scale size. One could regard Eq. (3.24) as a self-consistent equation for the effective coupling by replacing $x = 8\pi^2/g^2(\rho)$ under the integral by $\bar{x} = 8\pi^2/\bar{g}^2(\rho)$:

$$\bar{x}(\rho) = x(\rho) - \int_0^\rho \frac{d\rho'}{\rho'} \frac{4\pi^2}{\Delta(n)} \bar{x}^2(\rho) D_n(\bar{x}(\rho)), \qquad (3.25)$$

where $x(\rho)$ is taken to be the perturbative asymptotic freedom effective coupling. Differentiation with respect to $\ln\rho$ gives a renormalization-group equation for $\bar{x}$,

$$\frac{d\bar{x}}{d\ln(1/\rho\mu)} = C_n + \frac{4\pi^2}{\Delta(n)} \bar{x}^2 D_n(\bar{x}) \qquad (3.26)$$

(where $C_2 = \frac{22}{3}$, $C_3 = 11$), which we may integrate numerically. The results for SU(3) are displayed in Fig. 4 and for SU(2) in Fig. 5. In both cases the new effects turn on very sharply at scale sizes where our earlier estimates indicated that the integrated pseudoparticle density was rather small, say 10%. As soon as they are at all significant, the new effects are dominant.

A further point is that since $\bar{x}$ increases more rapidly with ρ than required by asymptotic freedom, the integrated density functions Eqs. (3.9) and (3.11) should be modified. Taking account of Eq. (3.25) we have

$$\bar{f}(x) = \pi^2 \int_x^\infty d\bar{x} \, \frac{D_n(\bar{x})}{C_n + [4\pi^2/\Delta(n)]\bar{x}^2 D_n(\bar{x})} \, . \qquad (3.27)$$

Now $\bar{f}(x)$ is less than $f(x)$ and, more importantly, always less than one for interesting values of x ($x \gtrsim 10$): One easily sees that $\bar{f}(x) < x\Delta(n)/4$. Therefore, the coupling-constant renormalization effects appear to reduce instanton densities to a manageable level and change our definition of the critical scale size (where a transition from vacuum physics to confinement physics occurs) to that scale ($\sim 0.1\mu^{-1}$) where the effective coupling begins to increase very rapidly. Therefore we must modify the picture we extracted from the behavior of the noninteracting instanton gas at the beginning of the section. There we said that the onset of new physics is associated with the passage of $f(x)$ through 1, identifying in that way a critical coupling and scale size. Once the effects of interactions are included, the density no longer rises dramatically, but there is still a well-defined scale size and coupling constant at which the renormalized coupling constant (and vacuum susceptibility) begin to rise rapidly. We now identify this transition as setting the hadron scale and find new critical couplings and scale sizes for SU(2) and SU(3) which do not, in fact, differ markedly from the original estimates.

These considerations are probably too crude to be taken very seriously since the rapid rise in $D(x)x^2$ means that most of the renormalization effect on an instanton of a given scale size is coming from instantons of nearly the same size. It probably should be renormalized by instantons of, say, half its size, or smaller, which would delay the onset of sizable renormalization effects to smaller x and integrated densities more nearly equal to one. At the moment we do not know how to translate this notion into manageable mathematics, but do not believe that it would materially change the qualitative conclusions we have reached. The most important of these qualitative effects, let us repeat, is the identification of a well-defined scale size and coupling constant (and a small coupling as well) at which there is a transition between asymptotic freedom behavior and confining behavior. This provides, we believe, the basic

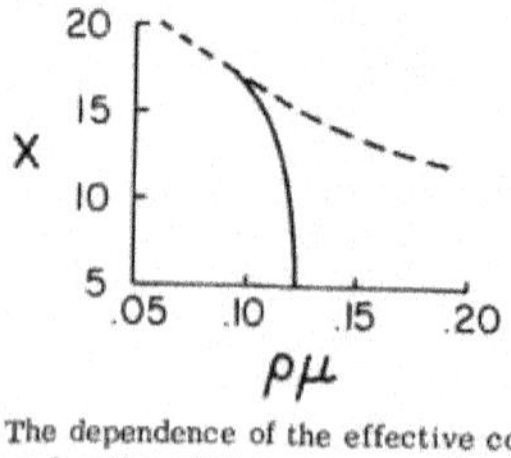

FIG. 5. The dependence of the effective coupling $x = 8\pi^2/g^2$ as a function of distance $\rho\mu$, taking the classical renormalization effects into account for SU(2) (no quarks). The dashed line represents the value of x using asymptotic freedom alone.

explanation for the apparent smallness of the Yang-Mills coupling on the scale of ordinary hadron sizes.

An alternate method of computing the "classical" coupling-constant renormalization is to compute the gauge field propagator $D^{ab}_{\mu\nu}(x-y)$. We now know that the effects of vacuum tunneling may not be negligible and we must include the non-Gaussian fluctuations associated with tunneling. The simplest way to do this is to write A^a_μ as a sum over instanton and anti-instanton fields (in singular gauge) and perform the average by integrating over instanton scale sizes, locations, and group orientations with the appropriate density function $D(\rho)$ [$D(\rho)$ summarizes the effects of *Gaussian* fluctuations about vacuum tunneling]. In the product $A^a_\mu(x)A^b_\nu(y)$, cross terms between different instantons vanish when we perform the independent group orientation averages and we are left with a sum over instantons of the correlation function $\langle A^a_\mu A^b_\nu\rangle$ for a single instanton.

It is best to evaluate the propagator in momentum space (momentum q) and we have ($A^{(0)}_\mu$ is the one-instanton field)

$$\int d^4y\, e^{iq\cdot y}\int d^4x\, e^{iq\cdot x}\int d^4z\, A^{a(0)}_\mu(x-z)A^{b(0)}_\nu(y-z)$$

$$=(2\pi)^4\delta(q'-q)\langle\bar{A}^{a(0)}_\mu(q)\bar{A}^b_\nu(-q)\rangle. \quad (3.28)$$

The Fourier transform of the singular gauge instanton field is easily seen to be

$$\bar{A}^{a(0)}_\mu(q)=\frac{i(4\pi)^2}{g(\rho)}\,\frac{\eta_{a\mu\nu}q_\nu}{q^4}F(\rho q)\,, \quad (3.29)$$

where ρ is the scale size, F has the property

$$F(x)\to 1,\quad x\to\infty$$

$$F(x)\to -\tfrac{1}{4}x^2,\quad x\to 0 \quad (3.30)$$

and $g(\rho)$ is the coupling appropriate to scale size ρ.

To construct the tunneling contribution to the propagator we must form $\bar{A}^a_\mu\bar{A}^b_\nu$, average over group orientations ($\eta_{a\mu\nu}\to R_{ab}\eta_{b\mu\nu}$, and average over R), sum over instantons and anti-instantons, and integrate over scale size. The result is

$$D^{ab}_{\mu\nu}(q)_{\text{tunnel}}=(4\pi)^2\delta_{ab}\,\frac{\delta_{\mu\nu}-q_\mu q_\nu q^{-2}}{q^6}\times\int\frac{d\rho}{\rho^5}\,\frac{D(\rho)}{3}\,\frac{F^2(\rho q)}{g^2(\rho)}\,. \quad (3.31)$$

The factor $\frac{1}{3}$ arises from group averaging if the group is SU(2). For SU(3) replace $\frac{1}{3}$ by $\frac{1}{8}$.

If there is an upper cutoff ρ_c on ρ and $q\rho_c<1$, then this result is simplified to

$$D^{ab}_{\mu\nu}(q)_{\text{tunnel}}=(2\pi)^4\delta_{ab}\,\frac{\delta_{\mu\nu}-q_\mu q_\nu/q^2}{q^2}\times\int_0^{\rho_c}\frac{d\rho}{\rho}\,\frac{1}{g^2(\rho)}\,\frac{D(\rho)}{3}\,. \quad (3.32)$$

This is just a numerical multiple of the free propagator, and the result of adding the two effects together is to produce a finite wave-function renormalization

$$Z=1+\frac{\pi^2}{3}\int_0^{\rho_c}\frac{d\rho}{\rho}\left(\frac{8\pi^2}{g^2(\rho)}\right)D(\rho)\,. \quad (3.33)$$

Since the effect of Z on g is $g^2\to g^2Z$ this classical wave-function renormalization increases the effective coupling. Our previous result was

$$\frac{1}{g^2}\to\frac{1}{g^2}\left(1-\frac{\pi^2}{3}\int_0^{\rho_c}\frac{d\rho}{\rho}\,\frac{8\pi^2}{g^2(\rho)}D(\rho)\right). \quad (3.34)$$

As long as the renormalization effect is small, the two results are identical. The previous calculation is in fact the more accurate. It was a self-consistent field type of calculation which takes into account *instanton-anti-instanton correlations* induced by the long-range R^{-4} interactions. Equation (3.27) does not take these correlations into account.

Finally, it is worth emphasizing that the instanton gas is a paramagnetic medium. It is magnetic rather than electric because a Euclidean gauge theory corresponds to static magnetism. It is paramagnetic because the coupling is renormalized upward.

*On leave from Princeton University 1976–77.

†On leave at Institute for Advanced Study 1977–78.

[1]M. Gell-Mann, Phys. Lett. 8, 214 (1964); G. Zweig, CERN Reports Nos. TH 401 and 412, 1964 (unpublished).

[2]O. W. Greenberg, Phys. Rev. Lett. 13, 598 (1964); O. W. Greenberg and M. Resnikoff, Phys. Rev. 163, 1844 (1968); M. Gell-Mann, in *Elementary Particle Physics*, proceedings of the XI Schladming conference, edited by P. Urban (Springer, Berlin, 1972) [Acta Phys. Austriaca Suppl. 9 (1972)], p. 733; W. A. Bardeen, H. Fritzsch, and M. Gell-Mann, in *Scale and Conformal Symmetry in Hadron Physics*, edited by R. Gatto (Wiley, New York, 1973).

[3]D. Gross and F. Wilczek, Phys. Rev. Lett. 30, 1343 (1973); H. Politzer, *ibid.* 30, 1346 (1973); S. Coleman and D. Gross, *ibid.* 31, 1343 (1973).

[4]D. Gross and F. Wilczek, Phys. Rev. D 8, 3633 (1973); 9, 980 (1974).

[5]H. Georgi and H. Politzer, Phys. Rev. D 9, 416 (1973).

[6]S. Weinberg, Phys. Rev. Lett. 31, 494 (1973).

[7]D. Gross and A. Neveu, Phys. Rev. D 10, 3235 (1974).

[8]H. Fritzsch, M. Gell-Mann, and H. Leutwyler, Phys. Lett. 47B, 365 (1973); P. Langacker and H. Pagels, Phys. Rev. D 9, 3413 (1974); J. Kogut and L. Susskind, *ibid.* 10, 3468 (1974); 11, 3594 (1975).

[9]S. Weinberg, Phys. Rev. D 11, 3583 (1975).

[10]K. Wilson, Phys. Rev. D 10, 2445 (1975).

[11]J. Kogut and L. Susskind, Phys. Rev. D 11, 395 (1975).

[12]A. Polyakov, Phys. Lett. 59B, 82 (1975); A. Belavin, A. Polyakov, A. Schwartz, and Y. Tyupkin, Phys. Lett. 59B, 85 (1975).

[13]G. 't Hooft, Phys. Rev. Lett. 37, 8 (1976).

[14]C. Callan, R. Dashen, and D. Gross, Phys. Lett. 63B, 334 (1976).

[15]R. Jackiw and C. Rebbi, Phys. Rev. Lett. 37, 172 (1976).

[16]S. Adler, Phys. Rev. 177, 2426 (1969); J. S. Bell and R. Jackiw, Nuovo Cimento 60A, 47 (1969).

[17]Y. Nambu and G. Jona-Lasinio, Phys. Rev. 122, 345 (1961).

[18]C. Callan, R. Dashen, and D. Gross, Phys. Rev. D 16, 2526 (1977).

[19]E. Witten, Phys. Rev. Lett. 38, 121 (1977); G. 't Hooft (unpublished); R. Jackiw, C. Nohl, and C. Rebbi, Phys. Rev. D 15, 1642 (1977); R. S. Ward, Phys. Lett. 61A, 81 (1977); M. Atiyah and R. S. Ward, Oxford report, 1977 (unpublished).

[20]When the gauge group is SU(n), $n \geq 3$, there are other ways of constructing instantons. In the case of SU(3) one can construct a solution of equations of motion with Pontryagin index 4 by embedding the one-instanton solution in the SO(3) subgroup of SU(3): F. Wilczek, *Quark Confinement and Field Theory* (Wiley-Interscience, N.Y., 1977), p. 211; W. Marciano, H. Pagels, and Z. Parsa, Phys. Rev. D 15, 1044 (1977). It is not known whether this is simply a special case of the complete four-instanton solution. If not, it is a new configuration that should be included. We shall, in this paper, ignore such configurations.

[21]G. 't Hooft, Phys. Rev. D 14, 3432 (1976).

[22]A. S. Schwartz, Phys. Lett. 67B, 172 (1977); R. Jackiw and C. Rebbi, *ibid.* 67B, 189 (1977); M. F. Atiyah, N. J. Hitching, and J. M. Singer, Proc. Natl. Acad. Sci. USA (to be published).

[23]R. Jackiw and C. Rebbi, Phys. Rev. D 14, 517 (1976).

[24]L. N. Lipatov, Zh. Eksp. Teor. Fiz. Pis'ma Red. 24, 179 (1976) [JETP Lett. 24, 157 (1976)]; E. Brezin, J. C. Le Guillou, and J. Zinn-Justin, Phys. Rev. D 15, 1544 (1977); 15, 1558 (1977).

[25]J. M. Kosterlitz and D. J. Thouless, J. Phys. C 6, 1181 (1973).

[26]D. Caldi, Phys. Rev. Lett. 39, 121 (1977).

[27]A. M. Polyakov, Nucl. Phys. B120, 429 (1977).

[28]V. De Alfaro, S. Fubini, and G. Furlan, Phys. Lett. 65B, 1631 (1977).

[29]S. Mandelstam, Phys. Rep. 23C, 245 (1975).

[30]C. Callan, R. Dashen, and D. Gross, Phys. Lett. 66B, 375 (1977).

[31]K. Rothe and J. Swieca (J. Swieca, private communication.)

[32]N. Nielsen and B. Schroer, Phys. Lett. 66B, 475 (1977).

[33]We are grateful to D. Laughton for help in calculating the finite part of the coupling renormalization. A detailed account of the calculation will be published by him.

II. COLLECTIVE COORDINATES AND THE INSTANTON MEASURE

INTRODUCTION

The classical Yang–Mills action is invariant with respect to a large class of transformations: (i) 15-parametric conformal group (4 translations, 4 special conformal transformations, 6 Lorentz rotations and 1 dilatation); (ii) global rotations in the color space (isospace) described by 3 parameters in the case of $G = SU(2)$. Some of these symmetries are broken for every given instanton solution. For instance, since the BPST instanton is a localized four-dimensional configuration, it has a center, and for any given position of the center, the translational invariance is of course absent. The symmetries of the action are restored only if one considers the family of all possible instanton solutions as a whole. Then, for any solution with the given center, there is another one, with the center shifted to a different point in the four-dimensional space.

As a matter of fact, the symmetry of the classical solutions can be higher than the symmetry of the classical action. This, for instance, is the case for the multi-instanton solutions, $|Q| > 1$.

Therefore, the first task is to characterize the general family of the instanton solutions by all relevant *collective coordinates* — parameters describing the response of the solutions to the symmetry transformations. The general strategy is quite straightforward. One starts from some known solution and transforms it by applying all symmetry transformations. The symmetry transformations which act nontrivially generate new solutions and require an introduction of the collective coordinates. The action of the symmetry generators prompts us on how these coordinates must be introduced.

The simplest example is the translation invariance mentioned above. $A^a_\mu = 2\eta^a_{\mu\nu}x_\nu(1+x^2)^{-1}$ is the BPST instanton centered at the origin. The translational symmetry tells us that substituting x by $x - x_0$, we generate another solution, with the center at x_0.

Apart from the instanton center, the generic $SU(2)$ instanton is characterized by its size ρ (sometimes called the instanton radius) and by three additional parameters $\vec{\omega}$ corresponding to global $SU(2)$ rotations. Altogether there are eight collective coordinates. All details relevant to the generic $SU(2)$ BPST solution are exhaustively discussed in the classic paper by 't Hooft presented in this section.

The question which immediately comes to mind concerns the fate of other symmetries. Why, say, do the special conformal transformations or Lorentz rotations not generate their own collective coordinates? The answer to this question can be found in Refs. [1–3].

The key point is the fact that the BPST instanton is actually invariant under certain combinations of the symmetry transformations (modulo gauge transformations).

Let us introduce the notations for the symmetry generators:[5]

[5]Concise reviews of the conformal group can be found in Ref. [4].

$$P_\mu \text{ (translations) ;}$$
$$M_{\mu\nu}\eta^a_{\mu\nu} \text{ (right-handed Lorentz rotations) ;}$$
$$M_{\mu\nu}\bar{\eta}^a_{\mu\nu} \text{ (left-handed Lorentz rotations) ;}$$
$$K_\mu \text{ (special conformal transformations) ; } D \text{ (dilatation) ;}$$
$$T^a \text{ (color space (isospace) rotations) ,}$$

where η and $\bar{\eta}$ are the 't Hooft symbols. Then the action of $P_\mu + 2\rho^2 K_\mu$, $M_{\alpha\beta}\bar{\eta}^a_{\alpha\beta}$ and $M_{\alpha\beta}\eta^a_{\alpha\beta} + T^a$ does not change the instanton solution up to irrelevant gauge transformations. Hence, out of 18 potential collective coordinates, only 8 survive.

A remark concerning the three collective coordinates associated with the $SU(2)$ rotations is in order here. One may say that these collective coordinates correspond to rotations in the color space (isospace). Alternatively, it is possible to interpret them as due to rotations in the $SU(2)_R$ subgroup of the $O(4)$ Euclidean Lorentz group. These two options are indistinguishable in pure gluodynamics. The introduction of the Higgs field, however, changes the situation. In the spontaneously broken Yang–Mills theory, with the nonvanishing vacuum expectation value of the Higgs field, only the second approach — where the chiral Lorentz rotations are used to generate the corresponding collective coordinates — is appropriate [3]. Rotations in the isospace are forbidden by the vacuum alignment of the Higgs field.

[Let us parenthetically note that the coordinate inversion $x_\mu \to x_\mu/x^2$ is also a symmetry of the conformally invariant action. As has been demonstrated in Ref. [2] (this paper is specifically devoted to the symmetries of the BPST instanton), the coordinate inversion can be used to generate the anti-instanton solution in the singular gauge starting from the instanton solution in the nonsingular gauge. Likewise, the anti-instanton in the nonsingular gauge becomes the singular-gauge instanton under the coordinate inversion.]

The number of collective coordinates grows with Q and with the rank of the gauge group considered. Thus, for $SU(N)$, the number of collective coordinates is $4N|Q|$. It is inevitable that some of them — actually, the majority — lose the transparent geometrical interpretation connecting the collective coordinates with the conformal and color (isotopic) symmetries of the classical action. "Superfluous" collective coordinates reflect the existence of a wider symmetry of classical solutions, a fact which has already been mentioned above.

After completing the description of the family of field configurations minimizing the action, one can proceed to the second stage, i.e. calculating the instanton contribution to the correlation functions which might be of physical interest. As a matter of fact, all such calculations contain a universal element, the so-called instanton measure $d\mu$ — the instanton contribution to the partition function Z.

The knowledge of the instanton action $S_{inst} = 8\pi^2/g^2$ tells us only that the instanton measure $d\mu \propto \exp(-8\pi^2/g^2)$, g being the bare coupling constant entering the QCD Lagrangian at the ultraviolet scale.

To calculate the instanton measure, it is necessary to include the effect of quantum fluctuations around the instanton solution,

$$Z_{inst} = \int \mathcal{D}a_\mu(x) \exp\{-S[a_\mu(x)]\} \tag{1}$$

where the gauge field is represented in the form

$$A_\mu = A_\mu^{inst}(\alpha) + a_\mu \ , \tag{2}$$

and α denotes generically the set of all relevant collective coordinates. Accounting for the quantum fluctuations will convert the bare coupling constant into the running one, $g^2(\rho)$, it will also fix pre-exponential factors, including the measure of integration over the collective coordinates. Only after the pre-exponential factors are established, the instanton contribution to the physical quantities becomes well defined.

Conceptually the procedure is quite standard and straightforward. One starts from the decomposition (2); the action is then expanded in powers of a_μ. The term linear in a_μ is absent due to the fact that A_μ^{inst} is the solution of the equation of motion. Hence,

$$S = S[A_\mu^{inst}(\alpha)] + \int d^4x a_\mu(x) L^{\mu\nu}(\alpha) a_\nu(x) + \dots \ , \tag{3}$$

where $L^{\mu\nu}$ is some linear differential operator of the second order. Limiting oneself to the leading (Gaussian) approximation means that all terms of the order $\mathcal{O}(a^3)$ and higher denoted by dots in Eq. (3) are neglected. Then the integral over $\mathcal{D}a_\mu(x)$ in Eq. (1) can be done provided that all modes of the operator $L^{\mu\nu}$ are known,

$$Z \sim \{Det L\}^{-1/2} \exp\{-8\pi^2/g^2\} \ .$$

The actual calculation is of course technically much more complicated than the outline given above. First, one should not forget about gauge fixing which entails, in turn, the appearance of the ghost sector. Second, one has to deal with the problem of the ultraviolet regularization. Third, all zero modes of the operator L have to be singled out and treated separately in order to avoid unphysical infinities in $\{Det L\}^{-1/2}$.

The program is carried out in full in Ref. [1], the work which essentially opened the instanton field for applications. This work (it opens this section) is on the desk of every instanton practitioner, and is recommended for thorough study.

The fact that the operator L must — and does — have zero modes is rather obvious. Each collective coordinate reflecting the existence of a nontrivially realized symmetry produces its own zero mode. Thus, for $SU(2)$ instanton we have 8 zero modes.

Indeed, the symmetry of the action implies that there is a variation of the classical field generated by a change in the corresponding collective coordinate which gives no variation of the action. For instance, $A_\mu^{inst}(\rho)$ and $A_\mu^{inst}(\rho+\delta\rho)$ have one and the same action, $8\pi^2/g^2$, and hence,

$$a_\mu^{(0)} \propto A_\mu^{inst}(\rho+\delta\rho) - A_\mu^{inst}(\rho)$$

has to be the zero mode of $L^{\mu\nu}$.

It is clear that the integrals over the zero modes in Eq. (1) are non-Gaussian, they have to be extracted from *Det L*; treating them properly, we arrive at the measure of integration over the collective coordinates.

The functional form of the instanton measure can be established from symmetry arguments alone, without explicit computations. As a matter of fact, as has been shown in Ref. [5], invoking some general properties of the Yang–Mills theory known in perturbation theory, one can readily fix $d\mu$ up to an overall constant factor.

First, it is rather obvious that $d\mu$ is proportional to d^4x_0 and $d\rho$: instantons can have all sizes and can be located everywhere in the four-dimensional space. Integration over the $SU(2)$ group orientations can be done in the closed form.[6] It gives just a constant factor since the $SU(2)$ group space is compact (sphere) and its volume is finite. On dimensional grounds, we then write

$$d\mu \propto d^4x_0 d\rho/\rho^5 \ .$$

Second, as explained in Ref. [5], each zero bosonic mode yields one power of the ultraviolet-regulator mass M_0 (following Ref. [1] we accept the Pauli–Villars regularization procedure, the only known procedure suitable for instanton calculations). This means that we must add the factor $(M_0\rho)^8$ where the appropriate power of ρ is inserted to ensure the fact that $d\mu$ is dimensionless. Now the instanton measure takes the form

$$\text{“}d\mu\text{”} = \text{const.}\ d^4x_0 \frac{d\rho}{\rho^5} \exp\left\{-\frac{8\pi^2}{g^2} + 4\ln(M_0\rho)^2\right\} , \tag{4}$$

where the quotation marks indicate that, so far, we have totally ignored the effects of nonzero modes.

Furthermore, from the renormalizability of the theory, one infers that $8\pi^2/g^2$ in the exponent is somehow transformed into $8\pi^2/g^2(\rho)$, the running coupling constant at the appropriate scale (which, in the case at hand, is the instanton size, of course). The coefficient 4 in front of the logarithm in Eq. (4) does not match the known expression for the first coefficient of the Gell–Mann–Low function — and it should not, of course, since “$d\mu$” does not take into account the nonzero modes.

Let us now recall that in $SU(N)$ gluodynamics, the first coefficient in the Gell–Mann–Low function is $(4-\frac{1}{3})N$, where $-1/3$ is a normal screening contribution of

[6] This statement, strictly speaking, refers to problems where one considers the instanton contribution in quantities which are color singlets (isosinglets); otherwise, we cannot integrate over the $SU(2)$ orientations beforehand and the corresponding collective coordinates should be kept in the measure, $d\mu \propto d^4x_0(d\rho/\rho^5)d\Omega$, where $d\Omega$ is the measure on the group space.

the transverse gauge quanta in the one-loop graphs, while 4 is an "abnormal" antiscreening part specific to non-Abelian theories and associated with the instantaneous Coulomb interaction [6] (for a review see Ref. [7]).

It is possible to show that, within instanton calculus, the antiscreening part is due to zero modes; and the net effect of the nonzero modes is to provide the missing screening part. This fact is demonstrated in Ref. [5] which the reader is referred to for further details.

If so, taking account of the nonzero modes merely shifts the coefficient in front of the logarithm in Eq. (4) and changes the overall constant which, anyway, cannot be established without explicit calculation,

$$\begin{aligned} d\mu &= \text{const.}\ d^4x_0 \frac{d\rho}{\rho^5} \exp\left\{-\frac{8\pi^2}{g^2} + (4-\frac{1}{3})\ln(M_0\rho)^2\right\} \\ &= \text{const.}\ d^4x_0 \frac{d\rho}{\rho^5} \exp\left\{-\frac{8\pi^2}{g^2(\rho)}\right\}\ . \end{aligned} \tag{5}$$

The 't Hooft calculation confirms Eq. (5) and provides the numerical factor in the pre-exponent. The striking feature of the instanton measure given above is the fact that at the one-loop level the expression becomes ρ-dependent. The scale dependence was absent at the classical level, in full accordance with the classical scale invariance of the Yang–Mills action. The occurrence of the scale dependence in the Gaussian approximation is of no surprise, of course. We know that the scale invariance is broken by the anomaly in the quantum corrections [8].

The instanton measure grows steeply with ρ; the expression (5) diverges in the infrared domain in a powerlike manner. The degree of divergence is even higher in $SU(3)$. This infrared instability is the reason why all naive instanton-based models of the QCD vacuum (in particular, the instanton gas model of Callan *et al.* [9]) turned out to be unsuccessful and were abandoned. The blow-up of the instanton measure at large ρ hinders quantitative applications of instantons in QCD, although their relevance at the qualitative level is beyond any doubt. At the same time, introducing the Higgs field with a large vacuum expectation value, we ensure a natural infrared cutoff in the theory which, obviously, must stabilize the instanton measure as well.

The issue of the instanton measure in the presence of the Higgs field has been first addressed by 't Hooft [1]. The cutoff is provided by an extra term in the action — the so-called 't Hooft term — originating from the mass term of the gauge field,

$$S_{inst} = \frac{8\pi^2}{g^2} + 2\pi^2 v^2 \rho^2\ , \tag{6}$$

where v is the vacuum expectation value of the Higgs field. Now, ρ dependence takes place at the classical level since v breaks the scale invariance right from the start. The exponential $\exp(-2\pi^2 v^2 \rho^2)$ renders all ρ integrations convergent. Strictly speaking, in the Higgs phase, there are no exact solutions of the classical equations of motion apart from the zero size instantons. In this case we rather speak abou the constrained instantons (see Sec. VI).

The 't Hooft term is absolutely crucial for modern development associated with instantons at high energies in the Standard Model in connection with the problem of the baryon number nonconservation [10,11] (for a review see Ref. [12]).

Embeddings of the $SU(2)$ instantons into higher groups are considered in Refs. [1 14]. The most studied case is that of $SU(N)$. A straightforward procedure of embedding suggested in Ref. [13] is usually referred to as standard. Within this procedure one places the $SU(2)$ instanton in the "upper left corner" of $SU(N)$, so that $A^a_\mu \neq 0$ if $a = 1,2,3$ and $A^a_\mu = 0$ if $a > 3$. The functional form of the BPST solution remains intact; new collective coordinates appear, though, reflecting the possibility of rotating the BPST instanton from the upper left corner to the outside, $4(N-2)$ new collective coordinates altogether. Thus, the total number of the collective coordinates is $4N$. Reference [13] presents a full calculation of instanton measure in the case of $SU(N)$.

The standard embedding is unique in the sense that it automatically generates $SU(N)$ self-dual fields with $|Q| = 1$ starting from the $SU(2)$ configurations with the same value of the topological charge Q. Needless to say that the $SU(2)$ solutions can be embedded into higher groups in a different way resulting in self-dual fields with $|Q| > 1$. A well-known example of this kind was found by Wilczek [14] who observed the following.

Assume that

$$A^a_\mu T^a = 2\frac{1}{x^2+\rho^2}\eta^b_{\mu\nu}x_\nu M_b \, , \tag{7}$$

where T^a are the generators of $SU(N)$ in the fundamental representation, [7] $\eta^b_{\mu\nu}$ are the 't Hooft symbols and M_b are some $N \times N$ matrices. Then the field strength tensor stemming from (7) will be automatically self-dual provided that

$$[M_a M_b] = i\epsilon_{abc}M_c, \tag{8}$$

i.e. the set of three matrices M_a forms an $SU(2)$ algebra.

This algebra is familiar to everybody from the courses of quantum mechanics. Its representations are $(2j+1)$-dimensional where $j = 1/2, 1, 3/2, \ldots$. In each representation one readily builds three $(2j+1)\times(2j+1)$ matrices of generators. For our purposes, one is free to choose any representation with the dimension $(2j+1) \leq N$. The $N \times N$ matrices M_b in Eq. (7) are obtained from the above generators by supplementing these $(2j+1)\times(2j+1)$ matrices by extra $N-(2j+1)$ columns and lines consisting of zeroes.

The standard embedding is obtained if we choose $j = 1/2$. The Wilczek embedding corresponds to $j = 1$. It is clear that one can go all the way up to $j = (N-1)/2$. It is not difficult to show that the topological charge of the field configuration obtained in this way is equal to $(2/3)j(j+1)(2j+1)$, so that $|Q|_{max} = (1/6)N(N^2-1)$.

[7] It is important that any traceless $N \times N$ matrix can be represented as a linear combination of T^a.

Thus, for $j \geq 1$, we actually get a degenerate multi-instanton solution, which means, in turn, that the paper in Ref. [14] could also have been reproduced in Sec. III, devoted to multi-instantons.

Finally, it is worth noting that the problem of embedding the $SU(2)$ instanton into higher groups in the case of the spontaneously broken Yang–Mills theory has its peculiarities. For $SU(3)$ the procedure is worked out in Ref. [3] where the analog of the 't Hooft term in the action is found.

References

[1] *G. 't Hooft, *Phys. Rev.* **D14** (1976) 3432; **D18** (1978) 2199.
[2] *R. Jackiw and C. Rebbi, *Phys. Rev.* **D14** (1976) 517.
[3] *M. Shifman and A. Vainshtein, *Nucl. Phys.* **B362** (1991) 21.
[4] R. Jackiw, *Field Theoretic Investigations in Current Algebra*, in S. B. Treiman *et al.*, *Current Algebra and Anomalies* (Princeton University Press, 1985); P. Fayet and S. Ferrara, *Phys. Rep.* **32** (1977) 249.
[5] *A. Vainshtein, V. Zakharov, V. Novikov and M. Shifman, *Sov. Phys. — Uspekhi* **25** (1982) 195.
[6] I. Khriplovich, *Yad. Fiz.* **10** (1969) 409 [*Sov. J. Nucl. Phys.*, **10** (1970) 235]; T. Appelquist, M. Dine and I. Muzinich, *Phys. Lett.*, **69B** (1977) 231.
[7] V. Novikov *et al.*, *Phys. Rep.* **41** (1978) 1, Sec. 1.3.
[8] J. Collins, A. Duncan and S. Joglekar, *Phys. Rev.* **D16** (1977) 438.
[9] *C. Callan, R. Dashen and D. Gross, *Phys. Rev.* **D17** (1978) 2717.
[10] A. Ringwald, *Nucl. Phys.* **B330** (1990) 1; O. Espinosa, *Nucl. Phys.* **B343** (1990) 310.
[11] L. McLerran, A. Vainshtein and M. Voloshin, *Phys. Rev.* **D42** (1990) 171; 180.
[12] M. Mattis, *Phys. Rep.* **214** (1992) 159.
[13] *C. Bernard, *Phys. Rev.* **D19** (1979) 3013.
[14] *F. Wilczek, *Phys. Lett.* **65B** (1976) 160.

PHYSICAL REVIEW D VOLUME 14, NUMBER 12 15 DECEMBER 1976

Computation of the quantum effects due to a four-dimensional pseudoparticle*

G. 't Hooft†
Physics Laboratories, Harvard University, Cambridge, Massachusetts 02138
(Received 28 June 1976)

A detailed quantitative calculation is carried out of the tunneling process described by the Belavin-Polyakov-Schwarz-Tyupkin field configuration. A certain chiral symmetry is violated as a consequence of the Adler-Bell-Jackiw anomaly. The collective motions of the pseudoparticle and all contributions from single loops of scalar, spinor, and vector fields are taken into account. The result is an effective interaction Lagrangian for the spinors.

I. INTRODUCTION

When one attempts to construct a realistic gauge theory for the observed weak, electromagnetic, and strong interactions, one is often confronted with the difficulty that most simple models have too much symmetry. In Nature, many symmetries are slightly broken, which leads to, for instance, the lepton masses, the quark masses, and *CP* violation. These symmetry violations, either explicit or spontaneous, have to be introduced artificially in the existing models.

There is one occasion where explicit symmetry violation is a necessary consequence of the laws of relativistic quantum theory: the Adler-Bell-Jackiw anomaly. The theory we consider is an SU(2) gauge theory with an arbitrary set of scalar fields and a number, N^f, of massless fermions. The apparent chiral symmetry of the form $U(N^f) \times U(N^f)$ is actually broken down to $SU(N^f) \times SU(N^f) \times U(1)$. This paper is devoted to a detailed computation of this effect.

The most essential ingredient in our theory is the localized classical solution of the field equations in Euclidean space-time, of the type found by Belavin *et al.*[1]

Although the main objective of this paper is the computation of the resulting effective symmetry-breaking Lagrangian in a weak-interaction theory, we present the calculations in such a way that they can also be used for possible color gauge theories of stong interactions based on the same classical field configurations. For such theories our intermediate expressions (12.5) and (12.8) will be applicable. Our final results are (15.1) together with the convergence factor (15.8).

Our general philosophy has been sketched in Ref. 2. We are dealing with amplitudes that depend on the coupling constant g in the following way:

$$g^{-C}\exp\left[-\frac{8\pi^2}{g^2}(1+a_1g^2+\cdots)\right]. \tag{1.1}$$

The coefficient a_1 involves one-loop quantum corrections, and it determines the scale of the amplitude. Clearly, then, to understand the main features of such an amplitude, complete understanding of all one-loop quantum effects is desired. For instance, if one changes from one renormalization subtraction procedure to another, so that $g^2 \to g^2 + O(g^4)$, then this leads to a change in (1.1) by an overall multiplicative constant. Thus, the renormalization subtraction point μ may enter as a dimensional parameter in front of our expressions. This is just one of the reasons to suggest that our results will also have interesting applications in strong-interaction color gauge theories.

The underlying classical solutions only exist in Euclidean space, but they give rise to a particular symmetry-breaking amplitude that can easily be continued analytically to Minkowski space. We interpret this amplitude as the result of a certain tunneling effect from one vacuum to a gauge-rotated vacuum. We recall that, indeed, tunneling through a barrier can sometimes be described by means of a classical solution of the equation of motion in the imaginary time direction.[2,3]

We compute in Euclidean space the vacuum-to-vacuum amplitude in the presence of external sources, thus obtaining full Green's functions. Of course, we must limit ourselves to gauge-invariant sources only, but that will be no problem. It turns out to be trivial to amputate the obtained Green's function and get the effective vertex.

The various calculational steps are the following. We first give in Sec. II the functional integral expression for the amplitude, first in a conventional Feynman gauge: $C_1 = \partial_\mu A_\mu$. Later, we go over to the so-called background gauge: $C_4 = D_\mu A_\mu^{qu}$. This is actually only correct up to an overall factor, as will be explained in Sec. XI. It is just for pedagogical reasons that we ignore this complication for a moment. It is in this gauge that the quantum excitations take a simple form: "Spin-orbit" couplings commute with the operator L^2

$= -\frac{1}{8} L_{\mu\nu} L_{\mu\nu}$, where

$$L_{\mu\nu} = x_\mu \frac{\partial}{\partial x_\nu} - x_\nu \frac{\partial}{\partial x_\mu}, \quad (1.2)$$

so that we can look at eigenstates of L^2. (In other gauges only total angular momentum $\vec{J} = \vec{L}$ + spin + isospin is conserved, not L^2.)

In Sec. III we consider the quantum fluctuations described by an eigenvalue equation,

$$\mathfrak{M}\psi = E\psi, \quad (1.3)$$

in order to compute det$\mathfrak{M}$. Now this looks like an ordinary scattering problem (in 4+1 dimensions), and, indeed, we show that the product of all nonzero eigenvalues E, in some large box, can be expressed in terms of the phase shift $\eta(k)$ as a function of the wave number k.

In Sec. IV we show that Eq. (1.3) is essentially the same for scalars, spinors, and vectors, from which we derive the important result that the product of all nonzero eigenvalues is the same for scalar fields as for each component of the spinor and vector fields. So, we turn to the (much easier) scalar case first.

But even for scalars Eq. (1.3) has no simple solutions in terms of well-known elementary functions. We decide not to compute det$\mathfrak{M}$ by solving (1.2), but we compute instead

$$\det[(1+x^2)\mathfrak{M}(1+x^2)] .$$

The factors $1+x^2$ drop out if we divide by the same determinant coming from the vacuum (i.e., the case $A^{cl}=0$). The equation

$$\mathfrak{M}\psi = \frac{\lambda}{(1+x^2)^2}\psi \quad (1.4)$$

is a simple hypergeometric equation that can be solved under the given boundary condition (Sec. V).

Now we must find the product of all eigenvalues λ, but that diverges badly even if we divide by the values they take in the vacuum. We must find a gauge-invariant regulator, and the regulator determinant must be calculable. Dimensional regularization is not applicable here, but we can use background Pauli-Villars regulators. They give messy equations unless they have a space-time-dependent mass:

$$\frac{M_i^2}{(1+x^2)^2} . \quad (1.5)$$

The rules are formulated in Sec. V.

In Sec. VI we compute the product of the eigenvalues using this regulator. In Sec. VII we make the transition to regulators with fixed mass by observing that a change toward fixed regulator mass must correspond to a local, space-time-dependent counterterm in the Lagrangian. The effect of this counterterm is computed.

Using the result of Sec. IV we now find also the contributions of all nonvanishing eigenvalues for the vectors and spinors. But there are also vanishing eigenvalues. They are listed in Sec. VIII. For the vector fields, we have eight zero eigenvectors in addition to the ones computed via the theorem of Sec. IV. They are to be interpreted as translations (Sec. IX), dilatation (Sec. X), and isospin rotations (Sec. XI). The last need special care and can only be interpreted correctly when different gauge choices are compared. This leads to the factor mentioned in the beginning.

In Sec. XII we combine the results so far obtained and add the fermions. This intermediate result may be useful to strong-interaction theories. In Sec. XIII we reexpress the result in terms of the dimensionally renormalized coupling constant g^D, as opposed to the previous coupling constant which was renormalized in a Pauli-Villars manner. In Sec. XIV the external sources for the fermions are considered and the amputation operation for the Green's function is performed. We obtain the desired effective Lagrangian, but there is still one divergence. So far, we only had massless particles, and as a consequence of that there is still a scale parameter ρ over which we must integrate. Asymptotic freedom gives a natural cutoff for this integral in the ultraviolet direction, but there is still an infrared divergence. In weak-interaction theories the Higgs field is expected to provide for the infrared cutoff. Section XV shows how to compute this cutoff.

The Appendix lists the properties of the symbols η, $\bar{\eta}$ which are used many times throughout these calculations.

II. FORMULATION OF THE PROBLEM

Let a field theory in four space-time dimensions be given by the Lagrangian

$$\mathfrak{L} = -\frac{1}{4} G^a_{\mu\nu} G^a_{\mu\nu} - D_\mu\Phi^* D_\mu\Phi - \bar{\psi}\gamma_\mu D_\mu\psi + \bar{\psi}_s \mathcal{J}_{st}\psi_t , \quad (2.1)$$

where the gauge group is SU(2):

$$G^a_{\mu\nu} = \partial_\mu A^a_\nu - \partial_\nu A^a_\mu + g\epsilon_{abc} A^b_\mu A^c_\nu . \quad (2.2)$$

The SU(2) indices will be called isospin indices. The scalars Φ, taken to be complex, may contain several multiplets of arbitrary isospin:

$$D_\mu\Phi = \partial_\mu\Phi - igT^a A^a_\mu\Phi , \quad [T^a, T^b] = i\epsilon_{abc}T^c \quad (2.3)$$

The spinors ψ are taken to be isospin-$\frac{1}{2}$ doublets.

The total number of doublets is N^f. Mass terms and interaction terms between scalars and spinors are irrelevant for the time being.

We inserted a source term in a gauge-invariant way, with respect to which we will expand, in order to obtain Green's functions. The indices $s,t=1,\ldots,N^f$, called flavor indices, label the different isospin multiplets. Isospin and Dirac indices have been suppressed. $\mathcal{J}_{st}$ must be diagonal in the isospin indices but may contain Dirac γ matrices.

The system (2.1) seems to have a chiral $U(N^f)\times U(N^f)$ global symmetry, but actually has an Adler-Bell-Jackiw anomaly[4] associated with the chiral U(1) current, breaking the symmetry down to $SU(N^f)\times SU(N^f)\times U(1)$. The aim of this paper is to find that part of the amplitude that violates the chiral U(1) conservation.

The functional integral expression for the amplitude is

$$W={}_{\text{out}}\langle 0\,|\,0\rangle_{\text{in}}$$
$$=\int \mathfrak{D}A\,\mathfrak{D}\psi\,\mathfrak{D}\Phi\,\mathfrak{D}\phi$$
$$\times\exp\left\{\int\left[\mathcal{L}-\tfrac{1}{2}C_1^2(A)+\mathcal{L}_1^{\text{ghost}}(\phi)\right]d^4x\right\}, \qquad (2.4)$$

to be expanded with respect to $\mathcal{J}$. Here $C_1(A)$ is a gauge-fixing term, and $\mathcal{L}_1^{\text{ghost}}$ are the corresponding ghost terms.

As is argued in Ref. 2, the U(1)-breaking part of this amplitude comes from that region of superspace where the A field approaches the solutions described in Ref. 1:

$$A_\mu^a(x)^{\text{cl}}=\frac{2}{g}\,\frac{\eta_{a\mu\nu}(x-z)^\nu}{(x-z)^2+\rho^2}\,, \qquad (2.5)$$

where z_μ and ρ are five free parameters associated with translation invariance and scale invariance. The coefficients η are studied in the Appendix. Conjugate to (2.5) we have its mirror image, described by the coefficients $\bar\eta$ (see Appendix).

Now these solutions form a local extremum of our functional integrand, and therefore it makes sense to consider separately that contribution to W in (2.4) that is obtained through a new perturbation expansion around these new solutions, taking the integrand there to be approximately Gaussian. The fields Φ, ϕ, and ψ all remain infinitesimal so that their mutual interactions may be neglected in the first approximation. Of course, we must also integrate over the values of z_μ and ρ. This will be done by means of the collective-coordinate formalism.[5] One writes

$$A_\mu^a=A_\mu^{a\,\text{cl}}+A_\mu^{a\,\text{qu}}, \qquad (2.6)$$

and those values of A^{qu} that correspond to translations or dilatations are replaced by collective coordinates.

The integrand in (2.4) now becomes

$$\mathcal{L}(A^{\text{cl}})-\tfrac{1}{2}(D_\mu A_\nu^{\text{qu}})^2+\tfrac{1}{2}(D_\mu A_\mu^{\text{qu}})^2-gA_\nu^{a\,\text{qu}}\epsilon_{abc}G_{\mu\nu}^{b\,\text{cl}}A_\mu^{c\,\text{qu}}$$
$$-D_\mu\Phi^*D_\mu\Phi-\bar\psi\gamma_\mu D_\mu\psi+\bar\psi\mathcal{J}\psi-\tfrac{1}{2}C_1{}^2+\mathcal{L}_1^{\text{ghost}}$$
$$+\mathcal{O}(A^{\text{qu}},\Phi,\psi)^3\,, \qquad (2.7)$$

where

$$S^{\text{cl}}=\int\mathcal{L}(A^{\text{cl}})d^4x$$
$$=-8\pi^2/g^2\,, \qquad (2.8)$$

and the "covariant derivative" D_μ only contains the background field A_μ^{cl}, for instance:

$$D_\mu A_\nu^{a\,\text{qu}}=\partial_\mu A_\nu^{a\,\text{qu}}+g\epsilon_{abc}A_\mu^{b\,\text{cl}}A_\nu^{c\,\text{qu}}\,, \qquad (2.9)$$

etc.

We abbreviate the integral over (2.7) by

$$S^{\text{cl}}-\tfrac{1}{2}A^{\text{qu}}\mathfrak{M}_A A^{\text{qu}}+\bar\psi\mathfrak{M}_\psi\psi-\Phi^*\mathfrak{M}_\Phi\Phi-\phi^*\mathfrak{M}_{\text{gh}}\phi\,. \qquad (2.10)$$

where the last term describes the Faddeev-Popov ghost. Thus, expression (2.4) is (ignoring temporarily the collective coordinates, and certain factors $\sqrt{\pi}$ from the Gaussian integration; see Sec. IX)

$$W=\exp(-8\pi^2/g^2)(\det\mathfrak{M}_A)^{-1/2}\det\mathfrak{M}_\psi(\det\mathfrak{M}_\Phi)^{-1}$$
$$\times\det\mathfrak{M}_{\text{gh}}\,. \qquad (2.11)$$

The determinants will be computed by diagonalization:

$$\mathfrak{M}_i\psi=E_i\psi\,, \qquad (2.12)$$

after which we multiply all eigenvalues E. Since there are infinitely many very large eigenvalues, this infinite product diverges very badly. There are two procedures that will make it converge:

(i) The vacuum-to-vacuum amplitude in the absence of sources must be normalized to 1, so that the vacuum state has norm 1. This implies that W must be divided by the same expression with $A^{\text{cl}}=0$.

(ii) We must regularize and renormalize. The dimensional procedure is not available here because the four-dimensionality of the classical solution is crucial. We will use the so-called background Pauli-Villars regulators (Secs. IV and V).

Taking a closer look at the eigenvalue equations (2.12) as they follow from (2.7), we notice that the background field in there gives rise to couplings between spin, isospin, and (the four-dimensional equivalent of) orbital angular momentum, through the coefficients $\eta_{a\mu\nu}$ in (2.5). Now these couplings simplify enormously if we go over to a new gauge that explicitly depends on the background field[6]

$$C_4^a(A^{qu}) = D_\mu A_\mu^{a\,qu}$$
$$= \partial_\mu A_\mu^{a\,qu} + g\epsilon_{abc} A_\mu^{b\,cl} A_\mu^{c\,qu} \tag{2.13}$$

Thus the third and eighth terms in (2.7) cancel.

This choice of gauge will lead to one complication, to be discussed in Sec. XI: The gauge for the vacuum-to-vacuum amplitude in the absence of sources, used for normalization, in the region $A_\mu \sim 0$, is usually invariant under global isospin rotations, but the classical solution (2.5) and the gauge (2.13) are not. Associated with this will be three spurious zero eigenvalues of $\mathfrak{M}_A$ that cannot be directly associated with global isospin rotations. The question is resolved in Sec. XI by careful comparison of the gauge C_1 with C_4 and some intermediate choices of gauge.

There will be five other zero eigenvalues of $\mathfrak{M}_A$ that of course must not be inserted in the product of eigenvalues directly, since they would render expression (2.11) infinite. They exactly correspond to the infinitesimal translations and dilatations of the classical solution, and, as discussed before, must be replaced by the corresponding collective coordinates (Secs. IX and X).

The matrices $\mathfrak{M}$ are now (ignoring temporarily the fermion source)

$$\mathfrak{M}_A A_\mu^{a\,qu} = -D^2 A_\mu^{a\,qu} - 2g\epsilon_{abc} G_{\mu\nu}^{b\,cl} A_\nu^{c\,qu},$$
$$-\mathfrak{M}_\psi^2 \psi = -D^2\psi + \tfrac{1}{4} i\tau^a G_{\mu\nu}^{a\,cl} \gamma_\mu\gamma_\nu\psi, \tag{2.14}$$
$$\mathfrak{M}_\Phi \Phi = -D^2\Phi,$$
$$\mathfrak{M}_{gh}\phi = -D^2\phi.$$

In order to substitute the classical solution (2.5) with $z=0$ and $\rho=1$ (generalization to other z and ρ will be straightforward), we introduce the space-time operators

$$L_1^a = -\tfrac{1}{2} i\eta_{a\mu\nu} x^\mu \frac{\partial}{\partial x^\nu},$$
$$L_2^a = -\tfrac{1}{2} i\bar{\eta}_{a\mu\nu} x^\mu \frac{\partial}{\partial x^\nu}, \tag{2.15}$$

with

$$[L_p^a, L_q^b] = i\delta_{pq}\epsilon_{abc} L_p^c,$$
$$L^2 = L_1^{\,2} = L_2^{\,2} = -\tfrac{1}{8}(x_\mu\partial_\nu - x_\nu\partial_\mu)^2. \tag{2.16}$$

They represent rotations in the two invariant SU(2) subgroups of the rotation group SO(4).

Isospin rotations will be generated by the operators T^a for the scalars, $T^a = \tfrac{1}{2}\tau^a$ for the spinors, and $T^b A_\mu^a = i\epsilon_{abc} A_\mu^c$ for the vectors. Then

$$D^2 = \left(\frac{\partial}{\partial r}\right)^2 + \frac{3}{r}\frac{\partial}{\partial r} - \frac{4}{r^2}L^2 - \frac{8}{r^2+1} T\cdot L_1 - \frac{4r^2}{(r^2+1)^2}T^2, \tag{2.17}$$

where $r^2 = (x-z)^2$. This clearly displays the isospin-orbit coupling.

The vector and spinor fields also have a spin-isospin coupling. For the spinors we define the spin operators

$$S_1^a\psi = -\tfrac{1}{8} i\eta_{a\mu\nu}\gamma_\mu\gamma_\nu\psi,$$
$$S_2^a\psi = -\tfrac{1}{8} i\bar{\eta}_{a\mu\nu}\gamma_\mu\gamma_\nu\psi, \tag{2.18}$$

satisfying

$$[S_p^a, S_q^b] = i\delta_{pq}\epsilon_{abc} S_p^c \tag{2.19}$$

and

$$S_1^{\,2} = \frac{3}{4}\,\frac{1-\gamma_5}{2},$$
$$S_2^{\,2} = \frac{3}{4}\,\frac{1+\gamma_5}{2}. \tag{2.20}$$

For the vector fields we define

$$S_1^a A_\mu^{qu} = -\tfrac{1}{2} i\eta_{a\mu\nu} A_\nu^{qu},$$
$$S_2^a A_\mu^{qu} = -\tfrac{1}{2} i\bar{\eta}_{a\mu\nu} A_\nu^{qu}, \tag{2.21}$$
$$S_1^{\,2} = S_2^{\,2} = \tfrac{3}{4}.$$

For the scalar fields $\vec{S}_1 = \vec{S}_2 = 0$. Thus right- and left-handed spinors are $(\tfrac{1}{2}, 0)$ and $(0, \tfrac{1}{2})$ representations of SO(4), and vectors are $(\tfrac{1}{2}, \tfrac{1}{2})$ representations. Scalars of course are $(0,0)$ representations.

In terms of the operators S and T, the spin-isospin couplings turn out to be universal for all particles. Substituting the classical value for $G_{\mu\nu}^{a\,cl}$ in (2.14) we find

$$\mathfrak{M} = -\left(\frac{\partial}{\partial r}\right)^2 - \frac{3}{r}\frac{\partial}{\partial r} + \frac{4}{r^2}L^2 + \frac{8}{1+r^2} T\cdot L_1 + \frac{4r^2}{(1+r^2)^2}T^2 + \frac{16}{(1+r^2)^2} T\cdot S_1, \tag{2.22}$$

with $\mathfrak{M} = \mathfrak{M}_A$ or $-\mathfrak{M}_\psi^{\,2}$ or $\mathfrak{M}_\Phi$ or $\mathfrak{M}_\phi$.

Observe the absence of spin-orbit and isospin-orbit couplings that contain x_μ or $\partial/\partial x_\mu$ explicitly. It all goes via the orbital angular momentum operator L_1 and that implies that L^2 commutes with $\mathfrak{M}$. This would not be so in other gauges. Further, $\mathfrak{M}$ commutes with $\vec{J}_1 = \vec{L}_1 + \vec{S}_1 + \vec{T}$ and $\vec{L}_2$ and $\vec{S}_2$. Eigenvectors of $\mathfrak{M}$ can thus be characterized by the quantum numbers

s_1 and s_2 (both either 0 or $\frac{1}{2}$),

t (total isospin, arbitrary for the scalars, $\frac{1}{2}$ for the spinors, 1 for the vector and the ghost),

$l = 0, \frac{1}{2}, 1, \ldots$,

$$j_1 = l - s_1 - t, l - s_1 - t + 1, \ldots, l + s_1 + t, \text{ as long as } j_1 \geq 0, \tag{2.23}$$

$j_1^3 = -j_1, \ldots, +j_1$,

$s_2^3 = -s_2, \ldots, +s_2$,

$l_2^3 = -l, \ldots, +l$.

For normalization we need the corresponding operator $\mathfrak{M}$ for the case that the background field is zero:

$$\mathfrak{M}_0 = -\left(\frac{\partial}{\partial r}\right)^2 - \frac{3}{r}\frac{\partial}{\partial r} + \frac{4}{r^2}L^2. \tag{2.24}$$

III. DETERMINANTS AND PHASE SHIFTS

The eigenvalue equation (2.12) with $\mathfrak{M}$ as in (2.22) differs in no essential way from an ordinary Schrödinger scattering problem. In this section we show the relation between the corresponding scattering matrix and the desired determinant.

Temporarily, we put the system in a large spherical box with radius R. At the edge we have some boundary condition: either $\Psi(R) = 0$, or $\Psi'(R) = 0$ (or a linear combination thereof). Here Ψ stands for any of the scalar, spinor, or vector fields. In the case $\Psi'(R) = 0$ the vacuum operator $\mathfrak{M}_0$ has a zero eigenvalue corresponding to Ψ = constant, and also the lowest eigenvalue of $\mathfrak{M}$ may go to zero more rapidly than $1/R^2$ when $R \to \infty$. Such eigenvalues have to be considered separately (negative eigenvalues can be proved not to exist).

We here consider all other eigenvalues of $\mathfrak{M}$. They approach the ones of $\mathfrak{M}_0$ if $R \to \infty$. We wish to compute the product

$$\prod_{n=1}^{\infty} \frac{E(n)}{E_0(n)}. \tag{3.1}$$

The scattering matrix $S(k) = e^{2i\eta(k)}$ will be defined by comparing the solution of

$$\mathfrak{M}\Psi = k^2\Psi \tag{3.2}$$

with

$$\mathfrak{M}_0\Psi_0 = k^2\Psi_0 \tag{3.3}$$

both with boundary condition $\Psi \to Cr^{2l}$ at $r = 0$. Let

$$\Psi_0(r) \sim Cr^{-3/2}(e^{-ik(r+a)} + e^{+ik(r+a)})$$
$$= 2Cr^{-3/2}\cos k(r+a) \quad \text{for large } r \tag{3.4}$$

and

$$\Psi(r) \propto Cr^{-3/2}[e^{-ik(r+a)} + S(k)e^{ik(r+a)}]$$
$$= 2C'r^{-3/2}\cos[k(r+a) + \eta(k)]. \tag{3.5}$$

If we require at $r = R$ the same boundary condition for Ψ_0 then we must solve

$$k(n)(R+a) + \eta(k(n)) = k_0(n)(R+a), \tag{3.6}$$

thus

$$\frac{k(n)}{k_0(n)} = 1 - \frac{\eta(k(n))}{(R+a)k(n)}. \tag{3.7}$$

The level distance $\Delta k = k(n+1) - k(n)$ is in both cases, asymptotically for large R,

$$\Delta k = \frac{\pi}{R} + O\left(\frac{1}{R^2}\right). \tag{3.8}$$

We find that

$$\prod_{n=1}^{\infty} \frac{E(n)}{E_0(n)} = \exp\left\{2\sum_1^{\infty} \ln[k(n)/k_0(n)]\right\}$$
$$= \exp\left\{2\frac{R}{\pi}\sum_1^{\infty} \Delta k\left[-\frac{\eta(k)}{Rk} + O\left(\frac{1}{R^2}\right)\right]\right\}$$
$$\to \exp\left[-\frac{2}{\pi}\int_0^{\infty}\frac{\eta(k)}{k}dk\right], \tag{3.9}$$

provided that the integral converges at both ends.

At $k \to 0$ the integral (3.9) converges provided that the interaction potential decreases faster than $1/r^2$ as $r \to \infty$; at $k \to \infty$ the integral converges if the interaction potential is less singular than $1/r^2$ as $r \to 0$. The latter condition is satisfied if we compare $\mathfrak{M}$ and $\mathfrak{M}_0$ at the same values for the quantum number l; the first condition is satisfied if $(L_1 + T)^2$ for the interacting matrix is set equal to L^2 for the vacuum matrix. If we consider the combined effect of all values for L^2 and $(L_1 + T)^2$ both for the vacuum and for the interacting case then we can split the integral (3.9) somewhere in the middle, and combine the $k \to \infty$ parts, so that we get convergence everywhere.

An easier way to get convergence is to regularize:

$$\prod_{n=1}^{\infty} \frac{E(n)[E_0(n)+M^2]}{E_0(n)[E(n)+M^2]} \to \exp\left[-\frac{2}{\pi}\int_0^\infty \frac{\eta(k)}{k}\frac{M^2}{(k^2+M^2)}dk\right]. \tag{3.10}$$

Regulators will be introduced anyhow, so we will not encounter difficulties due to non-convergence of the integral in (3.9).

IV. ELIMINATION OF THE SPIN DEPENDENCE

In Eq. (2.22) the operators $T\cdot L_1$ and $T\cdot S_1$ do not commute. Only in the case that

$$|j_1 - l| = s + t$$

(as defined in 2.23) do they simultaneously diagonalize. If

$$|j_1 - l| < s + t$$

and

$$j_1 \neq 0, \quad l \neq 0, \quad s = \tfrac{1}{2}, \quad t \neq 0,$$

then we have a set of coupled differential equations for two dependent variables.

In any other case there would be no hope of solving this set of equations analytically, but here we can make use of a unique property of the equation

$$\mathfrak{M}\Psi = E\Psi, \tag{4.1}$$

which enables us to diagonalize it completely. If $s_1 = 0$ the equation could describe a left-handed fermion with isospin t:

$$\mathfrak{M}\psi = -(\gamma\cdot D)^2\psi = E\psi, \qquad \gamma_5\psi = +\psi. \tag{4.2}$$

But then we can define, if $E\neq 0$,

$$\psi' = \gamma\cdot D\psi \tag{4.3}$$

with

$$\mathfrak{M}\psi' = E\psi', \qquad \gamma_5\psi' = -\psi' \tag{4.4}$$

Now ψ' has $s_1' = \frac{1}{2}$, and hence we found a solution for the set of coupled equations with $s_1' = \frac{1}{2}$ from a solution of the simpler equation with $s_1 = 0$. The operater $\gamma\cdot D$ in Eq. (4.3) does not commute with L^2, so if ψ has a given set of quantum numbers l, j_1, t then ψ' is a superposition of a state with $l' = l + \frac{1}{2}$ and one with $l' = l - \frac{1}{2}$. Now $\mathfrak{M}$ does commute with L^2, so if we project out the state with $l' = l + \frac{1}{2}$ or $l' = l - \frac{1}{2}$ then we get a new solution in both cases. Thus one solution with $s_1 = 0$ and quantum numbers l, j_1, t generates two solutions with $s_1' = \frac{1}{2}$, $l' = l \pm \frac{1}{2}$, $j_1' = j_1$, $t' = t$. In terms of the operators L, S, and T the new solutions to the coupled equations can be expressed in terms of the $S = 0$ solutions as follows:

$$L_a' + S_a' = L_a, \quad l' = l \pm \tfrac{1}{2}, \qquad T' = T, \quad J' = J, \tag{4.5}$$

$$\Psi' = \left[\frac{1}{r}(2L^2 - 2L'^2 + \tfrac{3}{2}) + \frac{4r}{1+r^2} S_1'\cdot T\right]\Psi + \frac{\partial}{\partial r}\Psi. \tag{4.6}$$

It is easy to check explicitly that if Ψ satisfies (2.22) with $S_1 = 0$, then the two wave functions Ψ' both satisfy (2.22) when the operators L, S, T are replaced by the primed ones.

Asymptotically, for large r, $\Psi' = (\partial/\partial r)\Psi$, and hence the phase shift $\eta(k)$ is the same for the primed case as for the original case. Consequently, the integral over the phase shifts as it occurs in (3.9) is the same for spinor and vector fields (with $s_1 = \frac{1}{2}$) as it is for scalar fields (with $s_1 = 0$).

The above procedure becomes more delicate if $E = 0$. Indeed, although scalar fields can easily be seen to have no zero-eigenvalue modes, spinor and vector fields do have them. In conclusion, the nonzero eigenvalues for the vector and spinor modes are the same as for the scalar modes, but the zero eigenvectors are different.

In the following sections we compute the universal value for the product. Note that also the regularized expressions (3.10) are equivalent because the $\eta(k)$ match for all k. The regulator of Eq. (3.10) corresponds to new fields with Lagrangians

$$\mathcal{L} = -\tfrac{1}{2}(D_\mu B_\nu)^2 - \tfrac{1}{2}M^2 B_\nu^{\,2} - gB_\nu^a\epsilon_{abc}G_{\mu\nu}^b B_\mu^c \tag{4.7}$$

for vectors,

$$\mathcal{L} = \bar{\chi}[-(\gamma\cdot D)^2 - M^2]\chi \text{ for spinors}, \tag{4.8}$$

and

$$\mathcal{L} = -(D_\mu\xi)^* D_\mu\xi - M^2\xi^*\xi \text{ for scalars}. \tag{4.9}$$

Within the background field procedure it is obvious that such regulator fields make the one-loop amplitudes finite. Later (Sec. XIII) we will make the link with the more conventional dimensional regulators.

V. A NEW EIGENVALUE EQUATION AND NEW REGULATORS

As stated in the Introduction, the solutions to the equations $\mathfrak{M}\Psi = E\Psi$ even in the scalar case cannot be expressed in terms of simple elementary functions. But eventually we only need $\det\mathfrak{M}/\det\mathfrak{M}_0$, and this can be obtained in another way.

We write

$$V=\tfrac{1}{4}(1+r^2)\mathfrak{M}(1+r^2)\,,$$
$$V_0=\tfrac{1}{4}(1+r^2)\mathfrak{M}_0(1+r^2)\,, \qquad (5.1)$$

and, formally,

$$\det(\mathfrak{M}/\mathfrak{M}_0)=\det(V/V_0)\,. \qquad (5.2)$$

The equation

$$V\Psi=\lambda\Psi \qquad (5.3)$$

corresponds to the expression

$$\left[\left(\frac{\partial}{\partial r}\right)^2+\frac{3}{r}\frac{\partial}{\partial r}-\frac{4}{r^2}L^2-\frac{4}{1+r^2}(J_1^{\,2}-L^2)+\frac{4(T^2+\lambda)}{(1+r^2)^2}\right]\Psi=0\,. \qquad (5.4)$$

Write

$$x=\frac{1}{1+r^2}\,,$$
$$\Psi=r^{2l}(1+r^2)^{-l-j_1-1}\Phi(x)\,, \qquad (5.5)$$

then

$$\left\{\left(\frac{\partial}{\partial x}\right)^2+\left[\frac{2(j_1+1)}{x}-\frac{2(l+1)}{1-x}\right]\frac{\partial}{\partial x}+\frac{1}{x(1-x)}\left[T^2+\lambda-(l+j_1+1)(l+j_1+2)\right]\right\}\Phi=0\,. \qquad (5.6)$$

This is a hypergeometric equation. The physical region is $1/(1+R^2)<x\le 1$. In the Hilbert space of square-integrable wave functions the spectrum is now discrete, which implies that we can safely take the limit $R\to\infty$. The solutions for Φ are just polynomials:

$$\Phi(x)=\sum_{\nu=0}^{\infty}a_\nu x^\nu\,, \qquad (5.7)$$

$$a_{\nu+1}=a_\nu\frac{(\nu-n)(\nu+n+2l+2j_1+3)}{(\nu+1)(\nu+2j_1+2)}\,, \qquad (5.8)$$

where n is defined by

$$(n+l+j_1+1)(n+l+j_1+2)=T^2+\lambda\,. \qquad (5.9)$$

If $n=$ integer ≥ 0 then the series (5.7) breaks off. Otherwise Φ is not square-integrable. So we find the eigenvalues

$$\lambda_n=(n+l+j_1+1-t)(n+l+j_1+2+t)\,, \qquad (5.10)$$
$$n=0,1,2,\ldots,\qquad T^2=t(t+1)\,.$$

The vacuum case, $V_0\Psi=\lambda_0\Psi$, is solved by the same equation, but with $j_1=l$, $t=0$.

The product of these eigenvalues, even when divided by the vacuum values, still badly diverges so we must regularize. The regulators of Sec. IV are not very attractive here because they spoil the hypergeometric nature of the equations. More convenient here is a set of regulator fields with masses that all depend on space-time in a certain way. They are given by the Lagrangians (4.7)–(4.9) but with M^2 replaced by

$$\frac{4M^2}{(1+r^2)^2}\,. \qquad (5.11)$$

We choose M^2 here so large that anywhere near the origin the regulator is heavy. Far from the origin the classical solution is expected to be close enough to the real vacuum, so that there the details of the regulators are irrelevant.

Of course the regulator procedure affects the definition of the subtracted coupling constant. In Sec. VII we link the regulator (5.11) with the more acceptable one of Sec. IV, and in Sec. XIII we make the link with the dimensional regulator.

The eigenvalues of the regulator are

$$\lambda_n^M=(n+l+j_1+1-t)(n+l+j_1+2+t)+M^2\,. \qquad (5.12)$$

The regulators M_i with $i=1,\ldots,R$ are as usual of alternating metric $e_i=\pm 1$. Consequently, $\det\mathfrak{M}$ is replaced by

$$(\det\mathfrak{M})\prod_{i=1}^{R}(\det\mathfrak{M}_i)^{e_i}\,. \qquad (5.13)$$

This converges rapidly if

$$\sum_1^R e_i=-1\,,$$
$$\sum_1^R e_iM_i=0\,, \qquad (5.14)$$
$$\sum_1^R e_iM_i^{\,2}=0\,,\ldots,$$

and

$$\sum_1^R e_i\ln M_i=-\ln M=\text{finite}. \qquad (5.15)$$

Let $i=0$ denote the physical field, then

$$e_0=1,\quad M_0=0,\quad \sum_{i=0}^{R}e_i=0,\quad \text{etc.} \qquad (5.16)$$

VI. THE REGULARIZED PRODUCT OF THE NONVANISHING EIGENVALUES

We now consider the logarithm of the regularized product of the nonvanishing eigenvalues, for a scalar field with total isospin t:

$$\ln\Pi(t)=\sum_{i=0}^{R} e_i \sum \ln\lambda^{M_i} \tag{6.1}$$

with

$$\lambda^{M_i}=(n+l+j_1+1-t)(n+l+j_1+2+t)+M_i^2 \tag{6.2}$$

(we imply that $e_0=1$ and $M_0=0$). The summation goes over the values of all quantum numbers. Now for given n,l,j_1 the degeneracy is $(2j_1+1)(2j_2+1)=(2j_1+1)(2l+1)$. The values of l, j_1, and n are restricted by

$$\sigma\equiv l+j_1-t\geq 0\,,\qquad \tau\equiv j_1-l+t\geq 0,\quad \tau\leq 2t\,,\quad n\geq 0\,. \tag{6.3}$$

[Later we will divide $\Pi(t)$ by the vacuum value $\Pi_0(t)$, which is obtained by the same formulas as above and the following, but with t replaced by zero, and the degeneracy will be $(2t+1)(2l+1)^2$.]

We go over to the variables σ and τ as given by (6.3) and s with

$$s=n+l+j_1+\tfrac{3}{2}\,,\quad s\geq t+\sigma+\tfrac{3}{2}\,. \tag{6.4}$$

We find that

$$\ln\Pi(t)=\sum_{s=t+3/2}^{\infty}\sum_i e_i\sum_{\tau=0}^{2t}\sum_{\sigma=0}^{s-t-3/2}(\sigma+\tau+1)(2t+\sigma-\tau+1)\times\ln[s^2+M_i^2-(t+\tfrac{1}{2})^2]\,. \tag{6.5}$$

The summation over σ and τ gives

$$\ln\Pi(t)=\frac{2t+1}{3}\sum_{s=t+3/2}^{\infty}\sum_i e_i[s^3-s(t+\tfrac{1}{2})^2]\times\ln[s^2+M_i^2-(t+\tfrac{1}{2})^2]\,. \tag{6.6}$$

The vacuum value $\Pi_0(t)$ is obtained from (6.6) by replacing t with zero and adding an additional multiplicity $2t+1$, thus

$$\ln\Pi_0(t)=\frac{2t+1}{3}\sum_{s=3/2}^{\infty}\sum_i e_i(s^3-\tfrac{1}{4}s)\times\ln(s^2+M_i^2-\tfrac{1}{4})\,. \tag{6.7}$$

Now we interchange the summation over s and i, letting first s go from $t+\frac{3}{2}$ to Λ and taking $\Lambda\to\infty$ in the end. We get

$$\ln[\Pi(t)/\Pi_0(t)]=\frac{2t+1}{3}\sum_i e_i(A^{M_i}(t+\tfrac{1}{2})-A^{M_i}(\tfrac{1}{2})) \tag{6.8}$$

with

$$A^{M_i}(\phi)=\sum_{s=\phi+1}^{\Lambda}(s^3-s\phi^2)\ln(s^2+M_i^2-\phi^2)\,. \tag{6.9}$$

Let us first consider the regulator contribution. Then M is large. We may consider the logarithm as a slowly varying function, and approximate the summation by means of the Euler-Maclaurin formula,

$$\sum_{s=p+1}^{\Lambda}f(s)=\int_p^{\Lambda}f(x)dx+[\tfrac{1}{2}f(x)+\tfrac{1}{12}f'(x)-\tfrac{1}{720}f'''(x)\cdots]\Big|_p^{\Lambda}\,, \tag{6.10}$$

and we obtain

$$A^M(\phi)=\text{indep}(\phi)+\phi^2(-\tfrac{1}{2}M^2-\Lambda^2\ln\Lambda-\Lambda\ln\Lambda-\tfrac{1}{2}\Lambda-\tfrac{1}{6}\ln\Lambda-\tfrac{1}{4}-\tfrac{1}{6}\ln M^2)+\tfrac{1}{4}\phi^4(2\ln\Lambda+1)+O\left(\frac{1}{M^2}\right)+O\left(\frac{1}{\Lambda}\right). \tag{6.11}$$

The first term stands for an array of expressions, all independent of ϕ, and is not needed because it cancels out in Eq. (6.8).

For $A^0(\phi)$ the series (6.10) will not converge at $x=p$ so it cannot be used. After some purely algebraic manipulations we find

$$\begin{aligned}A^0(\phi)&=\sum_{s=\phi+1}^{\Lambda}s(s+\phi)(s-\phi)[\ln(s+\phi)+\ln(s-\phi)]\\&=\text{indep}(\phi)+4\phi^2\sum_{s=1}^{\Lambda}s\ln s+\sum_{s=1}^{2\phi}s(2\phi-s)(s-\phi)\ln s\\&\quad+\phi^2(-3\Lambda^2-3\Lambda-\tfrac{1}{2})\ln\Lambda+\phi^2(\Lambda^2-\tfrac{1}{2}\Lambda-\tfrac{7}{12})+\phi^4(\tfrac{1}{2}\ln\Lambda-\tfrac{1}{12})+O\left(\frac{1}{\Lambda}\right).\end{aligned} \tag{6.12}$$

Now we insert (6.11) and (6.12) into (6.8):

$$\ln[\Pi(t)/\Pi_0(t)] = \frac{2t+1}{3}\left\{\sum_{s=1}^{2t+1} s(2t+1-s)\left(s-t-\tfrac{1}{2}\right)\ln s \right.$$
$$\left. + t(t+1)\left[4\sum_{1}^{\Lambda} s\ln s - 2\Lambda^2\ln\Lambda - 2\Lambda\ln\Lambda - \tfrac{1}{3}\ln\Lambda + \Lambda^2 + \tfrac{1}{3}\ln M - \tfrac{1}{3}t(t+1) - \tfrac{1}{2}\right]\right\}. \quad (6.13)$$

We made use of $\sum_0^R e_i = 0$, $\sum_0^R e_i M_i^2 = 0$, $\sum_1^R e_i \ln M_i^2 \equiv -\ln M$. The limit $\Lambda \to \infty$ exists. Defining

$$R = \lim_{\Lambda\to\infty}\left(\sum_{s=1}^{\Lambda} s\ln s - \tfrac{1}{2}\Lambda^2\ln\Lambda - \tfrac{1}{2}\Lambda\ln\Lambda - \tfrac{1}{12}\ln\Lambda + \tfrac{1}{4}\Lambda^2\right)$$
$$= 0.248\,754\,477\,, \quad (6.14)$$

we find that

$$\ln[\Pi(t)/\Pi_0(t)] = \frac{2t+1}{3}\left[t(t+1)\left(\frac{1}{3}\ln M + 4R - \frac{1}{3}t(t+1) - \frac{1}{2}\right) + \sum_{s=1}^{2t+1} s(2t+1-s)\left(s-t-\frac{1}{2}\right)\ln s\right]. \quad (6.15)$$

R is related to the Riemann zeta function $\zeta(z)$ as follows:

$$R = \tfrac{1}{12} - \zeta'(-1)$$
$$= \frac{\ln 2\pi + \gamma}{12} - \frac{\zeta'(2)}{2\pi^2}$$
$$= \tfrac{1}{12}(\ln 2\pi + \gamma) + \frac{1}{2\pi^2}\sum_{s=1}^{\infty}\frac{\ln s}{s^2}\,,$$

$\gamma = 0.577\,215\,664\,9$ is Euler's constant,

and

$$-\zeta'(2) = \sum \frac{\ln s}{s^2}$$
$$= 0.937\,548\,254\,315\,844\,.$$

VII. THE FIXED MASS REGULATOR

Equation (6.15) gives the regularized product of all nonvanishing eigenvalues of $\mathfrak{M}$. But the regulator used was a very unsatisfactory one, from a physical point of view, because the regulator mass μ depends on space-time:

$$\mu^2 = \frac{4M^2}{(1+r^2)^2}\,. \quad (7.1)$$

This μ must be interpreted as the subtraction point of the coupling constant g. Now g does not occur in $\Pi(t)/\Pi_0(t)$, but it does occur in the expression for the total action for the classical solution, and as we emphasized in the Introduction, any change in the subtraction procedure is important. The problem here is that we wish to make a space-time-dependent change in the subtraction point, from μ to a fixed μ_0. We solve that in the following way.

The effect of a change in the regulator mass can be absorbed by a counterterm in the Lagrangian, and hence is local in space-time. So we expect that, if we make a space-time-dependent change in the regulator mass, then this change can be absorbed by a space-time-dependent counterterm. Moreover, since our regulators are both gauge-invariant, this counterterm is gauge invariant. For space-time-independent regulators, this counterterm can be computed by totally conventional methods:

$$\Delta\mathcal{L} = \frac{-g^2}{32\pi^2} G_{\mu\nu}G_{\mu\nu} \times \frac{1}{9} t(t+1)(2t+1)\ln(\mu/\mu_0)\,. \quad (7.2)$$

From locality we deduce that the same formula must also be true for space-time-dependent regulator mass $\mu(x)$, simply because no other gauge-invariant, local expressions of the same dimensionality exist. Inserting the classical value for $G_{\mu\nu}$,

$$G^{a\,\mathrm{cl}}_{\mu\nu} = -\frac{4}{g}\frac{\eta_{a\mu\nu}}{(1+r^2)^2}\,, \quad (7.3)$$

and expression (7.1) for μ, we get

$$\Delta S^{\mathrm{cl}} = \int \Delta\mathcal{L}\, d^4x$$
$$= \frac{16\times 12\pi^2}{32\pi^2\times 9} t(t+1)(2t+1)$$
$$\times \int_0^\infty \frac{r^2 dr^2}{(1+r^2)^4}\ln\frac{\mu_0}{2M}(1+r^2)$$
$$= \tfrac{2}{3} t(t+1)(2t+1)\left(\tfrac{1}{6}\ln\frac{\mu_0}{2M} + \tfrac{5}{36}\right). \quad (7.4)$$

In the expression

$$\frac{\Pi_0(t)}{\Pi(t)} S^{cl},$$

with $\Pi(t)/\Pi_0(t)$ as computed in (6.15), we must correct S^{cl} with the above ΔS^{cl}, in order to get the corresponding expression with g subtracted with fixed mass regulators, as defined in (4.9). The regulator masses M_i in there must be such that

$$\sum_{1}^{R} e_i \ln M_i = -\ln \mu_0 . \qquad (7.5)$$

Expression (7.4) must be added to (6.15). Thus we get

$$\ln[\Pi(t)/\Pi_0(t)] = \frac{t(t+1)(2t+1)}{3}\left[\tfrac{1}{3}\ln\frac{\mu_0}{2} + 4R + \sum_{s=1}^{2t+1} s(2t+1-s)(s-t-\tfrac{1}{2})\ln s - \tfrac{1}{3}t(t+1) - \tfrac{2}{9}\right]. \qquad (7.6)$$

We note that the coefficient of the regulator term in (6.15) has the correct value. It matches the coefficient of (7.2) that has been computed independently. The regulator in this expression, (7.6), is the same as the one used in (3.10), and so we can use the result of Sec. IV to do the spinor and vector fields.

In Sec. IV we proved that the nonzero eigenvalues for vector and spinor fields are the same as for scalar fields, but we must take some multiplicity factors into account. Equation (7.6) holds for one complex scalar multiplet with isospin t. Fields with integer isospin may be real and then we have to multiply by $\frac{1}{2}$. The vector field has four components but is real, and hence its value for $\ln(\Pi/\Pi_0)$ is twice expression (7.6), with $t=1$. The complex Faddeev-Popov ghost has Fermi statistics and contributes with one unit, but opposite sign. Thus, altogether, the vector field contributes just like one complex scalar with $t=1$.

For fermions we must compute $\det\mathfrak{M}_\psi$, but the theorem of Sec. IV applies to $\mathfrak{M}_\psi^2$. The fermions have four Dirac components. So, altogether, fermions contribute just like two complex scalars, but the sign in $\ln(\Pi/\Pi_0)$ is opposite because of fermi statistics.

The above summarizes in words the complete contribution of all nonzero eigenstates to the functional determinants. But the spinor and vector fields have a few more modes, with $E=0$, and also the regulators have corresponding new modes, with $E=\mu_0^2$.

VIII. THE ZERO EIGENSTATES

First we consider the vector fields. We have $s_1=\frac{1}{2}$, $t=1$. Careful study of the operator $\mathfrak{M}$, Eq. (2.22), enables us to list the square-intergrable zero eigenstates as follows:

(i) $j_1=\frac{1}{2}$, $l=0$: $\Psi=(1+r^2)^{-2}$

$(j_2=s_2=\frac{1}{2})$, (8.1)

multiplicity $=(2j_1+1)(2j_2+1)=4$.

(ii) $j_1=0$, $l=\frac{1}{2}$: $\Psi=r(1+r^2)^{-2}$.

There are two possibilities for the other quantum numbers:

(a) $j_2=0$, multiplicity $=1$, (8.2)

(b) $j_2=1$, multiplicity $=3$. (8.3)

This completes the set of zero eigenstates. We interpret these as follows. States (i) have $j_1=j_2=\frac{1}{2}$, that is, the quantum numbers of an infinitesimal translation. The translations are considered in Sec. IX. State (iia) is the only singlet. It will correspond to the infinitesimal dilatation, Sec. X. State (iib) is just an anomaly. It will be discussed in Sec. XI. It is indirectly connected with infinitesimal global isospin rotations.

Spinors have similar sets of eigenstates, but their interpretation will be totally different. If $t=1$, then the eigenstates are essentially the same as the vector ones, but their multiplicity is half of that because $s_2=0$. In this paper we limit ourselves to $t=\frac{1}{2}$. Then there is just one zero eigenstate:

$$j_1=0,\quad l=j_2=0,\quad \Psi=(1+r^2)^{-3/2} \qquad (8.4)$$

Its multiplicity is of course N^f if there are N^f flavors. It leads to an N^f-fold zero in the amplitude [note that in (2.11) the amplitude is proportional to $\det\mathfrak{M}_\psi$ and thus is proportional to the product of the eigenvalues of $\mathfrak{M}_\psi$; if we have N^f zero eigenvalues then W has an N^f-fold zero]. But this zero will be removed if we switch on the fermion source $\mathcal{J}$ in the Lagrangian (2.1). In Sec. XIV we will construct the resulting N^f-point Green's function.

In strong-interaction theories the fermion mass will also remove this zero. The zero eigenstates must also be included in the regulator contributions. From Sec. VII on, our regulator mass is fixed and is essentially equal to μ_0. Every zero eigenvector of the operator $\mathfrak{M}$, Eq. (2.22), will be accompanied by a factor μ_0^{-2} for the regulator (a zero eigenvector of $\mathfrak{M}_\psi$ is accompanied by a factor μ_0^{-1}).

IX. COLLECTIVE COORDINATES: 1. TRANSLATIONS

Clearly, zero eigenvalues make no sense if they would be included in the products carefully computed in the previous sections. In the case of the vector fields, which we will now discuss, they would render the functional integral infinite because they are in the denominator. It merely means that the integration in those directions is not Gaussian.

Let us first consider the four modes (8.1). The angular dependence and index dependence can be read off from the quantum numbers. Written in full, the mode corresponds to the quantum field fluctuation (with arbitrarily chosen norm):

$$A_\mu^{a\,qu}(\nu) = 2\eta_{a\mu\nu}(1+r^2)^{-2}, \quad \nu = 1, \ldots, 4. \tag{9.1}$$

This can be seen to be the space-time derivative of the classical solution up to a gauge transformation:

$$A_\mu^{a\,qu}(\nu) = -\frac{g}{2}\frac{\partial}{\partial z^\nu}\left\{\frac{2\eta_{a\mu\lambda}(x-z)^\lambda}{g[1+(x-z)^2]}\right\}\bigg|_{z=0} + D_\mu \Lambda^a(\nu), \tag{9.2}$$

with

$$\Lambda^a(\nu) = -\eta_{a\nu\lambda}x^\lambda(1+x^2)^{-1}. \tag{9.3}$$

The gauge transformation is there because our gauge-fixing term depends on the background field.

If we want to replace the variable $\mathfrak{D}A^{qu}$ in this particular zero-mode direction by the collective variables dz^ν, then we must insert the corresponding Jacobian factor[5]

$$\int \mathfrak{D}A^{qu} \to \int \prod_\nu \left(\frac{2}{g}dz^\nu\right)\left\{\frac{1}{2\pi}\int [A_\mu^{a\,qu}(\nu)]^2 d^4x\right\}^{1/2}, \tag{9.4}$$

where $A_\mu^{a\,qu}(\nu)$ is the solution (9.1). The factor $2/g$ comes from the factor $g/2$ in (9.2). This way the result is independent of the normalization of $A_\mu^{a\,qu}(\nu)$ of (9.1). The factors $2\pi^{-1/2}$ arise from the fact that we compare this integral with Gaussian integrals of the form $\int dA \exp(-\frac{1}{2}A^2)$, and in these Gaussian integrals the factors $\sqrt{2\pi}$ that go with each eigenvalue had been suppressed previously. We could also have dragged along all factors $\sqrt{2\pi}$ at each of the eigenvalues of the matrices $\mathfrak{M}$, and then we would have noticed that the factors $\sqrt{2\pi}$ going with the corresponding modes of the regulators, which are still Gaussian, would have been left over. In (9.4) we just include these factors from the beginning.

The norm of the solution (9.1) is

$$\int A_\mu^{a\,qu}(\nu)A_\mu^{a\,qu}(\lambda)d^4x = 2\pi^2\delta_{\nu\lambda}, \tag{9.5}$$

so, together with the regulator, these four modes yield the factor

$$\left(\frac{2}{g}\right)^4\left(\frac{\mu_0^2}{2\pi}\right)^2(2\pi^2)^2 d^4z = 2^4\pi^2\mu_0^4 g^{-4}d^4z. \tag{9.6}$$

The integral over the collective coordinates z^μ will yield the total volume of space-time, if no massless fermions are present. If there are massless fermions, then we must include the sources $\mathcal{J}$, which break the translation invariance. In that case the z integration is rather like the integration over the location of an interaction vertex in a Feynman diagram in coordinate configuration, as we will see in Sec. XIV.

X. COLLECTIVE COORDINATES: 2. DILATATIONS

From the quantum numbers of the zero eigenstate (8.2) we deduce its angular and index dependence:

$$A_\mu^{a\,qu} = \eta_{a\mu\nu}x^\nu(1+x^2)^{-2}. \tag{10.1}$$

This is a pure infinitesimal dilatation of the classical solution:

$$A_\mu^{a\,qu} = \frac{g}{4}\frac{\partial}{\partial\rho}\left(\frac{2}{g}\frac{\eta_{a\mu\nu}x^\nu}{x^2+\rho^2}\right)\bigg|_{\rho=1}. \tag{10.2}$$

Thus, going from the integration variable A^{qu} in this direction to the collective variable ρ, we need a Jacobian factor:

$$\mathfrak{D}A_\mu^{a\,qu} \to \frac{4}{g}d\rho\left[\frac{1}{2\pi}\int (A_\mu^{a\,qu})^2 d^4x\right]^{1/2} \tag{10.3}$$

The norm of the solution (10.1) is

$$\int (A_\mu^{a\,qu})^2 d^4x = \pi^2. \tag{10.4}$$

Thus, from this mode we obtain the factor

$$\frac{4\pi}{g}\left(\frac{\mu_0^2}{2\pi}\right)^{1/2} d\rho = 2^{3/2}\pi^{1/2}\mu_0 g^{-1}d\rho \tag{10.5}$$

at $\rho = 1$. Our system is not scale invariant because of the nontrivial renormalization-group behavior. The complete ρ dependence for $\rho \neq 1$ will be deduced from simple dimensional arguments (including renormalization group) in Sec. XII.

XI. GLOBAL GAUGE ROTATIONS AND THE GAUGE CONDITION

Discussion of the legitimacy of the background gauge-fixing term has been deliberately postponed to this section, because we wanted to derive first the existence of the three anomalous zero eigenstates (8.3). They have the explicit form

(arbitrary normalization)

$$\psi^a_\mu(b) = 2\eta_{ak\mu}\bar{\eta}_{bk\lambda}x^\lambda(1+x^2)^{-2} \tag{11.1}$$

They are a pure gauge artifact

$$\psi^a_\mu(b) = D_\mu\psi^a(b)\,, \tag{11.2}$$

$$\psi^a(b) = \eta_{ak\nu}\bar{\eta}_{bk\lambda}x^\nu x^\lambda(1+x^2)^{-1}\,, \tag{11.3}$$

but $\psi^a(b)$ is not square-integrable. What is going on? Note that $\psi^a(b)$, since they are x dependent, do not generate a closed algebra of gauge rotations. They may not be replaced by a collective coordinate for global isospin rotations. To analyze this situation we first go back to a background-independent gauge-fixing term,

$$C^a_1(x) = \alpha\partial_\mu(A^{a\,\mathrm{cl}}_\mu + A^{a\,\mathrm{qu}}_\mu) = \alpha\partial_\mu A^{a\,\mathrm{qu}}_\mu\,, \tag{11.4}$$

where α is a free parameter. In this gauge we know exactly how to handle all zero eigenmodes: There are five for translations and dilatations and also three for global isospin rotations because global isospin is still an invariance in this gauge. To understand the latter we put the system in a large spherical box with volume V and assume that all (vector and ghost) fields vanish on the boundary. Let

$$\Lambda^a_1(b,x) = \delta^{ab} \tag{11.5}$$

generate an infinitesimal global isospin rotation. Then there is a zero eigenmode:

$$\psi^a_{1\mu}(b) = D_\mu\Lambda^a_1(b) = 2\epsilon_{acb}\eta_{c\mu\nu}x^\nu(1+x^2)^{-1}\,. \tag{11.6}$$

The subscript 1 is to remind us that this is a solution in the gauge C_1. Similarly as in the foregoing two sections, we can replace the integral over $\mathfrak{D}A^a_\mu$ by an integral over the collective coordinates $d\Lambda(b)$ by inserting the corresponding Jacobian factor

$$\mathfrak{D}A^a_\mu \to \int\prod_b d\Lambda(b)(\mathfrak{N}/2\pi)^{1/2} \tag{11.7}$$

with

$$\mathfrak{N} = \int[\psi^a_{1\mu}(b)]^2 d^4x\,, \tag{11.8}$$

which diverges as the volume V of space-time goes to infinity. The integral over the gauge rotation is just the volume of the group and yields

$$\int\prod_b d\Lambda^b = \frac{8\pi^2}{g^3}\,, \tag{11.9}$$

where the factor g^{-3} comes from our normalization of Λ_1 in (11.6). Thus, in this gauge, the zero eigenmode yields the factor

$$\frac{8\pi^2}{g^3}\left(\frac{\mathfrak{N}}{2\pi}\right)^{3/2} \tag{11.10}$$

Now we will study different gauges, and for that we need to change the boundary condition (we know from Sec. III that that will not affect the finite eigenvalues, but the zero eigenmodes change) into a gauge-invariant one: The ghosts ϕ and gauge generators Λ must satisfy

$$\frac{\partial\phi}{\partial r}(R) = \frac{\partial\Lambda}{\partial r}(R) = 0\,, \tag{11.11}$$

where R is the radius of the box, and the vector fields

$$A_r(R) = 0\,,\quad \frac{\partial}{\partial r}A_\parallel(R) = 0\,, \tag{11.12}$$

where $A_\parallel$ is the vector component parallel to the boundary.

Now observe the following. The gauge term C_1, Eq. (11.4), does not fix the gauge completely which can be seen in two ways: (a) Global isospin rotations are still an invariance; (b) one component of the gauge term C_1 is identically zero:

$$\int C^a_1(x)d^4x = 0\,. \tag{11.13}$$

This leaves us the possibility of adding a constant to C_1, which is orthogonal to it, with which we fix the remaining global gauge

$$C^a_2(x) = \alpha\partial_\mu A^{a\,\mathrm{qu}}_\mu - \kappa\sum_b\delta^{ab}\int\psi^c_{1\mu}(b,y)A^{c\,\mathrm{qu}}_\mu(y)d^4y \tag{11.14}$$

with α,κ free parameters and $\psi^a_{1\mu}(b,y) = D_\mu\delta^{ab}$, as in (11.6). The integral over group space is now replaced by a Gaussian integral. The Gaussian volume is corrected for by the Faddeev-Popov ghost,

$$\mathfrak{L}^{\mathrm{gh}}_2 = \phi^*_a(x)\left[\alpha\partial_\mu D_\mu\phi^a(x) - \kappa\sum_b\delta^{ab}\int\psi^c_{1\mu}(b,y)D_\mu\phi^c(y)d^4y\right], \tag{11.15}$$

so that the combined contribution of vector fields and ghosts is now independent of α and κ: The zero eigenvalues are replaced by

$$\lambda^{\mathrm{vector}}_2 = \kappa^2 V\mathfrak{N}$$

and (11.16)

$$\lambda^{\mathrm{ghost}}_2 = \kappa\mathfrak{N}\,.$$

(Remember that the ghost, with the new boundary

condition, has now an eigenstate ϕ = constant.) Thus, instead of (11.10), this gauge gives

$$(\lambda_2^{\text{ghost}})^3(\lambda_2^{\text{vector}})^{-3/2} = (\mathfrak{N}/V)^{3/2}. \tag{11.17}$$

Conclusion: If the redundant eigenmodes are fixed by an additional component in the gauge-fixing term, then a correction factor is needed: Equation (11.10) divided by (11.17):

$$\frac{8\pi^2}{g^3}\left(\frac{V}{2\pi}\right)^{3/2}. \tag{11.18}$$

Here V is the volume of the spherical box.

The background gauge $C_4 = D_\mu A_\mu^{a\,\text{qu}}$ has a problem similar to the gauge C_1. Both the ghost (under the new boundary condition) and the vector field have one eigenvalue that vanishes like $1/V$ as $V\to\infty$ (not $1/R^2$, as the other eigenvalues). Let $\psi_4^a(b)$ be the three ghost eigenstates and $\psi_{4\mu}^a(b) = D_\mu\psi_4^a(b)$ be the vector ones. Let λ_4 be the ghost eigenvalue

$$D^2\psi_4^a(b) = -\lambda_4\psi_4^a(b). \tag{11.19}$$

It is easy to see that

$$\lambda_4 = \frac{\int_V[\psi_{4\mu}^a(x)]^2d^4x}{\int_V[\psi_4^a(x)]^2d^4x}. \tag{11.20}$$

From (11.1) and (11.3) we see that this is $O(1/V)$. It is safer to have a gauge condition that fixes this gauge degree of freedom as $V\to\infty$,

$$C_3(x) = \alpha D_\mu A_\mu^{a\,\text{qu}}(x) - \kappa\sum_b \psi_4^a(b,x)\int\psi_{4\mu}^c(b,y)A_\mu^{c\,\text{qu}}(y)d^4y, \tag{11.21}$$

although the result, (11.25), will turn out to remain the same even if $\kappa = 0$, $\alpha = 1$. The ghost Lagrangian is

$$\mathcal{L}_3^{\text{gh}} = \phi_a^*(x)\left[\alpha D^2\phi_a(x) - \kappa\sum_b\psi_4^a(b,x)\int\psi_{4\mu}^c(b,y)D_\mu\phi^c(y)d^4y\right]. \tag{11.22}$$

In this gauge

$$\lambda_3^{\text{vector}} = \lambda_4\left(\alpha + \kappa\int[\psi_4^a(x)]^2d^4x\right)^2 \tag{11.23}$$

and

$$\lambda_3^{\text{ghost}} = \lambda_4\left(\alpha + \kappa\int[\psi_4^a(x)]^2d^4x\right), \tag{11.24}$$

where no summation over b is implied. In this gauge we find the contribution from the lowest eigenmodes:

$$(\lambda_3^{\text{ghost}})^3(\lambda_3^{\text{vector}})^{-3/2} = \lambda_4^{3/2}. \tag{11.25}$$

Using

$$\int[\psi^a(x)]^2d^4x = V, \qquad \int[\psi_\mu^a(x)]^2d^4x = 4\pi^2, \tag{11.26}$$

we find

$$\lambda_4 = 4\pi^2/V. \tag{11.27}$$

Thus, together with the correction factor (11.18), we find the correct contribution for the three eigenmodes (9.3), together with that of their regulator:

$$\frac{8\pi^2}{g^3}\left(\frac{V\mu_0^2}{2\pi}\right)^{3/2}\left(\frac{4\pi^2}{V}\right)^{3/2} = 2^{9/2}\pi^{7/2}\mu_0^3g^{-3} \tag{11.28}$$

XII. ASSEMBLING THE VECTOR, SCALAR, AND SPINOR TERMS

The eight zero-eigenvalue modes for the vector field give the factors (9.6), (10.5), and (11.28). Multiplying these gives

$$2^{10}\pi^6g^{-8}\mu_0^8d^4z\,d\rho \tag{12.1}$$

for $\rho = 1$ (later we will find the ρ dependence).

The contributions from the nonvanishing eigenmodes both for the scalar and for the vector fields are essentially contained in formula (7.6). As we saw before, the vector fields, combined with the Faddeev-Popov ghost, together count as two real, or one complex, scalar with $t = 1$.

Let there be $N^s(t)$ scalar multiplets for each isospin t, where each complex scalar multiplet counts as one, and each real scalar multiplet counts as one-half. Then from (7.6) we obtain the total contribution from vector and scalar nonzero modes:

$$\frac{\Pi_0}{\Pi} = \exp\left\{-\left[\tfrac{2}{3}+\tfrac{1}{6}\sum_t N^s(t)C(t)\right]\ln\mu_0 - \alpha(1) - \sum_t N^s(t)\alpha(t)\right\}. \tag{12.2}$$

Here

$$C(t) = \tfrac{2}{3}t(t+1)(2t+1) \tag{12.3}$$

and

$$\alpha(t) = C(t)\left[2R - \tfrac{1}{6}\ln 2 + \tfrac{1}{2}\sum_{s=1}^{2t+1} s(2t+1-s)(s-t-\tfrac{1}{2})\ln s - \tfrac{1}{6}t(t+1) - \tfrac{1}{9}\right]. \tag{12.4}$$

The numerical values for $C(t)$ and $\alpha(t)$ are listed in Table I. Combining (12.1), (12.2), and the classical action (2.8) gives the total amplitude in the absence of fermions:

$$2^{10}\pi^6 g^{-8}\frac{d^4z\,d\rho}{\rho^5}\exp\left\{-\frac{8\pi^2}{g^2(\mu_0)} + \ln(\mu_0\rho)\left[\tfrac{22}{3} - \tfrac{1}{6}\sum_t N^s(t)C(t)\right] - \alpha(1) - \sum_t N^s(t)\alpha(t)\right\}. \tag{12.5}$$

Note that the coefficient multiplying $\ln\mu_0$ coincides with the usual Callan-Symanzik β coefficient for $g^2(\mu_0)$ in such a way that (12.5) becomes independent of the subtraction point μ_0 if we choose $g^2(\mu_0)$ to obey the Gell-Mann–Low equation. We now also insert the ρ dependence if $\rho \neq 1$ by straightforward dimensional analysis.

The interpretation of (12.5) is best given in the language of path integrals: If $|0\rangle$ is the vacuum, and $|\bar{0}\rangle$ is the gauge-rotated vacuum, then (12.5) is the total contribution to $\langle\bar{0}|0\rangle$ from all paths in Euclidean space that have a pseudoparticle located at z within d^4z, having a scale between ρ and $\rho + d\rho$. The fermions can be introduced in two ways:

(i) If they have a mass $m \ll 1/\rho$ then only the lowest eigenvalue will depend critically on m.

(ii) If they are rigorously massless then the lowest eigenvalue depends critically on the external source $\mathcal{J}$.

In this paper we limit ourselves only to fermions with isospin $t = \frac{1}{2}$. In case (i) the contribution of the lowest modes will simply be

$$\left(\frac{m}{\mu_0}\right)^{N^f}. \tag{12.6}$$

The nonvanishing eigenmodes are again obtained from (7.6), which represents the eigenvalues of $\mathfrak{M}_\psi{}^2$. Now we wish to compute $\det\mathfrak{M}_\psi$ and we take into account that the Dirac field has four components. Thus the fermion nonvanishing eigenmodes will give

$$\exp\left[\frac{N^f}{3}C(\tfrac{1}{2})\ln\mu_0 + 2N^f\alpha(\tfrac{1}{2})\right]. \tag{12.7}$$

Together with (12.6) we find the total fermion factor that multiplies (12.5),

$$\rho^{N^f} m^{N^f}\exp[-\tfrac{2}{3}N^f\ln(\mu_0\rho) + 2N^f\alpha(\tfrac{1}{2})], \tag{12.8}$$

where we again inserted the factors ρ as they follow from dimensional arguments. Note that the well-known Callan-Symanzik β coefficient for $g^2(\mu_0)$ again matches the term in front of $\ln\mu_0$.

Equations (12.5) and (12.8) could be used as a starting point for a strong-interaction color theory where Euclidean pseudoparticles form a plasma-like statistical ensemble.[7] For that, one also needs to extend from SU(2) to SU(3). We will not do that in this paper.

In Sec. XIV we consider case (ii). In that case m must be replaced by the eigenvalue of the lowest mode as it is perturbed by the source insertion.

XIII. DIMENSIONAL RENORMALIZATION

The regulators used in Secs. VII–XII are what we call fixed mass Pauli-Villars regulators and they only make sense in the background-field formalism. They are given by the Lagrangians (4.7)–(4.9). In this section we wish to switch to another regulator scheme which is much more widely used in gauge theories: the dimensional method.[8] Let us emphasize again that if one switches to another regulator, then that affects the definition of $g(\mu)$ and that influences our calculation by an overall constant. We know that in the dimensional procedure the limit of large cutoff is replaced by a limit $n \to 4$, where n is the number of space-time dimensions, roughly in the following way:

$$\ln\Lambda \to \frac{1}{4-n} + \text{finite}. \tag{13.1}$$

In (12.5) and (12.7) the regulator mass μ_0 plays the role of the cutoff Λ. Clearly then, the finite part in (13.1) will be relevant. In this section we derive that finite part, in ordinary perturbation

TABLE I. Numerical values of the coefficients $C(t)$ and $\alpha(t)$ as they occur in the text.

t	$C(t)$	$\alpha(t)$	
0	0	0	
$\frac{1}{2}$	1	$2R - \frac{1}{6}\ln 2 - \frac{17}{12}$	$= 0.145\,873$
1	4	$8R + \frac{1}{3}\ln 2 - \frac{16}{9}$	$= 0.443\,307$
$\frac{3}{2}$	10	$20R + 4\ln 3 - \frac{5}{3}\ln 2 - \frac{265}{36}$	$= 0.853\,182$

$$R = \tfrac{1}{12}(\ln 2\pi + \gamma) + \frac{1}{2\pi^2}\sum_{2}^{\infty}\frac{\ln s}{s^2} = 0.248\,754\,477\,033\,784$$

theory. It corresponds to a finite counterterm in the Lagrangian. It is easy to compute this finite counterterm when, again, one makes use of the background fields. There are some diagrams to be computed and the rest is algebra. This algebra is identical to the algebra devised in Ref. 9. Symmetry arguments restrict the possible form of the finite counterterm in just two independent terms, X^2 and $Y_{\mu\nu}Y_{\mu\nu}$, in the language of Ref. 9. The first of these is obtained by comparing the integral

$$\frac{1}{(2\pi)^n}\int d^n k \frac{1}{(k^2+\mu_0{}^2)^2} \tag{13.2}$$

in the limits $\mu_0 \to \infty$ and $n \to 4$. For definiteness, we specify the theory at $n \neq 4$ dimensions: All trivial factors $(2\pi)^2$ must also be replaced by $(2\pi)^n$, which leads to the factor $(2\pi)^{-n}$ in (13.2).

The integral (13.2) is in this limit

$$\frac{1}{(4\pi)^2}\left[\frac{2}{4-n}-\gamma-2\ln\mu_0+\ln 4\pi+O(4-n)+O\left(\frac{1}{\mu_0{}^2}\right)\right], \tag{13.3}$$

where γ is again Euler's constant. So here

$$\ln\mu_0 \to \frac{1}{4-n}-\frac{1}{2}\gamma+\frac{1}{2}\ln 4\pi\,. \tag{13.4}$$

The coefficient in front of $Y_{\mu\nu}Y_{\mu\nu}$ is obtained in the same way by comparing the integral

$$\sum_i e_i \frac{1}{(2\pi)^2}\int d^n k \frac{k_\mu k_\nu k_\alpha k_\beta}{(k^2+\mu_i^2)^2} \tag{13.5}$$

in the two limits, where the signs e_i are defined as in Eqs. (5.15) and (5.16), replacing M by μ_0. This time we get

$$\sum_{i=0}^{R} e_i \frac{1}{(4\pi)^2 4!}\left[\frac{2}{4-n}-\gamma-2\ln\mu_i+\ln 4\pi+O(4-n) + O\left(\frac{1}{\mu_0^2}\right)\right]\left(\delta_{\mu\nu}\delta_{\alpha\beta}+\delta_{\mu\alpha}\delta_{\nu\beta}+\delta_{\mu\beta}\delta_{\nu\alpha}\right). \tag{13.6}$$

So we see that for both the X^2 terms and the $Y_{\mu\nu}Y_{\mu\nu}$ terms the substitution (13.4) is to be made. However, we have to remember that in dimensional regularization the number of the fields $A_m^{a\mu}$ is n rather than 4. For these fields one therefore must replace

$$\ln\mu_0 \to \frac{1}{4-n}-\frac{1}{2}\gamma+\frac{1}{2}\ln 4\pi-1\,. \tag{13.7}$$

In conclusion, (12.5) is to be replaced by

$$2^{10}\pi^6 g^{-8}\rho^{-5}d^4z\,d\rho\exp\left\{-\frac{8\pi^2}{g_B{}^2(n)}+\left(\ln\rho+\frac{1}{4-n}\right)\left[\tfrac{22}{3}-\tfrac{1}{6}\sum_t N^s(t)C(t)\right]+A-\sum_t N^s(t)A(t)\right\}, \tag{13.8}$$

with

$$A=-\alpha(1)+\frac{11}{3}(\ln 4\pi-\gamma)+\frac{1}{3}=7.05399103 \tag{13.9}$$

and

$$A(t)=-\alpha(t)+\frac{1}{12}(\ln 4\pi-\gamma)C(t).$$

Numerically,

$$A(0)=0,\ A(\tfrac{1}{2})=0.30869069,\ A(1)=1.09457662,$$
$$A(\tfrac{3}{2})=2.48135610 \tag{13.10}$$

Similarly for the fermion factor

$$\rho^{N^f}m^{N^f}\exp\left[-\tfrac{2}{3}N^f\left(\ln\rho+\frac{1}{4-n}\right)-N^f B\right] \tag{13.11}$$

with

$$B=-2\alpha(\tfrac{1}{2})+\frac{1}{3}(\ln 4\pi-\gamma)$$
$$=0.35952290\,. \tag{13.12}$$

If we define the subtracted coupling constant as in Ref. 10,

$$g_B(n)=\mu^{4-n}\left(g_R^D(\mu)+\frac{a_1}{n-4}+\cdots\right), \tag{13.13}$$

with a_1 depending only on g_R but not on n or μ, then we can make the following replacements in (13.8) and (13.11):

$$g_B(n)\to g_R^D(\mu)\,,$$
$$\ln\rho+\frac{1}{4-n}\to\ln(\rho\mu)\,. \tag{13.14}$$

Here the superscript D stands for the dimensional procedure which defines g_R^D. We see that the expression in terms of $g_R^D(\mu)$ differs slightly from the one in terms of $g(\mu_0)$.

XIV. THE FERMION SOURCE AND THE GREEN'S FUNCTION

We now consider the fermion zero eigenmode (9.4), and assume that the fermion mass (12.6) vanishes. In that case the source $\mathcal{J}_{st}$ in (2.1) must be taken into account, since the lowest eigenvalue will now mainly be determined by this source. We determine the lowest eigenvalues $E(i)$, $i=1,\ldots,N^f$ of the operator

$$\mathfrak{M}_\psi = -\gamma_\mu D_\mu \delta_{st} + \mathcal{J}_{st} \tag{14.1}$$

by perturbation theory, taking $\mathcal{J}$ as the small perturbation. The method is the standard one (the author thanks S. Coleman for an enlightening discussion on this point). The unperturbed, degenerate eigenmodes are (taking for simplicity $\rho=1$)

$$\psi_s^\alpha(t) = C(1+r^2)^{-3/2}u^\alpha\delta_{st}, \tag{14.2}$$

$$\alpha=1,2,\quad s,t=1,\ldots,N^f.$$

The coefficients u^α contain besides the isospin index α a Dirac index. They satisfy

$$J_1^a u^\alpha = S_1^a u^\alpha + T_{\alpha\beta}^a u^\beta = 0,$$

or

$$(\eta_{a\mu\nu}\gamma_\mu\gamma_\nu + 4i\tau_a)u = 0 \tag{14.3}$$

and

$$\gamma_5 u = -u. \tag{14.4}$$

The coefficient C is determined by normalizing ψ,

$$C^{-2} = u^* u\int d^4x(1+x^2)^{-3}$$

$$= \pi^2/2 \text{ if } u^*u = 1. \tag{14.5}$$

Let

$$H_{st} = \langle\psi(s)|\mathfrak{M}_\psi|\psi(t)\rangle$$

$$= \frac{2}{\pi^2}\int d^4x(1+x^2)^{-3}u_\alpha^*\mathcal{J}_{st}(x)u^\alpha, \tag{14.6}$$

then $E(i)$ are the N^f eigenvalues of H. We wish to compute

$$\prod_i E(i) = \det H$$

$$= \left(\frac{2}{\pi^2}\right)^{N^f}\det_{st}\int d^4x(1+x^2)^{-3}u_\alpha^*\mathcal{J}_{st}(x)u^\alpha. \tag{14.7}$$

For large x^2, this amplitude has exactly the space-time structure of an N^f-point Green's function, where each source point is connected to the origin by two fermion lines. The integral over the collective coordinate z [which is at the origin in Eq. (14.7)] will correspond to the integration in coordinate configuration over the vertex variable.

Equation (14.7) should really be considered as our final result for the space-time dependence of the fermion Green's function. But it would be enlightening if we could represent it in terms of an effective Lagrangian.

We found that the effective Lagrangian can best first be written in the form

$$\mathcal{L}^{\text{eff}} = C\prod_{s=1}^{N^f}(\bar\psi_s\omega)(\bar\omega\psi_s), \tag{14.8}$$

where ω is some fixed Dirac spinor with isospin $\frac{1}{2}$. Owing to Fermi statistics, the various terms of the determinant in (14.7) will arise with the appropriate minus signs so that we may limit ourselves to sources $\mathcal{J}_{st}$ that are diagonal in s and t:

$$\mathcal{J}_{st} = \mathcal{J}(x)\delta_{st}. \tag{14.9}$$

Here $\mathcal{J}$ may still contain Dirac matrices.

In the presence of this source, the amplitude from the effective interaction (14.8) would be

$$C\prod_s[-\bar\omega S_F(x)\mathcal{J}(x)S_F(-x)\omega], \tag{14.10}$$

where the minus sign comes from the Fermi statistics and the $S_F(x)$ are the Dirac propagators for massless fermions in coordinate configuration,

$$S_F(x) = \frac{\gamma x}{2\pi^2(x^2)^2}. \tag{14.11}$$

Comparing this with (14.7), at large x^2, we find that we must require (leaving aside temporarily the other contributions to the overall constant)

$$(\gamma x\omega_\alpha)(\bar\omega_\alpha\gamma x) = x^2u_\alpha\bar u_\alpha,$$
$$C = (8\pi^2)^{N^f}, \tag{14.12}$$

where we may sum over the isospin index α but not over the Dirac components. Now from (14.3) one can derive

$$\sum_\alpha u_\alpha\bar u_\alpha = \tfrac{1}{2}(1-\gamma_5), \tag{14.13}$$

so we must require the ω_α to be such that

$$\sum_\alpha \omega_\alpha\bar\omega_\alpha = \tfrac{1}{2}(1+\gamma_5). \tag{14.14}$$

Thus, the ω_α are some parity reflection of u_α.

There is clearly no gauge-invariant solution to (14.14), so our effective Lagrangian (14.8) is apparently not gauge invariant. But note that we only wish to reproduce the amplitude (14.7) for gauge-invariant currents $\mathcal{J}_{st}$. Thus, any gauge rotation of (14.8) does the same job. We get a gauge-invariant $\mathcal{L}^{\text{eff}}$ if we average over the whole group of gauge rotations:

$$\mathcal{L}^{\text{eff}} = C\left\langle \prod_{s=1}^{N^f} (\bar{\psi}_s \omega)(\bar{\omega}\psi_s) \right\rangle, \tag{14.15}$$

where the brackets $\langle\ \rangle$ denote the average for all gauge rotations of ω. We can then derive

$$\langle \omega_\alpha \bar{\omega}_\beta \rangle = \tfrac{1}{4}\delta_{\alpha\beta}(1+\gamma_5), \tag{14.16}$$

and, for instance,

$$\left\langle \prod_{s=1}^{2} (\bar{\psi}_s\omega)(\bar{\omega}\psi_s) \right\rangle = \tfrac{1}{24}\,(2\delta^{\beta_1}_{\alpha_1}\delta^{\beta_2}_{\alpha_2} - \delta^{\beta_2}_{\alpha_1}\delta^{\beta_1}_{\alpha_2})\epsilon^{st} \times \bar{\psi}_1^{\alpha_1}(1+\gamma_5)\psi_s^{\beta_1}\bar{\psi}_2^{\alpha_2}(1+\gamma_5)\psi_t^{\beta_2}. \tag{14.17}$$

The Lagrangian (14.15) only acts on the left-handed spinors. The parity-reflected Euclidean pseudoparticles will give a similar contribution acting on the right-handed spinors only. So in total we get $\mathcal{L}^{\text{eff}}$ of (14.15) plus its Hermitian conjugate.

Note that we obtain products of fermion fields, such as (14.17), that violate only chiral U(1) invariance. They have the chiral-symmetry properties of the determinant of an $N^f \times N^f$ matrix in flavor space and are therefore still invariant under chiral $SU(N^f)\times SU(N^f)$. The symmetry violation is associated with an arbitrary phase factor $e^{\pm i\omega}$ in front of the effective Lagrangians. If other mass terms or interaction terms occur in the Lagrangian that also violate chiral U(1), then they may have a phase factor different from these. We then find that our effective Lagrangian may violate P invariance, whereas C invariance is maintained. Thus we find that not only U(1) invariance but also PC invariance can be violated by our effect.

XV. CONVERGENCE OF THE ρ INTEGRATION

The entire expression that we now have for the effective Lagrangian is

$$\mathcal{L}^{\text{eff}}(z)d^4z = 2^{10+3N^f}\pi^{6+2N^f}g^{-8}d^4z \int \rho^{-5+3N^f}d\rho \exp\left\{-\frac{8\pi^2}{[g_R^D(\mu)]^2} + \ln(\mu\rho)\left[\tfrac{22}{3} - \tfrac{1}{6}\sum_t N^s(t)C(t) - \tfrac{2}{3}N^f\right] + A - \sum_t N^s(t)A(t) - N^f B\right\}\left\langle \prod_{s=1}^{N^f} (\bar{\psi}_s\omega)(\bar{\omega}\psi_s)\right\rangle + \text{H.c.}, \tag{15.1}$$

with

$$\langle \omega_\alpha \bar{\omega}_\beta \rangle = \tfrac{1}{4}\delta_{\alpha\beta}(1+\gamma_5), \text{ etc.},$$

and the numbers $A, A(t), B, C(t)$ as defined before.

The ρ dependence has been changed because the effective Lagrangian (14.8) is not dimensionless.

We see that this integral converges as $\rho \to 0$ (except when there are very many scalars). But there is an infrared divergence as $\rho \to \infty$. In an unbroken color gauge theory for strong interactions this is just one of the various infrared disasters of the theory to which we have no answer. But in a weak-interaction theory it is expected that the Higgs field provides for the cutoff. Let there be a Higgs field with isospin q and vacuum expectation value F. Let its contribution to the original Lagrangian be

$$\mathcal{L}^H = -D_\mu\Phi^* D_\mu\Phi - V(\Phi). \tag{15.2}$$

Formally, no classical solution exists now, because the Higgs Lagrangian tends to add to the total action of the pseudoparticle a contribution proportional to F^2, but this can always be reduced by scaling to smaller distances, until the action reaches the usual value $8\pi^2/g^2$ when the field configuration is singular.

On the other hand, it is clear that the quantum corrections, as can be seen in (15.1), act in the opposite way. There must be a region of values for ρ where the quantum effects compete with the effects due to the Higgs fields.

To handle this situation rigorously we alter slightly the philosophy of Sec. II. In Euclidean space it is not compulsory to consider only those classical fields for which the action is stationary. We will now look at approximate solutions of the classical equations, so that the total action is only a slowly varying function of one collective parameter, ρ.

We simply postulate the gauge field A to have the same configuration as before, with certain value for ρ, and now choose the Higgs field configuration in such a way that the total action is extreme. Only those infinitesimal variations that are pure scale transformations do not leave the action totally invariant, but nevertheless the parameter ρ gets the full treatment as a collective variable.

As will be verified explicitly, the dominant values for ρ will be those where the quantum effects and the Higgs contribution are equally important. Since the quantum effects are small we expect that there

$$\rho \ll 1/M_H\,, \tag{15.3}$$

which implies that the Higgs particle may be considered as approximately massless. Let us scale toward

$$\rho = 1\,, \quad |F| \sim 1\,, \quad M_H{}^2 \sim \lambda F^2 \ll 1\,.$$

The equation for this field will be approximately

$$\begin{aligned} &D^2\Phi = 0\,, \\ &\Phi^2(r \to \infty) = F^2 \end{aligned} \tag{15.4}$$

The solution to that is a zero-eigenvalue mode of the familiar operator (2.22):

$$\begin{aligned} &j_1 = 0\,, \; l = j_2 = q\,, \\ &|\Phi| = \left(\frac{x^2}{1+x^2}\right)^q |F|\,. \end{aligned} \tag{15.5}$$

The contribution to the classical action is

$$S^H = \int [-D_\mu\Phi^* D_\mu\Phi - V(\Phi)]d^4x\,. \tag{15.6}$$

The first term is (observing that $x_\mu A_\mu^{\rm cl} = 0$)

$$\begin{aligned} -\int_V \partial_\mu(\Phi^* D_\mu\Phi)d^4x &= -\int_S \Phi^*\partial_r\Phi\, d^3x \\ &= -4\pi^2 q F^2\,. \end{aligned} \tag{15.7}$$

The second term in (15.6) is of order λF^4, where λ is a small coupling constant. If we scale back to arbitrary ρ, then the Higgs field factor in the total expression is

$$\exp S^H = \exp[-4\pi^2 q F^2\rho^2 - O(\lambda F^4\rho^4)]\,. \tag{15.8}$$

We see that the second term in the exponent may be neglected at first approximation.

Thus (15.8) multiplies the integrand in (15.1) and the ρ integration is now completely convergent. The integration over ρ yields a factor

$$\tfrac{1}{2}(4\pi^2 q F^2)^{2-(3/2)N^f - C}\,\Gamma(\tfrac{3}{2}N^f + C - 2)\,, \tag{15.9}$$

where

$$C = \tfrac{11}{3} - \tfrac{1}{12}\sum_t N^s(t)C(t) - \tfrac{1}{3}N^f\,. \tag{15.10}$$

ACKNOWLEDGMENT

The author wishes to thank S. Coleman, R. Jackiw, C. Rebbi, and all other theorists at Harvard for their hospitality, encouragement, and discussions during the completion of this work.

APPENDIX: PROPERTIES OF THE η SYMBOLS

The group SO(4) is locally equivalent to SO(3) $\times$ SO(3). The antisymmetric tensors $A_{\mu\nu}$ in SO(4) having six components form a 3+3 representation of SO(3) $\times$ SO(3). The self-dual tensors

$$A_{\mu\nu} = \tfrac{1}{2}\epsilon_{\mu\nu\alpha\beta}A_{\alpha\beta} \tag{A1}$$

transform as 3-vectors of one SO(3) group. We now define the η symbols, in a way very similar to the Dirac γ matrices:

$$\begin{aligned} &A_{\mu\nu} = \eta_{a\mu\nu}A_a\,, \\ &a = 1,2,3, \quad \mu,\nu = 1,\ldots,4 \end{aligned} \tag{A2}$$

is a covariant mapping of SO(3) vectors on self-dual SO(4) tensors. A convenient representation is

$$\begin{aligned} &\eta_{a\mu\nu} = \epsilon_{a\mu\nu}\,, \text{ if } \mu,\nu = 1,2,3 \\ &\eta_{a4\nu} = -\delta_{a\nu}\,, \\ &\eta_{a\mu 4} = \delta_{a\mu}\,, \\ &\eta_{a44} = 0\,. \end{aligned} \tag{A3}$$

Let us also define

$$\bar\eta_{a\mu\nu} = (-1)^{\delta_{\mu4}+\delta_{\nu4}}\eta_{a\mu\nu}\,. \tag{A4}$$

The symbols $\bar\eta_{a\mu\nu}$ will then do the same with vectors of the other SO(3) group and tensors $B_{\mu\nu}$ that are minus their own dual.

We have the following identities:

$$\eta_{a\mu\nu} = \tfrac{1}{2}\epsilon_{\mu\nu\alpha\beta}\eta_{a\alpha\beta}\,, \quad \bar\eta_{a\mu\nu} = -\tfrac{1}{2}\epsilon_{\mu\nu\alpha\beta}\bar\eta_{a\alpha\beta}\,, \tag{A5}$$

$$\eta_{a\mu\nu} = -\eta_{a\nu\mu}\,, \tag{A6}$$

$$\eta_{a\mu\nu}\eta_{b\mu\nu} = 4\delta_{ab}\,, \tag{A7}$$

$$\eta_{a\mu\nu}\eta_{a\mu\lambda} = 3\delta_{\nu\lambda}\,, \tag{A8}$$

$$\eta_{a\mu\nu}\eta_{a\mu\nu} = 12\,, \tag{A9}$$

$$\eta_{a\mu\nu}\eta_{a\kappa\lambda} = \delta_{\mu\kappa}\delta_{\nu\lambda} - \delta_{\mu\lambda}\delta_{\nu\kappa} + \epsilon_{\mu\nu\kappa\lambda}\,, \tag{A10}$$

$$\delta_{\kappa\lambda}\eta_{a\mu\nu} + \delta_{\kappa\nu}\eta_{a\lambda\mu} + \delta_{\kappa\mu}\eta_{a\nu\lambda} + \eta_{a\sigma\kappa}\epsilon_{\lambda\mu\nu\sigma} = 0\,, \tag{A11}$$

$$\eta_{a\mu\nu}\eta_{b\mu\lambda} = \delta_{ab}\delta_{\nu\lambda} + \epsilon_{abc}\eta_{c\nu\lambda}\,, \tag{A12}$$

$$\epsilon_{abc}\eta_{b\mu\nu}\eta_{c\kappa\lambda} = \delta_{\mu\kappa}\eta_{a\nu\lambda} - \delta_{\mu\lambda}\eta_{a\nu\kappa} - \delta_{\nu\kappa}\eta_{a\mu\lambda} + \delta_{\nu\lambda}\eta_{a\mu\kappa}\,, \tag{A13}$$

$$\eta_{a\mu\nu}\bar\eta_{b\mu\nu} = 0\,, \tag{A14}$$

$$\eta_{a\kappa\mu}\bar\eta_{b\kappa\lambda} = \eta_{a\kappa\lambda}\bar\eta_{b\kappa\mu}\,. \tag{A15}$$

*Work supported in part by the National Science Foundation under Grant No. MPS 75-20427.

†On leave from the University of Utrecht.

[1]A. A. Belavin *et al.*, Phys. Lett. 59B, 85 (1975).

[2]G. 't Hooft, Phys. Rev. Lett. 37, 8 (1976); R. Jackiw and C. Rebbi, *ibid.* 37, 172 (1976); Phys. Rev. D 14, 517 (1976); C. Callan, R. Dashen, and D. Gross, Phys. Lett. 63B, 334 (1976). See also F. R. Ore, Jr., Phys. Rev. D (to be published).

[3]R. Jackiw and S. Coleman (private communication).

[4]J. S. Bell and R. Jackiw, Nuovo Cimento 51, 47 (1969); S. L. Adler, Phys. Rev. 177, 2426 (1969).

[5]J. L. Gervais and B. Sakita, Phys. Rev. D 11, 2943 (1975); E. Tomboulis, *ibid.* 12, 1678 (1975).

[6]J. Honerkamp, Nucl. Phys. B48, 269 (1972); J. Honerkamp, in *Proceedings of the Colloquium on Renormalization of Yang-Mills Fields and Applications to Particle Physics, 1972,* edited by C. P. Korthals-Altes (C.N.R.S., Marseille, France, 1972).

[7]A. M. Polyakov, Phys. Lett. 59B, 82 (1975).

[8]G. 't Hooft and M. Veltman, Nucl. Phys. B44, 189 (1972); C. G. Bollini and J. J. Giambiagi, Phys. Lett. 40B, 566 (1972); J. F. Ashmore, Lett. Nuovo Cimento 4, 289 (1972); G. 't Hooft and M. Veltman, Report No. CERN 73-9, 1973 (unpublished).

[9]G. 't Hooft, Nucl. Phys. B62, 444 (1973).

[10]G. 't Hooft, Nucl. Phys. B61, 455 (1973).

Phys. Rev. (E) **D18**, 2199 (1978)

Erratum: Computation of the quantum effects due to a four-dimensional pseudoparticle

[Phys. Rev. D 14, 3432 (1976)]

G. 't Hooft

1. In the transition towards collective coordinates, page 3442, we inserted a factor $1/\sqrt{\pi}$ [Eq. (9.4)] for each collective coordinate because these have to be normalized with a Gaussian integral. However, the relevant Gaussian integrals here are all of the type

$$\int \exp\left(-\frac{1}{2}x^2\right) dx = \sqrt{2\pi}\ ;$$

thus the expressions must be multiplied by a factor $1\sqrt{2}$ for each collective coordinate. We have eight of these. Equation (9.6) must be divided by 4 (both left and right), Eqs. (10.3) and (10.5) by $\sqrt{2}$, Eqs. (11.7), (11.10), (11.17), (11.18), and (11.28) by $2\sqrt{2}$, and in the final expressions (12.1), (12.5), (13.8), and (15.1) we must replace

2^{14} by 2^{10}.

2. In the transition from (6.13) to (6.15) the term

$$\sum_{s=1}^{2t+1} s(2t+1-s)(s-t-\frac{1}{2})\mathrm{Ins}$$

was erroneously multiplied by $t(t+1)$. This error propagates into Eq. (7.6) and Eq. (12.4) for $\alpha(t)$. The explicitly computed values for $\alpha(t)$ in Table I are free of this error. The author thanks F. R. Ore for making this observation.[1]

3. In Eq. (3.10) $1/\pi$ must be replaced by $2/\pi$. This has no further consequences.

In a report by Y. Iwasaki[2] it is suggested that certain zero modes of the ghost field in the background gauge could alter our conclusions. However, a careful reader of our paper will realize that these modes have been taken into account correctly.

[1] F. R. Ore, Phys. Rev. D 16, 2577 (1977).

[2] Y. Iwasaki, Princeton Institute for Advanced Studies report, 1978 (unpublished).

PHYSICAL REVIEW D VOLUME 14, NUMBER 2 15 JULY 1976

Conformal properties of a Yang-Mills pseudoparticle*

R. Jackiw and C. Rebbi
Laboratory for Nuclear Science and Department of Physics, Massachusetts Institute of Technology, Cambridge, Massachusetts 02139
(Received 12 April 1976)

The conformal transformation properties of the recently discovered pseudoparticle solution to a pure Yang-Mills theory are studied. It is shown that the solution is invariant under an O(5) subgroup of conformal transformations. A formalism is developed which renders this invariance explicit and which allows a very compact group-theoretical analysis of the propagation of fermions in the field of the pseudoparticle.

I. INTRODUCTION

A remarkable pseudoparticle solution to the SU(2) Yang-Mills theory in Euclidean four-space has been found by Belavin, Polyakov, Schwartz, and Tyupkin.[1] It has the property that although the gauge-covariant field strengths $F^{\mu\nu}=(\sigma^a/2i)F^{\mu\nu}_a$ vanish rapidly at large distance, the gauge potentials $A^\mu=(\sigma^a/2i)A^\mu_a$ decrease considerably more slowly—they tend to a pure gauge term

$$A^\mu \underset{r\to\infty}{\longrightarrow} g^{-1}\partial^\mu g. \tag{1.1}$$

As a consequence, the quantity

$$q=-\frac{1}{16\pi^2}\int d^4x\,\mathrm{Tr}\,{}^*F^{\mu\nu}F_{\mu\nu}, \tag{1.2}$$

where

$$F^{\mu\nu}=\partial^\mu A^\nu-\partial^\nu A^\mu+[A^\mu,A^\nu]\,,$$
$$^*F^{\mu\nu}=\tfrac{1}{2}\epsilon^{\mu\nu\alpha\beta}F_{\alpha\beta},$$

is nonvanishing, even though the integrand is a total derivative

$$\mathrm{Tr}\,{}^*F^{\mu\nu}F_{\mu\nu}=2\epsilon^{\mu\nu\alpha\beta}\mathrm{Tr}\,\partial_\mu(A_\nu\partial_\alpha A_\beta+\tfrac{2}{3}A_\nu A_\alpha A_\beta). \tag{1.3}$$

Since $(1/8\pi^2)\,\mathrm{Tr}\,{}^*F^{\mu\nu}F_{\mu\nu}$ is the anomalous divergence of the U(1) axial-vector current,[2] q also measures the anomalous violation of chirality. Such behavior in Minkowski-space quantum field theory would have far-reaching physical consequences,[3] especially as regards the well-known U(1) problem.[4] Indeed it has been suggested that the pseudoparticle solution be used to dominate the functional-integral description of quantum field theory, continued to Euclidean space.[5]

We discuss here the behavior of the pseudoparticle under general conformal transformations, which form an O(5, 1) invariance group for the field equation in Euclidean space[6]

$$\partial_\mu F^{\mu\nu}+[A_\mu,F^{\mu\nu}]=0, \tag{1.4}$$

We show in Sec. II that the solution respects a subgroup of the conformal group—the O(5) group generated by $M^{\mu\nu}$, the rotation generator, and by a combination of K^μ and P^μ, the conformal and translation generators, respectively. This group has been previously considered by Adler[7] and Fubini.[8] Moreover, coordinate inversion corresponds to pseudoparticle conjugation.

The O(5) invariance can be made explicit by formulating the theory in a manifestly O(5)-covariant fashion. This we do in Sec. III and find a truly elegant expression for the pseudoparticle field. Furthermore, the particularly simple form offers remarkable computational advantages, as we show in Sec. IV, where we study the interaction of fermions with the pseudoparticle.

II. CONFORMAL TRANSFORMATIONS

The form of the pseudoparticle solution[1] is

$$A^\mu=\frac{x^2}{1+x^2}g^{-1}\partial^\mu g, \tag{2.1}$$

with $g=(x_4-i\vec{x}\cdot\vec{\sigma})(x_\mu x^\mu)^{-1/2}$. Translations and dilatations do not leave (2.1) invariant; they give equivalant solutions with x shifted (translations) or 1 rescaled (dilatations). The field strength $F^{\mu\nu}$ is

$$F^{\mu\nu}=\frac{4}{(1+x^2)^2}i\sigma^{\mu\nu}, \tag{2.2}$$

where we have defined

$$\sigma^{ij}=\frac{1}{4i}[\sigma^i,\sigma^j],$$
$$\sigma^{i4}=\tfrac{1}{2}\sigma^i. \tag{2.3}$$

The matrices $\sigma^{\mu\nu}$ and the field strengths are self-dual, $^*F^{\mu\nu}=F^{\mu\nu}$ and $q=1$.

Another solution is obtained by replacing g with $g^\dagger=g^{-1}=(x_4+i\vec{x}\cdot\vec{\sigma})(x_\mu x^\mu)^{-1/2}$. In this case

$$F^{\mu\nu}=\frac{4}{(1+x^2)^2}i\bar{\sigma}^{\mu\nu} \tag{2.4}$$

and

$$\bar{\sigma}^{ij}=\sigma^{ij},$$
$$\bar{\sigma}^{i4}=-\sigma^{i4}. \tag{2.5}$$

The matrices $\bar{\sigma}^{\mu\nu}$ and the field strengths are now self-antidual, $*F^{\mu\nu}=-F^{\mu\nu}$, and $q=-1$.

We consider first a discrete transformation of the conformal group, coordinate inversion, which takes x^μ into $(1/x)^\mu \equiv x^\mu/x^2$. The vector field transforms under inversion into $\bar{A}^\mu(x)$, given by

$$A^\mu(x)\to\bar{A}^\mu(x)=\frac{1}{x^2}I^{\mu\nu}(x)A_\nu(1/x),$$
$$I^{\mu\nu}(x)=g^{\mu\nu}-2\frac{x^\mu x^\nu}{x^2}. \tag{2.6}$$

It follows that $F^{\mu\nu}$ transforms similarly:

$$F^{\mu\nu}(x)\to\bar{F}^{\mu\nu}(x)=\frac{1}{x^4}I^{\mu\alpha}(x)I^{\nu\beta}(x)F_{\alpha\beta}(1/x). \tag{2.7}$$

Suppose $F^{\mu\nu}$ is self-dual or self-antidual:

$$*F^{\mu\nu}=\pm F^{\mu\nu}. \tag{2.8a}$$

It is easy to verify that the inverted field strengths satisfy an inverted relation

$$*(\bar{F}^{\mu\nu})=\mp(\bar{F}^{\mu\nu}). \tag{2.8b}$$

Consequently,

$$q=-\frac{1}{16\pi^2}\int d^4x\,\mathrm{Tr}\,*F^{\mu\nu}F_{\mu\nu}$$
$$=\mp\frac{1}{16\pi^2}\int d^4x\,\mathrm{Tr}F^{\mu\nu}F_{\mu\nu}-\bar{q},$$
$$\bar{q}=-\frac{1}{16\pi^2}\int d^4x\,\mathrm{Tr}\,*(\bar{F}^{\mu\nu})(\bar{F}_{\mu\nu})$$
$$=\pm\frac{1}{16\pi^2}\int d^4x\,\mathrm{Tr}\bar{F}^{\mu\nu}\bar{F}_{\mu\nu}. \tag{2.9}$$

Since the action $-\frac{1}{8}\int d^4x\,\mathrm{Tr}F^{\mu\nu}F_{\mu\nu}$ is inversion-invariant, $\bar{q}=-q$. Hence coordinate inversion sends the pseudoparticle with $q=1$ into an antiparticle with $q=-1$.[9]

The explicit formula for the inverted potential is

$$A^\mu(x)\to\bar{A}^\mu(x)=\frac{1}{1+x^2}g^{-1}\partial^\mu g. \tag{2.10a}$$

This does not satisfy the boundary conditions (1.1) and is singular at the origin. But if $\bar{A}^\mu$ is subjected to a gauge transformation

$$\bar{A}^\mu\to U^{-1}\bar{A}^\mu U+U^{-1}\partial^\mu U,$$

with $U=g^{-1}$, then (2.10a) is seen to be equivalent to

$$\bar{A}^\mu=\frac{x^2}{1+x^2}g\partial^\mu g^{-1}, \tag{2.10b}$$

which is consistent with (1.1).

The particle and antiparticle solutions can be put together by extending the gauge group to SU(2) $\times$ SU(2) = O(4), which is a convenience already pointed out in Ref. 1, and is also adopted by us henceforth. The O(4) field strengths are now given by

$$F^{\mu\nu}=\frac{4}{(1+x^2)^2}i\Sigma^{\mu\nu}, \tag{2.11}$$

where $\Sigma^{\mu\nu}$ is constructed from the Dirac matrices

$$\Sigma^{\mu\nu}=\frac{1}{4i}[\alpha^\mu,\alpha^\nu]=\begin{pmatrix}\sigma^{\mu\nu} & 0\\ 0 & \bar{\sigma}^{\mu\nu}\end{pmatrix},$$
$$\alpha^i=\begin{pmatrix}0 & \sigma^i\\ \sigma^i & 0\end{pmatrix},\quad \alpha^4=i\begin{pmatrix}0 & -I\\ I & 0\end{pmatrix}. \tag{2.12}$$

These field strengths are obtained from the potentials

$$A^\mu=\frac{-2i}{1+x^2}\Sigma^{\mu\nu}x_\nu. \tag{2.13}$$

The matrices $\Sigma^{\mu\nu}$ are block diagonal, so that Eqs. (2.11) and (2.13) represent two separate solutions in the two SU(2)'s of the decomposition O(4) = SU(2) $\times$ SU(2). In terms of the O(4) formalism, we can say that the effect of the inversion is to produce solutions where, up to a gauge transformation, the upper and lower diagonal blocks of the matrices $\Sigma^{\mu\nu}$ are interchanged.

Next we consider the effect on the pseudoparticle of continuous transformations of the conformal group. Of course the solution does not respect translations and dilatations. An infinitesimal rotation with parameters $\omega^{\alpha\beta}=-\omega^{\beta\alpha}$ takes x^α into $x^\alpha-\omega^{\alpha\beta}x_\beta$, and

$$\delta F^{\alpha\beta}=\frac{4i}{(1+x^2)^2}(\omega^\alpha{}_\nu\Sigma^{\nu\beta}-\omega^\beta{}_\nu\Sigma^{\nu\alpha}). \tag{2.14}$$

The appraent noninvariance of $F^{\alpha\beta}$ can be compensated by an O(4) gauge transformation $U=e^{-i\Theta}$, $\Theta=\frac{1}{2}\theta_{\alpha\beta}\Sigma^{\alpha\beta}$. The response of $F^{\alpha\beta}$ to an infinitesimal gauge rotation is

$$\delta F^{\alpha\beta}=i[\Theta,F^{\alpha\beta}]$$
$$=-\frac{4}{(1+x^2)^2}[\Theta,\Sigma^{\alpha\beta}], \tag{2.15}$$

which compensates (2.14) provided $\theta^{\alpha\beta}=\omega^{\alpha\beta}$. Thus, as was already remarked in Ref. 1, the pseudoparticle is invariant under the combined space and isospace rotation generated by $J^{\mu\nu}$,

$$J^{\mu\nu}=M^{\mu\nu}+\Sigma^{\mu\nu}. \tag{2.16}$$

A conformal transformation, i.e., an inversion, a translation, and another inversion, will not leave the solution invariant, since translations do not. However, if we perform the infinitesimal transformation generated by[7,8]

$$R^\mu=\frac{1}{2}(K^\mu+P^\mu), \tag{2.17}$$

with parameters a_μ, we find

$$\delta F^{\alpha\beta} = \frac{4i}{(1+x^2)^2}(\Omega^\alpha{}_\nu \Sigma^{\nu\beta} - \Omega^\beta{}_\nu \Sigma^{\nu\alpha}),$$
$$\Omega^{\alpha\nu} = a^\alpha x^\nu - a^\nu x^\alpha. \tag{2.18}$$

Comparison with Eqs. (2.14) and (2.15) shows that the noninvariance may be compensated by a gauge transformation parametrized by Θ of the form

$$\Theta = \tfrac{1}{2}\Omega_{\mu\nu}\Sigma^{\mu\nu} = a_\mu x_\nu \Sigma^{\mu\nu}. \tag{2.19}$$

The invariance is therefore generated by $\mathcal{R}^\mu$:

$$\mathcal{R}^\mu = R^\mu + \Sigma^{\mu\nu} x_\nu. \tag{2.20}$$

When it is recalled that the action of R^μ on an arbitrary field Φ is[8]

$$\delta\Phi = \tfrac{1}{2}(1+x^2)^{1-d} a^\mu \partial_\mu[(1+x^2)^d \Phi]$$
$$+ a_\mu x_\nu (x^\mu \partial^\nu - x^\nu \partial^\mu + iS^{\mu\nu})\Phi, \tag{2.21}$$

where d is the scale dimension and $S^{\mu\nu}$ the spin matrix of Φ, it is recognized that $\mathcal{R}^\mu$ takes a similar form, except that $S^{\mu\nu}$ is replaced by $S^{\mu\nu} + \Sigma^{\mu\nu}$, just as in Eq. (2.16). The combination law for the modified generators $J^{\mu\nu}$ and $\mathcal{R}^\mu$ follows that of $M^{\mu\nu}$ and R^μ; the algebra closes on O(5).[7,8]

III. YANG-MILLS THEORY ON A HYPERSPHERE

The O(5) invariance of the pseudoparticle solution suggests that the theory be formulated in a manifestly O(5)-covariant fashion. This is achieved by projecting 4-dimensional Euclidean space onto the surface of a unit hypersphere embedded in 5-dimensional Euclidean space.[7,8] We introduce 5-dimensional coordinates r_a, $a=1,\ldots,5$, $r_a r_a = 1$ (in what follows Latin labels $a, b, c, \ldots$ run from 1 to 5; Greek labels $\mu, \nu, \zeta, \ldots$ run from 1 to 4):

$$r_\mu = \frac{2x_\mu}{1+x^2},$$
$$r_5 = \frac{1-x^2}{1+x^2}. \tag{3.1}$$

Rotations on the sphere are generated by M_{ab}, with $M_{5\mu} = R_\mu$, as is easily shown by defining the 5-dimensional orbital angular momentum tensor L_{ab}:

$$L_{ab} = -ir_a \frac{\partial}{\partial r_b} + ir_b \frac{\partial}{\partial r_a},$$
$$L_{\mu\nu} = -ix_\mu \frac{\partial}{\partial x^\nu} + ix_\nu \frac{\partial}{\partial x^\mu}, \tag{3.2}$$
$$L_{5\mu} = -ix_\mu x^\nu \frac{\partial}{\partial x^\nu} - \tfrac{1}{2}i(1-x^2)\frac{\partial}{\partial x^\mu}.$$

The formulation of an Abelian gauge theory on a hypersphere has been given by Adler.[7] To generalize his results to the non-Abelian case, one introduces gauge fields $\hat{A}_a$, which are anti-Hermitian matrices in the space of infinitesimal generators of the gauge group and obey the constraint $r_a \hat{A}_a = 0$. The relation between the 5-dimensional $\hat{A}_a$ and the conventional 4-dimensional A_μ is the following:

$$\frac{1+x^2}{2} A_\mu = \hat{A}_\mu - x_\mu \hat{A}_5,$$
$$\hat{A}_\mu = \frac{1+x^2}{2} A_\mu - x_\mu x_\nu A^\nu,$$
$$\hat{A}_5 = -x_\mu A^\mu. \tag{3.3}$$

Under a gauge transformation U, $\hat{A}_a$ is changed into $\hat{A}'_a$,

$$\hat{A}_a - \hat{A}'_a = U^{-1}\hat{A}_a U + U^{-1} i r_b L_{ba} U. \tag{3.4}$$

From the potentials $\hat{A}^a$ one constructs a totally antisymmetric field-strength tensor of rank three, $\hat{F}_{abc}$:

$$\hat{F}_{abc} = iL_{ab}\hat{A}_c + r_a[\hat{A}_b, \hat{A}_c]$$
$$+ \text{cyclic permutations of } a, b, c. \tag{3.5}$$

The invariant action is

$$I = -\tfrac{1}{48}\int d\Omega\, \mathrm{Tr}\hat{F}_{abc}\hat{F}_{abc}, \tag{3.6}$$

where the integral is over the angular hyperspherical variables. I is identical to the action $-\frac{1}{16}\int d^4x\, \mathrm{Tr} F^{\mu\nu}F_{\mu\nu}$ constructed with the 4-dimensional field strength. The variational principle $\delta I = 0$ then gives the equation satisfied by $\hat{F}_{abc}$,

$$iL_{ab}\hat{F}_{abc} + r_a[\hat{A}_b, \hat{F}_{abc}] - r_b[\hat{A}_a, \hat{F}_{abc}] = 0. \tag{3.7}$$

We have seen in Sec. II that O(5) invariance of the pseudoparticle is achieved by a combination of an O(5) rotation, $M_{ab} = \{M_{\mu\nu}, R_\mu\}$, and a gauge transformation. This suggests that in the present O(5) formalism, the invariance can be made explicit by adding to M_{ab} some suitable generators of a gauge transformation isomorphic to O(5). Of course, the gauge group we are dealing with is O(4), which is only a subgroup of O(5). But the matrix representation of O(4) provided by Eq. (2.12) can be immediately extended to a representation of O(5) by defining the matrices $\Sigma_{\mu 5}$,

$$\Sigma_{\mu 5} = \tfrac{1}{2}\alpha_\mu. \tag{3.8}$$

Then the 10 matrices Σ_{ab} are isomorphic to the infinitesimal generators of O(5). We stress that the embedding of the O(4) gauge group in an O(5) gauge group does not alter the theory; rather it is a convenient device for exposing invariance under the O(5) subgroup of conformal transformations.

Having extended the gauge group to O(5), we can

look for solutions which are explicitly invariant under the combined space and gauge rotations generated by $J_{ab}=M_{ab}+\Sigma_{ab}$. The most general form for $\hat{A}_a$ compatible with this invariance is

$$\hat{A}_a = i\alpha\Sigma_{ab}r_b, \tag{3.9}$$

where α is a constant. The field strength is determined from Eq. (3.5),

$$\hat{F}_{abc} = -i(\alpha^2+2\alpha)(r_a\Sigma_{bc}+r_b\Sigma_{ca}+r_c\Sigma_{ab}). \tag{3.10}$$

In order that the field equation (3.7) be satisfied, the following condition must be true:

$$-6i(\alpha+1)(\alpha^2+2\alpha)\Sigma_{ab}r_b=0, \tag{3.11}$$

which requires that $\alpha=0, -1, -2$. The two values $\alpha=0, -2$ lead to vanishing field strengths, and correspond therefore to a pure gauge ansatz for the potentials, while $\alpha=-1$ gives

$$\hat{A}_a = -i\Sigma_{ab}r_b, \tag{3.12a}$$

$$\hat{F}_{abc} = i(r_a\Sigma_{bc}+r_b\Sigma_{ca}+r_c\Sigma_{ab}). \tag{3.12b}$$

This is the most general nontrivial solution to an O(5) Yang-Mills field theory on the hypersphere which is O(5)-invariant.

The solution (3.12) is also equivalent to the pseudoparticle solution (2.11) and (2.13). To show this, we first eliminate the matrices $\Sigma_{\mu 5}$ from (3.12) with a gauge transformation (3.4), thereby rotating the fields into the O(4) subspace. Then we use Eq. (3.3) to construct A_μ from $\hat{A}_a$. The required gauge transformation is

$$U=\exp[if(r_5)\Sigma_{\mu 5}r^\mu]. \tag{3.13}$$

One verifies that the two choices for $f(r_5)$

$$f(r_5)=\frac{\cos^{-1}r_5}{(1-r_5^2)^{1/2}}, \tag{3.14a}$$

$$f(r_5)=\frac{\cos^{-1}r_5-\pi}{(1-r_5^2)^{1/2}} \tag{3.14b}$$

lead respectively to

$$A^\mu=\frac{-2i}{1+x^2}\Sigma^{\mu\nu}x_\nu, \tag{3.15a}$$

$$A^\mu=\frac{-2i}{(1+x^2)x^2}\Sigma^{\mu\nu}x_\nu, \tag{3.15b}$$

which are precisely the two gauge-equivalent forms of the potentials that characterize the pseudoparticle solution.

To summarize, the invariance of the pseudoparticle solution under the O(5) subgroup of conformal transformations generated by $M^{\mu\nu}$ and $R^\mu=\frac{1}{2}(K^\mu+P^\mu)$ can be made manifest by projecting Euclidean 4-space onto a hypersphere and embedding the O(4) gauge group in an O(5) gauge group. Using a gauge transformation, we can elegantly express the solution as in (3.12)—a formula invariant under the combined space and gauge transformations generated by $J_{ab}=M_{ab}+\Sigma_{ab}$. Note also that on the hypersphere the solution (3.12) is not concentrated around any definite point, and the action density $\mathcal{L}=-\frac{1}{48}\mathrm{Tr}\hat{F}_{abc}\hat{F}_{abc}$ is uniformly distributed, even though, by a dilatation transformation, it is of course possible to obtain solutions where $\mathcal{L}$ is not constant on the unit hypersphere.

IV. FERMIONS IN THE FIELD OF THE PSEUDOPARTICLE

In this section we analyze the fermion-pseudoparticle system. The power of the O(5) formalism allows for a complete solution of the equations. To describe Fermi fields on the hypersphere one must introduce a set of five anticommuting matrices Γ_a,

$$\{\Gamma_a,\Gamma_b\}=2\delta_{ab}. \tag{4.1}$$

We take these to be 4-dimensional,[10]

$$\Gamma_\mu=i\alpha_\mu\alpha_5,\quad \Gamma_5=\alpha_5,\quad \alpha_5=\alpha_1\alpha_2\alpha_3\alpha_4, \tag{4.2}$$

and we define

$$S_{ab}=\frac{1}{4i}[\Gamma_a,\Gamma_b]. \tag{4.3}$$

The Fermi fields $\hat{\psi}$ have four components[10]; they are related to the 4-dimensional, Euclidean Fermi fields ψ by

$$\hat{\psi}=\frac{(1+x^2)}{2}(1+i\alpha_\mu x^\mu)\psi. \tag{4.4}$$

It is possible to show that

$$\int d^4x\psi^\dagger i\alpha^\mu\frac{\partial}{\partial x^\mu}\psi=\int d\Omega\,\hat{\psi}^\dagger(S_{ab}L_{ab}+2)\hat{\psi}, \tag{4.5}$$

so that the free Dirac equation $i\alpha^\mu(\partial/\partial x^\mu)\psi=0$ is equivalent to the O(5)-covariant equation

$$(S_{ab}L_{ab}+2)\hat{\psi}=0. \tag{4.6}$$

In the presence of gauge fields, Eq. (4.6) becomes

$$[(S_{ab})_{ij}\delta_{mn}L_{ab}-2(S_{ab})_{ij}r_a\hat{A}_b^k T^k_{mn}+2\delta_{mn}\delta_{ij}]\hat{\psi}_{jn}=0, \tag{4.7}$$

where the matrices T^i_{mn} are the infinitesimal generators of the internal-symmetry group in the representation chosen for the fermions, and we have also put into evidence the spinor indices of $\hat{\psi}$. The equation may now be analyzed by expanding $\hat{\psi}$ in terms of O(5) harmonics appropriate to conservation of J_{ab}.

As an example of the computational simplification that our formalism effects, we consider the special case of fermions transforming as isospinors in the two SU(2)'s of the reduction O(4) = SU(2) × SU(2). We choose therefore for the fermions the 4-dimensional representation of O(4), where the infinitesimal generators are represented by the matrices $(\Sigma^{\mu\nu})_{mn}$. We notice then that the Fermi fields also span a representation of O(5), obtained simply by enlarging the set of infinitesimal generators to the matrices $(\Sigma_{ab})_{mn}$. We are thus led to consider the equation

$$[(S_{ab})_{ij}L_{ab}\delta_{mn} - 2(S_{ab})_{ij}r_a(\Sigma_{bc})_{mn}r_c + 2\delta_{mn}\delta_{ij}]\hat{\psi}_{jn} = 0, \tag{4.8}$$

which is obtained from Eq. (4.7) by substituting for $(1/i)\hat{A}^k_b T^k_{mn}$ the O(5)-invariant expression of Eq. (3.12a).

Equation (4.8) is the equation of motion for a Dirac field in the potential of the pseudoparticle, after a projection has been made from Euclidean space onto the surface of the hypersphere and a gauge transformation has been used to express the gauge fields in a convenient form. By means of an inverse gauge rotation and projection, it is always possible to transform a solution of Eq. (4.8) into a solution of the Dirac equation in Euclidean space with the SU(2) × SU(2) gauge group.

To analyze Eq. (4.8), it is convenient to perform a unitary transformation acting on internal-symmetry indices

$$\hat{\psi}_m \to \Psi_m = U_{mn}\hat{\psi}_n, \qquad U_{mn} = (i\alpha_1\alpha_3)_{mn}, \tag{4.9a}$$

with

$$U^\dagger\Sigma_{ab}U = -(\Sigma_{ab})^{tr}. \tag{4.9b}$$

The transformed Dirac equation becomes

$$[(S_{ab})_{ij}L_{ab}\delta_{mn} + 2(S_{ab})_{ij}r_a(\Sigma_{bc})_{nm}r_c + 2\delta_{ij}\delta_{mn}]\Psi_{jn} = 0, \tag{4.10a}$$

or, more compactly,

$$S_{ab}L_{ab}\Psi + 2r_a S_{ab}\Psi S_{bc}r_c + 2\Psi = 0, \tag{4.10b}$$

where we consider Ψ to be a 4 × 4 matrix, and since $\Sigma_{ab} = S_{ab}$ it is no longer necessary to distinguish between these two matrices in order to keep track of spin and isospin indices.[11]

The left-hand side of Eq. (4.10) is of the form $\mathfrak{D}\Psi$, where $\mathfrak{D}$ is a linear Hermitian operator with respect to the inner product $\int d\Omega\,\mathrm{Tr}\Psi_1^\dagger\Psi_2$. In the rest of this section we shall find eigenvalues and eigenfunctions of this operator, as well as its inverse, which gives the fermion propagator in the field of the pseudoparticle. The relevance of these quantities for an analysis of the field-theoretical implications of the pseudoparticle solution is obvious.[12]

It is important to notice that $\mathfrak{D}$ has definite symmetry properties under two unitary transformations of the field Ψ. These two transformations, which we shall call chiral transformation and chiral gauge transformation, are obtained by left-multiplying and right-multiplying the fields with the unitary matrix $\Gamma_a r_a$:

$$U_C:\ \Psi \to \Psi' = \Gamma_a r_a \Psi, \tag{4.11a}$$

$$U_{CG}:\ \Psi \to \Psi' = \Psi\Gamma_a r_a. \tag{4.11b}$$

[The chiral transformation U_C is the projection over the hypersphere of the chiral transformation $\psi \to \psi' = \alpha_5\psi$. The chiral gauge transformation is a useful addition symmetry of the system. It can be viewed as a gauge transformation that leaves the gauge fields unchanged, but to realize U_{CG} as a gauge transformation one must consider the gauge group O(5) further embedded into an SU(4) gauge group generated by the 15 matrices Σ_{ab} and $\frac{1}{2}\Gamma_a$.] The transformation properties of $\mathfrak{D}$ are the following:

$$U_C\mathfrak{D}U_C^{-1} = -\mathfrak{D}, \tag{4.12a}$$

$$U_{CG}\mathfrak{D}U_{CG}^{-1} = \mathfrak{D}. \tag{4.12b}$$

The eigenvalue equation which we solve is

$$\mathfrak{D}\Psi = \mu\Psi. \tag{4.13}$$

Note that U_C takes a solution of (4.13) into another solution with μ replaced by $-\mu$; this symmetry is analogous to Fermi-number conjugation. U_{CG} takes a solution of (4.13) into another solution with μ unchanged. Thus we expect to find doubly degenerate solutions both for $\mu > 0$ and $\mu < 0$, and a self-conjugate solution for $\mu = 0$.[13]

To solve (4.13) we expand Ψ into a complete set of 4 × 4 matrices,

$$\Psi = AI + B_a\Gamma_a + C_{ab}S_{ab}, \tag{4.14}$$

separate from B_a and C_{ab} components parallel and perpendicular to r^a, and obtain a set of coupled first-order differential equations, which converts easily to a set of uncoupled second-order differential equations. These equations are trivial to solve since they are simply free wave equations on the hypersphere, involving the wave operator L^2 whose eigenvectors are O(5) harmonics with eigenvalues $2n(n+3)$, $n = 0, 1, \ldots$.[7,8]

In this way the following solutions are found: $\mu = 0$,

$$\Psi_{(1)} = \tfrac{1}{2}Y_0 = \frac{1}{4\pi}\left(\tfrac{3}{2}\right)^{1/2}, \tag{4.15a}$$

$$\Psi_{(2)} = \Psi_{(1)}\Gamma_a r_a,$$

$$\mu^2 = n(n+3), \quad n = 1, 2, \ldots,$$

$$\Psi_{(1)} = \frac{1}{\sqrt{5}}\left(1 + \frac{1}{\mu} S_{ab} L_{ab}\right) Y_n, \tag{4.15b}$$

$$\Psi_{(2)} = \Psi_{(1)} \Gamma_a r_a,$$

$$\mu^2 = (n+1)(n+2), \quad n = 0, 1, \ldots,$$

$$\Psi_{(1)} = \frac{S_{cd}}{\sqrt{2}}\left[\left(1 - \frac{1}{\mu}\right) r_c + \frac{i}{\mu} r_a L_{ac}\right] \mathcal{Y}_{dn}, \tag{4.15c}$$

$$\Psi_{(2)} = \Psi_{(1)} \Gamma_a r_a,$$

where Y_n and $\mathcal{Y}_{an}$ are scalar and vector spherical harmonics. Their properties, already given by Adler,[7] are as follows. Each is an eigenfunction of L^2 with eigenvalue $2n(n+3)$. Y_n spans the $(n, 0)$ representation of O(5) with dimensionality $\frac{1}{6}(n+1) \times (n+2)(2n+3)$. [There is a "magnetic" label m, which we have suppressed, ranging from 1 to $\frac{1}{6}(n+1)(n+2)(2n+3)$.] The normalization is

$$\int d\Omega\, Y^*_{nm} Y_{n'm'} = \delta_{nn'} \delta_{mm'}.$$

Y_n may be constructed from

$$Y_n = r_{a_1} \cdots r_{a_n} M^{a_1 \cdots a_n},$$

where M is a constant, totally symmetric, and traceless tensor. $\mathcal{Y}_{an}$ is a vector harmonic, spanning the $(n, 1)$ representation of O(5) with dimensionality $\frac{1}{2}n(n+3)(2n+3)$. It satisfies the subsidiary conditions $r_a \mathcal{Y}_{an} = 0$, $iL_{ab}\mathcal{Y}_{bn} = \mathcal{Y}_{an}$, and is normalized by

$$\int d\Omega\, \mathcal{Y}^*_{anm} \mathcal{Y}_{an'm'} = \delta_{nn'} \delta_{mm'}.$$

One may construct $\mathcal{Y}_{an}$ from the overcomplete set of functions $(2L^2\delta_{ac} + 2L_{ab}L_{bc} + i6L_{ac})(u_I)_c Y_n$; $(u_I)_a$ is a constant unit vector with components δ_{Ia}.

The eigensolutions (4.15) are orthogonal to each other, and normalized to $\int d\Omega\, \mathrm{Tr}\Psi^\dagger\Psi = 1$. When nonvanishing, μ takes on positive and negative values of equal magnitude. There are the expected zero-eigenvalue solutions which can be arranged into eigenstates of $\Gamma^a r_a$. Evidently they also solve the Dirac equation (4.10).

We turn our attention now to the propagator associated with the operator $\mathfrak{D}$. It is obvious, from the existence of zero eigenvalues, that $\mathfrak{D}^{-1}$ does not exist. The propagator G should be rather defined through the equation

$$\mathfrak{D} G = I - P_0, \tag{4.16}$$

where I is the identity, and P_0 projects onto the space of eigenfunctions with $\mu = 0$. To make the expression of the propagator more explicit, we notice that it will be a matrix with indices i_1 (Dirac degrees of freedom), n_1 (internal degrees of freedom) referring to the initial configuration, and similarly i_2, n_2 for the final configuration. Furthermore, G depends on two position vectors r_1 and r_2. The symmetry properties of $\mathfrak{D}$ under chiral and chiral gauge transformations restrict G. It can be easily shown that these restrictions are satisfied when G is of the form

$$G = P^+_C P^+_{CG} \mathcal{G} P^-_C P^+_{CG} - P^+_C P^-_{CG} \mathcal{G} P^-_C P^-_{CG} + P^-_C P^+_{CG} \mathcal{G} P^+_C P^+_{CG} - P^-_C P^-_{CG} \mathcal{G} P^+_C P^-_{CG}, \tag{4.17}$$

$$P^\pm = \frac{1 \pm \Gamma_a r_a}{2}.$$

$P^\pm_C$ $(P^\pm{}_{CG})$ are the projection operators on the subspaces of positive and negative chirality (gauge chirality). The notation in (4.17) is schematic; matrix multiplication is not indicated. The projection operators standing to the left (right) of $\mathcal{G}$ refer to the first (second) configuration and are functions of r_1 (r_2). $P^\pm_C$ $(P^\pm_{CG})$ act on the spin (internal) indices. Further, $P^\pm_C$ $(P^\pm_{CG})$ of the first configuration act on the left (right), while those of the second configuration act on the right (left). For example, the first term on the right-hand side of (4.17) is explicitly

$$G^{(1)}_{i_1 n_1; n_2 i_2} = \tfrac{1}{2}(1 + \Gamma_a r_1^a)_{i_1 j_1} \tfrac{1}{2}(1 + \Gamma_a r_1^a)_{m_1 n_1} \mathcal{G}_{j_1 m_1; m_2 j_2} \times \tfrac{1}{2}(1 + \Gamma_a r_2^a)_{n_2 m_2} \tfrac{1}{2}(1 - \Gamma_a r_2^a)_{j_2 i_2}.$$

$\mathcal{G}$ can be expanded as

$$\mathcal{G}_{i_1 n_1; n_2 i_2}(r_1, r_2) = f(r_1^a r_2^a)(\delta_{i_1 n_1} r_1^b \Gamma^b_{n_2 i_2} + r_2^b \Gamma^b_{i_1 n_1} \delta_{n_2 i_2}) + ig(r_1^a r_2^a)(S^{ab}_{i_1 n_1} r_2^b \Gamma^a_{n_2 i_2} - \Gamma^a_{i_1 n_1} r_1^b S^{ab}_{n_2 i_2}), \tag{4.18}$$

with f and g to be determined. The form of G given by (4.17) and (4.18) satisfies a crossing property

$$G_{i_1 n_1; n_2 i_2}(r_1, r_2) = G^*_{i_2 n_2; n_1 i_1}(r_2, r_1). \tag{4.19}$$

In principle, f and g could be found from the eigenfunctions of $\mathfrak{D}$, Eqs. (4.15), but it is simpler to determine the propagator directly from (4.16). Inserting into this equation the expansion provided by (4.17) and (4.18), we obtain four equations for f and g

$$\begin{aligned}
&(x^2 - 1) f' + 4xf = -\frac{3}{16\pi^2}, \\
&(x - 1) g' + 2g = 0, \\
&-f' + \tfrac{1}{2} x g' + \tfrac{3}{2} g = 0, \\
&f + \frac{x^2 - 1}{2} g' + \tfrac{3}{2} x g = 0,
\end{aligned} \tag{4.20a}$$

where we have set $x = r_1^a r_2^a$. These equations are of course compatible and fix f and g completely:

$$\begin{aligned} f &= \frac{1}{16\pi^2}\frac{2-x}{(1-x)^2} \\ &= \frac{1}{4\pi^2}\frac{1}{|r_1-r_2|^4} + \frac{1}{8\pi^2}\frac{1}{|r_1-r_2|^2}, \\ g &= \frac{1}{8\pi^2}\frac{1}{(1-x)^2} \\ &= \frac{1}{2\pi^2}\frac{1}{|r_1-r_2|^4}. \end{aligned} \tag{4.20b}$$

Since we have thus satisfied (4.16), G is determined up to solutions of the homogeneous equation $\mathfrak{D}G_0=0$, that is, up to terms proportional to the zero-eigenvalue solutions (4.15a). Our propagator is the unique function which is orthogonal to these.

V. CONCLUSION

The Yang-Mills pseudoparticle is distinguished by its topological properties which give it a nonvanishing Pontryagin index q; this in turn makes it relevant to questions of anomalous nonconservation of symmetries. We have here demonstrated that the pseudoparticle is further distinguished by possessing a large kinematical invariance group, possibly important in future developments of the theory. We have already put into evidence the computational simplification that the O(5) formalism affords in analyzing the Dirac equation, which displays a peculiar zero-frequency mode, previously encountered in the spectrum of fermions in topologically interesting external fields, and having novel physical consequences.[5,11,12] Moreover, the O(5) invariance may possess further implications—in connection with spontaneous breakdown of space-time symmetries, as Fubini[8] has recently proposed.

ACKNOWLEDGMENT

We are grateful to G. 't Hooft for making available to us his results prior to their publication.

*This work is supported in part through funds provided by ERDA under Contract No. AT(11-1)-3069.

[1] A. Belavin, A. Polyakov, A. Schwartz, and Y. Tyupkin, Phys. Lett. 59B, 85 (1975).

[2] S. L. Adler, in *Lectures on Elementary Particles and Quantum Field Theory*, 1970 Brandeis Summer Institute in Theoretical Physics, edited by S. Deser, M. Grisaru, and H. Pendleton (MIT Press, Cambridge, Mass., 1970); S. Treiman, R. Jackiw, and D. Gross, *Lectures on Current Algebra and Its Applications* (Princeton Univ. Press, Princeton, New Jersey, 1972), p. 97.

[3] H. Pagels, Phys. Rev. D 13, 343 (1976).

[4] For a summary see S. Weinberg, Phys. Rev. D 12, 3583 (1975).

[5] A. Polyakov, Phys. Lett. 59B, 82 (1975); G. 't Hooft, Harvard report (unpublished).

[6] For a summary of the conformal group, see R. Jackiw, Ref. 2.

[7] S. L. Adler, Phys. Rev. D 6, 3445 (1972); 8, 2400 (1973).

[8] S. Fubini, CERN report (unpublished).

[9] It has been already remarked by S. Fubini, A. Hanson, and R. Jackiw, Phys. Rev. D 7, 1732 (1973), that, in an operator formulation of Euclidean field theory, coordinate inversion plays the role of conjugation.

[10] Here we differ from the method of Adler, Ref. 7, who uses 8-dimensional matrices and fields.

[11] This device was introduced by us in our recent study of the interaction of a Dirac isospinor with the SU(2) Yang-Mills monopole; R. Jackiw and C. Rebbi, Phys. Rev. D 13, 3398 (1976).

[12] G. 't Hooft (work in preparation) has also analyzed this problem using conventional methods.

[13] The structure is similar to that encountered by us previously; see Ref. 11.

Nuclear Physics B362 (1991) 21–32
North-Holland

COMMENTS ON THE SPACE-SYMMETRY INTERPRETATION OF THE GAUGE ORIENTATIONS OF THE INSTANTON IN THE HIGGS PHASE

M.A. SHIFMAN and A.I. VAINSHTEIN*

Theoretical Physics Institute, University of Minnesota, 116 Church St. SE, Minneapolis, MN 55455, USA

Received 5 March 1991
(Revised 16 May 1991)

We comment on introduction of the instanton collective coordinates in the theories with the spontaneously broken gauge symmetry. In distinction with the pure gauge theories some gauge-invariant quantities acquire explicit dependence on the collective coordinates associated with the gauge field orientation in the gauge group, which looks rather paradoxically. This dependence, however, is absolutely necessary for preserving the Lorentz invariance of the instanton-induced amplitudes.

1. Introduction

Recent works [1, 2] devoted to the baryon number violating processes at high energies have revived interest to instantons in non-abelian gauge theories with the spontaneously broken gauge symmetry. Instanton calculus [3–6] in such theories has certain peculiarities: although the number of the collective coordinates is the same as in the unbroken case their explicit realization in the Higgs sector is somewhat unusual. We are unaware of any discussion of the issue in the previous literature in spite of the fact that the correct introduction of the collective coordinates is absolutely crucial for preservation of the explicit Lorentz invariance of the instanton-induced interactions.

The main assertion to be presented below is as follows. Introduction of the Higgs field reveals the fact that actually there are no collective coordinates associated with the global orientations in the instantonic SU(2) gauge group. The corresponding three collective coordinates should be introduced through the Lorentz rotations in the $SU(2)_R$ subgroup of the O(4) euclidean space-time symmetry⋆. In the sector of the vector fields they cannot be distinguished from the

* Also at Institute for Nuclear Physics, 630090 Novosibirsk, USSR.

⋆ The instanton configuration considered refers to $SU(2)_L$ Lorentz sub-group. As a matter of fact, according to the established nomenclature we consider anti-instanton.

0550-3213/91/$03.50 © 1991 – Elsevier Science Publishers B.V. (North-Holland) standard ones. In the Higgs sector, however, the gauge rotations should be supplemented by "flavor" rotations in an additional group which is not explicit in the original lagrangian. The latter rather complicated assertion takes the most simple form in the unitary gauge where corresponding three coordinates do not appear in the Higgs sector at all.

The correct introduction of these collective coordinates is crucial for maintaining the Lorentz invariance of the instanton-induced amplitudes.

2. Description of the simplest model

To present the problem we will start from consideration of the truncated version of the standard model: SU(2) gauge fields plus the Higgs sector. We will omit the U(1) gauge group as well as the fermion fields which are irrelevant to the question to be discussed below. The lagrangian we will deal with is

$$L = -\frac{1}{4g^2} G^a_{\mu\nu} G^a_{\mu\nu} + \tfrac{1}{2} \operatorname{Tr} D_\mu \bar{\Phi} D_\mu \Phi - \lambda\left(\tfrac{1}{2} \operatorname{Tr} \bar{\Phi}\Phi - v^2\right)^2 . \tag{1}$$

Here $G^a_{\mu\nu} = \partial_\mu A^a_\nu - \partial_\nu A^a_\mu + \epsilon^{abc} A^b_\mu A^c_\nu$ is the gauge field strength tensor and SU(2) gauge group is implied so that $a = 1,2,3$.

Usually the Higgs field is described by the complex scalar field $\varphi^\alpha (\alpha = 1,2)$ transforming as a doublet of the gauge group (to be denoted below by $SU(2)_g$). For our purposes, however, it is more convenient to introduce a 2×2 matrix Φ connected with φ in the following way:

$$\Phi^{\alpha\beta} = \left\{\varphi^\alpha, \overline{\varphi^\lambda} \varepsilon^{\lambda\beta}\right\} = \begin{Bmatrix} \varphi^1 & -\overline{\varphi^2} \\ \varphi^2 & \overline{\varphi^1} \end{Bmatrix} . \tag{2}$$

The bar denotes hermitian conjugation. In this form it becomes explicit that the lagrangian (1), apart from the gauge $SU(2)_g$ symmetry, possesses a global SU(2) invariance [7] which will be called below $SU(2)_f$ ("flavor").

Below we will see that the extra flavor symmetry at the lagrangian level is not necessary for the solution of the problem under discussion. In the general case of the Higgs field in the arbitrary representation the flavor symmetry is absent in the lagrangian but it appears as a symmetry in the space of the classical solutions. The above-mentioned global $SU(2)_f$ invariance is responsible for the fact that all W-boson masses are equal in the truncated model under consideration.

Explicitly the symmetry transformations are of the form

$$\Phi'(x) = U(x)\Phi(x)V^{-1},$$

$$A'_\mu(x) = U(x)A_\mu(x)U^+(x) + iU(x)\partial_\mu U^+(x) . \tag{3}$$

Here the gauge transformations are presented by the x-dependent matrix $U(x)$ which multiplies the field $\Phi(x)$ from the left. The "flavor" transformations are presented by the x-independent matrix V^{-1} multiplying Φ from the right.

Summarizing, one can say that originally there are two symmetries in the lagrangian – one local $SU(2)_g$ and one global $SU(2)_f$ – which are both spontaneously broken by the vacuum expectation value of the Higgs field $\langle\Phi^{\alpha j}\rangle_{vac} \sim \delta^{\alpha j}$. In perturbation theory the vacuum configuration is

$$\langle\Phi^{\alpha j}\rangle_{vac} = v\delta^{\alpha j}, \qquad \langle A_\mu^a\rangle_{vac} = 0. \tag{4}$$

It is important to notice that this configuration is invariant under the action of the diagonal subgroup of $SU(2)_g \times SU(2)_f$. Indeed, in accordance with eq. (3) Φ_{vac} stays intact if $U = V$. This diagonal SU(2) subgroup will survive as a linearly realized symmetry, but not the gauge group itself; all three W-bosons become massive. Certainly, one could have chosen another – globally rotated – vacuum configuration,

$$\Phi_{vac}^{\alpha j} = v(U_0)^{\alpha j}, \qquad (A_\mu^a)_{vac} = 0, \tag{5}$$

U_0 being a fixed matrix belonging to SU(2). For any value of U_0 we get physically the same theory by just renaming the $W^\pm$- and W^0-bosons. What is important, however, is that U_0 should be *fixed* once and forever in order to specify the vacuum configuration. The instanton solution interpolates between the *given* vacuum configuration at $T = -\infty$ and its gauge copy (with the topologically nontrivial gauge matrix) at $T = +\infty$.

3. Instantons

The instanton trajectory connects in the imaginary time the point (4) with itself realizing one winding of a variable living on S^1 which exists [3] in the non-abelian gauge theories irrespectively of the presence (absence) of other fields. Technically the BPST instanton minimizes the action, being the solution of the classical euclidean equations of motion. Each particular instanton breaks spontaneously all symmetries of the model: gauge, flavor, Lorentz, translational, etc. One should take into account, however, that there exists a whole family of solutions corresponding to different orientations in the above-mentioned groups. The possibility of the different orientations is described by collective coordinates. Integration over the collective coordinates should (and does) restore all *unbroken* symmetries of the model.

Strictly speaking, introduction of the scalar field with the vacuum expectation value eliminates the exact solutions. The non-trivial topology (the existence of a variable living on S^1) remains intact, however, and therefore one should take into account the interpolating field configurations with the non-zero windings. In the weak coupling regime with the Higgs field instanton calculus survives in the form of the constrained instanton method [4, 8]. It is implied that the instanton action depends only weakly on some of the collective coordinates.

In order to show what kind of problems occur with the instanton collective coordinates in the case of the spontaneously broken gauge symmetry let us calculate in a standard way the interaction of the W-bosons with the instanton. The instanton configuration centered at the origin takes the form

$$A_\mu = \frac{ix^2}{x^2+\rho^2} g\partial_\mu g^+ , \qquad \Phi = v\left(\frac{x^2}{x^2+\rho^2}\right)^{1/2} g , \tag{6}$$

where

$$g = \frac{i\tau_\mu x_\mu}{\sqrt{x^2}} \equiv i\tau_\mu n_\mu , \qquad \tau_\mu = \{i, \boldsymbol{\tau}\} , \tag{7}$$

and it is implied that $x \ll M_W^{-1}, M_H^{-1}$.

As it is clear from the perturbative vacuum configuration (4) the physical W-boson can be defined in the gauge-invariant way as

$$W_\mu \equiv W_\mu^a(\tau^a/2) = -\frac{i}{2v^2}\bar{\Phi}\overleftrightarrow{D}_\mu \Phi . \tag{8}$$

According to the standard approach to the collective coordinates one should take into account the global gauge rotations of the configuration (6) by making in eq. (6) the following substitution:

$$g \to Gg , \tag{9}$$

where G is a constant SU(2) matrix. The definition of the W-boson field (8) is gauge invariant, however, and, hence, the matrix G drops out from W_μ. Explicitly,

$$(W_\mu)_{\text{inst}} = \frac{x^2}{x^2+\rho^2}\left(g^+ A_\mu^{\text{inst}} g + ig^+\partial_\mu g\right) = \frac{\rho^2}{(x^2+\rho^2)^2}\bar{\eta}_{a\mu\nu}\tau^a x_\nu , \tag{10}$$

where A_μ^{inst} is given in eq. (6) and $\bar{\eta}_{a\mu\nu}$ are the 't Hooft symbols. For $x \gg \rho$, eq. (10) coincides with A_μ^{inst} in the singular gauge except for the absence of the global gauge matrix G.

Moreover, using the standard procedure [9] one can convert eq. (10) into an effective lagrangian for the instanton-induced interactions. Performing the Fourier transformation in eq. (10) and amputating the pole propagator yields the vertices of emission of the W-bosons by the instanton. Concretely, from eq. (10) one obtains

$$S_{\text{inst}} = \exp\left(-\frac{2\pi^2\rho^2}{g^2}\bar{\eta}_{a\mu\nu}W^a_{\mu\nu}\right), \tag{11}$$

where

$$W^a_{\mu\nu} = \partial_\mu W^a_\nu - \partial_\nu W^a_\mu + \epsilon^{abc}W^b_\mu W^c_\nu \tag{12}$$

is the strength tensor of the W-boson field.

Thus, acting in a standard way we come to a rather paradoxical conclusion: the instanton-induced W-boson interaction (11) breaks the Lorentz invariance (as well as the flavor SU(2) symmetry). It is obvious that something is wrong with our understanding of the collective coordinates of the global gauge rotations.

Before we turn to a more detailed discussion of the collective coordinates let us discuss the asymptotic points connected by the instanton trajectory. To this end one should pass to the $A_0 = 0$ gauge. From eq. (6) it is obvious that the transition to the $A_0 = 0$ gauge is realized by the matrix V_0 satisfying the equation

$$\dot{V}_0 + \frac{x^2}{x^2+\rho^2}g^+\dot{g}V_0 = 0\,. \tag{13}$$

The solution of this equation is

$$V_0 = \exp\int_{-\infty}^{\tau}\left(-\frac{x^2}{x^2+\rho^2}g^+\dot{g}\right)\mathrm{d}\tau = \exp\int_{-\infty}^{\tau}\left(-\frac{i\sigma x}{x^2+\rho^2}\right)\mathrm{d}\tau\,. \tag{14}$$

Notice that $V_0(\tau = -\infty) = 1$ by construction while at $\tau = +\infty$ we have the famous "hedgehog" matrix

$$V_0(\tau = +\infty) = \exp\left(-\frac{i\pi \boldsymbol{x\sigma}}{(\boldsymbol{x}^2+\rho^2)^{1/2}}\right). \tag{15}$$

The gauge-transformed instanton configuration is

$$A^G_\mu = V_0^+ i\partial_\mu V_0 + \frac{ix^2}{x^2+\rho^2}V_0^+(g^+\partial_\mu g)V_0\,,$$

$$\Phi^G = V_0^+\Phi\,, \tag{16}$$

with the following asymptotics ($A_0 \equiv 0$):

$$\tau \to -\infty: \quad A_i^G = 0, \qquad \Phi^G = v \cdot I,$$

$$\tau \to +\infty: \quad A_i^G(x) = V_0^+(\tau = \infty, x) i\partial_i V_0(\tau = \infty, x), \qquad \Phi^G = -vV_0^+(\tau = \infty, x). \tag{17}$$

From eq. (17) it is quite clear that the instanton solution under consideration is one of the trajectories connecting the point (4) with itself in a topologically non-trivial way.

4. Collective coordinates

Given a single interpolating configuration (6) we can build a whole family by applying to the given solution the generators of all symmetries of the model which are unbroken. Thus, the translation invariance is resorted after shifting the center of the instanton from the origin to x_0,

$$x \to x - x_0$$

in eq. (6). The procedure is standard and is described in detail in the literature. The only new point which deserves attention is the introduction of the collective coordinates associated with $SU(2)_g$. Unlike the standard case to which we got used in QCD we *cannot* apply the generators of the gauge group to our solution because the vacuum configuration (4) does not allow us to do that. The gauge symmetry is spontaneously broken, and the orientation in $SU(2)_g$ is fixed by the choice of Φ_{vac}.

At first sight one might think that this fact kills the corresponding collective coordinates. This is not the case, however. These collective coordinates re-appear for the gauge-invariant quantities such as W_μ from eq. (8) in the form of the "flavor" rotations. Indeed, the vacuum configuration (4) preserves the symmetry with respect to the diagonal subgroup of $SU(2)_g \times SU(2)_f$. Therefore, we can (and must) act on our solution (6) by a sum of the generators, $T_f^a + T_g^a$ (three rotations altogether).

Technically, to describe the action of the generators of $SU(2)_g$ and $SU(2)_f$ it is convenient to pass from the vector notation in the instanton calculus to the spinor-index notation. All formulae then can be written in a much more economic way. Let us remind the spinor-index notation in application to the instantons.

The connection between the vector and spinor indices is given by the following expression:

$$V_{\alpha\dot{\alpha}} = (\sigma_\mu)_{\alpha\dot{\alpha}} V_\mu \qquad (\mu = 1, \ldots, 4, \alpha, \dot{\alpha} = 1, 2), \tag{18}$$

where $\sigma_\mu = (i, \boldsymbol{\sigma})$, $\boldsymbol{\sigma}$ are the Pauli matrices. The undotted α- and dotted $\dot{\alpha}$-indices are spinorial indices of the $SU(2)_L$ and $SU(2)_R$ subgroups of the $SU(2)_L \times SU(2)_R$ group of the Lorentz transformations continued to the euclidean space.

Using eq. (18) one finds the following decomposition for the antisymmetric tensor:

$$F_{\alpha\dot{\alpha},\beta\dot{\beta}} = (\sigma_\mu)_{\alpha\dot{\alpha}}(\sigma_\nu)_{\beta\dot{\beta}}F_{\mu\nu} = \varepsilon_{\dot{\alpha}\dot{\beta}}F_{\alpha\beta} + \varepsilon_{\alpha\beta}\bar{F}_{\dot{\alpha}\dot{\beta}}, \tag{19}$$

where symmetric tensors $F_{\alpha\beta}$ and $\bar{F}_{\dot{\alpha}\dot{\beta}}$ correspond to self-dual and anti-self-dual parts of $F_{\mu\nu}$.

In the spinor notation the instanton configuration (6) takes the form

$$A^{\eta\xi}_{\alpha\dot{\beta}} = -\frac{i}{x^2+\rho^2}\left(x^{\eta}_{\dot{\beta}}\delta^{\xi}_{\alpha} + x^{\xi}_{\dot{\beta}}\delta^{\eta}_{\alpha}\right),$$

$$\Phi^{\alpha f} = x^{\alpha f}\frac{iv}{(x^2+\rho^2)^{1/2}}, \tag{20}$$

where the triplet gauge indices are converted into the spinor ones,

$$A^{\eta\xi} = A^a(\tau^a/2)^{\eta}_{\delta}\varepsilon^{\xi\delta}. \tag{21}$$

Let us also give here the expressions for the gauge field strength tensor,

$$G^{\gamma\delta}_{\alpha\beta} = -\frac{2i\rho^2}{(x^2+\rho^2)^2}\left(\delta^{\gamma}_{\alpha}\delta^{\delta}_{\beta} + \delta^{\gamma}_{\beta}\delta^{\delta}_{\alpha}\right)$$

while $G^{\gamma\delta}_{\dot{\alpha}\dot{\beta}} = 0$.

After this digression we return to the construction of the family of solutions starting from eqs. (6) and (20). As was explained above we act on eq. (20) by the sum of the generators $T^a_f + T^a_g$. The flavor group does not act on the gauge sector, hence the three collective coordinates appear in the solution (20) in the standard way,

$$A^{\eta\xi}_{\alpha\dot{\beta}} = (M^+)^{\eta}_{\eta'}\left\{\frac{(-i)}{x^2+\rho^2}\left(x^{\eta'}_{\dot{\beta}}\delta^{\xi'}_{\alpha} + x^{\xi'}_{\dot{\beta}}\delta^{\eta'}_{\alpha}\right)\right\}M^{\xi}_{\xi'}, \tag{22}$$

where

$$M = \exp(i\omega^a\tau^a/2) \qquad (a = 1,2,3),$$

and ω^a are three collective coordinates. It does act in a non-trivial way in the

Higgs sector, so that

$$\Phi^{\alpha f} = \frac{iv}{\left(x^2+\rho^2\right)^{1/2}} (M^+)^{\alpha}_{\beta} x^{\beta\dot{\gamma}} M^f_{\dot{\gamma}}, \tag{23}$$

with the *same* matrix M.

In spite of the fact that Φ is originally in the fundamental representation of $SU(2)_g$ while the W-fields are in the adjoint representation, the collective coordinates residing in M transform these fields in one and the same way! Moreover, for the gauge-invariant combinations of eq. (8) the only rotations left are those associated with the flavor group.

This fact is directly related with the restoration of Lorentz symmetry. Indeed, under the Lorentz rotations the instanton fields are transformed as

$$A^{\eta\xi}_{\alpha\dot{\alpha}} \to \frac{(-i)}{x^2+\rho^2} (L^{-1})^{\xi}_{\alpha} (L)^{\eta}_{\eta'} x^{\eta'}_{\dot{\alpha}} + (\eta \leftrightarrow \xi),$$

$$\Phi^{\alpha f} \to \frac{iv}{\sqrt{x^2+\rho^2}} L^{\alpha}_{\beta} x^{\beta\dot{g}} (R^{-1})^f_{\dot{g}}, \tag{24}$$

where L and R are global matrices of $SU(2)_L$ and $SU(2)_R$ subgroups of the Lorentz group.

The first impression is that we get a new solution in this way, not reducible to eqs. (22) and (23). In order to stay in the family of eqs. (22) and (23) one should restrict oneself to $L = R$. To get the full Lorentz symmetry it is important to notice that the matrix L acts as a gauge rotation matrix. Therefore, it drops out from the gauge-invariant combinations. As a result, $SU(2)_L$ invariance is trivial for the gauge-invariant quantities while $SU(2)_R$ is ensured by averaging over orientations of the matrix M (see eqs. (22) and (23)). Simultaneously we will get the $SU(2)_f$ flavor symmetry. In particular, the expression (11) which presents a Lorentz non-invariant interaction will be corrected and instead of $\bar{\eta}_{a\mu\nu}$ in eq. (11) we will have

$$\bar{\eta}_{a\mu\nu} \to P_{ab}\bar{\eta}_{b\mu\nu}, \qquad P_{ab} = \tfrac{1}{2}\,\mathrm{Tr}\, M\tau^a M^+ \tau^b. \tag{25}$$

Thus, in the approximation considered we get the interaction which is identical to that in the unbroken theory (in spite of the fact that the W-boson fields are introduced as gauge singlets).

Three collective coordinates of the matrix M may potentially fail to restore the full $SU(2)_L \times SU(2)_R$ symmetry only in gauge non-invariant quantities. The lack of the explicit symmetries in such quantities can be considered, however, as a gauge artifact. If unphysical (gauge) degrees of freedom are eliminated the symmetry should become explicit.

When the gauge symmetry is spontaneously broken the unitary gauge is known to contain only the physical degrees of freedom. The gauge condition is

$$\Phi^{\alpha j}(x) = v(x)\delta^{\alpha j}. \tag{26}$$

Thus, the scalar sector contains one physical scalar field $v(x)$; the remaining three components of $\Phi^{\alpha j}$ are eaten up completely.

It is instructive to write down the instanton solution (20) in this gauge. Incidentally it turns out that the unitary gauge coincides (in the W-boson sector) with the singular gauge, well known in the instanton calculus. The matrix of the gauge transformations realizing the transition to the unitary gauge is

$$V_u = n_\mu \tau_\mu . \tag{27}$$

The instanton configuration in the unitary gauge takes the form

$$A^{\eta\xi}_{\alpha\beta} = -\frac{i\rho^2}{x^2(x^2+\rho^2)}(M^+)^{\eta}_{\dot\rho}\left(x^{\dot\rho}_{\alpha}\delta^{\dot\delta}_{\beta} + x^{\dot\delta}_{\alpha}\delta^{\dot\rho}_{\beta}\right)M^{\xi}_{\dot\delta},$$

$$\Phi^{\alpha j} = iv\left(\frac{x^2}{x^2+\rho^2}\right)^{1/2}\delta^{\alpha j}. \tag{28}$$

Notice the absence of any M factors in $\Phi^{\alpha j}_{\rm inst}$. This is another manifestation of the fact that M is not just pure global gauge rotations.

In the given form it is absolutely clear that the scalar sector respects $SU(2)_L \times SU(2)_R$ while the only non-invariance residing in A_μ is connected with the $SU(2)_R$ subgroup of the Lorentz group. After averaging over M the Lorentz symmetry of the instanton-induced amplitudes is restored automatically.

With the present understanding of the collective coordinates in hands we can answer subtle questions which might arise in connection with the instantons in the Higgs phase. As an example let us discuss the calculation of the instanton measure.

The subtle point here is the following. The integration over the collective coordinates residing in the matrix M (previously, one would say over orientations in the gauge group) gives rise to the volume of the SU(2) group in the instanton measure. More exactly, in the absence of the Higgs fields it is the volume of O(3) group which emerges in the instanton measure due to the fact that the gauge field is in the adjoint representation insensitive to the $\mathbb{Z}_2$ subgroup of SU(2). Naively, one might expect that introducing the Higgs field in the doublet representation increases by a factor of two the overall volume of the group integration (because $\mathbb{Z}_2$ acts non-trivially in the doublet representation).

This is not the case, however. The rotations corresponding to $SU(2)_f$ should be considered modulo gauge transformations, and this eats up $\mathbb{Z}_2$. Eq. (23) illustrates this fact. From this expression it is evident that M acts as if the Higgs field belonged to the adjoint representation.

5. General case and conclusions

Let us extend our discussion to include the case of arbitrary (not necessary irreducible) representation of the Higgs field in $SU(2)_g$. Consider the solutions of the equation

$$\left(\partial_\mu + iA_\mu^{\text{inst}}\right)^2 \varphi = 0, \qquad (29)$$

where φ is the Higgs field in the arbitrary representation R of $SU(2)_g$. It is known that in the singular gauge for the A_μ^{inst} the solutions take the form (see e.g. ref. [8])

$$\varphi^f = \left(\frac{x^2}{x^2+\rho^2}\right)^q c^f \qquad (f = 1, 2, \ldots, \dim \mathrm{R}), \qquad (30)$$

where q is $SU(2)_g$ isospin and c^f are arbitrary constants. One can form a basis of $\dim \mathrm{R}$ linearly independent solutions,

$$\varphi^{fi} = \left(\frac{x^2}{x^2+\rho^2}\right)^q v\delta^{fi} \qquad (f, i = 1, \ldots, \dim \mathrm{R}). \qquad (31)$$

In this form the $SU(2)_g \times SU(2)_f$ symmetry in the space of solutions is obvious. The global rotations in the $SU(2)_g$ can (and must) be completely compensated by that from $SU(2)_f$. As a result the proper configuration with the correct vacuum alignment includes a fixed φ-field together with the globally rotated A_μ^{inst}. This is fully equivalent to a global $SU(2)_R$ Lorentz rotation. Alternatively one can say that the Lorentz symmetry explains the flavor symmetry in the family of the solutions (see eqs. (30) and (31)).

A few remarks are in order concerning the instantons in the higher groups. Let us take the SU(3) gauge theory and a triplet Higgs field φ^α $(\alpha = 1, 2, 3)$ as an example. Assume for definiteness that the instanton in the singular gauge is placed in the SU(2) subgroup corresponding to $\alpha = 1, 2$. Then the basis of solutions of eq. (29) takes the form

$$\varphi^{\alpha 1} = v\sqrt{\frac{x^2}{x^2+\rho^2}}\begin{pmatrix}1\\0\\0\end{pmatrix}, \qquad \varphi^{\alpha 2} = v\sqrt{\frac{x^2}{x^2+\rho^2}}\begin{pmatrix}0\\1\\0\end{pmatrix}, \qquad \varphi^{\alpha 3} = v\begin{pmatrix}0\\0\\1\end{pmatrix}. \qquad (32)$$

Now the global gauge rotations of the Higgs field which involve the $\alpha = 3$ component lying outside the instantonic SU(2) subgroup cannot be exactly compensated by the flavor mixing in the family (32). Of course, we can do it for asymptotically large values of x, which is sufficient to preserve the vacuum alignment,

$$\varphi^{\text{vac}} = v\begin{pmatrix}1\\0\\0\end{pmatrix}. \tag{33}$$

In order to introduce the collective coordinates corresponding to SU(3) orientations let us consider the set of solutions (32) as a matrix $\{\varphi\} = \varphi^{\alpha j}$. Then the rotated solution for the vector and Higgs fields takes the form

$$A = M^{+}A_{\text{ins}}M\,, \qquad \varphi^{\beta} = (M^{+}\varphi M)^{\beta 1}\,, \tag{34}$$

where the 3×3 matrix M belongs to SU(3).

Unlike the SU(2) case the field φ is not aligned for finite values of x. Thus the standard singular gauge for the vector field does not coincide with the unitary gauge, where, by definition, the field $\varphi \sim (1,0,0)$.

It is clear that one can make an x-dependent gauge transformation (with the trivial asymptotics) which transforms eq. (34) into the unitary gauge changing simultaneously the vector fields. What is more important is the absence of the exact global symmetry in the family of the solutions which results in the fact that the instanton action becomes dependent on the orientation matrix M. This dependence is present only in the terms which are proportional to $v^2\rho^2$. Certainly those three collective coordinates which are connected with the Lorentz rotations do not appear in the instanton action.

It is not difficult to find explicitly the instanton action for the configuration (34). If we denote $M^{13} = \sin\theta/2$ the instanton action takes the form

$$S_{\text{inst}} = \frac{8\pi^2}{g^2} + 2\pi^2\rho^2 v^2 \cos^2\frac{\theta}{2}\,. \tag{35}$$

This is an obvious generalization of the 't Hooft result [4]. What is remarkable is the explicit dependence of the instanton action on the collective coordinates which were previously considered as gauge orientations!

Summarizing, we have shown that the collective coordinates considered previously as gauge orientations of the instanton actually should be introduced through "flavor" rotations in the space of solutions for the classical Higgs field (modulo appropriate gauge transformations). Generically, the classical action depends on these rotations and, moreover, they cannot be formulated as a symmetry of the theory. There is an SU(2) subgroup of these rotations, however, which is an exact symmetry. It is generated by $SU(2)_R$ subgroup of the Lorentz group of the space rotations.

We are grateful to M. Maggiore, L. McLerran, V. Novikov, M. Shaposhnikov and M. Voloshin for useful discussions.

References

[1] A. Ringwald, Nucl. Phys. B330 (1990) 1;
O. Espinosa, Nucl. Phys. B343 (1990) 310
[2] L. McLerran, A. Vainshtein and M. Voloshin, Phys. Rev. D42 (1990) 171, 180
[3] A. Belavin et al., Phys. Lett. B59 (1975) 85
[4] G. 't Hooft, Phys. Rev. D16 (1976) 3432 [Erratum: D18 (1978) 2199]
[5] R. Jackiw and C. Rebbi, Phys. Rev. D16 (1977) 1052
[6] V. Novikov, M. Shifman, A. Vainshtein and V. Zakharov, Nucl. Phys. B260 (1985) 157;
V. Novikov, Sov. J. Nucl. Phys. 46 (1987) 554
[7] A. Vainshtein and I. Khriplovich, Yad. Fiz. 13 (1971) 198;
A. Vainshtein, Lett. Nuovo Cimento 5 (1972) 680
[8] I. Affleck, Nucl. Phys. B191 (1981) 429
[9] M. Shifman, A Vainshtein and V. Zakharov, Nucl. Phys. B165 (1980) 45

PHYSICAL REVIEW D VOLUME 19, NUMBER 10 15 MAY 1979

Gauge zero modes, instanton determinants, and quantum-chromodynamic calculations

Claude Bernard
Department of Physics, University of California, Los Angeles, California 90024
(Received 2 February 1979)

A treatment of the gauge zero modes about an instanton in a singular gauge places them on the same footing as all other zero modes and simplifies the calculation of the collective-coordinate part of the instanton determinant. This determinant is calculated first for the gauge group SU(3) and then for general SU(N). The answers differ from previously published results: For SU(3), the reason for this difference is trivial [the inclusion of certain factors of $1/\sqrt{2}$ whose absence from 't Hooft's original SU(2) calculation was recently discovered] but the effects on quantum-chromodynamic calculations may be important; for large N, the reasons are more involved, but the usual conclusion that instantons are absent in the planar limit is unaffected.

I. INTRODUCTION

A necessary ingredient in the computation of all instanton effects in a gauge theory is the value of the quadratic functional integral about the instanton. This calculation was first performed by 't Hooft[1] for an SU(2) gauge group.

One of the more subtle aspects of 't Hooft's work was his treatment of the gauge zero modes which are due to the arbitrary orientation of the instanton within the gauge group. Because of the complicated space-time dependence of the gauge modes, it was necessary to place the system in a box and to resort to an unconventional form of the Faddeev-Popov ansatz in order to compute the contribution of these modes. Here, I show that the calculation may be greatly simplified by working with the instanton field in a singular gauge, rather than the regular gauge of Ref. 1. In a singular gauge, the infinitesimal gauge transformations which generate the zero modes approach constants at large distance (essentially because of the rapid falloff of the gauge potential) and may be easily identified with changes in the potential under variations in the collective coordinates which describe the gauge orientation of the instanton. I show in Sec. II that this allows one to treat the gauge modes on exactly the same footing as the other zero modes (translations, dilatations). For definiteness, I work first in the context of SU(3) and compute the one-loop functional integral about a single SU(3) instanton.

In Sec. III, I write down the answer for a general SU(N) theory, which requires only minor generalizations from the SU(3) computation. For reasons I explain, my results differ from the recently published work of Bashilov and Pokrovsky.[2] However, the prediction of vanishing instanton effects in the large-N limit[3,4] (the planar theory) is unaffected.

Because I take into account the recently discovered[5] numerical error in the original SU(2) calculation, my result for SU(3) differs by a predictable numerical factor from the previously published answer.[6] In Sec. IV, I make some comments about the effects of this new number on some recent instanton calculations[7-9] in quantum chromodynamics (QCD). I argue that while it is still possible to find significant instanton effects, to do so one is forced to go to larger values of the coupling constant, where other nonperturbative effects may be more important.

II. GAUGE ZERO MODES; SU(3)

We wish to calculate $W^{(1)}$, the one-loop vacuum-vacuum amplitude about a single instanton divided, for normalization, by the same amplitude about the ordinary vacuum. If the potential is expanded about the classical value,

$$A_\mu = A_\mu^{cl} + A_\mu^{qu}, \tag{1}$$

then the quadratic action about an instanton is

$$S = S^{cl} + \tfrac{1}{2} A^{qu} M_A A^{qu} + \phi^* M_{gh} \phi + \cdots . \tag{2}$$

where $S^{cl} = 8\pi^2/g^2$, and ϕ is the ghost field. Denoting the collective coordinates of the instanton by γ_i, $W^{(1)}$ is given by

$$W^{(1)} = \int \prod_i d\gamma_i J(\gamma) Q(\gamma) e^{-8\pi^2/g^2}, \tag{3a}$$

$$Q(\gamma) \equiv \frac{\det^{-1/2} M_A(\gamma) \det M_{gh}(\gamma)}{(\det^{-1/2} M_A \det M_{gh})_{A^{cl}=0}}, \tag{3b}$$

where $J(\gamma)$ is the collective-coordinate Jacobian, and where the determinants in (3b) are, of course, taken over nonzero modes only.

The focus here will be on the evaluation of the collective-coordinate Jacobian, in particular the contribution to it from the gauge zero modes. However, before evaluating $J(\gamma)$ in the case of interest, it is useful to review the usual method[10] for replacing zero modes with collective coordinates in a nongauge theory. For simplicity, we consider a scalar quantum field B which has a classical value $B=B^{cl}(\gamma)$, depending on a single collective coordinate γ. Let $M(\gamma)$ be the operator that appears in the expansion of the action to quadratic order about B^{cl}:

$$B=B^{cl}+B^{qu}\,,\qquad S=S^{cl}+\tfrac{1}{2}B^{qu}MB^{qu}+\cdots\,. \tag{4}$$

M has a complete set of orthogonal eigenfunctions χ_i with eigenvalues ϵ_i and norms $\sqrt{u_i}$:

$$u_i=\langle\chi_i|\chi_i\rangle. \tag{5}$$

There is, of course, a zero mode

$$\chi_0=\frac{\partial B^{cl}}{\partial\gamma}\,,\quad \epsilon_0=0\,. \tag{6}$$

If we expand B^{qu} as

$$B^{qu}=\sum_i \xi_i\chi_i\,, \tag{7}$$

then the measure for functional integration is

$$(dB)=(dB^{qu})=\prod_i\left(\frac{u_i}{2\pi}\right)^{1/2}d\xi_i\,. \tag{8}$$

That this is the correct measure can be seen by performing the integration over all the nonzero modes. The Gaussian integrations give

$$\int(dB)e^{-S}=\int\left(\frac{u_0}{2\pi}\right)^{1/2}d\xi_0 e^{-S^{cl}}\det{}^{-1/2}M+\cdots\,, \tag{9}$$

where $\cdots$ represents higher loops and the effects of other classical sectors. The fact that we get precisely $\det^{-1/2}M$ without some infinite multiplicative factor indicates that we have chosen the correct normalization.[5,2]

We may now insert a factor of unity which will require the quantum field to be orthogonal to the zero mode:

$$1=u_0\int d\gamma\,\delta(\langle B-B^{cl}(\gamma)|\chi_0(\gamma)\rangle)+\cdots\,, \tag{10}$$

where $\cdots$ represents terms of higher order in the quantum field. Inserting this factor into (9) and performing the ξ_0 integration results in

$$\int(dB)e^{-S}=\int d\gamma\left(\frac{u_0}{2\pi}\right)^{1/2}e^{-S^{cl}}\det{}^{-1/2}M+\cdots\,. \tag{11}$$

The case of the gauge theory is slightly more subtle. The difference lies in the requirement of fixing a gauge (taken here to be the usual background gauge with respect to the classical field). The derivative of the classical field with respect to a collective coordinate will not, in general, be in the background gauge. The ith zero mode is thus given by[1]

$$\psi_\mu^{(i)}=\frac{\partial A_\mu^{cl}}{\partial\gamma_i}+D_\mu^{cl}\Lambda^{(i)}\,,\qquad D_\mu^{cl}\psi_\mu^{(i)}=\partial_\mu\psi_\mu^{(i)}-ig[A_\mu^{cl},\psi_\mu^{(i)}]=0\,, \tag{12}$$

where D_μ^{cl} is the gauge-covariant derivative at the classical field and $\Lambda^{(i)}$ is the gauge transformation necessary to put the ith mode into the background gauge. Working in parallel with Eqs. (4) through (11), it is then easy to see that $J(\gamma)$, as defined in (3a), is given by

$$J(\gamma)=\left(\prod_i\frac{1}{\sqrt{2\pi}}\right)(\det V)(\det U)^{-1/2} \tag{13}$$

where the matrices V and U are defined by[11]

$$V_{ij}=\left\langle\frac{\partial A^{cl}}{\partial\gamma_i}\Big|\psi^{(j)}\right\rangle,\qquad U_{ij}=\langle\psi^{(i)}|\psi^{(j)}\rangle\,. \tag{14}$$

Now, if

$$\Lambda^{(i)}\psi_\mu^{(j)}<O\left(\frac{1}{r^3}\right) \tag{15}$$

at large distances r, then a simple integration by parts gives $V=U$ and the familiar result

$$J(\gamma)=\left(\prod_i\frac{1}{\sqrt{2\pi}}\right)(\det U)^{1/2}. \tag{16}$$

Thus, provided that we can express each zero mode as the derivative of the classical field with respect to a collective coordinate plus an additional gauge transformation [i.e., in the form of (12)] and provided that the gauge transformation $\Lambda^{(i)}$ vanishes sufficiently rapidly at large distances [i.e., (14) is obeyed], the calculation of the collective-coordinate Jacobian is straightforward. With the instanton in the regular gauge, only the translation and dilatation zero modes obey these conditions; however, in the singular gauge, the gauge modes are also well behaved (as we will see presently), and all modes may be treated on the same footing.

We now specialize the calculation to an SU(3) gauge theory. The general SU(N) case is only slightly more complicated and is presented in Sec. III. An SU(3) instanton can be obtained simply by embedding the SU(2) instanton into the "upper-left-hand corner" of the fundamental representation of SU(3).[12] Thus the singular gauge instanton has the form

$$A_\mu^{cl}(x)=\frac{2}{g}\,\frac{\bar\eta_{a\mu\nu}x_\nu}{x^2(x^2+\rho^2)}\,\frac{\lambda_a}{2}\,, \tag{17}$$

where λ_a $(a=1,2,3)$ are the first three Gell-Mann matrices and the symbols $\bar\eta_{a\mu\nu}$ are defined in Ref. 1. Under the action of this SU(2) subgroup, the generators of SU(3) form one triplet $(\lambda_1, \lambda_2, \lambda_3)$, two doublets (made from $\lambda_4, \lambda_5, \lambda_6, \lambda_7$), and one singlet (λ_8). Using this fact, it is easy to write down the twelve background gauge zero modes. First, there are the eight isospin-1 modes, which are just the ones given in Ref. 1, after conversion to the singular gauge of the instanton[13]:

$$\psi_\mu^{(\nu)}(x)=\left.\frac{\partial A_\mu^{cl}(x-z)}{\partial z^\nu}\right|_{z=0}+D_\mu^{cl}(A_\nu^{cl}(x))\,, \tag{18a}$$

$$\psi_\mu^{(\rho)}(x)=\frac{\partial A_\mu^{cl}(x)}{\partial\rho}\,, \tag{18b}$$

$$\psi_\mu^{(a)}(x)=D_\mu^{cl}\left[\frac{\lambda_a}{g}\,\frac{x^2}{x^2+\rho^2}\right], \tag{18c}$$

where $\psi_\mu^{(\nu)}$ $(\nu=1,\ldots,4)$ are the translation modes, $\psi_\mu^{(\rho)}$ is the dilatation mode, and $\psi_\mu^{(a)}$ $(a=1,2,3)$ are the gauge modes generated by $\lambda_1, \lambda_2, \lambda_3$. In addition, there are four modes which are members of isospin doublets and which can be obtained from the isospin-$\frac{1}{2}$ spinor modes of Ref. 1, since vectors and right-handed spinors obey the same equation. They are pure gauge modes, generated by λ_α $(\alpha=4,5,6,7)$:

$$\psi_\mu^{(\alpha)}=D_\mu^{cl}\left[\frac{\lambda_\alpha}{g}\left(\frac{x^2}{x^2+\rho^2}\right)^{1/2}\right]. \tag{19}$$

There are no normalizable isospin-0 zero modes (λ_8 does not generate a gauge mode since it commutes with the field of the instanton).

$\psi_\mu^{(\nu)}$ and $\psi_\mu^{(\rho)}$ are already in the form of (12), with $\Lambda^{(i)}$ vanishing rapidly at infinity. To put the gauge modes $\psi_\mu^{(k)}$ in this form, we must simply make more explicit the collective coordinates to which they correspond. The orientation of an instanton in SU(3) is described by a group element G:

$$A_\mu^{cl}[G]=G^{-1}A_\mu^{cl}G\,. \tag{20}$$

If we represent an infinitesimal change in G by

$$G+\delta G=(I-idt^i\lambda_i)G\,,\quad i=1,\ldots,8 \tag{21}$$

then seven gauge zero modes are given by

$$\chi_\mu^{(k)}[G]=\frac{\partial A_\mu^{cl}[G]}{\partial t^k}=-iG^{-1}[A_\mu^{cl},\lambda_k]G\,, \tag{22}$$

where $k=1,\ldots,7$. These modes are not in the background gauge. However, from (18c) and (19), we easily find the necessary gauge transformation[13] for $G=I$,[14]

$$\psi_\mu^{(k)}=\frac{\partial A_\mu^{cl}}{\partial t^k}+D_\mu\Lambda^{(k)}\,, \tag{23}$$

where

$$\Lambda^{(k)}=\begin{cases}\dfrac{1}{g}\left(\dfrac{-\rho^2}{x^2+\rho^2}\right), & k=1,2,3\\[2ex] \dfrac{1}{g}\left[\left(\dfrac{x^2}{x^2+\rho^2}\right)^{1/2}-1\right], & k=4,5,6,7.\end{cases} \tag{24}$$

Now all the modes are in the form of (12), with $\Lambda^{(i)}$ vanishing sufficiently rapidly at infinity so that (15) is obeyed. [This would not have been possible in the regular gauge since in that gauge the gauge transformations appearing in (18c) and (19) do not approach constants at infinity.] We can thus apply (16) to compute $J(\gamma)$. The modes are orthogonal and their normalization is easily calculated to be

$$\|\psi_\mu^{(\nu)}\|=\frac{1}{\sqrt2}\;\|\psi_\mu^{(\rho)}\|=\frac{1}{\rho\sqrt2}\;\|\psi_\mu^{(a)}\|=\frac{1}{\rho}\;\|\psi_\mu^{(\alpha)}\|=\frac{2\sqrt2\,\pi}{g}\,. \tag{25}$$

This implies

$$J(\gamma)=\frac{2^{14}\pi^6\rho^7}{g^{12}}\,. \tag{26}$$

To complete the calculation of $W^{(1)}$ [Eq. (3)], it is necessary to compute $Q(\gamma)$, which is the contribution of the nonzero modes. 't Hooft's calculation,[1] which has been verified by others,[15] gives the nonzero-mode determinants with Pauli-Villars regularization for arbitrary spin and isospin. Here we have vector (gauge) fields and scalar (ghost) fields each forming one isotriplet, two isodoublets, and one isoscalar. In addition, we must recall that the regulator fields contribute one factor of μ_0, the regulator mass, for each zero mode of the true fields. The results of Ref. 1 then immediately imply

$$Q(\gamma)=\mu_0^{12}\exp[-\ln(\mu_0\rho)-\alpha(1)-2\alpha(\tfrac12)]\,, \tag{27}$$

where the coefficients $\alpha(t)$ give the contribution of each isospin t and are tabulated in Ref. 1 [$\alpha(0)=0$].

We may now insert (26) and (27) into (3). Since the integrand in (3) is independent of the gauge orientation of the instanton, the integration over those collective coordinates may be performed. From (21), one learns that integration over all eight parameters t^i would simply give the volume of SU(3), calculated with the right-invariant Haar measure. However, only the seven t^k of (22) are collective coordinates; integration over them gives the volume of SU(3)/U(1), where the U(1) is generated by λ_8. SU(3)/U(1) is the set of equivalence classes on SU(3) given by

$$G'\propto G \text{ if } G'=e^{i\theta\lambda_8}G\,. \tag{28}$$

[In other words, two elements of SU(3) are counted

as equivalent if they produce the same instanton orientation.] The volume of SU(3)/U(1) is calculated in the Appendix to be $\pi^4/2$. Combining this information with (13), (26), and (27) gives the result

$$W^{(1)} = 2^{13}\pi^{10}e^{-\alpha(1)-2\alpha(1/2)}\int \frac{d^4z d\rho}{\rho^5}\frac{e^{-8\pi^2/g^2(\rho)}}{g^{12}}, \quad (29)$$

where z is the space-time location of the instanton, and

$$\frac{8\pi^2}{g^2(\rho)} = \frac{8\pi^2}{g^2} - 11\ln(\rho\mu_0) = -\ln(\rho\mu) \quad (30)$$

is just the usual renormalization-group result for $g(\rho)$. The second equality in (30) is simply a definition of the scale μ (the quantity that can be determined by electroproduction scaling violations).

III. SU(N)

The generalization from SU(3) to SU(N) is fairly straightforward. The only subtlety involves the integration over the collective coordinates which describe the gauge orientation of the instanton—this integration is performed in the Appendix.

If we embed the instanton in the standard way into the SU(2) in the "upper-left hand corner" of the fundamental representation of SU(N), the generators of SU(N) form one triplet (the analog of $\lambda_1, \lambda_2, \lambda_3$) and $2(N-2)$ doublets (the analogs of $\lambda_4, \ldots, \lambda_7$) under the action of this SU(2).[12] All other generators are singlets. This implies that, in addition to the eight zero modes of the form (18), there will be $4(N-2)$ zero modes of the form (19). Following the steps that led to (26), we now have

$$J(\gamma) = \frac{4}{\rho^5}\left(\frac{2\rho\sqrt{\pi}}{g}\right)^{4N}. \quad (31)$$

Similarly, following the steps leading to (27) gives

$$Q(\gamma) = \mu_0^{4N}\exp[-\tfrac{1}{3}N\ln(\mu_0\rho) - \alpha(1) - 2(N-2)\alpha(\tfrac{1}{2})]. \quad (32)$$

Equations (31) and (32) may then be inserted in (3a). The integral over the group orientation is defined in the Appendix as $V(C_N)$ and is given by (A14). The result is

$$W^{(1)} = \frac{4}{\pi^2}\frac{\exp[-\alpha(1) - 2(N-2)\alpha(\frac{1}{2})]}{(N-1)!(N-2)!} \times \int \frac{d^4z d\rho}{\rho^5}\left(\frac{4\pi^2}{g^2}\right)^{2N} e^{-8\pi^2/g^2(\rho)}, \quad (33)$$

where, according to the renormalization group,

$$\frac{8\pi^2}{g^2(\rho)} = \frac{8\pi^2}{g^2} - \frac{11N}{3}\ln(\mu_0\rho). \quad (34)$$

Equation (33) differs by a factor of $1/\sqrt{2N}$ from previously published results,[2] for reasons explained in the Appendix. However, the large-N limit is controlled by the factorial and power behavior of (33), and the conclusion that instantons are unimportant in the planar limit[3,4] ($N\to\infty$, Ng^2 fixed) is unchanged.

IV. COMMENTS ON QCD INSTANTON CALCULATIONS

We may rewrite our answer for SU(3), Eq. (29), in terms of the mean density of instanton of scale size ρ in the dilute-gas approximation:

$$\frac{d\rho}{\rho^5}D(\rho) = b\frac{d\rho}{\rho^5}x^6e^{-x(\rho)}, \qquad b = 0.0015, \quad (35)$$

where $x \equiv 8\pi^2/g^2$, $x(\rho) \equiv 8\pi^2/g^2(\rho)$. The number b in (35) differs by a factor of $\frac{1}{64}$ from the previously published[6] 0.1, which has been used in most QCD instanton calculations to date. The reason for this discrepancy is easily found: In calculating this number, we have taken cognizance of the recently discovered[5] error in 't Hooft's original calculation and have therefore been careful to normalize the functional measure correctly. [See Eq. (8) and the remarks following it.] Compared to the original incorrect normalization, this introduces a factor of $1/\sqrt{2}$ for each zero mode. The fact that the previous result must be corrected by a factor of $\frac{1}{64}$ is by now known to most specialists in this field[16]; however, a few comments are in order on the required modifications of some recent calculations[7-9] of physical instanton effects.

Examining (35) we see that a change in b can be absorbed into an additive constant in $x(\rho)$; from (30) this just implies a change in the scale $\mu\rho$. In fact, to absorb a factor of 64, we must change $x(\rho)$ by $\ln 64 \sim 4$, which means changing $\mu\rho$ by a factor of ~ 1.5. Of course, if we assume, in the usual way,[6] that the factor of x^6 in (35) is converted simply to $[x(\rho)]^6$ by the effect of higher loops, then the above argument is not strictly correct. Still, phenomena that occur for $x(\rho)$ considerably larger than 4 (for example, the interesting effects in Ref. 9 occur in the range $14 < x < 20$) are expected to take roughly the same form as before, but occur at somewhat larger values of the coupling constant [smaller $x(\rho)$] and correspondingly larger values of $\mu\rho$. We must keep in mind, though, that while pure dilute-gas instanton effects may be relatively unaffected, the larger coupling constants involved can bring other, nonperturbative effects into play, thus blurring some previous conclusions.

To be more specific, let us first consider the effects of instantons on the short-distance hadron currents that control e^+e^- annihilation.[7,8,17] In Ref. 7, the instanton correction to the photon self-energy $\Pi(q^2)$ for q^2 large and Euclidean is found to depend on all instanton sizes up to a maximum, cutoff scale. Thus we must ask whether we should change the cutoff with b. In the past, the cutoff could be taken, without significant difference, to be either the scale, ρ_D, at which the dilute-gas approximation breaks down [i.e., when the integrated instanton density is 1—which is given by $x(\rho_D)\simeq 14$ for $b=0.1$] or the scale, ρ_M, at which other nonperturbative effects become important [instantons are believed to ionize[8] into mesons at $x(\rho_M)\simeq 17$]. The meron calculation is purely a comparison of action and entropy and is not dependent on b (Ref. 18); however, ρ_D changes drastically with b. [In fact, a trival calculation gives $x(\rho_D)=0$ for $b=0.0015$—though of course one is hardly justified in using the renormalization group down to such values of $x(\rho)$.] Now one could take the point of view that the instanton dilute-gas approximation, while not quantitatively accurate below the point where merons appear, still gives a reasonably good qualitative picture of the effects of the whole nonperturbative sector. With this viewpoint, one may extend the cutoff to lower x, limited only by the validity of the renormalization-group calculation (as explained above, lack of diluteness is not a problem). In this way, one would find instanton effects of roughly the same numerical magnitude (within a factor of 2 or so) as those found by Andrei and Gross. On the other hand, with the point of view that instanton calculations cannot be trusted when instantons ionize into merons, we must divide by 64 the numbers R_1 and R_2 which compare instanton effects to perturbation theory—thereby making instanton effects much less important at the quoted values of q/μ. This difference can be made up by going to lower q/μ (larger coupling); however, one would again have to go beyond the point where merons appear for instanton corrections to be comparable to the perturbative ones.

In Ref. 8, Baulieu *et al.* argue that the contribution of instantons to the e^+e^- total cross section may be found by taking the imaginary part of the naive continuation of $\Pi(q^2)$ to timelike q^2. Their result depends only on instantons of scales $\rho\sim 1/q$, so there is no freedom to adjust this answer by changing a cutoff—numerical results must be divided by 64. Of course, in both Refs. 7 and 8, predictions for definite values of q (in GeV) depend on the identification of μ (in GeV), which is taken either from experiment (Ref. 8, $\mu\sim 300$–700 MeV) or from theory [Ref. 7, $\mu\sim\frac{1}{4}$ (hadron mass)—from the scale of meron ionization[8]]. To the extent that these numbers are uncertain, numerical predictions can change. (In Ref. 8, such high powers of momentum enter that our factor of 64 is lost in uncertainties in μ.)

We now turn to the recent work of Callan *et al.*[9] on the role of instantons in quark confinement and the formation of a hadron bag. These authors find that instantons act as permanent color magnetic dipoles which lead to a transition, at a critical value of an external color field, between a dilute phase with low paramagnetic permeability (the inside of the bag) and a dense phase of vacuum fluctuations with very high permeability (the outside of the bag). This phase transition is signalled by an instability in the phase diagram of color electric displacement D vs color electric field E—namely, a "nose" on the curve, where $\partial D/\partial E$ changes sign. In their calculations, with $b=0.1$, the "nose" occurs at a scale where $x(\rho)\simeq 19$. As explained at the beginning of this section, the effect of changing b should be to keep the form of this result essentially unchanged (i.e., to preserve the basic shape of the D vs E curve) but to displace the curve to somewhat larger scales and smaller values of $x(\rho)$. This is precisely what happens: When I repeat[19] their calculation using $b=0.0015$, I still find a "nose" in D vs E, but now at $x(\rho)\simeq 11$. The instanton gas is still dilute at this scale; furthermore, the coupling constant is still reasonably small so that ordinary perturbative corrections are not expected to be large. However, $x(\rho)=11$ is considerably below the point where instantons ionize into merons [recall that this occurs at $x(\rho)\simeq 17$, independent of b]. Thus the meaning we assign to this calculation will again depend on our philosophy of the nonperturbative sector. If instantons are believed to be representative of the whole sector, then we have a good qualitative understanding of an instability which leads to confinement. On the other hand, if we insist that we must at present stop calculating when instantons ionize, then our ability to see ends before things begin to look interesting. It is certainly true that no calculation has yet been done that indicates how a *meron* gas acts under an external color field. Thus, finding confinement in this picture is a tricky business. It is important to keep in mind, however, that another result of Ref. 9—that instanton corrections in the presence of large color fields inside hadrons are controlled and calculable—is unaffected.

ACKNOWLEDGMENTS

I would like to thank R. Blattner, C. Callan, J. M. Cornwall, R. Norton, E. J. Weinberg, and E. Witten for useful discussions. This work was supported in part by the National Science Foundation.

APPENDIX

Here we calculate the integral over the collective coordinates which describe the orientation of an instanton in the group SU(N). This is just the volume of the coset space C_N, defined by

$$C_N = \mathrm{SU}(N)/T_N\,, \tag{A1}$$

where T_N is the stability group of the instanton [the subgroup of SU(N) which leaves the instanton invariant].

As a preliminary, we compute the volume of SU(3). The calculation is simplified enormously[20] by considering the action of the fundamental representation on the vector

$$v = \begin{pmatrix} 0 \\ 0 \\ 1 \end{pmatrix}.$$

There is an SU(2) subgroup, generated by $\lambda_1, \lambda_2, \lambda_3$, which leaves v invariant; the rest of the group just takes v into an arbitrary complex 3-vector of length 1. Thus the set of equivalence classes of SU(3) under the above-mentioned SU(2) is in one-to-one correspondence with the points on the five-dimensional sphere S_5. However, the volume element of SU(3)/SU(2) is not numerically equal to the volume element on S_5; to get the relation between the two we consider how group elements near the identity act on v. Writing $I + \delta G = I - i\lambda_i dt^i$, we have

$$\delta G(v) = \begin{bmatrix} -idt^4 - dt^5 - \dfrac{i}{\sqrt{3}}\, dt^8 \\ -idt^6 - dt^7 - \dfrac{i}{\sqrt{3}}\, dt^8 \\ \dfrac{2idt^8}{\sqrt{3}} \end{bmatrix}. \tag{A2}$$

On the other hand, if we describe a point in S_5 as a complex 3-vector and denote the locally flat coordinates in the neighborhood of v by $x^1, \dots, x^5$, then the infinitesimal change in v under displacement by these coordinates is given by

$$\delta v = \begin{pmatrix} dx^1 + idx^2 \\ dx^3 + idx^4 \\ idx^5 \end{pmatrix}. \tag{A3}$$

Comparing (A2) and (A3) allows us to relate $dx^1, \dots, dx^5$ to $dt^4, \dots, dt^8$. We have, for the volume elements,

$$dt^4 dt^5 dt^6 dt^7 dt^8 = \frac{\sqrt{3}}{2}\, dx^1 dx^2 dx^3 dx^4 dx^5\,. \tag{A4}$$

Thus, the volume of SU(3) is given by[21]

$$V(\mathrm{SU}(3)) = \int \prod_i dt^i = \frac{\sqrt{3}}{2}\, V(S_5) V(\mathrm{SU}(2))\,. \tag{A5}$$

Using $V(S_5) = \pi^3$, $V(\mathrm{SU}(2)) = V(S_3) = 2\pi^2$, we have

$$V(\mathrm{SU}(3)) = \sqrt{3}\,\pi^5\,. \tag{A6}$$

We can now compute $V(C_3)$, the volume of C_3. This is just the volume of SU(3) divided by the volume of T_3, where T_3 is the U(1) generated by λ_8. If we write the elements of this U(1) as $e^{i\theta\lambda_8}$ then θ has the range $0 < \theta < 2\pi\sqrt{3}$. We thus have

$$V(C_3) = \frac{V(\mathrm{SU}(3))}{V(T_3)} = \frac{\pi^4}{2}\,. \tag{A7}$$

The calculation for general N follows the same lines. In parallel with (A2)–(A5), SU(N)/SU($N-1$) can be related to S_{2N-1}, giving[21]

$$V(\mathrm{SU}(N)) = \left(\frac{N}{2(N-1)}\right)^{1/2} V(S_{2N-1}) V(\mathrm{SU}(N-1))\,. \tag{A8}$$

Using

$$V(S_{2N-1}) = \frac{2\pi^N}{(N-1)!}\,, \tag{A9}$$

we have

$$V(\mathrm{SU}(N)) = \sqrt{N} \prod_{k=2}^{N} \frac{\sqrt{2}\,\pi^k}{(k-1)!}\,. \tag{A10}$$

The identification of T_N is slightly subtle. If we place the instanton in the upper-left-hand corner of the fundamental representation, then the generators of T_N are those which commute with that SU(2), namely the generators of the SU($N-2$) in the lower right and the generator

$$\lambda = \left(\frac{N-2}{N}\right)^{1/2} \begin{bmatrix} 1 & & & & & \\ & 1 & & & & \\ & & -2/(N-2) & & & \\ & & & -2/(N-2) & & \\ & & & & \ddots & \\ & & & & & -2/(N-2) \end{bmatrix}, \tag{A11}$$

which commutes with SU(2) and SU($N-2$). We may therefore parametrize T_N by writing $h \in T_N$ as

$$h = e^{i\theta\lambda} g\,, \tag{A12}$$

where $g \in \mathrm{SU}(N-2)$. The range of θ is $0 \leq \theta < 2\pi[N/(N-2)]^{1/2}$. (Note that for $N \geq 5$, this is less than the range of θ necessary for $e^{i\theta\lambda}$ to repeat; i.e., T_N is not the same as SU($N-2$) × U(1).

This is because $\theta_1 - \theta_2 = 2\pi[N/(N-2)]^{1/2}$ implies $e^{i\theta_1\lambda}$ and $e^{i\theta_2\lambda}$ differ by an element of SU($N-2$)—specifically, an element in the *center* of SU($N-2$).) We thus have

$$V(T_N) = 2\pi\left(\frac{N}{N-2}\right)^{1/2} V(\mathrm{SU}(N-2)), \qquad \text{(A13)}$$

and[22]

$$V(C_N) = \frac{V(\mathrm{SU}(N))}{V(T_N)} = \frac{\pi^{2N-2}}{(N-1)!(N-2)!}. \qquad \text{(A14)}$$

This differs from previously published results.[2,3] Aside from trivial differences in the normalization of the generators, my disagreement with these authors is based on their identification of T_N as SU($N-2$) (Ref. 3) or U($N-2$) (Ref. 2).

---

[1] G. 't Hooft, Phys. Rev. D 14, 3432 (1976).

[2] Yu. A. Bashilov and S. V. Pokrovsky, Nucl. Phys. B143, 431 (1978).

[3] J. Koplik, A. Neveu, and S. Nussinov, Nucl. Phys. B123, 109 (1977).

[4] E. Witten, Nucl. Phys. B149, 285 (1979).

[5] G. 't Hooft, Phys. Rev. D 18, 2199(E) (1978).

[6] C. Callan, R. Dashen, and D. Gross, Phys. Rev. D 17, 2717 (1978).

[7] N. Andrei and D. Gross, Phys. Rev. D 18, 468 (1978).

[8] L. Baulieu, J. Ellis, M. K. Gaillard, and R. J. Zakrzewski, Phys. Lett. 77B, 290 (1978).

[9] C. Callan, R. Dashen, and D. Gross, Phys. Rev. D 19, 1826 (1979).

[10] J. L. Gervais and B. Sakita, Phys. Rev. D 11, 2943 (1975); E. Tomboulis, *ibid.* 12, 1678 (1975).

[11] The scalar product, in conformity with that of Ref. 1, is defined by

$$\langle \psi^{(i)} | \psi^{(j)} \rangle = 2\int d^4x\, \mathrm{tr}(\psi^{(i)}_\mu(x)\,\psi^{(j)}_\mu(x))$$

$$= \int d^4x\, \psi^{(i)}_{\mu a}\, \psi^{(j)}_{\mu a},$$

where

$$\psi^{(i)}_\mu \equiv \tfrac{1}{2}\lambda_a \psi^{(i)}_{\mu a}$$

[12] For a description of instantons in a general gauge group, see C. Bernard, N. Christ, A. Guth, and E. Weinberg, Phys. Rev. D 15, 2967 (1977).

[13] C. Bernard, Phys. Rev. D 18, 2026 (1978). A^{cl}_ν is the correct gauge transformation parameter in (18a) since this gives $\psi^{(\nu)}_\mu = F^{cl}_{\mu\nu}$, which clearly obeys $D^{cl}_\mu \psi^{(\nu)}_\mu = 0$.

[14] For $G \neq I$ just rotate both sides of (23).

[15] F. R. Ore, Jr., Phys. Rev. D 16, 2577 (1977); S. Chadha, A. D'Adda, P. DiVecchia, and F. Nicodemi, Phys. Lett. 72B, 103 (1977).

[16] C. Callan (private communication).

[17] Other treatments, which we do not discuss here, are those of R. D. Carlitz and C. Lee, Phys. Rev. D 17, 3238 (1978); M. Suzuki, Phys. Lett. 76B, 466 (1978); and T. Appelquist and R. Shankar, Phys. Rev. D 18, 2952 (1978).

[18] While not dependent on b, the value of $x(\rho)$ at which merons ionize is of course dependent on the definition of the coupling constant in renormalization. Here, all numbers refer to the Pauli-Villars regularization scheme of Ref. 1.

[19] The calculation takes into account instanton interactions by placing each instanton in a spherical cavity of radius $R = 2.2\rho$ in the permeable medium formed by the other instantons. As in Ref. 9, other reasonable choices for R do not change the result significantly.

[20] I thank R. Blattner for explaining this method to me. A straightforward calculation of the volume by parametrizing the group and integrating yields the same answer.

[21] The normalization of the volume is determined by the normalization of the generators. Here, I use $\mathrm{tr}(\lambda_i\lambda_j) = 2\delta_{ij}$.

[22] While derived for $N > 3$, this formula is in fact true for $N = 2, 3$ also.

Volume 65B, number 2 PHYSICS LETTERS 8 November 1976

INEQUIVALENT EMBEDDINGS OF SU(2) AND INSTANTON INTERACTIONS *

Frank WILCZEK

Joseph Henry Laboratories, Princeton University, Princeton, New Jersey 08540, USA

Received 10 September 1976

The fact that SU(2) allows inequivalent embeddings inside larger groups is shown to have some consequences for the interaction of instantons.

In a recent Letter Belavin et al. [1] found a topological quantum number characterizing gauge fields in four dimensions. This number is given by

$$q = (1/8\pi^2)\,\mathrm{Tr} \int F_{\mu\nu}\tilde{F}_{\mu\nu}\,\mathrm{d}^4x \,, \tag{1}$$

where $\tilde{F}_{\mu\nu} \equiv \frac{1}{2}\epsilon_{\mu\nu\rho\sigma}F_{\rho\sigma}$ is the dual of $F_{\mu\nu}$ and we have used a matrix notation for gauge fields. It may be proved that q takes on integer values. There is an important inequality relating this change to the gauge field action, viz.

$$S = \frac{1}{4g^2}\,\mathrm{Tr}\int F_{\mu\nu}F_{\mu\nu}\,\mathrm{d}^4x \geqslant \frac{2\pi^2}{g^2}|q| \,. \tag{2}$$

The inequality is satisfied if and only if the field is self-dual, $F_{\mu\nu} = \pm\tilde{F}_{\mu\nu}$. In ref. [1] a gauge field was found which saturated the bound of eq. (2), for the group SU(2) and $q = \pm 1$. It is given by

$$A_\mu = \frac{x^2}{x^2+\lambda^2}\,g^{-1}\partial_\mu g\,, \qquad g = (x_0 + \mathrm{i}\boldsymbol{x}\cdot\boldsymbol{\sigma})/|x|\,, \tag{3}$$

where the σ^i are ordinary Pauli matrices. By a gauge transformation [2] this may be cast in a form more convenient for the following considerations:

$$A_\mu = \frac{-\lambda^2}{x^2+\lambda^2}(\partial_\mu g)g^{-1}\,. \tag{4}$$

If for a moment we imagine living in a five-dimensional world, then the field configuration just described corresponds to a sort of quasi-particle, called an instanton [3]. It is of some interest to know the interactions of such quasiparticles.

From the inequality (2) we see immediately that the interaction energy in a gas of instantons is always positive, suggesting repulsion. A charge 2 configuration is obtained by simply adding the potentials of two separated instantons of the form (4). For large separations r the interaction energy goes like c/r [4].

That the forces between instantons are not always repulsive follows from the following simple consideration. Let us write the potential of eq. (4) in the form

$$A_\mu = A_\mu^i\,\sigma^i/2\,, \tag{5}$$

leading to the field strength

$$F_{\mu\nu} = F_{\mu\nu}^i\,\sigma^i/2\,. \tag{6}$$

Now if the τ^i, $i = 1,2,3$ form *any* representation of SU(2), then the potential $A_\mu^i\tau^i$ will give the field strength

* Work supported in part by ERDA under contrace E(11-1)3072.

Volume 65B, number 2 PHYSICS LETTERS 8 November 1976

$F^i_{\mu\nu}\tau^i$, which is self-dual and thereby saturates the bound of eq. (2). For instance, in the physically interesting case of SU(3) we might take the spin-1 matrices

$$\tau_1 = \tfrac{1}{2}\begin{pmatrix} 0 & \sqrt{2} & 0 \\ \sqrt{2} & 0 & \sqrt{2} \\ 0 & \sqrt{2} & 0 \end{pmatrix}, \quad \tau_2 = \tfrac{1}{2}\begin{pmatrix} 0 & -\mathrm{i}\sqrt{2} & 0 \\ \mathrm{i}\sqrt{2} & 0 & -\mathrm{i}\sqrt{2} \\ 0 & \mathrm{i}\sqrt{2} & 0 \end{pmatrix}, \quad \tau_3 = \begin{pmatrix} 1 & 0 & 0 \\ 0 & 0 & 0 \\ 0 & 0 & -1 \end{pmatrix}. \tag{8}$$

Since the charge defined in eq. (2) is proportional to the square of the field strength, we readily see that this leads to *charge 4 instantons with no interaction energy*.

The charge 4 instanton can be reached by bringing together four separated charge 1 instantons as follows. We define four sets of spin-1/2 matrices embedded in different ways in SU(3):

$$\begin{aligned}
{}^{(1)}\sigma^1 &= \frac{\sqrt{2}}{2}\begin{pmatrix} 0 & 1+\mathrm{i} & 0 \\ 1-\mathrm{i} & 0 & 0 \\ 0 & 0 & 0 \end{pmatrix}, &
{}^{(1)}\sigma^2 &= \frac{\sqrt{2}}{2}\begin{pmatrix} 0 & 1-\mathrm{i} & 0 \\ 1+\mathrm{i} & 0 & 0 \\ 0 & 0 & 0 \end{pmatrix}, &
{}^{(1)}\sigma^3 &= \begin{pmatrix} 1 & 0 & 0 \\ 0 & -1 & 0 \\ 0 & 0 & 0 \end{pmatrix}, \\
{}^{(2)}\sigma^1 &= \frac{\sqrt{2}}{2}\begin{pmatrix} 0 & 1-\mathrm{i} & 0 \\ 1+\mathrm{i} & 0 & 0 \\ 0 & 0 & 0 \end{pmatrix}, &
{}^{(2)}\sigma^2 &= \frac{\sqrt{2}}{2}\begin{pmatrix} 0 & -1-\mathrm{i} & 0 \\ -1+\mathrm{i} & 0 & 0 \\ 0 & 0 & 0 \end{pmatrix}, &
{}^{(2)}\sigma^3 &= \begin{pmatrix} 1 & 0 & 0 \\ 0 & -1 & 0 \\ 0 & 0 & 0 \end{pmatrix}, \\
{}^{(3)}\sigma^1 &= \frac{\sqrt{2}}{2}\begin{pmatrix} 0 & 0 & 0 \\ 0 & 0 & 1+\mathrm{i} \\ 0 & 1-\mathrm{i} & 0 \end{pmatrix}, &
{}^{(3)}\sigma^2 &= \frac{\sqrt{2}}{2}\begin{pmatrix} 0 & 0 & 0 \\ 0 & 0 & 1-\mathrm{i} \\ 0 & 1+\mathrm{i} & 0 \end{pmatrix}, &
{}^{(3)}\sigma^3 &= \begin{pmatrix} 0 & 0 & 0 \\ 0 & 1 & 0 \\ 0 & 0 & -1 \end{pmatrix}, \\
{}^{(4)}\sigma^1 &= \frac{\sqrt{2}}{2}\begin{pmatrix} 0 & 0 & 0 \\ 0 & 0 & 1-\mathrm{i} \\ 0 & 1+\mathrm{i} & 0 \end{pmatrix}, &
{}^{(4)}\sigma^2 &= \frac{\sqrt{2}}{2}\begin{pmatrix} 0 & 0 & 0 \\ 0 & 0 & -1-\mathrm{i} \\ 0 & -1+\mathrm{i} & 0 \end{pmatrix}, &
{}^{(4)}\sigma^3 &= \begin{pmatrix} 0 & 0 & 0 \\ 0 & 1 & 0 \\ 0 & 0 & -1 \end{pmatrix},
\end{aligned} \tag{9}$$

Then the gauge field formed by adding the $A^i(x-r_a)^{(a)}\sigma^i/2$, ($a = 1$ to 4, not summed) where the r_a are four-vectors and A^i is given by eqs. (4) and (5), goes over continuously to the charge four instanton as the $r_a \to 0$ and to four separated charge one instantons as the $r_a \to \infty$. At large distances the two-body forces between instantons are dominant, and they are repulsive. Since, on the other hand, the interaction energy is zero when all four instantons sit on top of one another, we conclude that there must be a *domain of attraction*.

Since the representation theory of SU(2) is completely known, we can readily extend our construction to get higher charge instantons of minimal action in larger groups. For example, we have:

$$\text{in SU(3):}\quad \pm 1, 4\,, \tag{10}$$

$$\text{in SU(4):}\quad \pm 1, 2, 4, 10\,, \tag{11}$$

$$\text{in SU(8):}\quad \pm 1, 2, 3, 4\ \text{(two ways)}, 5, 6, 8, 9, 10, 11, 12, 14, 20\ \text{(two ways)}, 21, 24, 35, 36, 56, 84\,. \tag{12}$$

"Two ways" indicates that there are two inequivalent methods of achieving the charge: in SU(8) charge 4 can be reached by 4 × spin 1/2 or by spin 1; charge 20 by 2 × spin 3/2 or by spin 2.

From these simple remarks some interesting conclusions follow:

(a) Suppose we assume, as seems probable though not certain from the failure of many attempts to discover them, that no higher charge configurations besides the ones here discussed have no interaction energy. Then in SU(3) the two-distanton interactions always give a positive interaction energy. The action (2) is, however, quartic in the potentials, leading to the possibility of three or four-body forces. The considerations here indicate that these forces introduce qualitatively new effects.

(b) The fact that in SU(8) essentially different instantons exist with the same charge, indicates that the topological charge is not enough to classify solutions of the field equations, even after gauge, Poincaré, and scale transformations are accounted for.

Volume 65B, number 2 PHYSICS LETTERS 8 November 1976

Of course, the above considerations do not go very far in solving the general problem of interactions between instantons, and should be subsumed in a much larger theory. It seems worthwhile, however, to have some definite though modest results available.

Finally let us mention two directions in which this work is being carried further. Calculations of the fermion and gauge field propagators in the presence of the instantons in SU(3) are being done [4] along the lines of the SU(2) calculations [2, 3, 5]. It will be interesting to see if the quantum corrections to the charge 4 instanton raise or lower its action relative to four separated unit charges. Considerations analogous to the present ones are also relevant for monopoles and dyons in some models of broken gauge symmetry [4, 6].

References

[1] A. Belavin et al., Phys. Lett. 59B (1975) 85.
[2] R. Jackiw and C. Rebbi, M.I.T. preprints (1976).
[3] G. 't Hooft, Phys. Rev. Lett. 37 (1976) 8 and Harvard preprint (1976).
[4] D. Toussaint and F. Wilczek, in progress.
[5] F. Ore, Jr., M.I.T. preprint (1976).
[6] Related considerations are contained in: E. Corrigan et al., Nucl. Phys. B106 (1976) 475.

III. MULTI-INSTANTON SOLUTIONS

INTRODUCTION

As has been explained in Sec. I all finite action field configurations are classified according to their topological charge Q. If Q is fixed the minimum of the action in the given class is achieved on self-dual or anti-self-dual fields. So far we have discussed mainly the classical BPST instanton corresponding to $Q = \pm 1$. It is quite natural to ask how far one can go along the same lines in the case of $|Q| > 1$.

First of all let us note that a very diluted gas of instantons (i.e. n instantons widely separated from each other) is an example of almost a solution with $Q = n$. When all interinstanton distances become much larger than the instanton sizes, the solution becomes more and more exact, so that in the limit of infinite separations, we have an exact solution. This simple argument allows one to find the overall number of free parameters on which the general multi-instanton solution must depend. This number is just the number of the collective coordinates of the BPST instanton times the topological charge.

The question is whether it is possible to explicitly construct exact solutions with $|Q| > 1$ and finite interinstanton distances. Such multi-instanton ensembles are of importance in some two-dimensional field theories [1], to say nothing of the fact that they are, of course, very interesting on their own. Note that the action of the exact solution with $|Q| = n$ is equal to

$$S = \frac{8\pi^2}{g^2} n \ .$$

In other words individual instantons forming this ensemble do not interact with each other at the classical level.

A comprehensive answer to the above question is provided, in principle, in the work of Atiyah *et al.* [2], central to this section. These authors worked out a closed mathematical procedure allowing one to build self-dual fields with the arbitrary topological charge. Within their procedure the problem is reduced to an algebraic matrix equation. The solution obtained in this way depends on the maximal allowed number of free parameters and is thus a general solution. For instance, for $SU(2)$, the solution of Atiyah *et al.* contains $8|Q|$ collective coordinates, which is in full accordance with what one should expect from the consideration of $|Q|$ BPST instantons (anti-instantons) in the limiting case of a very diluted gas. Surprising as it is, the understanding of the fact that the general solution must depend on $8|Q|$ parameters did not come immediately, as the reader will see for himself/herself from the papers reproduced below.

Although, in principle, the construction by Atiyah *et al.* exhausts the problem of the multi-instantons, in practice, it is rather abstract and is far from being transparent. To get an insight and the necessary intuition, it is highly desirable to play with specific simple examples of multi-instanton solutions even though these examples will clearly lack the full generality and will depend on a fewer number of free parameters or will refer only to some specific values of the topological charge.

Several solutions with the closed form expressions for the gauge potentials A_μ^{inst} are known in the literature.

One of the first instructive examples has been found by Witten [3] — it corresponds to a sequence of instantons on a line. A very elegant approach allowing one to get a $5n$-parametric family of instantons was suggested by 't Hooft in a work which, unfortunately, remained unpublished. His starting point is as follows. Let us assume that

$$A_\mu^a = \frac{1}{g}\eta^a_{\mu\nu}\partial_\mu \ln\phi\ , \tag{1}$$

where ϕ is a scalar function of x and collective coordinates. It is rather easy to show that within this ansatz the duality condition for the field strength tensor results in an equation on ϕ which essentially reduces to that on the scalar potential in electrostatics,

$$\frac{1}{\phi}\Box\phi = 0. \tag{2}$$

The solution of Eq. (2) is given by a sum of (four-dimensional) Coulomb potentials from pointlike sources arbitrarily located in four-dimensional space,

$$\phi = 1 + \sum_{i=1}^{n}\frac{\rho_i^2}{(x-y_i)^2}. \tag{3}$$

The resulting field configuration describes n instantons in the singular gauge, with arbitrary positions and all with one and the same orientation in the isospace.

The 't Hooft solution is presented in more details in Ref. [4] where the authors also investigate its properties under conformal transformations.

The work of Christ *et al.* [5] deserves special attention. As a matter of fact, in the first reading, I would recommend that one starts the topic of multi-instantons from this paper. Among other things, it presents an extended commentary to the work by Atiyah *et al.* [2], hence providing a helpful background for those who would like to master this mathematical construction in full. In particular, the authors demonstrate how the general solution degenerates into the diluted gas in the limit of widely separated instantons (more exactly, the authors show how the general matrix equation by Atiyah *et al.* can be solved perturbatively and formulate an explicit limiting procedure reducing the field configuration of the general form to the dilute gas for large interinstanton distances). In this way all the free parameters introduced by Atiyah *et al.* acquire a graphic physical interpretation in terms of the collective coordinates of individual instantons.

Another problem addressed in Ref. [5] is the propagation of quanta in the background self-dual fields. The Green functions in the instanton external field are of practical importance in numerous applications — from the instanton-induced baryon number violation at high energies to models of the QCD vacuum. The next section will be fully devoted to this issue. Here I would like to note that Ref. [5] laid the foundation for further analyses. The Klein-Gordon equation for the massless (spatially) scalar field in the self-dual background was considered and the propagators were found in the cases of isospins 1/2 and 1 [for the gauge group $SU(2)$].

References [6, 7] conclude this section; they count the number of collective coordinates in the arbitrary solution of the duality equation for arbitrary gauge groups.

References

[1] V. Fateev, I. Frolov and A. Schwarz, *Nucl. Phys.* **B154** (1979) 1.
[2] *M. Atiyah, N. Hitchin, V. Drinfeld and Yu. Manin, *Phys. Lett.* **65A** (1978) 185.
[3] *E. Witten, *Phys. Rev. Lett.* **38** (1977) 121.
[4] *R. Jackiw, C. Nohl and C. Rebbi, *Phys. Rev.* **D15** (1977) 1642.
[5] *N. Christ, E. Weinberg and N. Stanton, *Phys. Rev.* **D18** (1978) 2013.
[6] *R. Jackiw and C. Rebbi, *Phys. Lett.* **67B** (1977) 189.
[7] *C. Bernard *et al.*, *Phys. Rev.* **D16** (1977) 2967.

VOLUME 38, NUMBER 3 PHYSICAL REVIEW LETTERS 17 JANUARY 1977

Some Exact Multipseudoparticle Solutions of Classical Yang-Mills Theory

Edward Witten*
Lyman Laboratory of Physics, Harvard University, Cambridge, Massachusetts 02138
(Received 2 November 1976)

I present some exact solutions of the Polyakov-Belavin-Schwartz-Tyupkin equation $F_{\mu\nu} = \tilde{F}_{\mu\nu}$ for an SU(2) gauge theory in Euclidean space. My solutions describe a system with an arbitrary number of pseudoparticles, with arbitrary scale parameters and arbitrary separations, arranged along a line. The action for an n-pseudoparticle solution is precisely n times the action for a single pseudoparticle.

Recently Polyakov[1] made the remarkable suggestion that localized, finite-action solutions of the classical Euclidean equations of motion may dominate the Euclidean path integrals of quantum field theory. Belavin, Polyakov, Schwartz, and Tyupkin (BPST) described such a localized, finite-action solution for non-Abelian gauge theories[2]; it has become known as the pseudoparticle.

Here I will describe a much more extensive class of exact, analytic solutions for a classical SU(2) gauge theory in Euclidean space. My solutions have arbitrary integral values of the topological charge discovered by Belavin, Polyakov, Schwartz, and Tyupkin. They describe an assembly of the BPST pseudoparticles, with arbitrary scale parameters and arbitrary separations, but arranged along a line. My solutions may help clarify many-pseudoparticle effects which, as Polyakov suggested, may play an important role in the strong interactions.

Belavin, Polyakov, Schwartz, and Tyupkin showed that the fields of minimum action for fixed boundary conditions are solutions of $F_{\mu\nu} = \tilde{F}_{\mu\nu}$. If $F_{\mu\nu} = \tilde{F}_{\mu\nu}$, then in view of the Bianchi identity $D_\mu \tilde{F}_{\mu\nu} = 0$, the field equation $D_\mu F_{\mu\nu} = 0$ is also satisfied. My solutions will all satisfy $F_{\mu\nu} = \tilde{F}_{\mu\nu}$.

I will seek solutions of $F_{\mu\nu} = \tilde{F}_{\mu\nu}$ that are invariant under three-dimensional rotations combined with gauge transformations. I will call this a cylindrical symmetry, because it determines the dependence of the fields on the three-dimensional polar angles and leaves unknown only the dependence on the three-dimensional radius r and the Euclidean time t. The most general gauge field with cylindrical symmetry can be written as follows:

$$A_j^{\,a} = \frac{(\varphi_2 + 1)}{r^2}\epsilon_{jak}x_k + \frac{\varphi_1}{r^3}[\delta_{ja}r^2 - x_j x_a] + A_1\frac{x_j x_a}{r^2},$$
$$A_0^{\,a} = \frac{A_0 x^a}{r}. \tag{1}$$

Here, j and k refer to the three spatial dimensions, and a is the isospin index. The precise definitions of φ_1, φ_2, A_0, and A_1 are chosen for future convenience. These functions depend only on r and t. I will find the most general solution of $F_{\mu\nu} = \tilde{F}_{\mu\nu}$ that can be written in the form (1).

The *Ansatz* (1) is consistent with gauge transformations generated by a unitary matrix $U(x, t) = \exp[if(r, t)\vec{x}\cdot\vec{T}]$, where f is arbitrary and T_i are the generators of SU(2). This is an Abelian subgroup of the full gauge group. For the moment I avoid a choice of gauge. Given (1), one readily calculates the field tensor $F_{\mu\nu}^{\,a} = \partial_\mu A_\nu^{\,a} - \partial_\nu A_\mu^{\,a} - \epsilon^{abc}A_\mu^{\,b}A_\nu^{\,c}$:

$$F_{0i}^{\,a} = (\partial_0\varphi_2 - A_0\varphi_1)\frac{\epsilon_{iak}x_k}{r^2} + (\partial_0\varphi_1 + A_0\varphi_2)\frac{(\delta_{ai}r^2 - x_a x_i)}{r^3} + r^2(\partial_0 A_1 - \partial_1 A_0)\frac{x_a x_i}{r^4},$$
$$\tfrac{1}{2}\epsilon_{ijk}F_{jk}^{\,a} = -\frac{\epsilon_{iak}x_k}{r^2}(\partial_1\varphi_1 + A_1\varphi_2) + \frac{(\delta_{ai}r^2 - x_a x_i)}{r^3}(\partial_1\varphi_2 - A_1\varphi_1) + \frac{x_a x_i}{r^4}(1 - \varphi_1^{\,2} - \varphi_2^{\,2}) \tag{2}$$

(where ∂_0 denotes $\partial/\partial t$ and ∂_1 denotes $\partial/\partial r$). The form of (2) suggests that I regard φ as a charged scalar interacting with the two-dimensional Abelian gauge field A_μ, with covariant derivative $D_\mu\varphi_i = \partial_\mu\varphi_i + \epsilon_{ij}A_\mu\varphi_j$. With integration over the polar angles, the action turns out to be

$$A = \tfrac{1}{4}\int d^3x \int dt\, F_{\mu\nu}^{\,a}F_{\mu\nu}^{\,a} = 8\pi\int_{-\infty}^{\infty}dt\int_0^\infty dr\left[\tfrac{1}{2}(D_\mu\varphi_i)^2 + \tfrac{1}{8}r^2F_{\mu\nu}^{\,2} + \tfrac{1}{4}r^{-2}(1 - \varphi_1^{\,2} - \varphi_2^{\,2})^2\right], \tag{3}$$

where $F_{\mu\nu}$ is of course $\partial_\mu A_\nu - \partial_\nu A_\mu$. This is very nearly the usual action for the two-dimensional

Abelian Higgs model; in fact, in curved space the action for the Abelian Higgs model is

$$\int d^2x\sqrt{g}\,[\tfrac{1}{2}g^{\mu\nu}D_\mu\varphi_i D_\nu\varphi_i+\tfrac{1}{8}g^{\mu\alpha}g^{\nu\beta}F_{\mu\nu}F_{\alpha\beta}+\tfrac{1}{4}(1-\varphi_1^{\,2}-\varphi_2^{\,2})^2],$$

which agrees with (3) if $g^{\mu\nu}=r^2\delta^{\mu\nu}$. This metric corresponds to a space of constant negative curvature.

I now consider the equation $F_{\mu\nu}=\tilde{F}_{\mu\nu}$ or, equivalently, $F_{0i}{}^a=\frac{1}{2}\epsilon_{ijk}F_{jk}{}^a$. Equating corresponding terms in (2), I find

$$\begin{aligned}&\partial_0\varphi_1+A_0\varphi_2=\partial_1\varphi_2-A_1\varphi_1,\\&\partial_1\varphi_1+A_1\varphi_2=-(\partial_0\varphi_2-A_0\varphi_1),\\&r^2(\partial_0A_1-\partial_1A_0)=1-\varphi_1^{\,2}-\varphi_2^{\,2}.\end{aligned}\tag{4}$$

I will find the general solution of these equations. The key is the choice of gauge. I set $\partial_\mu A_\mu=0$, so that $A_\mu=\epsilon_{\mu\nu}\partial_\nu\psi$ for some ψ. The first two equations in (4) now become

$$[\partial_0-(\partial_0\psi)]\varphi_1=[\partial_1-(\partial_1\psi)]\varphi_2,$$

$$[\partial_1-(\partial_1\psi)]\varphi_1=-[\partial_0-(\partial_0\psi)]\varphi_2.$$

If one lets $\varphi_1=e^\psi\chi_1$, $\varphi_2=e^\psi\chi_2$, one finds simply

$$\partial_0\chi_1=\partial_1\chi_2,$$

$$\partial_1\chi_1=\partial_0\chi_2.$$

These are the Cauchy-Riemann equations, which say that $f=\chi_1-i\chi_2$ is an analytic function of $z=r+it$.

It remains to consider the third equation in (4). It becomes

$$-r^2\nabla^2\psi=1-f^*fe^{2\psi}.\tag{5}$$

Let us first note that this equation possesses a remaining gauge invariance. Consider the transformation

$$\begin{aligned}&f\to fh,\\&\psi\to\psi-\tfrac{1}{2}\ln(h^*h),\end{aligned}\tag{6}$$

where $h(z)$ is an analytic function. Because $\nabla^2\ln h^*h=0$ for any analytic function h (as long as h has no zeroes), (5) is invariant under this transformation. This invariance exists because the gauge condition $\partial_\mu A_\mu=0$ permits transformations $A_\mu\to A_\mu+\partial_\mu\lambda$, where $\nabla^2\lambda=0$. If h does have zeroes for $r>0$, then (6) introduces isolated singularities at those zeroes.

In order to solve (5), let $\psi=\ln r-\frac{1}{2}\ln(f^*f)+\rho$, where ρ is a new unknown function. By using the fact that $\nabla^2\ln f^*f=0$ for any analytic function f, except for isolated singularities that I momentarily ignore, (5) becomes simply

$$\nabla^2\rho=e^{2\rho}.\tag{7}$$

Equation (7) is called Liouville's equation.[3] Its general solution can easily be found by using conformal invariance. Let $\rho_1(z)$ be any particular solution of Liouville's equation; for example, $\rho_1(z)=-\ln[\frac{1}{2}(1-z^*z)]$. Now, consider an arbitrary analytic function $\omega(z)$. The Laplacian with respect to ω is $\nabla_\omega^{\,2}=|dz/dw|^2\nabla_z^{\,2}$ and ρ_1, as a function of ω, satisfies

$$\nabla_\omega^{\,2}\rho_1(\omega)=|dz/dw|^2e^{2\rho_1(\omega)}.$$

This is Liouville's equation (7) except for the factor $|dz/dw|^2$. Letting $\rho(\omega)=\rho_1(\omega)-\frac{1}{2}\ln|dz/dw|^2$, and using the fact that $\nabla^2\ln|dz/dw|^2=0$, I find that $\rho(\omega)$ satisfies the Liouville equation $\nabla_\omega^{\,2}\rho=e^{2\rho}$. Thus, if g is any analytic function, $\rho(z)=-\ln[\frac{1}{2}(1-g^*g)]+\frac{1}{2}\ln|dg/dz|^2$ satisfies Liouville's equation. This is, in fact, known to be the general solution of Liouville's equation.

Returning now to (5), the various singularities cancel if and only if $(dg/dz)/f$ has neither zeroes nor poles in the right half plane. This means that, up to a gauge transformation of type (6), the most general nonsingular solution of (5) is

$$\psi=-\ln\left(\frac{1-g^*g}{2r}\right),\quad f=\frac{dg}{dz}.\tag{8}$$

For ψ to be nonsingular, I must require $|g|=1$ for $r=0$ and $|g|<1$ for $r>0$. The most general analytic function with these properties and smooth behavior for $z\to\infty$ is

$$g(z)=\prod_{i=1}^{k}\left(\frac{a_i-z}{a_i^*+z}\right),\tag{9}$$

where the a_i are an arbitrary set of complex numbers (some perhaps equal) constrained only to have $\mathrm{Re}\,a_i>0$. (8) and (9) provide the most general solution of $F_{\mu\nu}=\tilde{F}_{\mu\nu}$ with cylindrical symmetry and finite action.

Let us now consider the physical content of these solutions. In view of the gauge invariance (6), the only gauge-invariant property of f is the location of its zeroes in the right half plane. The zeroes of f in the right half plane may therefore play a central role.

If $k=1$ in (9), f has no zeroes and an easy calculation shows that this solution is a gauge transform of the vacuum. If $k=2$, f has precisely one zero in the right half plane. This field describes the BPST pseudoparticle (but in a different gauge). The imaginary part of the zero of f determines

VOLUME 38, NUMBER 3 PHYSICAL REVIEW LETTERS 17 JANUARY 1977

the location of the pseudoparticle along the time axis, while the real part determines the pseudoparticle scale.

For general k, the total multiplicity of the zeroes of f in the right half plane is always $k-1$. The natural generalization of the comments in the last paragraph would be that for general k, my solution describes $k-1$ pseudoparticles, with real and imaginary parts determined by the zeroes of f.

I will verify this, but first there is a counting problem to consider. For $k=2$ my solution (8) and (9) involves four real parameters—the real and imaginary parts of a_1 and a_2. But the BPST pseudoparticle, with cylindrical symmetry, has only two parameters—the position along the time axis and the scale. For general k my solution has $2k$ real parameters, but I expect the physics to involve only $k-1$ positions and $k-1$ scales. The explanation is that my solution (8) and (9) still possesses a remaining two-parameter gauge invariance. If one replaces f and g by

$$\tilde{g}=\frac{c+g}{c^*g+1}, \quad \tilde{f}=\frac{d\tilde{g}}{dz}, \tag{10}$$

where $|c|<1$, then $\tilde{g}$ is still of the form (9), and the transformation from f and g to $\tilde{f}$ and $\tilde{g}$ is a gauge transformation of type (6). Also, $\tilde{f}$ and f have the same zeroes. Consequently, the physics in (8) and (9) depends not on the k complex number a_i but only on $k-1$ complex functions of them—the zeroes of f.

I now wish to verify two facts: That the solution (8) and (9) always has a topological charge equal to $k-1$, and that, as the zeroes of f become widely separated, my solution describes $k-1$ widely speparated, BPST pseudoparticles.

I first consider what happens as the zeroes of f become widely separated. It is essential to choose the right gauge. Let us keep fixed one of the zeroes of f, α_0, which I assume to be a simple zero, and let the distance from α_0 to the other zeroes, $\alpha_1, \ldots, \alpha_m$, become large. In a general gauge, there is no simple relation between the zeroes of g and the zeroes of f, but by a gauge transformation of type (10), I may always arrange it so that α_0 is one of the zeroes of g. It will be a double zero, since $f=dg/dz$ has a simple zero at α_0. So g has the form

$$g=[(\alpha_0-z)/(\alpha_0^*+z)]^2\prod_{i=1}^{m}[(\beta_i-z)/(\beta_i^*+z)].$$

The β_i are complicated functions of the α_i. But as the differences $|\alpha_i-\alpha_0|$ become large, the differences $|\beta_i-\alpha_0|$ also become large. Then for z in the neighborhood of α_0, the entire factor $\pi[(\beta_i-z)/(\beta_i^*+z)]$ may be set equal to the constant $\pi(\beta_i/\beta_i^*)$. Dropping this phase factor I find that as the $|\alpha_i-\alpha_0|$ become large, with z fixed, I may approximate g by $[(\alpha_0-z)/(\alpha_0^*+z)]$. But this is the special case of (9) for $k=2$, which is already known to describe a single BPST pseudoparticle with scale and location controlled by α_0. Thus, in this limit, my solution describes a system of isolated BPST pseudoparticles, one for each zero of f. Hence, the topological charge $(\frac{1}{8}\pi^2)\int d^4xF_{\mu\nu}\tilde{F}_{\mu\nu}$ equals the number of zeroes of f, at least if those zeroes are widely separated. Since a solution with nearby zeroes of f can be reached continuously from a solution with widely separated zeroes, the topological charge equals the number of zeroes whether they are separated or not.

It is instructive to derive this result in a more direct way. From expression (2) for $F_{\mu\nu}$, I find that the four-dimensional topological charge $(\frac{1}{8}\pi^2)\int d^4xF_{\mu\nu}\tilde{F}_{\mu\nu}$ becomes

$$(1/2\pi)\int d^2x[\epsilon_{\mu\nu}\epsilon_{ij}D_\mu\varphi_iD_\nu\varphi_j+\tfrac{1}{2}\epsilon_{\mu\nu}F_{\mu\nu}(1-\varphi^2)].$$

By algebraic manipulations this can be rewritten

$$(1/2\pi)\int d^2x[\partial_\mu(\epsilon_{ij}\epsilon_{\mu\nu}\varphi_iD_\nu\varphi_j)+\tfrac{1}{2}\epsilon_{\mu\nu}F_{\mu\nu}].$$

The first term is a total divergence, and so can be written as a boundary integral which vanishes because for my solutions $D_\mu\varphi_i=0$ at the boundary of the space. The second term is also a total divergence, but its integral does not necessarily vanish—it is the usual topological charge of the Abelian Higgs model. Thus for cylindrically symmetric fields, I can identify the four-dimensional topological charge discovered by Belavin, Polyakov, Schwartz, and Tyupkin with the two-dimensional topological charge of the Higgs model.

For the Higgs model, $(1/4\pi)\int d^2x\epsilon_{\mu\nu}F_{\mu\nu}$ equal $1/2\pi$ times the change in phase of $\varphi=\varphi_1-i\varphi_2$ around a contour that encloses the region in which the fields differ significantly from the vacuum. With $\varphi=fe^{\psi}$, this is

$$\frac{1}{2\pi i}\oint ds\frac{d}{ds}\ln(fe^{\psi})=\frac{1}{2\pi i}\oint ds\frac{d}{ds}\ln f+\frac{1}{2\pi i}\oint ds\frac{d\psi}{ds},$$

where ds is the line element along the contour. The second term vanishes identically (since ψ, unlike $\ln f$, is a single-valued function); and the first term, by the argument principle of complex variable theory, is equal to the number of zeroes of f within the contour. This shows that the charge equals the number of zeroes of f.

A final remark is in order. My solutions possess finite action and finite $F_{\mu\nu}{}^a$, but in the gauge in which I am working, the four-dimensional gauge field $A_\mu{}^a$ is actually singular at $r=0$. This is obvious from (1), where I see that the field is nonsingular only if $\varphi_2=1$ and $\varphi_1=0$ at $r=0$. It is necessary to perform a gauge transformation on the solutions to satisfy these conditions; such a transformation always exists because I have $\varphi^2=1$ at $r=0$. In the language of (6), (8), and (9), a suitable gauge function is

$$h=-i\prod_{i=1}^{k}(a_i{}^*+z)^2.$$

Thus,

$$\psi=\ln\frac{2r}{(1-g^*g)(h^*h)^{1/2}}$$

and

$$\varphi_1-i\varphi_2=h\frac{dg}{dz}e^{\psi}$$

give nonsingular four-dimensional gauge fields that satisfy the equations of motion.

I wish to thank S. Coleman, R. Jackiw, C. Rebbi, B. Julia, and L. Dolan for thoughtful discussions. This work was initiated at the Aspen Center for Theoretical Physics.

---

*Research supported in part by the National Science Foundation under Grant No. MPS75-20427.

[1]A. M. Polyakov, Phys. Lett. 59B, 82 (1975).
[2]A. A. Belavin, A. M. Polyakov, A. Schwartz, and Y. Tyupkin, Phys. Lett. 59B, 85 (1975).
[3]H. Bateman, *Partial Differential Equations of Mathematical Physics* (Dover, New York, 1932).

PHYSICAL REVIEW D VOLUME 15, NUMBER 6 15 MARCH 1977

Conformal properties of pseudoparticle configurations*

R. Jackiw, C. Nohl, and C. Rebbi
Laboratory for Nuclear Science and Department of Physics, Massachusetts Institute of Technology, Cambridge, Massachusetts 02139
(Received 14 December 1976)

The known Euclidean Yang-Mills pseudoparticle solutions with Pontryagin index n are parametrized by $5n$ constants describing the size and location of each pseudoparticle. By insisting on conformal covariance of the solutions, we show that more general solutions exist—they are parametrized by $5n+4$ constants. We further demonstrate that the additional degrees of freedom are not gauge artifacts and correspond to a new degeneracy of pseudoparticle configurations.

I. INTRODUCTION

Recently Belavin, Polyakov, Schwartz, and Tyupkin[1] have shown that in the Euclidean domain the action functional of non-Abelian gauge theories possesses local minima different from the trivial absolute minimum corresponding to vanishing field strength $F^a_{\mu\nu}$. The implications of their discovery for the structure of the quantum theory are profound.[2]

The minima of the action are characterized by an integer n, the Pontryagin index, which labels topologically inequivalent classes of field configurations. Within each class, the action is bounded below by a constant multiple of $|n|$ and the bound is saturated by values of the potentials for which $F^a_{\mu\nu} = \pm {}^*F^a_{\mu\nu}$, where the dual of $F^a_{\mu\nu}$ is ${}^*F^a_{\mu\nu} = \frac{1}{2}\epsilon_{\mu\nu\rho\sigma}F^{a\rho\sigma}$.

A self-dual field configuration with unit Pontryagin index was exhibited in Ref. 1. This solution to the field equations is often called a pseudoparticle and depends on five parameters: the four coordinates of the pseudoparticle's position and a dimensional scale which measures the pseudoparticle's "size." The question of whether the bound on the action can be saturated also for values of the Pontryagin index different from unity was very recently answered in the affirmative by Witten,[3] who discovered a set of self-dual field configurations where arbitrary numbers of pseudoparticles appear aligned on a definite axis with arbitrary separations and sizes. Soon after, 't Hooft[4] was able to enlarge again the class of known exact solutions of the field equations by exhibiting self-dual field configurations with arbitrary n, described by $5n$ parameters, which may be interpreted as positions and sizes of the n pseudoparticles. 't Hooft's solution makes use of a previously proposed ansatz which reduces the condition of self-duality to the Laplace equation for a scalar "superpotential," which can be singular.[5] The positions and residues of the singularities are the parameters that specify the solutions.

In this note we wish to investigate the behavior under conformal transformations of the class of solutions discovered by 't Hooft. Our main result is that, in order to satisfy conformal covariance, the general n-pseudoparticle solution must depend on $5n+4$ parameters, rather than the $5n$ parameters one might expect. We show that the dependence on the additional four parameters does not generally correspond to a gauge freedom; consequently they must be interpreted as having physical significance.

II. SOLUTION OF SELF-DUALITY EQUATION

We begin by summarizing the construction of 't Hooft's solution. It is convenient to represent the potentials and field strengths as matrices in the space of infinitesimal generators of the internal-symmetry group. We consider an SU(2) gauge group and set

$$A_\mu = A^a_\mu \frac{\sigma^a}{2i}, \tag{2.1a}$$

$$F_{\mu\nu} = F^a_{\mu\nu}\frac{\sigma^a}{2i} = \partial_\mu A_\nu - \partial_\nu A_\mu + [A_\mu, A_\nu], \tag{2.1b}$$

where σ^a are Pauli matrices. The action density S and the Pontryagin density ${}^*S$ are

$$S = -\tfrac{1}{2}\,\mathrm{Tr}F_{\mu\nu}F^{\mu\nu}, \qquad {}^*S = -\tfrac{1}{2}\,\mathrm{Tr}\,{}^*F_{\mu\nu}F^{\mu\nu}. \tag{2.2}$$

The Pontryagin index is given by

$$q = \frac{1}{8\pi^2}\int d^4x\, {}^*S. \tag{2.3}$$

It is useful to define a set of antisymmetric matrices $\bar{\sigma}_{\mu\nu}$ (Ref. 6)

$$\bar{\sigma}_{ij} = \frac{1}{4i}[\sigma_i, \sigma_j], \quad \bar{\sigma}_{i4} = -\tfrac{1}{2}\sigma^i, \qquad i,j = 1,2,3. \tag{2.4}$$

These matrices are anti-self-dual $\bar{\sigma}_{\mu\nu} = -{}^*\bar{\sigma}_{\mu\nu}$, and the ansatz for the gauge field is

$$A_\mu = i\bar{\sigma}_{\mu\nu}a^\nu, \tag{2.5}$$

where a_ν is a vector field that will be further specified below. We see that the three potentials A_μ^a are expressed in terms of the single potential a_ν.

The self-duality condition $F_{\mu\nu} = {}^*F_{\mu\nu}$ reduces to equations for a_ν:

$$f_{\mu\nu} \equiv \partial_\mu a_\nu - \partial_\nu a_\mu = -{}^*f_{\mu\nu}, \tag{2.6a}$$

$$\partial_\mu a^\mu + a_\mu a^\mu = 0. \tag{2.6b}$$

Equation (2.6a) may be satisfied if a_μ is derived from a scalar superpotential ρ:

$$a_\mu = \partial_\mu \ln\rho. \tag{2.7}$$

Then Eq. (2.6b) becomes

$$\frac{1}{\rho}\Box\rho = 0, \tag{2.8}$$

and the action density, which now is equal to the Pontryagin density, may be expressed in terms of ρ (see also Ref. 7):

$$S = {}^*S = -\tfrac{1}{2}\Box\Box\ln\rho. \tag{2.9}$$

In order that S be integrable, ρ must never vanish, but singularities of the form $\rho(x) \approx \lambda^2/(x-y)^2$ are acceptable because S remains regular at $x = y$.[7]

't Hooft takes for the solution of (2.8)

$$\rho(x) = 1 + \sum_{i=1}^{n} \frac{\lambda_i^2}{(x-y_i)^2}. \tag{2.10}$$

It is clear that a more general solution, which as we show below is conformally covariant [a property not shared by (2.10)], can also be given by

$$\rho(x) = \sum_{i=1}^{N} \frac{\lambda_i^2}{(x-y_i)^2}, \tag{2.11}$$

and (2.10) may be regained for the case $y_N \to \infty$, $\lambda_N \to \infty$, with $\lambda_N^2/y_N^2 = 1$ and $n = N-1$.

It is important to evaluate the Pontryagin index. From (2.3) and (2.9) it follows that

$$q = -\frac{1}{16\pi^2}\int d^4x\,\Box\Box\ln\rho. \tag{2.12}$$

In (2.9) and (2.12) ρ may be multiplied by a constant factor or by $(x-y)^2$ without changing ${}^*S$ or q.[7] Hence when ρ is of the form (2.11), a manifestly nonsingular expression may be given for ${}^*S$:

$${}^*S = -\tfrac{1}{2}\Box\Box\ln P_{2N-2}. \tag{2.13}$$

Here P_{2N-2} is polynomial in x of degree $2N-2$:

$$P_{2N-2}(x) = \Big[\sum_{i=1}^{N}\lambda_i^2 \prod_{j\neq i}(x-y_j)^2\Big]\Big(\sum_{i=1}^{N}\lambda_i^2\Big)^{-1}. \tag{2.14}$$

From Eqs. (2.12), (2.13), and (2.14) we easily find the Pontryagin index of the solution (2.11):

$$\begin{aligned} q &= -\frac{1}{16\pi^2}\int d^4x\,\Box\Box\ln P_{2N-2} \\ &= -\lim_{R\to\infty}\int\frac{R^2 d\Omega}{16\pi^2}R_\mu\partial^\mu\Box\ln(R^{2N-2}+\cdots) \\ &= N-1. \end{aligned} \tag{2.15}$$

The use of Gauss's theorem in the evaluation of (2.15) is justified, since the integrand is nonsingular. Thus we see that although the superpotential ρ depends on N position parameters y_i^μ, the field configuration has a Pontryagin index appropriate to $N-1$ pseudoparticles. If the action density displays $N-1$ maxima at all, there is no obvious simple relation between their positions and the parameters y_i^μ, even in the limit of small λ_i's.

A more direct relation between the y_i^μ's and the positions of the pseudoparticles can be obtained when ρ is of the form (2.10). It is easy to show that for (2.10) the Pontryagin index is n, and the y_i^μ's and λ_i's can be interpreted as positions and sizes of the pseudoparticles, in the sense that for small λ_i^μ's the maxima of the action density are centered about the y_i^μ's. The more general expression, Eq. (2.11), contains precisely four more relevant parameters (a common rescaling of the λ_i's does not affect the expression of A_μ) than the $5(N-1)$ coordinates and sizes of the pseudoparticles. One may ask, then, whether the additional four parameters are physical, specifying a further degeneracy of the solution when two or more pseudoparticles are put together, or whether the more restrictive form (2.10) completely exhausts the multipseudoparticle solutions. In what follows we shall answer the question; we find that our additional four parameters are truly present—they are neither gauge phantoms nor can they be incorporated in reparametrizing Eq. (2.10).

III. EFFECTS OF CONFORMAL TRANSFORMATIONS

Let us first consider the behavior of the potentials A_μ under the infinitesimal special conformal transformation

$$x^\mu \to \tilde{x}^\mu = x^\mu + 2\epsilon\cdot x x^\mu - \epsilon^\mu x^2. \tag{3.1}$$

A field φ with scale dimension d transforms covariantly if

$$\delta_\epsilon\varphi = (2\epsilon\cdot x x^\alpha - \epsilon^\alpha x^2)\partial_\alpha\varphi + 2\epsilon_\alpha x_\beta(g^{\alpha\beta}d - \Sigma^{\alpha\beta})\varphi, \tag{3.2}$$

where $\Sigma^{\alpha\beta}$ is the spin matrix of the field.[8] It is apparent from Eq. (2.5) that an infinitesimal conformal transformation of A_μ will not correspond to any simple transformation of a_ν (indeed, in general it will not be compatible with the ansatz), because the spin matrix $\Sigma^{\alpha\beta}$ in Eq. (3.2) operates on the free index μ of $\bar{\sigma}_{\mu\nu}$ and not on the index which is contracted with a_ν. However, if together with the conformal transformation we perform a gauge transformation[6]

$$\delta A_\mu = \partial_\mu \chi - [\chi, A_\mu], \tag{3.3}$$

with

$$\chi = 2i\epsilon_\alpha x_\beta \bar{\sigma}^{\alpha\beta}, \tag{3.4}$$

then, because $\bar{\sigma}_{\mu\nu}$ and $\Sigma_{\mu\nu}$ have identical commutation relations, the net effect of the spin matrix in Eq. (3.2) and the commutator of the gauge transformation in Eq. (3.3) will be to transfer the action of the spin matrix from the free index to the second index of $\bar{\sigma}_{\mu\nu}$:

$$\begin{aligned}2\epsilon_\alpha x_\beta (\Sigma^{\alpha\beta})_{\mu\nu}\, i\bar{\sigma}^{\nu\omega} a_\omega + 2\epsilon_\alpha x_\beta [i\bar{\sigma}^{\alpha\beta}, i\bar{\sigma}_{\mu\nu}] a^\nu &= -2\epsilon_\alpha x_\beta (\Sigma^{\alpha\beta})_{\nu\omega} i\bar{\sigma}_\mu{}^\omega a^\nu \\ &= i\bar{\sigma}_\mu{}^\nu 2\epsilon_\alpha x_\beta (\Sigma^{\alpha\beta})_{\nu\omega} a^\omega \end{aligned} \tag{3.5}$$

Thus the variation of A_μ generated by a special conformal transformation followed by the gauge readjustment of Eqs. (3.3), (3.4) is identical to the variation induced by

$$\delta a_\mu = \delta_c a_\mu - 2\epsilon_\mu, \tag{3.6}$$

where the second term on the right-hand side comes from the gradient part of the gauge transformation. But this is precisely the variation of a_μ that follows from Eq. (2.7) if ρ transforms as a scalar density of scale dimension $d=1$, since

$$\delta\rho = (2\epsilon\cdot x x^\alpha - \epsilon^\alpha x^2)\partial_\alpha \rho - 2\epsilon\cdot x\rho \tag{3.7}$$

implies

$$\begin{aligned}\delta a_\mu &= \delta\partial_\mu \ln\rho \\ &= \partial_\mu (2\epsilon\cdot x x^\alpha - \epsilon^\alpha x^2)\partial_\alpha \ln\rho - 2\epsilon_\mu \\ &= \delta_c a_\mu - 2\epsilon_\mu .\end{aligned} \tag{3.8}$$

Summarizing, a conformal transformation of ρ as a scalar density of dimension 1 followed by a suitable gauge readjustment induces the correct conformal transformation of A_μ.

In a finite special conformal transformation

$$x^\mu \to \bar{x}^\mu = \frac{x^\mu - a^\mu x^2}{1 - 2a\cdot x + a^2 x^2} \tag{3.9}$$

a scalar density of dimension 1 transforms as

$$\rho(x) \to \bar{\rho}(x) = \frac{1}{1-2a\cdot x + a^2 x^2}\rho(\bar{x}). \tag{3.10}$$

With a little algebra one finds that (2.11) transforms as

$$\bar{\rho}(x) = \sum_{i=1}^{N} \frac{\bar{\lambda}_i^2}{(x-\bar{y}_i)^2}, \tag{3.11}$$

where

$$\bar{\lambda}_i^2 = \frac{\lambda_i^2}{1+2a\cdot y_i + a^2 y_i^2}, \tag{3.12a}$$

$$\bar{y}_i^\mu = \frac{y_i^\mu + a^\mu y_i^2}{1+2a\cdot y_i + a^2 y_i^2}. \tag{3.12b}$$

It is trivial to verify that Poincaré transformations and dilatations also leave invariant the form of Eq. (2.11). Thus, as anticipated, the class of solutions corresponding to Eqs. (2.5), (2.7), (2.8), and (2.11) is closed under the action of the full conformal group. On the other hand, conformal transformations take the solution (2.10) into the solution (2.11).

IV. RESIDUAL GAUGE FREEDOM

The results obtained in Sec. III still leave open the possibility that some variations of the parameters in Eq. (2.11) correspond to a pure gauge transformation of the fields, so that not all of the λ_i's and y_i^μ's would be physical parameters. We investigate in this section when such gauge transformations exist.

Under an infinitesimal gauge transformation, the gauge field transforms as

$$\delta A_\mu = \partial_\mu \chi - [\chi, A_\mu], \tag{4.1}$$

where χ is an anti-Hermitian matrix-valued field. We set

$$\chi = i\bar{\sigma}_{\alpha\beta}\omega^{\alpha\beta}, \tag{4.2}$$

with $\omega_{\alpha\beta}$ antisymmetric and anti-self-dual, and look for nontrivial solutions of the equation

$$\begin{aligned}\delta A_\mu &= i\bar{\sigma}_{\alpha\beta}\partial_\mu \omega^{\alpha\beta} + [\bar{\sigma}_{\alpha\beta}, \bar{\sigma}_{\mu\nu}]\omega^{\alpha\beta} a^\nu \\ &= i\bar{\sigma}_{\mu\nu}\delta a^\nu ,\end{aligned} \tag{4.3}$$

where $\delta a^\nu = \delta(\partial^\nu \rho/\rho)$ is the variation of a^ν induced by an infinitesimal change of the parameters in Eq. (2.11).

Equation (4.3) implies

$$\bar{\sigma}_{\alpha\beta}(\partial_\mu \omega^{\alpha\beta} - a_\mu \omega^{\alpha\beta} + 4g^\beta{}_\mu \omega^{\alpha\nu} a_\nu + g^\beta{}_\mu \delta a^\alpha) = 0, \tag{4.4}$$

where we have used the fact that both $\bar{\sigma}_{\alpha\beta}$ and $\omega_{\alpha\beta}$ are anti-self-dual to express

$$\bar{\sigma}^{\alpha\nu}\omega_{\mu\alpha} = -\bar{\sigma}^\alpha{}_\mu \omega^\nu{}_\alpha - \tfrac{1}{2} g^\nu{}_\mu \bar{\sigma}^{\alpha\beta}\omega_{\alpha\beta}. \tag{4.5}$$

It is convenient to define

$$\omega_{\alpha\beta} = \rho\tilde{\omega}_{\alpha\beta} \tag{4.6a}$$

and

$$4\omega_{\alpha\nu} a^\nu + \delta a_\alpha = \rho b_\alpha . \tag{4.6b}$$

Then Eq. (4.4) becomes

$$\bar{\sigma}_{\alpha\beta}(\partial_\mu \tilde{\omega}^{\alpha\beta} + g^\beta{}_\mu b^\alpha) = 0. \tag{4.7}$$

Multiplying with $\bar{\sigma}_{\alpha'\beta'}$ and taking the trace we find

$$4\partial_\mu \bar{\omega}_{\alpha\beta} + g_{\beta\mu} b_\alpha - g_{\alpha\mu} b_\beta + \epsilon_{\alpha\beta\mu\gamma} b^\gamma = 0. \tag{4.8}$$

This last equation poses formidable constraints on the possible forms of $\bar{\omega}_{\alpha\beta}$ and b_α. In particular, by using the symmetry properties of the various terms under permutations of the indices, one can show that $\bar{\omega}_{\alpha\beta}$ must obey an equation where b_α does not appear:

$$2\partial_\mu \bar{\omega}_{\alpha\beta} + \partial_\alpha \bar{\omega}_{\mu\beta} + \partial_\beta \bar{\omega}_{\alpha\mu} - g_{\alpha\mu} \partial_\gamma \bar{\omega}^\gamma{}_\beta + g_{\beta\mu} \partial_\gamma \bar{\omega}^\gamma{}_\alpha = 0. \tag{4.9}$$

(Notice the similarity to the equation satisfied by a Killing vector.) Once Eq. (4.9) is satisfied, Eq. (4.8) reduces to the condition

$$b_\alpha + \tfrac{4}{3} \partial^\beta \bar{\omega}_{\alpha\beta} = 0. \tag{4.10}$$

From the integrability conditions which follow from Eq. (4.9), after nontrivial algebra, one proves that the most general solution to Eq. (4.9) is

$$\bar{\omega}_{\alpha\beta}(x) = 2x_\alpha A_{\beta\gamma} x^\gamma - 2x_\beta A_{\alpha\gamma} x^\gamma + x^2 A_{\alpha\beta} + B_\alpha x_\beta - B_\beta x_\alpha - \epsilon_{\alpha\beta\gamma\delta} B^\gamma x^\delta + C_{\alpha\beta}, \tag{4.11}$$

with constant $A_{\alpha\beta}$, B_α, and $C_{\alpha\beta}$. Furthermore $A_{\alpha\beta}$ ($C_{\alpha\beta}$) must be anti-symmetric and self-dual (anti-self-dual).

Equations (4.11) and (4.6) now give

$$4(2x_\alpha A_{\beta\gamma} x^\gamma - 2x_\beta A_{\alpha\gamma} x^\gamma + x^2 A_{\alpha\beta} + B_\alpha x_\beta - B_\beta x_\alpha - \epsilon_{\alpha\beta\gamma\delta} B^\gamma x^\delta + C_{\alpha\beta}) \partial^\beta \rho(x) + \delta a_\alpha(x) - 4\rho(x)(2A_{\alpha\beta} x^\beta - B_\alpha) = 0. \tag{4.12}$$

The left-hand side of this equation develops singularities when x^μ approaches any of the y_i^μ. Requiring that the residues vanish, we find the conditions

$$2y_{i\alpha} A_{\beta\gamma} y_i^\gamma - 2y_{i\beta} A_{\alpha\gamma} y_i^\gamma + y_i^2 A_{\alpha\beta} + B_\alpha y_{i\beta} - B_\beta y_{i\alpha} - \epsilon_{\alpha\beta\gamma\delta} B^\gamma y_i^\delta + C_{\alpha\beta} = 0, \quad i = 1, \ldots, N, \tag{4.13}$$

and

$$\delta y_{i\alpha} + \lambda_i^2 (4A_{\alpha\gamma} y_i^\gamma - 2B_\alpha) = 0, \quad i = 1, \ldots, N. \tag{4.14}$$

Equations (4.13) represent a constraint on the positions of the singularities in ρ for the existence of a nontrivial gauge transformation preserving the ansatz of Eq. (2.5). For a general configuration of singularities the set of Eqs. (4.13) ($3N$ homogeneous linear equations for the ten independent components of $A_{\alpha\beta}, B_\alpha, C_{\alpha\beta}$) will admit no nonzero solutions and the ansatz completely fixes the gauge. If Eqs. (4.13) are compatible, then Eq. (4.14) specifies the changes of the position parameters that can be achieved with a gauge transformation.

A simple way to understand the geometrical meaning of conditions (4.13) and (4.14) is to observe that both equations are invariant under the fifteen-parameter group of general conformal transformations. With a rather straightforward computation one verifies that the conformally transformed variables $\bar{y}_i^\mu$ and $\delta \bar{y}_i^\mu$ satisfy equations analogous to (4.13) and (4.14), where $A_{\alpha\beta}$, B_α, and $C_{\alpha\beta}$ are replaced by new tensors $\bar{A}_{\alpha\beta}$, $\bar{B}_\alpha$, and $\bar{C}_{\alpha\beta}$. If Eqs. (4.13) admit a nontrivial solution, then by a suitable conformal transformation we can make $\bar{A}_{\alpha\beta} = \bar{C}_{\alpha\beta} = 0$, and the equations reduce to

$$\bar{B}_\alpha \bar{y}_{i\beta} - \bar{B}_\beta \bar{y}_{i\alpha} - \epsilon_{\alpha\beta\gamma\delta} \bar{B}^\gamma \bar{y}_i^\delta = 0, \quad i = 1, \ldots, N. \tag{4.15}$$

This implies

$$\bar{y}_{i\alpha} = K_i \bar{B}_\alpha, \quad i = 1, \ldots, N \tag{4.16}$$

and we see that the images $\bar{y}_i^\mu$ of the points y_i^μ lie on a straight line through the origin. We conclude that a nontrivial gauge transformation preserving the ansatz of Eq. (2.5) may exist only if the singularities in Eq. (2.11) lie on a circle (or on a straight line, as a circle through the point at infinity). If Eq. (4.16) is satisfied, the condition

$$\delta \bar{y}_{i\alpha} = 2\lambda_i^2 \bar{B}_\alpha, \quad i = 1, \ldots, N \tag{4.17}$$

further specifies that the effect of the gauge transformation may only be to move the images $\bar{y}_i^\mu$ on the same straight line, and therefore the points y_i^μ on the circle they determine, in a one-parameter group of transformations. If the points are on a straight line it is easy to prove that the gauge transformation actually exists, and by conformal covariance this extends also to the case where the points lie on a circle.

Summarizing, if the N points y_i^μ in Eq. (2.11) do not lie on a circle, the ansatz of Eqs. (2.5) and (2.7) completely fixes the gauge, and all the $5n+4$ parameters that specify a field configuration with Pontryagin index $n = N-1$ are relevant. For $N = 3$ (three points always lie on a circle) the two-pseudoparticle field configuration is characterized by thirteen physical parameters, since one of the fourteen parameters corresponds to a gauge transformation which moves the pseudoparticle around the circle. If $N = 2$ there is a threefold variety of circles passing through y_1^μ and y_2^μ, so that only five of the nine parameters in Eq. (2.11) are physical; thus a single *pseudoparticle is characterized* by position and size only.

It is intriguing that when two or more pseudoparticles are put together more parameters are necessary to specify the field configuration than the positions and sizes of the pseudoparticles. The results we have obtained call for a physical explanation of this additional degeneracy; one wonders whether there exist still further solutions to the self-duality equations beyond those discussed here. These problems are currently under investigation.

Note added in proof. Recently it has been possible to show that for an SU(2) gauge theory the general n-pseudoparticle solution is specified by at least $8n-3$ parameters.[9] The interpretation of the additional $3n-3$ parameters is that they describe the relative orientations of the pseudoparticles in group space.

---

*This work is supported in part through funds provided by ERDA under Contract No. EY-76-C-02-3079.

[1]A. Belavin, A. Polyakov, A. Schwartz, and Y. Tyupkin, Phys. Lett. 59B, 85 (1975).

[2]A. Polyakov, Phys. Lett. 59B, 82 (1975); G. 't Hooft, Phys. Rev. Lett. 37, 8 (1976); R. Jackiw and C. Rebbi, *ibid.* 37, 172 (1976); C. Callan, R. Dashen, and D. Gross, Phys. Lett. 63B, 334 (1976).

[3]E. Witten, Phys. Rev. Lett. 38, 121 (1977).

[4]G. 't Hooft (unpublished). We thank E. Witten for explaining 't Hooft's solution to us.

[5]F. Wilczek, Princeton University report (unpublished); F. Corrigan and D. Fairlie, Durham University report (unpublished).

[6]We use the notations and conventions adopted in our investigation of the conformal properties of the single pseudoparticle solution; R. Jackiw and C. Rebbi, Phys. Rev. D 14, 517 (1976). The anti-self-dual matrix $\bar{\sigma}_{\mu\nu}$ is appropriate to the self-dual field configurations which are discussed in the text. A parallel discussion may be given for anti-self-dual field configurations; for these the self-dual matrix $\sigma_{\mu\nu}$ is appropriate ($\sigma_{ij}=\bar{\sigma}_{ij}$, $\sigma_{i4}=-\bar{\sigma}_{i4}$).

[7]The equality $-\frac{1}{2}\mathrm{Tr}F^{\mu\nu}F_{\mu\nu}=-\frac{1}{2}\mathrm{Tr}\ {}^*F^{\mu\nu}F_{\mu\nu}=-\frac{1}{2}\Box\Box\ln\rho$ holds away from the singularities of ρ. At a singularity $\lambda^2/(x-y)^2$ the action density is regular but $\Box\Box\ln\rho$ acquires a $\delta^4(x-y)$ contribution. We avoid this difficulty by excluding from the integration of the action and Pontryagin densities infinitesimal neighborhoods around each singularity — a legitimate procedure since S and ${}^*S$ are nonsingular and continuous there. More simply we may define $\Box\Box\ln(x-y)^2=0$, a procedure we adopt henceforth.

[8]For a summary of the conformal group see S. Treiman, R. Jackiw, and D. Gross, *Lectures on Current Algebra and Its Applications* (Princeton Univ. Press, Princeton, New Jersey, 1972), p. 97.

[9]R. Jackiw and C. Rebbi, MIT report (unpublished); A. Schwartz, JINR report (unpublished).

Volume 65A, number 3 PHYSICS LETTERS 6 March 1978

CONSTRUCTION OF INSTANTONS

M.F. ATIYAH and N.J. HITCHIN
Mathematical Institute, Oxford University, UK

and

V.G. DRINFELD and Yu.I. MANIN
Steklov Institute of Mathematics, Academy of Sciences of the USSR, Moscow, USSR

Received 16 December 1977

A complete construction, involving only linear algebra, is given for all self-dual euclidean Yang–Mills fields.

In ref. [1] it was shown that "instantons", i.e. self-dual solutions of the Yang–Mills equations in the compactified euclidean 4-space S^4, corresponded in a precise way to certain real algebraic bundles on the complex projective 3-space $P_3(C)$. Using this correspondence and algebro-geometric results of Horrocks and Barth we shall show that, for the compact classical groups, all instantons have an essentially unique description in terms of linear algebra. For simplicity we will give details only for the case of SU(2), but we shall indicate briefly how to modify this for general G.

The construction starts from a complex linear map $A(z)$: $W \to V$, where $\dim W = k$, $\dim V = 2k+2$, z stands for a complex 4-vector (z_1, z_2, z_3, z_4) and $A(z)$ is itself linear in z, i.e. $A(z) = \Sigma_{i=1}^4 A_i z_i$. We assume V has a fixed bilinear form denoted by (v_1, v_2), which is skew and non-degenerate. For any subspace $U \subset V$ let U^0 be its annihilator, i.e. it consists of those v such that $(u, v) = 0$ for all $u \in U$. We now impose the following condition on $A(z)$:

for $z \neq 0$, $U_z = A(z) W \subset V$ has dimension k and is isotropic, i.e. $U_z \subset (U_z)^0$. (1)

Given condition (1) we take E_z to be the quotient space $(U_z)^0/U_z$. Since $\dim U_z = k$, $\dim (U_z)^0 = k+2$ and $\dim E_z = 2$. Moreover E_z inherits a non-degenerate skew form, and $E_z = E_{\lambda z}$ for all non-zero scalar λ. Hence the family of E_z defines an algebraic vector bundle over $P_3(C)$ with group SL(2, C). This construction was introduced and studied by Horrocks. The jumping lines of E [2,§2] are lines in $P_3(C)$ joining points (z) and (ξ) such that $(U_z)^0 \cap U_\xi \neq 0$.

To get a bundle corresponding to an SU(2)-instanton we need to impose reality conditions on $A(z)$ as explained in ref. [2]. We fix anti-linear maps σ on W, V, C^4 so that $\sigma^2 = 1$ on W, $\sigma^2 = -1$ on C^4, V. On C^4 define σ explicitly by $\sigma(z_1, z_2, z_3, z_4) = (-\bar{z}_2, \bar{z}_1, -\bar{z}_4, \bar{z}_3)$, and on V we require that σ be compatible with the skew form, i.e. $(\sigma v_1, \sigma v_2) = \overline{(v_1, v_2)}$. This gives V a hermitian form $\langle\,\rangle$ defined by $\langle v_1, \sigma v_2\rangle = (v_1, v_2)$, and we require this to be *positive definite*. Thus, as in ref. [2], C^4 can be identified with the 2-dimensional quarternion space H^2, while V can be identified with H^{k+1} (with its standard metric), σ being in both cases multiplication by the quaternion j.

Our reality condition is now that $A(z)$ be compatible with σ, i.e.

$$\sigma[A(z)w] = A(\sigma z)\sigma w \,. \tag{2}$$

The definition of the hermitian form shows that the orthogonal space $U^\perp$ of $U \subset V$ is $(\sigma U)^0$. Now condition (2) implies that $\sigma U_z = U_{\sigma z}$ (with U_z as in condition (1)). Hence $(U_z)^0 = (U_{\sigma z})^\perp$ and so, by the positive definiteness of the hermitian form, $(U_z)^0 \cap U_{\sigma z} = 0$. This shows that the "real" lines of $P_3(C)$ (i.e. lines joining (z) to (σz) – corresponding as in ref. [2] to points in the real 4-sphere S^4) are never jumping lines. Thus E satisfies the two conditions for an SU(2)-instanton described in ref. [2].

When $A(z)$ satisfies conditions (1) and (2) the resulting SU(2)-instanton, viewed differential-geometrically as a vector bundle with SU(2)-connection, can be constructed as follows. If $x \in S^4$ is represented by the line $(z, \sigma z)$ of $P_3(C)$, take F_x to be the 2-dimensional subspace of V which is orthogonal to both U_z and $U_{\sigma z}$ (note that, by condition (1), $U_z \subset U_z^0 = U_{\sigma z}^{\perp}$). As x varies we obtain a vector bundle F over S^4 as sub-bundle of the product $S^4 \times V$. We give F the connection induced, by orthogonal projection, from this product. It follows from the geometrical correspondence of ref. [2] that this connection is self-dual and the instanton number (or Pontrjagin index) is k. A direct verification is also not hard.

Our main assertion is that all SU(2)-instantons arise in this way and that gauge equivalence corresponds precisely to the obvious linear equivalence, i.e. replacing $A(z)$ by $PA(z)Q$, where P, Q act on V, W preserving their structures, i.e. $P \in \mathrm{Sp}(k+1)$, $Q \in \mathrm{GL}(k, \mathrm{R})$. A quick count of parameters (first made by Barth) yields $8k-3$ as the number of effective parameters of our construction.

In view of known results [1,7] this is encouraging but not conclusive evidence that our construction yields all instantons. For this we need a result of Barth [3] characterizing SL(2, C) bundles E on $P_3(C)$ which arise from the Horrocks construction (i.e. from $A(z)$ satisfying condition (1)). The essential condition is the vanishing of the sheaf cohomology group $H^1(P_3, E(-2))$, where $E(-2)$ denotes the sheaf of holomorphic sections of $E \otimes L^{-2}$ and L is the Hopf line-bundle over P_3. If E corresponds to a self-dual SU(2)-bundle F on S^4, then $H^1(P_3, E(-2))$ can be identified with the space of solutions of the linear differential equation $(\Delta + \frac{1}{6}R)u = 0$, where u is a section of F, Δ is the covariant laplacian of F and $R > 0$ is the scalar curvature of S^4. (For the case when E is trivial see Penrose [6] and [2,§4].) Since $\Delta \geqslant 0$ as an operator this equation has no global solutions, other than zero, and so Barth's criterion is fulfilled.

This vanishing theorem is analogous to the one used in refs. [1,7] in computing the dimension of the moduli space. It should be emphasized that vector bundles on $P_3(C)$ exist for which neither of these vanishing theorems hold (Hartshorne, unpublished). Fortunately the extra reality constraints arising from instantons exclude the unpleasant cases and simplify the problem, justifying the speculation in ref. [2,§5].

If E arises by the Horrocks construction from $A(z)$: $W \to V$, then W, V can be naturally identified with $H^2(P_3, E(-1))$ and $H^2(P_3, E(-4) \otimes T)$ respective[ly] (where T denotes the tangent bundle), while $A(z)$ is es[sentially] multiplication by $\partial/\partial z$, regarded as a section [of] $T(-1)$. If $A(z)$ also satisfies condition (2) so that E cor[responds] to a self-dual SU(2) bundle F over S^4, then W can be identified with the solutions of the Dirac equation for positive spinors coupled to F. The space V con[sists] of sections of E whose covariant derivatives are sums of products of solutions of the above Dirac equation and solutions of the "twistor equation". This show[s] that $A(z)$ can be reconstructed in an essentially unique way from F.

As an example, we may consider the 't Hooft instanton, depending on k real parameters λ_i and k points y_i in R^4. Writing $z \in C^4$ as a pair of quaternions (p, q), condition (2) implies that $A(z)$ is given by a $(k+1) \times k$ matrix of quaternions, which for the 't Hooft solution is

$$\begin{pmatrix} \lambda_1 p & \cdot & \cdot & \lambda_k p \\ y_1 p - q & \cdot & & \\ & & \cdot & \\ & & & y_k p - q \end{pmatrix},$$

where y_i is represented as a quaternion and $y_i p$ denote[s] quaternion product.

The potential A_μ may be obtained from the general $A(z)$ by a simple algebraic process which involves inverting a quaternionic matrix [4]. The diagonal nature of the above matrix leads to the particularly simple form of the 't Hooft potentials.

We now indicate how to modify the construction to deal with other groups than SU(2) = Sp(1). First observe that the construction works for any symplectic group Sp(n), the only change being that we now take $\dim V = 2k + 2n$. For the orthogonal group O(n) we take $\dim V = 2k + n$ to have a symmetric rather than a skew form and W becomes symplectic (i.e. $\sigma^2 = -1$). Since any compact Lie group G has a faithful representation as a subgroup of O(n) we can describe G-instantons as O(n)-instantons with additional structure. For $G = \mathrm{SU}(n) \subset \mathrm{O}(2n)$ this amounts to imposing compatibility with an orthogonal J with $J^2 = -1$ throughout. Note that the symplectic case can also be treated this way by the embedding $\mathrm{Sp}(n) \subset \mathrm{O}(4n)$.

In conclusion we should point out that although the

construction of G-instantons, a problem in non-linear differential equations, has now been reduced to linear algebra, certain points remain to be clarified. In particular, the geometric and topological nature of the space of moduli has still to be extracted from the linear algebra.

More detail on the results described here can be found in refs. [4,5] and will be developed further elsewhere.

We gratefully acknowledge our debt to G. Horrocks and W. Barth whose work on vector bundles has proved quite fundamental.

References

[1] M.F. Atiyah, N.J. Hitchin and I.M. Singer, Proc. Nat. Acad. Sci. USA 74 (1977).
[2] M.F. Atiyah and R.S. Ward, Commun. Math. Phys. 55 (1977) 117.
[3] W. Barth, to appear.
[4] V.G. Drinfeld and Yu.I. Manin, Funkts. Anal. Prilož., to appear.
[5] V.G. Drinfeld and Yu.I. Manin, Usp. Math. Nauk, to appear.
[6] R. Penrose, Rept. Math. Phys., to appear.
[7] A.S. Schwarz, Phys. Lett. 67B (1977) 172.

PHYSICAL REVIEW D VOLUME 18, NUMBER 6 15 SEPTEMBER 1978

General self-dual Yang-Mills solutions

Norman H. Christ and Erick J. Weinberg
Department of Physics, Columbia University, New York, New York

Nancy K. Stanton
Department of Mathematics, Columbia University, New York, New York

(Received 5 May 1978)

The recent work of Atiyah, Hitchin, Drinfeld, and Manin is used to discuss self-dual Yang-Mills solutions for the compact gauge groups O(n), SU(n), and Sp(n). It is shown that the resulting solutions contain the correct number of parameters for all values of the topological charge. Although explicit construction of a general self-dual field requires the solution of a finite-dimensional, nonlinear matrix equation, we show that for widely separated instantons this equation can be solved perturbatively, providing a systematic expansion about the dilute-gas limit and a physical interpretation of the independent parameters in this limit. Further, closed-form expressions can be obtained for the general SU(2) solutions with topological charge 2 or 3. Finally, explicit isospin-1/2 and isospin-1 propagators are derived for a massless scalar field in the presence of the general self-dual SU(2) solution.

I. INTRODUCTION

A significant advance in the study of self-dual Yang-Mills fields has recently been made by Atiyah, Hitchin, Drinfeld, and Manin.[1,2] These authors apply a construction due to Horrocks and Barth[3] to the complex formulation of self-dual Yang-Mills fields invented by Ward.[4,5] They reduce the nonlinear partial differential equations of self-duality to a nonlinear equation for a finite-dimensional matrix. The resulting SU(2) solutions contain the correct number of independent parameters and can be generalized in a straightforward way to Sp(n), SU(n), and O(n). Finally Atiyah, Hitchin, Drinfeld, and Manin argue that all self-dual fields can be obtained by this procedure.

In this paper we present the construction of the above authors as a remarkably simple *Ansatz* and show directly that the resulting gauge field is self-dual. This is done for SU(2) in Sec. II and explicitly generalized to O(n), SU(n), and Sp(n) in Sec. III. Next, in Sec. IV, we introduce a canonical form for the matrices appearing in the *Ansatz* and determine the number of independent parameters present in the O(n), SU(n), and Sp(n) solutions for arbitrary topological charge. The results agree with those of the previous analysis[6] based on the Atiyah-Singer index theorem. In Sec. V the above construction is applied to obtain both the familiar SU(2) 't Hooft k-instanton solution and an explicit, systematic expansion of the general SU(2) solution with topological charge k in the limit of large instanton separations. These results are generalized to O(n), SU(n), and Sp(n) in Sec. VI. Finally, in the concluding section we discuss the general self-dual SU(2) solutions for $k=2$ and 3 and present the isospin-$\frac{1}{2}$ and isospin-1 propagators for a scalar field in the presence of the general SU(2) solutions.

II. SU(2) CONSTRUCTION

We wish to construct a self-dual Yang-Mills field A_μ with topological charge

$$k=-\frac{1}{16\pi^2}\int d^4x\,\mathrm{tr}(F_{\mu\nu}F_{\mu\nu}) \tag{2.1}$$

and containing $8k-3$ independent parameters. Here the Yang-Mills field strength

$$F_{\mu\nu}=\partial_\mu A_\nu-\partial_\nu A_\mu+[A_\mu,A_\nu], \tag{2.2}$$

and the gauge field A_μ is viewed as a 2×2, anti-Hermitian, traceless matrix.

Such solutions can be constructed from a $(k+1)\times k$ matrix $M(x)$ made up of quarternions. Thus an element M_{ab} of the matrix M can be written in terms of four real numbers M^μ_{ab}, $0\leq\mu\leq 3$, and the unit quaternions $\hat{i}$, $\hat{j}$, and $\hat{k}$:

$$M_{ab}=M^0_{ab}+M^1_{ab}\hat{i}+M^2_{ab}\hat{j}+M^3_{ab}\hat{k}. \tag{2.3}$$

The quantities $\hat{i}$, $\hat{j}$, and $\hat{k}$ anticommute and obey

$$\hat{i}^2=\hat{j}^2=\hat{k}^2=-1,\quad \hat{i}\times\hat{j}=\hat{k}.$$

Occasionally it will be helpful to treat each quarternionic element M_{ab} as a 2×2 matrix

$$M_{ab}=\sum_{\mu=0}^{3}\sigma_\mu M^\mu_{ab}, \tag{2.5}$$

where σ_0 is the 2×2 identity matrix and $\sigma_j=-i\tau^j$, $1\leq j\leq 3$, τ^j being the jth Pauli matrix. The matrix M is chosen to be linear in x,

$$M=B-Cx, \tag{2.6}$$

where B and C are constant $(k+1)\times k$ quaternionic matrices of rank k (Ref. 7) and x is the quaternion

$$x = x_0 + x_1\hat{i} + x_2\hat{j} + x_3\hat{k} \tag{2.7}$$

formed from the four Euclidean position coordinates x_μ. Finally $M(x)$ is assumed to satisfy the nonlinear requirement

$$M^\dagger(x)M(x) = R\,, \tag{2.8}$$

a real, invertible $k\times k$ matrix. Here $M^\dagger$ is the transpose of the quaternionic conjugate of M: $(M^\dagger)^0_{ij} = M^0_{ji}$, $(M^\dagger)^a_{ij} = -M^a_{ji}$, $1\leq a\leq 3$.

In order to construct a self-dual gauge field $A_\mu(x)$ from $M(x)$ it is necessary to find a $(k+1)$-dimensional column vector $N(x)$ obeying

$$N^\dagger(x)M(x) = 0\,, \tag{2.9a}$$

$$N^\dagger(x)N(x) = 1\,. \tag{2.9b}$$

The linear equation (2.9a) can be viewed as k quaternionic conditions on the $k+1$ elements of N. Thus a solution $N(x)$ to Eq. (2.9a) can always be found and the requirement (2.9b) simply fixes its normalization. The gauge field $A_\mu(x)$ is now defined by

$$A_\mu(x) = N^\dagger(x)\partial_\mu N(x)\,, \tag{2.10}$$

where $A_\mu(x)$ can be identified as a 2×2 matrix if the quaternion on the right-hand side of Eq. (2.10) is considered to be a 2×2 matrix as is done in Eq. (2.5). Differentiating the normalization condition (2.9b), one easily sees that $A_\mu(x)$ is anti-Hermitian:

$$A_\mu(x)^\dagger = \partial_\mu N^\dagger(x)N(x) = -N^\dagger(x)\partial_\mu N(x)\,. \tag{2.11}$$

Furthermore, any anti-Hermitian 2×2 matrix which has the quaternionic form (2.5) must necessarily be traceless.

Thus Eq. (2.10) defines an SU(2) gauge potential. We will now show that the resulting field strength $F_{\mu\nu}$ is self-dual. Substituting into Eq. (2.2) one obtains

$$F_{\mu\nu} = (\partial_\mu N^\dagger)(\partial_\nu N) + (N^\dagger\partial_\mu N)(N^\dagger\partial_\nu N) - [\mu \leftrightarrow \nu] \tag{2.12}$$

or, using the right-hand part of Eq. (2.11) twice,

$$F_{\mu\nu} = \partial_\mu N^\dagger\{I - NN^\dagger\}\partial_\nu N - [\mu \leftrightarrow \nu]\,. \tag{2.13}$$

The quantity in curly brackets in Eq. (2.13) is simply the projection operator onto the k-dimensional quaternionic subspace orthogonal to N. Using Eqs. (2.9a) and (2.8) this can be written as the $(k+1)\times(k+1)$ matrix

$$I - NN^\dagger = M(x)R^{-1}(x)M^\dagger(x)\,, \tag{2.14}$$

where R^{-1} is the $k\times k$ matrix inverse of the real matrix R in Eq. (2.8). [It is easy to see that the matrix (2.14) is equal to its square.] Thus, using

$$(\partial_\mu N^\dagger)M = -N^\dagger\partial_\mu M \tag{2.15}$$

[a consequence of Eq. (2.9a)], $F_{\mu\nu}$ can be written

$$\begin{aligned} F_{\mu\nu} &= N^\dagger(\partial_\mu M)R^{-1}(\partial_\nu M^\dagger)N - [\mu \leftrightarrow \nu] \\ &= N^\dagger C(\sigma_\mu R^{-1}\sigma^\dagger_\nu - \sigma_\nu R^{-1}\sigma^\dagger_\mu)C^\dagger N\,, \end{aligned} \tag{2.16}$$

in the notation of Eqs. (2.5) and (2.6). Since the elements of the matrix R^{-1} are real, they commute with the "quaternions" σ_μ and $\sigma^\dagger_\nu$, and $F_{\mu\nu}$ explicitly contains the self-dual combination

$$\sigma_\mu\sigma^\dagger_\nu - \sigma_\nu\sigma^\dagger_\mu = 2i\tau^a\eta_{a\mu\nu}\,, \tag{2.17}$$

where $\eta_{a\mu\nu}$, $1\leq a\leq 3$ are similar to the three self-dual matrices defined by 't Hooft.[8]

As will be shown in Secs. IV and V, the solution (2.10) possesses topological charge k and contains precisely $8k-3$ independent parameters. Thus this surprisingly simple procedure has produced a general self-dual solution. Of course to obtain explicit solutions one must find matrices M obeying the quadratic condition (2.9a). This can be done easily for $k=1$, 2, and 3, and yields genuinely new $k=3$ solutions. In Sec. V we show how the familiar SU(2) solutions of 't Hooft arise using this procedure and derive a method for obtaining approximate solutions in the limit of small, widely separated instantons for arbitrary k.

III. O(n), SU(n), AND Sp(n) CONSTRUCTIONS

Let us now generalize the *Ansatz* of the previous section to give self-dual O(n) solutions. We will then impose restrictions on that *Ansatz* to reduce our O($2n$) solutions to SU(n) solutions and further conditions to specialize O($4n$) solutions to Sp(n). The Sp(1) case will then be the simple *Ansatz* described above [recall Sp(1) = SU(2)].

O(n). Let $M(x)$ be a $(4k+n)\times k$ matrix of quaternions linear in x,

$$M(x) = B - Cx\,, \tag{3.1}$$

where B and C are both $(4k+n)\times k$ constant quaternionic matrices of rank k. We require that the product $M^\dagger(x)qM(x)$ be a real $k\times k$ matrix for any choice of the quaternion q. Thus we can define a $k\times k$ quaternionic matrix R by the equation

$$\sum_{l=1}^{4k+n}[M(x)^\dagger]_{jl}\,qM(x)_{li} = 4\,\mathrm{Re}(R_{ij}q)\,, \tag{3.2}$$

valid for any quaternion q. The quantity $\mathrm{Re}(R_{ij}q)$ is the real part of the enclosed quaternion $R_{ij}q$, i.e., one-half of the trace of the quaternion $R_{ij}q$ when viewed as a 2×2 matrix. It follows from the definition (3.2) that $R = R^\dagger$; in addition, R is required to be invertible.

Next we introduce n real $(4k+n)$-dimensional vectors N_i, $1 \le i \le n$, with components N_{ji}, $1 \le j \le 4k+n$. These vectors are chosen to be orthonormal,

$$\sum_{l=1}^{4k+n} N_{li} N_{lj} = \delta_{ij}, \tag{3.3a}$$

and to satisfy the $4k$ real conditions

$$\sum_{l=1}^{4k+n} N_{li} M_{lj} = 0, \quad 1 \le j \le k. \tag{3.3b}$$

The resulting components N_{ij} can be viewed as a $(4k+n)\times n$ matrix obeying

$$N^t N = I, \tag{3.4a}$$

$$N^t M = 0, \tag{3.4b}$$

where I represents the $n \times n$ unit matrix. Finally the $O(n)$ vector potential $A_\mu(x)$ is defined by

$$A_\mu(x) = N^t \partial_\mu N. \tag{3.5}$$

The orthogonality condition (3.4a) directly implies that $A_\mu(x)$ is an antisymmetric matrix and therefore properly lies in the adjoint representation of O(n).

Just as in Sec II, we can demonstrate the self-duality of A_μ by studying the field strength $F_{\mu\nu}$ which, as before, can be written

$$F_{\mu\nu} = \partial_\mu N^t \{I - NN^t\} \partial_\nu N - [\mu \leftrightarrow \nu]. \tag{3.6}$$

Again the $(4k+n)\times(4k+n)$ matrix in curly brackets can be viewed as a projection operator onto the $4k$-dimensional subspace orthogonal to the n columns of N. This subspace and consequently the projection operator $I - NN^t$ can be directly constructed from the k independent quaternionic columns of M provided their normalization, Eq. (3.2), is taken into account. The result is

$$(I - NN^t)_{ll'} = \sum_{i,j=1}^{k} \mathrm{Re}(M^*_{l'i} R^{-1}_{ij} M_{lj}). \tag{3.7}$$

[Equation (3.2) can be used to show that the right-hand side of Eq. (3.7) is equal to its square. This calculation is most easily done by using the 2×2 matrix form for the quaternions.] In a more compact notation,

$$I - NN^t = \mathrm{Re}(M^* R^{-1} M^t), \tag{3.8}$$

where M^* is the $(4k+n)\times k$ matrix whose elements are the quaternionic conjugates of those in M. [Note that with noncommuting quaternions $(AB)^t \ne B^t A^t$.] Substituting the projection operator (3.8) into Eq. (3.6) for $F_{\mu\nu}$ and using $(\partial_\mu N^t)M = -N^t(\partial_\mu M)$, one obtains

$$\begin{aligned} F_{\mu\nu} &= N^t\,\mathrm{Re}(\partial_\nu M^* R^{-1} \partial_\mu M^t) N - [\mu \leftrightarrow \nu] \\ &= N^t\,\mathrm{Re}(\sigma_\nu^\dagger C^* R^{-1} C^t \sigma_\mu) N - [\mu \leftrightarrow \nu] \\ &= N^t\,\mathrm{Re}[C^* R^{-1} C^t (\sigma_\mu \sigma_\nu^\dagger - \sigma_\nu \sigma_\mu^\dagger)] N, \end{aligned} \tag{3.9}$$

which from Eq. (2.17) is explicitly self-dual. We have thus obtained a self-dual O(n) solution. It will be shown in Sec. IV that it in fact contains the right number of independent parameters.

SU(n). We can construct self-dual SU(n) solutions by beginning with the O($2n$) solutions found above and imposing a constraint. This constraint is designed to guarantee that the effect of the $2n \times 2n$ matrix A_μ is the same as that of an $n\times n$ Hermitian matrix A_μ^C acting on n-dimensional complex vectors. The real and imaginary parts of these complex vectors form the $2n$ real dimensions on which the original matrix A_μ operates.

Thus one begins with a $(4k+2n)\times k$ matrix of quaternions $M(x)$ linear in x and obeying Eq. (3.2). In addition, one requires

$$JM(x) = \hat{i} M(x), \tag{3.10}$$

where J is a $(4k+2n)\times(4k+2n)$ matrix made up of four $(2k+n)\times(2k+n)$ blocks

$$J = \begin{pmatrix} 0 & +I \\ -I & 0 \end{pmatrix}. \tag{3.11}$$

The condition (3.11) requires that the lower $2k+n$ rows of M be simply $\hat{i}$ times the upper $2k+n$ rows,

$$M(x) = \begin{pmatrix} M^C(x) \\ \hat{i} M^C(x) \end{pmatrix}, \tag{3.12}$$

where $M^C(x)$ is a $(2k+n)\times k$ quaternionic matrix.

Using the special form (3.12), the previous equations for the O(n) case can be significantly simplified. In terms of the matrix M^C, the condition (3.2) becomes

$$\sum_{l=1}^{2k+n} (M^C(x)^t)_{jl} (q - \hat{i} q \hat{i}) M^C_{li} = 4\,\mathrm{Re}(R_{ij} q). \tag{3.13}$$

Since the left-hand side of this equation automatically vanishes for $q = \hat{j}$ or $\hat{k}$, the number of conditions is reduced by a factor of 2, and the quaternionic matrix R_{ij} can contain only real and $\hat{i}$ parts and is thus essentially complex. Similarly, the condition (3.4b) on the $(4k+2n)$-dimensional real vector N becomes

$$\sum_{l=1}^{2k+n} (N_{l,a} + N_{l+2k+n,a}\hat{i}) M^C_{lb} = 0. \tag{3.14}$$

Given a real solution $N_{l,a}$ of Eq. (3.14) for $1 \le a \le n$, we can obtain a second solution, defined as $N_{l,a+n}$, by multiplying Eq. (3.14) on the left by $\hat{i}$ so that

$$\left.\begin{aligned} N_{l,a+n} &= -N_{l+2k+n,a} \\ N_{l+2k+n,a+n} &= N_{l,a} \end{aligned}\right\} \quad 1 \le l \le 2k+n\,. \tag{3.15}$$

If the original $N_{l,a}$, $1 \le a \le n$ are chosen so that the n complex vectors $(N_{l,a}+N_{l+2k+n,a}\hat{i})$, $1 \le a \le n$, are independent with respect to complex multiplication then the resulting $2n$ real vectors $N_{l,a}$, $1 \le a \le 2n$, will be linearly independent with respect to real multiplication. With this choice, the real $(4k+2n)\times 2n$ matrix N_{la} can be represented by a $(2k+n)\times n$ complex matrix

$$N^C_{l,a} = N_{l,a} - N_{l+2k+n,a}\hat{i}, \quad \begin{cases} 1 \le l \le 2k+n, \\ 1 \le a \le n, \end{cases} \tag{3.16}$$

where the last n columns $N_{l,a}$, $n < a \le 2n$, are not independent of the first n columns when complex multiplication is allowed. The conditions (3.4) become

$$\begin{aligned} N^{C\dagger}N^C &= I\,, \\ N^{C\dagger}M^C &= 0\,, \end{aligned} \tag{3.17}$$

while

$$A^C_\mu = N^{C\dagger}\partial_\mu N^C \tag{3.18a}$$

gives an $n\times n$ traceless[9] anti-Hermitian matrix whose real and imaginary parts make up the original $2n\times 2n$ matrix A of Eq. (3.5),

$$A_\mu = \begin{pmatrix} \mathrm{Re}A^C_\mu & \mathrm{Im}A^C_\mu \\ -\mathrm{Im}A^C_\mu & \mathrm{Re}A^C_\mu \end{pmatrix}. \tag{3.18b}$$

Finally, we note that the projection operator (3.8) is replaced by

$$I - N^C N^{C\dagger} = \mathrm{Re}(M^{C*}R^{-1}M^{C\dagger}) - \hat{i}\,\mathrm{Re}(M^{C*}\hat{i}R^{-1}M^{C\dagger})\,. \tag{3.19}$$

Thus, exploiting the extra symmetry that results from restricting O(2n) to SU(n), we obtain the following *Ansatz* for constructing SU(n) instantons: Begin with a $(2k+n)\times k$ matrix of quaternions $M^C(x)$ depending linearly on x and obeying

$$M^{C\dagger}(q-\hat{i}q\hat{i})M^C(x) = 4\,\mathrm{Re}(R^\dagger q)\,, \tag{3.20}$$

where R is invertible. Next find a $(2k+n)\times n$ complex matrix N^C obeying

$$N^{C\dagger}N^C = I\,, \tag{3.21a}$$

$$N^{C\dagger}M^C = 0\,. \tag{3.21b}$$

Then the equation

$$A_\mu = N^{C\dagger}\partial_\mu N^C \tag{3.22}$$

defines a traceless,[9] anti-Hermitian, self-dual gauge field which has topological charge k and, as we will see, depends on the proper number of parameters.

Sp(n). Instanton solutions for the symplectic groups can be constructed from our results for the orthogonal groups by treating Sp(n) as a subgroup of $O(4n)$ in a manner very similar to the above discussion of SU(n). First, recall that Sp(n) can be defined as the group of $n\times n$ matrices Q of quaternions obeying the "unitarity" condition.

$$Q^\dagger Q = I\,. \tag{3.23}$$

An $n\times n$ quaternionic matrix can be viewed as a transformation on an n-dimensional quaternionic vector space which is linear in the sense that

$$Q(vq) = (Qv)q\,, \tag{3.24}$$

where v is an n-dimensional quaternionic vector and q is a quaternion. However, such a matrix also describes a real transformation on the $4n$-dimensional real vector space formed from the $4n$ real numbers necessary to specify the n quaternionic components of v. Conversely, a $4n\times 4n$ real matrix L can be written as an $n\times n$ quaternionic matrix if the analog of the linearity conditions (3.24) is obeyed,

$$LJ_i = J_i L\,, \quad 1 \le i \le 3 \tag{3.25}$$

where the three real matrices $-J_i$ obey the algebra (2.4) of the unit quaternions $\hat{i}$, $\hat{j}$, and $\hat{k}$. If the matrix L obeying Eq. (3.25) is orthogonal, then the corresponding quaternionic matrix will by symplectic, obeying Eq. (3.23).

Adapting this point of view to the problem at hand, we begin with a $(4k+4n)\times k$ matrix $M(x)$, linear in x, satisfying the condition (3.2) and therefore generating an O(4n) self-dual solution. We then impose the additional constraints

$$J_1 M = \hat{i}M\,, \quad J_2 M = \hat{j}M\,, \tag{3.26}$$

where, if written in $(k+n)\times(k+n)$ blocks, the matrices J_1 and J_2 are given by

$$J_1 = \begin{pmatrix} 0 & +I & 0 & 0 \\ -I & 0 & 0 & 0 \\ 0 & 0 & 0 & +I \\ 0 & 0 & -I & 0 \end{pmatrix}, \quad J_2 = \begin{pmatrix} 0 & 0 & +I & 0 \\ 0 & 0 & 0 & -I \\ -I & 0 & 0 & 0 \\ 0 & +I & 0 & 0 \end{pmatrix}; \tag{3.27}$$

here I is the $(k+n)\times(k+n)$ identity matrix. Thus M can be written in terms of a single $(k+n)\times k$ matrix M^Q:

$$M = \begin{pmatrix} M^Q \\ \hat{i}M^Q \\ \hat{j}M^Q \\ \hat{k}M^Q \end{pmatrix}. \tag{3.28}$$

In terms of the matrix M^Q, the reality condition and definition of R, Eq. (3.2), becomes

$$M^Q(x)^\dagger(q-\hat{i}q\hat{i}-\hat{j}q\hat{j}-\hat{k}q\hat{k})M^Q(x)=4\,\mathrm{Re}(Rq)\,. \tag{3.29}$$

If $\mathrm{Re}(q)=0$, this condition is automatically obeyed and implies that R is real and thus symmetric. For $q=1$, it reduces to

$$M^{Q\dagger}(x)M^Q(x)=R\,. \tag{3.30}$$

Similarly, in terms of M^Q, the condition (3.3b) on the $(4k+4n)\times 4n$ real matrix N becomes

$$\sum_{l=1}^{k+1}(N_{l,a}+N_{l+k+n,a}\hat{i}+N_{l+2k+2n,a}\hat{j}+N_{l+3k+3n,a}\hat{k})M^Q_{l,b}=0\,,\quad 1\leq a\leq 4n\,,\quad 1\leq b\leq k\,. \tag{3.31}$$

We define a $(k+n)\times 4n$ quaternionic matrix η by

$$\eta_{l,a}=N_{l,a}-\hat{i}N_{l+k+n,a}-\hat{j}N_{l+2k+2n,a}-\hat{k}N_{l+3k+3n,a}\,. \tag{3.32}$$

The first n columns of the real matrix N are chosen so that the first n columns of η are independent with respect to quaternionic multiplication. The remaining $3n$ columns of N are then selected by requiring that η be made of four blocks of the form

$$\eta=(N^Q,-N^Q\hat{i},-N^Q\hat{j},-N^Q\hat{k})\,. \tag{3.33}$$

With this choice the matrix N obeys

$$J_iN=NJ'_i \tag{3.34}$$

and correspond to the $(k+n)\times n$ quaternionic matrix N^Q. Here J'_i is the obvious $4n\times 4n$ version of the J_i of Eq. (3.27). In terms of the matrices N^Q and M^Q, the projection operator (3.7) takes on the simple form

$$I-N^QN^{Q\dagger}=M^QR^{-1}M^{Q\dagger}\,. \tag{3.35}$$

An Sp(n), self-dual Yang-Mills field can thus be obtained from a $(k+n)\times k$ matrix M^Q of quaternions linear in x and satisfying the requirement

$$M^Q(x)^\dagger M^Q(x)=R\,, \tag{3.36}$$

a real, invertible $k\times k$ matrix. The gauge field is given as an $n\times n$ anti-Hermitian matrix of quaternions,

$$A^Q_\mu(x)=N^Q(x)^\dagger\partial_\mu N^Q(x)\,, \tag{3.37}$$

where the $(k+n)\times n$ matrix $N^Q(x)$ obeys

$$N^Q(x)^\dagger M^Q(x)=0\,,\qquad N^Q(x)^\dagger N^Q(x)=I\,. \tag{3.38}$$

For $n=1$, this is precisely the formulation of SU(2) given in Sec. II.

In this way it is possible to obtain Sp(n) as well as SU(n) self-dual solutions by suitable restriction of the more general O($4n$) and O($2n$) solutions obtained at the beginning of this section. In both cases, the necessary restrictions somewhat simplify the procedure, resulting in the simplified *Ansätze* (3.36)–(3.38) and (3.20)–(3.22) for Sp(n) and SU(n), respectively.

IV. CANONICAL FORM AND COUNTING OF PARAMETERS

In the preceding sections we have seen how a self-dual Yang-Mills field can be constructed quite simply from a matrix M of quaternions obeying a particular nonlinear constraint. It is natural to interpret the physical properties of the gauge field directly in terms of the matrix M. This is most easily done if we exploit the transformations that can be performed on M (but which leave A_μ invariant) in order to put M into a simple canonical form. Using this canonical form the number of parameters entering the k-instanton solution is readily determined. The O(n) case is considered first and treated in such a manner that the results will be applicable to SU(n) and Sp(n).

Let us begin by observing that the gauge field A_μ of Eq. (3.5) is not changed if N and M are transformed according to

$$M_{il}\to S_{ij}T_{kl}M_{jk}=M'_{il}\,,\qquad N_{il}\to S_{ij}N_{jl}=N'_{il}\,, \tag{4.1}$$

where S is a $(4k+n)\times(4k+n)$ real orthogonal matrix and T_{kl} is any invertible $k\times k$ matrix of quaternions. Furthermore, the reality constraint (3.2) on M and the orthogonality of N and M, Eq. (3.4), are unaffected by this transformation. Since the matrices M and M' are completely equivalent we can use the symmetry operations S and T to simplify M. First we choose an orthogonal matrix S which transforms to zero the first n rows of C:

$$S_{il}C_{lj}=0\,,\quad 1\leq j\leq k\,,\quad 1\leq i\leq n\,. \tag{4.2}$$

Viewing the columns of C as $4k$ real $(4k+n)$-component vectors, we see that such a matrix must exist.

Next, we write the reality condition (3.2) in terms of the matrices B and C:

$$B^\dagger qB=\text{real}\,, \tag{4.3a}$$

$$C^\dagger qC=\text{real}\,, \tag{4.3b}$$

$$B^\dagger qC=(B^\dagger q^\dagger C)^t\,, \tag{4.3c}$$

for any quaternion q. If we define a $k\times k$ quaternionic matrix $\mathcal{R}$ by

$$C^\dagger qC=4\,\mathrm{Re}q\mathcal{R}^t\,, \tag{4.4}$$

$$
C = \begin{pmatrix}
0 & & \cdot\ \cdot\ \cdot & 0 \\
\cdot & & & \cdot \\
\cdot & & & \cdot \\
\cdot & & & \cdot \\
0 & & \cdot\ \cdot\ \cdot & 0 \\
1 & 0 & & 0 \\
\hat{i} & 0 & & 0 \\
\hat{j} & 0 & \cdot\ \cdot\ \cdot & 0 \\
\hat{k} & 0 & & 0 \\
0 & 1 & & \\
0 & \hat{i} & & \\
0 & \hat{j} & & \\
0 & \hat{k} & \cdot & \\
\cdot & & & \\
\cdot & & \cdot & \\
\cdot & & & \\
0 & & \cdot & 1 \\
0 & & \cdot\ \cdot\ \cdot & \hat{i} \\
0 & & & \hat{j} \\
0 & & & \hat{k}
\end{pmatrix}
$$

FIG. 1. The canonical form for the $(4k+n)\times k$ matrix C.

then $\mathcal{R}=\mathcal{R}^\dagger$ and we can find a $k\times k$ quaternionic matrix T which diagonalizes $\mathcal{R}$,[10]

$$T^\dagger \mathcal{R} T = I . \tag{4.5}$$

Thus letting $C'_{il}=T^\dagger_{jl}C_{ij}$, Eq. (4.4) becomes

$$C'^\dagger q C' = 4I\,\mathrm{Re}q . \tag{4.6}$$

Or if we write $C'_{ij}=C'^{\mu}_{ij}\sigma_\mu$, Eq. (4.6) implies

$$\sum_{l=n+1}^{4k+n} C'^{\mu}_{lj}C'^{\nu}_{li}=\delta_{ji}\delta_{\mu\nu} , \tag{4.7}$$

so that C'^{μ}_{ij} itself can be viewed as a $4k\times 4k$ real orthogonal matrix. Our final operation is then an orthogonal transformation on the left of M by

$$
\begin{aligned}
&S_{ll}=\delta_{ll} , \quad 1\leq l\leq n \\
&\left.\begin{aligned} &S_{il}=0 \\ &S_{n+4i-3+\mu,\, l}=C'^{\mu}_{li} \end{aligned}\right\} 1\leq i\leq n , \quad n\leq l\leq 4k+n .
\end{aligned} \tag{4.8}
$$

After this transformation, C has the diagonal form

$$
\begin{aligned}
&C^{\mu}_{ij}=0 , \quad 1\leq i\leq n \\
&C^{\mu}_{n+4i-3+\nu,\, j}=\delta_{ij}\delta_{\mu\nu}
\end{aligned} \tag{4.9}
$$

(see Fig. 1). It should be noted that except for a reordering of the rows, this form is consistent with the O(n) description of SU(n) and Sp(n) solutions, having three quarters of the rows of C simple $\hat{i}$, $\hat{j}$, or $\hat{k}$ multiples of the first quarter.

The canonical form (4.9) for C can now be substituted into the reality condition (4.3c), which yields

$$\sum_{\nu=0}^{3}\left[(B^*_{n+4i-3+\nu,\, j})\sigma_\mu\sigma_\nu-(B^*_{n+4j-3+\nu,\, i})\sigma_\mu^\dagger\sigma_\nu\right]=0 , \tag{4.10}$$

for $0\leq\mu\leq 3$. These conditions require

$$
\begin{aligned}
&B_{ab}=v_{ab} , \quad 1\leq a\leq n \\
&\left.\begin{aligned}
&B_{n+4a-3,\, b}=b^{(0)}_{ab}+b^{(1)}_{ab}+b^{(2)}_{ab}+b^{(3)}_{ab} \\
&B_{n+4a-2,\, b}=\hat{i}(b^{(0)}_{ab}-b^{(1)}_{ab}) \\
&B_{n+4a-1,\, b}=\hat{j}(b^{(0)}_{ab}-b^{(2)}_{ab}) \\
&B_{n+4a,\, b}=\hat{k}(b^{(0)}_{ab}-b^{(3)}_{ab})
\end{aligned}\right\} , \quad 1\leq a\leq k
\end{aligned} \tag{4.11}
$$

where $b^{(0)}$ is symmetric and $b^{(1)}$, $b^{(2)}$, and $b^{(3)}$ are antisymmetric $k\times k$ quaterionic matrices, while v is an $n\times k$ quaternionic matrix. Equation (4.10) is obeyed for arbitrary values of the matrices $b^{(r)}$. However, the requirement (4.3a) that $B^\dagger qB$ be real restricts these four matrices so that further nonlinear matrix equations must be solved in order to obtain a satisfactory matrix M.

Requiring C to have the canonical form (4.9) does not completely eliminate the freedom to make the transformations (4.1). The choice (4.9) for C is not changed by the transformation (4.1) if the $k\times k$ quaternionic matrix T is symplectic and the O($4k+n$) matrix S has the block diagonal form

$$S=\begin{pmatrix} G & 0 \\ 0 & \tilde{T}^{-1}\end{pmatrix} , \tag{4.12}$$

where G is an element of O(n) and $\tilde{T}^{-1}$ is that $4k\times 4k$ orthogonal matrix corresponding to the $n\times n$ symplectic matrix of quaternions T^{-1}.

Having put M into canonical form, we next determine the number of parameters in the corresponding multi-instanton solution. The matrices v and $b^{(r)}$ have altogether $2k^2-k+nk$ quaternionic elements. However, Eq. (4.3a) requires the reality of one Hermitian (when $q=1$) and three anti-Hermitian matrices (for $q=\hat{i},\hat{j},\hat{k}$); this gives $3(2k^2+k)$ real conditions, so M depends on $2k^2-7k+4nk$ independent parameters. Of these, $\frac{1}{2}n(n-1)+k(2k+1)$ correspond to the O(n) $\times$ Sp(k) group of transformations which preserve the canonical form of C; subtracting these gives $4(n-2)k-\frac{1}{2}n(n-1)$

true parameters. There are two special cases for which this argument must be modified. The k columns of v may be viewed as $4k$ real n vectors which are acted on by the O(n) transformations. If $n>4k$, there will be an O($n-4k$) subgroup which leaves these vectors invariant, and thus has no effect on M, so we have oversubtracted by an amount equal to the dimension of this subgroup. Further, if $k=1$, one can transform v so that it has only four nonvanishing quaternionic components. Some algebra shows that these can be chosen to be c, $\hat{i}c$, $\hat{j}c$, and $\hat{k}c$, where c is some real number. Substituting for G in Eq. (4.12) the choice

$$G=\begin{pmatrix} T^{-1} & 0 \\ 0 & I \end{pmatrix} \tag{4.13}$$

[here I is the $(n-4)\times(n-4)$ unit matrix], then gives an Sp(1) subgroup which leaves M invariant, so we must add 3 to our answer. Thus, the number of parameters for the O(n) multi-instanton solution is

$$\begin{aligned} &4(n-2)k-\tfrac{1}{2}n(n-1)\,, \quad k\geq 4n \\ &8k^2-6k\,, \quad 4n\geq k\geq 1 \\ &5\,, \quad k=1\,. \end{aligned} \tag{4.14}$$

A similar analysis can be applied to the SU(n) and Sp(n) constructions. For the unitary case, the matrices S and T of Eq. (4.1) must not affect the SU(n) condition (3.10). Thus the $(4k+2n)\times(4k+2n)$ orthogonal matrix S must commute with J and consequently be equivalent to an element of U($2k+n$); likewise T must commute with $\hat{i}$ and hence be a complex $k\times k$ matrix. Of course, in the simplified formulation of Eqs. (3.20)–(3.22) the corresponding matrix S would be explicitly a $(2k+n)\times(2k+n)$ unitary matrix. Finally, if we consider Sp(n), the invariance of Eq. (3.26) under the transformation (4.1) requires that S be the real $(4k+4n)\times(4k+4n)$ version of an element of Sp(n) while T must be real.

In terms of the matrices $M^C(x)$ or $M^Q(x)$, the canonical form (4.8) and (4.11) adopted for O(n) becomes for SU(n)

$$\begin{aligned} &C^C_{ab}=0\,, \quad 1\leq a\leq n \\ &\left.\begin{aligned} &C^C_{n+2a-1,b}=\delta_{ab} \\ &C^C_{n+2a,b}=\hat{j}\delta_{ab} \end{aligned}\right\} 1\leq a\leq k\,, \\ &B^C_{ab}=v_{ab}\,, \quad 1\leq a\leq n \\ &\left.\begin{aligned} &B^C_{n+2a-1,b}=b^{(0)}_{ab}+b^{(2)}_{ab} \\ &B^C_{n+2a,b}=\hat{j}(b^{(0)}_{ab}-b^{(2)}_{ab}) \end{aligned}\right\} 1\leq a\leq k\,, \end{aligned} \tag{4.15}$$

while for Sp(n)

$$\begin{aligned} &C^Q_{ab}=0\,, \quad 1\leq a\leq n \\ &C^Q_{n+a,b}=\delta_{ab}\,, \quad 1\leq a\leq k \\ &B^Q_{ab}=v_{ab}\,, \quad 1\leq a\leq n \\ &B^Q_{n+a,b}=b^{(0)}_{ab}\,, \quad 1\leq a\leq k \end{aligned} \tag{4.16}$$

where, as before, $b^{(0)}$ is symmetric and $b^{(2)}$ antisymmetric. The group of transformations which preserves the canonical form of C is U(n)×U(k) for the unitary case and Sp(n)×O(k) for the symplectic case.

To count parameters for SU(n), we note that M has k^2+nk quaternionic elements. Equation (3.20) requires the reality of one Hermitian and one anti-Hermitian matrix, giving $3k^2$ real conditions. Subtracting the dimension of U(n)×U(k) would give $4nk-n^2$ parameters. However, M is invariant under the U(1) subgroup corresponding to $S=e^{-i\phi}$, $T=e^{-i\phi}$. Further, the columns of v give $2k$ complex vectors, so if $2k<n$ there is a U($n-2k$) subgroup which has no effect on M. Thus, the number of parameters is

$$\begin{aligned} &4nk-n^2+1\,, \quad 2k\geq n \\ &4k^2+1\,, \quad 2k<n\,. \end{aligned} \tag{4.17}$$

For Sp(n), M has $kn+\frac{1}{2}k^2+\frac{1}{2}k$ quaternionic components obeying $\frac{3}{2}k(k-1)$ real conditions. Subtracting the dimension of Sp(n)×O(k), and adding the dimension of Sp($n-k$) if $k<n$, we obtain

$$\begin{aligned} &4(n+1)k-n(2n+1)\,, \quad k\geq n \\ &2k^2+3k\,, \quad k<n\,. \end{aligned} \tag{4.18}$$

V. SU(2) SOLUTIONS

We now address the problem of explicitly finding matrices $M(x)$ satisfying the nonlinear constraints (4.3). In this section we restrict ourselves to SU(2) = Sp(1); the more general case will be considered in the next section. Let us begin with M in canonical form so that its first row is given by k constant quaternions

$$v_{1j}=M_{1j}=q_j\,, \tag{5.1}$$

while the remaining k rows of M form a square matrix $\hat{M}(x)$:

$$\hat{M}(x)_{ij}=M_{i+1,j}(x)=\delta_{ij}(y_i-x)+b_{ij}\,. \tag{5.2}$$

Here the quaternions y_i are the diagonal elements of $\hat{M}$, so that $b_{ii}=0$. The constraints (4.3) are obeyed if b_{ij} is symmetric and

$$\frac{1}{2}(q_i^*q_j-q_j^*q_i)+(y_i-y_j)^*b_{ij} +\frac{1}{2}\sum_{l=1}^{k}(b_{li}^*b_{lj}-b_{lj}^*b_{li})=r_{ij} \tag{5.3}$$

for some real $k\times k$ matrix r.

Finding the general solution to this set of quadratic equations is quite difficult. However, one solution can be obtained immediately. If the q_i are chosen to be real numbers λ_i, then Eq. (5.3) is satisfied if all the b_{ij} vanish. Before proceeding further, let us calculate the gauge field $A_\mu(x)$ corresponding to this simple choice of $M(x)$. We first must find a $(k+1)$-component quaternionic vector $N(x)$ satisfying

$$N^\dagger(x)M(x)=0\,, \tag{5.4a}$$

$$N^\dagger(x)N(x)=1\,. \tag{5.4b}$$

If we use Eq. (5.4a) to obtain the last k components of $N(x)$ in terms of the first, we find that the most general solution may be written as

$$N_i=\begin{cases}\dfrac{1}{\sqrt{\rho}}u, & i=1\\[2mm] -\dfrac{1}{\sqrt{\rho}}\left[(\hat{M}^\dagger)^{-1}v^\dagger\right]_{i-1}u, & 2\le i\le k+1\,,\end{cases} \tag{5.5a}$$

where

$$\rho=\left[1+v\hat{M}^{-1}(\hat{M}^\dagger)^{-1}v^\dagger\right] \tag{5.5b}$$

and u is an arbitrary, possibly x-dependent, unit quaternion; different choices for u will give gauge-equivalent Yang-Mills fields. For the simple case we are considering it is trivial to invert $\hat{M}$; we obtain for N

$$N_i(x)=\begin{cases}\dfrac{1}{\sqrt{\rho_0}}u, & i=1\\[2mm] -\dfrac{1}{\sqrt{\rho_0}}\dfrac{y_{i-1}-x}{|y_{i-1}-x|^2}q^*_{i-1}u, & 2\le i\le k+1\,,\end{cases} \tag{5.6a}$$

where

$$\rho_0=1+\sum_{l=1}^{k}\frac{\lambda_l^{\,2}}{|y_l-x|^2}\,. \tag{5.6b}$$

If we now make a particularly simple choice for u, namely $u=1$, we find

$$A_\mu(x)=N^\dagger(x)\partial_\mu N(x)$$

$$=-\frac{i}{\rho_0}\bar{\eta}_{a\mu\nu}\sum_{l=1}^{k}\frac{q_l\tau^a q_l^*(x-y_l)_\nu}{|x-y_l|^4}\,, \tag{5.7}$$

which is the familiar 't Hooft k-instanton solution in "singular gauge." Here $\bar{\eta}_{a\mu\nu}$ is the parity conjugate of $\eta_{a\mu\nu}$. (Note that this verifies that k is indeed the topological charge.[11]) The singularities in the gauge field at $x=v_l$ are not physical but are artifacts of our choice $u=1$.

In order to obtain a more general solution, we must relax the condition that the q_i be real. Since any quaternion q_i may be written as $(q^*q)^{1/2}U$ where U is some SU(2) matrix, it is natural to interpret the magnitudes of q_i (which we shall continue to denote by λ_i) as instanton scale parameters and the quaternionic phases of the q_i as specifying the SU(2) orientations of the instantons. An approximate solution to Eq. (5.3) may be obtained for general quaternions q_i if we let the scale parameters λ_i be of order β compared to the "instanton separations" $|y_i-y_j|$ and study the limit of small β. In such a limit the off-diagonal elements b_{ij} become very small and the quadratic piece of the equation can be treated perturbatively. This limit can be described more formally by treating the diagonal elements y_i as fixed and introducing a small parameter β into the q_i i.e., $q_i=\beta q_i'$, for fixed q_i'. We must then find the matrix $M(x)$ and the gauge potential $A_\mu(x)$ as a power series in β, choosing the particular branch to the solution of the nonlinear equation (5.3) with the property

$$b_{ij}\sim\beta^2 \tag{5.8}$$

as β tends to zero.

As a first step in this calculation, we exploit the freedom to conjugate $\hat{M}$ by a $k\times k$ orthogonal matrix T to simplify our nonlinear constraint (5.3). For example, if we use

$$T_{ij}=\delta_{ij}+t_{ij}+O(\beta^4)\,, \tag{5.9}$$

where t_{ij} is of order β^2, to trasnsform M, then to order β^2, Eq. (5.3) becomes

$$\tfrac{1}{2}(q_i^*q_j-q_j^*q_i)+(y_i-y_j)^*[b_{ij}+(y_i-y_j)t_{ij}]$$
$$+\frac{1}{2}\sum_{l=1}^{k}(b_{li}^*b_{lj}-b_{lj}^*b_{li})=r_{ij}\,. \tag{5.10}$$

If we choose

$$t_{ij}=\frac{r_{ij}}{|y_i-y_j|^2}\,, \tag{5.11}$$

then, to order β^2, r_{ij} drops out of the equation. This procedure can be repeated order by order in β^2 to define an orthogonal matrix T as a power series in β (convergent for β sufficiently small) so that

$$r_{ij}=0 \tag{5.12}$$

if the constraint (5.3) is written in terms of the elements of the transformed matrix $TM(x)T^{-1}$ With r_{ij} vanishing, Eq. (5.3) can be rewritten

$$b_{ij}=\frac{1}{2}\frac{y_i-y_j}{|y_i-y_j|^2}\left[q_j^*q_i-q_i^*q_j+\sum_l(b_{lj}^*b_{li}-b_{li}^*b_{lj})\right], \tag{5.13}$$

which immediately yields b_{ij} as a power series in β^2 by simple iteration in the quadratic term:

$$b_{ij} = \frac{1}{2}\frac{y_i - y_j}{|y_i - y_j|^2}(q_j^* q_i - q_i^* q_j) + \cdots . \tag{5.14}$$

The resulting series is convergent for sufficiently small β and gives a solution $M(x)$ containing the full $8k$-3 parameters [$4k$ positions y_i, and $4k$-3 scales and relative SU(2) orientations q_i] which corresponds to k widely separated instantons.

Given this series expansion for $M(x)$, we can again use Eq. (5.5) to explicitly calculate the corresponding gauge field $A_\mu(x)$ for the case of widely separated instantons. It is convenient to make the singular gauge choice $u = 1$. As can be seen from an examination of Eq. (5.5), $N_{j+1}(x)$ can be computed to leading order in β by using an approximate form for $\hat{M}(x)$ containing only diagonal terms so that the resulting gauge field is simply given by Eq. (5.7). Of course, the q_i in expression (5.7) are no longer real numbers and the terms in Eq. (5.7) nonleading in β are to be ignored.

It is not difficult to find the next correction to Eq. (5.7). In order to describe our approximation precisely, we will discuss the solution $A_\mu(x)$ for two nested regions in position space:

$$\begin{aligned} &|x - y_i| >> \lambda_i, \quad \text{region I} \\ &|x - y_i| >> \beta\lambda_i, \quad \text{region II}. \end{aligned} \tag{5.15}$$

Region I consists of points far outside the instantons where the lowest-order solution (5.7) is intrinsically quite small,

$$|A_\mu(x)| \sim \beta^2 . \tag{5.16}$$

If we substitute the order-β^2 value for b_{ij}, Eq. (5.14), into Eq. (5.5) for N_{j+1} and use the resulting corrections to N_{j+1} when computing $A_\mu(x)$, we obtain an explicit formula for the gauge potential in region I, including correctly the terms of order β^4:

$$\begin{aligned} A_\mu(x) = \frac{1}{\rho_0}\sum_{i=1}^{k}\Big\{ &-i\bar{\eta}_{a\mu\nu} Q_i(x)\tau^a Q_i(x)^* \frac{(x-y_i)_\nu}{|x-y_i|^4} \\ &+ \frac{1}{2}\frac{1}{|x-y_i|^2}[Q_i(x)\partial_\mu Q_i^*(x) \\ &- \partial_\mu Q_i(x) Q_i^*(x)]\Big\} \\ &+ O(\beta^6), \end{aligned} \tag{5.17}$$

where

$$Q_i(x) = q_i + \frac{1}{2}\sum_{j\neq i} q_j \frac{(x-y_j)^*}{|x-y_j|^2}\frac{(y_i - y_j)}{|x-y_j|^2}(q_j^* q_i - q_i^* q_j). \tag{5.18}$$

Region II includes all of position space, both inside and outside the instantons, except for very small regions around the singular points y_i. Inside these excluded regions the singular gauge field is so large that corrections of higher order than those being considered can have a significant effect. [Alternatively, our results, Eq. (5.17) and Eq. (5.19), differ from the exact solution by terms vanishing as β^6 and β^2, *uniformly* in regions I and II, respectively.] Since region II contains points where $|x - y_i| \sim \lambda_i$, the first approximation to the gauge potential, Eq. (5.7), can be of order β^{-1}. The order-β^2 corrections to $\hat{M}(x)$ can then be used to compute $A_\mu(x)$ accurate through order β everywhere in region II:

$$A_\mu(x) \simeq -\frac{i\bar{\eta}_{a\mu\nu}}{\rho(x)}\sum_{i=1}^{k} Q_i(y_i)\tau^a Q_i(y_i)^* \frac{(x-y_i)_\nu}{|x-y_i|^4} + O(\beta^2), \tag{5.19}$$

where

$$\rho(x) = 1 + \sum_{j=1}^{k}\frac{Q_j(y_j)Q_j(y_j)^*}{|x-y_j|^2}. \tag{5.20}$$

Thus through order β the gauge potential has precisely the form of a superposition of k single instanton solutions. The scale and group orientation of the ith instanton can be obtained by examining Eq. (5.19) in the region $|x - y_i| \sim |q_i|$. The result is simply the scale and SU(2) orientation of a 3rd quaternion $\bar{Q}_i$,

$$\bar{Q}_i = q_i - \frac{1}{2}\sum_{j\neq i}\frac{q_j q_i^* q_j}{|y_i - y_j|^2}, \tag{5.21}$$

which differs from q_i by a term of order β^3. If we express the q_j in terms of a gauge rotation relative to q_i,

$$q_j = \frac{\lambda_j}{\lambda_i} q_i e^{i\vec{\tau}\cdot\hat{n}_j\theta_j/2}, \tag{5.22}$$

then

$$\bar{Q}_i = q_i\left[1 - \frac{1}{2}\sum_{j}\frac{\lambda_j^2}{|y_i - y_j|^2}(\cos\theta_j + i\hat{n}_j\cdot\vec{\tau}\sin\theta_j)\right]. \tag{5.23}$$

Thus, if there are many randomly oriented instantons, their effects on instanton i will tend to cancel, at least to order β. By contrast, expansion of the 't Hooft solution ($\theta_j = 0$) would have suggested that as the number of instantons became very large the effective scale of each would tend toward zero.

VI. O(n), SU(n), AND Sp(n) SOLUTIONS

We now extend the methods of the previous section to other groups. We begin with the Sp(n) case, since it most closely resembles the SU(2) construction of Sec. V. M is a $(k+n)\times k$ quaternionic matrix, which we assume to be in canonical form. The first n rows form an $(n\times k)$ matrix v_{ij}, while the last k rows form a square matrix $\hat{M}$, which we continue to write in the form (5.2). The constraint (5.3) is replaced by the condition

$$\frac{1}{2}\sum_{l=1}^{n}(v_{li}^{*}v_{lj}-v_{lj}^{*}v_{li})+(y_i-y_j)^{*}b_{ij}+\frac{1}{2}\sum_{l=1}^{k}(b_{li}^{*}b_{lj}-b_{lj}^{*}b_{li})=r_{ij} \quad (6.1)$$

for some real $k\times k$ matrix r.

As with SU(2), one solution can be obtained immediately. If we view the columns of v as n-component vectors, and choose them to be real multiples of a fixed vector e

$$v_{ij}=\lambda_j e_i, \quad (6.2a)$$

then Eq. (6.1) is satisfied if all the b_{ij} vanish. Using the freedom to perform Sp(n) transformations, we can choose

$$e_i=\delta_{i1}. \quad (6.2b)$$

The gauge field $A_\mu(x)$ is then found by a straightforward calculation; it is simply an embedding into Sp(n) of the 't Hooft SU(2) k-instanton solution, with instanton positions y_i and scale parameters λ_i. Furthermore, the embedding is precisely that one which gives the minimum topological charge,[6] thus verifying that for Sp(n) k is equal to the topological charge.

To obtain more general solutions, we relax the condition that the columns of v be multiples of a fixed vector. Since any n-component quaternionic vector of unit length can be obtained by multiplying the vector e by a suitable Sp(n) matrix G, we can write

$$v_{ij}=\lambda_j(G^{(j)}e)_i, \quad (6.3)$$

where the $G^{(j)}$ are k distinct Sp(n) matrices. We thus obtain a solution with the correct number of physical parameters, namely the y_j, the λ_j, and the parameters needed to describe the relative orientation of the k vectors $G^{(j)}e$. Just as with SU(2), Eq. (6.1) can be analyzed in the limit of widely separated instantons, and these parameters are seen to correspond to the positions, scales and Sp(n) orientations of the instantons.

Let us now proceed to the O(n) and SU(n) cases. We begin by attempting to identify the matrix $M(x)$ corresponding to the embedding of the 't Hooft solution. The results for Sp(n) suggest that we should require [in the notation of Eq. (4.11)]

$$b_{ij}^{(0)}=y_i\delta_{ij},\quad b_{ij}^{(1)}=b_{ij}^{(2)}=b_{ij}^{(3)}=0,\quad v_{ij}=\lambda_j\bar{e}_i, \quad (6.4)$$

where $\bar{e}$ is fixed vector. Recall, however, that the O(n) and SU(n) constructions impose more conditions on $M(x)$ than does the Sp(n) construction. If we require (6.4), then the O(n) condition (4.3) will be satisfied if

$$\bar{e}^{\dagger}q\bar{e}=\text{real} \quad (6.5a)$$

for any quaternion q, while for SU(n) we need require only

$$\bar{e}^{\dagger}\hat{i}\bar{e}=\text{real}. \quad (6.5b)$$

Making use of those transformations which preserve the canonical form of M, we can write the solutions to these equations as

$$\bar{e}_i=\tfrac{1}{2}(\delta_{i1}+\hat{i}\delta_{i2}+\hat{j}\delta_{i3}+\hat{k}\delta_{i4}), \quad (6.6a)$$

for O(n) ($n\geq 4$) and

$$\bar{e}_i=\frac{1}{\sqrt{2}}(\delta_{i1}+\hat{j}\delta_{i2}) \quad (6.6b)$$

for SU(n). If we now determine the corresponding gauge field, we find that it is indeed an embedding of the 't Hooft SU(2) k-instanton solution with instanton positions y_i and scale parameters λ_i. Just as with Sp(n), the embedding is that which gives the minimum topological charge, thus verifying also for these groups that k is the topological charge.

To go beyond the 't Hooft solution we relax the conditions that the off-diagonal elements of $b^{(r)}$ vanish and that the columns of v be multiples of a fixed vector. In particular, we can consider the case of widely separated instantons and try to solve the analogs of Eq. (6.1) as a power series in β by requiring that the off-diagonal elements of B vanish as β^2 in this limit. Here, as in Sec. V, β parametrizes the ratio of the instanton scales λ_i to the instanton separations $|y_i-y_j|$. As can be seen by examining the terms of order β^2 in Eq. (4.3a), such a solution can exist only if the columns of v each obey a condition of the form (6.5) to order β^2. Consequently, we can write v in terms of k real numbers λ_j and k O(n) or SU(n) matrices $G^{(j)}$:

$$v_{ij}=\lambda_j(G^{(j)}\bar{e})_i+O(\beta^3). \quad (6.7)$$

Just as with Sp(n), the physical parameters are seen to correspond to the positions, scales, and group orientations of the instantons.

Note that the solution (6.6a) is not applicable to

O(3). In fact, once Eq. (6.4) is imposed there is no choice of $\vec{e}$ such that the O(n) condition (4.3) is satisfied. In particular, there is no acceptable $M(x)$ with $k=1$ for O(3), so k cannot be the topological charge. This result could have been anticipated from Eq. (4.14), which gives $4k$-3 as the number of paramters for O(3). This agrees with the SU(2) result only if the topological charge is $k/2$.

VII. CONCLUSION

We conclude with the following remarks:

1. The construction of Atiyah, Hitchin, Drinfeld, and Manin has a simple geometrical interpretation. For example, in the case of SU(2) the matrix $M(x)$ selects a k-dimensional, position-dependent subspace out of a fixed $(k+1)$-dimensional vector space (here all the dimensions are quaternionic). The vector $N(x)$ lies in the orthogonal complement of the subspace defined by $M(x)$. Given an isospin-$\frac{1}{2}$ field $\psi(x)$, $N(x)$ can be used to map it into this orthogonal complement:

$$\Psi_i(x) = N_i \bar{\psi}(x), \tag{7.1}$$

where $\bar{\psi}(x)$ is a quaternion formed from the two isotopic components ψ_1 and ψ_2 of ψ,

$$\bar{\psi} = \begin{pmatrix} \psi_1 & -\psi_2^* \\ \psi_2 & \psi_1^* \end{pmatrix}. \tag{7.2}$$

We can define a gauge-covariant derivative of the field Ψ by first taking the ordinary derivative of Ψ and then projecting orthogonally onto the subspace picked out by N. The resulting field $D_\mu \Psi$ then corresponds to the isospin-$\frac{1}{2}$ field $\partial_\mu \psi + A_\mu \psi$.

2. We have seen that the problem of finding self-dual Yang-Mills fields is reduced to that of finding a matrix M satisfying the nonlinear algebraic constraints (4.3). In Secs. V and VI we have solved this problem for the case of widely separated instantons. However, it is clearly desirable to obtain solutions for the general case. Although this is quite difficult for arbitrary topological charge, the general[12] $k=2$ and $k=3$ solutions for SU(2) (in fact for all the symplectic groups) can be found. For $k=2$ the SU(2) constraint Eq. (5.3) is linear in b_{ij}. For any choice of the q_i and y_i it is possible to find an orthogonal matrix T such that r_{12} vanishes; once this has been done, the unique solution to the constraint equation is

$$b_{12} = \frac{1}{2}\frac{(y_1 - y_2)}{|y_1 - y_2|^2}(q_2^* q_1 - q_1^* q_2). \tag{7.3}$$

The y_i and the magnitudes and relative phases of the q_i then give the $8k$-3 = 13 physical parameters. With $M(x)$ thus determined, substitution into Eq. (5.5) gives $N(x)$, and thus $A_\mu(x)$, explicitly. Furthermore, by comparing the SU(2) and Sp(n) constraints, Eqs. (5.3) and (6.1), the general $k=2$ result for Sp(n),

$$b_{12} = \frac{1}{2}\frac{(y_1 - y_2)}{|y_1 - y_2|^2}\sum_{l=1}^{n}(v_{l2}^* v_{l1} - v_{l1}^* v_{l2}), \tag{7.4}$$

is immediately obtained.

To obtain the general $k=3$ solution for SU(2), we first use the SU(2)×O(k) transformations which preserve the canonical form to require that q_1 be real and that the real parts of the b_{ij} vanish. We then take the y_i and the remaining components of the b_{ij} as the $8k-3=21$ physical parameters. Solving the constraint equation to obtain the q_i in terms of these parameters then gives

$$\begin{aligned} q_1^0 = \lambda_1 &= \frac{|\vec{W}_2 \times \vec{W}_3|}{[\vec{W}_1 \cdot (\vec{W}_2 \times \vec{W}_3)]^{1/2}} \\ q_2 = \lambda_1 &\frac{(\vec{W}_3 \times \vec{W}_2)\cdot(\vec{W}_3 \times \vec{W}_1)}{|\vec{W}_2 \times \vec{W}_3|^2} + i\vec{\tau}\cdot\frac{1}{\lambda_1}\vec{W}_3, \\ q_3 = \lambda_1 &\frac{(\vec{W}_3 \times \vec{W}_2)\cdot(\vec{W}_2 \times \vec{W}_1)}{|\vec{W}_2 \times \vec{W}_3|^2} - i\vec{\tau}\cdot\frac{1}{\lambda_1}\vec{W}_2, \end{aligned} \tag{7.5a}$$

where the vectors $\vec{W}_k$ are defined by

$$\vec{W}_k = \frac{i}{4}\epsilon_{ijk}\mathrm{tr}\left\{\vec{\tau}\left[(y_i - y_j)^* b_{ij} + \sum_{l=1}^{3}(b_{li}^* b_{lj})\right]\right\}. \tag{7.5b}$$

(Here we have represented the quaternions q_i as 2×2 matrices.) To extend the result to Sp(n) we use the Sp(n)×O(k) transformations to require that v_{lj} vanish if $l>j$ and be real if $l=j$, and that the real parts of the b_{ij} vanish. Just as with SU(2), the constraints can be used to obtain v_{lj} in terms of the physical parameters. Once $M(x)$ has been obtained, $A_\mu(x)$ can be found explicitly by substitution in Eqs. (5.5) and (2.10).

3. An important step in studying the quantum-mechanical implications of these solutions is to obtain the Green's functions for particles propagating in the presence of a classical self-dual Yang-Mills field. Brown, Carlitz, Creamer, and Lee[13] have shown that such propagators for massless spinor and vector fields are determined by the corresponding massless scalar propagator. Furthermore, they obtained explicit expressions for the isospin-$\frac{1}{2}$ and -1 massless scalar propagators in the presence of the $5k$-parameter SU(2) solution of 't Hooft. These expressions can be extended to the general SU(2) solution. For isospin $\frac{1}{2}$ the propagator takes the simple form[14]

$$\Delta^{(1/2)}(x,y) = \frac{N^\dagger(x)N(y)}{4\pi^2(x-y)^2}. \tag{7.4}$$

A proof of the formula, making use of the properties of $M(x)$, is given in the Appendix. Note that it is not necessary to specify the gauge in order to obtain the expression; the gauge choice is relevant only in specifying which solution for $N(x)$ is to be chosen. Using the results of the Appendix and following the method of Brown *et al.*, the isospin-1 propagator can also be obtained. It is given by

$$\Delta^{(1)}_{ab}(x,y)=\frac{1}{4\pi^2(x-y)^2}\,\mathrm{Re}[\tau^a N^\dagger(x)N(y)\tau^b N^\dagger(y)N(x)]$$
$$+\frac{1}{8\pi^2}\sum_{ijmn=1}^{k} f_{ij,mn}\,\mathrm{Re}\,[C^\dagger N(x)\tau^a N^\dagger(x)C]_{ij}$$
$$\times\mathrm{Re}[C^\dagger N(y)\tau^b N^\dagger(y)C]_{mn}, \tag{7.5}$$

where the constants $f_{ij,mn}$ are given by the matrix equation

$$f_{ij,mn}L_{mn,rs}=\delta_{ir}\delta_{js}-\delta_{jr}\delta_{is},$$
$$L_{mn,rs}=\mathrm{Re}\;2(C^\dagger B)_{mr}(B^\dagger C)_{sn}-(C^\dagger C)_{mr}(B^\dagger B)_{sn}$$
$$-(B^\dagger B)_{mr}(C^\dagger C)_{sn}]-(m\leftrightarrow n). \tag{7.6}$$

These results are quite encouraging, and suggest that the methods developed in this paper may be quite useful in the study of these nonperturbative effects in Yang-Mills theories.

Added note. After this manuscript was completed we received a report by E. J. Corrigan, D. B. Fairlie, S. Templeton, and P. Goddard containing some of these results.

ACKNOWLEDGMENTS

We wish to thank Claude Bernard, Anthony Duncan, Alfred Mueller, and Amarendra Sinha for helpful discussions. The work of N. H. C. and E. J. W. was supported in part by the U. S. Department of Energy, and the work of N. K. S. was supported in part by the National Science Foundation under Grant No. MCS 76-08478.

APPENDIX

We wish to show that the function

$$\Delta(x,y)=\frac{N^\dagger(x)N(y)}{4\pi^2(x-y)^2} \tag{A1}$$

is in fact the isospin-$\frac{1}{2}$ scalar propagator, i.e., that it satisfies

$$D_\mu D_\mu\Delta(x,y)=-\delta(x-y), \tag{A2}$$

where

$$D_\mu=\partial_\mu+A_\mu(x)=\partial_\mu+N^\dagger(x)\partial_\mu N(x). \tag{A3}$$

After substitution of the form (A1), Eq. (A2) becomes

$$\frac{1}{4\pi^2(x-y)^2}\left[-\frac{4(x-y)_\mu}{(x-y)^2}D_\mu N^\dagger(x)N(y)+D_\mu D_\mu N^\dagger(x)N(y)\right]=0. \tag{A4}$$

We first note that

$$D_\mu N^\dagger(x)=\partial_\mu N^\dagger(x)[I-N(x)N^\dagger(x)]$$
$$=N^\dagger(x)C\sigma_\mu R^{-1}(x)M^\dagger(x), \tag{A5}$$

where, as in the calculation of $F_{\mu\nu}$ in Sec. II, we have used Eq. (2.14) and the fact that M is linear in x. We next note that

$$M^\dagger(x)N(y)=[M^\dagger(y)-\sigma_\nu^\dagger(x-y)_\nu C^\dagger]N(y)$$
$$=-\sigma_\nu^\dagger(x-y)_\nu C^\dagger N(y). \tag{A6}$$

Then, using

$$\sigma_\mu\sigma_\nu^\dagger=\delta_{\mu\nu}+i\tau^a\eta_{a\mu\nu}, \tag{A7}$$

we obtain

$$-\frac{(x-y)_\mu}{(x-y)^2}D_\mu N^\dagger(x)N(y)=N^\dagger(x)CR^{-1}(x)C^\dagger N(y). \tag{A8}$$

Next, we obtain

$$D_\mu D_\mu N^\dagger(x)=N^\dagger(x)[\partial_\mu+C\sigma_\mu R^{-1}(x)M^\dagger(x)]\,[C\sigma_\mu R^{-1}(x)M^\dagger(x)]$$
$$=N^\dagger(x)\{-4CR^{-1}(x)C^\dagger+CR^{-1}(x)\;[4C^\dagger M(x)+2\sigma_\mu M^\dagger C\sigma_\mu]R^{-1}M^\dagger\}. \tag{A9}$$

For any quaternion q,

$$\sigma_\mu q\sigma_\mu=-2q^*, \tag{A10}$$

the quaternionic conjugate of q. Using this result and noting that the constraints (4.3) require that $M^\dagger C$ be symmetric, we see that the last two terms in Eq. (A9) cancel. Substitution of Eqs. (A8) and (A9) into Eq. (A4) then verifies that $\Delta(x,y)$ is indeed the desired propagator.

[1]V. G. Drinfeld and Yu. I. Manin (unpublished); V. G. Drinfeld and Yu. I. Manin (unpublished).

[2]M. F. Atiyah, N. J. Hitchin, V. G. Drinfeld, and Yu. I. Manin, Phys. Lett. 65A, 185 (1978).

[3]W. Barth, Inventiones Math. 42, 63 (1977).

[4]R. S. Ward, Phys. Lett. 61A, 81 (1977).

[5]M. F. Atiyah and R. S. Ward, Commun. Math. Phys. 55, 117 (1977).

[6]C. W. Bernard, N. H. Christ, A. H. Guth, and E. J. Weinberg, Phys. Rev. D 16, 2967 (1977), and references therein.

[7]If B is not of rank k the matrix R defined below is not invertible at $x=0$ and the construction fails. If C has rank less than k, one simply obtains a solution with lower topological charge.

[8]G. 't Hooft, Phys. Rev. D 14, 3432 (1976).

[9]The gauge field A_μ given by Eq. (3.18) or (3.22) is not automatically traceless. However, it is possible to eliminate the trace of A_μ by a gauge transformation (equivalent to a position-dependent change in the overall phase of N^{C}) provided $\mathrm{tr}(F_{\mu\nu})=0$. Furthermore, $\mathrm{tr}(F_{\mu\nu})=\partial_\mu(\mathrm{tr}\,A_\nu)-\partial_\nu(\mathrm{tr}\,A_\mu)$ corresponds to the field strength of a self-dual, finite action Abelian gauge field and must therefore vanish.

[10]This congruence of the Hermitian, quaternionic matrix $\mathcal{R}$ with I is the quaternionic version of a standard theorem in linear algebra for the case of real symmetric or Hermitian matrices. See for example, theorem 9.4 in D. Finkbeiner, *Introduction to Matrices and Linear Transformations* (Freeman, San Francisco, 1960).

[11]Having shown that k is the topological charge for the 't Hooft solution, we can argue that k must be the topological charge for all solutions constructed using our canonical form with M of rank k since these can be obtained from the 't Hooft solutions by continuous deformation of B.

[12]A different form for the general $k=2$, SU (2) solution has been previously obtained by conformal transformation of the special 't Hooft $k=2$ solution: R. Jackiw, C. Nohl, and C. Rebbi, Phys. Rev. D 15, 1642 (1977).

[13]L. S. Brown, R. D. Carlitz, D. B. Creamer, and C. Lee, Phys. Rev. D 17, 1583 (1978).

[14]We would like to thank Amarendra Sinha for pointing out to us that the expression found in Ref. 13 could be generalized to this form.

Volume 67B, number 2 PHYSICS LETTERS 28 March 1977

DEGREES OF FREEDOM IN PSEUDOPARTICLE SYSTEMS*

R. JACKIW and C. REBBI
Laboratory for Nuclear Science and Department of Physics, Massachusetts Institute of Technology, Cambridge, Massachusetts 02139, USA

Received 1 February 1977

We show that at least $8n-3$ parameters are required to specify an n-pseudoparticle solution in Euclidean SU(2) Yang-Mills theory.

1. Introduction. A very important property of the theory of non-Abelian gauge fields is that the action functional has local minima in the Euclidean domain with non-vanishing field strength $F^a_{\mu\nu}$. The corresponding field configurations, which are often called pseudoparticles, have self-dual or anti-self-dual field strength, i.e.

$$^*F^a_{\mu\nu} \equiv \tfrac{1}{2}\epsilon_{\mu\nu\rho\sigma} F^{a\rho\sigma} = \pm F^a_{\mu\nu},$$

and fall into topologically inequivalent classes labelled by an integer n, the Pontryagin index.

The existence of these non-local minima was first pointed out by Belavin et al. [1] who also exhibited the solution of the self-duality equation with $n = 1$ for an SU(2) gauge group. It is very relevant in the context of this article that the single pseudoparticle solution depends on five parameters, a vector y_μ and a scale λ, which specify position and size of the pseudoparticle.

Solutions of the self-duality equations with an arbitrary number of pseudoparticles were discovered by Witten [2] and 't Hooft [3]. The class of self-dual field configurations found by 't Hooft contains $5n$ free parameters for a Pontryagin index equal to n, which may be interpreted as positions and sizes of the n pseudoparticles. This number of degrees of freedom appears at first sight satisfactory, because it accounts precisely for the number of degrees of freedom of n individual pseudoparticles at extremely large separation. However, it was shown in ref. [4] that the solution must be further generalized to make it covariant under transformations of the conformal group. The surprising conclusion was that, whereas a single pseudoparticle is described by five parameters, at least thirteen parameters are necessary to describe two pseudoparticles at finite separation, at least $5n + 4$ to describe n pseudoparticles for $n \geqslant 3$.

The value of thirteen, found for the number of degrees of freedom for two pseudoparticles, suggests that a relative orientation in group space may come into play in the specification of the solution. These three additional degrees of freedom would not be seen in the case of a single pseudoparticle, because of gauge invariance in the system, but for two pseudoparticles it is no longer possible to perform separate gauge transformations on them, since their fields overlap. If this indeed is what is happening, then one should expect that a system of n pseudoparticles be described by $8n-3$ *independent* parameters, and in particular there should be a wider class of solutions than those discovered thus far.

In this article we do not attempt to enlarge the class of known exact pseudoparticle configurations, but approach the problem of the degrees of freedom of the system by studying its infinitesimal perturbations. For small oscillations about an n-pseudoparticle solution the self-duality constraint can be represented as $\mathcal{L}\delta A_\mu = 0$, where $\mathcal{L}$ is a linear differential operator. Any degeneracy of the solution implies the existence of an eigenmode of $\mathcal{L}$ with zero eigenvalue, and, vice versa, every independent zero-eigenmode, which is not a pure gauge deformation, signals the existence of a degree of freedom. In the next section we shall study the solutions of the equation $\mathcal{L}\delta A_\mu = 0$ and we shall show that $\mathcal{L}$ has more zero-eigenmodes than those associated with the degeneracy of the known pseudoparticle solutions. In sect. 2 we shall demonstrate that after removal of the gauge freedom the number of independent eigenmodes for Pontryagin index n is indeed $8n-3$.

* This work is supported in part through funds provided by ERDA under Contract EY-76-C-02-3069.*000.

Volume 67B, number 2 PHYSICS LETTERS 28 March 1977

2. *Small deformations of pseudoparticle configurations.* We recall the form of the known pseudoparticle solutions. We consider throughout this paper a gauge group SU(2). It is convenient to represent the Yang-Mills potentials and field strength by anti-hermitian matrices

$$A_\mu = A_\mu{}^a\,(\sigma^a/2\mathrm{i}), \tag{2.1a}$$

$$F_{\mu\nu} = F^a_{\mu\nu}(\sigma^a/2\mathrm{i}) = \partial_\mu A_\nu - \partial_\nu A_\mu + [A_\mu, A_\nu], \tag{2.1b}$$

where σ^a are the Pauli matrices.

The self-duality equation is

$$^*F_{\mu\nu} \equiv \tfrac{1}{2}\epsilon_{\mu\nu\rho\sigma}\,F^{\rho\sigma} = F_{\mu\nu}. \tag{2.2}$$

One can show [1] that when this equation is satisfied the action of the system

$$S = -\frac{1}{2}\int \mathrm{d}^4x\,\mathrm{Tr}\,F_{\mu\nu}F^{\mu\nu}, \tag{2.3}$$

is minimum and equals $8\pi^2 n$.

Eq. (2.2) can be solved by making the *ansatz* [3]

$$A_\mu = \mathrm{i}\bar{\sigma}_{\mu\nu}a^\nu, \tag{2.4}$$

where the matrices $\bar{\sigma}_{\mu\nu}$,

$$\bar{\sigma}_{ij} = (1/4\mathrm{i})[\sigma_i, \sigma_j], \quad \bar{\sigma}_{i4} = -\tfrac{1}{2}\sigma_i, \tag{2.5}$$

are anti-self-dual, and expressing the vector field a_μ in terms of a scalar superpotential [3, 5]

$$a_\mu = \partial_\mu \ln \rho. \tag{2.6}$$

It is straightforward then to verify that eq. (2.2) reduces to the condition

$$(1/\rho)\,\Box\,\rho = 0. \tag{2.7}$$

The solution of eq. (2.7) [4],

$$\rho = \sum_{i=1}^{n+1}\frac{\lambda_i^2}{(x-y_i)^2}, \tag{2.8}$$

which is the most general with positive definite ρ, when inserted back in eqs. (2.6) and (2.4) produces a self-dual field configuration with Pontryagin index n. Although the right hand side of eq. (2.8) contains $5(n+1)$ parameters, at most $5n+4$ enter in the expression of A_μ, because a common rescaling of all λ_i's leaves a_μ unchanged.

If we perform a small variation of the potentials

$$A_\mu \to A_\mu + \delta A_\mu, \tag{2.9}$$

this induces a variation of $F_{\mu\nu}$

$$\delta F_{\mu\nu} = D^A{}_\mu\,\delta A_\nu - D^A_\nu\,\delta A_\mu, \tag{2.10}$$

with

$$D^A{}_\mu\,\delta A_\nu = \partial_\mu\delta A_\nu + [A_\mu, \delta A_\nu]. \tag{2.11}$$

The constraint

$$\delta F_{\mu\nu} = {}^*\delta F_{\mu\nu}, \tag{2.12}$$

becomes then a first order linear differential equation for δA_μ.

The most general variation of A_μ can be expressed as

$$\delta A_\mu = \mathrm{i}\bar{\sigma}_{\alpha\beta}X_\mu{}^{\alpha\beta}, \tag{2.13}$$

where the tensor field $X_{\mu\alpha\beta}$ is anti-symmetric and anti-self-dual in the indices $\alpha\beta$. It is not convenient however to analyze eq. (2.12) in terms of the twelve independent components of $X_{\mu\alpha\beta}$, because one is bound to find as solutions all infinitesimal gauge transformations of A_μ. These are given by

$$\delta^g A_\mu = D^A{}_\mu\,\mathrm{i}\bar{\sigma}_{\alpha\beta}\omega^{\alpha\beta}, \tag{2.14}$$

where the three independent components of $\omega^{\alpha\beta}$ are the infinitesimal generators of the transformation. We make therefore an *ansatz* for $X_{\mu\alpha\beta}$, which reduces the number of independent components to nine.

We set

$$\delta A_\mu = \mathrm{i}\bar{\sigma}_{\mu\nu}\partial_\nu(\delta\rho/\rho) + \mathrm{i}\bar{\sigma}_{\alpha\beta}\,\rho\,\partial_\nu Y_\mu{}^{\nu\alpha\beta}, \tag{2.15}$$

where $Y_{\mu\nu\alpha\beta}$ is a tensor field anti-symmetric and anti-self-dual both pairs of indices $\mu\nu$, $\alpha\beta$ and subject to the condition

$$Y_{\mu\nu}{}^{\mu\nu} = 0. \tag{2.16}$$

The inclusion of the first term in the right hand side of eq. (2.15) is dictated by the fact that we want to find among the zero-modes the variation of A_μ induced by a change of the parameters of ρ, but we do not have at present a better justification for the particular form of the second term other than that it leads to remarkable simplifications in the algebra. The number of independent functions in the *ansatz* of eq. (2.15) is ten ($\delta\rho$ and the nine independent components of $Y_{\mu\nu\alpha\beta}$); eq. (2.16) removes one of the components of $Y_{\mu\nu\alpha\beta}$ and we are left with nine functions, so that we may consider eq. (2.15) as a gauge specification (although not a full gauge specification as we shall see in the next section).

The tensor $Y_{\mu\nu\alpha\beta}$ can be decomposed into a symmetric, traceless and an anti-symmetric part as follows.

$$Y_{\mu\nu\alpha\beta} = S_{\mu\nu\alpha\beta} + A_{\mu\nu\alpha\beta}, \qquad (2.17)$$

with

$$S_{\mu\nu\alpha\beta} = S_{\alpha\beta\mu\nu}, \qquad (2.18a)$$

$$A_{\mu\nu\alpha\beta} = -A_{\alpha\beta\mu\nu} = \tfrac{1}{4}(g_{\mu\alpha}V_{\nu\beta} - g_{\nu\alpha}V_{\mu\beta} - g_{\mu\beta}V_{\nu\alpha} + g_{\nu\beta}V_{\mu\alpha}). \qquad (2.18b)$$

$V_{\alpha\beta}$ is anti-symmetric and anti-self-dual. After some non-trivial algebra and using many of the properties of anti-self-dual quantities one finds that eq. (2.12) implies the Laplace equation for the symmetric part of $Y_{\mu\nu\alpha\beta}$.

$$\Box S_{\mu\nu\alpha\beta} = 0. \qquad (2.19)$$

All non-trivial solutions of this equation have singularities which cannot be removed from $\delta F_{\mu\nu}$ by a gauge transformation. Therefore we set $S_{\mu\nu\alpha\beta} = 0$.

When the symmetric part of $Y_{\mu\nu\alpha\beta}$ vanishes, the anti-self-duality condition implies that $\delta\rho$ and $V_{\mu\nu}$ must satisfy the equations

$$\Box \rho V_{\mu\nu} = 0 \qquad (2.20)$$

and

$$\Box\delta\rho + 2\rho\partial_\mu V^\mu{}_\nu \partial^\nu\rho = 0. \qquad (2.21)$$

These are solved by

$$V_{\mu\nu} = \frac{1}{\rho}\sum_{i=1}^{n+1} \frac{k^i_{\mu\nu}}{(x-y_i)^2}, \qquad (2.22)$$

$$\delta\rho(x) = \int d^4x' \frac{1}{4\pi^2(x-x')^2}(2\rho\partial_\mu V^\mu{}_\nu\partial^\nu\rho)(x') + \delta\tilde{\rho}(x), \qquad (2.23)$$

where the $k^i_{\mu\nu}$'s are constant anti-self-dual matrices, and $\delta\tilde{\rho}(x)$ is the harmonic variation of $\rho(x)$ induced by a change of its parameters which we shall disregard in the rest of this section.

Although $V_{\mu\nu}$ and $\delta\rho/\rho$, as given in eqs. (2.22) and (2.23) are regular, substituting into eq. (2.15) we find a δA_μ which behaves as $|x - y_i|^{-1}$ near the poles of ρ. We must check therefore whether this singularity can be removed by a gauge transformation or if it is a true singularity of $\delta F_{\mu\nu}$, which would make the solution physically unacceptable. Assuming without loss of generality that $y^\mu_i = 0$ and $\lambda_i = 1$, we find in the neighborhood of the origin

$$\delta A_\mu = \sigma_\alpha{}^\beta \frac{x_\gamma}{x^2}(g_\mu{}^\alpha g^{\nu\gamma} - g_\mu{}^\gamma g^{\nu\alpha}) \times (ak_{\beta\nu} - l_{\beta\nu} + b_\rho x^\rho k_{\beta\nu} - m_{\beta\nu\rho}x^\rho) + \tfrac{1}{4}\sigma_{\alpha\beta}(k^{\alpha\beta}b_\mu - m^{\alpha\beta}{}_\mu) + O(x^2), \qquad (2.24)$$

where we have expanded

$$\rho(x) = (1/x^2) + a + b_\rho x^\rho + O(x^2), \qquad (2.25a)$$

$$V_{\alpha\beta} = (k_{\alpha\beta}/x^2) + l_{\alpha\beta} + m_{\alpha\beta\rho}x^\rho + O(x^2). \qquad (2.25b)$$

Simple algebra shows that, if we add to δA_μ the field $\delta^g A_\mu$ obtained from A_μ by a gauge transformation

$$\delta^g A_\mu = D^A{}_\mu\, \mathrm{i}\,\bar{\sigma}_{\alpha\beta}\omega^{\alpha\beta}, \qquad (2.14)$$

$\delta A_\mu + \delta^g A_\mu = O(x)$ provided that

$$\omega_{\alpha\beta} = -\tfrac{1}{4}(ak_{\alpha\beta} - l_{\alpha\beta} + b_\rho x^\rho k_{\alpha\beta} - m_{\alpha\beta\rho}x^\rho) + O(x^2). \qquad (2.26)$$

It follows that $\delta F_{\mu\nu}$ is finite at the singularities of ρ, and that eqs. (2.22) and (2.23) give an acceptable solution to the equations for first-order deformations.

3. Removal of the gauge degrees of freedom. When the traceless symmetric part of $Y_{\mu\nu\alpha\beta}$ vanishes in eq. (2.15), δA_μ can be brought to a more compact form by a gauge transformation. By choosing

$$\omega_{\alpha\beta} = \rho(V_{\alpha\beta}/4) \qquad (3.1)$$

in eq. (2.14), we find

$$\delta A'_\mu = \delta A_\mu + \delta^g A_\mu = \mathrm{i}\,\bar{\sigma}_{\mu\nu}(\partial^\nu(\delta\rho/\rho) + \partial_\lambda\rho V^{\lambda\nu}). \qquad (3.2)$$

The zero-mode solution appears thus as a first order variation of the *ansatz* of eq. (2.4), where δa_ν consists of a term derived from a potential and a term with zero divergence. The gauge transformation that leads to eq. (3.2) is singular, and $\delta A'_\mu$ behaves now as $|x - y_i|^{-3}$ near the poles of $V_{\alpha\beta}$. Because of this singular behavior, it is not clear to us whether eq. (3.2) can be a useful starting point for the construction of a finite

Volume 67B, number 2 PHYSICS LETTERS 28 March 1977

solution; it is however very convenient to study the residual gauge freedom.

If we consider any further gauge transformation, the demand that

$$\delta A''_\mu = \delta A'_\mu + \delta^{g'} A_\mu, \tag{3.3}$$

be still of the form

$$\delta A''_\mu = \mathrm{i}\, \bar\sigma_{\mu\nu} \delta a^{\nu'} = \mathrm{i}\, \bar\sigma_{\mu\nu}(\delta a^\nu + \delta^{g'} a^\nu), \tag{3.4}$$

poses very severe constraints on the possible forms of the generator $\omega'_{\alpha\beta}$. In fact, one can repeat the analysis made in ref. [4], sect. 3, to find that $\omega'_{\alpha\beta}$ must be of the form

$$\omega'_{\alpha\beta} = \rho\widetilde{\omega}_{\alpha\beta}, \tag{3.5}$$

with

$$\widetilde{\omega}_{\alpha\beta} = 2x_\alpha A_{\beta\gamma} x^\gamma - 2x_\beta A_{\alpha\gamma} x^\gamma + x^2 A_{\alpha\beta}$$
$$+ B_\alpha x_\beta - B_\beta x_\alpha - \epsilon_{\alpha\beta\gamma\delta} B^\gamma x^\delta + C_{\alpha\beta}, \tag{3.6}$$

where B_α is a constant vector, $A_{\alpha\beta}$ and $C_{\alpha\beta}$ are constant self-dual and anti-self-dual tensors. The change in δa_ν is then given by

$$\delta^{g'} a_\nu = -4\widetilde{\omega}_{\nu\alpha}\partial^\alpha\rho - \tfrac{4}{3}\rho\partial^\alpha\widetilde{\omega}_{\nu\alpha}. \tag{3.7}$$

With a little algebra we find

$$\delta^{g'}\rho V_{\alpha\beta} = 4\sum_{i=1}^{n+1} \frac{\lambda_i^2}{(x-y_i)^2}\, \widetilde{\omega}_{\alpha\beta}(y_i) \tag{3.8}$$

and

$$\frac{\delta^{g'}\rho}{\rho} = \tfrac{4}{3}\sum_{i=1}^{n+1} \frac{\lambda_i^2}{(x-y_i)^2}\,(x_\alpha - y^i_\alpha)\,\partial_\beta\widetilde{\omega}^{\alpha\beta}|_{x=y_i}, \tag{3.9}$$

and we see that, since $\widetilde{\omega}_{\alpha\beta}$ depends on ten independent parameters, ten components of the residual $k^i_{\alpha\beta}$ of the poles of $V_{\alpha\beta}$ can be removed by a gauge transformation, all the others represent physical degrees of freedom.

In conclusion, we find that, given any solution of the self-duality equations, there are $8(n+1) - 1 - 10 = 8n - 3$ physical zero-modes, which signal an $(8n-3)$-fold degeneracy of the solution. $8(n+1)$ is the total number of parameters (y^i_μ, λ_i and $k^i_{\alpha\beta}$) which can be varied, one parameter is subtracted because only ratios of λ_i's enter in the expression of the fields, ten more because of the residual gauge freedom.

Notice that eqs. (3.8) and (3.9) also clarify one of the results obtained in ref. [4]. It was found there that a variation of the y^i_μ and λ_i parameters can be a pure gauge transformation only if the points y^i_μ obey the equation $\widetilde{\omega}_{\alpha\beta}(y^i_\mu) = 0$, which constrains them to lie on a circle. We understand now that $\widetilde{\omega}_{\alpha\beta}(y^i_\mu) = 0$ is precisely the condition necessary to preserve $V_{\alpha\beta} = 0$.

4. Conclusion. The results which we have presented in this article strongly suggest that the number of degrees of freedom of a system of n pseudoparticles is $8n-3$ for an SU(2) gauge group. Barring unexpected pathologies, indeed, in principle it should be possible to extend the infinitesimal deformations of the solutions (the physical zero-modes) to finite variations.

The problem of actually finding the full set of solutions as well as of understanding all the implications that they have for the theory remains open. We shall not dwell on this, but would like to conclude with the following observation. The class of solutions with unit Pontryagin index has a group of covariance, the four-dimensional conformal group, which acts in a transitive way, in the sense that all solutions can be obtained from any one by a finite group transformation. The class of solutions discovered by Witten has an analogous property, being covariant under the conformal group in two dimensions, with an infinite number of generators. The conformal group in four-dimensions is still a group of covariance of the multi-pseudoparticle system, but obviously it cannot generate all solutions from any definite one. It is fascinating to think that a larger group of covariance, yet to be discovered, might exist, with the property that all pseudoparticle configurations can be generated through the action of its transformations. The existence of such a group, which at the present stage is only a speculative hypothesis, would be far-reaching for the theory.

References

[1] A. Belavin et al., Phys. Lett. 59B (1975) 85.
[2] E. Witten, Harvard University, preprint.
[3] G. 't Hooft, unpublished.
[4] R. Jackiw, C. Nohl and C. Rebbi, Phys. Rev. D, in print.
[5] F. Wilczek, Princeton University preprint;
F. Corrigan and D. Fairlie, Durham University preprint.

PHYSICAL REVIEW D VOLUME 16, NUMBER 10 15 NOVEMBER 1977

Pseudoparticle parameters for arbitrary gauge groups*

Claude W. Bernard, Norman H. Christ, Alan H. Guth, and Erick J. Weinberg
Department of Physics, Columbia University, New York, New York 10027
(Received 6 July 1977)

The number of parameters entering a Euclidean Yang-Mills solution with topological charge k is determined for a theory constructed from an arbitrary Lie group G. It is shown that 'his number is precisely that required to specify the position, scale, and relative group orientation of k independent solutions each with minimum topological charge 1. Such minimal single-pseudoparticle solutions can be obtained by embedding the familiar SU_2 pseudoparticle of Belavin *et al.* into the general Lie group.

I. INTRODUCTION

A considerable amount of information is now known about self-dual solutions[1-4] to the Euclidean, Yang-Mills field equations. Of particular interest is the recent application of the Atiyah-Singer index theorem which determines the number of parameters entering the general solution with given topological charge k.[5-7] For SU_2 the number of parameters, $8k-3$, can be readily interpreted as resulting from the combination of k "elementary" SU_2 pseudoparticle solutions with topological charge 1—each with a particular size, space-time location, and SU_2 orientation.

In this paper we ask whether a similar interpretation is possible for self-dual solutions in a Yang-Mills theory with arbitrary gauge group G. This question is answered in three steps: First, it is observed that a solution with minimum topological charge (normalized to $k=1$) can be obtained by a particular embedding of the SU_2 solution of Belavin, Polyakov, Schwartz, and Tyupkin[1] into the general gauge group G. Second, for a solution with arbitrary k, we apply the Atiyah-Singer[8] index theorem to determine the number of parameters on which such a solution depends. Finally, we show that this number can be interpreted as that required to describe the scale, position, and group orientation of k examples of the SU_2 embedding found in the first step. Thus it may well be possible to view an arbitrary *self-dual* Yang-Mills solution, even for a general Lie group, as an appropriate combination of familiar SU_2 pseudoparticles.

We begin by considering a Yang-Mills theory with simple gauge group G (Ref. 9) of dimension $d(G)$ and with action

$$S=\tfrac{1}{4}\int (F^i_{\mu\nu})^2 d^4x\,, \tag{1.1}$$

where

$$F^i_{\mu\nu}=\partial_\mu A^i_\nu-\partial_\nu A^i_\mu+f_{ijk}A^j_\mu A^k_\nu\,. \tag{1.2}$$

Repeated group indices are summed from 1 to $d(G)$, and the structure constants f_{ijk} are chosen completely antisymmetric—for SU_2, $f_{ijk}=\epsilon_{ijk}$. The topological charge k is defined as

$$k=\frac{1}{32\pi^2}\int F^i_{\mu\nu}\tilde{F}^i_{\mu\nu}d^4x\,, \tag{1.3}$$

where the dual of $F_{\mu\nu}$ is

$$\tilde{F}_{\mu\nu}=\tfrac{1}{2}\epsilon_{\mu\nu\rho\sigma}F_{\rho\sigma}\,. \tag{1.4}$$

For the case of SU_2, a self-dual solution with topological charge k depending on $5k$ parameters has been given by 't Hooft[10]:

$$A^i_\mu(x)=-\bar{\eta}^i_{\mu\nu}\partial_\nu \ln\left[1+\sum_{j=1}^{k}\frac{\lambda_j^2}{(x-x_j)^2}\right]. \tag{1.5}$$

The singularities of this solution at the points $x_1,\ldots,x_n$ are not physical and can be removed by a gauge transformation. Since all non-Abelian groups contain SU_2 as a subgroup, these SU_2 solutions can be used to generate self-dual solutions with various topological charges for a Yang-Mills theory with an arbitrary group. More can be learned about the space of self-dual solutions by considering small fluctuations $A^i_\mu+\delta A^i_\mu$ about a particular solution A^i_μ and asking that the resulting field strength continue to be self-dual. If expanded to first order in δA_μ, this requirement can be written

$$\delta F_{\mu\nu}=\delta\tilde{F}_{\mu\nu}\,, \tag{1.6}$$

where

$$\delta F_{\mu\nu}=D_\mu\delta A_\nu-D_\nu\delta A_\mu\,. \tag{1.7}$$

The gauge-covariant derivative D depends on the initial solution A^i_μ and is defined by

$$(D_\mu\delta A_\nu)^i=\partial_\mu\delta A^i_\nu+f_{ijk}A^j_\mu\delta A^k_\nu\,. \tag{1.8}$$

In addition, one must require that the modified solution $A^i_\mu+\delta A^i_\mu$ represents a new solution and not simply a gauge transformation

$$\delta A^i_\mu=\delta\Lambda^j f_{ijk}A^k_\mu-\partial_\mu\delta\Lambda^i=-(D_\mu\delta\Lambda)^i \tag{1.9}$$

of the original A^i_μ. This can be done by requiring

that δA^i_μ be orthogonal to all functions of the form (1.9), i.e.,

$$\int d^4x(D_\mu \delta\Lambda)^i \delta A^i_\mu = 0 \qquad (1.10)$$

for all functions $\delta\Lambda^i(x)$. If integration by parts in Eq. (1.10) is allowed, this orthogonality requirement is equivalent to the usual background field gauge condition

$$D_\mu \delta A_\mu = 0 . \qquad (1.11)$$

For the case of SU_2, the analysis of small fluctuations about the solution (1.5) has been approached in three different ways. First, Jackiw and Rebbi[4] have found $8k-3$ solutions to Eq. (1.6) which are not gauge transformations of A^i_μ. Second, Schwartz[5] and Atiyah, Hitchin and Singer[7] transform the SU_2 Yang-Mills theory to the four-dimensional sphere S^4, apply the Atiyah-Singer index theory to the simultaneous linear differential equations (1.6) and (1.11), and show that there are precisely $8k-3$ parameters appearing in the general solution. Third, Brown, Carlitz, and Lee[6] combine equations (1.6) and (1.11), writing them as a single spinor equation. They then employ a variant of a method suggested by Coleman[11] to show directly that in Euclidean space, E^4, Eqs. (1.6) and (1.11) possess exactly $8k$ simultaneous solutions. This result does agree with the S^4 application of the Atiyah-Singer theorem. On S^4 the integration by parts relating Eq. (1.11) and the condition (1.10) is always permitted so that in the background gauge on S^4 all gauge freedom is eliminated. However, on E^4 there is a three-parameter family of gauge transformations which generate a δA^i_μ satisfying Eq. (1.11). Thus, as Brown *et al.* observe, there are only $8k-3$ physical modes for E^4.

Let us now consider self-dual solutions for an arbitrary simple compact Lie group G. Just as in the SU_2 case, each such solution (in a nonsingular gauge) approaches a direction-dependent gauge transformation at infinity

$$\lim_{|x|\to\infty} A_\mu(x) = g(\hat{x})\partial_\mu g^{-1}(\hat{x}) \qquad (1.12)$$

with $\hat{x}_\mu = x_\mu/(x^2)^{1/2}$ Hence to each solution there corresponds a mapping of directions in four dimensions (i.e., the three-dimensional sphere S^3) into the gauge group. These mappings fall naturally into equivalence classes, with elements in each class being continuously deformable into one another. The topological charge k is determined by the equivalence class to which the mapping belongs. The group of all such equivalence classes, $\Pi_3(G)$, has been thoroughly analyzed in the mathematical literature. In particular, the familiar result that $\Pi_3(SU_2)$ is isomorphic to the integers is valid for any simple Lie group. Furthermore, there is a particular minimal embedding of SU_2 into an arbitrary simple Lie group G such that each equivalence class of mappings in $\Pi_3(G)$ contains representatives obtained by mapping S^3 into that particular SU_2 subgroup.[12] Thus each topologically distinct set of boundary conditions (1.12) for an arbitrary simple group G can be obtained by embedding one of the SU_2 solutions (1.5), transformed to a regular gauge, into G. We have normalized the definition (1.3) of the topological charge so that for an arbitrary field configuration in G, k takes on the value of the corresponding topologically equivalent SU_2 embedding.

The remainder of this paper is arranged as follows. In Sec. II we analyze a general embedding of the SU_2 pseudoparticle of Belavin *et al.* in an arbitrary simple Lie group G. We show that the minimum topological charge k for such an embedding is obtained when one uses an SU_2 subgroup of G generated by E_α, $E_{-\alpha}$, and $[E_\alpha, E_{-\alpha}]$, where α is a root of maximum length. These $k=1$ solutions correspond to the minimal SU_2 embeddings referred to in the paragraph above.

In Sec. III the Atiyah-Singer index theorem is introduced. This theorem relates the index $\mathscr{I}$,

$$\mathscr{I} = h^0 - h^1 + h^2 , \qquad (1.13)$$

of the simultaneous equations (1.6) and (1.11) to the topological charge k. Here h^0 is the number of linearly independent solutions to the equation

$$D_\mu \delta\Lambda = 0 , \qquad (1.14)$$

h^1 is the number of linearly independent simultaneous solutions of Eqs. (1.6) and (1.11), and $h^2=0$ for S^4. Following Schwartz,[5] we do not evaluate $\mathscr{I}$ directly but instead use its linear dependence on k and explicit evaluation of h^0 and h^1 for $k=0$ and 1 to determine it in general. The result is a formula determining $h^1 - h^0$ as a function of k.

Throughout this section and the remainder of the paper we are specifically discussing the Yang-Mills equations on the four-dimensional sphere. This eliminates the problem of surface terms and is necessary for the validity of the index theorem. It is well known that the conformal invariance of the Yang-Mills equations ensures that when any solution on S^4 is stereographically projected onto Euclidean space it will also solve the Euclidean space equations. Conversely, all known Yang-Mills solutions in Euclidean space are sufficiently regular at infinity that, when appropriately gauge transformed, they can be mapped onto S^4.

Thus we have a formula for $h^1 - h^0$, where the value of h^0 depends on the particular configuration $A^i_\mu(x)$. Two configurations with the same value of k may have different values of h^0. In Sec. IV

TABLE I. The rank, quadratic Casimir operator $C(G)$, dimension $d(G)$, and the quantities $M(G)$ and $I(G,k)$ are listed for all simple compact Lie groups. The number of parameters necessary to describe a configuration of topological charge k is given by $N(G,k)=4C(G)k-d(G)$ if $k\geq M(G)$, and $N(G,k)=I(G,k)$ if $k<M(G)$. For any group, $I(G,1)=5$. (Note that $SO_3\cong SU_2$, $SO_5\cong Sp_4$, $SO_6\cong SU_4$, and SO_4 is not simple).

Group	Rank	$C(G)$	$d(G)$	$M(G)$	$I(G,k)$
SU_n	$n-1$	n	n^2-1	$\frac{1}{2}n$	$4k^2+1$
SO_n, n odd, $n\geq 7$	$\frac{1}{2}(n-1)$	$n-2$	$\frac{1}{2}n(n-1)$	$\frac{1}{4}(n-1)$	$8k^2-6k$ $(k\neq 1)$
SO_n, n even, $n\geq 8$	$\frac{1}{2}n$	$n-2$	$\frac{1}{2}n(n-1)$	$\frac{1}{4}n$	$8k^2-6k$ $(k\neq 1)$
Sp_{2n}	n	$n+1$	$n(2n+1)$	n	$2k^2+3k$
G_2	2	4	14	2	...
F_4	4	9	52	2	...
E_6	6	12	78	3	$I(2)=20$
E_7	7	18	133	3	$I(2)=20$
E_8	8	30	248	3	$I(2)=20$

we show that if a configuration has a certain regularity property, then its value of h^0 leads to a value of h^1 which gives the true number of parameters $N(G,k)$ necessary to specify the general self-dual solution of topological charge k. Furthermore, we show how to calculate these values of h^0. In Table I we display our final results for $N(G,k)$ for all simple compact groups G. Finally in Sec. V it is shown that the number of parameters we have obtained is precisely equal to the number necessary to describe the positions, scales, and relative group orientations of k independent pseudoparticles.

II. THE MINIMAL SU_2 EMBEDDING

In this section we explain how to embed the SU_2 pseudoparticle into an arbitrary, compact, simple Lie group to give the minimum topological charge

$$k=\frac{1}{32\pi^2}\int_M F^i_{\mu\nu}\tilde{F}^{i,\mu\nu}dx. \tag{2.1}$$

Here the $F^i_{\mu\nu}$ are the components of the field strength $F_{\mu\nu}$:

$$F_{\mu\nu}=F^i_{\mu\nu}T_i. \tag{2.2}$$

The T_i form a basis for the Lie algebra $\mathfrak{G}$ of the compact simple group G, and will be chosen to belong to the adjoint representation. The basis vectors T_i are chosen orthonormal with respect to the Cartan invariant inner product

$$\langle T_i,T_j\rangle=\frac{1}{C(G)}\operatorname{tr}(T_iT_j)=\delta_{ij}, \tag{2.3}$$

where $C(G)$ is a normalization constant that will be specified later. (Recall that it is for such a choice of basis that the structure constants f_{ijk} become completely antisymmetric.)

Given three matrices $\{J^i\}, 1\leq i\leq 3$ in the adjoint representation of G which obey the commutation relations of angular momenta

$$[J^i,J^j]=i\epsilon_{ijk}J^k \tag{2.4}$$

or

$$[J^3,J^\pm]=\pm J^\pm,\quad [J^+,J^-]=J^3$$

with

$$J^\pm=\frac{1}{\sqrt{2}}(J^1\pm iJ^2), \tag{2.5}$$

we can easily obtain a pseudoparticle solution in G,

$$F_{\mu\nu}=F^{I,i}_{\mu\nu}J^i, \tag{2.6}$$

where $F^{I,i}_{\mu\nu}$, $1\leq i\leq 3$ are the SU_2 components of the single-pseudoparticle solution obtained by setting $k=1$ in Eq. (1.5). The topological charge k of the solution (2.6) is simply the length of any one of the matrices J^i

$$k=\langle J^i,J^i\rangle=\frac{1}{C(G)}\operatorname{tr}(J^iJ^i)\ \text{(no sum on } i). \tag{2.7}$$

Our problem then is to find an SU_2 subalgebra $\{J^i\}, 1\leq i\leq 3$ of $\mathfrak{G}$, whose generators J^i have minimum length.

Let us first recall some properties of the root diagram of a simple Lie algebra $\mathfrak{G}$.[13,14] We choose a regular Abelian subalgebra H of $\mathfrak{G}$ which contains the maximum number of commuting generators. The dimension of H is called the rank r of $\mathfrak{G}$. We choose our basis T_i so that the first r elements, written as $\hat{h}_i$, $1 \le i \le r$ form a basis for H. Next, raising and lowering operators, E_α, are constructed out of the remaining elements of $\mathfrak{G}$. The action of the E_α on $\hat{h}_i$ can be represented by a root diagram which is a vector diagram in a space of dimension r. The vectors are labeled by α and their components α_i are the amounts by which E_α changes $\hat{h}_i$. Thus we have n

$$[\hat{h}_i, \hat{h}_j] = 0, \qquad [\hat{h}_i, E_\alpha] = \alpha_i E_\alpha. \tag{2.8}$$

The E_α, with suitable normalization, obey

$$[E_\alpha, E_{-\alpha}] = \sum_{i=1}^{r} \alpha_i \hat{h}_i, \qquad [E_\alpha, E_\beta] = N_{\alpha\beta} E_{\alpha+\beta}. \tag{2.9}$$

The constant $N_{\alpha\beta} = 0$ whenever $\alpha + \beta$ is not another root vector. Root vectors for all the compact simple Lie groups are listed by Racah.[13]

We conclude these preliminaries by discussing the normalization of the inner product (2.3) and hence of the basis T_i. The normalization of the T_i is fixed by the requirement that the maximal length of any root vector is one. Thus for such a root α^0

$$\sum_i (\alpha_i^0)^2 = 1. \tag{2.10}$$

With this choice of normalization, the constant $C(G)$ of Eq. (2.3),

$$C(G) = \mathrm{tr}(T_j^2) = \frac{1}{d(G)} \sum_{i=1}^{d(G)} \mathrm{tr}(T_i)^2 \qquad \sum_{i=1}^{d(G)} T_i^2, \tag{2.11}$$

becomes the usual quadratic Casimir operator for the adjoint representation of G.

We can now determine the embedding of SU_2 in G which has the minimum topological charge. Suppose we have any embedding $\{J^i\}$, $1 \le i \le 3$ obeying Eq. (2.4). It is always possible to pick a regular Abelian subalgebra H so that it contains any given element of $\mathfrak{G}$;[15] in particular we can choose H so that J^3 is an element of H. Thus we have

$$J^3 = \sum_{i=1}^{r} \beta_i \hat{h}_i \tag{2.12}$$

for some set of coefficients β_i and seek an embedding with a J^3 whose length $|\beta| \equiv (\sum_i \beta_i^2)^{1/2}$ is a minimum. It follows that J_+ and J_- have the form

$$J^\pm = \sum_\alpha f_\alpha^\pm E_\alpha, \tag{2.13}$$

where f_α^+ and f_α^- are two sets of coefficients. Thus one of the commutation relations (2.5) becomes

$$\left[\sum_i \beta_i \hat{h}_i, \sum_\alpha f_\alpha^+ E_\alpha\right] = \sum_\alpha f_\alpha^+ E_\alpha. \tag{2.14}$$

Equation (2.8) and the linear independence of the E_α now yield

$$1 = \sum_i \beta_i \alpha_i = |\beta|\,|\alpha| \cos\theta, \tag{2.15}$$

whenever $f_\alpha^+ \ne 0$, where θ is the angle between α and β. To minimize $|\beta|$ we should, according to Eq. (2.15), choose β parallel to a root vector, α^0, of maximum length ($|\alpha^0| = 1$) and then choose $f_\alpha^+ = 0$ for all other α. Thus if α^0 is a root vector of maximum length, we have a minimum SU_2 subgroup given by

$$J^+ = E_{\alpha^0}, \quad J^- = E_{-\alpha^0}, \qquad J^3 = \sum_i \alpha_i^0 \hat{h}_i. \tag{2.16}$$

Consequently the generators J_i of this subgroup also have length 1 and the minimal SU_2 pseudoparticle constructed from them, Eq. (2.6), will have topological charge $k = 1$.

A minimal SU_2 subgroup obtained in this manner is easily described for the series SU_n, Sp_{2n}, and SO_n. For SU_n and SP_{2n}, it is just the obvious "upper-left-hand-corner" embedding of the two-dimensional representation of $SU_2 = Sp_2$ into the n-dimensional representation of SU_n or the $2n$-dimensional representation of Sp_{2n}. For SO_n, $n \ge 5$, a minimal SU_2 subgroup is obtained by embedding the four-dimensional representation of SO_4 into the n-dimensional representation of SO_n and then using one of the factors $SO_4 = SU_2 \times SU_2$. (The more obvious subgroup obtained by embedding the three-dimensional representation of SO_3 gives a J^i whose length is the $\sqrt{2}$ times that obtained by this method.

The embedding (2.16) has a simple property that will be extremely useful to us later on. First note that if we let the generators J^i of any SU_2 subgroup act on all the generators of the group T_a by

$$[J^i, T_a] = L^i_{ab} T_b, \tag{2.17}$$

then the matrices L^i_{ab} form a representation of SU_2. Now by (2.8) for the particular subalgebra (2.16) we have

$$[J^3, \hat{h}_j] = 0, \qquad [J^3, E_\alpha] = m_\alpha E_\alpha, \tag{2.18}$$

where

$$m_\alpha = \sum_{i=1}^{r} \alpha_i^0 \alpha_i.$$

Thus

$$|m_\alpha| \le 1. \tag{2.19}$$

This inequality implies that when the generators of the group are arranged into standard angular momentum representations under the action of J^i, these representations can have only $j=0, \frac{1}{2}$, or 1 and that the only $j=1$ piece consists of the original generators J^i. Further, if $p(G)$ is defined as the number of generators which belong to doublets and $s(G)$ the number which are singlets, then Eq. (2.11) and Eq. (2.18) imply the relation

$$C(G)=\mathrm{tr}(J_3{}^2)=2+\tfrac{1}{4}p(G)$$
$$=\tfrac{1}{4}[5+d(G)-s(G)], \qquad (2.20)$$

which will be referred to in Sec. III. The value of $C(G)$ for each of the simple Lie groups is listed in Table I.

III. APPLICATION OF THE INDEX THEOREM

We now apply the Atiyah-Singer index theorem[8] to our problem. Since the index theorem is valid only on compact manifolds, it cannot be applied directly to the Yang-Mills theory in Euclidean space. However, as explained in Sec. I, the conformal invariance of the Yang-Mills action makes it possible to project the theory onto the four-dimensional hypersphere, where the index theorem is applicable. Throughout this section we will understand the theory to be in this projected form.

Given a self-dual field strength $F_{\mu\nu}$ arising from a vector potential A_μ, we wish to investigate infinitesimal variations δA_μ which preserve to first order the self-duality of $F_{\mu\nu}$. The condition that the first-order change in $F_{\mu\nu}$ be self-dual may be written as

$$(D_1\delta A)_{\alpha\beta}\equiv \Pi_{\alpha\beta}{}^{\mu\nu}(D_\mu\delta A_\nu-D_\nu\delta A_\mu)=0, \qquad (3.1)$$

where $D_\mu\delta A_\nu$, defined in Eq. (1.8), is the gauge-covariant derivative using the unperturbed A_μ, and $\Pi_{\alpha\beta}{}^{\mu\nu}$ is a projection matrix which picks out the anti-self-dual part of a tensor. Among the solutions to Eq. (3.1) are those arising from infinitesimal gauge transformations of the unperturbed A_μ; these are of the form

$$(D_0\delta\Lambda)_\mu\equiv D_\mu\delta\Lambda. \qquad (3.2)$$

Two solutions of Eq. (3.1) are gauge equivalent if they differ by a field of the form of Eq. (3.2). Our problem is to find the number of linearly independent gauge-inequivalent solutions of Eq. (3.1).

A convenient device for formulating our problem and for applying the index theorem is the sequence of mappings[7]

$$0\xrightarrow{D_{-1}}M^0\xrightarrow{D_0}M^1\xrightarrow{D_1}M^2_-\xrightarrow{D_2}0. \qquad (3.3)$$

M^0, M^1, and M^2_- are the spaces of scalar, vector, and anti-self-dual antisymmetric rank-two tensors, respectively, all transforming under the adjoint representation of the group G. D_{-1} takes 0 to the scalar field which is identically 0, D_0 and D_1 are the differential operators defined by Eqs. (3.1) and (3.2), and D_2 takes all of M^2_- to 0. At each step in the sequence, the image of D_{l-1} is contained in the kernel of D_l. (The only nontrivial case is the application of D_1D_0 to a scalar field. This vanishes because a gauge transformation cannot change the self-duality of $F_{\mu\nu}$.) We may define equivalence classes of elements in the kernel of D_l by defining two elements to be equivalent if they differ by an element in the image of D_{l-1}. These equivalence classes form a vector space

$$H^l=\frac{\text{kernel } D_l}{\text{image } D_{l-1}} \qquad (3.4)$$

whose dimension we shall denote by h^l. In particular, H^1 consists of the classes of gauge-equivalent solutions to Eq. (3.1); its dimension, h^1, is precisely the quantity in which we are interested.

Since the image of D_{-1} is the field which is identically zero, H^0 is just the space of scalar fields in the adjoint representation with vanishing covariant derivative. This space may be simply described in terms of the holonomy group of the vector potential A_μ. (The holonomy group at a point x_0 is defined as follows: Given a vector potential, the operation of parallel transport along a closed path beginning and ending at x_0 determines an element of the group G. The holonomy group at x_0 is the subgroup of G obtained by considering all possible paths. For a connected manifold, the holonomy groups at different points are easily seen to be isomorphic.) Since the fields in H^0 have vanishing covariant derivative, they are unchanged by parallel transport about a closed path and thus at every point are left unchanged by the holonomy group. Conversely, any element left unchanged by the holonomy group at a point x_0 determines, by parallel transport, an element at every point, thus giving a well-defined field $\phi(x)$ with vanishing covariant derivative. Thus h^0 is equal to the dimension of the subspace of the adjoint representation which is left unchanged by the holonomy group; this is equal to the dimension of the largest subgroup of G commuting with the holonomy group.

Since the kernel of D_2 is all of M^2_-, h^2 is equal to the dimension of the subspace of M^2_- orthogonal to the image of D_1. But this subspace is just the kernel of D_1^*, where D_1^* is the adjoint of D_1, so h^2 is the number of linearly independent solutions to $D_1^*T=0$. Any tensor field satisfying this equation also satisfies $D_1D_1^*T=0$; in Appendix B, we show that on a sphere (with the usual metric) $D_1D_1^*$ is a positive-definite operator, so $h^2=0$.

We now define an elliptic differential operator $\mathfrak{D}$, which takes ordered pairs of scalar and anti-self-dual tensor fields to vector fields, by

$$(\mathfrak{D}(S,T))_\mu = (D_0 S)_\mu + (D_1^* T)_\mu . \qquad (3.5)$$

Using some linear algebra, one can show that its index, defined by

$$\mathcal{I}(\mathfrak{D}) = \dim(\text{kernel } \mathfrak{D}) - \dim(\text{kernel } \mathfrak{D}^*), \qquad (3.6)$$

is related to the h^i defined above by

$$\mathcal{I}(\mathfrak{D}) = h^0 - h^1 + h^2. \qquad (3.7)$$

(Keep in mind that for a hypersphere, the case in which we are interested, $h^2 = 0$.) On the other hand, the Atiyah-Singer index theorem gives an expression for $\mathcal{I}(\mathfrak{D})$ in terms of the topological charge k, the topological invariants of the manifold on which the fields are defined, and constants which depend on the group G. For a four-dimensional manifold, this expression will have the form

$$\mathcal{I}(\mathfrak{D}) = ak + b. \qquad (3.8)$$

The constants a and b can be calculated by purely topological methods; instead, we shall obtain them by analytically determining $\mathcal{I}(\mathfrak{D})$ for $k=0$ and $k=1$.[16]

For $k=0$, any self-dual configuration must have zero action, and therefore $F_{\mu\nu}=0$. The holonomy group then consists of only the unit element, so h^0 is equal to the dimension of the group, $d(G)$. Since there are no solutions to Eq. (3.1) other than infinitesimal gauge transformations, $h^1=0$, and so $\mathcal{I}(\mathfrak{D}) = d(G)$.

For an example with $k=1$ we embed the one-pseudoparticle solution of Belavin *et al.*[1] into the group G via the minimal SU_2 subgroup described in Sec. II. It was shown in Sec. II that with respect to this subgroup the generators of G belong to one triplet (the generators of the SU_2 itself), $\frac{1}{2}p(G)$ doublets, and $s(G)$ singlets; clearly h^0 is equal to $s(G)$. We can obtain h^1 by a simple extension of 't Hooft's analysis[17] of the fluctuations about an SU_2 pseudoparticle. For the SU_2 case 't Hooft found eight modes corresponding to solutions of Eq. (3.1). Three of these correspond to gauge transformations, so only the remaining five, corresponding to translations and dilatation, contribute to h^1. Extending the analysis to vector fields belonging to doublets yields two modes per doublet, but both correspond to gauge transformations (see Appendix C). For singlet vector fields there are no modes. Thus for any group we obtain $h^1=5$, and $\mathcal{I}(\mathfrak{D}) = s(G) - 5$. The constants in Eq. (3.8) are now determined; using Eq. (3.7), we obtain

$$h^1 = [5 + d(G) - s(G)]k - [d(G) - h^0]. \qquad (3.9)$$

Using Eq. (2.20) we may rewrite this as

$$h^1 = 4C(G)k - d(G) + h^0. \qquad (3.10)$$

It should be noted that this result can also be obtained by using the method of Brown *et al.*[6] on E^4.

IV. DETERMINATION OF h^0

Given a group G and a value of the topological charge k, our goal is to find the dimension $N(G,k)$ of the manifold of self-dual configurations (modulo the action of the gauge group).[18] In this section we will show how to use the results of Sec. III, along with some knowledge of the structure of the Lie groups, to determine $N(G,k)$.

Given any initial self-dual configuration, the index theorem allows us to calculate h^1, the number of linearly independent gauge-inequivalent solutions to the infinitesimal variation problem. On the right-hand side of Eq. (3.10), however, appears the quantity h^0, the dimension of the largest subgroup of G which commutes with the holonomy group of the initial configuration. Two different configurations with the same value of k may have different values of h^0. Thus, the central problem is to find the appropriate value of h^0 such that $h^1 = N(G,k)$.

We begin by establishing some facts about the holonomy group. Given a configuration of Yang-Mills fields, let $G'(x)$ denote the holonomy group associated with the point x. Given two points x and y, $G'(x)$ and $G'(y)$ are equivalent in the following sense: There is an element $g_{xy} \in G$ which generates an isomorphism between the two groups of the form $g_x = g_{xy} g_y g_{xy}^{-1}$, where $g_x \in G'(x)$ and $g_y \in G'(y)$. (g_{xy} can be taken as the element of G defined by parallel transport along some fixed path between the two points.) We now show that it is possible to choose a gauge in which $G'(x)$ is identical for all points x, and in which $A_\mu^i(x)$ has nonzero values only within the holonomy group. On R^4 one imposes the following gauge conditions:

$$\begin{aligned} &A_1^i = 0 \text{ everywhere},\\ &A_2^i = 0 \text{ if } x_1 = 0,\\ &A_3^i = 0 \text{ if } x_1 = x_2 = 0,\\ &A_4^i = 0 \text{ if } x_1 = x_2 = x_3 = 0. \end{aligned} \qquad (4.1)$$

These conditions can be achieved (one at a time) by performing gauge transformations which are determined by simple first-order differential equations. (S^4 can be covered by two overlapping coordinate systems, and within each system one can impose the above gauge conditions.) In this gauge an arbitrary point x can be connected to the origin by a path along which $A_\mu dx^\mu = 0$, and hence parallel transport is trivial. Thus, $G'(x) = G'(x=0) \equiv G'$. One then chooses a basis for the Lie algebra $\mathfrak{G}$ such that G' is generated by the first m generators, where $m = \dim(G')$. Then $F_{\mu\nu}^i(x) = 0$ unless $i \le m$, since G' contains the transformations obtained by parallel transport around infinitesimal loops. Finally, one notes that these gauge conditions allow one to express $A^i_\mu(x)$ in terms of integrals which are linear in $F^i_{\mu\nu}$. Thus,

$$A^i_\mu(x)=0 \text{ unless } i \le m. \tag{4.2}$$

Within the manifold of self-dual configurations, we will call a particular configuration "regular" if it is contained in an open region of configurations which all have equivalent holonomy groups. By a gauge transformation, it is possible to write this entire family of configurations in a form consistent with Eq. (4.2). We will see that such regular configurations are exactly the ones which give us the desired value for h^0, and hence h^1.

Suppose we consider a regular initial configuration with a given value of k. Using the index theorem, we found in Sec. III that $h^1=4C(G)k-d(G)+h^0$. These infinitesimal variations can then be iterated to obtain a family of self-dual solutions with h^1 parameters, with the initial configuration taken as the origin of parameter space. If $h^2=0$, such an iteration exists to all orders and the series has a nonzero radius of convergence.[7,19] Since the initial configuration is regular, there exists a region about the origin of parameter space in which all the configurations can be written to obey Eq. (4.2). Within this region, it is clear that the configurations are gauge inequivalent. Recall that the infinitesimal variation problem was formulated to remove at the outset the possibility of equivalence by infinitesimal gauge transformations. One must, however, also consider the possibility that there is a class of finite gauge transformations which leave the initial configuration invariant, but which lead to an equivalence among the infinitesimal variations. With our choice of gauge, the class of gauge transformations which leave the initial configuration invariant is simply the h^0-parameter class of global gauge transformations which commute with G'. These gauge transformations, however, also leave invariant any configuration obeying Eq. (4.2), and thus the gauge inequivalence of the family of configurations is established. Thus, we have constructed a local manifold of dimension h^1. (If the initial configuration were not regular, only the argument about gauge inequivalence would break down.) If we let $L(G,k)$ be the value of h^0 corresponding to a regular configuration of topological change k, then

$$N(G,k)=4C(G)k-d(G)+L(G,k). \tag{4.3}$$

[Since $N(G,k)$ has a well-defined value, the above relation implies that h^0 has the same value for all regular configurations.]

We are now prepared to derive $N(G,k)$ for all compact simple Lie groups G. We begin with the series $G=SU_n$.

Our proof will be based on mathematical induction, so we begin by stating the answer:

$$N(SU_n,k)=\begin{cases}4k^2+1 & \text{if } k\le \frac{1}{2}n,\\ 4nk-(n^2-1) & \text{if } k\ge \frac{1}{2}n.\end{cases} \tag{4.4}$$

[The lower expression is simply $4C(G)k-d(G)$. The upper expression is the maximum value of the lower expression for fixed k. The crossover point is the value of n for which this maximum occurs.] One first verifies the solution for SU_2: Clearly $L(SU_2,k)=0$, since there are no non-Abelian proper subgroups. We now assume that the formula holds for $SU_2,\ldots,SU_{n-1}$, and consider the case of SU_n. By Eq. (4.3),

$$N(SU_n,k)=4nk-(n^2-1)+L(SU_n,k). \tag{4.5}$$

Since any SU_{n-1} configuration can be embedded into an SU_n theory, it follows that

$$N(SU_n,k)\ge N(SU_{n-1},k)\,. \tag{4.6}$$

The problem will then be solved by proving one more relation:

$$N(SU_n,k)=N(SU_{n-1},k) \text{ if } L(SU_n,k)\ne 0. \tag{4.7}$$

To prove Eq. (4.7), imagine constructing a family of self-dual configurations with the same holonomy group G'. Then $L(SU_n,k)\ne 0$ implies that there is at least one generator τ_0 which commutes with G'. One can diagonalize τ_0 (in the fundamental representation). Suppose that each repeated eigenvalue λ_i, $i=1,\ldots,r$, occurs with multiplicity p_i; let s be the number of unrepeated eigenvalues. Then

$$s+\sum_i p_i=n,\quad p_i<n. \tag{4.8}$$

If G'' is defined as the group generated by all elements of $\mathfrak{G}$ which commute with τ_0, then

$$G'\subset G''=SU_{p_1}\times\cdots\times SU_{p_r}\times(U_1)^{r+s-1}. \tag{4.9}$$

One can then write the entire family of configurations in a gauge satisfying Eq. (4.2), and one can further choose a basis for $\mathfrak{G}$ in which the generators of the subgroups $SU_{p_1},\ldots,SU_{p_r}$ occur as elements. Each configuration then decomposes into a superposition of r mutually independent Yang-Mills configurations. (Self-duality implies that the U_1 fields must vanish.) Within each subgroup SU_{p_i} the configuration will have a topological charge k_i, with $k=\sum k_i$. It follows that

$$N(SU_n,k)\le \sum_i N(SU_{p_i},k_i). \tag{4.10}$$

Regardless of the values of the k_i's and p_i's, the above inequality will hold if the right-hand side is replaced by its maximum possible value. It is a straightforward exercise to verify from Eq. (4.4) that this maximum value is $N(SU_{n-1}, k)$. Equation (4.7) is then obtained by recalling the inequality (4.6).

One can now carry out the induction in two steps First, suppose $k \geq \frac{1}{2}n$. Then

$$N(SU_n, k) \geq 4nk - (n^2 - 1) > N(SU_{n-1}, k). \qquad (4.11)$$

This contradicts Eq. (4.7), so one must have $L(SU_n, k) = 0$. Then suppose $k < \frac{1}{2}n$. It follows that

$$4nk - (n^2 - 1) < N(SU_{n-1}, k). \qquad (4.12)$$

Equations (4.5), (4.6), and (4.12) imply that $L(SU_n, k) > 0$, and then Eq. (4.7) determines $N(SU_n, k)$. Thus, the induction hypothesis has been shown for SU_n.

For the other groups, the answer can again be stated and then proved by induction:

$$N(G,k) = \begin{cases} I(G,k) & \text{if } k \leq M(G), \\ 4C(G)k - d(G) & \text{if } k \geq M(G), \end{cases} \qquad (4.13)$$

where all of the quantities on the right are listed in Table I.

The proof of Eq. (4.13) for the symplectic groups follows the same pattern as the proof given for the unitary groups. The induction begins at $Sp_2 = SU_2$. Equation (4.9) must of course be revised. Here the analysis is facilitated by using the root diagrams rather than the fundamental representations and choosing a regular subalgebra containing τ_0.[15] For Sp_{2n} one can show that

$$G' \subset SU_p \times Sp_{2r} \times (U_1)^{n+1-p-r}, \qquad (4.14)$$

where

$$p + r \leq n, \quad r < n. \qquad (4.15)$$

There is one further complication: When the SU_p root diagram is looked at within the root diagram of Sp_{2n}, its maximum root length is only $1/\sqrt{2}$. This means that a given configuration, when viewed in the Sp_{2n} theory, will have some $k = k_{SU} + k_{Sp}$. However, when the SU_p fields are viewed as a configuration in an SU_p theory, they have topological charge $k' = \frac{1}{2}k_{SU}$. This fact is necessary to show that

$$N(Sp_{2n}, k) = N(Sp_{2n-2}, k) \text{ if } L(Sp_{2n}, k) \neq 0. \qquad (4.16)$$

The proof for the orthogonal groups is then a straightforward exercise. The induction begins at $SO_5 \cong Sp_4$ for the odd orthogonal groups, and $SO_6 \cong SU_4$ for the even orthogonal groups. For SO_{2n+1} one can show that

$$G' \subset SU_p \times SO_{2r+1} \times (U_1)^{n+1-p-r} \qquad (4.17)$$

where again (4.15) holds. For SO_{2n} one can show

$$G' \subset SU_p \times SO_{2r} \times (U_1)^{n+1-p-r}, \qquad (4.18)$$

where (4.15) also holds. Using these relations, one can show that for $n \geq 8$

$$N(SO_n, k) = N(SO_{n-2}, k) \text{ if } L(SO_n, k) \neq 0. \qquad (4.19)$$

For $n = 7$ the right-hand side of the above equation is replaced by $N(SU_3, k)$.

Only the exceptional groups remain, and fortunately it is necessary to know only a few facts to determine the answer. Using the fact that $D_5 \subset E_6 \subset E_7 \subset E_8$, Eq. (4.6) may be replaced by

$$N(E_8, k) \geq N(E_7, k) \geq N(E_6, k) \geq N(D_5, k). \qquad (4.20)$$

Since SU_2 is always a subgroup,

$$N(G_2, k) \geq N(SU_2, k), \qquad (4.21)$$

$$N(F_4, k) \geq N(SU_2, k). \qquad (4.22)$$

(One must of course check that the above embeddings do not involve a rescaling of the topological charge.) Equation (4.9) may be replaced by the general statement that if $L(G, k) \neq 0$, then G' is contianed in a group G'' which, apart from U_1 factors must have a rank which is less than the rank of G. The decomposition of G'' may involve a rescaling of topological charge, but general arguments guarantee that such a rescaling must always be in the same direction as it was for the symplectic groups. In these cases such a rescaling has no effect on the derivation.

V. CONCLUSION

A self-dual configuration of topological charge k is often thought of as a kind of nonlinear superposition of k single pseudoparticles. (In Appendix A we discuss two examples which support this hypothesis.) This superposition interpretation makes a definite prediction for the number of parameters in the general solution: For each pseudoparticle there should be one scale parameter, four position parameters, and some number of parameters to specify the relative orientation of the pseudoparticle in group space. We now show that the number of parameters which we have calculated is in agreement with this interpretation.

To clarify the basis for this agreement, we summarize some of the logic used in Sec. III. Using the Atiyah-Singer index theorem, one asserts that

$$h^1 = -ak - b + h^0. \qquad (5.1)$$

For $k = 0$ one knows that $h^1 = 0$, $h^0 = d(G)$. For $k = 1$ one must show that there exists an embedding of the elementary SU_2 pseudoparticle into the group G for which $h^1 = 5$. Using the basis for the Lie algebra adopted in Sec. II, the value of h^0 for this embedding is simply equal to the number $s(G)$ of singlet generators (those which commute with the J_i used in the embedding). These two cases determine the constants a and b, leading to

$$h^1 = [5 + d(G) - s(G)]k - [d(G) - h^0] . \tag{5.2}$$

As discussed in Sec. IV, h^1, the number of solutions to the linearized equations, is equal to the number of parameters in the general solution, $N(G,k)$, provided we are expanding about a regular initial configuration, in which case $h^0 = L(G,k)$.

The terms in Eq. (5.2) can be easily recognized if such a configuration is viewed as a superposition of k, SU_2 pseudoparticles. Since $s(G)$ is the number of independent generators which commute with an arbitrary minimal SU_2 embedding $\{J_i\}$, $1 \le i \le 3$, the quantity $d(G) - s(G)$ in Eq. (5.2) is precisely the number of parameters necessary to specify the orientation of the J_i within the group G.

Thus, the first term on the right-hand side of Eq. (5.2) is just the number of parameters required to fix the scale, position, and group orientation (relative to a fixed basis) of each pseudoparticle. The second term simply subtracts out the number of nontrivial global gauge transformations of the entire configuration—thus, only the relative group orientations are counted.

In fact the value of the second term in Eq. (5.2), $d(G) - L(G,k)$, determined explicitly in Sec. IV, is also correctly given by the hypothesis that the solution is made up of k, SU_2 pseudoparticles. Consider a subgroup G_k of G generated by k minimal embeddings of SU_2. Again, such a subgroup G_k might be called regular if it commutes with the minimum number of independent generators of G. A simple rewording of the arguments in Sec. IV shows that the number of independent generators which commute with such a regular subgroup G_k is the same quantity $L(G,k)$ appearing above. Consequently $d(G) - L(G,k)$ is also the number of parameters entering a global gauge transformation which rotates a regular embedding of k minimal SU_2 subgroups.

Thus the superposition interpretation is in precise agreement with the number of parameters necessary to describe a self-dual configuration of any topological charge k, in any gauge group G. This agreement certainly suggests that it may be possible to parametrize the general configuration of topological charge k by the variables appropriate to this superposition interpretation.

ACKNOWLEDGMENTS

We thank Professor D. Burns, Professor I. M. Singer, and Professor R. T. Smith for helpful discussions.

APPENDIX A

We will now describe two examples which illustrate the hypothesis that a self-dual solution with $k>1$ can be viewed as a combination of $k=1$, SU_2 pseudoparticles. We first consider the one family of SU_2 multipseudoparticle solutions which is completely known[3]—the $k=2$ solution obtained by conformal transformation of Eq. (1.5):

$$A^i_\mu(x) = -\bar{\eta}^i_{\mu\nu}\partial_\nu \ln\left[\sum_{i=0}^{2} \frac{\lambda_i^2}{(x-x_i)^2}\right] . \tag{A1}$$

Consider the configuration in which the separations $|x_i - x_j|$ are comparable and the ratios λ_1/λ_0 and λ_2/λ_0 are very small. With this choice of parameters the solution (A1) looks much like two isolated pseudoparticles located at points x_1 and x_2. In particular, for x very near x_1, i.e.,

$$|x - x_1| \sim \frac{\lambda_1}{\lambda_0} |x_1 - x_0| ,$$

the argument of the logarithm becomes

$$\frac{\lambda_0^2}{(x_1-x_0)^2} + \frac{\lambda_1^2}{(x-x_1)^2} + O\left(\frac{\lambda_0\lambda_1}{(x_1-x_0)^2}\right) + O\left(\frac{\lambda_2^2}{(x_1-x_2)^2}\right), \tag{A2}$$

which approximates a $k=1$ solution at x_1 with scale $(\lambda_1/\lambda_0)|x_1 - x_0|$. On the other hand, for x outside two small regions about x_1 and x_2, i.e.,

$$|x - x_i| \gg \frac{\lambda_i}{\lambda_0} |x - x_0| ,$$

the argument of the logarithm is approximately the single term $\lambda_0^2/(x-x_0)^2$ which corresponds to a pure gauge transformation

$$A^i_\mu \frac{\sigma^i}{2} = -g\partial_\mu g^{-1} \tag{A3}$$

with

$$g(x) = \frac{x^4 - x_0^4 - i(\vec{x} - \vec{x}_0)\cdot\vec{\sigma}}{[(x-x_0)^2]^{1/2}} . \tag{A4}$$

If we perform the inverse of that gauge transformation, the solution looks precisely like the superposition of two $k=1$ pseudoparticles of the type (1.5) at the positions x_i, with scales $\lambda_i' = (\lambda_i/\lambda_0)|x_i - x_0|$ but each with a different SU_2 orientation,

$$\sigma^j A^j_\mu \simeq \sum_{i=1}^{2} -g^{-1}(x_i)\sigma^j g(x_i)\bar{\eta}^j_{\mu\nu}\partial_\nu \ln\left(1 + \frac{(\lambda_i')^2}{(x-x_i)^2}\right). \tag{A5}$$

The terms omitted from the right-hand side of Eq. (A5) are smaller than those retained by at least a factor of λ_i/λ_0 for $i = 1$ or 2.

A complementary test of this interpretation of self-dual solutions with $k>1$ can be found by generating such solutions from nonminimal embeddings of SU_2 into larger groups. Consider for example SU_3, generated by the standard eight matrices λ_i,

$$[\lambda_i, \lambda_j] = 2if_{ijk}\lambda_k \tag{A6}$$

with f_{ijk} normalized according to our convention. Since f_{ijk} vanishes when only two of its indices lie between one and three and since $f_{123} = +1$ we can immediately construct an SU_3 solution using the $k = 1$, SU_2 solution of Eq. (1.5),

$$A_\mu^i = \begin{cases} -\bar{\eta}^i_{\mu\nu}\partial_\nu \ln(1+\lambda^2/x^2), & 1 \le i \le 3 \\ 0, & 3 < i \end{cases} \tag{A7}$$

which, from Eq. (1.3), must also have topological charge 1. The three matrices λ^2, λ^5, and λ^7 also form an SU_2 subalgebra but with $f_{257} = \frac{1}{2}$. Thus a second SU_3 solution[20,21] can be written

$$A_\mu^{\prime i} = -2\rho^i_{\mu\nu}\partial_\nu \ln(1+\lambda^2/x^2), \tag{A8}$$

where

$$\rho^7_{\mu\nu} = \eta^1_{\mu\nu}, \quad \rho^5_{\mu\nu} = -\eta^2_{\mu\nu}, \quad \rho^2_{\mu\nu} = \eta^3_{\mu\nu},$$

and

$$\rho^i_{\mu\nu} = 0 \text{ for } i \ne 2,\ 5, \text{ or } 7. \tag{A9}$$

Because of the factor 2 in Eq. (A8), this solution has topological charge 4.

We now observe[20] that this $k = 4$, SU_3 solution can be written as a simple superposition of four minimal solutions of the type (A7) with various gauge orientations:

$$\sum_{i=1}^{8} A^{\prime i}\lambda^i = \sum_{l=1}^{4}\sum_{i=1}^{3} \tau_l^i A_\mu^i. \tag{A10}$$

Here the four sets of three matrices τ_l^i obey the SU_2 commutation relations

$$[\tau_l^i, \tau_l^j] = 2i\epsilon_{ijk}\tau_l^k \tag{A11}$$

and are given by

$$\tau_1^i = M^{-1}\begin{pmatrix} u\sigma^i u^{-1} & 0 \\ & 0 \\ 0\ \ 0 & 0 \end{pmatrix} M,$$

$$\tau_2^i = M^{-1}\begin{pmatrix} u^{-1}\sigma^i u & 0 \\ & 0 \\ 0\ \ 0 & 0 \end{pmatrix} M,$$

$$\tau_3^i = M^{-1}\begin{pmatrix} 0 & 0\ \ 0 \\ 0 & \\ 0 & u\sigma^i u^{-1} \end{pmatrix} M, \tag{A12}$$

$$\tau_4^i = M^{-1}\begin{pmatrix} 0 & 0\ \ 0 \\ 0 & \\ 0 & u^{-1}\sigma^i u \end{pmatrix} M,$$

where σ^i, $1 \le i \le 3$ are the Pauli matrices, $u = e^{(i\pi\sigma^3/}$ and

$$M = \begin{pmatrix} -\frac{1}{\sqrt{2}} & +\frac{i}{\sqrt{2}} & 0 \\ 0 & 0 & 1 \\ \frac{1}{\sqrt{2}} & \frac{i}{\sqrt{2}} & 0 \end{pmatrix} \tag{A13}$$

is a matrix which diagonalizes λ^2.

APPENDIX B

In Sec. III it was stated that $D_1D_1^*$ is a positive-definite operator on the four-dimensional hypersphere; in this appendix we prove this statement. In a curved space we may write ($\tilde{D}_\mu$ denotes the generally covariant and gauge covariant derivative)

$$(D_1 V)_{\alpha\beta} = \tfrac{1}{2}\left(g^\mu_\alpha g^\nu_\beta - g^\nu_\alpha g^\mu_\beta - \frac{1}{\sqrt{g}}\epsilon_{\alpha\beta}{}^{\mu\nu}\right)D_\mu V_\nu$$
$$= \tfrac{1}{2}\left(g^\mu_\alpha g^\nu_\beta - g^\nu_\alpha g^\mu_\beta - \frac{1}{\sqrt{g}}\epsilon_{\alpha\beta}{}^{\mu\nu}\right)\tilde{D}_\mu V_\nu. \tag{B1}$$

Intergrating by parts to find the effect of D_1^* on an anti-self-dual tensor $T_{\mu\nu}$, we obtain

$$(D_1^* T)_\alpha = -2\tilde{D}_\beta T^\beta{}_\alpha. \tag{B2}$$

Combining Eqs. (B1) and (B2), and using the anti-self-duality of $T_{\mu\nu}$, we find after some manipulation

$$(D_1D_1^*T)_{\mu\nu} = -g^{\alpha\beta}\tilde{D}_\alpha\tilde{D}_\beta T_{\mu\nu} + R_{\nu\lambda}T^\lambda{}_\mu - R_{\mu\lambda}T^\lambda{}_\nu$$
$$+ 2R_{\mu\beta\nu\lambda}T^{\beta\lambda} + F_{\beta\mu}T^\beta{}_\nu - F_{\beta\nu}T^\beta{}_\mu. \tag{B3}$$

Because $F_{\mu\nu}$ is assumed to be self-dual, the last two terms cancel. The Riemann tensor $R_{\mu\beta\nu\lambda}$ may be decomposed as[22]

$$R_{\mu\beta\nu\lambda} = \tfrac{1}{2}(g_{\mu\nu}R_{\beta\lambda} - g_{\mu\lambda}R_{\beta\nu} - g_{\beta\nu}R_{\mu\lambda} + g_{\beta\lambda}R_{\mu\nu})$$
$$- \tfrac{1}{6}R(g_{\mu\nu}g_{\beta\lambda} - g_{\mu\lambda}g_{\beta\nu}) + C_{\mu\beta\nu\lambda}. \tag{B4}$$

The tensor $C_{\mu\beta\nu\lambda}$ is called the Weyl, or conformal, tensor and vanishes whenever it is possible to choose a coordinate system in which $g_{\mu\nu}$ is proportional to a constant matrix throughout the manifold. Since the sphere has this property, we may set $C_{\mu\beta\nu\lambda} = 0$ and substitute Eq. (B4) into Eq. (B3), obtaining

$$(D_1D_1^*T)_{\mu\nu} = -g^{\alpha\beta}\tilde{D}_\alpha\tilde{D}_\beta T_{\mu\nu} - \tfrac{1}{3}RT_{\mu\nu}. \tag{B5}$$

Since $-g^{\alpha\beta}\tilde{D}_\alpha\tilde{D}_\beta$ is a positive operator, and R is everywhere negative on the hypersphere, $D_1D_1^*$ is positive-definite.

APPENDIX C

The small oscillations about the SU_2 one-pseudoparticle solution have been investigated by 't

Hooft.[17] Working in a background gauge, he showed that for either scalar, spinor, or vector fields the normal modes are eigenmodes of the operator

$$\mathfrak{M} = -\left(\frac{\partial}{\partial r}\right)^2 - \frac{3}{r}\frac{\partial}{\partial r} + \frac{4}{r^2}L^2 + \frac{8}{(1+r^2)}\vec{T}\cdot\vec{L}_1 + \frac{4r^2}{(1+r^2)^2}\vec{T}^2 + \frac{16}{(1+r^2)^2}\vec{T}\cdot\vec{S}_1\,, \qquad (C1)$$

where $\vec{L}_1$, $\vec{L}_2$, $\vec{S}_1$, and $\vec{S}_2$ are orbital and spin angular momentum operators for the two SU_2 components of SO_4 and $\vec{T}$ is the isospin operator. The orbital angular momenta satisfy $\vec{L}_1^{\,2} = \vec{L}_2^{\,2} \equiv L^2$. For scalars, $\vec{S}_1 = \vec{S}_2 = 0$, while (s_1, s_2) is $(\frac{1}{2}, 0)$ and $(0, \frac{1}{2})$ for right- and left-handed spinors and $(\frac{1}{2}, \frac{1}{2})$ for vectors. The multiplicity of a mode is $(2j_1+1)\times(2j_2+1)$, where $\vec{J}_1 = \vec{L}_1 + \vec{S}_1 + \vec{T}$ and $\vec{J}_2 = \vec{L}_2 + \vec{S}_2$.

This analysis may be used to study the small oscillations about self-dual solutions in a theory with a larger gauge group, G, which is obtained by embedding the SU_2 one-pseudoparticle solution. The generators of G can be classified into multiplets according to their transformation properties under the SU_2 of the embedding. In particular, if the minimal SU_2 is used, there will be one triplet, with the remaining generators belonging to doublets and singlets; thus we must consider vector field small oscillations with $t = 1, \frac{1}{2}$, or 0.

The modes which preserve the self-duality of $F_{\mu\nu}$ are precisely those with zero eigenvalue. Thus, to calculate h^1 for the $k=1$ configuration considered in Sec. III, we must determine the number of such modes, excluding those which correspond to gauge transformations. The $t=1$ modes are just those considered by 't Hooft; there are eight modes with zero eigenvalue, of which three correspond to gauge transformations. The $t=\frac{1}{2}$ case follows immediately from 't Hooft's result for right-handed spinors. [Since Eq. (C1) does not involve $\vec{S}_2$, the vector and right-handed spinor eigenvalues are the same, except for their multiplicity.] These modes have $j_1 = l_1 = l_2 = 0$, $j_2 = s_2 = \frac{1}{2}$, so there are two modes per doublet. However, these modes may be written in the form $D_\mu \delta\Lambda$, where $\delta\Lambda$ is an isospinor given by

$$\delta\Lambda = \left[\frac{x_4 - i\vec{x}\cdot\vec{\tau}}{(1+x^2)^{1/2}}\right] v \qquad (C2)$$

with v an arbitrary isospinor. Thus, when viewed in terms of the gauge group G, these modes correspond to gauge transformations and do not contribute to h^1. Finally, for $t=0$ there are no normalizable solutions of $\mathfrak{M}\psi = 0$.

---

*This research was supported in part by the U. S. Energy Research and Development Administration.

[1]A. Belavin, A. Polyakov, A. Schwartz, and Y. Tyupkin, Phys. Lett. 59B, 85 (1975).

[2]E. Witten, Phys. Rev. Lett. 38, 121 (1977).

[3]R. Jackiw, C. Nohl. and C. Rebbi, Phys. Rev. D 15, 1642 (1977).

[4]R. Jackiw and C. Rebbi, Phys. Lett. 67B, 189 (1977).

[5]A. Schwartz, Phys. Lett. 67B, 172 (1977).

[6]L. Brown, R. Carlitz, and C. Lee, Phys. Rev. D 16, 417 (1977).

[7]M. Atiyah, N. Hitchin, and I. Singer, Proc. Natl. Acad. Sci. USA 74, 2662 (1977).

[8]M. Atiyah and I. Singer, Ann. Math. 87, 484 (1968); M. Atiyah, R. Bott, and V. Patodi, Invent. Math. 19, 279 (1973).

[9]Semisimple algebras can of course be written as the tensor product of simple algebras. The Yang-Mills field for the semisimple theory is then just the superposition of the corresponding mutually independent fields of the simple Yang-Mills theories.

[10]G. 't Hooft (unpublished). The matrix $\bar{\eta}_{i\mu\nu}$ is defined by G. 't Hooft, Phys. Rev. D 14, 3432 (1976).

[11]S. Coleman (unpublished). See also J. Kiskis, Phys. Rev. D 15, 2329 (1977).

[12]We have been able to find a proof of this statement for all simple compact Lie groups except E_6, E_7, and E_8. A. Borel, Bull. Am. Math. Soc. 61, 397 (1955). See also N. Steenrod, *The Topology of Fibre Bundles* (Princeton Univ. Press, Princeton, New Jersey, 1951).

[13]G. Racah, in *Ergebnisse der Exakten Naturwissenshaften*, edited by G. Höhler (Springer, Berlin, 1965), Vol. 37.

[14]L. Pontryagin, *Topological Groups* (Gordon and Breach, New York, 1960); V. Varadarajan, *Lie Groups, Lie Algebras and Their Representations* (Prentice-Hall, Englewood Cliffs, New Jersey, 1974).

[15]The possibility of this choice is a consequence of theorem 4.1.6. of V. Varadarajan (Ref. 14) when restricted to compact Lie algebras.

[16]This method was used by A. Schwartz (Ref. 5) to calculate $\mathcal{G}(\mathfrak{D})$ for SU_2.

[17]G. 't Hooft, Phys. Rev. D 14, 3432 (1976).

[18]It should be noted that the manifold may not be connected, and that different components of the manifold could conceivably have different dimensions. If this should occur, we let $N(G,k)$ refer to the largest of the dimensions of the various components. The arguments to be used in the text would then be valid provided that one applies them only to the components of maximum dimension. One such maximal component contains the mimimal embedding of an SU_2 solution of topological charge k. In fact, by using methods similar to those of this section, we have been able to show, for all simple compact Lie groups except E_8, that all components have the same dimension.

[19]M. Kuranishi, in *Proceedings of the Conference on Complex Analysis*, edited by H. Röhrl (Springer, Berlin, 1965).

[20]F. A. Wilczek, Phys. Lett. 65B, 160 (1976).

[21]W. Marciano, H. Pagels, and Z. Parsa, Phys. Rev. D 15, 1044 (1977).

[22]S. Weinberg, *Gravitation and Cosmology* (Wiley, New York, 1972).

IV. GREEN FUNCTIONS IN THE INSTANTON BACKGROUND

INTRODUCTION

The calculation of propagators in the arbitrary self-dual background corresponding to k instantons of a generic form was started by Christ *et al.* (see paper in Sec. III). These authors considered the $SU(2)$ gauge group and the scalar field of isospin 1/2 and 1; they followed the method of Ref. [2]. The analysis has been extended to arbitrary gauge groups and different representations of the scalar field in Ref. [1].

The issue of propagators for higher spins was treated in Ref. [2]. This work presents a very elegant procedure which teaches us to build the Green functions for (spatially) spinor and vector fields in terms of those for the scalar field in the self-dual background. A subtlety which deserves mention here is the fact that the corresponding Dirac and Klein–Gordon equations have zero modes (in the former case this is the 't Hooft zero mode). In the presence of the zero modes the equation defining the Green functions has to be modified in a well-known way. We encounter no trouble in the spinor propagator. At the same time, in the vector propagator, an infrared divergence emerges whose fate is not completely clear at the moment. The question of how these infrared divergent terms in the vector propagator might be eliminated is addressed in Ref. [3]. Some additional comments on this divergence can be found in Ref. [4].

References

[1] *E. Corrigan, P. Goddard and S. Templeton, *Nucl. Phys.* **B151** (1979) 93.
[2] *L. Brown *et al.*, *Phys. Rev.* **D17** (1978) 1583.
[3] H. Levine and L. Yaffe, *Phys. Rev.* **D19** (1979) 1225.
[4] M. Voloshin, *Nucl. Phys.* **B359** (1991) 301.

PHYSICAL REVIEW D VOLUME 17, NUMBER 6 15 MARCH 1978

Propagation functions in pseudoparticle fields

Lowell S. Brown, Robert D. Carlitz,* Dennis B. Creamer,* and Choonkyu Lee†
Department of Physics, University of Washington, Seattle, Washington 98195
(Received 6 October 1977)

The Green's functions for massless spinor and vector particles propagating in a self-dual but otherwise arbitrary non-Abelian gauge field are shown to be completely determined by the corrresponding Green's functions of scalar particles. Simple, explicit algebraic expressions are constructed for the scalar Green's functions of isospin-1/2 and isospin-1 particles in the self-dual field of a configuration of n pseudoparticles described by $5n$ arbitrary parameters.

I. INTRODUCTION

The existence of classical, pseudoparticle solutions[1] in non-Abelian gauge theory has profound implications for the structure and physical consequences of this theory. The classical pseudoparticle solution in Euclidean space provides a tunneling path which yields a finite quantum transition amplitude between different would-be vacuum states characterized by a vanishing field-strength tensor but with a nontrivial gauge field. Thus the true vacuum state in non-Abelian gauge theory has a rich structure.[2-4] The pseudoparticle solution may provide a resolution[2,3] of the "U(1) problem" by removing the unwanted Goldstone boson from a theory which has an apparent chiral-phase symmetry. Pseudoparticle solutions may also provide a mechanism for quark confinement.[5,6]

Clearly, it is important to develop the dynamical theory of fields quantized about classical pseudoparticle solutions. The first part of this program entails the determination of the nature of the small fluctuations of such quantum fields. This involves the calculation of propagators (Green's functions) for particles moving in the external field of a pseudoparticle. The small fluctuation of the non-Abelian gauge field and its associated "ghost" field correspond to the motion of spin-1 and spin-0, massless particles. Hadronic matter is presumably described by the interaction of these fields with those of spin-$\frac{1}{2}$, massless particles (quantum chromodynamics). In this paper, we shall present explicit and simple algebraic formulas for the Euclidean propagation functions of massless particles with spin 0, $\frac{1}{2}$, and 1 moving in the external, classical, non-Abelian gauge field of any pseudoparticle solution. A brief account of some of our results has already appeared.[7]

The pseudoparticle solutions that concern us are characterized by field-strength tensors which are either self-dual or anti-self-dual. We begin our development by showing that the propagation functions for massless spin-$\frac{1}{2}$ and spin-1 particles moving in a self-dual or anti-self-dual but otherwise arbitrary gauge field are determined explicitly by the propagation functions of the corresponding massless, spin-0 particles. This is a completely general result which holds for any gauge group; the only restrictions are that the field strength be self-dual (or anti-self-dual) and that the particles be massless. The spin-$\frac{1}{2}$ case will be worked out below in Sec. II and the spin-1 case in Sec. III. Thus we will need explicit expressions only for the propagation functions of massless, spin-0 particles moving in pseudoparticle fields.

The pseudoparticle solutions, and the propagation functions in these external fields, are defined in Euclidean space-time. We shall work entirely in Euclidean space in this paper. Moreover, we shall restrict our discussion of the spin-0 propagators to those in a SU(2) gauge field. The original pseudoparticle solution[1] approaches a pure SU(2) gauge transformation at infinity, with the gauge transformation covering the SU(2) group once as the field point at infinity covers the S_3 hypersphere once. The gauge field has a topological character described by a winding number (or Pontryagin index) 1. Thus the solution is referred to as the field of one pseudoparticle. Its field-strength tensor is self-dual. There is another solution giving a similar mapping of the S_3 hypersphere once onto the SU(2) group, but with the points on S_3 mapped to the inverse group elements. This is the antipseudoparticle with winding number -1; its field-strength tensor is anti-self-dual. The general solution with n pseudoparticles or n antipseudoparticles covers the SU(2) group n times as the field point at infinity covers the hypersphere S_3 once. It has winding number $\pm n$ (with a self-dual or anti-self-dual field-strength tensor). The general solution is determined by $8n$ parameters[8] (three of which designate a global gauge orientation). Of these $8n$ parameters, $4n$ determine the positions of the n pseudoparticles, n describe their sizes, and $3n$ fix their orientations in the

SU(2) gauge space. Explicit n-pseudoparticle solutions have been constructed[9] in terms of $5n$ parameters[10]; these solutions do not contain the $3n$ parameters needed to fix the gauge orientations of the pseudoparticles. *We shall construct explicit and simple algebraic expressions for spin-0 propagators in this somewhat restricted n-pseudoparticle (or antipseudoparticle) field.* The construction for isospin $\frac{1}{2}$ will be carried out in Sec. IV and that for isospin 1 in Sec. V.

We discuss our results in Sec. VI. Here we note that the spin-$\frac{1}{2}$ and spin-1 propagation functions decrease more slowly at large distances than do free propagators. At large distances the spin-$\frac{1}{2}$ propagator is of order $1/x^2$ in contrast to the $O(1/x^3)$ behavior of the free propagator. Similarly, at large distances the spin-1 propagation function is of order $1/x$ while the free propagator is $O(1/x^2)$. This slowly vanishing character of the propagators in pseudoparticle fields is a gauge-independent property; it may significantly affect physical processes occurring in pseudoparticle fields.

II. SPIN $\frac{1}{2}$

We turn now to the construction of the massless spin-$\frac{1}{2}$ propagator in a self-dual or anti-self-dual but otherwise arbitrary non-Abelian gauge field. We shall use simple operator techniques to express this propagator in terms of the corresponding spin-0 propagator. Before considering the spin-$\frac{1}{2}$ propagator, let us first establish our notation and conventions. We work in Euclidean space-time, and use skew-Hermitian Dirac matrices $\gamma_1, \gamma_2, \gamma_3, \gamma_4 = i\gamma^0$ obeying the anticommutator condition

$$\{\gamma_\mu, \gamma_\nu\} = -2\delta_{\mu\nu}. \tag{2.1}$$

Hermitian Dirac matrices γ_5 and $\sigma_{\mu\nu}$ are defined by

$$\gamma_5 = \gamma_1\gamma_2\gamma_3\gamma_4 \tag{2.2}$$

(with $\gamma_5^2 = +1$) and

$$\sigma_{\mu\nu} = \tfrac{1}{2}i[\gamma_\mu, \gamma_\nu]. \tag{2.3}$$

The dual of a skew-symmetrical tensor $f_{\mu\nu} = -f_{\nu\mu}$ is defined by

$${}^d f_{\mu\nu} = \tfrac{1}{2}\epsilon_{\mu\nu\lambda\kappa} f_{\lambda\kappa}, \tag{2.4}$$

where $\epsilon_{\mu\nu\lambda\kappa}$ is the completely antisymmetrical tensor with $\epsilon_{1234} = +1$. A simple exercise in the Dirac matrix algebra shows that

$$\sigma_{\mu\nu}\left(\frac{1 \pm \gamma_5}{2}\right) = \mp {}^d\sigma_{\mu\nu}\left(\frac{1 \pm \gamma_5}{2}\right). \tag{2.5}$$

The matrices $\sigma_{\mu\nu}$ are generators of the Euclidean O(4) rotation group while the matrices $\frac{1}{2}(1 \pm \gamma_5)$ project into spaces of definite chirality, $\gamma_5' = \pm 1$. The formula (2.5) shows that the spaces of definite chirality reduce the representation of the O(4) algebra generated by $\sigma_{\mu\nu}$ into the direct product, $O(4) = SU(2) \otimes SU(2)$.

We shall consider the spin-$\frac{1}{2}$ propagation function in the non-Abelian field of an arbitrary gauge group specified by the real, completely antisymmetrical structure constants f_{abc}. The spin-$\frac{1}{2}$ field belongs to some representation of the gauge group with generators specified by Hermitian matrices T_a which obey the commutation relations

$$[T_a, T_b] = i f_{abc} T_c, \tag{2.6}$$

where a sum over repeated indices will always be implicit. The gauge-covariant derivative D_μ is defined by

$$D_\mu = \partial_\mu - iT_a A_{\mu a}(x), \tag{2.7}$$

where $A_{\mu a}(x)$ is the non-Abelian gauge field. By virtue of Eq. (2.6), this derivative obeys

$$[D_\mu, D_\nu] = -iT_a F_{\mu\nu a}, \tag{2.8}$$

where

$$F_{\mu\nu a}(x) = \partial_\mu A_{\nu a}(x) - \partial_\nu A_{\mu a}(x) + f_{abc} A_{\mu b}(x) A_{\nu c}(x) \tag{2.9}$$

is the non-Abelian field-strength tensor. Now, using Eqs. (2.1), (2.3), and (2.8), we find that

$$-(\gamma D)^2 = D^2 + \tfrac{1}{2}\sigma_{\mu\nu} T_a F_{\mu\nu a}. \tag{2.10}$$

Let us suppose that $F_{\mu\nu a}$ is self-dual,

$${}^d F_{\mu\nu a} = +F_{\mu\nu a}. \tag{2.11}$$

Then, according to Eqs. (2.5) and (2.10), we have

$$-(\gamma D)^2\left(\frac{1+\gamma_5}{2}\right) = D^2\left(\frac{1+\gamma_5}{2}\right), \tag{2.12}$$

since the $\sigma_{\mu\nu}F_{\mu\nu}$ term arising from Eq. (2.10), when acting on the chiral projection matrix $\frac{1}{2}(1+\gamma_5)$, is equal to $-{}^d\sigma_{\mu\nu}F_{\mu\nu} = -\sigma_{\mu\nu}{}^dF_{\mu\nu} = -\sigma_{\mu\nu}F_{\mu\nu}$ and thus vanishes. It is this simple statement [Eq. (2.12)] which will enable us to relate the massless spin-$\frac{1}{2}$ propagator to a massless spin-0 propagator in a self-dual, non-Abelian gauge field. In order to keep our notation clear, we shall write out the development for the case of a self-dual field. The anti-self-dual case is obtained simply by changing the sign of γ_5, $\gamma_5 \to -\gamma_5$.

The massless spin-$\frac{1}{2}$ propagator $S(x, y)$ has a formal representation as sum over normal modes,

$$S(x, y) \sim \sum_n \frac{\psi_n(x)\psi_n^\dagger(y)}{\lambda_n}, \tag{2.13}$$

with mode functions ψ_n of eigenvalues λ_n,

$$\gamma D\psi_n = \lambda_n \psi_n. \tag{2.14}$$

There are, however, a finite number N of zero-mode functions[11] $\psi_n^{(0)}$, satisfying

$$\gamma D\psi_n^{(0)} = 0 . \tag{2.15}$$

Since the matrix γ_5 anticommutes with γD, the zero-mode functions $\psi_n^{(0)}$ can be chosen to simultaneously diagonalize γ_5; i.e., they can be chosen to be chirality eigenstates. Now if we multiply the zero-mode equation (2.15) by γD we get, by Eq. (2.10),

$$(-D^2 - \tfrac{1}{2}\sigma_{\mu\nu}T_a F_{\mu\nu a})\psi_n^{(0)} = 0 . \tag{2.16}$$

If $\psi_n^{(0)}$ has positive chirality, $\gamma_5\psi_n^{(0)} = +\psi_n^{(0)}$, then, according to the discussion following Eq. (2.12), $\sigma_{\mu\nu}F_{\mu\nu}\psi_n^{(0)}$ vanishes. In this case there can be no zero-mode solution to Eq. (2.16) because the remaining operator, $-D^2$, is essentially a positive operator. (See Brown *et al.*, Ref. 8.) Hence the zero-mode functions must have negative chirality,

$$\gamma_5\psi_n^{(0)} = -\psi_n^{(0)} . \tag{2.17}$$

In this case, the $-\frac{1}{2}\sigma_{\mu\nu}T_a F_{\mu\nu a}$ contribution in Eq. (2.16) no longer vanishes and, in fact, presents a negative potential energy in appropriate spin-isospin states. Thus, Eq. (2.16) can appear as a Schrödinger equation at zero energy for a particle moving in a negative potential, an equation that may possess square-integrable solutions. As shown rigorously in Ref. 11, there are indeed a finite number of these solutions.

None of these zero modes can appear in the mode sum representation of the propagator, Eq. (2.13), if the propagator is to exist. Note, however, that the quantum transition amplitude with spin-$\frac{1}{2}$ fermion fields involves a factor of Det $\gamma D = \prod \lambda_n$ times the propagation function for some number, k, of spin-$\frac{1}{2}$ Fermi particles. The latter, k-particle propagation function is formed as an antisymmetric product of k two-point propagators, $\det_k S(x_m, y_n)$, corresponding to the Fermi statistics of spin-$\frac{1}{2}$ particles. On account of the factor Det γD, the quantum amplitude will vanish unless $k \geq N$ and all N zero-mode states are included in the k-particle propagation function. By virtue of the complete antisymmetry of the determinant (the Pauli principle), the remaining k-N particles in $\det_k S(x_m, y_n)$ cannot be in zero-mode states. Thus the propagation of these remaining particles, the particles with which we are concerned, is described by a propagator

$$S(x, y) = \sum_n{}' \frac{\psi_n(x)\psi_n^\dagger(y)}{\lambda_n} , \tag{2.18}$$

where the prime on the summation sign indicates that the zero modes (states with $\lambda_n = 0$) are to be deleted.

It follows from Eq. (2.18) that the spin-$\frac{1}{2}$ propagator obeys the Green's function equation

$$\gamma D S(x, y) = Q(x, y) , \tag{2.19}$$

where

$$Q(x, y) = \delta(x - y) - \sum_n \psi_n^{(0)}(x)\psi_n^{(0)\dagger}(y) , \tag{2.20}$$

with the summation running over all the zero-mode functions. The quantity $Q(x, y)$ represents the projection operator into the subspace of all nonzero modes, the complement of the zero-mode subspace. It also follows from Eq. (2.18) that the spin-$\frac{1}{2}$ propagator is orthogonal to all the zero-mode functions,

$$\int (d_E{}^4x)\psi_n^{(0)\dagger}(x)S(x, y) = 0 . \tag{2.21}$$

The Green's function equation (2.19) and the orthogonality constraint (2.21) serve to define the spin-$\frac{1}{2}$ propagation function $S(x, y)$.

A construction of this propagator is easily achieved with operator techniques. We write the function $S(x, y)$ as the matrix element of an operator S,

$$S(x, y) = \langle x|S|y\rangle . \tag{2.22}$$

Similarly, we write the corresponding spin-0 propagation function $\Delta(x, y)$, which is defined by

$$-D^2\Delta(x, y) = \delta(x - y) , \tag{2.23}$$

as the matrix element of an operator $1/-D^2$,

$$\Delta(x, y) = \left\langle x \left| \frac{1}{-D^2} \right| y \right\rangle . \tag{2.24}$$

We now assert that the operator expression of the spin-$\frac{1}{2}$ propagator is

$$S = \gamma D \frac{1}{-D^2}\left(\frac{1+\gamma_5}{2}\right) + \frac{1}{-D^2}\gamma D\left(\frac{1-\gamma_5}{2}\right) \tag{2.25}$$

The proof of this assertion is quick. We multiply Eq. (2.25) by γD and use Eq. (2.12) to get

$$\gamma D S = Q , \tag{2.26}$$

where

$$Q = \frac{1+\gamma_5}{2} + \gamma D \frac{1}{-D^2}\gamma D\left(\frac{1-\gamma_5}{2}\right) . \tag{2.27}$$

Now Eq. (2.26) implies that Q contains no zero modes since these modes are annihilated by γD. On the other hand, using Eq. (2.12) again, we find that

$$\gamma D Q = \gamma D . \tag{2.28}$$

Hence Q is the operator which projects into the subspace of all the nonzero modes, and we have

$$Q(x, y) = \langle x|Q|y\rangle , \tag{2.29}$$

where $Q(x, y)$ is the function defined in Eq. (2.20). The matrix elements of Eq. (2.26) thus reproduce the Green's function equation (2.19), and therefore the matrix elements of Eq. (2.25) produce a spin-$\frac{1}{2}$ propagation function which obeys this Green's function equation. [Incidentally, it is a simple matter to use Eq. (2.12) to check directly that the definition (2.27) does obey the projection property $Q^2 = Q$.] It remains to be shown that the assertion (2.25) gives a propagator which is orthogonal to all the zero modes, i.e., a propagator which obeys the constraints (2.21). This, however, is immediate, for Eq. (2.12) implies that

$$QS = S , \tag{2.30}$$

which is tantamount to the constraints (2.21).

The matrix elements of Eq. (2.25) express the massless spin-$\frac{1}{2}$ propagation function in a self-dual but otherwise arbitrary non-Abelian gauge field in terms of the corresponding massless, spin-0 propagation function,

$$S(x, y) = \gamma D^{(x)} \Delta(x, y)\left(\frac{1+\gamma_5}{2}\right) + \Delta(x, y)\gamma \overleftarrow{D}^{(y)}\left(\frac{1-\gamma_5}{2}\right) . \tag{2.31}$$

(Note that when the symbol D_μ acts to the left, $\overleftarrow{D}_\mu$, it involves the derivative ∂_μ with its sign reversed, $-\overleftarrow{\partial}_\mu$.) Equation (2.31) expresses the principal result of this section. The spin-$\frac{1}{2}$ propagator in an arbitrary anti-self-dual field is obtained from Eq. (2.31) by changing the sign of the γ_5 matrices and using the appropriate spin-0 propagator.

III. SPIN 1

The non-Abelian gauge field is governed by the action

$$g^2 W = -\int (d_E{}^4 x)\left[\tfrac{1}{4}(F_{\mu\nu a})^2 + \frac{1}{2\xi}(D^{\rm cl}_{\mu ab}A_{\mu b})^2\right] . \tag{3.1}$$

With our normalization of the gauge field [cf. Eq. (2.9)] the coupling constant g appears as an overall factor. Here we have chosen a "background gauge" specified by the parameter ξ: The operator $D^{\rm cl}_{\mu ab}$ is the gauge-covariant derivative

$$D^{\rm cl}_{\mu ab} = \partial_\mu \delta_{ab} + f_{acb} A^{\rm cl}_{\mu c} , \tag{3.2}$$

where the vector potential $A^{\rm cl}_{\mu a}$ describes a classical solution of the non-Abelian field equations and is fixed in all variations of the action. We shall study the small fluctuations $\phi_{\mu a}$ of the gauge field about the classical solution,

$$A_{\mu a} = A^{\rm cl}_{\mu a} + \phi_{\mu a} . \tag{3.3}$$

Inserting this decomposition of the vector potential into the field-strength tensor [Eq. (2.9)] and extracting pieces quadratic in ϕ_μ from the resulting action [Eq. (3.1)] yields the small-fluctuation, vector-field action

$$g^2 W_2 = -\tfrac{1}{2}\int (d_E{}^4 x)\phi_\mu[-D^2\delta_{\mu\nu} - 2F_{\mu\nu} + (1 - 1/\xi)D_\mu D_\nu]\phi_\nu . \tag{3.4}$$

Here and henceforth we omit the superscript cl on the classical fields $A^{\rm cl}_\mu$, $F^{\rm cl}_{\mu\nu}$, and on the corresponding gauge-covariant derivative $D^{\rm cl}_\mu$. We have also adopted a matrix notation with regard to group indices $a, b, \ldots$, defined

$$(F_{\mu\nu})_{ab} = f_{acb}F_{\mu\nu c} , \tag{3.5}$$

and used the commutation relation

$$[D_\mu , D_\nu] = F_{\mu\nu} , \tag{3.6}$$

which follows from the fact that the structure-constant matrices

$$(f_c)_{ab} = f_{acb} \tag{3.7}$$

obey the relation

$$[f_a, f_b] = f_{abc} f_c . \tag{3.8}$$

From the definition [Eq. (2.9)] of the field-strength tensor, it follows algebraically that the dual tensor ${}^dF_{\mu\nu,a}$ obeys

$$D_{\mu ab}\,{}^dF_{\mu\nu,b} = 0 . \tag{3.9}$$

Hence any gauge field with a self-dual (or anti-self-dual) field-strength tensor ${}^dF_{\mu\nu,a} = \pm F_{\mu\nu,a}$ provides automatically a solution of the field equations

$$D_{\mu ab}F_{\mu\nu b} = 0 . \tag{3.10}$$

Moreover, by a suitable gauge transformation, we may impose the background gauge condition

$$D_{\mu ab}A_{\mu b} = \partial_\mu A_{\mu a} = 0 . \tag{3.11}$$

In general, a continuously connected family of self-dual (or anti-self-dual) fields exists, labeled by some set of continuously varying parameters. [For example, if $A_\mu(x)$ yields a self-dual field-strength tensor, then so does the translated field $A_\mu(x - z)$, where z_μ are four constant parameters.] Thus, given a self-dual (or anti-self-dual) field, we can take its derivative with respect to one of its parameters to get a small-fluctuation field which is also self-dual (or anti-self-dual). Moreover, an infinitesimal gauge transformation can be added to this small-fluctuation field to bring it into the background gauge. We find that any self-dual (or anti-self-dual) field will support some number of zero-mode fluctuation fields $\phi_{\mu,a}^{(s)}$, $s=1,2,\ldots,N$, satisfying

$$D_\mu \phi_\mu^{(s)}=0, \tag{3.12}$$

and the field equation which follows from the action (3.4),

$$(D^2\delta_{\mu\nu}+2F_{\mu\nu})\phi_\nu^{(s)}=0. \tag{3.13}$$

In the background gauge, the fluctuation fields $\phi_\mu^{(s)}$ are square integrable,[12] and their contribution to the spin-1 propagator must be deleted just as the zero modes were deleted from the spin-$\frac{1}{2}$ propagator in the preceding section. Thus, according to the action (3.4), the spin-1 propagator $G_{\mu\nu}(x,y)$ in an arbitrary self-dual (or anti-self-dual) field obeys the Green's function equation

$$-[D^2\delta_{\mu\lambda}+2F_{\mu\lambda}-(1-1/\xi)D_\mu D_\lambda]G_{\lambda\nu}(x,y)=Q_{\mu\nu}(x,y), \tag{3.14}$$

where

$$Q_{\mu\nu}(x,y)=\delta_{\mu\nu}\delta(x-y)$$

$$\sum_{S=1}^{N}\phi_\mu^{(s)}(x)\phi_\nu^{(s)}(y) \tag{3.15}$$

projects onto the space of the nonzero-mode functions. (We should note that the zero modes play a different role here from that in the spin-$\frac{1}{2}$ case. The zero modes in the spin-1 field are accounted for by the introduction of appropriate collective coordinates.[13])

The vector propagation function in an arbitrary self-dual (or anti-self-dual) field can be constructed in terms of the corresponding scalar-field propagator using simple operator techniques akin to those employed in the preceding section for the spin-$\frac{1}{2}$ propagator. To proceed with this construction, we note that

$$p_{\mu\nu\lambda\kappa}^{(\pm)}=\tfrac{1}{4}(\delta_{\mu\lambda}\delta_{\nu\kappa}-\delta_{\mu\kappa}\delta_{\nu\lambda}\pm\epsilon_{\mu\nu\lambda\kappa}) \tag{3.16}$$

projects out the self-dual part ($p^{(+)}$) or anti-self-dual part ($p^{(-)}$) of an antisymmetrical tensor. We now define

$$q_{\mu\nu\lambda\kappa}^{(\pm)}=\delta_{\mu\nu}\delta_{\lambda\kappa}+4p_{\mu\nu\lambda\kappa}^{(\pm)} \tag{3.17}$$

and, for an arbitrary operator X,

$$\{X\}_{\mu\nu}^{(\pm)}=q_{\mu\nu\lambda\kappa}^{(\pm)}D_\lambda X D_\kappa. \tag{3.18}$$

(Some motivation for the introduction of the bracket operation is presented in the Appendix.) This bracket operation has several useful properties when the field $F_{\mu\nu}$ is either self-dual [for the (+) brackets] or anti-self-dual [for the (−) brackets]. First, we note that

$$p_{\mu\nu\lambda\kappa}^{(\pm)}D_\lambda D_\kappa=p_{\mu\nu\lambda\kappa}^{(\pm)}\tfrac{1}{2}[D_\lambda,D_\kappa]=\tfrac{1}{2}F_{\mu\nu}. \tag{3.19}$$

Hence

$$\{1\}_{\mu\nu}^{(\pm)}=D^2\delta_{\mu\nu}+2F_{\mu\nu}, \tag{3.20}$$

which expresses the field equation operator (with $\xi=1$) as the bracket operation on the identity [cf. Eq. (3.4)]. Second, we use the commutation relation of the gauge-covariant derivatives [Eq. (3.6)] and the self-duality property

$$\tfrac{1}{2}\epsilon_{\mu\nu\lambda\kappa}F_{\lambda\kappa}=\pm F_{\mu\nu} \tag{3.21}$$

to establish that

$$D_\mu\{X\}_{\mu\nu}^{(\pm)}=D^2 X D_\nu \tag{3.22a}$$

and

$$\{X\}_{\mu\nu}^{(\pm)}D_\nu=D_\mu X D^2. \tag{3.22b}$$

Third, we make use of the algebraic relation proved in the Appendix:

$$q_{\mu\sigma\alpha\beta}^{(\pm)}q_{\sigma\nu\lambda\kappa}^{(\pm)}=\delta_{\beta\lambda}q_{\mu\nu\alpha\kappa}^{(\pm)}+r_{\mu\nu\alpha\kappa\beta\lambda}^{(\pm)}. \tag{3.23}$$

Here $r_{\mu\nu\alpha\kappa\beta\lambda}^{(\pm)}$ is a numerical tensor whose detailed structure need not concern us; we need only the fact that it is antisymmetrical in its last pair of indices, and that its duality character in this last index pair is reversed:

$$\tfrac{1}{2}r_{\mu\nu\alpha\kappa\rho\sigma}^{(\pm)}\epsilon_{\rho\sigma\beta\lambda}=\mp r_{\mu\nu\alpha\kappa\beta\lambda}^{(\pm)}. \tag{3.24}$$

Hence,

$$r_{\mu\nu\alpha\kappa\beta\lambda}^{(\pm)}D_\beta D_\lambda=r_{\mu\nu\alpha\kappa\beta\lambda}^{(\pm)}\tfrac{1}{2}F_{\beta\lambda}=0, \tag{3.25}$$

and we secure the bracket composition law

$$\{X\}_{\mu\sigma}^{(\pm)}\{Y\}_{\sigma\nu}^{(\pm)}=\{XD^2Y\}_{\mu\nu}^{(\pm)}. \tag{3.26}$$

Henceforth, so as to achieve a simpler notation, we shall consider only self-dual fields and delete the superscript (±). (The treatment of the anti-self-dual case is obvious.)

The construction of the spin-1 propagation function $G_{\mu\nu}(x,y)$ can now be quickly performed with the aid of these operator techniques. We assert that $G_{\mu\nu}(x,y)$ has the formal operator realization

$$G_{\mu\nu}=-\left\{\left(\frac{1}{D^2}\right)^2\right\}_{\mu\nu}+(1-\xi)D_\mu\left(\frac{1}{D^2}\right)^2 D_\nu. \tag{3.27}$$

(An alternative derivation of this result is sketched in the Appendix.) To prove this assertion, we first observe that

$$\begin{aligned}[-D^2\delta_{\mu\lambda}-2F_{\mu\lambda}+(1-1/\xi)D_\mu D_\lambda]D_\lambda &= -\{1\}_{\mu\lambda}D_\lambda+(1-1/\xi)D_\mu D^2\\ &= -(1/\xi)D_\mu D^2,\end{aligned} \tag{3.28}$$

whence

$$[-D^2\delta_{\mu\lambda} - 2F_{\mu\lambda} + (1-1/\xi)D_\mu D_\lambda]G_{\lambda\nu} = \{1\}_{\mu\lambda}\left\{\left(\frac{1}{D^2}\right)^2\right\}_{\lambda\nu} = Q_{\mu\nu}, \quad (3.29)$$

where

$$Q_{\mu\nu} = \{1/D^2\}_{\mu\nu}. \quad (3.30)$$

The quantity in square brackets on the left-hand side of Eq. (3.29) is the field-equation operator which appears on the left-hand side of the Green's function, Eq. (3.14). Thus, we will have proved that the operator $G_{\mu\nu}$ given by Eq. (3.27) is a formal operator realization of the spin-1 propagation function $G_{\mu\nu}(x, y)$ if we prove that $Q_{\mu\nu}$ [cf. Eqs. (3.29) and (3.30)] is an operator realization of the projector $Q_{\mu\nu}(x, y)$ onto the nonzero modes, Eq. (3.15). The proof that $Q_{\mu\nu}$ is the correct projection operator proceeds exactly as in the spin-$\frac{1}{2}$ case discussed in the preceding section. Since Eq. (3.29) expresses $Q_{\mu\nu}$ as the field equation applied to some operator, $Q_{\mu\nu}$ itself can contain no zero modes. On the other hand, it follows from Eqs. (3.20), (3.26), and (3.30) that

$$\begin{aligned}(D^2\delta_{\mu\lambda} + 2F_{\mu\lambda})Q_{\lambda\nu} &= \{1\}_{\mu\lambda}\{1/D^2\}_{\lambda\nu} \\ &= \{1\}_{\mu\nu} \\ &= D^2\delta_{\mu\nu} + 2F_{\mu\nu}, \end{aligned} \quad (3.31)$$

which implies that $Q_{\mu\nu}$ contains all the nonzero modes. Hence $Q_{\mu\nu}$ is indeed the correct projection operator. [Incidentally, the fact that $Q_{\mu\nu}$ is a projection operator, $Q_{\mu\lambda}Q_{\lambda\nu} = Q_{\mu\nu}$, follows immediately from Eqs. (3.26) and (3.30)].

There are problems with the formal operator construction (3.27) for the spin-1 propagators. These problems arise from the convolution integral that defines the matrix element of the operator $(1/D^2)^2$,

$$\left\langle x\left|\left(\frac{1}{D^2}\right)^2\right|y\right\rangle = \int (d_E{}^4 z)\Delta(x,z)\Delta(z,y). \quad (3.32)$$

At large distances ($z^2 \to \infty$) the spin-0 propagator $\Delta(x, z)$ behaves as $1/z^2$ and the integral in Eq. (3.32) diverges logarithmically. We shall discuss this difficulty in Sec. VI after we have derived explicit forms for the spin-0 propagators $\Delta(x, y)$ in specific pseudoparticle fields.

IV. ISOSPIN-$\frac{1}{2}$ SCALAR PROPAGATORS

Here we shall construct explicitly the massless, spin-0 propagation function with isospin $\frac{1}{2}$ in the self-dual (or anti-self-dual) SU(2) gauge field of n pseudoparticles. First we shall review some properties of the pseudoparticle fields. The vector potential of a single pseudoparticle is given by[1]

$$A_{\mu a}^{(\pm)}(x) = \frac{2\eta_{\mu\nu a}^{(\pm)}(x-z)_\nu}{(x-z)^2+\rho^2}, \quad (4.1)$$

where $a = 1, 2, 3$ is the SU(2) group index and $\eta_{\mu\nu a}^{(\pm)}$ is antisymmetrical in $\mu\nu$,

$$\eta_{\mu\nu a}^{(\pm)} = -\eta_{\nu\mu a}^{(\pm)}, \quad (4.2a)$$

with

$$\eta_{kla}^{(\pm)} = \epsilon_{kla} \quad (4.2b)$$

and

$$\eta_{k4a}^{(\pm)} = \pm\delta_{ka}. \quad (4.2c)$$

The (+) superscript denotes a pseudoparticle solution which has a self-dual field-strength tensor while the (−) superscript denotes an antipseudoparticle solution which has an anti-self-dual field-strength tensor. (These self-duality properties will become evident in the following.) The four constants z_λ parametrize the position of the pseudoparticle and the fifth constant ρ parametrizes its size.

It is convenient to introduce the Hermitian 2×2 Pauli spin matrices τ_a which obey

$$\tau_a\tau_b = \delta_{ab} + i\epsilon_{abc}\tau_c, \quad (4.3)$$

and define the matrix field

$$A_\mu^{(\pm)} = A_{\mu a}^{(\pm)}\tfrac{1}{2}\tau_a. \quad (4.4)$$

It is also convenient to introduce the four-vector symbols

$$\tau_\mu = (\vec{\tau}, i), \quad \tau_\mu^\dagger = (\vec{\tau}, -i), \quad (4.5)$$

which have the useful properties

$$\tau_\mu^\dagger\tau_\nu = \delta_{\mu\nu} + i\eta_{\mu\nu a}^{(+)}\tau_a \quad (4.6a)$$

and

$$\tau_\mu\tau_\nu^\dagger = \delta_{\mu\nu} + i\eta_{\mu\nu a}^{(-)}\tau_a. \quad (4.6b)$$

The coordinate-dependent matrix

$$\Omega(x) = \frac{-i\tau_\mu x_\mu}{(x^2)^{1/2}} \quad (4.7)$$

is a unitary matrix

$$\Omega(x)^\dagger = \Omega^{-1}(x), \quad (4.8)$$

which connects each space-time point with an element of the SU(2) gauge group in a particular way: As the coordinate x_μ ranges once over the S_3 hypersphere $x^2 = \text{const}$, the matrix $\Omega(x)$ covers the SU(2) group space once; this mapping thus has winding number +1. It follows from Eqs. (4.6) that

$$\Omega(x)^{\mp 1} i\partial_\mu \Omega(x)^{\pm 1} = \eta_{\mu\nu a}^{(\pm)}\frac{x_\nu}{x^2}\tau_a. \quad (4.9)$$

Hence, the matrix form (4.4) of the pseudoparticle

solution (4.1) can be expressed as

$$A_\mu^{(\pm)}(x) = \frac{(x-z)^2}{(x-z)^2+\rho^2}\,\Omega(x-z)^{\mp 1} \times i\partial_\mu \Omega(x-z)^{\pm 1}. \tag{4.10}$$

This form shows immediately that at large distances the vector potential approaches a pure gauge transformation,

$$A_\mu^{(\pm)}(x) \xrightarrow[x^2\to\infty]{} \Omega(x)^{\mp 1} i\partial_\mu \Omega(x)^{\pm 1}. \tag{4.11}$$

Thus, the pseudoparticle potential $A_\mu^{(+)}(x)$ provides a mapping with winding number +1, and the antipseudoparticle potential $A_\mu^{(-)}(x)$ provides a mapping with winding number −1.

In order to motivate the construction of the n-pseudoparticle solution, we first gauge transform the single pseudoparticle solution [Eq. (4.10)] so that it vanishes more rapidly at infinity,

$$\bar{A}_\mu^{(\pm)}(x) = \Omega(x-z)^{\pm 1}[i\partial_\mu + A_\mu^{(\pm)}(x)]\Omega(x-z)^{\mp 1}$$

$$= \left[1 - \frac{(x-z)^2}{(x-z)^2+\rho^2}\right]\Omega(x-z)^{\pm 1} i\partial_\mu \Omega(x-z)^{\mp 1} \tag{4.12}$$

We use Eqs. (4.4) and (4.9) to write this as

$$\bar{A}_{\mu a}^{(\pm)}(x) = \frac{2\rho^2}{(x-z)^2+\rho^2}\,\eta_{\mu\nu a}^{(\mp)}\,\frac{(x-z)_\nu}{(x-z)^2}$$

$$= -\eta_{\mu\nu a}^{(\mp)}\partial_\nu \ln\left[1+\frac{\rho^2}{(x-z)^2}\right]. \tag{4.13}$$

The solution in this gauge is singular at $x_\mu = z_\mu$. This singularity arises because the gauge transformation matrix $\Omega(x-z)$ is singular at $x_\mu = z_\mu$ in the sense that its value at this point depends upon the direction in which the limit $x_\mu = z_\mu$ was approached. Nonetheless, since the singular solution [Eq. (4.13)] can be gauge transformed into a regular solution [Eq. (4.10)], the singular solution is an acceptable one. In view of the structure of Eq. (4.13), it is natural to try the form[9]

$$\bar{A}_{\mu a}^{(\pm)}(x) = -\eta_{\mu\nu a}^{(\mp)}\partial_\nu \ln \Pi(x), \tag{4.14}$$

for a general n-pseudoparticle solution. This will be a solution to the field equations if the field-strength tensor

$$\bar{F}_{\mu\nu a}^{(\pm)} = \partial_\mu \bar{A}_{\nu a}^{(\pm)} - \partial_\nu \bar{A}_{\mu a}^{(\pm)} + \epsilon_{abc}\bar{A}_{\mu b}^{(\pm)}\bar{A}_{\nu c}^{(\pm)} \tag{4.15}$$

is self-dual (or anti-self-dual), i.e.,

$$p_{\mu\nu\lambda\kappa}^{(\mp)}\bar{F}_{\lambda\kappa a}^{(\pm)} = 0, \tag{4.16}$$

where the projection operators $p^{(\pm)}$ are defined in Eq. (3.16). The symbols $\eta_{\mu\nu a}^{(\pm)}$ define transformation matrices that take the three independent components of an antisymmetrical, self-dual (or anti-self-dual) tensor into a three-vector labeled by the index a. Thus they decompose the O(4) rotation group into its SU(2)⊗SU(2) subgroups. Indeed

$$p_{\mu\nu\lambda\kappa}^{(\pm)} = \tfrac{1}{4}\eta_{\mu\nu a}^{(\pm)}\eta_{\lambda\kappa a}^{(\pm)}, \tag{4.17}$$

and the conditions (4.16) are tantamount to the conditions

$$\eta_{\mu\nu b}^{(\mp)}\bar{F}_{\mu\nu a}^{(\pm)} = 0. \tag{4.18}$$

The $\eta_{\mu\nu a}^{(\pm)}$ symbols play another role. They are isomorphic to the generators $i\tau_a$ of the SU(2) group since

$$\eta_{\mu\lambda a}^{(\pm)}\eta_{\nu\lambda b}^{(\pm)} = \delta_{ab}\delta_{\mu\nu} + \epsilon_{abc}\eta_{\mu\nu c}^{(\pm)}. \tag{4.19}$$

Inserting Eqs. (4.14) and (4.15) into the self-duality condition (4.18) and using Eq. (4.19), we find that it is satisfied if

$$\Pi(x)^{-1}\partial^2\Pi(x) = 0. \tag{4.20}$$

If the function $\Pi(x)$ approaches unity at large distances, then the vector potential (4.14) will vanish at infinity. Hence, a general solution[9] of Eq. (4.20) is given by[10]

$$\Pi(x) = 1 + \sum_{s=1}^{n}\frac{\rho_s^2}{(x-z_s)^2} \tag{4.21}$$

This describes an n-pseudoparticle configuration specified by $5n$ parameters: $4n$ position variables $z_{s\mu}$ and n sizes ρ_s. It is not the most general n-pseudoparticle solution, for the latter involves[8] $8n$ parameters, with $3n$ additional parameters specifying the gauge orientation of each pseudoparticle. [Specializing to $n=1$ we recover the solution (4.13) and hence, by the gauge transformation (4.12), verify that the vector potential (4.1) is the regular, single-pseudoparticle solution.]

The n-pseudoparticle solution (4.14) is singular at each of the pseudoparticle positions, i.e., as $x\to z_s$,

$$\bar{A}_{\mu a}^{(\pm)}(x) \to 2\eta_{\mu\nu a}^{(\mp)}\frac{(x-z_s)_\nu}{(x-z_s)^2}. \tag{4.22}$$

If we write the vector potential in a matrix notation and use Eq. (4.9), we see that this singularity has the structure of a pure gauge transformation,

$$\bar{A}_\mu^{(\pm)}(x) \xrightarrow[x\to z_s]{} \Omega(x-z_s)^{\pm 1} i\partial_\mu \Omega(x-z_s)^{\mp 1}. \tag{4.23}$$

Hence, these singularities of $\bar{A}_{\mu a}^{(\pm)}(x)$ can be removed by a gauge transformation[14]

$$\tilde{A}_\mu^{(\pm)}(x) = U^{(\pm)}(x)^{-1}[i\partial_\mu + \bar{A}_\mu^{(\pm)}(x)]U^{(\pm)}(x), \tag{4.24}$$

if a unitary matrix $U^{(\pm)}(x)$ can be found which obeys for all z_s, $s=1,\ldots,n$,

$$U^{(\pm)}(x) \xrightarrow[x\to z_s]{} \Omega(x-z_s)^{\pm 1}R_s^{(\pm)}(x), \tag{4.25}$$

with $R_s^{(\pm)}(x)$ a unitary matrix that is regular at x

$= z_s$. Such a matrix $U^{(\pm)}(x)$ obeying Eq. (4.25) can be constructed, and thus the divergences in $\bar{A}_\mu^{(\pm)}(x)$ can be removed by a gauge transformation. Indeed, for a single pseudoparticle, $n = \pm 1$, we have

$$U^{(\pm)}(x) = \Omega(x - z_1)^{\pm 1}, \tag{4.26a}$$

for two pseudoparticles, $n = \pm 2$, we have

$$U^{(\pm)}(x) = \Omega(x - z_1)^{\pm 1}\Omega(z_2 - z_1)^{\mp 1}\Omega(x - z_2)^{\pm 1}, \tag{4.26b}$$

while for three pseudoparticles, $n = \pm 3$, we have

$$\begin{aligned} U^{(\pm)}(x) = {} & \Omega(x - z_1)^{\pm 1}\Omega(z_2 - z_1)^{\mp 1}\Omega(x - z_2)^{\pm 1} \\ & \times \Omega(z_3 - z_2)^{\mp 1}\Omega(z_2 - z_1)^{\pm 1}\Omega(z_3 - z_1)^{\mp 1} \\ & \times \Omega(x - z_3)^{\pm 1}, \end{aligned} \tag{4.26c}$$

and so forth. Note that this construction shows that the $U^{(\pm)}(x)$ matrix for $\pm n$ pseudoparticles covers the SU(2) group $\pm n$ times as the space-time coordinate x ranges once over the surface of a large hypersphere S_3 that encloses all of the pseudoparticle positions z_s. Since as $x^2 \to \infty$,

$$\bar{A}_\mu^{(\pm)}(x) \to U^{(\pm)}(x)^{-1} i\partial_\mu U^{(\pm)}(x), \tag{4.27}$$

we find explicitly that the vector potential has a winding number $\pm n$. We should remark that although the gauge transformation (4.24) does remove all the divergent pieces of the vector potentail, there remain "singularities" in the transformed potential $\bar{A}_\mu^{(\pm)}(x)$ involving ill-defined quantities of the form $(x - z)_\nu[(x - z_s)^2]^{-1/2}$. These weaker singularities, however, cause no trouble in that they do not give rise to singularities in the action.

With this lengthy introduction completed, we now proceed with the construction of the massless, spin-0, isospin-$\frac{1}{2}$ propagation function $\bar{\Delta}^{(\pm)}(x - y)$ in the presence of the pseudoparticle field $\bar{A}_\mu^{(\pm)}(x)$ [Eqs. (4.14) and (4.21)]. This propagator is defined by

$$-\bar{D}^{(\pm)2}\bar{\Delta}^{(\pm)}(x, y) = \delta(x - y), \tag{4.28}$$

where

$$\bar{D}_\mu^{(\pm)} = \partial_\mu - \tfrac{1}{2} i\tau_a \bar{A}_{\mu a}^{(\pm)}(x). \tag{4.29}$$

Utilizing Eqs. (4.3), (4.19), and (4.14), we find that

$$\bar{D}^{(\pm)2} = \partial^2 - \tfrac{3}{4}(\partial_\mu \ln\Pi)^2 + i\tau_a \eta_{\mu\nu a}^{(\mp)}(\partial_\nu \ln\Pi)\partial_\mu. \tag{4.30}$$

The propagator can be easily constructed because $\bar{D}^{(\pm)2}$ can be factored. We use Eq. (4.6a) or Eq. (4.6b) and Eq. (4.20) to secure

$$\bar{D}^{(+)2} = \Pi^{1/2}\tau\partial\Pi^{-1}\tau^\dagger\partial\Pi^{1/2} \tag{4.31a}$$

and

$$\bar{D}^{(-)2} = \Pi^{1/2}\tau^\dagger\partial\Pi^{-1}\tau\partial\Pi^{1/2}. \tag{4.31b}$$

(Here and henceforth we write scalar products such as $\tau_\mu\partial_\mu$ simply as $\tau\partial$.)

The propagation function must have the same short-distance singularity as the free propagation function,

$$\bar{\Delta}^{(\pm)}(x, y) \underset{x \to y}{\sim} \frac{1}{4\pi^2(x - y)^2}, \tag{4.32}$$

to produce the inhomogeneous term $\delta(x - y)$ in the Green's function equation (4.28). On the other hand, the appearance of the factors of $\Pi^{1/2}$ on the right-hand side of Eqs. (4.31) suggests that $\bar{\Delta}^{(\pm)}(x - y)$ should contain a factor of $\Pi(x)^{-1/2}\Pi(y)^{-1/2}$. Thus, we are led to write

$$\bar{\Delta}^{(\pm)}(x, y) = \Pi(x)^{-1/2}\frac{F^{(\pm)}(x, y)}{4\pi^2(x - y)^2}\Pi(y)^{-1/2}, \tag{4.33}$$

where the function $F^{(\pm)}(x, y)$ must obey the boundary condition

$$F^{(\pm)}(x, x) = \Pi(x). \tag{4.34}$$

Inserting the decomposition (4.33) into the Green's function equation (4.28) and using Eq. (4.31a) and the fact that

$$\Pi(x)^{-1/2}\delta(x - y)\Pi(y)^{1/2} = \delta(x - y), \tag{4.35}$$

we get

$$\tau\partial[\Pi(x)4\pi^2(x - y)^2]^{-1}\left[\tau^\dagger\partial F^{(+)}(x, y) - \frac{2\tau^\dagger(x - y)}{(x - y)^2}F^{(+)}(x, y)\right] = \delta(x - y). \tag{4.36}$$

Now according to Eq. (4.6b) we have

$$\tau\partial\tau^\dagger\partial = \partial^2, \tag{4.37}$$

and hence

$$\tau\partial\frac{\tau^\dagger(x - y)}{2\pi^2[(x - y)^2]^2} = \delta(x - y). \tag{4.38}$$

This allows us to write Eq. (4.36) as

$$\tau\partial[\Pi(x)(x - y)^2]^{-1}\left\{\tau^\dagger\partial F^{(+)}(x, y) - 2\frac{\tau^\dagger(x - y)}{(x - y)^2}[F^{(+)}(x, y) - \Pi(x)]\right\} = 0. \tag{4.39}$$

Since $\tau\partial$ has no zero eigenvalues [cf. Eq. (4.37)], we conclude that the quantity in curly brackets in Eq. (4.39) must vanish,

$$\tau^\dagger\partial F^{(+)}(x, y) - 2\frac{\tau^\dagger(x - y)}{(x - y)^2}[F^{(+)}(x, y) - \Pi(x)] = 0. \tag{4.40}$$

The same first-order differential equation holds for $F^{(-)}(x, y)$, but with $\tau_\nu^\dagger$ replaced by τ_ν.

We recall that

$$\Pi(x) = 1 + \sum_{s=1}^{n}\frac{\rho_s^2}{(x - z_s)^2}, \tag{4.21}$$

and assert that Eq. (4.39) is satisfied if $F^{(+)}(x,y)$ is given by

$$F^{(+)}(x,y)=1+\sum_{s=1}^{n}\rho_s^{\,2}\,\frac{\tau(x-z_s)}{(x-z_s)^2}\,\frac{\tau^\dagger(y-z_s)}{(y-z_s)^2}\,. \tag{4.41}$$

The proof rests on the repeated use of Eqs. (4.6). They imply that the boundary condition (4.34) is obeyed [as is necessary for any solution of Eq. (4.40)]. They also imply that

$$\tau^\dagger\partial F^{(+)}(x,y)=2\sum_{s=1}^{n}\frac{\rho_s^{\,2}}{(x-z_s)^2}\,\frac{\tau^\dagger(y-z_s)}{(y-z_s)^2}\,, \tag{4.42}$$

and the relations

$$\tau^\dagger(x-z_s)\tau(x-z_s)=(x-z_s)^2\,, \tag{4.43}$$

$$\tau^\dagger(y-z_s)\tau(x-z_s)\tau^\dagger(y-z_s)$$
$$=-(y-z_s)^2\tau^\dagger(x-z_s)+2(x-z_s)(y-z_s)\tau^\dagger(y-z_s)\,. \tag{4.44}$$

Hence, on writing

$$(x-y)=(x-z_s)-(y-z_s)\,, \tag{4.45}$$

we get

$$2\,\frac{\tau^\dagger(x-y)}{(x-y)^2}[F^{(+)}(x,y)-\Pi(x)]=\frac{2}{(x-y)^2}\sum_{s=1}^{n}\rho_s^{\,2}\left\{\frac{\tau^\dagger(y-z_s)}{(y-z_s)^2}+\frac{\tau^\dagger(x-z_s)}{(x-z_s)^2}-2(x-z_s)(y-z_s)\,\frac{\tau^\dagger(y-z_s)}{(x-z_s)^2(y-z_s)^2}-\frac{\tau^\dagger[(x-z_s)-(y-z_s)]}{(x-z_s)^2}\right\}$$
$$=2\sum_{s=1}^{n}\frac{\rho_s^{\,2}}{(x-z_s)^2}\,\frac{\tau^\dagger(y-z_s)}{(y-z_s)^2}\,, \tag{4.46}$$

which, in view of Eq. (4.42), proves that the structure (4.41) does indeed satisfy the first-order differential equation (4.39). The antipseudoparticle solution is given by

$$F^{(-)}(x,y)=1+\sum_{s=1}^{n}\rho_s^{\,2}\,\frac{\tau^\dagger(x-z_s)}{(x-z_s)^2}\,\frac{\tau(y-z_s)}{(y-z_s)^2}\,. \tag{4.47}$$

Equations (4.33), (4.41), and (4.47) provide a simple algebraic expression for the massless, spin-0, isospin-$\frac{1}{2}$ propagators in the field of an arbitrary configuration of n pseudoparticles or antipseudoparticles.

We have calculated the propagators $\bar\Delta^{(\pm)}(x,y)$ in a singular vector potential $\bar A_\mu^{(\pm)}(x)$ [Eq. (4.14)]. A regular vector potential $\tilde A_\mu^{(\pm)}(x)$ can be obtained by the gauge transformation $U^{(\pm)}(x)$ [Eq. (4.24)]. The propagator $\tilde\Delta^{(\pm)}(x,y)$ in this (regular) vector potentail is given by a gauge rotation of $\bar\Delta^{(\pm)}(x,y)$:

$$\tilde\Delta^{(\pm)}(x,y)=U^{(\pm)}(x)^{-1}\bar\Delta^{(\pm)}(x,y)U^{(\pm)}(y). \tag{4.48}$$

For a single pseudoparticle (or antipseudoparticle) it is a simple matter to perform this transformation explicitly. Using Eq. (4.7), we may write

$$F^{(\pm)}(x,y)=1+\frac{\rho^2}{(x^2y^2)^{1/2}}\Omega(x)^{\pm1}\Omega(y)^{\mp1}, \tag{4.49}$$

where, for simplicity, we have located the pseudoparticle at the origin, $z_1=0$. Thus, using Eq. (4.26a), we get

$$\tilde\Delta^{(\pm)}(x,y)=\frac{\Omega(x)^{\mp1}\Omega(y)^{\pm1}+\rho^2(x^2y^2)^{-1/2}}{\Pi^{1/2}(x)\Pi^{1/2}(y)4\pi^2(x-y)^2}\,, \tag{4.50}$$

and, employing Eqs. (4.6),

$$\tilde\Delta^{(\pm)}(x,y)=\frac{\rho^2+xy+i\eta^{(\pm)}_{\mu\nu a}x_\mu y_\nu\tau_a}{(\rho^2+x^2)^{1/2}(\rho^2+y^2)^{1/2}4\pi^2(x-y)^2}\,. \tag{4.51}$$

This solution (4.51) was obtained previously[15] by exploiting the invariance of the single pseudoparticle field under an O(5) group of conformal coordinate transformations.

V. ISOSPIN-1 SCALAR PROPAGATORS

We turn now to the construction of the massless, spin-0, isospin-1 propagation function $\Delta_{ab}(x,y)$ in a general pseudoparticle field. This propagator obeys the Green's function equation

$$-D^2_{ac}\Delta_{cb}(x,y)=\delta_{ab}\delta(x-y)\,, \tag{5.1}$$

where

$$D_{\mu ab}=\partial_\mu\delta_{ab}+\epsilon_{acb}A_{\mu c}(x). \tag{5.2}$$

In order to simplify the notation, we shall work out explicitly only the propagation function for the n-pseudoparticle field in the singular gauge, $\bar A^{(+)}_{\mu a}(x)$, and delete the various superscripts that refer to this field.

To proceed with the construction, we note that the isospin-$\frac{1}{2}$ propagator has the form

$$\Delta(x,y)=\frac{M(x,y)}{4\pi^2(x-y)^2}\,, \tag{5.3}$$

with

$$M(x,y)=\Pi^{-1/2}(x)F(x,y)\Pi^{-1/2}(y), \tag{5.4}$$

and $F(x,y)=F^{(\ )}(x,y)$ [Eq. (4.41)]. The matrix $M(x,y)$ is Hermitian in the sense that

$$M(x,y)^\dagger=M(y,x). \tag{5.5}$$

By inserting the form (5.3) into the Green's function equation (4.28) for the isospin-$\frac{1}{2}$ propagator, we find that

$$D^2M(x,y)=4\,\frac{(x-y)_\mu}{(x-y)^2}D_\mu M(x,y), \tag{5.6}$$

where here the gauge-covariant derivative refers to isospin $\frac{1}{2}$, the derivative displayed in Eq. (4.29). The symmetry (5.5) implies that we also have

$$M(y,x)D^2=-4M(y,x)\overleftarrow{D}_\mu\,\frac{(x-y)_\mu}{(x-y)^2}, \tag{5.7}$$

where it should be remembered that the derivative $\overleftarrow{D}_\mu$ acting to the left involves $-\overleftarrow{\partial}_\mu$.

We combine two isospin-$\frac{1}{2}$ matrices $M(x,y)$ to form an isospin-1 matrix

$$W_{ab}(x,y)=\tfrac{1}{2}\operatorname{tr}\tau_a M(x,y)\tau_b M(y,x) \tag{5.8}$$

and express the isospin-1 propagator in terms of another matrix $C_{ab}(x,y)$:

$$\Delta_{ab}(x,y)=\frac{W_{ab}(x,y)}{4\pi^2(x,y)^2}+\frac{C_{ab}(x,y)}{4\pi^2\Pi(x)\Pi(y)}. \tag{5.9}$$

Since

$$W_{ab}(x,x)=\delta_{ab}, \tag{5.10}$$

the decomposition (5.9) will produce the inhomogeneous term $\delta_{ab}\delta(x-y)$ in the Green's function equation if $C_{ab}(x,y)$ is regular when $x\to y$. We insert the decomposition (5.9) into the Green's function equation (5.1); note that the isospin-1 covariant derivatives acting on $W_{ab}(x,y)$ are converted to isospin-$\frac{1}{2}$ covariant derivatives acting on $M(x,y)$ by the commutation relation

$$[\tau_a,\tau_b]=2i\epsilon_{abc}\tau_c, \tag{5.11}$$

and use Eqs. (5.6) and (5.7) to secure

$$D^2_{ac}\frac{C_{cb}(x,y)}{\Pi(x)\Pi(y)}=\frac{1}{(x-y)^2}\operatorname{tr}\tau_a D_\mu M(x,y)\tau_b M(y,x)\overleftarrow{D}_\mu. \tag{5.12}$$

The covariant derivatives on the left-hand side of this equation refer to isospin 1 [Eq. (5.2)], while those on the right-hand side refer to isospin $\frac{1}{2}$ [Eq. (4.29)]. Specializing to the particular pseudoparticle vector potential

$$A_{\mu a}(x)=-\eta^{(-)}_{\mu\nu a}\,\partial_\nu\ln\Pi(x), \tag{5.13}$$

writing $M(x,y)$ in terms of $F(x,y)$ [Eq. (5.4)], and commuting factors of $\Pi(x)$ with covariant derivatives, we obtain an explicit differential equation for the unknown function $C_{ab}(x,y)$:

$$\left\{\partial^2\delta_{ac}-2\left[\eta^{(-)}_{\mu\nu d}\epsilon_{adc}\frac{\partial_\nu\Pi(x)}{\Pi(x)}+\delta_{ac}\frac{\partial_\mu\Pi(x)}{\Pi(x)}\right]\partial_\mu\right\}C_{cb}(x,y)=K_{ab}(x,y), \tag{5.14}$$

in which

$$K_{ab}(x,y)=\frac{1}{(x-y)^2}\operatorname{tr}\tau_a\left[\partial_\mu-\frac{1}{2}\frac{\tau\partial\Pi(x)}{\Pi(x)}\tau^\dagger_\mu\right]F(x,y)\tau_b F(y,x)\left[\overleftarrow{\partial}_\mu-\tau_\mu\frac{1}{2}\frac{\tau^\dagger\partial\Pi(x)}{\Pi(x)}\right]. \tag{5.15}$$

Here we have also used Eqs. (4.6) to introduce the matrices τ_μ and $\tau^\dagger_\mu$. We shall often make use of Eqs. (4.6) in the algebraic reductions that follow.

The quantity $K_{ab}(x,y)$ can be expressed in terms of fairly simple, explicit formulas. First, we note that since

$$\tau_\mu\tau^\dagger\partial\Pi\,\tau_a\tau\partial\Pi\,\tau^\dagger_\mu=0, \tag{5.16}$$

we can write

$$K_{ab}(x,y)=K^{(1)}_{ab}(x,y)+K^{(2)}_{ab}(x,y), \tag{5.17}$$

with

$$K^{(1)}_{ab}(x,y)=\frac{1}{(x-y)^2}\operatorname{tr}\tau_a[\partial_\mu F(x,y)-\tfrac{1}{2}\tau\partial\Pi(x)\tau^\dagger_\mu]\tau_b[F(y,x)\overleftarrow{\partial}_\mu-\tfrac{1}{2}\tau_\mu\tau^\dagger\partial\Pi(x)] \tag{5.18}$$

and

$$K^{(2)}_{ab}(x,y)=\frac{-1}{2(x-y)^2}\frac{1}{\Pi(x)}\operatorname{tr}\{\tau_a\tau\partial\Pi(x)\tau^\dagger_\mu[F(x,y)-\Pi(x)]\tau_b[F(y,x)\overleftarrow{\partial}_\mu-\tfrac{1}{2}\tau_\mu\tau^\dagger\partial\Pi(x)]$$
$$+\tau_a[\partial_\mu F(x,y)-\tfrac{1}{2}\tau\partial\Pi(x)\tau^\dagger_\mu]\tau_b[F(y,x)-\Pi(x)]\tau_\mu\tau^\dagger\partial\Pi(x)\}. \tag{5.19}$$

In order to remove some of the notational clutter, we define

$$x_s = x - z_s\,, \quad y_s = y - z_s\,, \tag{5.20}$$

so that

$$\Pi(x) = 1 + \sum_s \frac{\rho_s^2}{x_s^2} \tag{5.21}$$

and

$$F(x,y) = 1 + \sum_s \rho_s^2 \frac{\tau x_s}{x_s^2}\frac{\tau^\dagger y_s}{y_s^2}\,. \tag{5.22}$$

Manipulations similar to those done before [cf. Eqs. (4.43)–(4.45)] give

$$\partial_\mu F(x,y) - \tfrac{1}{2}\tau\partial\Pi(x)\tau_\mu^\dagger = -\sum_s \frac{\rho_s^2}{(x_s^2)^2 y_s^2}\,\tau x_s \tau_\mu^\dagger \tau(x-y)\tau^\dagger y_s \tag{5.23a}$$

and

$$F(y,x)\overleftarrow{\partial}_\mu - \tfrac{1}{2}\tau_\mu\tau^\dagger\partial\Pi(x) = -\sum_t \frac{\rho_t^2}{(x_t^2)^2 y_t^2}\,\tau y_t \tau^\dagger(x-y)\tau_\mu \tau^\dagger x_t\,. \tag{5.23b}$$

Therefore, the quantity $K^{(1)}_{ab}(x,y)$ involves

$$\tau_\mu^\dagger \tau(x-y)\tau^\dagger y_s \tau_b \tau y_t \tau^\dagger(x-y)\tau_\mu = -4i\eta^{(-)}_{\mu t b} y_{s\mu} y_{t\nu}(x-y)^2\,, \tag{5.24}$$

and we get

$$K^{(1)}_{ab}(x,y) = 8\sum_{s,t} \frac{\rho_s^2}{(x_s^2)^2 y_s^2}\frac{\rho_t^2}{(x_t^2)^2 y_t^2} \times \eta^{(-)}_{\lambda\kappa a} x_{s\lambda} x_{t\kappa}\eta^{(-)}_{\mu\nu b} y_{s\mu} y_{t\nu}\,. \tag{5.25}$$

Using

$$F(x,y) - \Pi(x) = \sum_s \frac{\rho_s^2}{x_s^2 y_s^2}\,\tau(x-y)\tau^\dagger y_s\,, \tag{5.26a}$$

and

$$F(y,x) - \Pi(x) = \sum_t \frac{\rho_t^2}{x_t^2 y_t^2}\,\tau y_t \tau^\dagger(x-y)\,, \tag{5.26b}$$

together with Eqs. (5.23), we find that the quantity $K^{(2)}_{ab}(x,y)$ also involves the expression displayed in Eq. (5.24), and we get

$$K^{(2)}_{ab}(x,y) = \frac{-16}{\Pi(x)}\sum_{s,t,u} \frac{\rho_s^2}{x_s^2 y_s^2}\frac{\rho_t^2}{(x_t^2)^2 y_t^2}\frac{\rho_u^2}{(x_u^2)^2} \times \eta^{(-)}_{\lambda\kappa a} x_{u\lambda} x_{t\kappa}\eta^{(-)}_{\mu\nu b} y_{s\mu} y_{t\nu}\,. \tag{5.27}$$

The y coordinate enters into both $K^{(1)}_{ab}(x,y)$ and $K^{(2)}_{ab}(x,y)$ only through the functions

$$\Phi^{(+)}_{stb}(y) = \frac{\rho_s\rho_t}{y_s^2 y_t^2}\,\eta^{(-)}_{\mu\nu b} y_{s\mu} y_{t\nu}\,, \tag{5.28}$$

and in terms of these functions

$$K^{(1)}_{ab}(x,y) = 8\sum_{s,t} \frac{1}{x_s^2 x_t^2}\,\Phi^{(+)}_{sta}(x)\Phi^{(+)}_{stb}(y)\,, \tag{5.29}$$

$$K^{(2)}_{ab}(x,y) = -\frac{16}{\Pi(x)}\sum_{s,t,u} \frac{\rho_s\rho_u}{x_s^2 x_t^2 x_u^2}\,\Phi^{(+)}_{uta}(x)\Phi^{(+)}_{stb}(y)\,. \tag{5.30}$$

The propagation function is symmetrical under the interchange of its coordinates and isospin indices and this symmetry must be shared by the function $C_{ab}(x,y)$,

$$C_{ab}(x,y) = C_{ba}(y,x)\,. \tag{5.31}$$

The differential equation (5.14) for $C_{ab}(x,y)$ has a driving term $K_{ab}(x,y)$ which depends upon the y coordinate only through functions $\Phi^{(+)}_{stb}(y)$ which are the antisymmetrical in their indices s and t. Hence, by virtue of the symmetry (5.31) of $C_{ab}(x,y)$ it can only involve these functions,

$$C_{ab}(x,y) = \sum_{r,s,t,u} \Phi^{(+)}_{rsa}(x)\,c_{rs,tu}\,\Phi^{(+)}_{tub}(y)\,, \tag{5.32}$$

with coefficients $c_{rs,tu}$ that do not involve the coordinates x or y and have the symmetries

$$c_{rs,tu} = c_{tu,rs}\,, \tag{5.33}$$

$$c_{rs,tu} = -c_{sr,tu} = -c_{rs,ut}\,. \tag{5.34}$$

After a little calculation, we find that

$$\left\{\partial^2\delta_{ac} - 2\left[\eta^{(-)}_{\mu\nu d}\epsilon_{adc}\frac{\partial_\nu\Pi(x)}{\Pi(x)} + \frac{\partial_\mu\Pi(x)}{\Pi(x)}\delta_{ac}\right]\partial_\mu\right\}\Phi^{(+)}_{rsc}(x)$$
$$= -4\frac{(z_r - z_s)^2}{x_r^2 x_s^2}\Phi^{(+)}_{rsa}(x) + \frac{4}{\Pi(x)}\sum_t \frac{1}{x_r^2 x_s^2 x_t^2}\{\rho_t\rho_r[(z_r - z_s)^2 - x_r^2]\Phi^{(+)}_{tsa}(x) + \rho_s\rho_t[(z_r - z_s)^2 - x_s^2]\Phi^{(+)}_{rta}(x)\}\,. \tag{5.35}$$

We now insert the expansion (5.32) of $C_{ab}(x,y)$ in terms of the Φ functions into the differential equation (5.14) obeyed by $C_{ab}(x,y)$, use the explicit forms Eqs. (5.17), (5.29), and (5.30) for the driving term $K_{ab}(x,y)$ in this differential equation, and identify the coefficients of $\Phi^{(+)}_{rsa}(x)\Phi^{(+)}_{uvb}(y)$ to obtain a matrix equation for the constants $c_{rs,uv}$:

$$\Pi(x)(z_r - z_s)^2 c_{rs,uv} - \sum_p\left\{\frac{\rho_r\rho_p}{x_p^2}[(z_p - z_s)^2 - x_p^2]c_{ps,uv} + \frac{\rho_s\rho_p}{x_p^2}[(z_r - z_p)^2 - x_p^2]c_{rp,uv}\right\}$$
$$= \Pi(x)(\delta_{ru}\delta_{sv} - \delta_{rv}\delta_{su}) - \left(\frac{\rho_r\rho_u}{x_u^2}\delta_{sv} - \frac{\rho_s\rho_u}{x_u^2}\delta_{rv} - \frac{\rho_r\rho_v}{x_v^2}\delta_{su} + \frac{\rho_s\rho_v}{x_v^2}\delta_{ru}\right). \tag{5.36}$$

Although this equation involves the coordinate x explicitly, it is solved by constant coefficients $c_{rs,uv}$. The solution is of the form

$$c_{rs,uv} = \frac{\delta_{ru}\delta_{sv} - \delta_{rv}\delta_{su}}{(z_r - z_s)^2} - \frac{1}{(z_r - z_s)^2}\left(\rho_r\rho_u f_{sv} - \rho_r\rho_v f_{su} - \rho_s\rho_u f_{rv} - + \rho_s\rho_v f_{rv}\right)\frac{1}{(z_u - z_v)^2} \,. \tag{5.37}$$

On substituting this structure into Eq. (5.36), we find that all the x-dependent terms cancel independent of the values of the constants f_{sv} and that these constants are determined by

$$\sum_t g_{st} f_{tv} = \delta_{sv} \,, \tag{5.38}$$

in which

$$g_{st} = \left[1 + \sum_{r\neq s}\frac{\rho_r^2}{(z_r - z_s)^2}\right]\delta_{st} - \frac{\rho_s\rho_t}{(z_s - z_t)^2}(1 - \delta_{st}) \,. \tag{5.39}$$

Thus, the constants f_{st} are the elements of the matrix which is the inverse of the matrix g_{st}. Since the latter is symmetrical, so is f_{st}. This symmetry ensures that the structure (5.37) obeys the necessary symmetry conditions on $c_{rs,uv}$ given in Eqs. (5.33) and (5.34).

We have now completed our construction of the isospin-1 propagation function. It is given by Eq. (5.9) with the functions $C_{ab}(x,y)$ determined by Eqs. (5.32), (5.28), and (5.37)–(5.39). Our development has been for the case of an n-pseudoparticle field, but the result for n antipseudoparticles is obvious: One need only replace the function $F = F^{(+)}$ by $F^{(-)}$ and the functions $\Phi^{(+)}$ by $\Phi^{(-)}$, where $\Phi^{(-)}$ is constructed with $\eta^{(+)}$ symbols rather than with the $\eta^{(-)}$ symbols that appear in $\Phi^{(+)}$ [Eq. (5.28)].

The result for a single pseudoparticle (or antipseudoparticle) field is particularly simple. In this case the function $C_{ab}(x,y)$ vanishes and the isospin-1 propagation function is determined entirely in terms of the isospin-0 function. Our result agrees, when properly gauge transformed, with that obtained from an O(5) group theory construction.[15] The result for two pseudoparticles is also quite simple. In this case the function $C_{ab}(x,y)$ no longer vanishes but is determined by the single constant

$$c_{12,12} = [(z_1 - z_2)^2 + \rho_1^2 + \rho_2^2]^{-1} \tag{5.40}$$

VI. DISCUSSION

We have shown that the propagation functions for massless spin-$\frac{1}{2}$ and spin-1 particles moving in a self-dual but otherwise arbitrary non-Abelian gauge field are determined by the corresponding spin-0 propagators. The spin-0 propagators were constructed for particles of isospin $\frac{1}{2}$ and 1 moving in the self-dual field of n pseudoparticles. These scalar propagators have a remarkably simple algebraic form. This simplicity suggests the existence of some deeper underlying principles which, however, we have not yet been able to fathom.

The field of a single pseudoparticle is invariant under an O(5) group of conformal coordinate transformations.[16] The corresponding massless, spin-0 propagators which we have constructed for isospin-$\frac{1}{2}$ and isospin-1 particles are covariant under this O(5) group and, in fact, coincide with the functions obtained directly from an O(5) group theory analysis.[15] The situation is different for massless particles with spin $\frac{1}{2}$ or 1. The propagation functions for these particles do not transform covariantly under the O(5) group. This lack of covariance results because the propagators obey Green's function equations [Eqs. (2.19) and (2.20) and Eqs. (3.14) and (3.15)] from which the zero-mode components of the inhomogeneous δ function have been subtracted. The products of zero-mode functions which thus appear on the right-hand side of these equations have the same conformal weight as does the propagator (-3 for spin $\frac{1}{2}$ and -2 for spin 1). This weight differs from the conformal weight of the δ function (-4). Hence the inhomogeneous, driving terms in the spin-$\frac{1}{2}$ and spin-1 Green's function equations do not have a single, pure conformal weight. These equations—and hence the propagators they define—are thus not covariant under the transformations of the O(5) conformal group. Accordingly, the propagators that we have constructed for spin $\frac{1}{2}$ and spin 1 do not agree with those obtained from an O(5) conformal group analysis.[16]

The Green's function equations [Eqs. (2.19) and (2.20) for spin $\frac{1}{2}$ and Eqs. (3.14) and (3.15) for spin 1] dictate the asymptotic behavior of the propagators S and $G_{\mu\nu}$. For large x, the dominant zero-mode functions in Eqs. (2.20) and (3.15) fall as x^{-3}. The term $\delta(x-y)$ in Eqs. (2.19) and (3.15) is irrelevant in this limit, and S and $G_{\mu\nu}$ must therefore fall as x^{-2} and x^{-1}, respectively. This behavior should be contrasted with the large-x behavior of the free propagators for spin-$\frac{1}{2}$ and spin-

1 particles, namely x^{-3} and x^{-2}, respectively. This feature of our results is gauge independent and may be of considerable importance for physical processes occurring in pseudoparticle fields.

Note that this large-distance behavior conflicts with that derived from an O(5) conformal group analysis. The O(5) results are physically relevant only if they are supplemented by a corresponding O(5)-covariant treatment of the various collective coordinates which enter any pseudoparticle calculation. An O(5)-covariant approach to the collective coordinates has not been formulated and may in fact be an impossibility. It may also be true that such a formulation is possible and that it resolves the discrepancies between O(5)-covariant propagators and those derived in the present paper.

There are, as noted in Sec. III, some problems with the spin-1 propagation function that we have constructed. Now that we have found an explicit expression for the scalar, isospin-1 propagator,

$$\Delta_{ab}(x,y)=\frac{W_{ab}(x,y)}{4\pi^2(x-y)^2}+\frac{C_{ab}(x,y)}{4\pi^2\Pi(x)\Pi(y)}, \tag{5.9}$$

we can discuss these problems more fully. The matrix $C_{ab}(x,y)$ is a bilinear combination [Eq. (5.32)] of functions $\Phi^{(+)}_{rsa}(x)\,\Phi^{(+)}_{rsb}(y)$ [Eq. (5.28)] which vanish sufficiently rapidly at infinity so as to be square integrable:

$$\Phi^{(+)}_{rsa}(x)\underset{x^2\to\infty}{\sim}O(x^{-3}). \tag{6.1}$$

The matrix $W_{ab}(x,y)$, on the other hand, does not vanish when one of its coordinates becomes large:

$$W_{ab}(x,y)\underset{x^2\to\infty}{\longrightarrow}\delta_{ab}\Pi(y)^{-1} \tag{6.2}$$

[see Eqs. (5.8) and (5.4)]. Thus, in the formal operator construction of the spin-1 propagator

$$G^{(\pm)}_{\mu\nu}=-\left\{\left(\frac{1}{D^2}\right)^2\right\}^{(\pm)}_{\mu\nu}+(1-\xi)D_\mu\left(\frac{1}{D^2}\right)^2D_\nu, \tag{3.27}$$

the convolution integral which defines $(1/D^2)^2$ is divergent. In a coordinate representation we have

$$\left\langle x\left|\left(\frac{1}{D^2}\right)^2_{ab}\right|y\right\rangle=\int(d_E{}^4z)\Delta_{ac}(x,z)\Delta_{cb}(z,y), \tag{6.3}$$

and the z integration diverges logarithmically at large z, giving a contribution

$$\left\langle x\left|\left(\frac{1}{D^2}\right)^2_{ab}\right|y\right\rangle_{\rm div}=\frac{\ln\infty}{8\pi^2}\delta_{ab}\Pi(x)^{-1}\Pi(y)^{-1}. \tag{6.4}$$

The bracket operation appearing in the formal operator expression for the propagator is defined by

$$\{X\}^{(\pm)}_{\mu\nu}=q^{(\pm)}_{\mu\nu\lambda\kappa}D_\lambda XD_\kappa, \tag{3.18}$$

where, according to Eqs. (3.17) and (3.16),

$$q^{(\pm)}_{\mu\nu\lambda\kappa}=\delta_{\mu\nu}\delta_{\lambda\kappa}+\delta_{\mu\lambda}\delta_{\nu\kappa}-\delta_{\mu\kappa}\delta_{\nu\lambda}\pm\epsilon_{\mu\nu\lambda\kappa}. \tag{6.5}$$

The character of the divergent piece in the spin-1 propagator is revealed when we rewrite $q^{(\pm)}_{\mu\nu\lambda\kappa}$ in terms of an appropriate combination of the symbols $\eta^{(\mp)}$. To this end, we recall that these symbols project out the self-dual (or anti-self-dual) part of an antisymmetrical tensor [Eq. (4.17)], giving

$$\eta^{(\pm)}_{\alpha\beta a}\eta^{(\pm)}_{\gamma\delta a}=\delta_{\alpha\gamma}\delta_{\beta\delta}-\delta_{\alpha\delta}\delta_{\beta\gamma}\pm\epsilon_{\alpha\beta\gamma\delta}. \tag{6.6}$$

Thus we have

$$q^{(\pm)}_{\mu\nu\lambda\kappa}=\delta_{\mu\lambda}\delta_{\nu\kappa}+\eta^{(\pm)}_{\mu\lambda a}\eta^{(\pm)}_{\nu\kappa a}, \tag{6.7}$$

and we can write the divergent piece of the spin-1 propagator as

$$G^{(\pm)}_{\mu\nu ab}(x,y)_{\rm div}=-[\eta^{(\pm)}_{\mu\lambda d}D_{\lambda ac}\Pi(x)^{-1}\eta^{(\pm)}_{\nu\kappa d}\Pi(y)^{-1}\overleftarrow{D}_{\kappa cb}+\xi D_{\mu ac}\Pi(x)^{-1}\Pi(y)^{-1}\overleftarrow{D}_{\nu cb}]\frac{\ln\infty}{8\pi^2}. \tag{6.8}$$

The functions

$$\begin{aligned}\phi^{(\pm)c}_{\mu a}(x)&=D_{\mu ac}\Pi(x)^{-1}\\&=-\Pi(x)^{-2}[\partial_\mu\Pi(x)\delta_{ac}+\eta^{(\mp)}_{\mu\nu d}\partial_\nu\Pi(x)\epsilon_{adc}]\end{aligned} \tag{6.9}$$

are three $(c=1,2,3)$ zero-mode functions of the small-fluctuation, vector field. They correspond to an overall, global gauge rotation of the small-fluctuation vector field.[17] In general,[12] a zero-mode function $\phi^{(\pm)}_{\mu a}(x)$ obeys the background gauge constraint

$$D_{\mu ab}\phi^{(\pm)}_{\mu b}=0 \tag{6.10}$$

and produces a small-fluctuation field-strength tensor

$$f^{(\pm)}_{\mu\nu a}(x)=D_{\mu ab}\phi^{(\pm)}_{\nu b}(x)-D_{\nu ab}\phi^{(\pm)}_{\mu b}(x), \tag{6.11}$$

which is self-dual (or anti-self-dual), i.e.,

$$\eta^{(\mp)}_{\mu\nu c}D_{\mu ab}\phi^{(\pm)}_{\nu b}(x)=0, \tag{6.12}$$

for $c=1,2,3$. It is a simple matter to show that the gauge-rotation functions (6.9) do obey Eqs. (6.10) and (6.12) and thus prove that they are indeed zero-mode functions. Since the η symbols obey the Pauli-matrix algebra

$$\eta^{(\pm)}_{\mu\lambda a}\eta^{(\pm)}_{\nu\lambda b}=\delta_{\mu\nu}\delta_{ab}+\epsilon_{abc}\eta^{(\pm)}_{\mu\nu c}, \tag{4.19}$$

we see that if $\phi^{(\pm)}_{\nu a}(x)$ is a zero-mode solution of Eqs. (6.10) and (6.12), then so are the three functions $\overline{\phi}^{(\pm)d}_{\mu a}(x)$ $(d=1,2,3)$ defined by

$$\overline{\phi}^{(\pm)d}_{\mu a}(x)=\eta^{(\mp)}_{\mu\nu d}\phi^{(\pm)}_{\nu a}. \tag{6.13}$$

Thus the other functions

$$\eta^{(\mp)}_{\mu\lambda d} D_{\lambda ac}\Pi(x)^{-1} = \eta^{(\mp)}_{\mu\lambda d}\phi^{(\pm)c}_{\lambda a}(x) \qquad (6.14)$$

appearing in the divergent piece (6.8) of the propagator are also zero-mode functions. Now, using the explicit form for the gauge-rotation mode displayed in Eq. (6.9), we find that

$$\eta^{(\mp)}_{\mu\lambda d}\phi^{(\pm)c}_{\lambda a}(x) = -\epsilon_{dce}\phi^{(\pm)e}_{\mu a}(x) + \delta_{dc}\psi^{(\pm)}_{\mu a}(x), \qquad (6.15)$$

where

$$\psi^{(\pm)}_{\mu a}(x) = -\Pi(x)^{-2}\eta^{(\mp)}_{\mu\nu a}\partial_\nu\Pi(x). \qquad (6.16)$$

Thus there is a fourth zero-mode function $\psi^{(\pm)}_{\mu a}(x)$ attached to the three gauge-rotation, zero-mode functions $\phi^{(\pm)}_{\mu a}(x)$. Recalling that

$$A^{(\pm)}_{\mu a}(x) = -\eta^{(\mp)}_{\mu\nu a}\partial_\nu \ln\Pi(x), \qquad (4.14)$$

with

$$\Pi(x) = 1 + \sum_{s=1}^{n}\frac{\rho_s^{\,2}}{(x-z_s)^2}, \qquad (4.21)$$

we see that

$$\psi^{(\pm)}_{\mu a}(x) = \sum_{s=1}^{n}\rho_s^{\,2}\frac{\partial}{\partial\rho_s^{\,2}}A^{(\pm)}_{\mu a}(x), \qquad (6.17)$$

which identifies $\psi^{(\pm)}_{\mu a}(x)$ as a zero mode corresponding to an overall scale change.[17]

Collecting these results, we can rewrite the divergent piece (6.8) of the propagator as

$$G^{(\pm)}_{\mu\nu ab}(x,y)_{\rm div} = [3\psi^{(\pm)}_{\mu a}(x)\psi^{(\pm)}_{\nu b}(y) + (2+\xi)\,\phi^{(\pm)c}_{\mu a}(x)\,\phi^{(\pm)c}_{\nu b}(y)]\,\frac{\ln\infty}{8\pi^2}. \qquad (6.18)$$

Since zero-mode functions can be subtracted from the vector propagator without any change in the Green's function equation, this propagator can be rendered finite by simply deleting its infinite terms. There is, however, no unique way in which to perform this subtraction. The propagator behaves asymptotically as $O(x^{-1})$ as dictated by the asymptotic behavior[11] $[O(x^{-3})]$ of the product of vector zero-mode functions which appear on the right-hand side of the Green's function equation. Thus, one cannot impose the constraint that the propagator be orthogonal to the zero-mode functions; the inner product integral diverges logarithmically. We believe that this ambiguity may be resolved by an appropriate redefinition of the collective coordinates associated with the zero modes, but of this we are not yet sure.

ACKNOWLEDGMENTS

S. D. Ellis collaborated in the early stages of some of the work presented here. Part of the research and writing of this paper was performed when one of the authors (L.S.B.) was a visitor at the Los Alamos Scientific Laboratory and the Aspen Center for Physics. Another author (R.D.C.) would like to acknowledge the support of the Alfred P. Sloan Foundation. This work was supported, in part, by the U. S. Energy Research and Development Administration and by the National Science Foundation.

APPENDIX

The construction of the spin-1 propagator in Sec. III can be motivated as follows. We decompose the propagator into a sum of various mode functions $\chi_{\mu a}(x)$. We set the gauge-fixing parameter $\xi = 1$ so that, in view of Eq. (3.14), the mode functions obey

$$-[D^2\delta_{\mu\lambda} + 2F_{\mu\lambda}]\chi_\lambda = \kappa^2\chi_\mu. \qquad (A1)$$

Making use of the commutator of the gauge-covariant derivatives, Eq. (3.6), we find that

$$\chi^0_\mu = D_\mu\phi \qquad (A2)$$

satisfies the vector mode equation (A1) if ϕ is a scalar mode function,

$$-D^2\phi = \kappa^2\phi. \qquad (A3)$$

As we have remarked in the text, the $\eta^{(\pm)}$ symbols defined in Eq. (4.2) obey the algebra of the SU(2) generators $i\tau_a$,

$$\eta^{(\pm)}_{\mu\lambda a}\eta^{(\pm)}_{\nu\lambda b} = \delta_{\mu\nu}\delta_{ab} + \epsilon_{abc}\eta^{(\pm)}_{\mu\nu c}. \qquad (4.19)$$

Moreover, the two $\eta^{(\pm)}$ symbols correspond to the two SU(2) spins in the decomposition O(4) = SU(2) ⊗ SU(2), and they commute,

$$\eta^{(\pm)}_{\mu\lambda a}\eta^{(\mp)}_{\lambda\nu b} = \eta^{(\mp)}_{\mu\lambda b}\eta^{(\pm)}_{\lambda\nu a}. \qquad (A4)$$

Now if the field-strength tensor $F_{\mu\nu}$ is self-dual (or anti-self-dual), we can write [cf. Eq. (4.17)]

$$F^{(\pm)}_{\mu\nu} = \eta^{(\pm)}_{\mu\nu a}F_a \qquad (A5)$$

and thus conclude that $\eta^{(\mp)}$ commutes with $F^{(\pm)}$,

$$F^{(\pm)}_{\mu\lambda}\eta^{(\mp)}_{\lambda\nu a} = \eta^{(\mp)}_{\mu\lambda a}F^{(\pm)}_{\lambda\nu}. \qquad (A6)$$

In this case, $\eta^{(\mp)}$ commutes with the operator on the left-hand side of the vector mode equation (A1), so that if χ_μ is a solution of this equation, then so also are the three functions $\eta^{(\mp)}_{\mu\nu a}\chi_\nu$. In particular, starting from the scalar mode function ϕ, we can construct three more vector mode functions

$$\chi^a_\mu = \eta^{(\mp)}_{\mu\nu a}D_\nu\phi. \qquad (A7)$$

It follows from Eq. (4.19) that all four vector mode functions χ^a_μ, $a = 0, 1, 2, 3$, are orthogonal to one another and from Eq. (A3) that their normalization differs from that of the scalar mode function φ by a factor of the eigenvalue κ^2,

$$\int (d_g{}^4x)\chi^a_\mu\chi^b_\mu = \delta_{ab}\kappa^2\int (d_g{}^4x)\phi^2. \qquad (A8)$$

Thus the set of all χ_μ^a is a complete set of orthogonal vector mode functions, and with the scalar mode functions normalized to unity, the vector propagator can be written as

$$G_{\mu\nu}^{(\pm)}(x,y) = \sum_\chi \frac{\chi_\mu^a(x)\chi_\nu^a(y)}{\kappa^4}$$

$$= -(\delta_{\mu\lambda}\delta_{\nu\kappa} + \eta_{\mu\lambda a}^{(\mp)}\eta_{\nu\kappa a}^{(\mp)}) D_\lambda \sum_\phi \frac{\varphi(x)\phi(y)}{\kappa^4} \bar{D}_\kappa \,. \tag{A9}$$

The tensor quantity in parentheses here is precisely the quantity $q_{\mu\nu\lambda\kappa}^{(\pm)}$ [cf. Eq. (6.7)]. The scalar mode sum in Eq. (A9) represents the operator $(1/D^2)^2$, and Eqs. (A9) and (3.27) are thus equivalent.

Finally, let us prove, as was asserted in Sec. III, that

$$q_{\mu\alpha\alpha\beta}^{(\pm)} q_{\alpha\nu\lambda\kappa}^{(\pm)} = \delta_{\beta\lambda} q_{\mu\nu\alpha\kappa}^{(\pm)} + r_{\mu\nu\alpha\kappa\beta\lambda}^{(\pm)} \,, \tag{3.23}$$

where $r_{\mu\nu\alpha\kappa\beta\lambda}^{(\pm)}$ is antisymmetrical in its last index pair with a reversed duality character,

$$\tfrac{1}{2} r_{\mu\nu\alpha\kappa\rho\sigma}^{(\pm)} \epsilon_{\rho\sigma\beta\lambda} = \mp r_{\mu\nu\alpha\kappa\beta\lambda}^{(\pm)} \,. \tag{3.24}$$

The proof is brief if the form (6.7) is employed, for the substitution of Eq. (6.7) in Eq. (3.23) and the use of Eq. (4.19) give

$$r_{\mu\nu\alpha\kappa\beta\lambda}^{(\pm)} = (\delta_{\mu\alpha}\eta_{\nu\kappa c}^{(\mp)} - \delta_{\nu\kappa}\eta_{\mu\alpha c}^{(\mp)} + \eta_{\mu\alpha a}^{(\mp)}\eta_{\nu\kappa b}^{(\mp)}\epsilon_{abc})\,\eta_{\beta\lambda c}^{(\pm)} \,. \tag{A10}$$

which is manifestly antisymmetric in $\beta\lambda$ with the opposite duality character (3.24).

---

*Now at the Department of Physics and Astronomy, University of Pittsburgh, Pittsburgh, Pa. 15260.

†Now at the Department of Physics, University of Michigan, Ann Arbor, Michigan 48109.

[1]A. Belavin, A. Polyakov, A. Schwartz, and Y. Tyupkin, Phys. Lett. 59B, 85 (1975).

[2]G. 't Hooft, Phys. Rev. Lett. 37, 8 (1976).

[3]G. 't Hooft, Phys. Rev. D 14, 3432 (1976).

[4]C. G. Callan, R. Dashen, and D. J. Gross, Phys. Lett. 63B, 334 (1976); R. Jackiw and C. Rebbi, Phys. Rev. Lett. 37, 172 (1976).

[5]A. M. Polyakov, Phys. Lett. 59B, 82 (1975).

[6]C. G. Callan, R. Dashen, and D. J. Gross, Phys. Lett. 66B, 375 (1977).

[7]L. S. Brown, R. D. Carlitz, D. B. Creamer, and C. Lee, Phys. Lett. 70B, 180 (1977) or 71B, 103 (1977).

[8]A. S. Schwartz, Phys. Lett. 67B, 172 (1977); R. Jackiw and C. Rebbi, *ibid.* 67B, 189 (1977); L. S. Brown, R. D. Carlitz, and C. Lee, Phys. Rev. D 16, 417 (1977); M. Atiyah, N. Hitchin, and I. Singer, Proc. Natl. Acad. Sci. U.S.A. 74, 2662 (1977).

[9]E. Corrigan and D. B. Fairlie, Phys. Lett. 67B, 69 (1977). These authors dismissed these solutions as unphysically singular, but G. 't Hooft (private communication) has emphasized that they are, in fact, physically acceptable.

[10]A slightly more general pseudoparticle solution involving $5n+4$ parameters has been constructed by R. Jackiw, C. Nohl, and C. Rebbi, Phys. Rev. D 15, 1642 (1977). The results of our paper are easily generalized to accommodate solutions of this form.

[11]The spin-$\frac{1}{2}$ zero modes have been discussed by G. 't Hooft, Ref. 3 above. The number of these modes in a general self-dual field has been determined by J. Kiskis, Phys. Rev. D 15, 2329 (1977) and by L. S. Brown, R. D. Carlitz, and C. Lee, Ref. 8 above. The spin-$\frac{1}{2}$ zero mode functions in the $5n$-parameter pseudoparticle field have been explicitly computed for isospin $\frac{1}{2}$ by B. Grossman, Phys. Lett. 61A, 86 (1977) and by R. D. Carlitz (unpublished) and for isospin 1 by R. Jackiw and C. Rebbi, Phys. Rev. D 16, 1052 (1977).

[12]A full discussion of the square-integrable, zero-mode wave functions is given in L. S. Brown, R. D. Carlitz, and C. Lee, Ref. 8.

[13]J. L. Gervais and B. Sakita, Phys. Rev. D 11, 2943 (1975) and G. 't Hooft, Ref. 3.

[14]Similar observations have been made independently by J. J. Giambiagi and K. D. Rothe, Nucl. Phys. B129, 111 (1977), and by S. Sciuto (unpublished).

[15]D. B. Creamer, Phys. Rev. D 16, 3496 (1977).

[16]R. Jackiw and C. Rebbi, Phys. Rev. D 14, 517 (1976); F. R. Ore, Jr., *ibid.* 15, 470 (1977); 16, 1041 (1977). Ore works in a gauge which differs from the one employed in this paper, so that even his results for the "ghost" (scalar) propagator are not directly comparable to those given here. For a single pseudoparticle he calculates not $(-D^2)^{-1}$ but rather $\{-D^2 + [4x_\mu/(\rho^2+x^2)]\partial_\mu\}^{-1}$

[17]These results differ slightly from the speculation offered in our letter, Ref. 7.

Nuclear Physics B151 (1979) 93–117

INSTANTON GREEN FUNCTIONS AND TENSOR PRODUCTS

E. CORRIGAN *
Laboratoire de Physique Théorique de l'Ecole Normale Supérieure, Paris **, *France*

P. GODDARD ‡
CERN, Geneva, Switzerland

S. TEMPLETON
Department of Mathematics, University of Durham, Durham, UK

Received 20 November 1978

The structure of the Green function for a scalar field, transforming under the adjoint representation of a gauge group, in the background field of an arbitrary self-dual instanton field, is considered. It is shown that it can be written in the same elegant form as the Green function for the fundamental vector representation of the group, given the introduction of a certain conformally invariant matrix. The same method solves the more general problem of finding the Green function for a field transforming under the direct product of two groups. Simple expressions for the solutions of the corresponding massless Dirac equation are obtained. The results on the products of representations may be applied, iteratively, to construct the Green function and massless solutions of the Dirac equation for any higher representation of the group.

1. Introduction

Our understanding of the self-dual solutions to the Euclidean Yang-Mills field equations (self-dual instantons) has been completely transformed by the work of Atiyah, Drinfeld, Hitchin and Manin (ADHM) [1,2]. They have given a construction for the general self-dual solution in terms of constant parameters constrained by certain quadratic equations. This construction may be used, without any essential variations in the formalism, for any compact classical group, involves only elemen-

* On leave from the Department of Mathematics, University of Durham, Durham, UK; present address: CALTECH, Pasadena, California, USA.
** Laboratoire propre du CNRS, associé à l'Ecole Normale Supérieure et à l'Université de Paris-Sud.
‡ On leave from DAMTP, University of Cambridge, Cambridge, UK.

tary matrix algebra, and leads to expressions for the gauge potentials which are rational functions of the spatial coordinates (in a suitable gauge). Further, it provides a framework within which the previously accumulated results on problems associated with self-dual solutions may be understood and their structure elucidated and simplified. In particular, using the techniques implicit in the construction of Atiyah et al., simple and elegant forms have been found for the Green function for a (spatially) scalar field transforming under the fundamental vector representation of the gauge group [3,4], and for the massless solutions of the corresponding Dirac equation [3,5]. (By the "fundamental vector representation" we mean that representation with the same dimension as the matrices forming the group, e.g., isospin-$\frac{1}{2}$ for SU(2); we shall simply call it the fundamental representation henceforth.) These results not only generalize the results found previously [6,7], for the special case of the 't Hooft SU(2) solutions [8], but clarify their essentially simple structure.

In addition to their intrinsic mathematical interest, such Green functions are of practical importance in heuristic estimations of instanton effects in quantum chromodynamics. For example, work has begun on calculating instanton effects in electron-positron annihilation [9,10] and on studying effects of instantons in bound-state mass calculations [11]. The latter problem requires knowledge of the adjoint representation Green function, as does the former, once one proceeds beyond the lowest order. Once the scalar Green function is known for a given representation, the corresponding (spatially) spinor and vector Green functions may be found easily using the techniques of Brown et al. [6]. However, a good knowledge of the corresponding zero-frequency modes is desirable [12].

Brown et al. [6] showed how to construct the adjoint representaion (isospin-1) Green function for the 't Hooft SU(2) solutions. Christ et al. [4], pointed out that these methods could be applied immediately to the general self-dual SU(2) solution. Superficially, the resulting expression shows little of the elegance so striking in the isospin-$\frac{1}{2}$ case [3,4], but we shall see that this appearance is rather deceptive. The results described in this present paper have grown out of an attempt to understand this expression for the adjoint Green function, its structure and generality. We feel that these attributes are somewhat obscured in ref. [4] by the notation and explicit symmetrization. Suitably recast, the expression reveals an intricate elegance intimately connected with the conformal invariance of the theory, and is completely general: it works for any compact classical group. Further, it solves the wider problem of constructing the Green function for a scalar field transforming under the direct product of two groups, $G_1 \times G_2$, with respect to the fundamental representation of each, in the background field obtained by taking a self-dual solution for each of the groups. In other words, it constructs the Green function for the fundamental representation of a non-simple Lie group consisting of two simple compact factors. (We shall refer to the solution for $G_1 \times G_2$ obtained by taking particular (self-dual) solutions for G_1 and G_2 as the *tensor product* of those solutions. The adjoint representation occurs as a special case when we take $G_1 = G_2^*$ and the solutions to be conjugate.)

From another point of view it would seem that working out the Green function for the tensor product and, hence, the adjoint representation should be easy. We can always regard $G_1 \times G_2$ as a subgroup of some larger simple group, G. The tensor product of solutions for G_1 and G_2 provides us with a solution for G and, if G is suitably chosen, the Green function for the fundamental representation of G will be what is required. The reason that this approach is not as trivial as it sounds is that to implement it we have to describe the solution for G in the terms of the linear algebra of the ADHM construction [1] for G, given the corresponding descriptions for G_1 and G_2. This is not easy because taking tensor products destroys linearity. In fact the additional structure in the adjoint representation, or tensor product, Green function is really there to solve this problem. By elaborating this structure we have discovered how to describe the tensor product of instantons within the framework of the linear ADHM construction, thus unifying the descriptions of the fundamental and adjoint representation Green functions. In addition, this enables us to immediately extend results obtained for the fundamental representation (for example, the massless solution of the Dirac equation) to tensor products and so, in particular, the adjoint representation. Going further, we may use this new description of the adjoint representation, or any tensor product, as an input and calculate the Green function for *its* tensor product with some other instanton. In this way we may, at least in principle, obtain the Green function, etc., for any representation of the group by an iterative process of taking tensor products of representations and reducing.

Our paper is arranged as follows. In sect. 2 we review the formalism we shall use and summarize the results we have obtained. In sect. 3 we establish our results on the Green function for the adjoint representation and, more generally, the tensor product of instantons. Sect. 4 is devoted to understanding the tensor product of instantons within the ADHM framework. As an illustration of the power of this point of view, in sect. 5 we show how the expression for the massless solutions of the Dirac equation in the fundamental representation yields those for the adjoint representation. Finally, in sect. 6, we discuss the significance of our results and further problems.

2. A summary of results

We shall begin this section by reviewing briefly the construction of Atiyah et al. [1] and the results on Green functions for the fundamental representation [3,4]. Throughout this paper we shall describe our results with specific reference to Sp(n) which may conveniently be regarded as the group of $n \times n$ unitary matrices with quaternion entries. In particular $\mathrm{Sp}(1) \simeq \mathrm{SU}(2)$, the group of quaternions with unit modulus. Any compact classical group may be thought of as a subgroup of Sp(n) for sufficiently large n, and it is simple to deduce the corresponding results for the orthogonal and unitary groups in this way. To be quite explicit we shall restate some of our results for SU(n).

In the ADHM construction for Sp(n), the vector potential, A_α, represented as a traceless $n \times n$ anti-Hermitian matrix of quaternions, is given in the form

$$A_\alpha = v^\dagger \partial_\alpha v \,. \tag{2.1}$$

where $v \equiv v(x)$ is a $(k+n) \times n$ matrix of quaternions which must satisfy the equations,

$$v^\dagger v = 1 \,, \tag{2.2}$$

$$v(x)^\dagger \, \Delta(x) = 0 \,. \tag{2.3}$$

Here $\Delta(x)$ is an $(n+k) \times k$ matrix of quaternions, linear in the position coordinates x, which has the form

$$\Delta_{\lambda i}(x) = a_{\lambda i} + b_{\lambda i} x \,, \qquad 1 \leqslant i \leqslant k \,, \qquad 1 \leqslant \lambda \leqslant n+k \,, \tag{2.4}$$

with x being the quaternionic representation of the position coordinates,

$$x = x_0 - i\boldsymbol{x} \cdot \boldsymbol{\sigma} = \begin{pmatrix} x_0 - ix_3 & -x_2 - ix_1 \\ x_2 - ix_1 & x_0 + ix_3 \end{pmatrix} . \tag{2.5}$$

The quaternionic matrices a, b are constants, independent of x. Provided they satisfy the constraints that $a^\dagger a$, $b^\dagger b$ and $a^\dagger b$ all be symmetric as $k \times k$ matrices of quaternions, eq. (2.1) yields a self-dual solution of the field equations, determined up to gauge equivalence by eqs. (2.2), (2.3). Further this solution is non-singular at points where $\Delta(x)$ has maximal rank. All these conditions may be summarized by saying that one obtains a solution which is non-singular everywhere if the $k \times k$ matrix of quaternions $\Delta(x)^\dagger \, \Delta(x)$ is actually real and invertible for all x. (We shall denote its inverse by $f(x)$.)

It is not difficult to give simple and elementary demonstrations that the resulting field strength,

$$F_{\alpha\beta} = \partial_\alpha A_\beta - \partial_\beta A_\alpha + [A_\alpha, A_\beta] \,, \tag{2.6}$$

is self-dual,

$$F_{\alpha\beta} = {}^*F_{\alpha\beta} = \tfrac{1}{2}\epsilon_{\alpha\beta\gamma\delta} F_{\gamma\delta} \,, \tag{2.7}$$

with instanton number k,

$$k = \frac{1}{16\pi^2} \int d^4x \, \mathrm{Tr}\{ F_{\alpha\beta} \, {}^*F_{\alpha\beta} \} \,, \tag{2.8}$$

see refs. [2], [3] or [4]. (For this last statement it is also necessary that the solution be regular at infinity, which is ensured by $b^\dagger b$ being non-singular.) A deeper statement [1] is that the ADHM construction yields all the self-dual solutions which are regular everywhere including infinity.

For further details such as the form of a and b for the 't Hooft SU(2) solutions see ref. [3].

The Green functions we shall be concerned with satsify equations of the form

$$\mathcal{D}^2 G(x,y) = -\delta(x-y)\,, \tag{2.9}$$

where, in the case of the fundamental representation

$$\mathcal{D}_\alpha = \partial_\alpha + A_\alpha = \partial_\alpha + v^\dagger \partial_\alpha v\,. \tag{2.10}$$

In the case of the fundamental representation, a form has been found for G which is valid for any self-dual solution for any compact-classical group [3,4]

$$G(x,y) = \frac{v(x)^\dagger\, v(y)}{4\pi^2 (x-y)^2}\,. \tag{2.11}$$

This could hardly be simpler. The Green function for the ordinary Laplacian, corresponding to putting $A_\alpha = 0$ in eq. (2.9), is just $[4\pi^2(x-y)^2]^{-1}$. Thus 1 has just been replaced by $v(x)^\dagger\, v(y)$, or δ_{ab} by $[v(x)^\dagger\, v(y)]_{ab}$ to be more explicit.

This result might tempt one into hoping that this prescription would work for other representations too: that the Green functions for products of the fundamental representation, q, could be obtained by judiciously replacing 1 by $v(x)^\dagger\, v(y)$ in appropriate places. In particular the adjoint representation can be obtained by decomposing $\mathrm{q} \otimes \bar{\mathrm{q}}$, for which

$$\mathcal{D}_\alpha = \partial_\alpha + A_\alpha \otimes 1 + 1 \otimes A_\alpha^*\,, \tag{2.12}$$

A_α^* being the complex conjugate of A_α. The prescription would lead to a result of the form,

$$G_{ab,cd}(x,y) = \frac{[v(x)^\dagger\, v(y)]_{ac} [v(x)^\dagger\, v(y)]^*_{bd}}{4\pi^2 (x-y)^2} + \frac{1}{4\pi^2} C_{ab,cd}(x,y)\,, \tag{2.13}$$

in which $C = 0$, and the labels a, b, c, d take $2n$ values. Unfortunately this does not quite work; the C term really is necessary, as was first pointed out by Brown et al. in the case of the 't Hooft SU(2) solutions [6]. The tempting prescription provides the correct δ-function singularity as $x \to y$, so that $C(x,y)$ is non-singular in this limit. In their pioneering papers, Brown et al., found a form for $C(x,y)$ which was extended to the general self-dual SU(2) solution by Christ et al. [4]. Both these sets of authors explicitly performed the simple reduction from $\mathrm{q} \otimes \bar{\mathrm{q}}$ to the adjoint representation, which tends to obscure the structure of $C(x,y)$. Leaving this undone, it becomes apparent that this structure is independent of the group and actually solves a wider class of problems.

The structure of $C(x, y)$ is contained in the general formula,

$$C_{ab,cd}(x,y) = M_{ij,lm}[v_1(x)^\dagger b_1]_{a,iI}[v_2(x)^\dagger b_2]_{b,jJ}\epsilon_{IJ}[b_1^\dagger v_1(y)]_{lL,c} \times [b_2^\dagger v_2(y)]_{mM,d}\,\epsilon_{LM}\,, \tag{2.14}$$

where $M_{ij,lm}$ is defined by the equation

$$M_{rs,ij}\{[a_1^\dagger a_1]_{il}[b_2^\dagger b_2]_{jm} + [b_1^\dagger b_1]_{il}[a_2^\dagger a_2]_{jm} - [a_2^\dagger b_1]_{iI,lL}[a_2^\dagger b_2]_{jJ,mM} \cdot \times \epsilon_{IJ}\epsilon_{LM}\} = \delta_{rl}\delta_{sm}\,. \tag{2.15}$$

(See eq. (3.20) for a diagrammatic representation of M^{-1}). In eqs. (2.14), (2.15), ϵ is the usual two-dimensional antisymmetric tensor,

$$\epsilon_{12} = -\epsilon_{21} = 1\,, \qquad \epsilon_{11} = \epsilon_{22} = 0\,. \tag{2.16}$$

The upper case labels I, J, L, M run over the values 1, 2 and label the rows and columns of quaternions; we shall refer to these as the quaternionic labels.

To construct the Green function for $\mathrm{q} \otimes \bar{\mathrm{q}}$ we take

$$a_1 = a\,, \qquad b_1 = b\,, \qquad a_2 = a^*\epsilon\,, \qquad b_2 = b^*\epsilon\,, \tag{2.17}$$

where a, b are the matrices used to define the potential A_a *via* eqs. (2.1)–(2.4). In this case i, j, l, m, r, s run over k values. (The multiplication by ϵ in eq. (2.17) is on the quaternionic labels, of course.)

To extract the adjoint representation Green function, $G_{\rho\tau}(x, y)$, from that for $\mathrm{q} \otimes \bar{\mathrm{q}}$, we multiply eq. (2.13) by Λ^ρ_{ab}, Λ^τ_{cd}, and sum over a, b, c, d, where $\{\Lambda^\rho_{ab}, p = 1, ..., n(2n + 1)\}$ is a suitably normalized basis of Hermitian generators for Sp(n), i.e., for Hermitian $2n \times 2n$ matrices satisfying $\epsilon\Lambda\epsilon^{-1} = -\Lambda^{\mathrm{T}}$. It follows from this property of the generators that only the antisymmetric part of $M_{ij,lm}$, $M_{[ij],[lm]}$, contributes in the adjoint representation. For Sp(1) $\simeq$ SU(2) the $\{\Lambda^\rho\}$ are just the Pauli σ matrices.

The motivation for writing eqs. (2.14), (2.15) as we have, is that with $C(x, y)$ so defined, eq. (2.13) solves a more general problem. Consider a non-simple gauge group Sp(n_1) $\times$ Sp(n_2) and a scalar field $\phi_{ab}(x)$ which transforms under the fundamental representation of each factor. For such a field the covariant derivative takes the form

$$\mathcal{D}_\alpha = \partial_\alpha + A_{1\alpha} \otimes 1 + 1 \otimes A_{2\alpha}\,. \tag{2.18}$$

The self-dual solutions for Sp(n_1) $\times$ Sp(n_2) are given by taking self-dual solutions for each of the factors. These can be specified by equations of the form of eqs. (2.1)–(2.4):

$$A_{m\alpha} = v_m^\dagger \partial_\alpha v_m\,, \qquad m = 1, 2\,, \tag{2.19}$$

where $v_m^\dagger v_m = 1$, $v_m^\dagger \Delta_m = 0$ and $\Delta_m = a_m + b_m x$, $m = 1, 2$. The quaternionic matrices a_m, b_m have dimensions $(n_m + k_m) \times k_m$ for $m = 1, 2$, respectively. The solution has two instanton numbers k_1, k_2. Eq. (2.19) will provide a self-dual solution if $\Delta_m(x)^\dagger \Delta_m(x)$ is everywhere a real and invertible $k_m \times k_m$ matrix, $f_m(x)^{-1}$, for $m = 1, 2$. Then

$$G_{ab,cd}(x,y) = \frac{[v_1(x)^\dagger v_1(y)]_{ac}[v_2(x)^\dagger v_2(y)]_{bd}}{4\pi^2(x-y)^2} + \frac{1}{4\pi^2} C_{ab,cd}(x,y)\,, \tag{2.20}$$

with $C(x, y)$ given by eqs. (2.14). (2.15), solves the Green function equation (2.9) in which the covariant derivative is given by eq. (2.18). In other words, eqs. (2.13)–(2.15) solve the Green function problem for the direct product of two compact simple classical groups. This our first main result which will be proved in sect. 3, where we shall also introduce a convenient diagrammatic notation to make the structure of equations like eqs. (2.14), (2.15) clearer.

We can use eqs. (2.13)–(2.15) and (2.20) to go further and consider non-simple groups with more than two factors, for example, $G_1 \times G_2 \times G_3$, if we can find a description of the solution for $G_1 \times G_2$ which is strictly analogous to eq. (2.1), and which would mean we could amalgamate the two terms in eq. (2.20) to give a precise analogue of eq. (2.11). Having achieved this we may use the solution for $G_1 \times G_2$ as an input, in the places we used the solution for G_1, and replace G_2 by G_3, to obtain the Green function for the fundamental representation of $G_1 \times G_2 \times G_3$ and so on. Making identifications of the sort made in eq. (2.17), we may particularize, obtaining Green functions for the higher dimensional representations of a given group, working iteratively.

The main task of this paper is to demonstrate how the program described in the last paragraph may be carried out. The first step is to rewrite eq. (2.20) in the form of eq. (2.11). In sect. 3 we shall show that we can rewrite eq. (2.20) as

$$G(x,y) = \frac{[v_1(x) \otimes v_2(x)]^\dagger (1 - \mathcal{M})[v_1(y) \otimes v_2(y)]}{4\pi^2(x-y)^2}\,, \tag{2.21}$$

where

$$\begin{aligned}\mathcal{M}_{ab,cd} = \{&a_{1a,iI} b_{2b,jJ} a^\dagger_{1lI,c} b^\dagger_{2mJ,d} + b_{1a,iI} a_{2b,jJ} b^\dagger_{1lI,c} a^\dagger_{2mJ,d} \\ &- a_{1a,iI} b_{2b,jJ} b^\dagger_{1lJ,c} a^\dagger_{2mI,d} - b_{1a,iI} a_{2b,jJ} a^\dagger_{1lJ,c} b^\dagger_{2mI,d} \\ &- [a_{1a,iI} a_{2b,jJ} b^\dagger_{1lL,c} b^\dagger_{2mM,d} + b_{1a,iI} b_{2b,jJ} a^\dagger_{1lL,c} a^\dagger_{2mM,d}]\, \epsilon_{IJ} \epsilon_{LM}\} \, M_{ij,lm}\,. \end{aligned} \tag{2.22}$$

Whilst a diagrammatic notation is helpful for discerning the structure of eqs. (2.14), (2.15) it is essential for eq. (2.22); the reader should compare this equation with eq. (3.29).

The matrix $1 - \mathcal{M}$ has a number of significant properties which we shall establish and use in the remainder of this paper. Briefly, it is conformally invariant, its eigenvalues are non-negative and at least $4k_1k_2$ of them are zero. As a consequence of the second statement we may define a positive square root $(1 - \mathcal{M})^{1/2}$, which is

Hermitian, and set

$$\tilde{v}(x) = (1 - \mathcal{M})^{1/2}\, v_1(x) \otimes v_2(x)\,. \tag{2.23}$$

The non-singularity of $C(x, y)$ as $x \to y$ implies that $\tilde{v}(x)^\dagger\, \tilde{v}(x) = 1$. Consequently, if we can find a $\tilde{\Delta}(x)$, linear in x such that $\tilde{v}(x)$ is defined by $\tilde{v}(x)^\dagger\, \tilde{\Delta}(x) = 0$ and $\tilde{\Delta}(x)^\dagger \tilde{\Delta}(x)$ satisfies the usual cónstraints, we shall have achieved our objective. In sect. 4 we show how to define such a $\tilde{\Delta}(x)$. Note that the Green functions of eq. (2.20) can now be obtained by replacing v by $\tilde{v}$ in eq. (2.11); the validity of this expression then follows merely from the existence of some $\tilde{\Delta}(x)$ with the properties just described and the previous proof of the formula for the Green function [3,4].

Having thus unified the treatments of the product of the adjoint and fundamental representations, we may immediately take results from one to the other. In particular we may use the formula for the massless solutions of the Dirac equation for the fundamental representation [3,5] to write them down for the adjoint representation, or, more generally, a tensor product of instantons. In a sense this is a slightly circular point of view, because the construction of $\tilde{\Delta}(x)$, is intimately linked with these solutions of the Dirac equation. However, in sect. 5, we show how they may be obtained immediately from the simple result for the fundamental representation, given $\tilde{\Delta}(x)$. Here we state the results in their simplest form; for further details of the formalism for the Dirac equation see sect. 5.

Consider the massless Dirac equation corresponding to the covariant derivative of eq. (2.18). It has only negative chirality solutions described by two spinors $\psi_{\rm L} \equiv \psi_{ab,I}$ satisfying

$$e_\alpha^\dagger \mathcal{D}_\alpha \psi_{\rm L} = 0\,. \tag{2.24}$$

Here $e_\alpha^\dagger = \partial_\alpha x^\dagger$ acts on the spinor index, I. There are $2(n_1 k_2 + n_2 k_1)$ linearly independent solutions to eq. (2.24) and these are of the form,

$$\psi_{ab,I} = [v_1^\dagger b_1 \epsilon f_1]_{a,iI} [v_2^\dagger d]_{b,i} + [v_1^\dagger c]_{a,i} [v_2^\dagger b_2 \epsilon f_2]_{b,iI}\,, \tag{2.25}$$

where c, d are (constant) $2(k_1 + n_1) \times k_2$ and $2(k_2 + n_2) \times k_1$ dimensional matrices, respectively, satisfying the linear equations

$$[a_1^\dagger c]_{iI,j} = [a_2^\dagger d]_{jI,i}\,,$$

$$[b_1^\dagger c]_{iI,j} = [b_2^\dagger d]_{jI,i}\,, \tag{2.26}$$

for $1 \leqslant i \leqslant k$, $1 \leqslant j \leqslant k$, $I = 1, 2$. Again we may particularize to the $q \otimes \bar{q}$ representation by using eq. (2.17), and extract the adjoint representation solutions, $\psi_{\rho,I}$, by multiplying by Λ^ρ_{ab} and summing. In the latter case the symmetry of Λ^ρ implies that only those c, d with $d = \epsilon c$ contribute.

We first obtained eq. (2.26) by studying the so called "supersymmetric" solutions [13,14] to the massless Dirac equation for the adjoint representation and generalising. These solutions, which in general have the form

$$\psi_L = (e_\alpha e_\beta^\dagger - e_\beta e_\alpha^\dagger)\,\chi F_{\alpha\beta}\,,$$

$$\psi_L = (e_\alpha e_\beta^\dagger - e_\beta e_\alpha^\dagger)\,x\chi F_{\alpha\beta}\,, \tag{2.27}$$

where χ is a fixed two-spinor, here correspond to $c = b\chi$ and $c = a\chi$, respectively. Eqs. (2.25), (2.26) generalize the results obtained by Jackiw and Rebbi [14] for the 't Hooft SU(2) solutions in the adjoint representation.

The results we have described may be transcribed immediately so that they apply directly to SU(n). The ADHM construction for SU(n) may be described [3] using eqs. (2.1)–(2.3), where now v is a $(2k+n)\times n$ complex matrix and $\Delta_{\lambda,iI} = \alpha_{\lambda,iI} + b_{\lambda,iJ}x_{Ji}$, $1 \leqslant \lambda \leqslant 2k+n$, $1 \leqslant i \leqslant k$, $I, J = 1, 2$, is a $(2k+n)\times 2k$ complex matrix, such that $\{\Delta(x)^\dagger\,\Delta(x)\}_{iJ,jJ}$ is proportional to δ_{IJ} and invertible for all x (including, in an appropriate sense, infinity). Given solutions for SU(n_m) with topological charge k_m, described by $\Delta_m(x)$, $m = 1, 2$, we may use eq. (2.15) to construct an Hermitian k_1k_2 dimensional matrix M, hence obtaining the n_1n_2 dimensional complex matrix $c(x, y)$ from eq. (2.14), and the Green function for the tensor product. Similarly we may define $\mathcal{M}$, which has dimensions $(2k_1+n_1)(2k_2+n_2)$ and $4k_1k_2$ unit eigenvalues, using eq. (2.22). (The instanton number of the tensor product, regarded as a solution for SU(n_1n_2) is now $k_1n_2 + k_2n_1$.) Subsequent formulae and equations may be reinterpreted similarly.

3. Green functions

As was emphasised in ref. [3], the covariant derivative is most easily understood and calculated with, if it is thought of, in a larger space, as differentiation followed by orthogonal projection [1]. We define a map between $2n$-dimensional fields, ϕ, in the fundamental representation and $2(n+k)$ dimensional fields, $\hat{\phi}$, using matrix multiplication by $v(x)$.

$$\hat{\phi}(x) = v(x)\,\phi(x)\,. \tag{3.1}$$

Then the covariant derivative takes the form

$$\mathcal{D}_\alpha\hat{\phi} \equiv \widehat{\mathcal{D}_\alpha\phi} = P\partial_\alpha\hat{\phi}\,, \tag{3.2}$$

where the orthogonal projection operator

$$P(x) = v(x)\,v(x)^\dagger\,. \tag{3.3}$$

Because of the normalization condition on v, eq. (2.2), there is one to one correspondence between $2n$-dimensional fields ϕ and $2(n+k)$ dimensional fields $\hat{\phi}$ satisfying $P\hat{\phi} = \hat{\phi}$. As in ref. [3] we shall employ this correspondence to lift equations into the larger space and use the simple form for the covariant derivative.

For a field in the $q \otimes \bar{q}$ representation $\phi_{ab}(x)$, we may define

$$\hat{\phi}(x) = v(x)\, \phi(x)\, v(x)^\dagger \,, \tag{3.4}$$

regarding ϕ as a $2n \times 2n$ matrix, and the covariant derivative of eq. (2.12) then takes the form

$$\mathcal{D}_\alpha \hat{\phi} \equiv \mathcal{D}_\alpha \phi = P(\partial_\alpha \hat{\phi})\, P \,. \tag{3.5}$$

More generally, for the tensor product case of eq. (2.18) we define

$$\hat{\phi}_{\lambda\mu}(x) = v_1(x)_{\lambda a}\, v_2(x)_{\mu b}\, \phi_{ab}(x) \,, \tag{3.6}$$

and the covariant derivative is given by

$$\mathcal{D}_\alpha \hat{\phi} = (P_1 \otimes P_2)\, \partial_\alpha \hat{\phi} \,, \tag{3.7}$$

where $P_m(x) = v_m(x)\, v_m(x)^\dagger$, $m = 1, 2$.

The Green function equation for the tensor product may thus be written in the form,

$$\mathcal{D}^2 \hat{G}_{\lambda\mu,\nu\rho}(x, y) = -P_{1\lambda\nu}(x)\, P_{2\mu\rho}(x)\, \delta(x - y) \,. \tag{3.8}$$

The first guess at a solution, to which eq. (2.11) might tempt us, is $\hat{G} = \hat{H}$, where

$$\hat{H}_{\lambda\mu,\nu\rho}(x, y) = \frac{[P_1(x)\, P_1(y)]_{\lambda\nu} [P_2(x)\, P_2(y)]_{\mu\rho}}{4\pi^2 (x - y)^2} \,. \tag{3.9}$$

Before discussing the extent to which eq. (3.9) fails to yield a solution to eq. (3.8) and how eqs. (2.14), (2.15) and (2.20) rectify this, we shall introduce a diagrammatic notation which we find convenient for representing and manipulating such expressions. We shall use lines to indicate the contraction of indices. The matrices being multiplied typically have dimensions $2k_m$ or $2(k_m + n_m)$, $m = 1, 2$; we shall treat the two-dimensional quaternionic labels separately from those of dimension k_m or $k_m + n_m$, which label the different quaternions. We shall use broken lines to stand for sums over the former and solid lines for sums over the latter. The matrix ϵ occurs so frequently that it is convenient to have a special symbol for it, and we shall use an arrow. Thus,

$$i \;\text{———————}\; j = \delta_{ij} \,,$$

$$I \cdots\cdots J = \delta_{IJ} \,,$$

$$I \cdots \rightarrow \cdots J = \epsilon_{IJ} \,,$$

$$I \cdots \leftarrow \cdots J = -\epsilon_{IJ} = \epsilon_{JI} \,, \tag{3.10}$$

where $1 \leqslant i, j \leqslant k_m$ or $n_m + k_m$ and $I, J = 1, 2$. We shall write those operators P_m, a_m, b_m, etc., with a suffix 2 below those with suffix 1, and omit the suffices for brevity. In this notation eq. (3.9) becomes

$$\hat{H} = \frac{1}{4\pi(x-y)^2} \frac{[P(x)]\text{-----}[P(y)]}{[P(x)]\text{-----}[P(y)]} , \tag{3.11}$$

The suffices λ, μ, ν, ρ are left implicit at the four corners.

In subsequent calculations we shall often need to use the identity,

$$\epsilon_{AB}\epsilon_{CD} = \delta_{AC}\delta_{BD} - \delta_{AD}\delta_{CB} , \tag{3.12}$$

which diagrammatically appears as

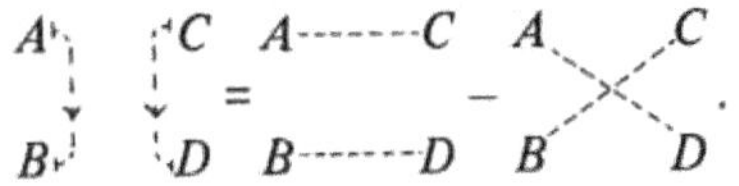

$$
\begin{array}{cc}
A' \quad \quad C \\
\vdots \quad \vdots \quad = \\
B \quad \quad D
\end{array}
\quad
\begin{array}{c}
A\text{------}C \\
\\
B\text{------}D
\end{array}
\quad - \quad
\begin{array}{c}
A \qquad C \\
\times \\
B \qquad D
\end{array}
. \tag{3.13}
$$

Another useful identity is

$$[e_\alpha]_{AC}[e_\alpha]_{BD} = 2\epsilon_{AB}\epsilon_{CD} , \tag{3.14}$$

which appears as

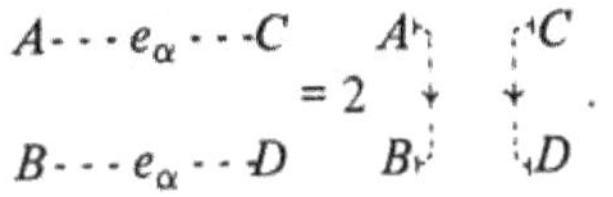

$$
\begin{array}{c}
A\text{---}e_\alpha\text{---}C \\
= 2 \\
B\text{---}e_\alpha\text{---}D
\end{array}
\quad
\begin{array}{cc}
A' \quad \quad C \\
\downarrow \quad \downarrow \\
B \quad \quad D
\end{array}
. \tag{3.15}
$$

In evaluating $\mathcal{D}^2\hat{H}$ we may divide the terms into four groups. We have a term involving $\partial^2\{(x-y)^{-2}\}$ which gives the desired δ-function as on the right-hand side of eq. (3.8). We have a term involving

$$P_1\partial_\alpha\{P_1\partial_\alpha P_1\}(x-y)^{-2} + 2\{P_1\partial_\alpha P_1\}\,\partial_\alpha\{(x-y)^{-2}\} ,$$

which vanishes as in the proof of the Green function for the fundamental representation [3]. We have a similar term involving P_2 instead of P_1. Finally we have a term involving $P_1\partial_\alpha P_1$ and $P_2\partial_\alpha P_2$, which we might like to vanish, but does not. Thus,

$$\mathcal{D}^2\hat{H} = -\frac{[P(x)]}{[P(x)]}\,\delta(x-y) + \frac{2}{4\pi(x-y)^2} \frac{[P(x)\,be_\alpha f(x)\,\Delta(x)^\dagger\,P(y)]}{[P(x)\,be_\alpha f(x)\,\Delta(x)^\dagger\,P(y)]} . \tag{3.16}$$

We use $\Delta(x)^\dagger P(y) = (x-y)^\dagger\,b^\dagger P(y)$ and eq. (3.15) to rewrite the last term in eq. (3.16), obtaining that $\hat{G} = \hat{H} + \hat{C}$ will solve eq. (3.8) provided that

$$\mathcal{D}^2\hat{C} = -\frac{1}{\pi^2} \frac{[P(x)\,bf(x)]\text{----}[b^\dagger P(y)]}{[P(x)\,bf(x)]\text{----}[b^\dagger P(y)]} . \tag{3.17}$$

If we set

$$4\pi^2\hat{C} = \frac{[P(x)\,b]\text{----}}{[P(x)\,b]\text{----}} M \frac{\text{----}[b^\dagger P(y)]}{\text{----}[b^\dagger P(y)]} , \tag{3.18}$$

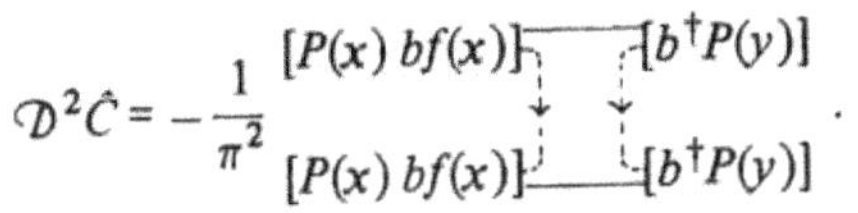

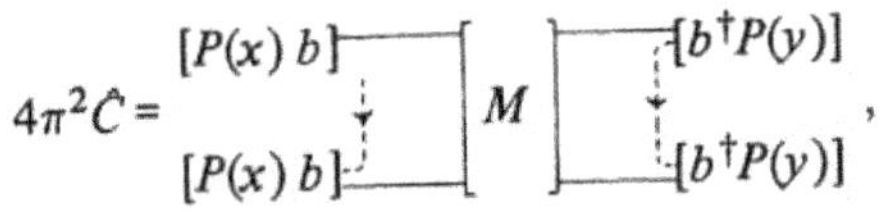

where $M_{ij,lm}$ is a constant $k_1k_2 \times k_1k_2$ matrix, $\hat{C}$ will satisfy eq. (3.17) provided that M satisfies the equation,

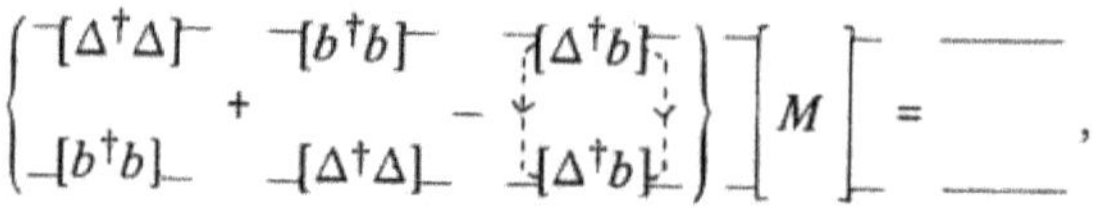

(3.19)

where $\Delta \equiv \Delta(x)$ and we have used $P\partial_\alpha\{P\partial_\alpha P\} = - \; Pbfb^\dagger$. At first sight eq. (3.19) offers no hope for a solution since it appears to state that the constant matrix M is the inverse of an x-dependent matrix. But, as was first noted by Brown et al. [6] in what is effectively a special case of this equation, the appearance is deceptive because the x-dependence cancels out; eq. (3.19) merely states that M is the inverse of

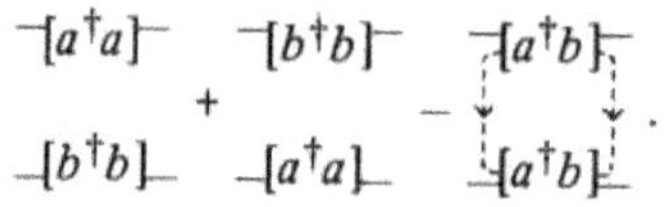

(3.20)

To show this, note that

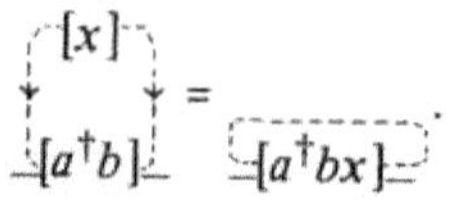

(3.21)

We conclude that, defining M as the inverse of expression (3.20), which is equivalent to eq. (2.15), the solution to eq. (3.8) is $\hat{G} = \hat{H} + \hat{C}$. This constitutes a generalization of a result obtained by Christ et al. [4] for the adjoint representation of SU(2).

We shall now discuss the properties of M. Firstly the expression (3.20), and so M, is invariant under the interchange of a and b. This is obvious from the symmetry of $a^\dagger b$ once one notes that this can be expressed as

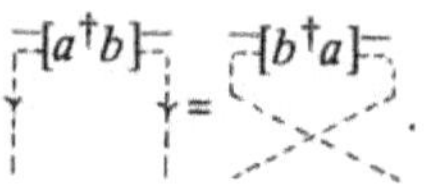

(3.22)

The x-independence of eq. (3.19) is equivalent to the statement that expression (3.20) is invariant under the transformations

$$a_m \to a_m + b_m x\,, \qquad b_m \to b_m\,, \qquad m = 1, 2\,. \tag{3.23}$$

Because of the symmetry under interchange of a and b it is also invariant under the transformations

$$a_m \to a_m\,, \qquad b_m \to b_m + a_m x\,, \qquad m = 1, 2\,, \tag{3.24}$$

whilst under the transformations

$$a_m \to a_m x\,, \qquad b_m \to b_m y\,, \qquad m = 1, 2\,, \tag{3.25}$$

it changes by a factor $|x|^2|y^2|$ for any quaternions x, y. Taken together transformations (3.23)–(2.25) generate a sixteeen parameter group

$$a_m \to a_m \delta + b_m \beta , \qquad b_m \to b_m \alpha + a_m \gamma , \tag{3.26}$$

where $\alpha, \beta, \gamma, \delta$ are quaternions such that

$$\kappa = \det \begin{pmatrix} \alpha & \beta \\ \gamma & \delta \end{pmatrix} \neq 0 . \tag{3.27}$$

We deduce that under these transformations expression (3.20) is scaled by $|\kappa|^2$ so that M is merely multiplied by $|\kappa|^{-2}$. Actually the transformation (3.26) corresponds to performing a conformal transformation,

$$x \to (\alpha x + \beta)(\gamma x + \delta)^{-1} , \tag{3.28}$$

on the solutions specified by $\Delta_1 \Delta_2$. In such a transformation we may as well take $\kappa = 1$ and, with this restriction, M is conformally invariant.

We have implicitly assumed the real $k_1 k_2 \times k_1 k_2$ matrix (3.20) is invertible. We have no precise set of conditions under which this is true, but it is easily seen to be true for particular tensor products or $q \otimes \bar{q}$ representations, using the 't Hooft solutions. Therefore, because the whole construction is algebraic, it can only be singular for certain special tensor products, or representations q, which form a manifold of smaller dimension within the whole space of solutions. In other words typically our procedure will work, though we have not proved rigorously that it will work everywhere.

To conclude this section, we shall discuss how we may amalgamate the two parts of G given by eqs. (3.11) and (3.18). Consider the square matrix $\mathcal{M}$ of dimension $4(n_1 + k_1)(n_2 + k_2)$ defined by

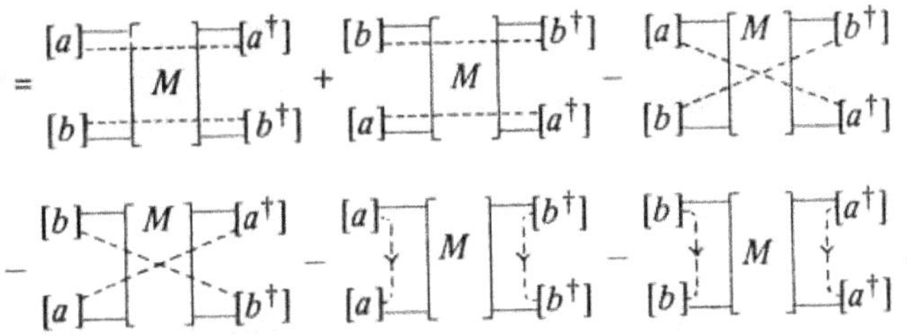

$$\mathcal{M} = \begin{array}{c}[a]\text{---}[a^\dagger] \\ [b]\text{---}[b^\dagger]\end{array}\boxed{M} + \begin{array}{c}[b]\text{---}[b^\dagger] \\ [a]\text{---}[a^\dagger]\end{array}\boxed{M} - \begin{array}{c}[a]\quad[b^\dagger] \\ [b]\quad[a^\dagger]\end{array}\boxed{M}$$

$$- \begin{array}{c}[b]\quad[a^\dagger] \\ [a]\quad[b^\dagger]\end{array}\boxed{M} - \begin{array}{c}[a]\quad[b^\dagger] \\ [a]\quad[b^\dagger]\end{array}\boxed{M} - \begin{array}{c}[b]\quad[a^\dagger] \\ [b]\quad[a^\dagger]\end{array}\boxed{M} . \tag{3.29}$$

The matrix $\mathcal{M}$ is conformally invariant. The simplest way to prove this is by checking invariance under transformations (3.23)–(3.25), using the invariance of M, and eq. (3.13). In particular we may replace a everywhere by $\Delta(x)$ in eq. (3.29) and, thus,

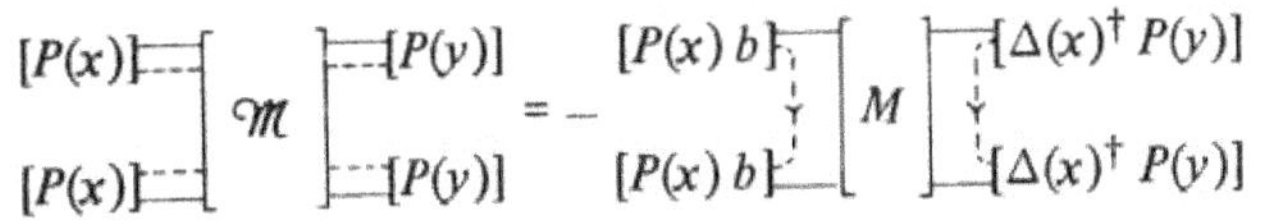

$$\begin{array}{c}[P(x)]\text{---}[P(y)] \\ [P(x)]\text{---}[P(y)]\end{array}\mathcal{M} = - \begin{array}{c}[P(x)\,b]\quad[\Delta(x)^\dagger P(y)] \\ [P(x)\,b]\quad[\Delta(x)^\dagger P(y)]\end{array}\boxed{M} \tag{3.30}$$

$$= -4\pi^2 \hat{C}(x, y)(x - y)^2 , \tag{3.31}$$

since only the last term of eq. (3.29) contributes after the conformal transformation, and using $\Delta(x)^\dagger P(y) = (x-y)^\dagger b^\dagger P(y)$ again. As a result we have the compact formula

$$\hat{G}(x,y) = \frac{\{P_1(x)\otimes P_2(x)\}\{1-\mathcal{M}\}\{P_1(y)\otimes P_2(y)\}}{4\pi^2(x-y)^2}. \tag{3.32}$$

Eq. (3.32) goes someway towards unifying the tensor product with the fundamental representation and we shall pursue this further in sect. 4.

4. The tensor products of instantons

As we have indicated in sects. 2 and 3, the existence of the form for the tensor product Green function given in eq. (3.32) is not really surprising. The self-dual $\mathrm{Sp}(n_1)\times\mathrm{Sp}(n_2)$ solution specified by eq. (2.18) may be considered as a solution for a larger simple group, e.g., $\mathrm{SU}(4n_1n_2)\supset\mathrm{Sp}(n_1)\times\mathrm{Sp}(n_2)$. Because of the completeness of their results, there must exist a $\Delta(x)$, $\tilde{v}(x)$, etc., which will give us this solution directly through the ADHM construction. The complex dimension of $\tilde{\Delta}$ must be $(4n_1n_2+2\tilde{k})\times 2\tilde{k}$, where $\tilde{k}$ is the instanton number of the field tensor

$$F_{\alpha\beta} = F_{1\alpha\beta}\otimes 1_{2n_2} + 1_{2n_1}\otimes F_{2\alpha\beta}, \tag{4.1}$$

regarded as an $\mathrm{SU}(4n_1n_2)$ solution. Using eq. (2.8) we find

$$\tilde{k} = 2(n_1k_2+n_2k_1).$$

This shows that we may not identify $\tilde{v}$ with $v_1\otimes v_2$, since the former must have dimension $4(n_1n_2+k_1n_2+n_1k_2)\times 4n_1n_2$ whilst the latter has dimension $4(n_1+k_1)(n_2+k_2)\times 4n_1n_2$. This fact is enough in itself to tell us that we need the extra (x,y) term in the tensor product Green function because this Green function would be given by eq. (2.13) with $C=0$ if and only if $\tilde{v}=v_1\otimes v_2$. Indeed, looking at the form we have found, we see that

$$\tilde{v}(x)^\dagger\tilde{v}(y) = \{v_1(x)\otimes v_2(x)\}^\dagger\{1-\mathcal{M}\}\{v_1(y)\otimes v_2(y)\}, \tag{4.3}$$

for all values of x and y. Now, at least for a typical tensor product, the columns of $v_1(x)\otimes v_2(x)$ will span the whole $4(n_1+k_1)(n_2+k_2)$ dimensional complex space. In consequence, the rank of $1-\mathcal{M}$ cannot exceed that of the left-hand side, i.e., $4(n_1n_2+k_1n_2+k_2n_1)$, and this must be true for all tensor products since the rank can only decrease at atypical points. For the same reason $1-\mathcal{M}$ must be non-negative. Thus $\mathcal{M}$ has at least $4k_1k_2$ unit eigenvalues and the rest must be less than or equal to unity.

The discussion would be simpler if $\mathcal{M}^2=\mathcal{M}$, so that $1-\mathcal{M}$ was a projection operator. This is possible *a priori* since eqs. (3.19) and (3.29) imply that $\mathrm{tr}\,\mathcal{M} = 4k_1k_2$, consistent with all its eigenvalues being zero or unity. We have studied the eigenvalues of $\mathcal{M}$ numerically on a computer for a large number of particular cases (all corresponding to 't Hooft solutions for ease of computation) and explicitly checked the statement we have made about the eigenvalues and that $\mathcal{M} \neq \mathcal{M}^2$ in general. In appendix A, we explicitly find $4k_1k_2$ linearly independent unit eigenvectors of $\mathcal{M}$.

Because of the properties of its spectrum, $1 - \mathcal{M}$ has a unique positive square root and, as we indicated in eq. (2.23), we may identify $\tilde{v}$ with $(1 - \mathcal{M})^{1/2}\, v_1 \otimes v_2$. We may see that this identification and eq. (4.3) are consistent with the normalization of $\tilde{v}$ either from the non-singularity of $C(x, y)$ or, more directly, from the conformal invariance of $\mathcal{M}$; if we use the latter property to change all the a's in eq. (3.29) to Δ's, we see that none of the terms in $\mathcal{M}$ contribute in eq. (4.3) when we put $x = y$. Further we may see directly that

$$\tilde{A}_\alpha = A_{1\alpha} \otimes 1 + 1 \otimes A_{2\alpha} = \tilde{v}^\dagger \partial_\alpha \tilde{v}\,, \tag{4.4}$$

by differentiating

$$\{v_1(y) \otimes v_2(y)\}^\dagger \{v_1(x) \otimes v_2(x)\} = \tilde{v}(y)^\dagger\, \tilde{v}(x) + O(|x - y|^2) \tag{4.5}$$

with respect to x and then setting $x = y$.

Eqs. (2.23) and (4.4), (4.5) together with the normalization condition $\tilde{v}^\dagger \tilde{v} = 1$ would be enough in themselves to establish the formula for the tensor product Green function if we could also find a matrix $\tilde{\Delta}(x)$, linear in x such that

$$\tilde{v}(x)^\dagger\, \tilde{\Delta}(x) = 0 \tag{4.6}$$

determines $\tilde{v}$ up to normalization, and $\tilde{\Delta}(x)^\dagger\, \tilde{\Delta}(x)$ is an invertible matrix, proportional to the unit matrix in the quaternionic labels. One might look for a $\tilde{\Delta}(x)$ of the form $(1 - \mathcal{M})^{1/2}\{\Delta_1(x) \otimes \Delta_2(x)\}$ but this is quadratic in x. It seems that there is no obvious candidate for $\tilde{\Delta}(x)$. One would expect an arbitrariness corresponding to $GL(\tilde{k}, C)$ for reasons which are explained, for example, in ref. [3]. At first sight it will appear that the arbitrariness is even greater but this appearance is deceptive.

In attempting to construct a suitable $\tilde{\Delta}(x)$ we can be guided by the fact that it can be viewed as a mapping from the space of massless solutions of the Dirac equation in the background field in question. In eqs. (2.25), (2.26) we summarized the form we found (by inductive guess work) for these solutions. The essential new ingredients needed to construct these solutions are matrices c and d satisfying the linear eqs. (2.26) which have the diagrammatic form,

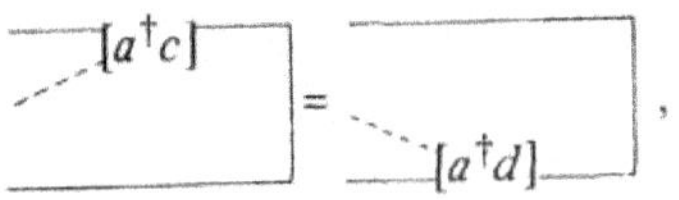

(4.7a)

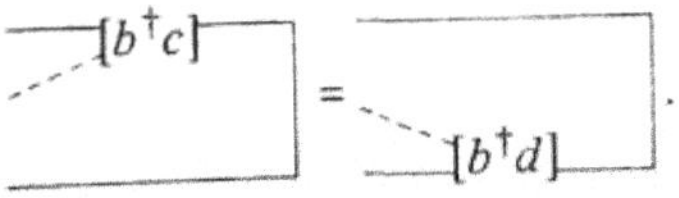

(4.7b)

Eqs. (4.7) from the $4k_1k_2$ linear equations for the $2(n_1 + k_1)\,k_2 + 2(n_2 + k_2)\,k_1$ unknown elements of c and d. Thus we expect $\tilde{k} = 2n_1k_2 + 2n_2k_1$ linearly independent solutions which we label (c_r, d_r). (There must be at least $\tilde{k}$ independent solutions and we shall argue in appendix B that there can be no more if M exists.)

In constructing $\tilde{\Delta}(x)$, we first attempted to find as general as possible $\tilde{\Delta}(x)$, linear in x satisfying eq. (4.6). In doing so we were led to introduce a $4(n_1 + k_1)(n_2 + k_2) \times k_1k_2$ dimensional matrix $\mathcal{N}$ of the form

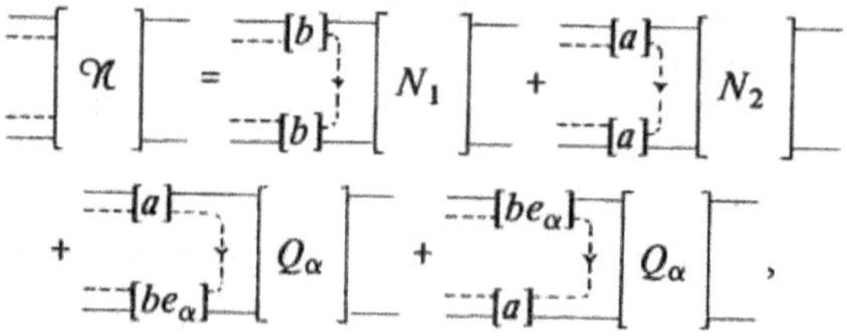

(4.8)

and define $\tilde{\Delta}(x)$ by

$$\tilde{\Delta}(x) = (1 - \mathcal{M})^{1/2}\, \tilde{\Gamma}(x)\ , \tag{4.9}$$

where $\Gamma_{\lambda\mu,rR}(x)$, $1 \leqslant \lambda \leqslant 2(k_1 + n_1)$, $1 \leqslant \mu \leqslant 2(k_2 + n_2)$, $1 \leqslant v \leqslant k$, $1 \leqslant R \leqslant 2$, is defined by

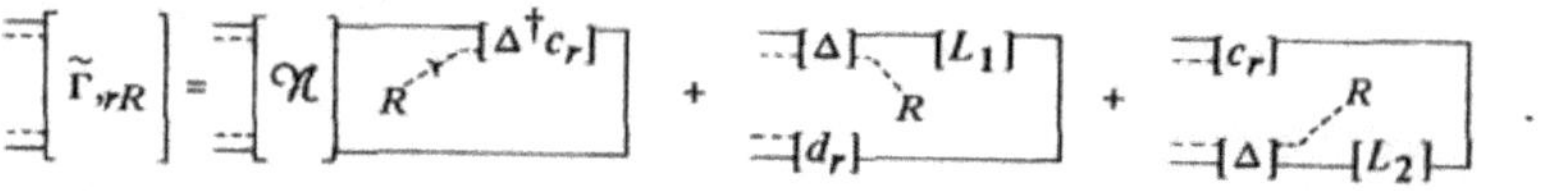

(4.10)

In eqs. (4.8)–(4.10) N_1, N_2 and Q_α, $\alpha = 0, \ldots 3$, are square matrices of dimension k_1k_2 and L_1, L_2 are square matrices of dimension k_1 and k_2, respectively.

It is reasonably straightforward to show, using the definitions and conformal invariance of $\mathcal{M}$ and M, that $\tilde{\Delta}(x)$, as defined by eqs. (4.8)–(4.10), satisfies eq. (4.6), that is

$$\{v_1(x) \otimes v_2(x)\}^\dagger \{1 - \mathcal{M}\}\, \tilde{\Gamma}(x) = 0\ , \tag{4.11}$$

provided that $N_1, N_2, Q_\alpha, L_1, L_2$ satisfy

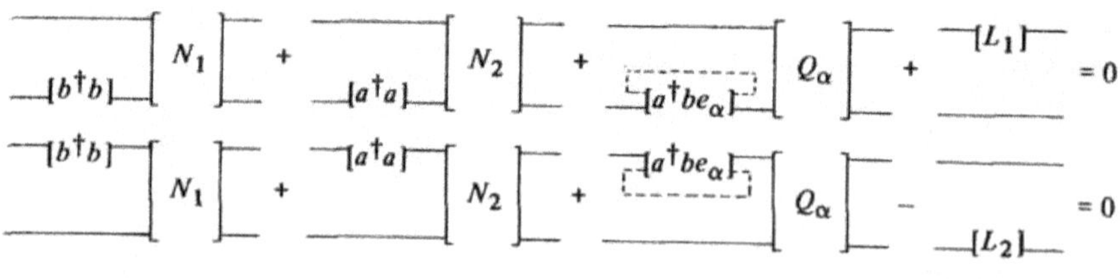

(4.12)

Further details of the proof of this statement are given in appendix B. It is for deriving statements like this that a diagrammatic notation seems almost obligatory to us. Merely relying on suffices obscures the essential simplicity of the algebra; in any case one quickly runs out of letters.

Although constrained by eqs. (4.12) a vast amount of arbitrariness remains in $\tilde{\Delta}$. It would appear to have $6k_1^2k_2^2 + k_1^2 + k_2^2$ degrees of freedom subject to $2k_1^2k_2^2$ constraints, leaving, apparently, $4k_1^2k_2^2 + k_1^2 + k_2^2$. Some of this freedom may be quickly shown to be illusory; clearly we may change $\tilde{\Gamma}$, using the $4k_1k_2$ linearly independent with vectors of $1 - \mathcal{M}$, without affecting $\tilde{\Delta}$ or the validity of eqs. (4.6) or (4.11). (See appendix A for further details.) In fact this accounts for $4k_1^2k_2^2$ of the degrees of freedom and we may think of the remaining $k_1^2 + k_2^2$ effective degrees of freedom as corresponding to the ability to choose L_1 and L_2 arbitrarily. This last point is reinforced by the explicit calculations of $(1 - \mathcal{M})\,\tilde{\Gamma}(x)$ and

$$\tilde{\Delta}(x)^\dagger\,\tilde{\Delta}(x) = \tilde{\Gamma}(x)^\dagger(1 - \mathcal{M})\,\tilde{\Gamma}(x)\,, \tag{4.13}$$

which are reported in appendix B, and show that these quantities depend only on L_1 and L_2, the dependence on N_1, N_2 and Q_α cancelling out. This is important because whilst *in general* eqs. (4.12) have solutions for N_1, N_2, Q_α for any given L_1, L_2, for specific tensor products (in particular, for the $\mathrm{q}\otimes\bar{\mathrm{q}}$ representation) we cannot choose L_1, L_2 freely and obtain a solution. On the other hand what we need to construct is $\tilde{\Delta}(x) = (1 - \mathcal{M})^{1/2}\,\tilde{\Gamma}(x)$ which will depend only on L_1, L_2 because $(1 - \mathcal{M})\,\tilde{\Gamma}(x)$ does. We may regard the calculations of appendix B as being performed for a typical tensor but use the results, which depend only on L_1, L_2, for an arbitrary tensor product. Having realized this, we see that N_1, N_2, Q_α and eqs. (4.12) are really irrelevant, their only function having been to enable us to calculate $\tilde{\Delta}(x)^\dagger\,\tilde{\Delta}(x)$.

The calculations of appendix B also show that $\tilde{\Delta}(x)^\dagger\,\tilde{\Delta}(x)$ is diagonal in and independent of the quaternionic labels, and so may be regarded as an hermitian matrix of dimension $\tilde{k}$ (see the ADHM formalism for SU(n) as described in sect. 2 and ref. [3]). Further it is established there, using properties of the c's and d's, that it has the following simple expression

$$[\tilde{\Delta}(x)^\dagger\,\tilde{\Delta}(x)]_{rR,sS} = [Z^\dagger\Omega^{-1}Z]_{rs}\,\delta_{RS}\,, \tag{4.14}$$

where

$$Z_{rs} = \begin{bmatrix} [c_r^\dagger c_s] & \\ & [L_2] \end{bmatrix} + \begin{bmatrix} [L_1] & \\ & [d_r^\dagger d_s] \end{bmatrix}\,, \tag{4.15}$$

and

$$\Omega_{rs} = \begin{bmatrix} [c_r^\dagger c_s] & \\ & [f] \end{bmatrix} + \begin{bmatrix} [f] & \\ & [d_r^\dagger d_s] \end{bmatrix} - \begin{bmatrix} [c_r^\dagger \Delta] & [f] \\ [f] & [\Delta^\dagger d_s] \end{bmatrix} \tag{4.16}$$

Note that the last term in eq. (4.16) may be written as

$-$ 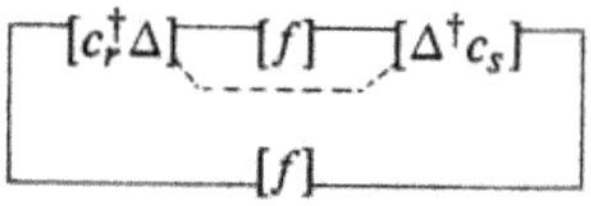 , (4.17)

or one of two other symmetrical possibilities.

The non-singularity of Ω follows from that of f_1, f_2 using eqs. (4.16), (4.17). Also Z is clearly non-singular for suitable choices of L_1, L_2, for example, $L_1 = L_2 = 1$. (It is here that the irrelevance of N_1, N_2, Q_α becomes important because they do not always exist in this case. In the obvious cases in which they do exist like $L_1 = -(a_1^\dagger a_1)^{-1}, L_2 = (a_2^\dagger a_2)^{-1}, N_1 = Q_\alpha = 0, (N_2)_{ij,lm} = (a_1^\dagger a_1)^{-1}_{il}(a_2^\dagger a_2)^{-1}_{jm}$, Z is not always non-singular.)

Eq. (4.14) shows completely explicitly that the $\tilde{f}(x) = [\tilde{\Delta}(x)^\dagger \tilde{\Delta}(x)]^{-1}$ for the various $\tilde{\Delta}(x)$ we have found, i.e., for the various L_1, L_2, are related to $\Omega(x)$ by the *constant* matrices $Z \in GL(\tilde{k}, C)$, and so to one another, as had to be the case from the nature of the ADHM construction.

Because of the non-singularity of $\tilde{\Delta}(x)^\dagger \tilde{\Delta}(x)$, eq. (4.6) defines $\tilde{v}(x)$ and we have found a construction of the ADHM type for the tensor product. The form for the tensor product Green function now follows from the argument used for the fundamental representation [3,4] and in sect. 5 we shall apply the formula for the massless solutions of the Dirac equation [3,5] to obtain eq. (2.25).

5. The massless solutions of the Dirac equation

In this section we shall consider the massless Dirac equation for a tensor product of instantons,

$$\gamma_\alpha \mathcal{D}_\alpha \psi = 0\,, \tag{5.1}$$

where the covariant derivative, $\mathcal{D}_\alpha$ is as in eq. (2.18) and, as in ref. [3], we take as a representation of the γ matrices

$$\gamma_\alpha = \begin{pmatrix} 0 & e_\alpha \\ e_\alpha^\dagger & 0 \end{pmatrix}. \tag{5.2}$$

Eq. (5.1) possesses no normalizable positive chirality ($\gamma_5\psi \equiv \gamma_0\gamma_1\gamma_2\gamma_3\psi = +\psi$) solutions and exactly $\tilde{k}$ negative chirality solutions. For these

$$\psi = \begin{pmatrix} \psi_L \\ 0 \end{pmatrix}, \tag{5.3}$$

with

$$e_\alpha^\dagger \mathcal{D}_\alpha \psi_L = 0\,. \tag{5.4}$$

For the corresponding Dirac equation for the fundamental representation, the k

independent solutions are given by [3,5]

$$(\psi_L)_{a,I} = (v^\dagger b f \epsilon)_{a,iI} ,\tag{5.5}$$

for each fixed i, $1 \leqslant i \leqslant k$, I being the spinor index. Following the formulation of sect. 4 the solution to eq. (5.4) will be given by

$$\tilde{v}^\dagger \widetilde{bf}\epsilon = (v_1 \otimes v_2)(1 - \mathcal{M}) b\epsilon Z^{-1}\Omega Z^{\dagger -1}\tag{5.6}$$

where $\tilde{\Gamma}(x) = \hat{a} + \hat{b}x$. Eq. (5.6) looks formidable but simplifies greatly. In fact eq. (5.6) reduces to

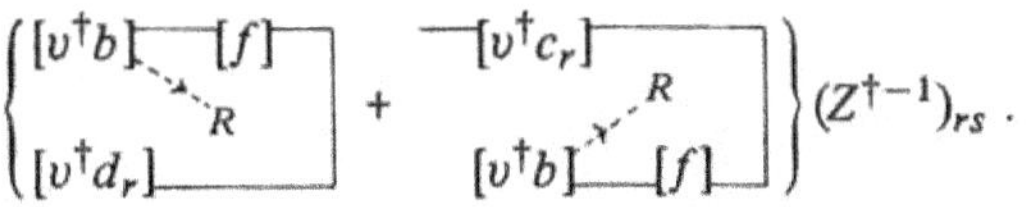

$$\left\{ \begin{array}{c} \boxed{\begin{array}{c} [v^\dagger b] \quad [f] \\ {}_{\searrow} R \\ [v^\dagger d_r] \end{array}} + \boxed{\begin{array}{c} [v^\dagger c_r] \\ R \\ [v^\dagger b] \quad [f] \end{array}} \right\} (Z^{\dagger -1})_{rs} .\tag{5.7}$$

To establish this we first note, from appendix B, that $(v_1 \otimes v_2)(1 - \mathcal{M}) b\epsilon$ simplifies to

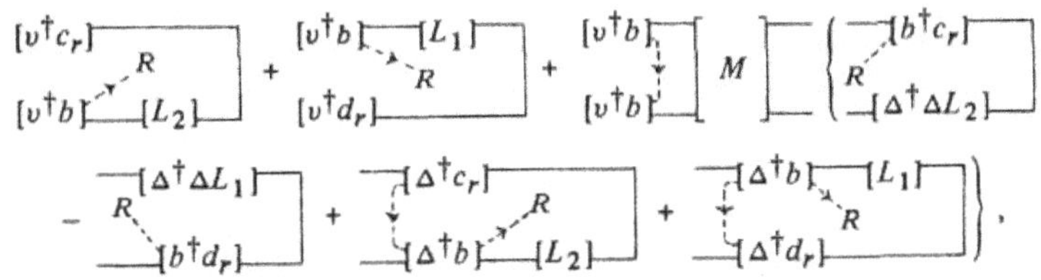

$$\boxed{\begin{array}{c} [v^\dagger c_r] \\ R \\ [v^\dagger b] \quad [L_2] \end{array}} + \boxed{\begin{array}{c} [v^\dagger b] \quad [L_1] \\ {}_{\searrow} R \\ [v^\dagger d_r] \end{array}} + \boxed{\begin{array}{c} [v^\dagger b] \\ [v^\dagger b] \end{array}} M \boxed{\left\{ \begin{array}{c} [b^\dagger c_r] \\ R \\ [\Delta^\dagger \Delta L_2] \end{array} \right.}$$
$$- \boxed{\begin{array}{c} [\Delta^\dagger \Delta L_1] \\ R \\ [b^\dagger d_r] \end{array}} + \boxed{\begin{array}{c} [\Delta^\dagger c_r] \\ R \\ [\Delta^\dagger b] \quad [L_2] \end{array}} + \boxed{\begin{array}{c} [\Delta^\dagger b] \quad [L_1] \\ R \\ [\Delta^\dagger d_r] \end{array}} \right\},\tag{5.8}$$

It remains to show that we obtain expression (5.8) if we multiply expression (5.7) by $Z^\dagger \Omega^{-1} Z$. To do this we use the relations for $c_r(\Omega^{-1})_{rs} c_s^\dagger$, $c_r(\Omega^{-1})_{rs} d_s^\dagger$, etc. proved in appendix B, together with eq. (3.13).

The form for the Dirac equation solutions given is just that we claimed in eq. (2.25) apart from the (irrelevant) constant matrix $Z^{\dagger -1}$. It is straightforward to check directly that these satisfy eq. (5.4) using the results for the fundamental representation and eqs. (4.7). Indeed this was the way we established this result first, but now we see it fitting into the same framework as the previous results.

6. Conclusions and outlook

Our aim in beginning this study was to understand the structure of the adjoint representation Green function, whose form at first seems rather complicated and enigmatic compared to that for the fundamental representation. We have seen that this structure enables us to solve a much wider class of problems, involving tensor products of instantons. Further, we have seen how to unify our treatment of these problems with the treatment of the fundamental representation. Results proved for the fundamental representation can be immediately extended to this more general situation. In principle, we can obtain results for higher dimensional representations iteratively.

On the other hand, since we have often obtained elegant results at the end of extensive and tedious algebra, we conclude that there must be better concepts in terms of which to frame the discussion. Presumably there should be some natural geometrical way to approach these problems.

We hope that our results go some small way to indicate the way the work of Atiyah et al. [1,2] has increased one's computational power in dealing with self-dual gauge fields. The solutions to the problems of the Green function and the massless Dirac wave functions for the fundamental and adjoint representation, which were obtained before their work with considerable ingenuity for the special case of the 't Hooft SU(2) solutions, can now be unified and completely generalized using a few simple formulae.

The surprising simplicity of what has so far been discovered leads one to hope that there may be comparable results, for example, for the determinants of the covariant Laplacian and the covariant Dirac operator. Another related problem which is of considerable interest is the possible relation of functional integrals over gauge fields to the (gauge invariant) description of solutions, and other fields, provided by the ADHM construction.

Edward Corrigan thanks the CERN Theory Division for hospitality whilst part of this work was done and Peter Goddard thanks the Theoretical Physics Group at the Ecole Normale Supérieure similarly. Stephen Templeton is grateful to the UK Science Research Council for a studentship.

Appendix A

Zero eigenvectors of $1 - \mathcal{M}$

In this appendix we construct $4k_1k_2$ linearly independent eigenvectors of with unit eigenvalues and discuss their relation to the construction of $\tilde{\Delta}(x)$. We consider vectors of the form

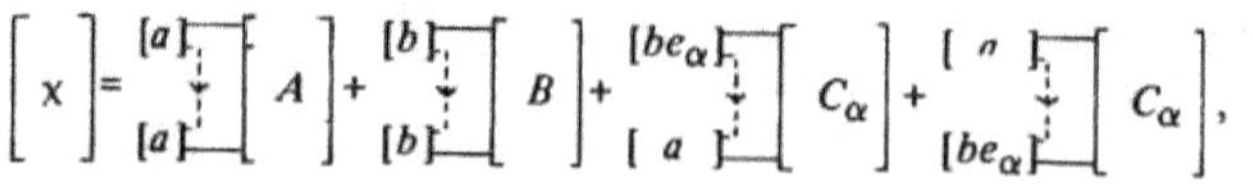

$$
\left[x \right] = \left[\begin{array}{c} [a] \\ \vdots \\ [a] \end{array} \right] A + \left[\begin{array}{c} [b] \\ \vdots \\ [b] \end{array} \right] B + \left[\begin{array}{c} [be_\alpha] \\ \vdots \\ [a] \end{array} \right] C_\alpha + \left[\begin{array}{c} [a] \\ \vdots \\ [be_\alpha] \end{array} \right] C_\alpha , \tag{A.1}
$$

where $A, B, C_\alpha, \alpha = 0, ..., 3$ are arbitrary vectors of dimension $k_1 k_2$. Using the definitions of $\mathcal{M}, M$ and eq. (3.13) one may show by straightforward algebra (facilitated by the diagrammatic notation) that

$$
\mathcal{M}\chi = \chi , \tag{A.2}
$$

provided that

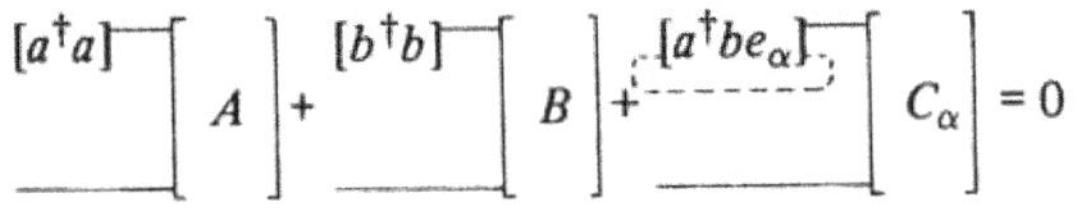

$$
\left[\begin{array}{c} [a^\dagger a] \\ \quad\quad A \end{array} \right] + \left[\begin{array}{c} [b^\dagger b] \\ \quad\quad B \end{array} \right] + \left[\begin{array}{c} [a^\dagger b e_\alpha] \\ \quad\quad\quad C_\alpha \end{array} \right] = 0 \tag{A.3}
$$

and

$$
\left[\begin{array}{c} \quad\quad A \\ [a^\dagger a] \end{array} \right] + \left[\begin{array}{c} \quad\quad B \\ [b^\dagger b] \end{array} \right] + \left[\begin{array}{c} \quad\quad C_\alpha \\ [a^\dagger b e_\alpha] \end{array} \right] = 0 . \tag{A.4}
$$

Since A, B, C_α possess $6k_1k_2$ degrees of freedom and eqs. (A.3), (A.4) provide $2k_1k_2$ equations we are left with (at least) $4k_1k_2$ linearly independent solutions for A, B, C_α. We should show that these yield linearly independent solutions for χ. This will follow if we show that $\chi = 0$ implies A, B, C_α vanish. To show this consider multiplying eq. (A.1) by

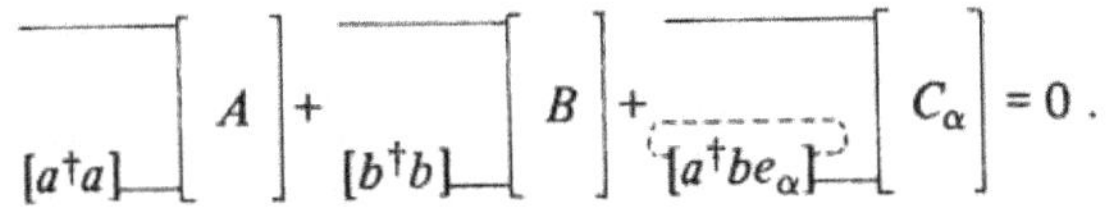

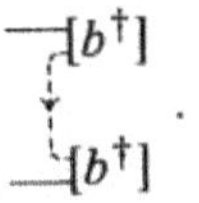

$$
\begin{array}{c} [b^\dagger] \\ \\ [b^\dagger] \end{array} . \tag{A.5}
$$

Using eqs. (A.3), (A.4) we thus see that $\chi = 0$ implies $(M^{-1}) A$ vanishes and so that A vanishes. It may be shown similarly that B, C_α vanish if $\chi = 0$.

Comparing eq. (A.1) with eqs. (4.8)–(4.10) we see that given one solution of eqs. (4.12) for given L_1, L_2, we may use the vectors χ to construct new N_1, N_2, Q_α, and hence a new $\tilde\Gamma$, which also solve the equation. In fact it is clear that given one such solution we may obtain $4k_1^2k_2^2$ linearly independent solutions in this way. But changing $\tilde\Gamma$ in this way does not effect $\tilde\Delta = (1 - \mathcal{M})^{1/2} \tilde\Gamma$ because the vectors χ are null vectors of $1 - \mathcal{M}$. Thus we have understood $4k_1^2k_2^2$ of the $4k_1^2k_2^2 + k_1^2 + k_2^2$ degrees of freedom of $\tilde\Delta$; they are not effective. Really everything is fixed by the choice of L_1, L_2 as we show in appendix B.

Appendix B

Properties of $\tilde\Delta(x)$

We begin this appendix by discussing the derivation of eqs. (4.12). We begin by noting that $\mathcal{M}$, as defined by eq. (4.8) is unchanged if we replace a by Δ and replace N_1 by N_1', N_2 by N_2' and Q_α by Q_α', where

$$
N_1' = N_1 - 2x_\alpha Q_\alpha + x^2 N_2 ,
$$

$$
N_2' = N_2 ,
$$

$$
Q_\alpha' = Q_\alpha - x_\alpha N_1 . \tag{B.1}
$$

Further, eqs. (4.12) are equivalent to the same equations for N_1', N_2', Q_α' with a

everywhere replaced by Δ. Using this new expression for $\mathcal{N}$ and the new form of eqs. (4.12) it is straightforward to show, using the definition of M, that $\{v_1 \otimes v_2\}$ $\{1 - \mathcal{M}\}\, \mathcal{N}$ is given by

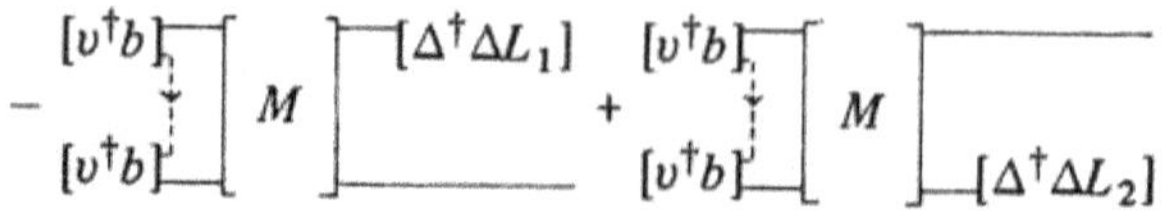

(B.2)

Eq. (4.11) now follows immediately from eq. (4.10).

Using similar techniques (together with eq. (3.13)), but working rather harder, one may express $\{1 - \mathcal{M}\}\, \mathcal{N}$ entirely in terms of L_1 and L_2 obtaining

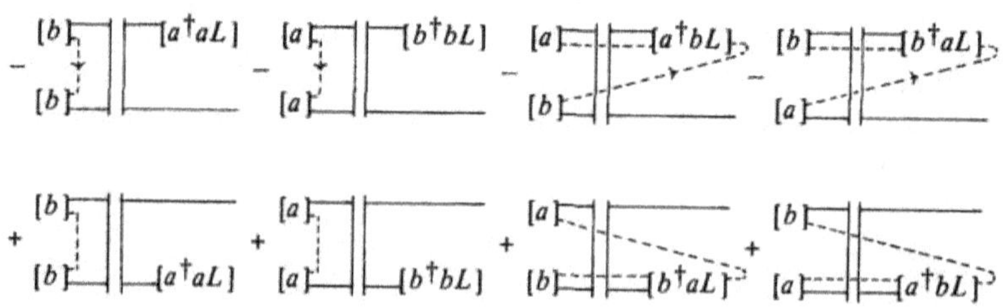

(B.3)

where we have further developed our diagrammatic notation by abbreviating $[M]$ to two vertical lines: ‖ and by suppressing the indices on L_1 and L_2. Note that (B.3) is conformally invariant. Going further one may calculate $\mathcal{N}^\dagger(1 - \mathcal{M})\, \mathcal{N}$ to obtain

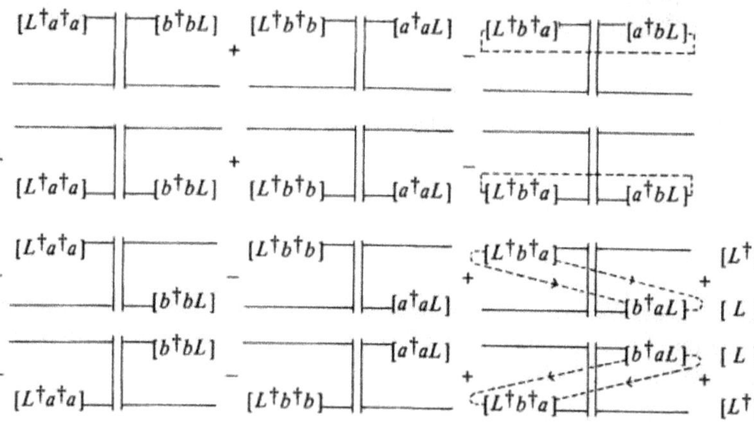

(B.4)

Working even harder one may proceed to calculate $(\tilde{\Delta}^\dagger\tilde{\Delta})_{rR,sS}$; using (B.4) we obtain the following expression

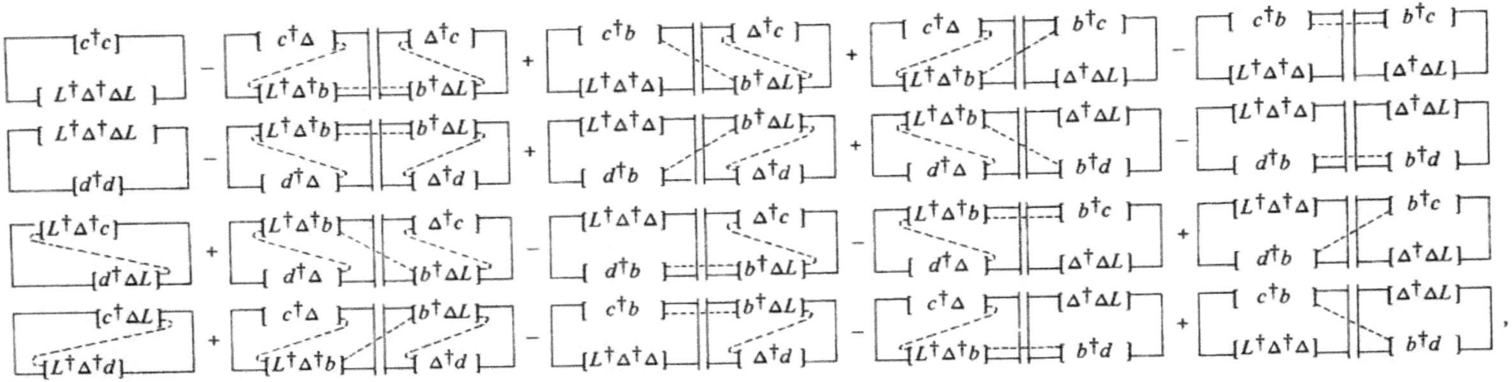

(B.5)

115

where we have suppressed the labels r, s on the c and d matrices, multiplied by δ_{RS}. This expression has a certain elegance but is rather unwieldy. It is important to realise that it may be simplified by writing the "central parts" of each line in terms of c and d. Let us define

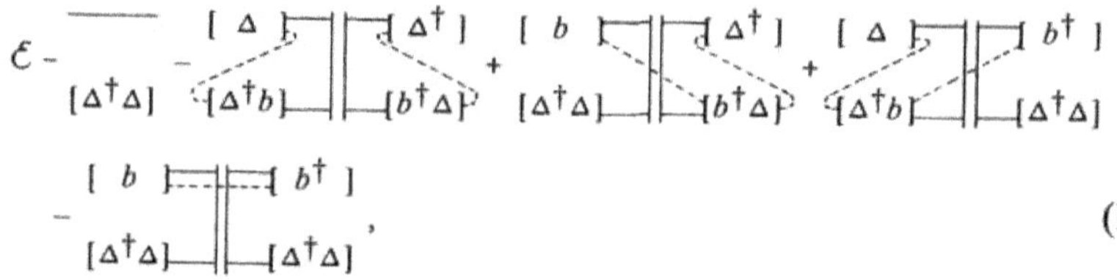

(B.6)

and define $\mathcal{F}$, $\mathcal{G}$, $\mathcal{H}$ similarly using the second, third and fourth rows of (B.5), respectively, in the same way we have used the first row to define $\mathcal{E}$. Now it is straightforward to show that

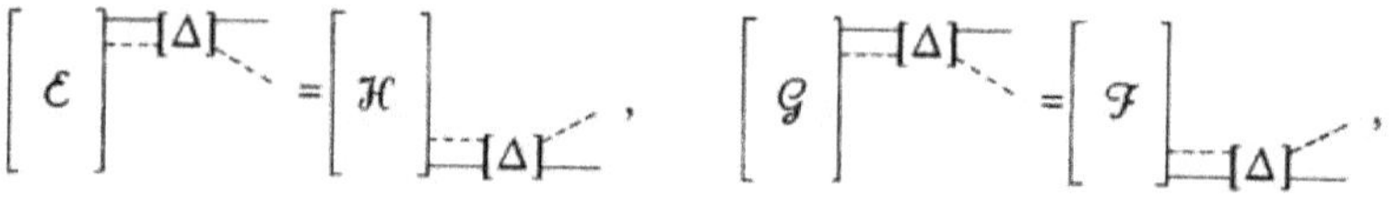

(B.7a)

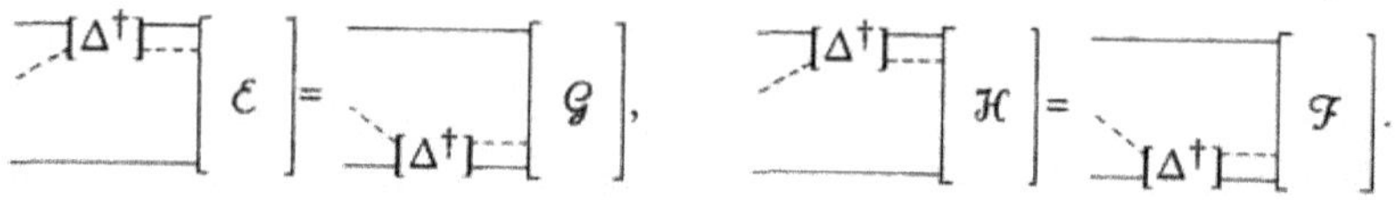

(B.7b)

Using the definition of M we can show that eqs. (B.7) hold with Δ replaced by b. It follows from this and the fact that the matrices (c_r, d_r) form a complete set of solutions to eqs. (4.7) that we must be able to write

$$\mathcal{E} = c_r x_{rs} c_s^\dagger\ , \qquad \mathcal{F} = d_r x_{rs} d_s^\dagger\ , \qquad \mathcal{G} = d_r x_{rs} c_s^\dagger\ , \qquad \mathcal{H} = c_r x_{rs} d_s^\dagger\ , \quad \text{(B.8)}$$

for some matrix X. It is not difficult to determine X; consider

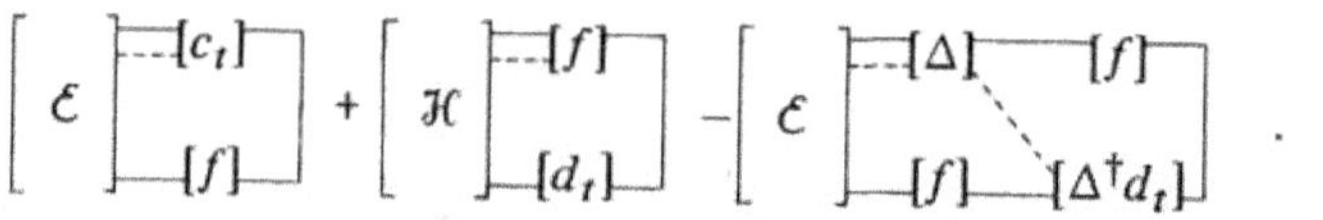

(B.9)

This simplifies to give just c_t. Since the c_r's are linearly independent this implies

$$X_{rs}\Omega_{sr} = \delta_{rt}\ ,$$

with Ω defined as in eq. (4.16). So $X = \Omega^{-1}$ and using this in expression (B.5) gives eq. (4.14). (We first found the relations of eq. (B.8) by studying the relationship of the Dirac equation Green function to the complete sum over the solutions of the Dirac equation; compare the calculation of ref. [5].)

The manifest positive definiteness of $\tilde{\Delta}^\dagger\tilde{\Delta}$ for suitable L_1, L_2 shows that $\tilde{\Delta}$ may have as many linearly independent columns as there are independent solutions for (c, d), multiplied by two for quaternionic label. Since the number of columns, being each orthogonal to the $4k_1k_2$ columns of $\tilde{v}$ in a space of dimension $4k_1k_2 + 4n_1k_2 + 4n_2k_1$, can not exceed $4n_1k_2 + 4k_1n_2$, we conclude that there are precisely $\tilde{k}$ linearly independent solutions (c, d).

References

[1] M.F. Atiyah, V.G. Drinfeld, N.J. Hitchin and Yu.I. Manin, Phys. Lett. 65A (1978) 185.
[2] V.G. Drinfeld and Yu.I. Manin, ITEP, Moscow preprint (1978).
[3] E.F. Corrigan, D.B. Fairlie, P. Goddard and S. Templeton, Nucl. Phys. B140 (1978) 31.
[4] N.H. Christ, E.J. Weinberg and N.K. Stanton, Columbia preprint (1978).
[5] H. Osborn, Nucl. Phys. B140 (1978) 45.
[6] L.S. Brown, R.D. Carlitz, D.B. Creamer and C. Lee, Phys. Lett. 70B (1977) 180; 71B (1977) 103; Phys. Rev. D17 (1978) 1583.
[7] B. Grossmann, Phys. Lett. 61A (1977) 86.
[8] G. 't Hooft, unpublished;
R. Jackiw, C. Nohl and C. Rebbi, Phys. Rev. D15 (1977) 1642.
[9] N. Andrei and D.J. Gross, Phys. Rev., to be published.
[10] L. Baulieu, J. Ellis, M.K. Gaillard and W.J. Zakrzewski, Phys. Lett. 77B (1978) 290.
[11] A. Duncan, Columbia preprint (1978).
[12] D. Amati and A. Rouet, CERN preprint TH.2468 (1978).
[13] S. Chadha, A. D'Adda, P. Di Vecchia and F. Nicodemi, Phys. Lett. 67B (1977) 103.
[14] R. Jackiw and C. Rebbi, Phys. Rev. D16 (1977) 1052.

V. FERMIONS

INTRODUCTION

The fermion sector is an important part of both QCD and the Standard Model. Although the very existence of instantons is due to the nontrivial topology in the space of the *gauge* fields, introducing the massless fermion fields in the Yang–Mills Lagrangian and considering them in the instanton background leads to drastic consequences — this observation was first made by 't Hooft in his pioneering paper [1]. The point is that instantons make explicit the nonconservation of certain fermion quantum numbers associated with the triangle anomaly in the axial-vector current [2].

Let us elucidate the above statement with a simple example. Consider a toy version of QCD where the fermion sector consists of one massless quark. As is well known, one Dirac field Ψ is the same as two chiral (Weyl) fields, ψ_L and ψ_R. In the absence of the mass term, the latter two fields decouple from each other in the Lagrangian,

$$\bar{\Psi} i \not{D} \Psi = \bar{\psi}_L i \not{D} \psi_L + \bar{\psi}_R i \not{D} \psi_R \, .$$

Hence, any Feynman graph of any finite order conserves the number of the ψ_L's and ψ_R's separately. In other words, our toy model possesses (in perturbation theory) two conserved charges, Q and Q_5, associated with the vector current $\bar{\Psi}\gamma_\mu\Psi$ and the axial-vector current $\bar{\Psi}\gamma_\mu\gamma_5\Psi$, respectively. The first charge counts the total number of ψ_L's and ψ_R's while the second one counts the difference between these two numbers.

Furthermore, the axial-vector current is actually anomalous [2],

$$\partial_\mu \bar{\Psi}\gamma_\mu\gamma_5\Psi = \frac{\alpha_s}{4\pi} G^a_{\mu\nu}\tilde{G}^a_{\mu\nu} \, .$$

The anomaly in the divergence of the axial-vector current does not affect the conservation of Q_5 in perturbation theory since the right hand side is a full derivative, $G^a_{\mu\nu}\tilde{G}^a_{\mu\nu} \propto \partial_\mu K_\mu$, where K_μ is the Chern–Simons current (see Sec. I). For topologically trivial field configurations one deals with in the conventional perturbation theory, $\mathcal{K} \equiv \int K_0 d^3x = 0$, and hence, $\dot{Q}_5 = 0$.

In the instanton transition, however, $\Delta\mathcal{K} \neq 0$; thus the conservation of Q_5 is lost in the instanton background field and a coupling between ψ_L and ψ_R is generated.

In the real QCD we have three (almost) massless flavors, not one. The essence of the phenomenon remains the same, however: instantons do break a would-be chiral invariance and help solve, at least at the qualitative level, the famous $U(1)$ problem of QCD [3]. Let us remind the reader of the essence of the problem, briefly and in a simplified form. (The reader interested in subtleties is referred to the original publications.)

In the chiral limit, when all quark masses are set equal to zero, the quark sector of the QCD Lagrangian, with three flavors, is invariant under $U(3)_V \times U(3)_A$ (here the subscripts V and A denote the vector and axial-vector symmetries). As is well known, the axial-vector symmetry cannot be realized linearly. The spontaneous breaking of this symmetry leads to the occurrence of the pseudoscalar Goldstone bosons whose number is equal to that of the broken generators. Thus, naively, one should expect to get nine Goldstone bosons. The quark mass term (it breaks the chiral invariance explicitly) shifts the masses of the Goldstone bosons from zero but does not create a gap between the mass of the ninth boson (flavor-singlet η') and that of the pions [3].

The octet of the pseudoscalar Goldstone bosons associated with the spontaneous breaking of $SU(3)_A$ is familiar to everybody. The ninth Goldstone boson, corresponding to $U(1)_A$, is definitely absent in nature — η' does not belong to the Goldstone family. The vacuum tunnelings described by the instanton field solve the puzzle of the missing ninth Goldstone boson. The instanton-induced fermion interaction explicitly violates the axial-vector $U(1)$ so that no ninth Goldstone meson is to be expected [1].

In the Standard Model of electroweak interactions the instantons make explicit the baryon number nonconservation. There is a close parallel between Q_5 in QCD and the baryon charge in the Standard Model. The baryon number is conserved in any finite order of perturbation theory but is broken nonperturbatively [1]. Namely, at low energies, the baryon number violating cross section $\sigma_{bnv} \propto \exp\{-4\pi/\alpha_2\}$ where α_2 is the $SU(2)_{weak}$ gauge coupling.[8] It is remarkable that this cross section grows with energy incredibly fast — exponentially [4]. At the sphaleron mass (~ 10 TeV in the Standard Model), the 't Hooft suppression factor is partly lifted and the cross section reaches its maximum,

$$\sigma_{bnv}(E \sim M_{sph}) \propto \exp\{-C4\pi/\alpha_2\} ,$$

where C is a constant strictly less than 1 (but of order 1). The difference between σ_{bnv} in the GeV range and its maximal value attained near the sphaleron mass is enormous since $4\pi/\alpha_2 \approx 356$!

Even more radical is the impact of the topologically distinct vacua on the baryon number violation at high temperatures. In this case the exponential growth of the baryon number violation with temperature is not frozen. As the temperature T approaches the sphaleron mass M_{sph}, the 't Hooft suppression associated with tunnelings under the barrier is completely lifted — first a conjecture [5, 6, 7, 8] which was later worked out and at present is a well-established fact [9, 10, 11]. The disappearance of the suppression factor means that there are no traces of the baryon number conservation at high temperatures, $T \sim M_{sph}$. This result [9, 10] entails, in turn, a complete revision of the scenarios of the baryon asymmetry generation in the early universe, known for years. Whatever asymmetry is generated at $T >> M_{sph}$ it is likely to be swept out by the above mechanism. As a result one has to invent a novel scenario [12, 13] in which the baryon asymmetry appears at later stages when the universe cools down to temperatures $T < M_{sph}$.

Technically the $U(1)$-violating fermion interaction, mentioned above more than once, is generated because of the occurrence of the zero modes in the Dirac equation

[8]The above exponential is usually referred to in this context as the 't Hooft suppression factor.

in the instanton background field [1],

$$\not{D}\Psi = 0. \tag{1}$$

Here $\mathcal{D}_\mu$ is the covariant derivative. The number of the fermion zero modes is related to the topological charge of the gauge field by the so-called Atiyah–Singer (or the index) theorem [14], which was derived in the instanton context in Ref. [15] (see also Refs. [16, 17, 18]). Specifically, if the number of the normalizable zero modes of positive (negative) chirality is n_+ (n_-), then

$$n_+ - n_- = Q \tag{2}$$

for each Dirac fermion field Ψ, in the fundamental representation [since the operator $\not{D}$ is hermitean, Eq. (1) implies that the equation on $\bar{\Psi}$ has a zero-eigenvalue solution as well]. A brief but illuminating discussion of the derivation of Eq. (2) can be found in the review paper by Coleman [17]. For our present purposes, it suffices to mention only that this theorem is perfectly equivalent to the statement of the triangle anomaly in the axial-vector current [2]. A reinterpretation of this anomaly within the instanton calculus as the index theorem [15, 16, 17] shows that the phenomenon has two faces and can be discussed in two complementary languages. On the one hand, in perturbation theory, the anomaly is related to a regularization of the fermion triangular graph. On the other hand, it reflects the occurrence of the zero modes of the corresponding Dirac operator in the topologically nontrivial gauge fields.

In the case of the BPST instanton ($Q = 1$), the index theorem (2) is realized in the simplest way: the Dirac equation has one solution in the right-handed sector and none in the left-handed one. For $Q = -1$ (anti-instanton), the situation is reversed.

The zero eigenvalue is required, of course, by the index theorem. Still, this fact seems rather surprising. Indeed, as a rule, the vanishing eigenvalues appear in the spectral problems as a manifestation of some symmetry. For instance, the zero modes of the gauge field correspond to translational and scale invariances of the action and to its symmetry with respect to isorotations (see Sec. 2). By analogy one could expect that the zero eigenvalue in the Dirac spectral problem might also be caused by a symmetry of the Yang–Mills action including the fermion fields. No such symmetry is visible.

As was realized much later [19], there exists a roundabout line of reasoning providing an argument which explains the existence of the fermion zero mode. To reveal a hidden symmetry in the family of classical solutions, one has to supersymmetrize this family! Proceeding to the supersymmetric version of the Yang–Mills theory with the matter fields, one can obtain the 't Hooft zero mode from symmetry considerations. Further details are given in Sec. VII.

The effect of the fermion zero modes can be summarized, in a concise form, by an effective Lagrangian whose particular structure depends on the fermion sector of the theory (generically called the 't Hooft Lagrangians [1, 20]). Thus, with N_f Dirac fermions in the fundamental representation, the effective Lagrangian will contain $2N_f$ fermion lines. It explicitly exhibits the absence of the axial-vector $U(1)$ invariance inherent to the classical action in the massless limit. In the general case the number of fermion lines is determined by the coefficient in the triangle $U(1)$ anomaly. If the Dirac fermion field belongs to the representation R of the gauge group, and the corresponding generators are denoted by T^a, then the number of lines is equal to $4T(R)$, where $T(R)$ is defined as

$$\mathrm{Tr}(T^aT^b) = T(R)\delta^{ab} .$$

A simple derivation of the 't Hooft effective Lagrangian for gauge group $SU(3)$ with two and three quark flavors is given in Ref. [21].

The inclusion of fermions in the instanton calculus might seem a rather straightforward exercise. It requires some technical skills though. The most concise and convenient formalism is obtained if one consistently uses the spinorial representations of $O(4) = SU(2) \times SU(2)$ for *all* fields. The formalism of the Dirac bispinors, quite standard elsewhere, turns out to be awkward in instanton problems. All algebraic manipulations become much more transparent if one uses the Weyl spinors. Not only are they quite natural for massless spinors, they also allow one to express in the simplest form the "hedgehog" structure of the instanton — the fact that the Lorentz $SU(2)$ indices are entangled with the color $SU(2)$.

In particular, the solution of the Dirac equation $\not{D}\Psi = 0$ (the 't Hooft zero mode) has the form

$$\Psi^\alpha_\gamma \propto \delta^\alpha_\gamma \frac{\rho}{(x^2+\rho^2)^{3/2}} , \tag{3}$$

where the lower (undotted) subscript refers to the left-handed $SU(2)$ subgroup of the Lorentz $O(4)$ while the upper one is the spinorial index of the color $SU(2)$. I would like to remind the reader that the solution for the gauge field in these notations is

$$A^{\gamma\delta}_{\alpha\dot\alpha} = -i(\delta^\gamma_\alpha x^\gamma_{\dot\alpha} + \delta^\delta_\alpha x^\gamma_{\dot\alpha})\frac{1}{x^2+\rho^2} , \tag{4}$$

where, instead of the vector A_μ, we introduce an object with two spinorial indices, a $\{\frac{1}{2},\frac{1}{2}\}$ representation with respect to $SU(2) \times SU(2)$,

$$A^a_\mu \to A^{\gamma\delta}_{\alpha\dot\alpha} = (\sigma_\mu)_{\alpha\dot\alpha} A^a_\mu (T^a)^\gamma_\rho \epsilon^{\delta\rho} , \quad T^a = \sigma^a/2. \tag{5}$$

The self-dual (anti-self-dual) field strength tensor is then $\{1,0\}$ ($\{0,1\}$). Expressions in Eqs. (3) and (4) correspond to the nonsingular gauge.

[Warning: in my notation, the spinors with the undotted indices are left-handed. The zero mode presented in Eq. (3) is thus left-handed. Traditionally, the name "instantons" is applied to gauge field solutions whose fermion zero modes are right-handed. This means that Eq. (4) actually gives the anti-instanton solution.]

A nice introduction to the theory of spinors in four dimensions is given in Ref. [22] in Sec. III [where, however, the left-handed spinors are denoted by dotted indices while those referring to the right-handed $SU(2)$ are marked by undotted ones — a convention opposite to ours]. Note that, if, in the Minkowski space, the complex conjugate of the dotted spinor transforms in the same way as the undotted one, there is no such connection in the Euclidean space. The dotted and undotted spinors belong to different $SU(2)$ subgroups of $O(4)$ and transform independently. A systematic adaptation of the spinorial formalism for the instanton calculus was worked out first in Ref. [23] which opens this section. A brief discussion can also be found in Ref. [19], devoted to applications of supersymmetric instantons. In supersymmetric Yang–Mills theories, the spinor language becomes practically indispensable.

A remark concerning Witten's paper [24], the last in this section, is in order here. It is devoted to a question that puzzled instanton practitioners for years, namely, what happens if one considers the Weyl, not Dirac spinor fields, in the instanton background. Many people (including myself) knew that in the $SU(2)$ gauge theory with one Weyl fermion field in the fundamental representation, the (analog of the) 't Hooft effective Lagrangian contains *one* fermion line, which seems very surprising. At the same time, nobody could figure out the origin of the pathology — the $SU(2)$ theory with one Weyl fermion doublet is anomaly-free and seemed perfectly well defined. Witten pointed out that although this theory is free from the standard Adler–Bell–Jackiw anomaly, it is actually nonexistent as a consistent gauge-invariant theory. He discovered a new phenomenon, the so-called global anomaly, which makes the theory "sick". Instanton calculus makes the existence of this anomaly explicit.

References

[1] *G. 't Hooft, *Phys. Rev. Lett.* **37** (1976) 8.

[2] S. Adler, *Phys. Rev.* **177** (1969) 2426; J. Bell and R. Jackiw, *Nuov. Cim.* **60A** (1969) 47; For a review see S. Treiman *et al.*, *Current Algebra and Anomalies* (Princeton University Press, 1985).

[3] S. Weinberg, *Phys. Rev.* **D11** (1975) 3583.

[4] A. Ringwald, *Nucl. Phys.* **B330** (1990) 1; O. Espinosa, *Nucl. Phys.* **B343** (1990) 310; L. McLerran, A. Vainshtein and M. Voloshin, *Phys. Rev.* **D42** (1990) 171.

[5] A. Linde, *Phys. Lett.* **70B** (1977) 306.

[6] S. Dimopoulos and L. Susskind, *Phys. Rev.* **D18** (1978) 4500.

[7] A. Polyakov, in *Proc. 1979 Int. Symp. on Lepton and Photon Interactions at High Energies*, eds. T. Kirk and H. Abarbanel, p. 520.

[8] F. Klinkhammer and N. Manton, *Phys. Rev.* **D30** (1984) 2212.

[9] V. Kuzmin, V. Rubakov and M. Shaposhnikov, *Phys. Lett.* **155B** (1985) 36.

[10] P. Arnold and L. McLerran, *Phys. Rev.* **D36** (1987) 581.

[11] P. Arnold and L. McLerran, *Phys. Rev.* **D37** (1988) 1020; M. Dine *et al.*, *Nucl. Phys.* **B342** (1990) 381; D. Grigoriev, V. Rubakov and M. Shaposhnikov, *Nucl. Phys.* **B32** (1989) 737; J. Ambjorn *et al.*, *Nucl. Phys.* **B353** (1991) 346.
[12] M. Shaposhnikov, *Nucl. Phys.* **B287** (1987) 757; **B299** (1988) 797; A. Bochkarev, S Khlebnikov and M. Shaposhnikov, *Nucl. Phys.* **B329** (1990) 493.
[13] L. McLerran, *Phys. Rev. Lett.* **62** (1989) 1075.
[14] M. Atiyah and I. Singer, *Ann. Math.* **87** (1968) 484; **93** (1971) 119.
[15] A. Schwarz, *Phys. Lett.* **67B** (1977) 172.
[16] L. Brown, R. Carlitz and C. Lee, *Phys. Rev.* **D16** (1977) 417.
[17] *S. Coleman, "The uses of instantons," in *The Whys of Subnuclear Physics*, ed. A. Zichichi (Plenum Press, 1977) [reprinted in S. Coleman, *Aspects of Symmetry* (Cambridge University Press, 1985), p. 265].
[18] D. Friedan and P. Windey, *Nucl. Phys.* **B235** (1984) 395 [reprinted in *Supersymmetry*, ed. S. Ferrara (North-Holland/World Scientific, 1987), p. 572].
[19] V. Novikov *et al.*, *Nucl. Phys.* **B260** (1985) 157.
[20] *G. 't Hooft, *Phys. Rev.* **D14** (1976) 3432.
[21] M. Shifman, A. Vainshtein and V. Zakharov, *Nucl. Phys.* **B163** (1980) 46.
[22] V. Berestetskii, E. Lifshitz and L. Pitaevskii, *Quantum Electrodynamics* (Pergamon Press, 1982).
[23] *R. Jackiw and C. Rebbi, *Phys. Rev.* **D16** (1977) 1052.
[24] *E. Witten, *Phys. Lett.* **117B** (1982) 324 [reprinted in S. Treiman *et al.*, *Current Algebra and Anomalies* (Princeton University Press, 1985), p. 429].

PHYSICAL REVIEW D VOLUME 16, NUMBER 4 15 AUGUST 1977

Spinor analysis of Yang-Mills theory*

R. Jackiw and C. Rebbi†

Laboratory for Nuclear Science and Department of Physics, Massachusetts Institute of Technology, Cambridge, Massachusetts 02139

(Received 13 April 1977)

We formulate the Euclidean Yang-Mills gauge theory for isospin in terms of multispinors of $SU(2)\times SU(2)[=O(4)]\times SU(2)$. The Dirac equation for fermions with arbitrary isospin interacting with the self-dual, conformally covariant Yang-Mills field is analyzed and completely solved for the isovector case. The relevance for this problem of the Atiyah-Singer index theory and its relation to the anomalous divergence of the axial-vector current are also explained.

I. INTRODUCTION

In recent times we have succeeded in solving many differential equations which arise in the study of pseudoparticles—classical solutions in Euclidean space-time to Yang-Mills field equations. For the original pseudoparticle with unit Pontryagin index,[1] we showed that an O(5)-covariant formalism leads to a dramatic simplification of the equations.[2] Group theory replaces much analysis; all that had to be solved were harmonic equations. For the conformal pseudoparticle solution with Pontryagin index N and $5N+4$ parameters,[3] the O(5) formalism is not applicable. In this paper we present a spinorial formulation of Yang-Mills theory with which it is again possible to recognize harmonic equations. With the help of this formalism, we have reduced the Dirac equation for zero-eigenvalue modes of a fermion with arbitrary isospin to a simple differential equation. We have also completely solved the equation for isovector fermions, and have found $4N$ eigenfunctions. (The isospinor case has been previously analyzed.[4])

The spinorial formulation begins with the observation that the O(4) covariants with which we are concerned in Euclidean four-space may be designated by $SU(2)\times SU(2)$ representation labels.[5] Also, the internal SU(2) gauge group gives rise to such labels. Hence all objects of interest are SU(2) multispinors, and equations become simplified when the various SU(2) groups are cunningly coupled to each other.

The spinorial method makes use of 2×2 matrices that are Euclidean analogs of the 2×2 SL(2,C) matrices encountered in the study of the Lorentz group. These matrices have been employed on previous occasions to give a compact presentation of the theory, but their role in a spinorial description was not made explicit.[6] Indeed, it is possible to carry through the analysis of the zero-eigenvalue isovector fermion modes with the help of our matrices, without reference to the underlying spinorial content. This we do in the second section of the paper, where we also summarize some results of Atiyah-Singer index theory—a body of mathematical knowledge which gives precise information on the number and nature of zero-eigenvalue modes.[7] In a local version, the index theorem has already appeared in the physics literature as the anomalous divergence of the axial-vector current.[8] In view of the mathematical results, we now appreciate that this phenomenon is not only an "anomaly," arising from the necessary regularization of a field theory with Fermi operators, but also that it reflects topological properties of gauge-field configurations.

The derivation in Sec. II, though self-contained and complete, employs various matrix manipulations which are unmotivated and *ad hoc*. In Sec. III we present the spinorial formulation of Yang-Mills theory. In that framework, we recognize the manipulations of Sec. II as very natural spin-recoupling procedures, and we successfully analyze the problem with arbitrary isospin.

II. DIRAC EQUATIONS

A. Preliminaries

We are interested in Euclidean 4-space solutions to

$$i\gamma^\mu(\partial_\mu + A_\mu)\psi = 0\,. \tag{2.1}$$

In an infinitesimal group transformation with generators θ_a, ψ transforms according to some definite representation of SU(2),

$$\delta\psi = iT^a\psi\theta_a\,, \tag{2.2a}$$
$$[T^a, T^b] = i\epsilon_{abc}T^c\,,$$

$a=1,2,3$, and iA^μ is the Yang-Mills potential, in a Hermitian matrix representation:

$$iA^\mu = A^\mu_a T_a\,. \tag{2.2b}$$

We assume that A^μ leads to a sufficiently well-behaved field strength

$$F^{\mu\nu}=\partial^\mu A^\nu-\partial^\nu A^\mu+[A^\mu,A^\nu] \tag{2.3}$$

so that the action is finite. In a familiar fashion, this implies that the Pontryagin index

$$q=\frac{1}{32\pi^2}\int d^4x\,{}^*F_a^{\mu\nu}F_{a\mu\nu}\,,\qquad {}^*F_a^{\mu\nu}=\tfrac{1}{2}\epsilon^{\mu\nu\alpha\beta}F^a_{\alpha\beta}\quad(\epsilon_{1234}=1)\,, \tag{2.4}$$

is an integer, characterizing the gauge-field configuration.[1]

ψ also is a 4-component spinor in Euclidean space. The 4×4 Dirac matrices γ^μ satisfy Euclidean anticommutation relations,

$$\{\gamma^\mu,\gamma^\nu\}=2\delta^{\mu\nu}. \tag{*2.5a}$$

(There is no distinction between upper and lower indices.) A convenient realization is

$$\vec{\gamma}=\begin{pmatrix}0 & -i\vec{\sigma}\\ i\vec{\sigma} & 0\end{pmatrix},\quad \gamma_4=\begin{pmatrix}0 & I\\ I & 0\end{pmatrix}, \tag{2.5b}$$

where $\vec{\sigma}$'s are Pauli matrices. The choice (2.5b) diagonalizes γ_5,

$$\gamma_5=\gamma_1\gamma_2\gamma_3\gamma_4=\begin{pmatrix}I & 0\\ 0 & -I\end{pmatrix}, \tag{2.5c}$$

which anticommutes with the Hermitian Dirac operator $i\gamma^\mu(\partial_\mu+A_\mu)$. Consequently, if we consider the full eigenvalue spectrum

$$i\gamma^\mu(\partial_\mu+A_\mu)\psi_E=E\psi_E\,, \tag{2.6}$$

we recognize that γ_5 takes eigenfunctions ψ_E into ψ_{-E}. On the other hand, zero-eigenvalue modes can be chosen to be eigenstates of γ_5; they have either positive or negative chirality.

From (2.5b), we see that the γ matrices may also be written as

$$\gamma^\mu=\begin{pmatrix}0 & \alpha^\mu\\ \bar{\alpha}^\mu & 0\end{pmatrix}, \tag{2.7}$$

where

$$\alpha^\mu=(-i\vec{\sigma},1)\,, \tag{2.8a}$$

$$\bar{\alpha}^\mu=(i\vec{\sigma},1)=(\alpha^\mu)^\dagger\,. \tag{2.8b}$$

These 2×2 matrices will be used in the subsequent discussions, and we record some of their properties:

$$\bar{\alpha}^\mu\alpha^\nu+\bar{\alpha}^\nu\alpha^\mu=2\delta^{\mu\nu}\,, \tag{2.9a}$$

$$\alpha^\mu\bar{\alpha}^\nu+\alpha^\nu\bar{\alpha}^\mu=2\delta^{\mu\nu} \tag{2.9b}$$

The Hermitian spin matrices defined by

$$\sigma^{\mu\nu}=\frac{1}{4i}(\bar{\alpha}^\mu\alpha^\nu-\bar{\alpha}^\nu\alpha^\mu)\,, \tag{2.10a}$$

$$\bar{\sigma}^{\mu\nu}=\frac{1}{4i}(\alpha^\mu\bar{\alpha}^\nu-\alpha^\nu\bar{\alpha}^\mu) \tag{2.10b}$$

appear in a reduction of the product of two α matrices:

$$\bar{\alpha}^\mu\alpha^\nu=\delta^{\mu\nu}+2i\sigma^{\mu\nu}\,, \tag{2.11a}$$

$$\alpha^\mu\bar{\alpha}^\nu=\delta^{\mu\nu}+2i\bar{\sigma}^{\mu\nu}\,. \tag{2.11b}$$

The spin matrices behave simply under the duality transformation,

$$^*\sigma^{\mu\nu}=\sigma^{\mu\nu}\,, \tag{2.12a}$$

$$^*\bar{\sigma}^{\mu\nu}=-\bar{\sigma}^{\mu\nu}\,. \tag{2.12b}$$

They can be expressed by Pauli matrices

$$\sigma^{i4}=-\bar{\sigma}^{i4}=\tfrac{1}{2}\sigma^i,\qquad \sigma^{ij}=\bar{\sigma}^{ij}=\tfrac{1}{2}\epsilon^{ijk}\sigma^k\,. \tag{2.13}$$

and satisfy $O(4)$ commutation relations:

$$i[\sigma^{\mu\alpha},\sigma^{\nu\beta}]=\delta^{\alpha\nu}\sigma^{\mu\beta}-\delta^{\mu\nu}\sigma^{\alpha\beta}+\delta^{\alpha\beta}\sigma^{\nu\mu}-\delta^{\mu\beta}\sigma^{\nu\alpha}\,, \tag{2.14a}$$

$$i[\bar{\sigma}^{\mu\alpha},\bar{\sigma}^{\nu\beta}]=\delta^{\alpha\nu}\bar{\sigma}^{\mu\beta}-\delta^{\mu\nu}\bar{\sigma}^{\alpha\beta}+\delta^{\alpha\beta}\bar{\sigma}^{\nu\mu}-\delta^{\mu\beta}\bar{\sigma}^{\nu\alpha}\,. \tag{2.14b}$$

Products of three α matrices are equal to linear combinations of α matrices:

$$\bar{\sigma}^{\mu\nu}\alpha^\alpha=(\bar{\alpha}^\alpha\bar{\sigma}^{\mu\nu})^\dagger=\frac{1}{2i}\delta^{\nu\alpha}\alpha^\mu-\frac{1}{2i}\delta^{\mu\alpha}\alpha^\nu+\frac{1}{2i}\epsilon^{\mu\nu\alpha\beta}\alpha_\beta\,, \tag{2.15a}$$

$$\sigma^{\mu\nu}\bar{\alpha}^\alpha=(\alpha^\alpha\sigma^{\mu\nu})^\dagger=\frac{1}{2i}\delta^{\nu\alpha}\bar{\alpha}^\mu-\frac{1}{2i}\delta^{\mu\alpha}\bar{\alpha}^\nu-\frac{1}{2i}\epsilon^{\mu\nu\alpha\beta}\bar{\alpha}_\beta\,. \tag{2.15b}$$

Finally, we note the following identites involving the Pauli matrices σ^a:

$$\sigma^a\sigma^{\mu\nu}\sigma^a=-\sigma^{\mu\nu}\,, \tag{2.16a}$$

$$\sigma^a\bar{\sigma}^{\mu\nu}\sigma^a=-\bar{\sigma}^{\mu\nu}\,, \tag{2.16b}$$

$$\bar{\alpha}^\mu\sigma^a\alpha_\mu=0\,. \tag{2.16c}$$

The eigenvalue problem (2.6) may be written in a form which makes explicit the chiral structure and uses the α matrices:

$$\begin{pmatrix}0 & L\\ L^\dagger & 0\end{pmatrix}\begin{pmatrix}\psi_E^+\\ \psi_E^-\end{pmatrix}=E\begin{pmatrix}\psi_E^+\\ \psi_E^-\end{pmatrix},\qquad L=i\alpha^\mu(\partial_\mu+A_\mu)\,,\quad L_*=i\bar{\alpha}^\mu(\partial_\mu+A_\mu)\,. \tag{2.17}$$

The zero-eigenvalue modes satisfy

$$\bar{\alpha}^\mu(\partial_\mu+A_\mu)\psi^+=0\,, \tag{2.18a}$$

$$\alpha^\mu(\partial_\mu+A_\mu)\psi^-=0\,. \tag{2.18b}$$

Here $\psi^\pm$ are two-component spinors, which also carry an isospin label appropriate to the representation (2.2a).

B. Index theory analysis of zero-eigenvalue modes

When the Atiyah-Singer index theorem is applied to our problem, the following result emerges: If the number of normalizable zero-eigenvalue modes of positive (negative) chirality is n_+ (n_-), then[8]

$$n_+ - n_- = \frac{1}{16\pi^2}\int d^4x \,\mathrm{tr}\, {}^*F^{\mu\nu}F_{\mu\nu} . \tag{2.19}$$

For fermions with total isospin T and gauge fields with Pontryagin index N, the above may be evaluated with the help of

$$\mathrm{tr} T_a T_b = \tfrac{1}{3} T(T+1)(2T+1)\delta_{ab} .$$

So we find

$$n_- - n_+ = \tfrac{2}{3} T(T+1)(2T+1)N . \tag{2.20}$$

Although the index theorem should be applied only to compact manifolds, it may, nevertheless, also be used in our Euclidean 4-space problem, since the conformal invariance of the theory and the assumption that the gauge fields decrease rapidly at infinity allow our problem to be mapped onto the surface of a 5-dimensional hypersphere which is compact.[2] (This introduces the following subtlety: The normalizablility condition for the Dirac equation in an O(5)-covariant formulation requires only that $\int [d^4x/(1+x^2)]\psi^\dagger(x)\psi(x)$ converge, while $\int d^4x\, \psi^\dagger(x)\psi(x)$ may diverge. However, we have not encountered a situtation in 4 dimensions where this distinction makes a difference.)

There is also a local version of (2.19).[7] A derivation of that formula begins by considering the resolvent of (2.6),

$$R(x,y;\mu) = \sum_E \frac{\psi_E(x)\psi_E^\dagger(y)}{E+i\mu} . \tag{2.21a}$$

We shall want to make x and y coincident, which in general will produce infinities that must be regulated. A convenient, gauge-invariant regularization involves subtracting from(2.21a) the same expression with μ replaced by M, and at the end of the calculation M is passed to infinity,

$$R_{\mathrm{reg}}(x,y;\mu) = \lim_{M\to\infty} [R(x,y;\mu) - R(x,y;M)] . \tag{2.21b}$$

Next we form the "axial-vector current," and its divergence,

$$J_5^\mu(x) = \mathrm{tr} i\gamma^\mu\gamma_5 R_{\mathrm{reg}}(x,x;\mu) . \tag{2.22}$$

A simple calculation based on (2.6) gives

$$\partial_\mu J_5^\mu(x) = 2i\mu \sum_E \frac{\psi_E^\dagger(x)\gamma_5\psi_E(x)}{E+i\mu} - \lim_{M\to\infty}\left(2iM\sum_E \frac{\psi_E^\dagger(x)\gamma_5\psi_E(x)}{E+iM}\right) . \tag{2.23a}$$

The second term is known to produce the axial-vector current anomaly and the result, familiar to physicists as the anomalous divergence of the axial-vector current, is also the local form of the index theorem[7,8]:

$$\partial_\mu J_5^\mu(x) = 2\mu J_5(x) - \frac{1}{8\pi^2}\mathrm{tr}\, {}^*F^{\mu\nu}F_{\mu\nu} ,$$
$$J_5(x) = \sum_E \frac{\psi_E^\dagger(x)\gamma_5\psi_E(x)}{E+i\mu} . \tag{2.23b}$$

Upon integration of this equation over x, (2.19) is regained if the left-hand-side in (2.23b) is assumed to produce no surface terms, and if the series on the right-hand side is integrated term by term. [Recall that $\int d^4x\, \psi_E^\dagger(x)\gamma_5\psi_E(x)$ vanishes for $E\neq 0$, and equals +1 (−1) for zero-eigenvalue modes of positive (negative) chirality.][9,10(a),10(b)]

The above discussion also exposes circumstances which may modify the simple, integrated expression (2.19): The surface term for the integral of $\partial_\mu J_5^\mu$ need not vanish; the term-by-term integration of J_5 may be illegitimate.[10(c)] In that case, an additional contribution is present in (2.19); it is called the "signature defect".[7] We expect that such pathologies occur when long-range potentials are present in the Dirac equation. In our example, the gauge potential can be long range, but gauge-invariant quantities see only the short-range gauge field, and in all known 4-dimensional examples the simple result holds. (However, we have encountered Dirac equations in dimensions less than 4, where no known anomaly exists, yet there are zero-eigenvalue modes. These examples involve soliton-monopole potentials which include a long-range Higgs field; they provide physically interesting applications of the signature defect.[11])

The index theorem may be strengthened by showing that only n_- or n_+ is nonzero; a "vanishing theorem" may sometimes hold.[7] The vanishing theorem can be always established when the gauge field is self-dual or anti-self-dual: Apply L^+ to (2.18a) and L to (2.18b) to get

$$[(\partial_\mu + A_\mu)^2 + 2i\bar\sigma_{\mu\nu}F^{\mu\nu}]\psi^+ = 0 , \tag{2.24a}$$
$$[(\partial_\mu + A_\mu)^2 + 2i\sigma_{\mu\nu}F^{\mu\nu}]\psi^- = 0 . \tag{2.24b}$$

The duality properties of $\bar\sigma_{\mu\nu}$ ($\sigma_{\mu\nu}$) [see (2.12)] assure that $\bar\sigma_{\mu\nu}F^{\mu\nu}$ ($\sigma_{\mu\nu}F^{\mu\nu}$) vanishes for self-dual (anti-self-dual) gauge fields. Since $(\partial_\mu + A_\mu)^2$ is a positive-definite operator, the differential equation without the gauge-field term has no normalizable solution. All known Yang-Mills solutions with finite action are self-dual or anti-self-dual; hence in those instances there are precisely $\frac{2}{3}T(T+1)(2T+1)N$ zero-eigenvalue modes with chirality determined by the gauge field's duality properties.[5,10(b),12] Whether a vanishing theorem can be established for general gauge fields remains an open question. Note also that $\psi^\dagger T^a\gamma_\mu\psi$ vanishes when ψ has definite chirality. Hence the functions A^μ and ψ are solutions of the coupled Yang-Mills fermion equations, when A^μ has definite duality properties and ψ solves the Dirac equation.

C. Isospinor Fermi fields coupled to pseudoparticle potentials

We solve Eq. (2.1) for an isovector fermion field, belonging to the adjoint representation of SU(2) ($T^a_{nm}=i\epsilon_{nam}$), with the conformally covariant Yang-Mills potential of the N-pseudoparticle self-dual field configuration[3]:

$$A^\mu = A^\mu_a \frac{\sigma^a}{2i} = i\bar\sigma^{\mu\nu}\partial_\nu \ln\rho\,, \qquad (2.25\text{a})$$

$$\rho = \sum_{i=1}^{N+1} \frac{\lambda_i^2}{(x-x_i)^2}\,. \qquad (2.25\text{b})$$

According to the index theory described in Sec. II B, we expect to find $4N$ zero-eigenvalue modes of negative chirality. Thus we seek $4N$ linearly independent solutions to

$$\alpha_\mu(\partial^\mu\delta_{ac}+\epsilon_{abc}A^\mu_b)\psi^-_c=0\,. \qquad (2.26)$$

We are interested in isovector Fermi fields for several reasons. Firstly, our formulas add to an expanding assembly of explicit solutions for Dirac equations in a pseudoparticle field.[13] Secondly, it is known that the isovector Dirac equation is equivalent to the equation for small oscillations of a self-dual Yang-Mills field, in the background gauge, in the sense that if ψ^-_a solves the Dirac equation with a self-dual potential A^μ_a, then $A^\mu_a + u^\dagger\alpha^\mu\psi^-_a$ is self-dual to first order.[5,10(b)] (Here u is a constant spinor.) Hence the present solution provides explicit formulas for Yang-Mills zero-frequency modes which complement the ones found previously by us.[14] (Our earlier solution made use of an unfamiliar gauge condition.) Thirdly, the Yang-Mills isovector fermion system has peculiar supersymmetric properties, even when all fields are treated as commuting c numbers. The supersymmetric transformation of a Fermi field is $\delta\psi_a = F_a^{\mu\nu}\gamma_\mu\gamma_\nu(c_1+c_2\gamma\cdot x)u$, where the c_i's are constants and u is an x-independent, anticommuting spinor. The curiosity is that $F^{\mu\nu}_a\gamma_\mu\gamma_\nu u$ and $F^{\mu\nu}_a\gamma_\mu\gamma_\nu\gamma\cdot xu$ also solve the Dirac equation, provided only that $F^{\mu\nu}_a$ satisfies the Yang-Mills equation, and u is a constant c-number spinor. In other words, we assert that $F^{\mu\nu}_a\bar\sigma_{\mu\nu}u$ and $F^{\mu\nu}_a\bar\sigma_{\mu\nu}\alpha\cdot xu$ solve (2.18a), while $F^{\mu\nu}_a\sigma_{\mu\nu}u$ and $F^{\mu\nu}_a\sigma_{\mu\nu}\bar\alpha\cdot xu$ solve (2.18b). [This is easy easy to verify by simply applying the appropriate differential operator, L or $L^\dagger$, reducing products of matrices according to (2.15), and using the fact that both $F^{\mu\nu}_a$ and $^*F^{\mu\nu}_a$ satisfy the Yang-Mills equation.][15] Of course, when $F^{\mu\nu}_a$ is self-dual (anti self-dual), positive chirality (negative chirality) solutions vanish. In that case these "supersymmetric" solutions provide four of the total $4N$ eigenmodes. (Each spinor u has two arbitrary components.) Moreover, the small-oscillation solution to the self-duality equation mentioned above, $u^\dagger\alpha^\mu\psi^-_a$, is just the supersymmetric transformation of a gauge potential A^μ_a. Fourthly, and lastly, our analysis of the Dirac equation well illustrates the usefulness of the spinorial formalism presented in the next section.

The solution of (2.26) is made possible by the following identity, satisfied by the gauge field A^μ_a, when $A^\mu = A^\mu_a(\sigma^a/2i)$ is given by (2.25a):

$$\alpha_\mu A^\mu_a = -i\sigma^a\alpha\cdot\partial\ln\rho\,, \qquad (2.27\text{a})$$

$$\sigma_{\mu\nu}F^{\mu\nu}_a = \tfrac12\rho\bar\alpha\cdot\partial\sigma^a\alpha\cdot\partial\frac{1}{\rho}\,. \qquad (2.27\text{b})$$

Upon substituting (2.27a) into (2.26) and also redefining the eigenfunction by

$$\psi^-_a = \rho\bar\alpha\cdot\partial\eta_a\,, \qquad (2.28\text{a})$$

we arrive at

$$0=\rho[\Box\eta_a + (\delta_{ac}-i\epsilon_{abc}\sigma^b)(\alpha\cdot\partial\ln\rho)\bar\alpha\cdot\partial\eta_c]$$
$$=\rho[\Box\eta_a+\sigma_a\sigma_c(\alpha\cdot\partial\ln\rho)\alpha\cdot\partial\eta_c]\,. \qquad (2.28\text{b})$$

We multiply this equation by σ^a and sum on a to find

$$\rho\sigma_a(\alpha\cdot\partial\ln\rho)\bar\alpha\cdot\partial\eta_a = -\tfrac13\rho\Box\sigma^a\eta_a\,. \qquad (2.28\text{c})$$

Combining with (2.28b) results in a purely kinematical equation for a projection of η_a,

$$\rho\Box\left(\delta_{ab}-\frac{\sigma_a\sigma_b}{3}\right)\eta_b=0\,. \qquad (2.28\text{d})$$

This equation states that if η_a is decomposed according to

$$\eta_a=\sigma_a\eta+\eta^\perp_a\,,$$
$$\eta=\tfrac13\sigma_a\eta^a\,,$$
$$\eta^\perp_a=\left(\delta_{ab}-\frac{\sigma_a\sigma_b}{3}\right)\eta_b\,, \qquad (2.29)$$
$$\sigma^a\eta^\perp_a=0\,,$$

then the transverse part is harmonic, which in turn forces it to vanish, for otherwise ψ^-_a would be irregular and non-normalizable. We thus learn that ψ^-_a has the form

$$\psi^-_a=\rho\bar\alpha\cdot\partial\sigma_a\eta\,, \qquad (2.30\text{a})$$

where, according to (2.28c), η satisfies

$$\rho[\Box\eta+\sigma^a(\alpha\cdot\partial\ln\rho)\bar\alpha\cdot\partial\sigma^a\eta]=0\,. \qquad (2.30\text{b})$$

The remaining sum over a is evaluated with the help of (2.11b) and (2.16b). The result may be written as

$$\alpha\cdot\partial\frac{1}{\rho}\bar{\alpha}\cdot\partial\rho^2\eta=0. \tag{2.30c}$$

(We have set $\Box\rho=0$; we ignore the singularities of ρ. This is justified, for as we show below, ψ_a^- is nonsingular.) Equation (2.30c) may also be written as

$$\begin{aligned}0&=\rho^{-1/2}(\partial^\mu+i\bar{\sigma}^{\mu\alpha}\partial_\alpha\ln\rho)\\&\quad\times(\partial_\mu+i\bar{\sigma}_{\mu\beta}\partial^\beta\ln\rho)\rho^{3/2}\eta\\&=\rho^{-1/2}(\partial^\mu+A^\mu)(\partial_\mu+A_\mu)\rho^{3/2}\eta\end{aligned} \tag{2.30d}$$

which demonstrates that in the spinor ψ_a^-, with spin $\frac{1}{2}$ and isospin 1, spin and isospin have combined to form an isospin-$\frac{1}{2}$ quantity $\rho^{3/2}\eta$ which satisfies (2.30d).

The conclusion therefore is that the formula

$$\psi_1^-=\rho\bar{\alpha}\cdot\partial\sigma_a\frac{1}{\rho^2}\chi \tag{2.31a}$$

reduces the Dirac equation to

$$\alpha\cdot\partial\frac{1}{\rho}\bar{\alpha}\cdot\partial\chi=0 \tag{2.31b}$$

away from the singularities of ρ. We need $4N$ linearly independent solutions to (2.31b); if we set $\chi=Mu$, where M is a 2×2 matrix and u an arbitrary x-independent 2-component spinor, then we need $2N$ linearly independent solutions to

$$\alpha\cdot\partial\frac{1}{\rho}\bar{\alpha}\cdot\partial M=0. \tag{2.31c}$$

Equation (2.31c) holds when M is proportional to $\alpha\cdot\partial h$, or $(1/\rho)\bar{\alpha}\cdot\partial M$ is proportional to $\bar{\alpha}\cdot\partial h$, with h harmonic. In order to avoid singularities in ψ_a^-, we must match the singularities of h with the poles of ρ. Thus there are $N+1$ choices for h,

$$h_i=\frac{\lambda_i^2}{(x-x_i)^2},\quad i=1,\ldots,N+1. \tag{2.32}$$

Thereby we get $N+1$ expressions for M_i:

$$M_i^{(1)}=\frac{\lambda_i^2}{(x-x_i)^4}\alpha\cdot(x-x_i). \tag{2.33a}$$

Additional expressions are found by solving the equation

$$\bar{\alpha}\cdot\partial M_i^{(2)}=-4\rho\frac{\lambda_i^2}{(x-x_i)^4}\bar{\alpha}\cdot(x-x_i).$$

One readily verifies that a solution is

$$\begin{aligned}M_i^{(2)}&=\rho\frac{\lambda_i^2}{(x-x_i)^2}\\&\quad+\sum_{j\neq i}\frac{\lambda_i^2\lambda_j^2}{(x_i-x_j)^2}\left[\frac{\alpha\cdot(x-x_i)+\alpha(x-x_j)}{(x-x_i)^2(x-x_j)^2}\right]\\&\quad\times\bar{\alpha}\cdot(x_i-x_j),\end{aligned} \tag{2.33b}$$

which produces $N+1$ further M_i's.

A tedious, but straightforward, calculation establishes that the corresponding eigenfunctions behave near the poles as $\bar{\alpha}\cdot(x-x_i)\sigma^a\alpha\cdot(x-x_i)/(x-x_i)^2$; such a discontinuity contributes a constant to the gauge-invariant norm density $\psi_a^\dagger\psi_a$. The large-x behavior is also well tempered so that all $4N+4$ eigenfunctions are normalizable. We still must exhibit four linear dependences among the solutions, to reduce the number to $4N$. Clearly it sufficies to show that two combinations of M_i's give no contribution to (2.31a). One such combination is the sum of the $M_i^{(2)}$'s. Since

$$\sum_i M_i^{(2)}=\rho^2, \tag{2.34a}$$

it is seen from (2.31a) that the corresponding ψ_a^- vanishes. To exhibit the second combination, we form the sum

$$\sum_i\lambda_i^2M_i^{(1)}+\sum_i M_i^{(2)}\alpha\cdot x_i=\rho^2\alpha\cdot x. \tag{2.34b}$$

By virtue of (2.16c), this combination also leads to a vanishing ψ_a^-.

Where among these solutions are the "supersymmetric" eigenfunctions? One is found by summing the $M_i^{(1)}$'s:

$$\sum_i M_i^{(1)}=-\tfrac{1}{2}\alpha\cdot\partial\rho. \tag{2.35a}$$

The corresponding eigenfunction becomes

$$\begin{aligned}\psi_a^-&=-\tfrac{1}{2}\rho\bar{\alpha}\cdot\partial\sigma^a\frac{1}{\rho^2}\alpha\cdot\partial\rho u\\&=\tfrac{1}{2}\rho\bar{\alpha}\cdot\partial\sigma^a\alpha\cdot\partial\frac{1}{\rho}u.\end{aligned} \tag{2.35b}$$

According to (2.27b) this is just $F_a^{\mu\nu}\sigma_{\mu\nu}u$, one of the supersymmetric solutions. For the other, we form the combination

$$\sum_i M_i^{(1)}\bar{\alpha}\cdot x_i=-\rho-\tfrac{1}{2}(\alpha\cdot\partial\rho)\bar{\alpha}\cdot x. \tag{2.36a}$$

The corresponding eigenfunction is evaluated with the help of (2.16c) and we find

$$\psi_a^-=\left(\tfrac{1}{2}\rho\bar{\alpha}\cdot\partial\sigma^a\alpha\cdot\partial\frac{1}{\rho}\right)\bar{\alpha}\cdot xu, \tag{2.36b}$$

which is $F_a^{\mu\nu}\sigma_{\mu\nu}\bar{\alpha}\cdot xu$, the second supersymmetric solution.

III. SPINOR FORMULATION OF YANG-MILLS THEORY

A. Definitions

In this section we present a spinorial formulation of Euclidean Yang-Mills theory for the SU(2) gauge group. All objects carry spinor labels $A, B, C\ldots$, which take on two values, and describe the spin and isospin degrees of freedom. An antisymmetric metric tensor with two upper indices is defined by

$$\epsilon^{AB} = \begin{pmatrix} 0 & 1 \\ -1 & 0 \end{pmatrix} = i\sigma^2 . \tag{3.1a}$$

The negative inverse of this matrix is a metric tensor with lower indices,

$$\epsilon_{AB} = \begin{pmatrix} 0 & 1 \\ -1 & 0 \end{pmatrix} = \epsilon^{AB} \tag{3.1b}$$

A spinor may have lower or upper indices, which can be raised or lowered with the metric tensors according to the following rules (repeated indices are summed):

$$\xi^A = \epsilon^{AB}\xi_B , \quad \xi_B = \xi^A \epsilon_{AB} . \tag{3.2}$$

Covariant summations always involve one upper and one lower index. From (3.1) and (3.2) it follows that $\xi^A{}_A = -\xi_A{}^A$. For every pair of indices we may define a symmetric and antisymmetric part,

$$\xi_{AB} = \tfrac{1}{2}\epsilon_{AB}\xi_C{}^C + \tfrac{1}{2}\xi_{\overline{AB}} . \tag{3.3}$$

The bar over the indices A and B indicates that a symmetric sum should be performed, $\xi_{\overline{AB}} = \xi_{AB} + \xi_{BA}$. More generally for a multiindexed object $\xi_{A_1A_2\cdots A_n}$, already symmetric in $A_2 \cdots A_n$, we define $\xi_{\overline{A}_1\overline{A}_2\cdots\overline{A}_n} = \xi_{A_1A_2\cdots A_n} + \xi_{A_2A_1\cdots A_n} + \cdots + \xi_{A_NA_2\ldots A_1}$ (n terms)

An O(4), two-component spinor is described by a spinor with one index. To each O(4) tensor with indices $\mu, \nu, \ldots$, there corresponds a spinor with index pairs $AA', BB', \ldots$. The rule of association is given through the α matrices, defined in Sec. II A:

$$\xi_\mu(\alpha^\mu)_{AA'} = \xi_{AA'} , \quad \xi_\mu(\overline{\alpha}^\mu)_{A'A} = \xi^{AA'} \tag{3.4}$$

Consistency of this definition is established by noting that $(\overline{\alpha}^\mu)_{B'B}\epsilon_{BA}\epsilon_{B'A'} = -(i\sigma^2\overline{\alpha}^\mu i\sigma^2)_{A'A} = (\alpha^\mu)_{AA'}$. From (3.3) follows

$$\xi_{AA'}\xi_B{}^{A'} = \tfrac{1}{2}\epsilon_{AB}\xi_{CA'}\xi^{CA'} + \tfrac{1}{2}\xi_{\overline{A}A'}\xi_{\overline{B}}{}^{A'} , \tag{3.5a}$$

while comparison with (3.4) gives

$$\tfrac{1}{2}\xi_{CA'}\xi^{CA'} = \xi_\mu\xi^\mu . \tag{3.5b}$$

We note that the above correspondence also holds for derivatives, $\partial_\mu - \partial_{AA'}$, and that O(4) covariants may be regained from spinors by multiplying by the appropriate α matrix and taking the trace,

$$\xi^\mu = \tfrac{1}{2}(\overline{\alpha}^\mu)_{A'A}\xi_{AA'} = \tfrac{1}{2}(\alpha^\mu)_{AA'}\xi^{AA'} . \tag{3.6}$$

Isospinor indices are represented as follows. Isospin-$\frac{1}{2}$ objects are described by one-index spinors. For isospin 1 a two-index spinor, symmetric in the two indices, is used. In general, an isospin-T quantity is described by a spinor with $2T$ indices, symmetric in all of them, so that there are $2T+1$ independent components. The correspondence between the conventional description and the present spinorial one is immediate for isospin $\frac{1}{2}$ —the two descriptions coincide. For units isospin the correspondence is achieved by contracting an isovector ξ^a, $a=1,2,3$, with the Pauli matrices and defining

$$\xi^a \frac{(\sigma^a)_{UV}}{2i} = -\xi^V{}_U . \tag{3.7a}$$

Since $\sigma^2\sigma^a$ and $\sigma^a\sigma^2$ are symmetric matrices one sees that both $\xi_{VU} = \xi^W{}_U\epsilon_{WV}$ and $\xi^{VU} = \epsilon^{UW}\xi^V{}_W$ are symmetric in $U - V$. Again we may regain ξ^a from the spinor by

$$\xi^a = \frac{1}{i}(\sigma^a)_{VU}\xi^V{}_U . \tag{3.7b}$$

This definition has the pleasant consequence that the quantity $\epsilon_{abc}\xi^b\overline{\xi}^c$ corresponds to $\xi_{\overline{U}W}\overline{\xi}^W{}_{\overline{V}}$, as the following equations show:

$$\begin{aligned} -C^V{}_U &= \epsilon_{abc}\xi^b\overline{\xi}^c\frac{(\sigma^a)_{UV}}{2i} \\ &= \left[\xi^a\frac{\sigma^a}{2i}, \xi^b\frac{\sigma^b}{2i}\right]_{UV} \\ &= \xi^W{}_U\overline{\xi}^V{}_W - \xi^V{}_W\overline{\xi}^W{}_U , \end{aligned}$$

$$C_{VU} = \xi_{VW}\overline{\xi}^W{}_U + \xi_{WU}\overline{\xi}_V{}^W .$$

But $\xi_{WU} = \xi_{UW}$ and $\overline{\xi}_V{}^W = \overline{\xi}^W{}_V$, so

$$C_{VU} = \xi_{\overline{U}W}\overline{\xi}^W{}_{\overline{V}} .$$

The correspondences for higher isospin are more complicated, and will not be needed here.

B. Gauge fields

The gauge potential A_a^μ is described by $A_{AA';UV}$, and the gauge field $F_a^{\mu\nu}$ is described by $F_{AA',BB';UV}$, which is antisymmetric under the interchange $A - B$, $A' - B'$; both expressions are symmetric in $U - V$. The formula relating the two is

$$F_{AA',BB';UV} = \partial_{AA'}A_{BB';UV} - \partial_{BB'}A_{AA';UV} + A_{AA';\overline{U}W}A_{BB';}{}^W{}_{\overline{V}} . \tag{3.8}$$

Owing to its antisymmetry properties, F may be split into two parts:

$$F_{AA',BB';UV} = \tfrac{1}{2}\epsilon_{AB}F^+_{A'B';UV} + \tfrac{1}{2}\epsilon_{A'B'}F^-_{AB,UV} , \tag{3.9a}$$

where $F^+_{A'B';UV}$ is symmetric in $A' - B'$, and

$F^-_{AB;UV}$ is symmetric in $A-B$. From (2.8a) it follows that

$$F^+_{A'B';UV}=\partial_{A\bar{X}'}A^A{}_{\bar{B}';UV}+A_{AA';\bar{U}W}A^A{}_{B';}{}^W{}_{\bar{V}}, \tag{3.9b}$$

$$F^-_{AB;UV}=\partial_{\bar{X}A'}A_{\bar{B}}{}^{A'}{}_{;UV}+A_{AA';\bar{U}W}A_B{}^{A'}{}_{;}{}^W{}_{\bar{V}}. \tag{3.9c}$$

Under the duality transformation one finds

$$^*F_{AA',BB';UV}=F_{AB',BA';UV} \tag{3.10a}$$

or

$$^*F^+_{A'B';UV}=F^+_{A'B';UV}, \tag{3.10b}$$

$$^*F^-_{AB;UV}=-F^-_{AB;UV}. \tag{3.10c}$$

Hence for self-dual fields (3.9c) must vanish, while for anti-self-dual fields (3.9b) is zero.

We solve the self-duality equation

$$\partial_{\bar{X}A'}A_{\bar{B}}{}^{A'}{}_{;UV}+A_{AA';\bar{U}W}A_B{}^{A'}{}_{;}{}^W{}_{\bar{V}}=0 \tag{3.11}$$

with the ansatz

$$A_{AA';UV}=\tfrac{1}{2}\epsilon_{AU}\partial_{VA'}\ln\rho+\tfrac{1}{2}\epsilon_{AV}\partial_{UA'}\ln\rho. \tag{3.12}$$

Substitution of this into (3.11) yields the familiar result[16]

$$\frac{1}{\rho}\Box\rho=0 \tag{3.13}$$

It is straightforward to verify that (3.12) is equivalent to (2.25a) and (2.27a).

It is also instructive to exhibit the equation for small deformations of a self-dual gauge potential $A^0_{AA';UV}$. Writing

$$A_{AA';UV}=A^0_{AA';UV}+\chi_{AA';UV} \tag{3.14a}$$

and substituting this into (3.11) gives to first order in $\chi_{AA';UV}$

$$\partial_{\bar{X}A'}\chi_{\bar{B}}{}^{A'}{}_{;UV}+A^0_{\bar{X}A';UW}\chi_{\bar{B}}{}^{A'}{}_{;}{}^W{}_{V}+A^0_{\bar{X}A';VW}\chi_{\bar{B};U}=0. \tag{3.14b}$$

The background gauge condition is imposed on $\chi_{AA';UV}$,

$$\partial_{CC'}\chi^{CC'}{}_{;UV}+A^0_{CC';\bar{U}W}\chi^{CC'}{}_{;}{}^W{}_{\bar{V}}=0. \tag{3.14c}$$

We multiply (3.14c) by ϵ_{AB}, add that to (3.14b), and obtain from (3.3)

$$\partial_{AA'}\chi_B{}^{A'}{}_{;UV}+A^0_{AA';\bar{U}W}\chi_B{}^{A'}{}_{;}{}^W{}_{\bar{V}}=0. \tag{3.14d}$$

Note that the B index is free; hence (3.14d) describes two uncoupled equations.

C. Dirac equations

Equation (2.18b), transcribed to the spinorial formalism, reads

$$\partial_{AA'}\psi^{A'}{}_{;U_1\cdots U_{2T}}+A_{AA';\bar{U}_1V}\psi^{A'}{}_{;}{}^V{}_{\bar{U}_2\cdots\bar{U}_{2T}}. \tag{3.15}$$

Here the spinor carries the index A', describing its two spatial components; also, it is entirely symmetric in the $2T$ indices U_i, which refer to the $2T+1$ components of isospin. [Observe that for the adjoint representation, $T=1$, (3.15) is identical with each of the two equations of (3.14d) obtained by setting $B=1, 2$. This exhibits the relation between the small deformation problem and the isovector Dirac equation.[5,10(b)]]

Substituting (3.12) into (3.15) gives

$$\frac{1}{2T}\partial_{AA'}\psi^{A'}{}_{;U_1\cdots U_{2T}}+\tfrac{1}{2}\epsilon_{AU_1}(\partial_{BB'}\ln\rho)\psi^{B'}{}_{;}{}^{B}{}_{U_2\cdots U_{2T}}-\tfrac{1}{2}(\partial_{U_1A'}\ln\rho)\psi^{A'}{}_{;AU_2\cdots U_{2T}}+\text{permutations}=0 \tag{3.16a}$$

Next using (3.3), this becomes

$$\frac{1}{4T}\epsilon_{AU_1}\partial_{BB'}\psi^{B'}{}_{;}{}^{B}{}_{U_2\cdots U_{2T}}+\frac{1}{4T}\partial_{\bar{X}A'}\psi^{A'}{}_{;\bar{U}_1\cdots U_{2T}}+\tfrac{1}{2}\epsilon_{AU_1}(\partial_{BB'}\ln\rho)\psi^{B'}{}_{;}{}^{B}{}_{U_2\cdots U_{2T}}$$
$$+\tfrac{1}{4}\epsilon_{AU_1}(\partial_{BB'}\ln\rho)\psi^{B'}{}_{;}{}^{B}{}_{U_2\cdots U_{2T}}-\tfrac{1}{4}(\partial_{\bar{X}A'}\ln\rho)\psi^{A'}{}_{;\bar{U}_1\cdots U_{2T}}+\text{permutations}=0. \tag{3.16b}$$

The last term may be eliminated by redefining $\psi=\rho^T f$ [compare (2.28)]. Thus we get

$$\frac{1}{4T}\epsilon_{AU_1}\partial_{BB'}f^{B'}{}_{;}{}^{B}{}_{U_2\cdots U_{2T}}+\frac{1}{4T}\partial_{\bar{X}A'}f^{A'}{}_{;\bar{U}_1\cdots U_{2T}}+\epsilon_{AU_1}(\partial_{BB'}\ln\rho)f^{B'}{}_{;}{}^{B}{}_{U_2\cdots U_{2T}}+\text{permutations}=0, \tag{3.16c}$$

which is the same as

$$\partial_{AA'}f^{A'}{}_{;U_1\cdots U_{2T}}+\epsilon_{A\bar{U}_1}(\partial_{BB'}\ln\rho)f^{B'}{}_{;}{}^{B}{}_{\bar{U}_2\cdots\bar{U}_{2T}}=0. \tag{3.16d}$$

From this equation we may extract a purely kinematical relation which must be satisfied by f. Multiply (3.16d) by ϵ^{AU_1} and sum on those two indices,

$$\partial_{BB'}f^{B'}{}_{;}{}^{B}{}_{U_2\cdots U_{2T}}+(2T+1)(\partial_{BB'}\ln\rho)f^{B'}{}_{;}{}^{B}{}_{U_2\cdots U_{2T}}=0. \tag{3.17a}$$

Substituting this back in (3.16d) gives the desired

kinematical result [compare (2.28)]

$$\partial_{AA'}f^{A'}{}_{;U_1\cdots U_{2T}} - \frac{1}{2T+1}\epsilon_{A\bar{U}_1}\partial_{BB'}f^{B';B}{}_{\bar{U}_2\cdots\bar{U}_{2T}} = 0. \tag{3.17b}$$

To solve this with nonsingular functions we must set

$$f_{A';U_1\cdots U_{2T}} = \partial_{\bar{U}_1A'}\eta_{\bar{U}_2\cdots\bar{U}_{2T}}. \tag{3.18}$$

Then (3.17) imply that η satisfies

$$\Box\eta_{U_2\cdots U_{2T}} + (\partial^{BB'}\ln\rho)\partial_{BB'}\eta_{U_2\cdots U_{2T}} + (\partial^{BB'}\ln\rho)\partial_{\bar{U}_2B'}\eta_{B\bar{U}_3\cdots\bar{U}_{2T}} = 0. \tag{3.19}$$

A last change of variables [compare (2.31a)]

$$\eta = \frac{1}{\rho^{2T}}\chi \tag{3.20}$$

gives

$$\Box\chi_{U_2\cdots U_{2T}} - (2T-1)(\partial^{BB'}\ln\rho)\partial_{BB'}\chi_{U_2\cdots U_{2T}} + (\partial^{BB'}\ln\rho)\partial_{\bar{U}_2B'}\chi_{B\bar{U}_3\cdots\bar{U}_{2T}} = 0, \tag{3.21a}$$

Finally, we make use of the identity

$$\partial_{U_2B'}\chi_{BU_3\cdots U_{2T}} = \partial_{BB'}\chi_{U_2\cdots U_{2T}} + \epsilon_{U_2B}\partial_{CB'}\chi^{C}{}_{U_3\cdots U_{2T}}$$

to rewrite (3.21a) as

$$\Box\chi_{U_2\cdots U_{2T}} - (\partial_{\bar{U}_2B'}\ln\rho)\partial^{CB'}\chi_{C\bar{U}_3\cdots\bar{U}_{2T}} = 0 \tag{3.21b}$$

or

$$\frac{1}{2T-1}\partial_{U_2B'}\partial^{CB'}\chi_{CU_3\cdots U_{2T}} - (\partial_{U_2B'}\ln\rho)\partial^{CB'}\chi_{CU_3\cdots U_{2T}} + \text{permutations} = 0, \tag{3.21c}$$

which is the same as

$$\partial_{U_2B'}\frac{1}{\rho^{2T-1}}\partial^{CB'}\chi_{CU_3\cdots U_{2T}} + \text{permutations} = 0. \tag{3.21d}$$

To summarize, the zero-eigenvalue modes $\psi_{A';U_1\cdots U_{2T}}$ of the Dirac equation for fermions with isospin T have the following form:

$$\psi_{A';U_1\cdots U_{2T}} = \rho^{T}\partial_{U_1A'}\rho^{-2T}\chi_{U_2\cdots U_{2T}} + \text{permutations}, \tag{3.22a}$$

where χ satisfies

$$\partial_{U_2B'}\rho^{-2T+1}\partial^{CB'}\chi_{CU_3\cdots U_{2T}} + \text{permutations} = 0. \tag{3.22b}$$

For isospin $\frac{1}{2}$, this reduces to

$$\psi_{A';U} = \rho^{1/2}\partial_{UA'}\rho^{-1}\chi, \quad \Box\chi = 0 \tag{3.23}$$

which has already been found by conventional methods.[4] For isospin 1, Eqs. (3.22) are equivalent to those of Sec. II. For higher isospins, one needs to solve (3.22b).

ACKNOWLEDGMENT

We were aided enormously in gaining understanding of the general theory by conversations with M. Atiyah and I. Singer, who patiently explained to us their beautiful results.

*This work is supported in part through funds provided by ERDA under Contract No. EY-76-C-02-3069.

† Present address: Department of Physics, Brookhaven National Laboratory, Upton, New York 11973.

[1] A. Belavin, A. Polyakov, A. Schwartz, and Y. Tyupkin, Phys. Lett. 59B, 85 (1975).

[2] R. Jackiw and C. Rebbi, Phys. Rev. D 14, 517 (1976). For further developments, see F. Ore, *ibid*. 15, 470 (1977); S. Chadha, A. D'Adda, P. DiVecchia, and F. Nicodemi, Phys. Lett. 67B, 103 (1977); A. Belavin and A. Polyakov (unpublished); F. Ore, preceding paper, Phys. Rev. D 16, (1977).

[3] R. Jackiw, C. Nohl, and C. Rebbi, Phys. Rev. D 15, 1642 (1977).

[4] B. Grossman, Phys. Lett. A (to be published).

[5] See also L. S. Brown, R. Carlitz, and C. Lee, Phys. Rev. D 16, 417 (1977).

[6] R. Jackiw and C. Rebbi, Phys. Rev. D 14, 517 (1976). F. Ore, *ibid*. 15, 470 (1977); R. Jackiw, C. Nohl, and C. Rebbi, *ibid*. 15, 1642 (1977); B. Grossman, Phys. Lett. A (to be published); R. Jackiw and C. Rebbi, Phys. Lett. 67B, 189 (1977); F. Ore, preceding paper, Phys. Rev. D 16, (1977).

[7] For a recent account and guide to tne earlier mathematical literature, see M. Atiyah, V. Patodi, and I. Singer, Math. Proc. Camb. Philos. Soc. 77, 43 (1975); 78, 405 (1975); 79, 71 (1976).

[8] S. Adler, in *Lectures on Elementary Particles and Quantum Field Theory*, 1970 Brandeis Summer Institute in Theoretical Physics, edited by S. Deser, M. Grisaru, and H. Pendleton (MIT Press, Cambridge, Mass., 1970); S. Treiman, R. Jackiw, and D. Gross, *Lectures on Current Algebra and Its Applications* (Princeton Univ. Press, Princeton, 1972), p. 97.

[9] In the mathematical literature, the resolvent operator is not used in the derivation of the local index theorem, rather the evolution operator is analyzed. In order to see the relation between the two approaches, let us observe that in the equality $2(n_+ - n_-) = \int d^4x\, 2i\mu J_5(x)$ the integral may also be written as Tr $2i\mu\gamma_5 R$, where the Tr symbol refers to a functional trace of the resolvent (2.21a) over all labels, including coordinates. In the notation (2.17) this is also

$$\mathrm{Tr}\left(\frac{2\mu^2}{\mu^2 + LL^\dagger} - \frac{2\mu^2}{\mu^2 + L^\dagger L}\right).$$

The above equality is an example of the more general formula $f(0)\,(n_+ - n_-) = \mathrm{Tr}[f(LL^\dagger) - f(L^\dagger L)]$. Mathematicians choose for f the exponential, so $n_+ - n_- = \mathrm{Tr}[e^{-tLL^\dagger} - e^{-tL^\dagger L}]$, where t is a parameter. Since the left-hand side is independent of t, the right-hand side is evaluated at $t = 0$, with the help of known asymptotic expansions for $\mathrm{Tr}\, e^{-tLL^\dagger}$ at small t (see

Ref. 7). Physicists observe that

$$n_+ - n_- = \mathrm{Tr}\left(\frac{\mu^2}{\mu^2 + LL^\dagger} - \frac{\mu^2}{\mu^2 + L^\dagger L}\right)$$

is independent of μ^2 and evaluate the trace at $\mu^2 = \infty$, where the result is known from the position-space derivation of the anomaly, see Ref. 8. Of course the two approaches are identical, as the following sequence of equations demonstrates:

$$\lim_{\mu^2 \to \infty} \mathrm{Tr}\left(\frac{\mu^2}{\mu^2 + LL^\dagger} - \frac{\mu^2}{\mu^2 + L^\dagger L}\right)$$

$$= \lim_{\mu^2 \to \infty} \mathrm{Tr} \int_0^\infty dT\, e^{-T}\left(e^{-(T/\mu^2)LL^\dagger} - e^{-(T/\mu^2)L^\dagger L}\right)$$

$$= \lim_{t \to 0} \mathrm{Tr}\left(e^{-tLL^\dagger} - e^{-tL^\dagger L}\right).$$

[10] The connection between the anomaly and the zero-eigenvalue modes is implicit in the work of G. 't Hooft, Phys. Rev. Lett. 37, 8 (1976); Phys. Rev. D 14, 3432 (1976). The connection was made explicit by (a) S. Coleman (unpublished), who integrated the anomaly equation (2.23b) over all space to obtain (2.19). For further discussion see (b) A. Schwartz, Phys. Lett. 67B, 172 (1977); (c) J. Kiskis, Phys. Rev. D 15, 2329 (1977); and (d) N. Nielsen and B. Schroer (unpublished).

[11] R. Jackiw and C. Rebbi, Phys. Rev. D 13, 3398 (1976).

[12] M. Atiyah, N. Hitchin, and I. Singer (unpublished).

[13] R. Jackiw and C. Rebbi, Phys. Rev. D 14, 517 (1976); G. 't Hooft, Phys. Rev. D 14, 3432 (1976); S. Chadha, A. D'Adda, P. DiVecchia, and F. Nicodemi, Phys. Lett. 67B, 103 (1977); B. Grossman, Phys. Lett. A (to be published).

[14] R. Jackiw and C. Rebbi, Phys. Lett. 67B, 189 (1977).

[15] D. Freedman and D. Gross (unpublished); S. Chadha, A. D'Adda, P. DiVecchia, and F. Nicodemi, Phys. Lett. 67B, 103 (1977).

[16] G. 't Hooft, talk given at the 1977 Coral Gables Conference (unpublished).

VOLUME 37, NUMBER 1 PHYSICAL REVIEW LETTERS 5 JULY 1976

Symmetry Breaking through Bell-Jackiw Anomalies*

G. 't Hooft†
Department of Physics, Harvard University, Cambridge, Massachusetts 02138
(Received 22 March 1976)

In models of fermions coupled to gauge fields certain current-conservation laws are violated by Bell-Jackiw anomalies. In perturbation theory the total charge corresponding to such currents seems to be still conserved, but here it is shown that nonperturbative effects can give rise to interactions that violate the charge conservation. One consequence is baryon and lepton number nonconservation in $V-A$ gauge theories with charm. Another is the nonvanishing mass squared of the η.

When one attempts to construct a realistic model of nature one is often confronted with the difficulty that most simple models have too much symmetry. Many symmetries in nature are slightly broken, which leads to, for instance, the lepton and quark masses, and *CP* violation. Here I propose to consider a new source of symmetry breaking: the Bell-Jackiw anomaly.

My starting point is the solution of classical field equations given by Belavin *et al.*[1] in four-dimensional (4D) Euclidean gauge-field theories. The solution is obtained from the vacuum by mapping SU(2) gauge transformations onto a large sphere in Euclidean space. Taking the new, gauge-rotated, vacuum as a boundary condition, they obtain a nontrivial solution inside the sphere, characterized by a topological quantum number. If the Lagrangian is

$$\mathcal{L}^{\rm YM} = -\tfrac{1}{4} G_{\mu\nu}{}^a G_{\mu\nu}{}^a, \quad a = 1,2,3,$$
$$G_{\mu\nu}{}^a = \partial_\mu A_\nu{}^a - \partial_\nu A_\mu{}^a + g\epsilon_{abc} A_\mu{}^b A_\nu{}^c, \tag{1}$$

then the topological quantum number is

$$n = (g^2/32\pi^2)\int G_{\mu\nu}{}^a \tilde{G}_{\mu\nu}{}^a \, d^4x, \tag{2}$$

with

$$\tilde{G}_{\mu\nu}{}^a = \tfrac{1}{2}\epsilon_{\mu\nu\alpha\beta} G_{\alpha\beta}{}^a. \tag{3}$$

n is an integer for all field configurations in Euclidean space that have the vacuum (or a gauge transformation thereof) at the boundary. In Minkowsky space n would be i times an integer (if we take d^4x and ϵ_{1234} to be real and A_4, $\partial/\partial x_4$ imaginary).

The solution with $n=1$ in Euclidean space is

$$A_\mu{}^a(x)^{\rm cl} = \frac{2}{g}\frac{\eta_{a\mu\nu}(x-x_0)^\nu}{(x-x_0)^2+\lambda^2}. \tag{4}$$

Here, x_0 is free because of translation invariance and λ is a free scale parameter; η is a tensor that maps antisymmetric representations of SO(4) onto vectors of one of its two invariant subgroups SO(3):

$$\eta_{a\mu\nu} = \epsilon_{a\mu\nu} \quad (a,\mu,\nu = 1,2,3),$$
$$\eta_{a4\nu} = -\delta_{a\nu} \quad (a,\nu = 1,2,3),$$
$$\eta_{a\mu4} = \delta_{a\mu} \quad (a,\mu = 1,2,3), \tag{5}$$
$$\eta_{a44} = 0.$$

Thus isospin is linked to one of the SO(3) subgroups of SO(4). The solution has

$$S = \int \mathcal{L}\{A^{\rm cl}\}\, d^4x = -8\pi^2/g^2. \tag{6}$$

Since we have a 4D rotational symmetry, the solution is not only localized in three-space, but also instantaneous in time. I shall refer to such objects as "Euclidean-gauge solitons,"[2] EGS for short.

There is a simple heuristic argument that explains why these solutions of the Euclidean field equations are relevant for describing a tunneling mechanism in real (Minkowsky) space-time, from one vacuum state to a gauge-rotated vacuum (a gauge rotation that cannot be obtained via a series of infinitesimal gauge rotations). Consider an ordinary quantum mechanical system with a potential barrier V larger than the available energy E, which I put equal to zero. Then the leading exponential of the tunneling amplitude is $\exp(-\int p\,dx)$, with

$$p^2/2m = V - E.$$

This corresponds to the classical equations of motion, except for a sign difference. Thus the leading exponential is obtained by replacing in the equations of motion t by it and computing the action S for a path from one to the other vacuum. [Note that both in Euclidean and in Minkowsky space the gauge group is the compact group SU(2).]

Suppose now that we have in addition N massless fermion doublets coupled to the gauge field:

$$\mathcal{L}^{\rm fermion} = -\sum_{i=1}^{N} \bar{\psi}^i \gamma_\mu D_\mu \psi^i, \tag{7}$$

where

$$D_\mu\psi_i{}^t = \partial_\mu\psi_i{}^t - \tfrac{1}{2}i\tau_{ij}{}^a A_\mu{}^a\psi^t. \qquad (8)$$

I will call the SU(2) index i "color" and the index $t=1,\ldots,N$ "flavor." The vector currents

$$J_\mu{}^{st} = i\bar\psi^s\gamma_\mu\psi^t, \qquad (9)$$

and the traceless part of the axial vector current

$$J_\mu{}^{5st} = i\bar\psi^s\gamma_\mu\gamma_5\psi^t, \qquad (10)$$

are all conserved without anomalies. Thus we have the exact chiral flavor symmetry $SU(N)_L \otimes SU(N)_R \otimes U(1)$. But the current

$$J_\mu{}^5 = \textstyle\sum_t J_\mu{}^{5tt}$$

has an anomaly[3]

$$\partial_\mu J_\mu{}^5 = -i(Ng^2/16\pi^2)G_{\mu\nu}{}^a\tilde G_{\mu\nu}{}^a. \qquad (11)$$

Let us now compare this with Eq. (2).

A configuration in Minkowsky space with $n=1$ would be associated with a violation of axial charge conservation:

$$\Delta Q^5 = 2N. \qquad (12)$$

To calculate the amplitude for such an event directly in Minkowsky space one needs more understanding of the quantum mechanical tunneling from one vacuum to the gauge-rotated vacuum. In practice it is much easier to make use of the explicit solution in Euclidean space. Let us assume then that all Green's functions in Minkowsky space can simply be obtained from the Euclidean ones by analytic continuation.

Let us consider the vacuum-to-vacuum amplitude in Euclidean space, first without, and then with source insertions in the Lagrangian[4]:

$$\langle 0|0\rangle = \int \mathfrak{D}A\,\mathfrak{D}\psi\,\mathfrak{D}\varphi \exp\int\{\mathcal{L}^{\text{gauge}}(A) + \mathcal{L}^{\text{fermion}}(A,\psi) + \mathcal{L}^{\text{fix}}(A) + \mathcal{L}^{\text{ghost}}(A,\varphi)\}d^4x, \qquad (13)$$

where $\mathcal{L}^{\text{fix}}$ fixes the gauge and $\mathcal{L}^{\text{ghost}}$ is the corresponding Faddeev-Popov ghost term; φ is the ghost field. We perform the perturbation expansion around those values of the fields where the exponent is stationary. The solution of Eq. (3) is such a stationary point. Collective coordinates[5] must be introduced for x_0 and λ. The first will lead to energy-momentum conservation in an obvious way; the second might at first sight lead to infinities at both ends of the scale, but there are natural cutoffs, as we will see later.

The arguments that follow now must be considered as a summary of a series of mathematical manipulations needed to compute the wanted amplitudes. Let us expand

$$A_\mu{}^a = A_\mu{}^{a\text{cl}} + A_\mu{}^{a\text{qu}}, \quad \mathcal{L}^{\text{gauge}+\text{fermion}+\text{ghost}} = \mathcal{L}\{A^{\text{cl}}\} - A^{\text{qu}}M_1A^{\text{qu}} - \bar\psi M_2\psi - \varphi^* M_3\varphi$$
$$+ \text{higher orders in the quantum fields.} \qquad (14)$$

It will be very convenient[6] to use the so-called "background gauge":

$$\mathcal{L}^{\text{fix}} = -\tfrac{1}{2}(D_\mu{}^{\text{cl}}A_\mu{}^{qa})^2, \quad D_\mu{}^{ac,\text{cl}} = \partial_\mu\delta^{ac} + g\epsilon_{abc}A_\mu{}^{b,\text{cl}} \qquad (15)$$

Because we introduced collective coordinates, we may restrict the quantum fields to be orthogonal to those values that generate pure translations or dilatations of the classical solution. The amplitude (in the one-EGS sector) is now formally

$$\langle 0|0\rangle = \int d^4x_0\int d\lambda\,(\det J)(\det M_1)^{-1/2}(\det M_2)(\det M_3)\exp\int\mathcal{L}(A^{\text{cl}})d^4x. \qquad (16)$$

Here $(\det J)$ is the Jacobian following from our transition to collective coordinates. If the background gauge is used it turns out to be finite and proportional to λ^{-1}. From Eq. (6) it follows that the exponent equals

$$\exp(-8\pi^2/g^2), \qquad (17)$$

which explains why we get results that are unobtainable through ordinary perturbation expansions with respect to g^2.

The other determinants are in principle obtained by solving the equations

$$M_1A^{\text{qu}} = E_1A^{\text{qu}}, \quad M_2\psi = E_2\psi, \quad M_3\varphi = E_3\varphi. \qquad (18)$$

Now M_1 and M_3 have some zero eigenvalues that neatly cancel. But there are also solutions to

$$M_2\psi = 0, \quad \psi = (1+r^2)^{-3/2}u, \qquad (19)$$

where u is a fixed tensor with Dirac and isospin indices. There is one such solution for each of the N flavors. They are chiral solutions, very much like the fermion bound states described by Jackiw and Rebbi[7] in one and three spacelike dimensions (but stationary in time). These zero eigenvalues are not canceled by anything, so

$\det M_2 = 0$, and the amplitude (13) vanishes. I interpret this result as being a consequence of Eq. (12): We must not sandwich the functional integral expression between two vacuum states, because the initial and final chiral charges Q^5 should be different.

Let us now insert a source term $\bar{\psi}_i^{s} J_{(x)}^{st} \psi_i^{t}$ into the Lagrangian, where $J(x)$ may contain flavor indices and γ matrices, but must be gauge invariant. Now the lowest eigenvalues will become different from zero:

$$M_2 \psi_{(i)}^{s} + J^{st}(x) \psi_{(i)}^{t} = E_{2(i)} \psi_{(i)}^{s}, \quad i = 1, \ldots, N. \tag{20}$$

Using lowest-order perturbation expansion we find that

$$\int d^4x\, \psi_{(i)}^{*s} J^{st}(x) \psi_{(j)}^{t} = E_{(j)} \int \psi_{(i)}^{*s} \psi_{(j)}^{s} d^4x$$
$$= E_{(i)} \delta_{ij}, \quad i, j = 1, \ldots, N,$$

where $\psi_{(i)}^{s} = \psi_0 a_{(i)}^{s}$, and ψ_0 is the zero-"energy" solution for any flavor, and $a_{(i)}^{s}$ are coefficients. Thus

$$\det(M_2 + J) = \prod_{(i)} E_{(i)} \simeq \det_{st} \int \psi_0^* J^{st}(x) \psi_0 d^4x$$
$$\times \{\int \psi_0^* \psi_0 d^4x\}^{-1}. \tag{21}$$

Substituting the known form of ψ_0, Eq. (19), we find that Eq. (21) goes like

$$\prod_{i=1}^{N} (x_i - x_0)^{-6} J(x_i)$$

for large distances, and only the $1 - \gamma_5$ part of J^{st} is selected out. We ask now for an effective vertex that could mimic the same amplitude (neglecting the finite size of the EGS) and we find

$$\mathcal{L}^{\text{eff}} = C g^{-8} \exp(-8\pi^2/g^2) \mathcal{L}^f + \text{H.c.}, \tag{22}$$

where $\mathcal{L}^f$ is a $2N$ fermion interaction that has the chiral transformation properties of

$$\det_{st} \bar{\psi}^s (1 + \gamma_5) \psi^t$$

but the isospin indices are arranged in a more general way. The factor

$$\prod_{i=1}^{N} (x_i - x_0)^{-6},$$

is exactly reproduced by the $2N$ propagators that connect the sources with the EGS (Fig. 1).

Note that the sources have to switch chirality. This explains why the instanton gives $\Delta Q^5 = 2N$. I have found that the constant in Eq. (22) can be computed analytically to zeroth order in g. The calculation is lengthy and will be discussed in a separate publication.

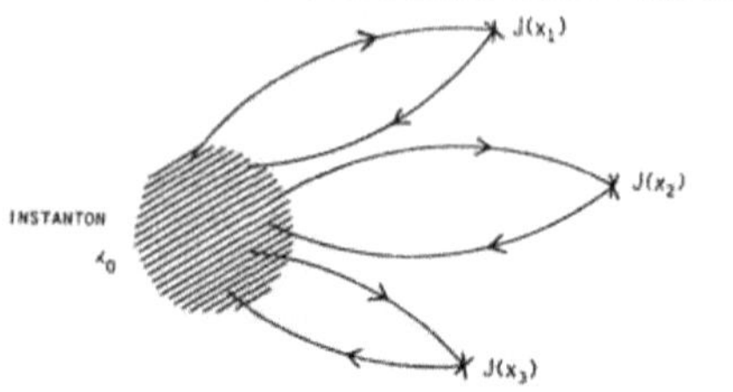

FIG. 1. The case of three flavors. The sources J turn the axial charge Q^5 into $-Q^5$ for each flavor. The amplitude goes like J^3.

The integration over the collective coordinate $x_{0\mu}$ in Eq. (16) simply implies that the effective action is Eq. (22) integrated over space and time, so that we get energy-momentum conservation. The integration over λ is finite if the Callan-Symanzik β function for g is negative and if also the ir divergence is cut off by the Higgs mechanism.

A notable application is the case of the Weinberg-Salam SU(2)⊗U(1) model in an often cited form. The leptons are e_R, (ν_L^e, e_L), μ_R, (ν_L^μ, μ_L); and the quarks are

$$\begin{matrix} u_R^r, & u_R'^r, & (u_L, d_L^C)^r \\ u_R^r, & s_R^r, & (u_L', s_L^C)^r \end{matrix} \tag{23}$$

where r denotes red, yellow, or blue, and C denotes Cabibbo rotation. All currents that are coupled to gauge fields are anomaly free, hence the model is renormalizable. But the baryon and lepton currents have anomalies. We find that one Euclidean-gauge soliton gives

$$\begin{aligned} \Delta E &= \Delta M = 1, \\ \Delta u + \Delta d^C &= 3, \\ \Delta u' + \Delta s^C &= 3, \end{aligned} \tag{24}$$

where E is the electron number, M is the muon number, and u, u', d^C, and s^C are the numbers of the corresponding (Cabibbo rotated) quarks. Thus, because of the Cabibbo rotation, a proton and a neutron (two baryons equal six quarks) may annihilate to form two antileptons, one of electron and one of muon type.

The factors $\exp(-16\pi^2/g^2) = \exp(-4\pi \times 137 \times \sin^2\theta_W)$ in the cross sections and lifetimes will make none of the predicted effects observable if the Weinberg angle is not very small.

In a color gauge theory for strong interactions with two massless quark triplets, Eq. (22) is an effective four-fermion interaction with exactly the chiral quantum numbers of a mass term for the η particle. Comparison with the actual value of the η mass is not yet possible because of the infrared divergences in that theory.

The author would like to thank S. Coleman, R. Jackiw, S. L. Glashow, and all other physicists of the Harvard theory group for discussions.

Note added.—In Eq. (22) the isospin indices have been suppressed. The isospin structure of this expression, however, is more complicated than the compact notation suggests. There is also a power of the coupling constant g in front of the exponent. The full details will be published soon.

---

*Work supported in part by the National Science Foundation under Grant No. MPS75-20427.

†On leave from the University of Utrecht, Utrecht, The Netherlands.

[1]A. A. Belavin *et al.*, Phys. Lett. 59B, 85 (1975).

[2]The term "instanton" has been occasionally used elsewhere, and is used in Fig. 1 here.

[3]J. S. Bell and R. Jackiw, Nuovo Cimento 60A, 47 (1969); S. L. Adler, Phys. Rev. 117, 2426 (1969).

[4]L. D. Faddeev and V. N. Popov, Phys. Lett. 25B, 29 (1967); G. 't Hooft and M. Veltman, Nucl. Phys. B50, 318 (1972).

[5]J. L. Gervais and B. Sakita, Phys. Rev. D 11, 2943 (1975); E. Tomboulis, Phys. Rev. D 12, 1678 (1975).

[6]J. Honerkamp, Nucl. Phys. B48, 269 (1972), and in *Renormalization of Yang-Mills Fields and Applications to Particle Physics, Marseille, 19–23 June 1972*, edited by C. P. Korthals Altes (Centre de Physique Théorique, Centre Nationale de la Recherche Scientifique, Marseille, 1972).

[7]R. Jackiw and C. Rebbi, Phys. Rev. 13, 3398 (1976).

Volume 117B, number 5 PHYSICS LETTERS 18 November 1982

AN SU(2) ANOMALY

Edward WITTEN [1]

Joseph Henry Laboratories, Princeton University, Princeton, NJ 08544, USA

Received 14 July 1982

A new restriction on fermion quantum numbers in gauge theories is derived. For instance, it is shown that an SU(2) gauge theory with an odd number of left-handed fermion doublets (and no other representations) is mathematically inconsistent.

It has been a long-standing puzzle to elucidate the properties of an SU(2) gauge theory with a single doublet of left-handed (Weyl) fermions. This theory defies simple phenomenological descriptions. There is no obvious attractive channel in which a fermion condensate could form, consistent with Fermi statistics and Lorentz invariance. But it is hard to believe that the fermions could remain massless in the presence of strong SU(2) gauge forces at long distances.

This puzzle persists (in the absence of other representations) whenever the number of elementary fermion doublets is odd. An even number of doublets, even if they have zero bare mass, could pair up and become massive Dirac fermions through spontaneous chiral symmetry breaking. With an odd number of elementary doublets, however, there would always be one massless doublet left over after any assumed chiral symmetry breaking, as long as the SU(2) gauge symmetry remains unbroken.

Of course, there is no real paradox here. Perhaps our heuristic pictures of strongly interacting gauge theories are inadequate. However, the facts noted above do suggest that something is strange about an SU(2) gauge theory with an odd number of elementary doublets. The purpose of this paper is to determine precisely what is strange about these theories; we will see that they are mathematically inconsistent! The inconsistency arises from a problem somewhat analogous to the Adler–Bell–Jackiw anomaly.

[1] Supported in part by the National Science Foundation under Grant No. PHY80-19754.

Although a hamiltonian approach exists, let us first look at this problem from the point of view of euclidean functional integrals. The starting point is the fact [1] that the fourth homotopy group of SU(2) is nontrivial,

$$\pi^4(\mathrm{SU}(2)) = Z_2. \tag{1}$$

[Note that we are dealing with the *fourth* homotopy group, while the *third* homotopy group, $\pi^3(\mathrm{SU}(2)) = Z$, has entered in instanton studies [2]. The analogue of π^4 has entered in some recent studies [3] of 2 + 1 dimensional models.] Eq. (1) means that in four-dimensional euclidean space, there is a gauge transformation $U(x)$ such that $U(x) \to 1$ as $|x| \to \infty$, and $U(x)$ "wraps" around the gauge group in such a way that it cannot be continuously deformed to the identity. The fact that the homotopy group is Z_2 means that a gauge transformation that wraps twice around SU(2) in this way can be deformed to the identity. We will not need the detailed form of $U(x)$.

The existence of the topologically non-trivial mapping $U = U(x)$ means that when we carry out the euclidean path integral

$$\int (\mathrm{d}A_\mu) \exp\left(-\frac{1}{2g^2}\int \mathrm{d}^4x \,\mathrm{tr}\, F_{\mu\nu}F^{\mu\nu}\right), \tag{2}$$

we are actually double counting. For every gauge field A_μ, there is a conjugate gauge field

$$A_\mu^U = U^{-1}A_\mu U - \mathrm{i}U^{-1}\partial_\mu U,$$

which makes exactly the same contribution to the functional integral. There is no way to eliminate this double counting because A_μ and A_μ^U lie in the same sector of field space; A_μ^U can be reached continuously from A_μ without passing through singularities or infinite action barriers. But, in the absence of fermions, the double counting is harmless and cancels out when one calculates vacuum expectation values.

Now let us include fermions. Introducing, say, a single doublet of left-handed fermions, we now must deal with

$$Z = \int d\psi\, d\bar{\psi} \int dA_\mu$$

$$\times \exp\left(-\int d^4x\,[(1/2g^2)\,\mathrm{tr}\,F_{\mu\nu}^2 + \bar{\psi} i \not{D}\psi]\right). \quad (3)$$

We would like to integrate out the fermions and discuss the effective theory with the fermions eliminated.

As is well known, for a theory with a doublet of *Dirac* fermions, the basic integral is

$$\int (d\psi\, d\bar{\psi})_{\mathrm{Dirac}} \exp(\bar{\psi} i\not{D}\psi) = \det i\not{D}. \quad (4)$$

Here the right-hand side is, formally, the infinite product of all eigenvalues of the hermitian operator $i\not{D} = i\gamma_\mu D^\mu$. Certain theories – those that are afflicted with Adler–Bell–Jackiw anomalies – are ill-defined because it is impossible to renormalize this formal product so as to get a gauge invariant answer. However, a doublet of Dirac fermions could have a gauge invariant bare mass; this means that Pauli–Villars regularization is available, and hence that the determinant in (4) can be defined satisfactorily. This determinant is completely gauge invariant – invariant both under infinitessimal gauge transformations and under the topologically non-trivial gauge transformation U discussed earlier.

Now, with the gauge group SU(2), a doublet of Dirac fermions is exactly the same as two left-handed or Weyl doublets. Hence the fermion integration with a single Weyl doublet would give precisely the square root of (4):

$$\int (d\psi\, d\bar{\psi})_{\mathrm{Weyl}} \exp(\bar{\psi} i\not{D}\psi) = (\det i\not{D})^{1/2}. \quad (5)$$

But an ambiguity arises here; the square root has two signs. As we will see, this ambiguity leads to trouble.

Picking a particular gauge field A_μ, we are free to define in an arbitrary way the sign of $(\det i\not{D})^{1/2}$ for this field. Once this is done, there is no further freedom; to satisfy the Schwinger–Dyson equations we must define the fermion integral $(\det i\not{D})^{1/2}$ to vary smoothly as A_μ is varied.

Defined in this way $(\det i\not{D})^{1/2}$ is certainly invariant under infinitessimal gauge transformations – since the sign does not change abruptly. But nothing guarantees that $(\det i\not{D})^{1/2}$ is invariant under the topologically non-trivial gauge transformation U. In fact, as we will see, $(\det i\not{D})^{1/2}$ is odd under U. We will see that for any gauge field A_μ,

$$[\det i\not{D}(A_\mu)]^{1/2} = -[\det i\not{D}(A_\mu^U)]^{1/2}. \quad (6)$$

In other words, if one continuously varies the gauge field from A_μ to A_μ^U, one ends up with the opposite sign of the square root.

Before explaining why eq. (6) is valid, let us first discuss why it results in the mathematical inconsistency of the SU(2) theory with a single left-handed doublet. The partition function would be

$$Z = \int dA_\mu (\det i\not{D})^{1/2} \exp\left(-\frac{1}{2g^2}\int d^4x\,\mathrm{tr}\,F_{\mu\nu}^2\right). \quad (7)$$

But this vanishes identically, because the contribution of any gauge field A_μ is exactly cancelled by the equal and opposite contribution of A_μ^U! Likewise the path integral Z_X with insertion of any gauge invariant operator X is identically zero. So expectation values are indeterminate, $\langle X\rangle = Z_X/Z = 0/0$. For this reason, the theory is ill-defined.

One cannot avoid this problem by taking the absolute value of $(\det i\not{D})^{1/2}$; the resulting theory would not obey the Schwinger–Dyson equations. Nor can one consistently integrate over only "half" of field space, since A_μ and A_μ^U are continuously connected.

It remains to explain eq. (6). For convenience, take space–time to be a sphere of large volume so that the spectrum of $i\not{D}$ is discrete. We may as well assume there are no zero eigenvalues since otherwise $\det i\not{D}(A_\mu)$ vanishes and (6) is certainly true. The eigenvalues of $i\not{D}$ are real and (fig. 1) for every eigenvalue λ there is an eigenvalue $-\lambda$, since if $i\not{D}\psi = \lambda\psi$, then $i\not{D}(\gamma_5\psi) = -\lambda(\gamma_5\psi)$.

Taking the square root of $\det i\not{D}$ means that we want the product of only half of the eigenvalues, not all of them. We may suppose that for every pair of eigenvalues $(\lambda, -\lambda)$ we pick one or the other, but not both. For instance, for a particular gauge field A_μ we

Volume 117B, number 5 PHYSICS LETTERS 18 November 1982

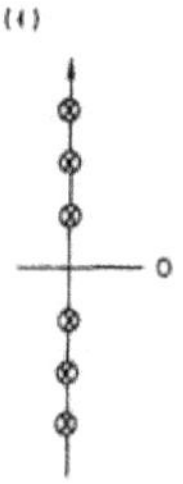

Fig. 1. The spectrum of the Dirac operator for a particular gauge field A_μ. The square root of the determinant may be defined – for this particular gauge field – as the product of the positive eigenvalues.

may define $(\det i\not{D})^{1/2}$ to be the product of the positive eigenvalues (fig. 1).

Now imagine varying the gauge field along a continuous path in field space from A_μ to A_μ^U. For instance, one may consider the gauge field $A_\mu^t = (1-t)A_\mu + tA_\mu^U$, with t varied smoothly from zero to one. Let us follow the flow of the eigenvalues as a function of t. The spectrum of $i\not{D}$ is precisely the same at $t = 1$ as it is at $t = 0$. However, the individual eigenvalues may rearrange themselves as t is varied from zero to one.

As will be explained, the Atiyah–Singer index theorem predicts that such a rearrangement occurs. The simplest rearrangement allowed by the theorem is indicated in fig. 2. A single pair of eigenvalues $\{\lambda(t), -\lambda(t)\}$ cross at zero and change places as t is varied form zero to one.

In particular, one of the eigenvalues which was pos-

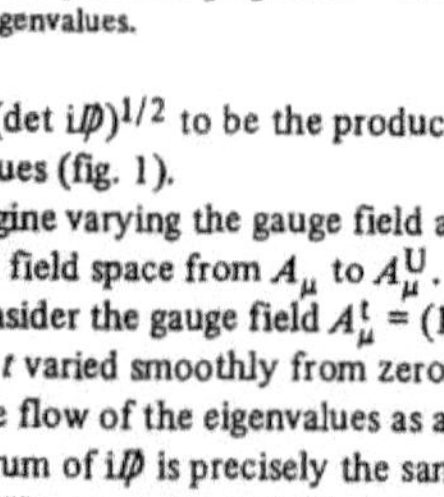

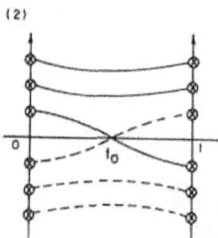

Fig. 2. The flow of eigenvalues as the gauge field is varied from A_μ (left of drawing) to A_μ^U (right of drawing). The square root of the determinant may be defined as the product of the eigenvalues indicated by solid lines; it vanishes and changes sign at $t = t_0$.

itive at $t = 0$ is negative by the time $t = 1$. If at $t = 0$, $(\det i\not{D})^{1/2}$ was defined as the product of the positive eigenvalues, then, following the eigenvalues continuously, by the time we reach $t = 1$ $(\det i\not{D})^{1/2}$ is the product of many positive eigenvalues and a single negative one. This means that $(\det i\not{D})^{1/2}$ has the opposite sign at $t = 1$ from its value at $t = 0$. The square root vanishes when the eigenvalue pair passes through zero ($t = t_0$ in fig. 2) and is negative for $t > t_0$.

The Atiyah–Singer theorem permits more complicated rearrangements of eigenvalues, but the number of positive eigenvalues that become negative as t is varied from 0 to 1 is always *odd*. This is the basis for eq. (6).

The connection between the index theorem and the flow of eigenvalues is well known in mathematics [4] and has been discussed in the physics literature [5]. What is relevant for our problem is a slightly exotic form of the index theorem, namely the mod two index theorem for a certain *five*-dimensional Dirac operator [6].

Consider the five-dimensional Dirac equation for an SU(2) doublet of fermions,

$$\not{D}^{(5)}\Psi = \sum_{i=1}^{5} \gamma^i \left(\partial_i + \sum_{a=1}^{3} A_i^a T^a\right)\Psi = 0. \tag{8}$$

The spinor Ψ has eight components because the spinor representation of O(5) is four dimensional while an SU(2) doublet has two components.

The spinor representation of O(5) is pseudo-real, rather than real, and the doublet of SU(2) is likewise pseudo-real. But the tensor product of the spinor representation of O(5) with the doublet of SU(2) is a *real* representation of O(5) × SU(2). This means that in (8), we can take the gamma matrices γ^i to be real, symmetric 8 × 8 matrices while the anti-hermitian generators T^a of SU(2) are real, anti-symmetric matrices [‡1].

The five-dimensional Dirac operator $\not{D}^5$ for an SU(2) doublet is therefore a real, antisymmetric opera-

[‡1] In fact, one can arrange Ψ as a two-component column vector of quaternions $\Psi = \binom{\psi_1}{\psi_2}$ – which would have eight real components. Such a column vector can be acted on from the left by a 2 × 2 unitary matrix of quaternions [making the group Sp(2) or O(5)] and on the right by a unitary quaternion [the group Sp(1) or SU(2)]. This is the desired eight-dimensional real representation of O(5) × SU(2).

Volume 117B, number 5 PHYSICS LETTERS 18 November 1982

(3)

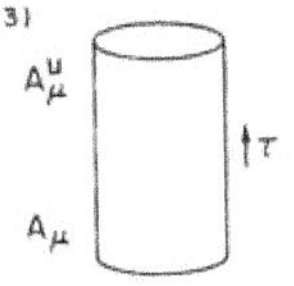

Fig. 3. A five-dimensional cylinder, $S^4 \times R$, on which an instanton-like gauge field is defined.

tor, acting on an infinite dimensional space. The eigenvalues of such a real, antisymmetric operator either vanish or are imaginary and occur in complex conjugate pairs. When the gauge field A_μ is varied, the number of zero eigenvalues of $\not{D}^5$ can change only if a complex conjugate pair of eigenvalues moves to – or away from – the origin. The number of zero eigenvalues of $\not{D}^5$ modulo two is therefore a topological invariant. It is known as the mod two index of the Dirac operator.

Now, consider a five dimensional cylinder $S^4 \times R$ (fig. 3). Let $x^\mu, \mu = 1, ..., 4$, be coordinates for S^4 while the position in the "time" direction (position in R) is called τ. Consider the following five-dimensional instanton-like SU(2) gauge field. For all x^μ and τ, $A_\tau = 0$. But $A_\mu(X^\sigma, \tau), \mu = 1, ..., 4$, is – for each τ – a four-dimensional gauge field described as follows. For $\tau \to -\infty$ $A_\mu(x^\sigma, \tau)$ approaches the four-dimensional gauge field A_μ of our previous discussion ($t = 0$ in fig. 2). For $\tau \to +\infty$ $A_\mu(x^\sigma, \tau)$ approaches what we previously called A^U_μ ($t = 1$ in fig. 2). As τ varies from $-\infty$ to $+\infty$, $A_\mu(X^\sigma, \tau)$ varies *adiabatically* from A_μ to A^U_μ, along the same path in field space considered in fig. 2.

The mod two Atiyah–Singer index theorem predicts that the number of zero modes in this five-dimensional gauge field is odd – equal to one modulo two [‡2]

On the other hand, the number of zero modes of $\not{D}^5$, modulo two, can be calculated in terms of the eigenvalue flow of fig. 2. The Dirac equation $\not{D}^5\Psi = 0$ can be written

$$d\Psi/d\tau = -\gamma^\tau \not{D}^4 \Psi, \tag{9}$$

[‡2] Actually, in a special case one can easily find the zero mode. If one conformally compactifies $S^4 \times R$ to the five-sphere S^5, then on S^5 one can choose the instanton field to be invariant (up to a gauge transformation) under an SU(3) subgroup of the symmetry group O(6) of the five-sphere. The fermion zero mode is then the unique SU(3) invariant spinor field that can be defined.

where $\not{D}^4 = \Sigma^4_{i=1}\gamma^\mu D_\mu$ is – at each τ – a four-dimensional Dirac operator.

Since $A_\mu(x^\sigma, \tau)$ evolves adiabatically in τ, (9) can be solved in the adiabatic approximation. We write $\Psi(x^\mu, \tau) = F(\tau)\,\phi^\tau(x^\mu)$, where $\phi^\tau(x^\mu)$ is a smoothly evolving solution of the eigenvalue equation

$$\gamma^\tau \not{D}^4 \phi^\tau(x^\mu) = \lambda(\tau)\phi^\tau(x^\mu). \tag{10}$$

The eigenvalues $\lambda(\tau)$ evolve on the curves of fig. 2 ($i\not{D}^4$ and $\gamma^\tau \not{D}^4$ have the same spectrum). In the adiabatic limit, eq. (9) now reduces to $dF/d\tau = -\lambda(\tau)F(\tau)$, and the solution is

$$F(\tau) = F(0)\exp\left(-\int_0^\tau d\tau'\,\lambda(\tau')\right). \tag{11}$$

This is normalizable only if λ is positive for $\tau \to +\infty$, and negative for $\tau \to -\infty$.

In the adiabatic approximation, the number of zero eigenvalues of $\not{D}^5$ is therefore equal to the number of eigenvalue curves in fig. 2 which pass from negative to positive values (or from positive to negative values) between $t = 0$ and $t = 1$. When corrections to the adiabatic approximation are considered, this gives the number of zero eigenvalues modulo two.

The Atiyah–Singer theorem, which requires that $\not{D}^5$ has an odd number of zero eigenvalues, therefore implies that the number of eigenvalue curves that pass from positive to negative values in fig. 2 is odd. This is precisely what we needed to show that $(\det i\not{D})^{1/2}$ is odd under the topologically non-trivial gauge transformation U.

Now let us consider some generalizations. With n left-handed fermion doublets, the fermion integration would give $(\det i\not{D})^{n/2}$. If n is even, the sign of the square root does not matter, but if n is odd, the theory suffers from the same inconsistency as before.

This persists even if additional gauge or Yukawa couplings are added to an SU(2) gauge theory. Since the fermion integration is necessarily either even or odd under U, if it is odd in a pure SU(2) gauge theory, it remains odd if additional gauge or Yukawa couplings are smoothly switched on. In particular, the standard $SU(3) \times SU(2) \times U(1)$ model of strong, weak, and electromagnetic interaction would be inconsistent if the number of left-handed fermion doublets were odd.

If one considers theories with SU(2) representations of isospin bigger than one half, the Atiyah–Singer

Volume 117B, number 5 PHYSICS LETTERS 18 November 1982

theorem gives the following result. If one normalizes the SU(2) generators conventionally so that tr $T_3^2 = 1/2$ in the doublet representation, then the fermion integration is even or odd depending on whether tr T_3^2, evaluated among all the left-handed fermions, is an integer or half-integer. The inconsistent theories are those where tr T_3^2 is a half-integer. (In an ordinary instanton field, the number of fermion zero modes is 2 tr T_3^2, so the inconsistent theories are precisely those with an odd number of fermion zero modes in an instanton field.)

Considering gauge groups other than SU(2), we have

$$\pi^4(\mathrm{SU}(N)) = 0, \quad N > 2,$$

$$\pi^4(\mathrm{O}(N)) = 0, \quad N > 5,$$

$$\pi^4(\mathrm{Sp}(N)) = Z_2, \quad \text{any } N. \tag{12}$$

Thus non-trivial conditions arise only for Sp(N) groups.

Finally, let us note how this appears in a hamiltonian framework. Space permits only a brief statement of results.

From a hamiltonian viewpoint, one introduces the group G consisting of all gauge transformations $U(x, y, z)$ defined in *three*-dimensional space such that $U(x) \to 1$ as $|x| \to \infty$.

The fact that $\pi^4(\mathrm{SU}(2)) = Z_2$ means that $\pi^1(\mathrm{G}) = Z_2$. For the topologically non-trivial gauge transformation $U = U(x, y, z, t)$ that we have discussed is – at each t – an element of G. At $t \to -\infty$ or $t \to +\infty$ it is the identity in G; varying t from $-\infty$ to $+\infty$, U describes a loop in G which cannot be deformed away.

In canonical quantization, one encounters operators

$$Q^a(x) = g^{-2} D_i F^a_{0i}(x) - \bar{\psi}\gamma^0 T^a \psi, \tag{13}$$

which are generators of the Lie algebra of G. However, when a group – in this case G – is not simply connected, a representation of the Lie algebra does not necessarily provide a representation of the group. In general one gets a representation only of the simply connected covering group $\bar{\mathrm{G}}$. Since $\pi^1(\mathrm{G}) = Z_2$, the center of $\bar{\mathrm{G}}$ has a single non-trivial element P.

In quantum field theory, P – being in the center of $\bar{\mathrm{G}}$ – commutes with all fields and therefore is a c-number. Since $P^2 = 1$, we must have $P = +1$ or $P = -1$ (as an operator statement) in any given field theory. The theories in which the fermion integration is odd under U are the theories in which $P = -1$.

Theories with $P = -1$ are inconsistent for the following reason. According to Gauss's law, physical states $|\psi\rangle$ must be gauge invariant, obeying $Q^a|\psi\rangle = 0$ and hence $P|\psi\rangle = |\psi\rangle$. If $P = -1$, there are no states in the entire Hilbert space that obey Gauss's law.

Similar behavior can be seen in the models of ref. [3] by means of canonical quantization. This was one motivation for the present work.

I would like to thank S. Coleman and H. Georgi for discussions of the SU(2) theory with an odd number of doublets, I. Affleck and J. Harvey for discussions of fermion integration, and R. Jackiw for discussions about the models of ref. [3]. I also wish to thank W. Browder and J. Milnor for discussions about topology.

References

[1] S.T. Hu, Homotopy theory (Academic Press, New York, 1959) Ch. 11;
H. Toda, Composition methods in homotopy groups of spheres (Princeton U.P., Princeton, 1962).

[2] A.A. Belavin, A.M. Polyakov, A. Schwarz and Y. Tyupkin, Phys. Lett. 59B (1975) 85;
C. Callan R. Dashen and D.J. Gross, Phys. Lett. 63B (1976) 334;
R. Jackiw and C. Rebbi, Phys. Rev. Lett. 37 (1976) 334.

[3] S. Deser, R. Jackiw and S. Templeton, MIT preprint (1982).

[4] M.F. Atiyah, V. Patodi and I. Singer, Math. Proc. Camb. Philos. Soc. 79 (1976) 71.

[5] C.G. Callan, R. Dashen and D.J. Gross, Phys. Rev. D17 (1978) 2717;
J. Kiskis, Phys. Rev. D18 (1978) 3061.

[6] M.F. Atiyah and I.M. Singer, Ann. of Math. 93 (1971) 119.

VI. CONSTRAINED INSTANTONS, I–A INTERACTION AND THE VALLEY METHOD

INTRODUCTION

In numerous applications one often encounters situations in which the field configuration considered is *not* the exact solution of the equations of motion. Historically, one of the first examples of this type is an analog ensemble of the I/A "atoms" introduced in Ref. [1], the so-called dilute instanton gas (I stands for the instanton and A for the anti-instanton). Each given pseudoparticle in the ensemble is affected by the presence of others, and it is quite obvious that the topologically neutral gas cannot, strictly speaking, correspond to a minimum of the action and, hence, is not the exact solution. In other words the action of n pseudoparticles in the I/A ensemble $S_n \neq (8\pi^2 n/g^2)$; instead,

$$S_n = \frac{8\pi^2}{g^2} n - S_{int} \ ,$$

where generically the interaction "energy" $S_{int} > 0$. (From now on we will use the terminology of a static problem in four-dimensional space; the action will be reinterpreted as the energy, and the word "pseudoparticle" becomes appropriate.)

Another example of a similar type is the instanton in theories with the spontaneously broken gauge symmetry [2]. In the Higgs phase, the vacuum expectation value of the Higgs field v generates the mass term for the gauge field. As a result, the BPST solution is not the exact solution any more. It remains approximately valid only at distances $x \ll m_W^{-1}$, where we also assume that $\rho \ll m_W^{-1}$; at $x > m_W^{-1}$, all fields in the massive theory must decay exponentially, of course, unlike powers of x^{-2} characteristic to the BPST expression.

Thus, the question arises as to whether one can make quantitative the notion of proximity of the given field configuration to the exact solution and, if yes, what small parameter measures this proximity. A related crucial question is as follows: "If the selection criteria are relaxed and the functional integral is not represented exclusively by the stationary points of the classical action, how far can one distance oneself from the exact solutions?"

It is intuitively clear that, say, the small size instanton, $\rho \ll m_W^{-1}$, is a legitimate contribution. Likewise, in pure gluodynamics, well-separated I–A pairs should be included in the partition function. Unfortunately this intuitive feeling of what is relevant and what is not is not easy to formalize. The understanding existing in current literature is heuristic at best, and since we agreed to avoid controversial issues in this volume, we will focus only on the first question. Namely, this section presents a discussion of the approximate solutions when they are arbitrarily close to the exact ones; the question of their fate with the worsening of the proximity parameter is left aside. However, some marginal remarks on the relevance of "nearly" solutions will be presented later on.

The approach allowing one to deal with approximate solutions is usually referred to in literature as "constrained instantons". Assume we have a field configuration $\phi(\beta)$ continuously depending on some parameter(s) β; ϕ is a generic notation for the set of all relevant fields. Assume that at $\beta \to \beta_0$, the field ϕ tends to become an exact solution. Let us call $\beta = \beta_0$ the limiting point. In the vicinity of the limiting point, $\phi(\beta)$ is almost an exact solution. The fact that it is still not the exact solution means that there exist such deformations of ϕ that decrease the action Typically, such "decreasing" deformations are characteristic to one or at most, to several directions in the functional space; we call them destabilizing directions. Deformations along all other directions increase the action ("energy"). The basic idea of the constrained instantons is as follows. One introduces a constraint in the functional integration measure in such a way as to lock up all destabilizing directions. Then one minimizes the action subject to this constraint. Only those variations of ϕ which go in the directions perpendicular to the destabilizing ones are allowed. In this way one arrives at the constrained instanton. Dynamics in the destabilizing directions is studied separately.

To make graphic a simple physical picture lying behind the program, let us turn to the example of the instanton in the Higgs phase. At $\rho \to 0$, the BPST instanton becomes the exact solution. This is the limiting point which one can approach arbitrarily closely. If $\rho \neq 0$, there exists one direction in the functional space along which the action ("energy") slowly decreases. This direction corresponds to rescaling the instanton solution as a whole to smaller radii. By imposing a constraint, we forbid ourselves to move in the functional space along this direction. In the orthogonal subspace, any deformation of the field configuration only increases the action, so it is possible to find one which minimizes the action, the constrained instanton. We then calculate the contribution of the constrained instanton in physically observable quantities. At the last stage we eliminate the constraint by integrating the result over all possible values of ρ.

The idea of the constrained instanton was first put forward in Ref. [3]. The method was later developed by Affleck [4] whose work is reproduced here. It is worth noting that there is no unambiguous prescription as to how one can choose the constraint. Usually, in each particular problem, the most convenient and adequate choice is pretty clear from the physical context. For instance, in the example above, the appropriate constraint must fix the size of the instanton. The classical equations of motion are changed, of course, once the constraint is introduced. New equations do have a solution which at small x, behaves like that of Belavin *et al.*, while at large distances, it decays exponentially, $\mathcal{O}(\exp(-m_W x))$ for the gauge field and $\mathcal{O}(\exp(-m_H x))$ for the Higgs field. The corresponding action now becomes ρ-dependent and is readily calculable in the form of expansion in $\rho^2 v^2$,

$$S = \frac{8\pi^2}{g^2} + 2\pi^2\rho^2 v^2 + \dots \,. \tag{1}$$

It is important that the ρ-dependent terms are $\mathcal{O}(g^2)$ compared to the BPST term, so that the ρ dependence is indeed weak provided that $\rho \ll m_W^{-1}$.

As a matter of fact the term $\rho^2 v^2$ in Eq. (1) was obtained by 't Hooft [2] without explicitly invoking the constrained instanton technique, a fact explanable by some specific features of the instanton expression for the scalar field. The heavy artillery of the constrained instantons becomes relevant at the next-to-leading order.

A version of the constrained instanton technique receives much attention in current literature in connection with approximate solutions, the valley or streamline method [5] (see also Ref. [6]; for recent reviews referring to the Yang–Mills theory, see e.g. Refs. [7, 8]). The valley method is a variant of the constrained instanton approach, with a specific prescription as to how one should choose the constraints. It is most suitable to the problem of I–A pairs. Let us sketch a transparent physical picture lying behind the valley method in the simplest example, one I–A pair.

We start from a pair of pseudoparticles at very large separations, large compared to their sizes which for simplicity are assumed to coincide for both pseudoparticles. This configuration — our boundary condition — is not an exact solution but is arbitrarily close to that. The field equations will experience a force tending to change the field in the direction of lowering the total energy of two pseudoparticles. Let us do a *gedankenexperiment* — introduce an auxiliary fifth coordinate, "fifth time", and trace the evolution of the original configuration in the fifth time. In other words, our I and A atoms are set free to move as they wish at $t_5 = 0$.

To visualize the picture further it is instructive to discretize the fifth time. Then at step zero, we have $\phi_0 = \phi_I^{R/2} + \phi_A^{-R/2}$, where the superscripts indicate the position of the I (A) centers.

At step one the field is deformed. The variation of ϕ is obtained in the following way. Let us try to change ϕ along different directions in the functional space. Almost every such attempt will lead to a higher energy, with an exception of, say, one direction [9] where the energy of the field configuration $\phi_1 = \phi_0 + \delta\phi_1$ is smaller than that of ϕ_0. At step two we take ϕ_1 as an input and repeat the procedure. In this way we get a chain of field configurations ϕ_i which becomes, in the limit of the continuous fifth time, a one-parametric family $\phi(\beta)$, the bottom of the valley.

In the quantitative terms the bottom of the valley is defined by the following requirement: as we move along the bottom $\phi(\beta)$ at each point the variation of the coordinate of the bottom must be proportional to the force at the given point,

$$\frac{\partial\phi(\beta)}{\partial\beta} \propto \frac{\delta S}{\delta\phi}\Big|_{\phi=\phi(\beta)} \tag{2}$$

The proportionality coefficient is, generally speaking, a function of β sensitive to particular parametrizations of the bottom of the valley. (The coordinates along the bottom can be introduced in different ways.)

If further clarification is needed it may be instructive to turn from the infinite-dimensional functional space to the mechanical motion of a stream or a brook (with large friction) on a two-dimensional surface with a trough (Fig. 1) in the gravitational field. The bottom of the valley in this case is a one-dimensional curve

[9] It may well happen that the number of such directions is more than one; for clarity, we confine ourselves to the simplest case.

in the three-dimensional space, $\vec{x}(\beta)$. The force at the bottom is the gradient of the potential energy V, $F_i = \partial_i V(\vec{x}) \mid_{\vec{x}=\vec{x}(\beta)}$. The velocity of the stream in the given point at the bottom $\dot{\vec{x}}$ is proportional to the force,

$$\dot{\vec{x}} = \frac{\partial \vec{x}}{\partial \beta} \dot{\beta} \propto \partial_i V \ ,$$

full analog of Eq. (2).

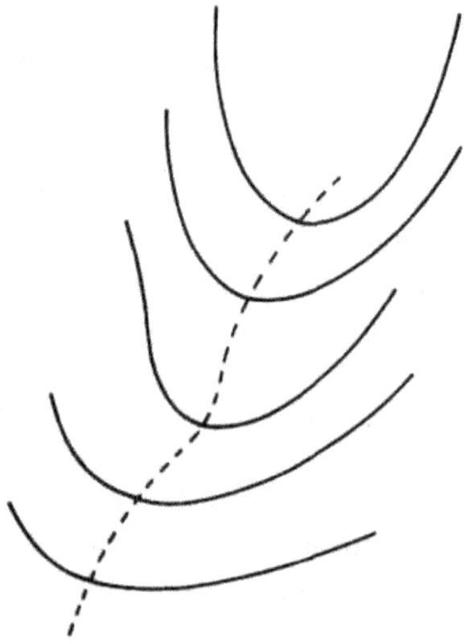

Fig. 1. Mechanical motion of a steam (- - - -) on the bottom of a two-dimensional surface, a trough. The surface is steep in the direction perpendicular to the stream trajectory.

The valley method has been applied to calculate the I–A interaction in the form of the series in ρ^2/R^2 In pure Yang–Mills theory this interaction starts from R^{-4} — it is of the dipole–dipole type. The leading R^{-4} term is universal and was extracted long ago [9, 1] from a simple additive ansatz, $A = A_I + A_A$, without submerging into the subtleties of the valley construction which was, of course, unknown at that time. The valley method becomes operative in full at the level $\mathcal{O}(R^{-6})$. The form of the I–A interaction energy in this order was obtained in Ref. [10] (see also Ref. [11]) by combining the valley approach with arguments of the conformal invariance. The valley method is also useful in the theory with the vacuum expectation value of the Higgs field [12, 13, 8]. In this case the leading term in the I–A interaction is R^{-2}.

It is worth emphasizing that the valley calculation of the higher order terms in the R^{-2} expansion requires a very careful and thoughtful attitude. Indeed, the pseudoparticles are extended objects. When they are not infinitely far apart from each other, they are deformed, and the question of what is to be called the instanton size ρ and the I–A separation R requires a special investigation. The parameters appearing in this or that *ansatz* are not physically measurable, and different *ansätze* can produce expansions which may look differently but be physically equivalent. The only way out of this ambiguity is to use a procedure relating ρ and R to physical quantities. The present state of the art is such that practically we have to invoke the effective Lagrangian method (see below).

Before we proceed to the instanton-induced effective Lagrangians, I would like to make one last remark on the general aspects of approximate solutions.

If the beginning of the valley is a well-defined construction, and one can hardly doubt that, say, a well-separated I–A pair contributes to the observable effects, the question of where and how the valley ends up is rather obscure. Indeed, when the I and A "atoms" approach each other, so that the defect of the action becomes comparable to $8\pi^2/g^2$, they tend to annihilate each other. Continuing the journey along the bottom of the valley, we smoothly interpolate to a point where the field is weak, the action is $\mathcal{O}(1)$; this point, of course, belongs to ordinary perturbation theory. To avoid double-counting, this part of the valley (a flat part) should be definitely excluded from consideration based on instantons. The quasiclassical approximation here, the basis of the whole picture, certainly fails. We must probably stop earlier. The collapse of the instanton-based approximations might manifest itself as a phase transition taking place when we approach a critical point from the "other" side of the valley (the one corresponding to large separations). Another option is the occurrence of a bifurcation in the valley [14, 15], or even the appearance of new dimensions (the latter seems natural from the mechanical analogy considered above).

Attempts to circumvent all unclear questions concerning the end of the valley are known in the literature from the early eighties when a few problems requiring a precise quantitative analysis of the I–A interaction were addressed. If only one could turn the I–A attraction into the I–A repulsion, the field configurations with close pseudoparticles would be suppressed and the dominant contribution would be determined by a well-separated I–A pair (the well-defined beginning of the valley). A trick allowing one to achieve this goal was suggested in Ref. [16] in the context of a toy model — quantum mechanics of a particle in the double-well potential,

$$\mathcal{L} = \frac{1}{2}\dot{x}^2 + \frac{1}{2}x^2(1-\lambda x)^2 \ .$$

The idea is to make an analytic continuation in the coupling constant λ^2 If $\lambda^2 \to -\lambda^2$, then the I–A attraction becomes the I–A repulsion. The I–A contribution then becomes well defined and determines the divergence of the perturbative series in the model at hand. The supersymmetric Yang–Mills theory with matter fields is another example whereby an analytic continuation in the coupling constant helps define the I–A contribution [6]. I hasten to add that the physical meaning of this procedure — analytic continuation in the coupling constant — is as obscure now as it was 10 years ago; it produces more questions than answers and, therefore, I will not touch on it any more. The interested reader is referred to the original publications [16, 6].

The only modest achievement the theory can claim in recent years in the issue of the evolution of approximate solutions away from the exact stationary points, is the understanding of how one could formulate the problem to make it well defined and, simultaneously, not to lose the physical clarity.

For exact solutions with nontrivial topology, there is no doubt: they minimize the action under certain boundary conditions which are, in turn, dictated by the topology of the space of gauge fields; their contribution to the partition function is not contained in the ordinary perturbation theory and has to be included additionally. As for the I–A pair (or pairs), the corresponding topological charge is zero – the same as for topologically trivial fields. The latter, however, are somehow reflected in ordinary perturbation theory, and we face the problem of double-counting. To make the notion of such "almost" solutions quantitative and well defined one must invent something which would unambiguously single out the topologically neutral I–A ensemble from the perturbative sector of the theory. This "something" may be the fermion charge nonconservation (baryon charge in realistic models).

As we already know from Sec. V, switching on the massless fermion fields results in a remarkable phenomenon: the axial $U(1)$ symmetry explicitly present in any Feynman graph ceases to exist in the instanton background. The corresponding fermion quantum number conserved to any finite order in perturbation theory is broken by instantons. Moreover, the inclusive cross section describing the fermion number violation is related, by means of unitarity, to the problem of the I–A interaction (see below), a crucial observation [17] giving some hope that the questions

— what happens in the middle of the valley?

— what is the minimal action of the "minimal" I–A molecule still to be included?

may find answers some day in the future.

Further discussion of this topic would lead us further astray from the domain of clean results, the subject of this volume. The latter reason explains why, from a wealth of works treating approximate solutions, only three are reproduced in this section.

Let us return to the issue of the I–A interaction at large separations in the form of expansion in ρ^2/R^2. In the above I noted that this is a perfectly defined procedure but did not explain why. The most convenient line of reasoning demonstrating that this is indeed the case is that based on effective Lagrangians.

Instanton-induced effective Lagrangians date back to the work of 't Hooft [2] who presented the multifermion interaction due to the fermion zero modes in the form of a local Lagrangian. Later on it was realized that, from the point of view of effective Lagrangians, there is nothing specific in fermion vertices and they should be supplemented by an infinite set of instanton-induced boson vertices [18]. The general formulation of the problem is as follows. Let us consider the vacuum-to-vacuum transition generated by the instanton trajectory in the presence of additional background fields which are assumed to be (i) weak compared to a typical instanton field and (ii) slowly varying at the scale set by the instanton size ρ. A bosonic or fermionic plane wave with the frequency $\omega \ll \rho^{-1}$ might be a good example of such an additional background field. (To preserve the standard definition of the plane wave at large distances, we must choose the instanton field in the singular gauge so that it falls off sufficiently fast at infinity. In the nonsingular gauge the large distance tails of the instanton distort the plane wave much in the same way as the scattering wave function of the Schrödinger problem in the Coulomb field is distorted.)

Then we calculate the transition amplitude as a functional of the background field. This functional is a sum of all amplitudes, with an arbitrary number of quanta produced or annihilated in the instanton transition. In the leading in g approximation this can be viewed as an effective Lagrangian by itself. In the general case we have to make one step further in order to proceed from amplitudes to operators. Using the fact that the external lines carry low frequencies (in the scale ρ^{-1}) we expand the result in the derivatives of the fields. In this way we arrive at an infinite sum of local operators.

The physical meaning of the procedure is quite transparent. The small size instanton is a source which absorbs or emits locally any number of large wavelength quanta. Instead of explicitly considering all these processes in the instanton transition we substitute the effect of the instanton by a series of local operators. Once this is done we can forget about instantons altogether in the consideration of low energy processes.

The simplest effective Lagrangian of this type was obtained in Ref. [18] in the pure Yang–Mills theory in the leading approximation — a prototype for all Lagrangians constructed later and including the Higgs fields and/or higher order effects (see e.g. Refs. [12, 13, 8]). The specific expression takes the form

$$\mathcal{L}_\rho(x_0) \sim \exp\left(-\frac{2\pi^2\rho^2}{g} O^{ab}\bar{\eta}_{b\mu\nu} G^a_{\mu\nu}(x_0)\right) + \text{h.c.}\,, \tag{3}$$

where O^{ab} is a global color rotation matrix, x_0 is the coordinate of the instanton center, the subscript ρ reminds us of its radius, and $\bar{\eta}_{b\mu\nu}$ is the 't Hooft symbol. $G^a_{\mu\nu}(x_0)$ is the operator of the gluon field strength tensor.[10] In principle, beyond the leading approximation, the exponent in Eq. (3) will contain other operators, say with derivatives $D_\alpha G_{\mu\nu}$ or with two or more G's, along with a series in g.

If the instanton-induced effective Lagrangian is known, one can proceed to determine the I–A interaction as a systematic double expansion, in the ratio ρ/R and in the coupling constant. Thus, the leading dipole–dipole term is obtained from Eq. (3) by substituting the operator $G^a_{\mu\nu}(x_0)$ by the anti-instanton field at x_0. The anti-instanton centered at y_0 should be taken in the singular gauge, and it is assumed that $R = |x_0 - y_0| \gg \rho$.

A somewhat different version (which I like more) does not use the language of the classical fields at all. It operates with vertices and particles and allows us to connect the classical problem of the interaction energy with the quantum problem

[10] Equation (3) implies that $G_{\mu\nu}$ is normalized as follows: $L = (1/4)G^a_{\mu\nu}G^a_{\mu\nu}$.

of the instanton-induced cross sections. Let us illustrate in more details this lin of reasoning in the calculation of the I–A interaction in the leading approximation Relevant digrams are depicted in Fig. 2. The instanton with size ρ_1 is placed at x and the anti-instanton with size ρ_2 is placed at the origin. It is assumed, of course that $|x| \gg \rho_{1,2}$ so that the effective Lagrangian makes sense.

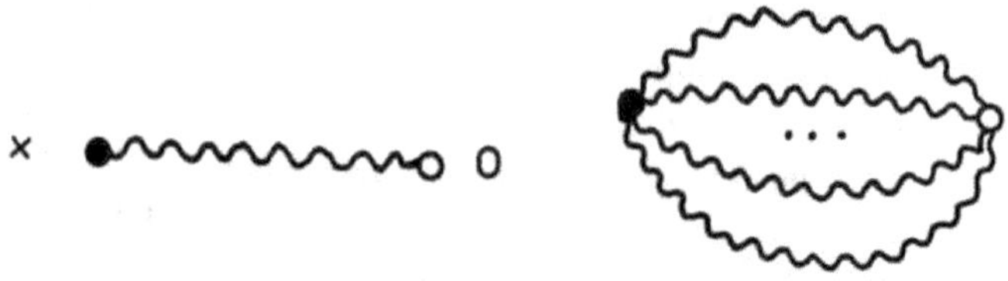

(a) One-gluon exchange. (b) Multigluon exchange.

Fig. 2. The I–A interaction from the effective Lagrangian. The instanton(•) is at point x while the anti-instanton (o) is at the origin. The vertices in diagrams (a) and (b) correspond to those in Eq. (3) and a similar Lagrangian for the anti-instanton.

The diagram of Fig. 2(a) is the basic element of the calculation. We expand the exponent in Eq. (3) and a similar one for the anti-instanton, keep the linear in $G^a_{\mu\nu}$ terms and contract $G(x)$ and $G(0)$ to get

$$\left(\frac{4\pi^2}{g^2}\rho_1^2\rho_2^2\right) O_I^{ab}\bar{\eta}_{b\mu\nu}O_A^{cd}\eta_{d\alpha\beta} < G^a_{\mu\nu}(x)G^c_{\alpha\beta}(0) > , \tag{4}$$

where $< G^a_{\mu\nu}(x)G^c_{\alpha\beta}(0) >$ is the free Green function of the gauge field. Moreover, in the Green function $< A_\mu(x)A_\nu(0) >$, we can keep only the $g_{\mu\nu}$ part, since the part $x_\mu x_\nu$ drops out,

$$< G^a_{\mu\nu}(x)G^c_{\alpha\beta}(0) >= \frac{2\delta^{ab}}{\pi^2}(g_{\nu\alpha}x_\mu x_\beta+g_{\mu\beta}x_\nu x_\alpha-g_{\nu\beta}x_\mu x_\alpha-g_{\mu\alpha}x_\nu x_\beta)\frac{1}{x^6}+\dots , \tag{5}$$

where the dots denote the terms which do not contribute due to the fact that $\eta_{a\mu\nu}\bar{\eta}_{b\mu\nu} = 0$.

Now, it is not difficult to see that Fig. 2(b) merely exponentiates the result of Eq. (4) $[1/(n!)^2$ from the expansion of the effective Lagrangians is supplemented by $n!$ coming from combinatorics] and we finally get for the I–A amplitude

$$\exp\left[-\frac{8\pi^2}{g^2} - \frac{32\pi^2}{g^2}\rho_1^2\rho_2^2\eta_{a\lambda\mu}\bar{\eta}_{b\lambda\nu}\Omega^{ab}\frac{x_\mu x_\nu}{x^6}\right] , \tag{6}$$

where Ω^{ab} is the matrix of the relative orientation of the pseudoparticles, $\Omega^{ab} = O_I^{cb}O_A^{ca}$.

The x^{-4} term above is the famous dipole–dipole interaction obtained in Refs. [9, 1] within a totally different approach — a purely classical analysis of superimposed I–A fields. At the level of x^{-4}, the classical analysis of this type is a well-defined procedure. As was explained above, it becomes ambiguous in higher orders:[11] even if the prescription of how instantons should be superimposed is fixed, the definition of, say, the instanton center in such a superposition, is not so clear. The centers can experience a *zitterbewegung* affecting the x^{-8} terms.

The effective Lagrangian approach fixes the definition of ρ and R and allows one to build a systematic expansion of the interaction energy in ρ/R. Each term in the effective Lagrangian is immediately translatable in the corresponding term in the interaction energy. For instance, the term given in Eq. (3) generates the $g^{-2}R^{-4}$ part and, through the diagrams of Fig. 3, a part $\propto g^{-2}R^{-6}$. [In Fig. 3(b), one of the vertices is due to the quadratic term in $G^a_{\mu\nu}$, if both vertices in Fig. 3(b) are due to the quadratic term in $G^a_{\mu\nu}$, this graph generates a piece $\propto R^{-4}$ in the interaction energy.] If the Higgs field is included, it generates the terms of order v^2R^{-2}, v^2R^{-4}, etc. through the diagrams depicted in Fig. 4. The last term which has been explicitly computed is of order $g^{-2}R^{-8}\ln R$ corresponding to the diagram of Fig. 5.

(a) (b)

Fig. 3. Higher order terms in the I–A interaction due to quantum corrections in the gluon exchanges.

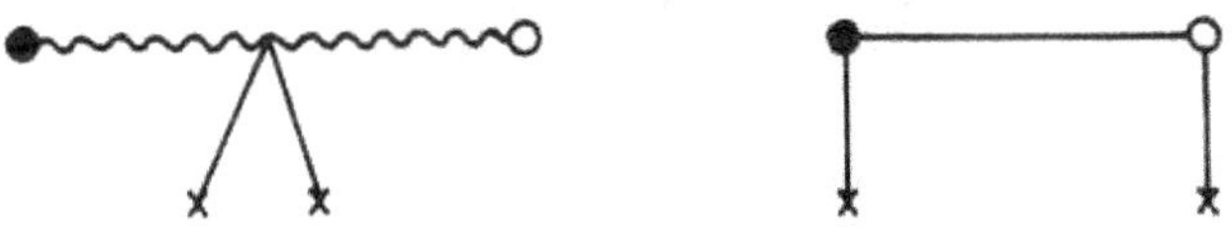

(a) Mass term of the gauge boson. (b) Higgs exchange.

Fig. 4. Higher order terms in the I–A interaction due to the Higgs field. The crosses denote the vacuum expectation value of the Higgs field.

[11]The conformal invariance of the classical Yang–Mills theory allows one to fix unambiguously also the x^{-6} term [10]. It is still unambiguously predicted by the valley method since it turns out to be rigidly connected with the x^{-4} term. The conformal invariance implies that the interaction energy depends on the conformal parameter [10]

$$\zeta = \frac{|x|^2 + \rho_1^2 + \rho_2^2}{\rho_1\rho_2}$$

and the x^{-6} term appears as a correction to x^{-4} due to the expansion of this parameter.

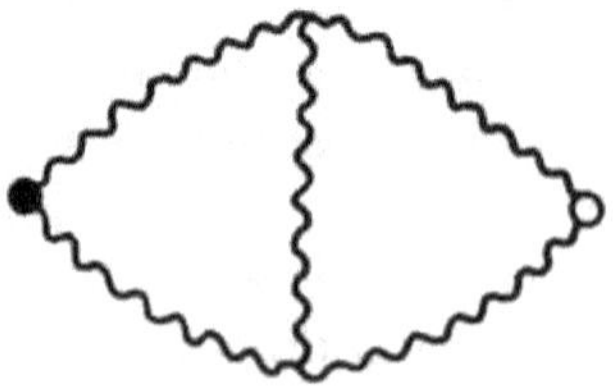

Fig. 5. An example of the two-loop diagram contributing to the I–A interaction at the level $g^{-2}R^{-8}\ln R$.

It is instructive to apply the very same approach to see how other known features of the instanton interaction are recovered. For instance, Ref. [12] nicely demonstrates the vanishing of the instanton–instanton interaction, a consequence of the fact that two instantons are, of course, the exact solution of the duality equation.

References

[1] *C. Callan, R. Dashen and D. Gross, *Phys. Rev.* **D17** (1978) 2717.
[2] *G. 't Hooft, *Phys. Rev.* **D14** (1976) 3432.
[3] Y. Frishman and S. Yankielowicz, *Phys. Rev.* **D19** (1979) 540.
[4] *I. Affleck, *Nucl. Phys. Rev.* **B191** (1981) 429.
[5] *I. Balitsky and A. Yung, *Phys. Lett.* **B168** (1986) 113.
[6] I. Balitsky and A. Yung, *Nucl. Phys.* **B274** (1986) 475.
[7] A. Yung, "Valley Method for Instanton-Induced Effects in Quantum Field Theory," in *Proc. ICTP 1991 Summer School on High Energy Physics and Cosmology*, p. 580.
[8] I. Balitsky and A. Schäfer, *Nucl. Phys.* **B404** (1993) 639.
[9] D. Förster, *Phys. Lett.* **66B** (1977) 279.
[10] A. Yung, *Nucl. Phys.* **B297** (1988) 47.
[11] V. V. Khoze and A. Ringwald, "Valley trajectories in Gauge Theories," preprint CERN-TH-60282/91, 1991, unpublished.
[12] A. Yung, "Instanton-Induced Effective Lagrangian in the Gauge–Higgs Theory," preprint SISSA 181/90/EP, 1990, unpublished.
[13] D. Dyakonov and M. Polyakov, *Nucl. Phys.* **B389** (1993) 109.
[14] J. Grandey and M. Mattis, *Phys. Lett.* **B307** (1993) 140.
[15] P. Provero, *Phys. Lett.* **B321** (1994) 95.
[16] E. B. Bogomolny, *Phys. Lett.* **B91** (1980) 431 [reprinted in *Large Order Behaviour of Perturbation Theory*, eds. J. C. Le Guillou and J. Zinn-Justin (North-Holland, 1990), p. 175].
[17] V. I. Zakharov, *Nucl. Phys.* **B371** (1992) 637; *Nucl. Phys.* **B383** (1992) 218.
[18] *M. Shifman, A. Vainshtein and V. Zakharov, *Nucl. Phys.* **B165** (1980) 45.

Nuclear Physics B191 (1981) 429–444

ON CONSTRAINED INSTANTONS

Ian AFFLECK[1]

Lyman Laboratory of Physics, Harvard University, Cambridge, Massachusetts 02138, USA

Received 5 December 1980
(Revised 13 April 1981)

A simple method is presented for doing systematic constrained instanton calculations in models such as ϕ^4 or Higgs theories where the presence of a mass term prevents the existence of a classical solution. As an application, instanton estimates of the large-order behavior of the perturbation series in massive ϕ_4^4 theory are derived. (These estimates agree with those of Frishman and Yankielowicz.)

1. Introduction

If a classically scale-invariant theory has instantons, they necessarily have arbitrary size. This leads to the well-known infrared problem in Yang-Mills theory [1]. One way to overcome this problem is to break scale invariance by adding Higgs fields. Instanton solutions then cease to exist but some sort of approximate instanton solution may still dominate the functional integral. As was first pointed out by 't Hooft [1], instantons larger than the inverse Higgs expectation value, $\langle\phi\rangle^{-1}$, are then suppressed and the resulting effect is exponentially small. An approximate calculation of this effect was presented by 't Hooft. However, it has never been quite clear (at least to the present author) what small parameter justified the approximations, nor how one could calculate corrections. Furthermore, one is left with the impression that calculations can be performed using regular Yang-Mills instantons (with a cut-off on their scale). These instantons produce long-range (but exponentially small) effects since they behave asymptotically like an inverse power of x. On the other hand, one expects physical quantities to go like $e^{-\langle\phi\rangle|x|}$ at large x due to the Higgs phenomenon. These problems may be of some practical importance in the context of grand unified theories where an instanton effect, although containing a small exponential, may also be proportional to a large mass ratio. With this justification we present a systematic method for doing calculations in such a context. The basic idea is to introduce a constraint into the functional integration measure.

[1] Research supported in part by the National Science Foundation under grant no. PHY77-22864, and the Harvard Society of Fellows.

One must then minimize the action subject to the constraint. This procedure was discussed by Frishman and Yankielowicz [2] who constrained the integrals of fields over localized regions of space-time. We observe here that a somewhat simpler procedure is to constrain the integral over all space-time of some local operator not already appearing in the lagrangian. The resulting constrained instanton looks like the Yang-Mills instanton at short distances but decays exponentially at large distances.

A very similar situation arises in ϕ_4^4 theory (with a negative ϕ^4 term). Instanton solutions exist only in the massless theory. In sect. 2 the constrained instanton method is developed in this somewhat simpler case. We use these results to rederive the instanton estimates of Frishman and Yankielowicz for the large-order behavior of the perturbation series in massive ϕ_4^4 theory, in sect. 3. These estimates differ in various ways from those of the massless theory [3, 4]; we show that at least some of these differences can be explained in terms of large-order diagrams. However, these estimates do not include renormalon effects [5]. In sect. 4 the constrained instanton method is extended to Higgs theories.

2. The constrained instanton method (ϕ_4^4 theory)

Consider ϕ_4^4 theory with euclidean lagrangian

$$\mathcal{L}=\frac{1}{g}\left[\left(\partial_\mu\phi\right)^2+\tfrac{1}{2}\mu^2\phi^2-\frac{1}{4!}\phi^4\right]. \tag{2.1}$$

We have chosen the "wrong" sign for the ϕ^4 term. This is the relevant sign for performing asymptotic estimates in the theory with a positive ϕ^4 term (sect. 3). Also this theory may be of some intrinsic interest since it has a metastable ground state ($\phi=0$). The dimensionless coupling, $g_{\text{eff}}(\mu)$, will be chosen to be small.

The massless theory has an instanton:

$$\partial^2\phi_0=-\tfrac{1}{6}\phi_0^3, \tag{2.2}$$

$$\phi_0(x)=\frac{4\sqrt{3}\,\rho}{\rho^2+x^2}, \tag{2.3}$$

ρ being the (arbitrary) scale parameter. In the massive theory no instanton solution exists as can be seen by considering the scale transformation $\phi(x)\to a\phi(ax)$, under which

$$gS\to\int d^4x\left[\tfrac{1}{2}(\partial\phi)^2-\frac{1}{4!}\phi^4\right]+\tfrac{1}{2}\mu^2a^{-2}\int d^4x\,\phi^2. \tag{2.4}$$

If $\phi(x)$ is a stationary point then $dS/da|_1=0$; thus the only solution is $\phi=0$. In terms of the instantons of the massless theory, the scale transformation maps ρ into $a^{-1}\rho$; the action is minimized by shrinking the instanton to a point.

If we take the mass to be small compared to the inverse instanton scale ($\rho\mu \ll 1$), then it is somewhat surprising that a solution doesn't exist. Normally, when one adds a small perturbation to a soluble equation, a new solution can be computed as an expansion in the perturbation. If we attempted this here then the equation for the leading correction, $\delta\phi$, would be

$$\left(\partial^2 + \tfrac{1}{2}\phi_0^2\right)\delta\phi = \mu^2\phi_0. \tag{2.5}$$

Let us consider a more general source term, $J(x)$, on the right-hand side of eq. (2.5). We may solve it using the Green function,

$$\delta\phi(x) = \int d^4y\left(\partial^2 + \tfrac{1}{2}\phi_0^2\right)^{-1}(x, y)J(y). \tag{2.6}$$

[Actually, $(\partial^2 + \frac{1}{2}\phi_0^2)$ has four normalizable zero-modes, $\partial_\mu\phi_0(x)$. This is irrelevant since we take $J(x)$ to be a scalar.] The problem arises when we attempt to impose boundary conditions on $\delta\phi$ at large x (e.g. $\delta\phi \to 0$). Due to the existence of a (non-normalizable) scalar zero mode, $\partial\phi_0/\partial\rho$, the asymptotic behavior of $\delta\phi$ is already determined by J. To see this, multiply both sides of eq. (2.5) by $\partial\phi_0/\partial\rho$ and integrate over x. Upon integration by parts we find

$$\int_{|x|=\infty} dS_\mu \frac{\partial\phi_0}{\partial\rho}\overleftrightarrow{\partial}_\mu\delta\phi = \int d^4x\frac{\partial\phi_0}{\partial\rho}J. \tag{2.7}$$

This implies that $\delta\phi$ approaches a constant at infinity,

$$\delta\phi(x) \to \sqrt{\tfrac{1}{3}}\,\pi^2\int d^4x\frac{\partial\phi_0}{\partial\rho}J. \tag{2.8}$$

We can't require $\delta\phi \to 0$ unless the perturbation, J, has vanishing projection onto the zero mode. A similar problem would arise at each order in perturbation theory, requiring order by order adjustment of J to ensure correct asymptotic behavior of ϕ.

The present situation is slightly more complicated than this, because the perturbation, $\mu^2\phi_0$, vanishes at infinity less rapidly than the unperturbed term, ϕ_0^3. A finite action solution, if it existed, would, at $x \to \infty$, become proportional to $G_\mu(x)$, the solution of

$$\left(-\partial^2 + \mu^2\right)G_\mu(x) = \delta(x), \tag{2.9}$$

which vanishes exponentially at large $x, G_\mu(x) \sim e^{-\mu|x|}$. This has the small μ expansion

$$G_\mu(x) = \frac{1}{4\pi^2}\left[\frac{1}{x^2} + \tfrac{1}{2}\mu^2 \ln c\mu|x| + O\left[\mu^4 x^2 \ln \mu|x|\right]\right],$$

c being a numerical constant. Thus, if a finite action solution to eq. (2.5) existed, it would have an expansion of the form

$$\phi(x) = \phi_0(\rho, x) + (\rho\mu)^2 \ln \rho\mu\phi_1(\rho, x) + (\rho\mu)^2\phi_2(\rho, x) + (\rho\mu)^4 \ln \rho\mu\phi_3(\rho, x) + \cdots, \tag{2.10}$$

with boundary conditions

$$\phi_0(\rho, x) \underset{|x|\to\infty}{\to} \frac{4\sqrt{3}\,\rho}{x^2},$$
$$\phi_1(\rho, x) \to \frac{2\sqrt{3}}{\rho},$$
$$\phi_2(\rho, x) \to \frac{2\sqrt{3}}{\rho} \ln c|x|/\rho. \tag{2.11}$$

[The overall constant of proportionality has been fixed by the asymptotic behavior of ϕ_0.] Alas, we again cannot enforce these boundary conditions because of the zero mode.

As we shall see, only field configurations with scale $\rho \lesssim \sqrt{g}\,\mu^{-1}$ make important contributions to the functional integral of the field theory. At these short distances one feels that the instantons of the massless theory should dominate the functional integral. A way to realize this notion is to break the functional integral up into sectors labeled by a scale parameter, ρ. For each value of ρ the integral may be dominated by an instanton-like configuration; at the end the integral over ρ is performed. A method must be derived for fixing the scale at a particular value, ρ. A simple object to hold fixed is the integral over space-time of some local operator, O. If the operator has dimension d, then we may require

$$\int d^4x\, O = c\rho^{4-d}, \tag{2.12}$$

for a conveniently chosen constant, c. [Clearly we must choose $d \neq 4$.] Can we find a stationary point subject to this condition (with $\rho\mu \ll 1$)? The constrained variational problem is equivalent to stationarizing

$$S+\frac{\sigma}{g}\left[\int d^4x\,O-c\rho^{4-d}\right],$$

with respect to σ as well as $\phi(x)$. This adds a term to the Euler-Lagrange equation,

$$\partial^2\phi=-\tfrac{1}{6}\phi^3+\mu^2\phi+\sigma\frac{\delta O}{\delta\phi}\,. \tag{2.13}$$

For a rather wide class of operators, O, we can now contrast a perturbative solution in the manner discussed earlier. We choose O to vanish more rapidly at large x than the mass term, so that the perturbative expansion and the boundary conditions are as in eqs. (2.10) and (2.11). The Lagrange multiplier, σ, can now be adjusted order by order in perturbation theory to enforce the boundary conditions. [If we choose $0=\phi^2$ then the only solution is $\sigma=\frac{1}{2}\mu^2$ and we get back the instanton ϕ_0. However we then find $\int d^4x\,\phi^2=\infty$ so the procedure doesn't work.] Some suitable operators, O are $\phi^3,\phi^6,(\partial_\mu\phi)^4,\ldots$. [For the ϕ^6 case we construct an explicit proof of the existence of an instanton for $\rho\mu$ less than some critical value in the appendix.] To see how the method works consider the equation for $\phi_1(\rho,x)$:

$$\left[\partial^2+\tfrac{1}{2}\phi_0^2\right]\phi_1=\sigma_1\frac{\delta O}{\delta\phi}\,, \tag{2.14}$$

where $\sigma=(\mu\rho)^2\ln\rho\mu\sigma_1+\cdots$. Note that the mass term doesn't appear in this equation since it is logarithmically suppressed. As observed above [see eq. (2.8)], σ_1 is determined by the boundary condition $\phi_1\to2\sqrt{3}/\rho$:

$$\sigma_1=\frac{6}{\pi^2}\left[\frac{\partial}{\partial\rho}\int d^4x\,O\right]^{-1}. \tag{2.15}$$

What does the constrained instanton look like? It is characterized by two length scales, ρ [which we shall see is $\lesssim\sqrt{g}\,\mu^{-1}$] and μ^{-1}. Near the origin it resembles the massless instanton, ϕ_0. At a distance large compared to ρ (but still small compared to μ^{-1}) it goes over smoothly to the function $G_\mu(x)$. In this region it is characterized by the scale μ^{-1}, decreasing as $e^{-\mu|x|}$ at infinity. As $g\to0$ it looks like $G_\mu(x)$ everywhere except inside a very small core (of radius $\sim\sqrt{g}\,\mu^{-1}$) where the $1/x^2$ singularity of G_μ gets smoothed into an instanton $\sim1/(x^2+\rho^2)$. Corrections to the constrained instanton inside the core may be calculated as explained above. We can also calculate corrections to the form $G_\mu(x)$ outside the core by writing an expansion:

$$\phi(x,\rho,\mu)=\rho\left[16\pi^2\sqrt{3}\,G_\mu(x)+(\rho\mu)^2\phi^1(\mu,x)+(\rho\mu)^4\phi^2(\partial,x)+\cdots\right]. \tag{2.16}$$

All the ϕ^i's vanish exponentially at $x \to \infty$ but blow up as $x \to 0$ with boundary conditions determined by matching onto $\phi_0(x)$ at intermediate length scales:

$$\phi_0(x) = \frac{4\sqrt{3}\,\rho}{x^2}\left[1 - \frac{\rho^2}{x^2} + \frac{\rho^4}{x^4}\cdots\right], \tag{2.17}$$

so

$$\phi^1(\mu, x) \underset{|x|\to 0}{\to} -\frac{4\sqrt{3}}{\mu^3 x^4},$$

$$\phi^2(\mu, x) \to \frac{4\sqrt{3}}{\mu^5 x^6}, \qquad \ldots. \tag{2.18}$$

The action of this instanton can be computed as an expansion in the small parameter $\rho\mu$ by using the two expansions for ϕ in the appropriate (overlapping) regions of validity. As shown by Frishman and Yankielowicz [2], this expansion contains logarithms. This lends some amusing features to the asymptotic estimates of sect. 3. Consider the first two terms in the expansion of gS. It turns out that most of the action comes from the core where the first expansion is valid. To leading order we simply get the action of the massless theory,

$$\int d^4x\left[\tfrac{1}{2}(\partial\phi_0)^2 - \frac{1}{4!}\phi_0^4\right] = 16\pi^2. \tag{2.19}$$

To next order we should consider the (boundary) term linear in ϕ_1:

$$(\rho\mu)^2 \ln \rho\mu \int dS_\mu \phi_1 \partial_\mu \phi_0 = -24\pi^2 (\rho\mu)^2 \ln \rho\mu. \tag{2.20}$$

(The sceptical reader may be concerned that the mass term appears to be infrared divergent:

$$\tfrac{1}{2}\mu^2 \int_{|x|<L} d^4x\, \phi_0^2 = 24\pi^2(\rho\mu)^2 \ln(L/\rho). \tag{2.21}$$

He will be relieved to note that this cancels with the term linear, in ϕ_2: $(\mu\rho)^2 \int_{|x|=L} dS_\mu \phi_2 \partial_\mu \phi_0$ [by eq. (2.11)].) We see that the expansion of S involves both powers and ln's of $\rho\mu$.

Having discussed the classical properties of the constrained instanton let us now consider how to employ it semiclassically. We start with the massless theory. The instanton gives an imaginary part to the vacuum energy which is (formally)

$$\mathrm{Im}\, E_0 \sim e^{-16\pi^2/g} \frac{1}{g^{5/2}} V \int_0^\infty \frac{d\rho}{\rho^5} (\rho\Lambda)^3. \tag{2.22}$$

We remind the reader how the various factors arise [4]. $e^{-16\pi^2/g}$ is e^{-S}. We introduce collective coordinates for the five translational and conformal zero modes. This gives the factor $g^{-5/2}$ and the integrals $\int d^4x = V$ and $\int d\rho$. The ultraviolet divergence of the gaussian integral produces the factor Λ^3 (Λ is an ultraviolet cut-off) which renormalizes g:

$$\Lambda^3 e^{-16\pi^2/g} = M^3 e^{-16\pi^2/g_R}, \tag{2.23}$$

where M is a renormalization scale, g_R a renormalized coupling. The remaining powers of ρ in eq. (2.22) can be determined by dimensional analysis leaving only a numerical constant to be found by explicit calculation. Now consider the imaginary part of the n-point function

$$\begin{aligned} \operatorname{Im} G^{(n)}(x_1,\ldots,x_n) &\sim g_R^{-(n+5)/2} e^{-16\pi^2/g_R} \int_0^\infty \frac{d\rho}{\rho^5}(\rho M)^3 \\ &\quad \times \int d^4x_0 \phi_0(\rho, x_1 - x_0)\cdots\phi_0(\rho, x_n - x_0). \end{aligned} \tag{2.24}$$

Fourier transforming neatly undoes the integral over instanton location,

$$\operatorname{Im} G^{(n)}(p_1,\ldots,p_n) \sim g_R^{-(n+5)/2} e^{-16\pi^2/g_R} \int_0^\infty \frac{d\rho}{\rho^5}(\rho M)^3 \tilde{\phi}_0(\rho, p_1)\cdots\tilde{\phi}_0(\rho, p_n). \tag{2.25}$$

Here

$$\begin{aligned} \tilde{\phi}_0(\rho, p) \equiv \int d^4x\, e^{ip\cdot x}\phi_0(\rho, x) &\to \frac{16\sqrt{3}\pi^2\rho}{p^2} \qquad |p|\rho \to 0 \\ &\sim e^{-|p|\rho} \qquad |p|\rho \to \infty. \end{aligned} \tag{2.26}$$

The ρ-integral is now convergent at $\rho \to 0$. It is also convergent at $\rho \to \infty$, being cut off at scales $\sim 1/p$. We expect perturbation theory about the instanton to give a power series in $g_{\rm eff}(\rho^{-1})$, so we can only trust this one-loop calculation at large p [where $g_{\rm eff}(p) \ll 1$].

Now consider the massive theory. We use the identity

$$\int (d\phi) \propto \int_0^\infty \frac{d\rho\,\rho^{d-3}}{|d-4|} \int (d\phi)\,\delta\left[\int d^4x\, O - c\rho^{4-d}\right], \tag{2.27}$$

and integrate out ϕ by the saddle-point method. This procedure may introduce extra ultraviolet divergences if the interaction O is non-renormalizable. However, these must cancel out at the end of the calculation. To leading order in $\rho\mu$, this procedure amounts to introducing ρ as a collective coordinate so we find

$$\mathrm{Im}\,G^{(n)}(p_1\cdots p_n)\sim g_{\mathrm{R}}^{-(n+5)/2}\int\frac{\mathrm{d}\rho}{\rho^5}(\rho M)^3$$

$$\times\mathrm{e}^{-(1/g_{\mathrm{R}})[16\pi^2-24\pi^2(\rho\mu)^2\ln\rho\mu]}\tilde{\phi}(\rho,p_1)\cdots\tilde{\phi}(\rho,p_n), \quad (2.28)$$

$\tilde{\phi}$ being the Fourier transform of the constrained instanton. [We have kept the normalization point, M, different than the physical mass, μ, to facilitate comparison with the zero-mass case.] As we shall see very shortly, the ρ-integral gets cut off (by the instanton action) at small $\rho\mu$, $\rho\mu\sim\sqrt{g_{\mathrm{R}}}$. If we take the p_i to be $\mathrm{O}(\mu)$, that is, hold p_i fixed as $g_{\mathrm{R}}\to 0$, then we only need $\tilde{\phi}(\rho,p)$ for $\rho p\ll 1$. This is given by the Fourier transform of $\phi(\rho,x)$ for $\rho/|x|\ll 1$,

$$\phi(\rho,x)=16\pi^2\sqrt{3}\,\rho G_\mu(x)\Rightarrow\tilde{\phi}(\rho,p)=\frac{16\pi^2\sqrt{3}\,\rho}{p^2+\mu^2}. \quad (2.29)$$

The p-dependence simply supplies propagators for the external legs and the amputated Green function is local:

$$\mathrm{Im}\,\Gamma^{(n)}(p_1\cdots p_n)\sim\mathrm{e}^{-16\pi^2/g_{\mathrm{R}}}g_{\mathrm{R}}^{-(n+5)/2}\int_0^\infty\mathrm{d}\rho\,M^3\rho^{n-2}\mathrm{e}^{(24\pi^2/g_{\mathrm{R}})(\rho\mu)^2\ln\rho\mu}. \quad (2.30)$$

The integrand goes through a maximum at $(\rho\mu)^2\approx g_{\mathrm{R}}/(96\pi^2(n-1)\ln g_{\mathrm{R}}^{-1})\ll 1$. It eventually goes through a minimum at $\rho\mu=\mathrm{e}^{-1/2}$ and then increases without bound at $\rho\mu\to\infty$. However, this occurs at $\rho\mu\gtrsim\mathrm{O}(1)$, well outside the region of validity of our approximate classical calculations ($\rho\mu\ll 1$), and, what is more essential, outside the range of validity of the semiclassical approximation [$g_{\mathrm{eff}}(\rho^{-1})\ll 1$]. We will assume that the dominant contribution is from $\rho\mu\ll 1$ and that the large ρ region is suppressed by the presence of the mass term. Thus we have

$$\mathrm{Im}\,\Gamma^{(n)}\sim g_{\mathrm{R}}^{-(n+5)/2}\mathrm{e}^{-16\pi^2/g_{\mathrm{R}}}M^3\int_0^{(\mu\sqrt{\varepsilon})^{-1}}\mathrm{d}\rho\,\rho^{n-2}\mathrm{e}^{(24\pi^2/g_{\mathrm{R}})(\rho\mu)^2\ln\rho\mu}. \quad (2.31)$$

Making the change of variables

$$(\rho\mu)^2\equiv\frac{2g_{\mathrm{R}}x^2}{\ln g_{\mathrm{R}}^{-1}}, \quad (2.32)$$

we find

$$\mathrm{Im}\,\Gamma^{(n)} \sim g_R^{-3}(\ln g_R^{-1})^{(1-n)/2} M^3\mu^{1-n}\int_0^\infty \mathrm{d}x\, x^{n-2}\mathrm{e}^{-24\pi^2x^2}. \tag{2.33}$$

Some comments on this expression are in order. It was obtained in the normal weak-coupling limit, $g_R \to 0$ with μ and p_i held fixed. The relevant constrained instantons have size $\rho = \mu^{-1}\sqrt{g_R/\ln g_R^{-1}}$. Thus at $g_R \to 0$ the cores of the instantons shrink to zero and they simply behave like free propagators, $\phi \sim G_\mu(x)$. This is why the instanton gives a local n-point vertex. One could also consider the limit $g_{\text{eff}}(p_i)$ small but fixed, and $\mu/|p_i| \to 0$: In this limit we simply get back the result of the massless theory, $S \to 16\pi^2/g_R, \phi \to \phi_0$. Note that the massless and weak-coupling limits don't commute. If we expand the ρ integral in μ, we obtain factors of $(1/g)(\mu/p_i)^2\ln(\mu/p_i)$ from expanding e^{-S}.

3. Asymptotic estimates

It is a trivial extension of the results of sect. 2 to obtain estimates of the large-order behavior of the perturbation expansion. We again start by reviewing the massless case [3, 4]. Writing

$$G^{(n)}(p_1\cdots p_n) \sim g^{-n/2}\int(\mathrm{d}\phi)\mathrm{e}^{-S}\tilde{\phi}(p_1)\cdots\tilde{\phi}(p_n) \equiv \sum_{N=0}^{\infty} A_N g^N, \tag{3.1}$$

we (naively) apply the residue theorem to obtain

$$A_N = \frac{1}{2\pi i}\oint\frac{\mathrm{d}g}{g^{N+1+n/2}}\int(\mathrm{d}\phi)\mathrm{e}^{-S}\tilde{\phi}(p_1)\cdots\tilde{\phi}(p_n). \tag{3.2}$$

The g integral is around a contour encircling the origin. For large N we seek a joint stationary point with respect to both the ϕ and g integrals. That is, we stationarize $(S + N\ln g)$. The two conditions are

$$\partial^2\phi = \tfrac{1}{6}\phi^3, \tag{3.3}$$

$$\int \mathrm{d}^4x\left[\tfrac{1}{2}(\partial\phi)^2 + \frac{1}{4!}\phi^4\right] = Ng. \tag{3.4}$$

(We've written the theory with positive ϕ^4 term.) The solution is

$$\phi = i\phi_0, \qquad g = \frac{-16\pi^2}{N}. \tag{3.5}$$

Thus

$$e^{-(S+N\ln g)}=(-)^N e^{N[\ln(N/16\pi^2)-1]}. \tag{3.6}$$

Doing the gaussian integrals over g and ϕ gives the result of sect. 2 with the replacement $g\to 16\pi^2/N$ and an extra factor of $N^{-3/2}$ from the g integral:

$$A_N\sim(-)^N N^{n/2-1}e^{N[\ln(N/16\pi^2)-1]}\int_0^\infty\frac{d\rho}{\rho^5}(\rho M)^3\tilde{\phi}_0(\rho,p_1)\cdots\tilde{\phi}_0(\rho,p_n). \tag{3.7}$$

Higher-order corrections are suppressed by $1/N$ and are thus small if the p_i are large enough so that, in the theory with negative ϕ^4 term, $g_{\rm eff}(p_i)\ll 1$. This is somewhat ironic since it is at *small* momentum that perturbation theory is useful ($g_{\rm eff}$ small) in the actual theory of interest (positive ϕ^4 term). Massless ϕ_4^4 theory has infrared singularities in the Borel transform that are not given by the instanton [6]. The potential infrared problem with the instanton prediction may be related to them.

Now let us consider the massive theory. We proceed in the same way, using the constrained instanton. The stationary point for the g integral is now at

$$g=-\frac{1}{N}\left[16\pi^2-24\pi^2(\mu\rho)^2\ln\mu\rho\right], \tag{3.8}$$

$$\frac{1}{g^N}e^{-S}=(-)^N e^{N[\ln(N/16\pi^2)-1+(3/2)(\mu\rho)^2\ln\mu\rho]}, \tag{3.9}$$

and thus

$$A_N\sim(-)^N N^{(n-2)/2}e^{N[\ln(N/16\pi^2)-1]}M^3\int^\infty\frac{d\rho}{\rho^2}e^{(3/2)N(\mu\rho)^2\ln\mu\rho}\tilde{\phi}(\rho,p_1)\cdots\tilde{\phi}(\rho,p_n). \tag{3.10}$$

This result was also obtained by Frishman and Yankielowicz. We have a choice of taking the large-N limit with μ fixed or else the massless limit. The former gives

$$A_N\sim\prod_{i=1}^n\left(p_i^2+\mu^2\right)^{-1}e^{N[\ln(N/16\pi^2)-1]}N^{1/2}(\ln N)^{(1-n)/2}M^3\mu^{1-n}. \tag{3.11}$$

For large N, the Nth term in the perturbative expansion for the amputated Green function becomes momentum independent. Furthermore, we now trust the calculation right down to zero momentum since S itself cuts off the ρ-integral. Note that, A_N is smaller than in the massless case by a factor $(N\ln N)^{(1-n)/2}$. Both these facts may reflect the absence of infrared Borel singularities in the massive theory. If we take $\mu/p_i \to 0$ for N fixed, then we get back the estimate eq. (3.7). A small mass expansion involves factors of $N(\mu/p_i)^2\ln(\mu/p_i)$. Thus the large N and small mass limits don't commute. In other words, if μ/p_i is small but fixed and we start increasing N, then the estimate of eq. (3.7) would be valid at first but at a large value of $N[N\sim(p_i/\mu)^2/\ln(p_i/\mu)]$ cross-over to the estimate of eq. (3.11) would occur. This result can easily be understood in terms of large-order Feynman graphs. Imagine attempting to expand the sum of all Nth order graphs about the zero-mass limit. Multiple mass insertions on the same propagator produce infrared divergences. As is well known, this problem can be overcome by a resummation procedure which changes the expansion parameter from $(\mu/p_i)^2$ to $(\mu/p_i)^2\ln(\mu/p_i)$. The number of propagators in an Nth order graph is $O(N)$, so there are $O(N)$ places to make mass insertions. Thus the correct expansion involves $N(\mu/p_i)^2\ln(\mu/p_i)$, in agreement with the instanton estimate. It would be interesting to see if the other features of these estimates (locality, suppression compared to massless case) could be derived from a statistical treatment [7] of Feynman graphs.

Of course we have been neglecting ultraviolet Borel singularities [5], "renormalons". There are single diagrams which are larger than A_N as predicted above. Probably, the best one can hope is that these estimates, in some sense, describe the sum of all diagrams not including the ones responsible for "renormalons".

4. Higgs theories

Consider an SU(2) gauge field coupled to a Higgs scalar of "isospin" q:

$$\mathcal{L} = \operatorname{tr}\left\{\frac{1}{4g^2}F_{\mu\nu}^2 + \frac{1}{\lambda}\left[\tfrac{1}{2}(D_\mu\phi)^2 + \tfrac{1}{8}(\phi^2 - \langle\phi\rangle^2)^2\right]\right\}. \tag{4.1}$$

If $\langle\phi\rangle = 0$ there is an instanton

$$A_{\mu,0} = \frac{\sigma^a 2\rho^2\eta_{a\mu\nu}x_\nu}{x^2(x^2+\rho^2)}, \qquad \phi = 0. \tag{4.2}$$

For $\langle\phi\rangle^2 > 0$ no non-trivial solution exists as can be seen from the transformation

$$A_\mu(x) \to aA_\mu(ax), \qquad \phi(x) \to \phi(ax) \tag{4.3}$$

(which preserves the finite action boundary condition $\phi \to \langle\phi\rangle$).

$$S \to \operatorname{tr}\int d^4x\left\{\frac{1}{4g^2}F^2 + \frac{1}{\lambda}\left[a^{-2}\tfrac{1}{2}(D\phi)^2 + a^{-4}\tfrac{1}{8}(\phi^2 - \langle\phi\rangle^2)^2\right]\right\}. \tag{4.4}$$

The action is reduced by shrinking the scale to zero. In what follows we shall be interested in scales, ρ, such that $\rho\langle\phi\rangle \lesssim \sqrt{\lambda} \ll 1$. In this limit both Euler-Lagrange equations can be solved:

$$\frac{1}{g^2} D_\mu F_{\mu\nu} = \frac{1}{\lambda}\frac{1}{2}\phi \overleftrightarrow{D}_\nu \phi \to 0, \tag{4.5}$$

$$D^2\phi = \tfrac{1}{2}\phi(\phi^2 - \langle\phi\rangle^2) \to 0. \tag{4.6}$$

The solutions are $A_{\mu,0}$ and [1]

$$\phi_0(x) = \left(\frac{x^2}{x^2+\rho^2}\right)^q \langle\phi\rangle. \tag{4.7}$$

As in sect. 2 we would naively expect to be able to find solutions perturbatively in $\langle\phi\rangle^2$. Again as in sect. 2 the desired boundary conditions can't be enforced due to zero modes. The equations for the leading corrections are

$$\frac{1}{g^2}\left[-\delta_{\mu\nu} D^2 + 2F_{\mu\nu}\right]\delta A_\nu = \frac{1}{\lambda}\phi_0 \overleftrightarrow{D}_\mu \phi_0, \tag{4.8}$$

$$D^2\delta\phi = \tfrac{1}{2}\phi_0(\phi_0^2 - \langle\phi\rangle^2). \tag{4.9}$$

The operators $(-\delta_{\mu\nu}D^2 + 2F_{\mu\nu})$ and D^2 have respectively the zero modes $\partial A_{\mu,0}/\partial\rho$ and ϕ_0. Hence we are not free to impose the desired boundary conditions on δA_ν and $\delta\phi$. Perturbative solutions can only be constructed if we add extra terms to the right-hand sides of both eqs. (4.8) and (4.9) with coefficients that can be adjusted order by order in perturbation theory to obtain the desired boundary conditions. Let us find out what the desired boundary conditions are. At $x \to \infty$ both equations linearize:

$$\left[-\delta_{\mu\nu}\partial^2 + \partial_\mu\partial_\nu + \frac{g^2}{\lambda}\langle\phi\rangle^2\right]\delta A_\nu = 0, \tag{4.10}$$

$$[-\partial^2 + \langle\phi\rangle^2](\phi - \langle\phi\rangle) = 0. \tag{4.11}$$

The solutions that vanish at infinity are

$$\phi^0 - \langle\phi\rangle \propto G_{\langle\phi\rangle}(x), \tag{4.12}$$

$$A_\mu^0(x) \propto \sigma^a \eta_{a\mu\nu}\partial_\nu G_{(g\langle\phi\rangle/\sqrt{\lambda})}(x). \tag{4.13}$$

We must expand in $\langle\phi\rangle$ to obtain the correct boundary conditions:

$$\phi^0(x)=\langle\phi\rangle-\frac{q\rho^2\langle\phi\rangle}{x^2}-\tfrac{1}{2}q\rho^2\langle\phi\rangle^3\ln\frac{c|x|}{\langle\phi\rangle}+\cdots, \tag{4.14}$$

$$A_\mu^0(x)=\sigma^a\eta_{a\mu\nu}\left[\frac{2x_\nu}{x^4}-\frac{1}{2}\frac{g^2\langle\phi\rangle^2}{\lambda}\frac{x_\nu}{x^2}+\cdots\right]. \tag{4.15}$$

The overall constants have again been determined by matching onto ϕ_0, $A_{\mu,0}$ at $\rho\ll x\ll\langle\phi\rangle^{-1}$. Thus the correct small $\langle\phi\rangle$ expansion is

$$\phi(x)=\phi_0(x)+\langle\phi\rangle^3\rho^3\ln\langle\phi\rangle\rho\phi_1(x)+\cdots, \tag{4.16}$$

$$A_\mu(x)=A_{\mu,0}(x)+\langle\phi\rangle^2\rho^2A_{\mu,1}(x)+\cdots, \tag{4.17}$$

where

$$\phi_1(x)\to-\tfrac{1}{2}q/\rho,$$

$$A_{\mu,1}(x)\to-\frac{1}{2}\frac{g^2}{\lambda\rho^2}\sigma^a\eta_{a\mu\nu}\frac{x_\nu}{x^2},\qquad\cdots. \tag{4.18}$$

As before, the required extra terms can be obtained by imposing a constraint. The constraint must now be a sum of two terms in order to have independently adjustable parameters in each equation (the relative weight of the two terms being a function of $\langle\phi\rangle\rho$ which is adjusted order by order in perturbation theory). The simplest such sum of operators consists of one depending only on A_μ and the other only on ϕ. The gauge field operator must have dimension other than four in order to give non-zero projection onto $\partial A_\mu/\partial\rho$ (e.g. $\operatorname{tr}F^3,\operatorname{tr}(D^2F)^2,\ldots$). It also must vanish at infinity rapidly enough that eq. (4.10) is still correct asymptotically. Similar conditions apply to the scalar operator, suitable candidates being $\operatorname{tr}(\phi^2-\langle\phi\rangle^2)^3,\ldots$.

Proceeding in this way we obtain a constrained instanton which looks like the pure Yang-Mills instanton (and $\phi=\phi_0$) inside a small core of radius $\rho(\lesssim\langle\phi\rangle^{-1}\sqrt{\lambda})$ and is given by $-4\pi^2\sigma^a\eta_{a\mu\nu}\partial_\nu G_{(g\langle\phi\rangle/\sqrt{\lambda})}(x)$, with $\phi=\langle\phi\rangle(1-4\pi^2q\rho^2G_{\langle\phi\rangle}(r))$, elsewhere. It decays exponentially, $A_\mu\sim e^{-(g\langle\phi\rangle|x|/\sqrt{\lambda})}$, $\phi-\langle\phi\rangle\sim e^{-\langle\phi\rangle|x|}$. Substituting this expansion into S, we find

$$\operatorname{tr}\int d^4x\,F^2=8\pi^2+O[\langle\phi\rangle^4\rho^4], \tag{4.19}$$

$$\operatorname{tr}\int d^4x(D_\mu\phi)^2=\int dS_\mu\phi_0\partial_\mu\phi_0=4\pi^2q\rho^2\langle\phi\rangle^2+O[\langle\phi\rangle^4\rho^4\ln\langle\phi\rangle\rho], \tag{4.20}$$

$$\operatorname{tr}\int d^4x(\phi^2-\langle\phi\rangle^2)^2=O[\langle\phi\rangle^4\rho^4\ln\rho\langle\phi\rangle], \tag{4.21}$$

$$S=\frac{8\pi^2}{g^2}+\frac{4\pi^2q\rho^2\langle\phi\rangle^2}{\lambda}+O\left[\frac{(\rho\langle\phi\rangle)^4\ln\langle\phi\rangle\rho}{\lambda}\right]. \tag{4.22}$$

The instanton contribution to the vacuum energy is then

$$\frac{E_0}{V} \sim \frac{1}{g^8} \int \frac{d\rho}{\rho^5} (\rho\Lambda)^{22/3-\beta(q)} e^{-(8\pi^2/g^2)-(4\pi^2 q\rho^2\langle\phi\rangle^2/\lambda)}$$

$$\sim \frac{\lambda^{(5/3-\beta(q)/2)}}{g^8} e^{-8\pi^2/g^2_{\rm eff}(\langle\phi\rangle)} . \qquad (4.23)$$

[$\beta(q)$ is proportional to the scalar contribution to the β function.] As promised, the integral is cut off at scales $\rho \sim \langle\phi\rangle^{-1}\sqrt{\lambda}$ so that our approximations were justified. Including corrections to S would give a series in powers and logs of λ and g^2. [The leading correction is down by $\lambda \ln \lambda$.]

This leading result was obtained by 't Hooft [1]. We have simply rederived it in a way which indicates the limit in which it is correct [$g^2_{\rm eff}(\langle\phi\rangle), \lambda_{\rm eff}(\langle\phi\rangle) \ll 1$] and shows the nature of the corrections. This derivation also makes it clear that the constrained instanton decays exponentially at large $|x|$ and thus does not affect the long-range behavior of the Higgs theory. If we were naively to calculate the gauge field propagator using the pure Yang-Mills instanton we would find

$$\langle A_\mu(x) A_\nu(y)\rangle \underset{|x-y|\to\infty}{\sim} e^{-8\pi^2/g^2(\langle\phi\rangle)} \int d^4x_0 A_{\mu,0}(x-x_0) A_{\nu,0}(y-x_0)$$

$$\sim \frac{e^{-8\pi^2/g^2(\langle\phi\rangle)}}{(x-y)^2} . \qquad (4.24)$$

[An equivalent, gauge-invariant result would be obtained from the Wilson loop average.] Thus we would arrive at the absurd conclusion that an exponentially small instanton effect is potent enough to prevent the Higgs phenomenon from occurring (a tree-level effect). This paradox is avoided by the exponential decay of the constrained instanton. (Another method of avoiding this paradox was pointed out by Levine [8]. If we do not introduce a constraint, but simply expand the functional integral about $A_{\mu,0}$ and $\langle\phi\rangle$, then the action contains terms linear in the "quantum fluctuations", in particular a term $(1/2\lambda)\int d^4x\, \mathrm{tr}\, A_\mu \delta A_\mu \langle\phi\rangle^2$. Summing over insertions of this vertex is equivalent to summing over insertions of the mass $m^2 = (g^2/\lambda)\langle\phi\rangle^2$. The constrained instanton incorporates this infinite set of "quantum corrections" at the classical level.)

I would like to thank E. Witten for interesting me in this subject, and S. Coleman for some very useful questions and comments.

Appendix

EXISTENCE PROOF

In this appendix we give a proof of the existence of a constrained instanton solution, with the constraint $\int d^4x\phi^6 = c\rho^{-2}$, for $\rho\mu$ less than some critical value. We will look for a spherically symmetric solution. For a scalar potential $U(\phi)$, the euclidean equation of motion is

$$\frac{d^2\phi}{dr^2} = -\frac{3}{r}\frac{d\phi}{dr} + U'(\phi). \tag{A.1}$$

This is the equation for a particle moving in the potential $-U$ with position ϕ at time r and a frictional force with time-dependent coefficient $3/r$. Consider first the potential $U = -(1/4!)\phi^4$. Then the solution [eq. (2.3)] represents the particle starting at $\phi(0)$ with zero velocity, rolling down to the bottom of the well (fig. 1) and stopping (due to friction) without oscillation. If we add a mass term (fig. 2) it is obvious why there is no longer a solution. The particle doesn't have enough energy to climb the hill near $\phi = 0$. If we add a $\sigma\phi^6$ term then $-U$ is as in fig. 3 for sufficiently small σ and a solution exists. To see this, simply observe that if the particle starts very close to the top of the higher hill then it can stay there a very long time, losing essentially no energy, until the frictional coefficient is essentially zero,

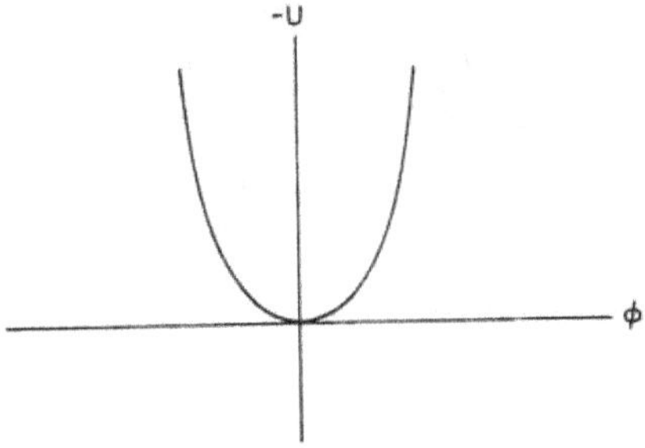

Fig. 1. The potential, $-U$, for the analogue classical particle problem in the case $-U = (1/4!)\phi^4$.

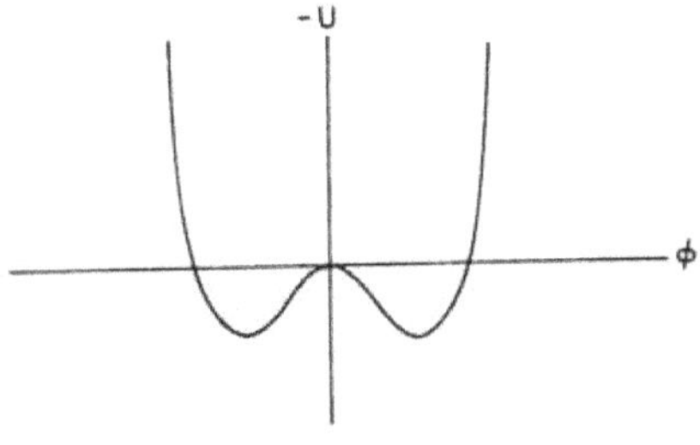

Fig. 2. The same for $-U = -\frac{1}{2}\mu^2\phi^2 + (1/4!)\phi^4$.

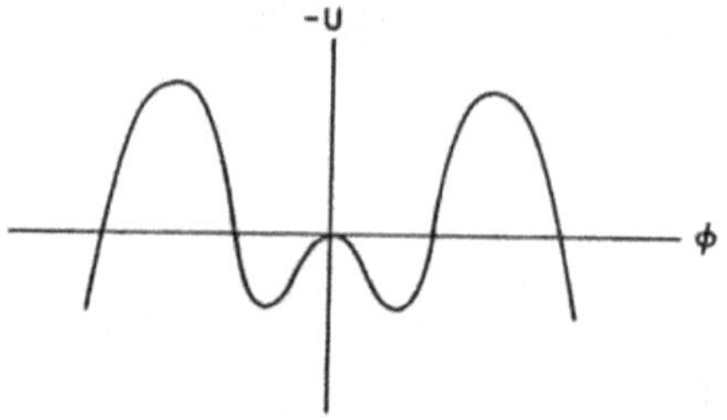

Fig. 3. The same for $-U = -\frac{1}{2}\mu^2\phi^2 + (1/4!)\phi^4 - \sigma\phi^6$.

whereafter it can travel through the well with virtually no energy loss and easily overshoot $\phi = 0$. On the other hand undershoot can easily be arranged for $\phi(0)$ close enough to zero. By continuity there must exist an intermediate choice of $\phi(0)$ such that the particle just comes to rest at $\phi = 0$ as $r \to \infty$. This argument applies for all σ small enough that the double hills are higher than the hill at $\phi = 0$. This range of σ corresponds to a finite range of $\int d^4x\,\phi^6 \equiv c\rho^{-2}$. Our perturbative arguments showed that this range includes $\mu\rho \to 0$ $(\sigma \to 0)$. In terms of the particle analogy, when the hill at $\phi = 0$ becomes very small the larger hill can move very far away and still give the particle a slightly smaller initial acceleration allowing it to lose less energy and hence reach the top of the hill at $\phi = 0$.

References

[1] G. 't Hooft, Phys. Rev. D14 (1976) 3432

[2] Y. Frishman and S. Yankielowicz, Phys. Rev. D19 (1979) 540

[3] L. Lipatov, JETP (Sov. Phys.) 72 (1977) 411

[4] E. Brézin, J. Le Guillou and J. Zinn-Justin, Phys. Rev. D15 (1977) 1544

[5] G. 't Hooft, 1977 Ericè Lecture Notes, *in* Whys of subnuclear physics, ed. A. Zichichi (Plenum Press, N.Y., 1979);
B. Lautrup, Phys. Lett. 69B (1977) 109;
P. Olesen, Phys. Lett. 73B (1977) 327

[6] G. Parisi, Nucl. Phys. B150 (1979) 163

[7] C. Bender and T.T. Wu, Phys. Rev. Lett. 27 (1971) 461; Phys. Rev. D7 (1972) 1620

[8] H. Levine, Nucl. Phys. B172 (1980) 119

Nuclear Physics B165 (1980) 45–54

INSTANTONS IN NON-PERTURBATIVE QCD VACUUM

M.A. SHIFMAN, A.I. VAINSHTEIN* and V.I. ZAKHAROV
Institute of Theoretical and Experimental Physics, Moscow 117259, *USSR*

Received 19 September 1979

The effects of large-scale vacuum fluctuations on instantons of a small size ρ are considered. We find the instanton density with account of the gluon vacuum condensate. It is shown that the instanton gas approximation fails at unexpectedly small ρ. The instanton interaction with the gluon condensate already becomes strong at $\rho \approx (1.1\ \mathrm{GeV})^{-1}$, while the absolute bound for the use of quasiclassical methods lies at $\rho \approx (0.5\ \mathrm{GeV})^{-1}$.

There is no theory of the physical vacuum state in QCD. Still, it seems safe to say that fluctuations of the gluon field which go beyond mere perturbation theory are most important. Moreover, rather large-scale fluctuations of size R, $R \sim (200\ \mathrm{MeV})^{-1}$, dominate. The lack of understanding of the quark-gluon interaction at such distances hampers the construction of explicit vacuum wave function.

It is clear then that any particular example of non-perturbative vacuum fields is very interesting and therefore it is not surprising that the celebrated BPST instanton [1] is one of the most widely discussed objects nowadays. Instantons of small size are described reliably and must be present in the vacuum state. However, the probability of finding them is very low and in the limit of the vanishing size instantons are irrelevant. On the other hand, in the limit of large instanton size the one-instanton field is greatly modified by interactions. Nevertheless, it is tempting to assume [2] that the instanton gas approximation makes sense up to $\rho \sim R$ so that the vacuum can be considered as an ensemble of non-interacting (or weakly interacting) instantons in a rough but valuable approximation.

In this paper we investigate the range of validity of the instanton gas approximation. To this end we consider interaction of a small-size instanton with large-scale vacuum fluctuations. The correction to the standard instanton density [3]** due to this interaction vanishes for $\rho \to 0$, but grows rapidly with ρ.

* Permanent address: Institute for Nuclear Physics, Novosibirsk 90, USSR.

** The expression for the instanton density quoted below for $SU(3)_c$ differs by a factor 64 from what was first proposed in ref. [2]. The difference is apparently due to the literal use by the authors of this paper of the original 't Hooft result [3] for the $SU(2)_c$ instantons which omits the factor $2^{-k/2}$, k being the number of the zero modes (see the erratum [4]).

In other words, we view an instanton as developing in a physical vacuum densely populated by large-scale fluctuations, while the standard derivation describes the instanton evolution in the mathematical vacuum of perturbation theory. In this respect the present work is a direct extension of our previous paper [5] where we accounted in a similar fashion for the effect of the quark condensate in the vacuum. Here we address ourselves to the role of the vacuum gluon field.

The central point is that "external" vacuum fields are much weaker than the field of the small-size instanton considered and therefore the effect of the large-scale vacuum fields can be treated perturbatively. More precisely, we use the operator expansion technique [6] modified to incorporate non-perturbative terms as well [5].

We do not assume any particular vacuum structure and are able to reduce the effect to certain phenomenological quantities. The most important one is the vacuum expectation value of the gluon field strength $G^a_{\mu\nu}$ squared:

$$\langle 0|\frac{\alpha_s}{\pi} G^a_{\mu\nu}G^a_{\mu\nu}|0\rangle \approx 0.012 \text{ GeV}^4 , \tag{1}$$

where the number is borrowed from the analysis [7–9] of QCD sum rules sensitive to the vacuum fluctuations.

It is essential that the small-size instanton already interacts with the physical vacuum at the classical level. This might explain our result that the vacuum "medium" strongly affects instantons of rather small size, $\rho \approx (1.1 \text{ GeV})^{-1}$, which is very far from the expected crucial size $\rho_c \approx (200 \text{ MeV})^{-1}$. (When comparing the numbers one must keep in mind that the ρ dependence peaks very sharply at the upper end.)

We find that the effect of the vacuum fields on instantons is of a definite sign so that the vacuum stimulates instantons.

For $\rho > (1.1 \text{ GeV})^{-1}$ the standard expression for the instanton density [3, 2] should be abandoned if the physical vacuum is considered. At the price of an additional, although plausible, assumption we are able to consider instantons of larger size, up to $\rho \sim (500 \text{ MeV})^{-1}$. In the interval of $\rho \sim (1.1 \text{ GeV})^{-1}$–$(0.5 \text{ GeV})^{-1}$ we account for the medium effects quasiclassically, but in all orders in the vacuum field. Here the instanton density grows exponentially and at $\rho = (0.5 \text{ GeV})^{-1}$ is about six orders of magnitude higher than the well-known 't Hooft result. At $\rho > (0.5 \text{ GeV})^{-1}$ the quasiclassical approach is not valid any longer and one must invent some other method which goes far beyond the quasiclassical approximation based picture.

Thus, the quasiclassical approximation is not self-consistent. Accounting for the instanton interaction with the vacuum does not introduce any intrinsic cut-off on the instanton scale but, to the contrary, results in an exponential blow-up at $\rho \approx (0.5 \text{ GeV})^{-1}$. It seems, therefore, that it is quantum effects which enter the game and provide a closed description. Unfortunately, at present nothing is known on this phenomenon.

An important question is whether fluctuations of the size $\rho<(0.5\text{ GeV})^{-1}$ — for which one has a quantitative description — control hadron physics. In other words, the problem is whether the instanton approximation, bounded by the condition $\rho<(0.5\text{ GeV})^{-1}$, can be relevant to the hadronic states.

Althtough there might be no universal answer to this question, it seems reasonable to approach the problem in the following way. Consider the two-point function

$$i\int \mathrm{d}x\, \mathrm{e}^{iqx}\langle 0|\mathrm{T}\{A(x), B(0)\}|0\rangle\,, \tag{2}$$

where A and B are some local operators. Its imaginary part can be expressed in terms of observable quantities (at least in some cases, say, for $A, B = j_{\text{e.m.}}$).

At euclidean $q\sim 1$ GeV, the two-point function (2) becomes sensitive to the contribution of low-lying resonances. The problem, therefore, is whether the $\rho<(0.5\text{ GeV})^{-1}$ instantons are relevant to $q\sim 1$ GeV. The answer is a flat no.

Really, at large (euclidean) momenta, q, the contribution of instantons with (variable) size $\rho\sim 1/q$ is readily distinguishable from that of the fixed-size vacuum fluctuations. The point is that the former yield the piece which is proportional to $(1/q^2)^\gamma$, where γ is not necessarily an integer, while the latter give rise to terms regular in $1/q^2$.

Diminishing Q we move towards larger ρ and the question is whether or not the instanton approximation is valid for $Q\sim 1$ GeV. The crucial number then is the proportionality coefficient C between $1/Q$ and ρ_{eff}, $\rho_{\text{eff}}=C/Q$ (ρ_{eff} is an effective value of ρ in the integral over ρ; $Q^2=-q^2$). For the standard 't Hooft instanton density $C\approx 5$. If the modification due to the interaction with the vacuum field is taken into account, then C becomes even larger. Thus, the standard instanton density can only be used for $Q^2\gtrsim(5.5\text{ GeV})^2$, where the contribution of the instantons of size $\sim 1/Q$ is extremely small, both compared to the ordinary perturbation theory and to terms regular in $1/Q^2$. Relying on the modified instanton density extends the (modified) instanton approximation up to $\sim(3\text{ GeV})^2$, but it is still far enough from the really interesting region $Q^2\sim 1\text{ GeV}^2$ sensitive to a single hadronic state.

We proceed now to a systematic derivation of the results announced and start for simplicity with pure gluodynamics.

Since we consider fields of two distinct types, namely, small-size instantons and fixed-size vacuum fluctuations it is convenient to introduce an effective lagrangian. As usual, one integrates explicitly over rapidly varying fields and includes their contribution into the coefficients in front of various operators which act in the space of the slowly varying fields (dominant in the vacuum state).

Thus, if one considers a particular instanton of size ρ centered at x_0, then its effect reduces to the following lagrangian:

$$\Delta\mathscr{L}(x_0)=\frac{\mathrm{d}\rho}{\rho^5}\sum_n C_n(\rho)O_n(x_0)\,, \tag{3}$$

where $C_n(\rho)$ are c-number coefficients and $O_n(x_0)$ are the local operators constructed from the gluon field (all the physical quantities do not depend, of course, on x_0; the integration over x_0 resores the uniformity of the vacuum state).

The probability of finding the instanton considered, or in other words the instanton density, is given by lagrangian (3) averaged over the physical vacuum state.

To find the coefficients C_n explicitly use the following trick. In the quasiclassical approximation evaluate the transition between states a and b in the presence of the instanton (a and b belong to the set of perturbation theory states and correspond, say, to free gluons). Then, compare the result with the matrix element

$$\langle \mathrm{b}|\frac{\mathrm{d}\rho}{\rho^5}\sum_n C_n(\rho)O_n(x_0)|\mathrm{a}\rangle .$$

Choosing the states a and b in an appropriate way picks up a certain coefficient C_n. It is worth emphasizing once more that the effective lagrangian is meaningful only if the external momenta are much smaller than ρ^{-1}, in particular, we take $p_{\mathrm{a,b}} \ll \rho^{-1}$.

In the limit of vanishing ρ, the expansion (3) is dominated by the unit operator I. The corresponding coefficient C_I has been found by 't Hooft and for $\mathrm{SU(3)_{color}}$ is equal to

$$C_I(\rho) \equiv d(\rho) \simeq 1.7 \cdot 10^{-3}\left(\frac{2\pi}{\alpha_s(\rho)}\right)^6 \mathrm{e}^{-2\pi/\alpha_s(\rho)}[1+\mathrm{O}(\alpha_s(\rho))], \tag{4}$$

$$\frac{2\pi}{\alpha_s(\rho)} = \frac{2\pi}{\alpha_s^{(0)}} - b\ln M_0\rho + \frac{b'}{b}\ln\left(1-\frac{\alpha_s^{(0)}}{2\pi}b\ln M_0\rho\right) + \mathrm{O}(\alpha_s(\rho)),$$

where $\alpha_s^{(0)}$ is the bare coupling constant regularized according to the Pauli–Villars prescription; M_0 is the regulator mass, and b, b' are the coefficients in the Gell-Mann–Low function, which for pure gluodynamics are [10, 11]

$$b = 11, \qquad b' = 51 .$$

The coefficient $C_I(\rho)$ is singled out by averaging the lagrangian (3) over the vacuum state of the perturbation theory. To find the other coefficients consider the transition of the vacuum state into n gluons with momenta much smaller than ρ^{-1} and use the reduction formula for the corresponding amplitude:

$$M(\mathrm{vac} \to n \text{ gluons}) = \mathrm{T}\langle \mathrm{vac}|\int \prod_{k=1}^{n} i\,\mathrm{d}x_k\, \mathrm{e}^{iq_kx_k}\epsilon_{\mu_k}^{a_k}(q_k)q_k^2\mathscr{A}_{\mu_k}^{a_k}(x_k)|\mathrm{vac}\rangle , \tag{5}$$

where $\epsilon_{\mu_k}^{a_k}(q_k)$ and q_k are the polarization and 4-momentum of the kth gluon, respectively, and $\mathscr{A}_{\mu_k}^{a_k}(x)$ are the operators of the gluon field.

To find the Green function (5) in the quasiclassical approximation it is sufficient to proceed to euclidean space-time, $(x_k)_0 \to -i(x_k)_4$ and to substitute $\mathscr{A}_{\mu_k}^{a_k}(x_k)$ by the classical one-instanton solution $A^{a_k}_{\mu_k}(x_k - x_0)$,

$$A^a_\mu(x-x_0) = -\frac{2}{g}\frac{\bar\eta_{a\mu\nu}(x-x_0)_\nu\rho^2}{[(x-x_0)^2+\rho^2](x-x_0)^2}. \tag{6}$$

Moreover, the vacuum-to-vacuum amplitude $C_I(\rho)$ enters as an overall factor so that the final result looks as

$$M(\text{vac}\to n \text{ gluons}) = \frac{d\rho}{\rho^5} C_I(\rho)\, e^{-i\Sigma q_k x_0}\{dx\, e^{-iqx} q^2 \epsilon^a_\mu A^a_\mu(x)\}^n, \tag{7}$$

where all the quantities on the right-hand side are euclidean.

It is worth emphasizing that the instanton field is taken in the singular gauge. The advantage of this gauge is that the potential $A^a_\mu(x-x_0)$ falls off rapidly at large distances and the inverse propagators entering the reduction formula (5) have the standard form, as for the free fields. For non-singular gauges the potential falls too slowly at infinity and q_k^2 in eq. (5) must be replaced by some more complicated expression.

The Fourier transform of the instanton solution in the limit of small ρ is readily found:

$$\int dx\, e^{-iqx} q^2 A^a_\mu(x) \underset{q\rho\to 0}{=} \frac{4\pi^2 i}{g}\bar\eta_{a\mu\nu} q_\nu \rho^2. \tag{8}$$

Substituting (8) into the amplitude (7) and comparing the result with the matrix element $\langle n \text{ gluons}|\Delta\mathcal{L}|\text{vac}\rangle$ we find $\Delta\mathcal{L}$:

$$\Delta\mathcal{L}(x_0) = \frac{d\rho}{\rho^5} C_I(\rho) \exp\left\{-\frac{2\pi^2}{g}\rho^2 \bar\eta_{a\mu\nu} G^a_{\mu\nu}(x_0)\right\}. \tag{9}$$

where $G^a_{\mu\nu}$ is the operator of the gluon field strength tensor. Note that the factorial-like factors characteristic to the exponential arise due to combinatorics when one converts the amplitude found into the effective lagrangian. Note also that eq. (9) assumes the color orientation of the instanton considered to be fixed so that the whole expression is to be multiplied, in fact, by the integration measure in color space, $d\mu$, normalized to unity.

Expression (9) for the case of an instanton interacting with an external field was first derived by Callan, Dashen and Gross [2] in an alternative way. The derivation presented above might be more transparent than the original one. What is most important, unlike in refs. [2, 12] we do not fix external field $G^a_{\mu\nu}$ "by hand" but ascribe it to the large-scale vacuum fluctuations.

The next step in calculating the instanton density is the averaging of lagrangian (9) over the physical vacuum. Then the term linear in $G^a_{\mu\nu}$ evidently drops off and the first non-vanishing correction is proportional to G^2:

$$d_{\rm eff}(\rho)\,d\rho/\rho^5 \equiv \langle 0|\Delta\mathscr{L}|0\rangle_{\rm phys}$$

$$=\frac{d\rho}{\rho^5}\,d(\rho)\left\{1+\frac{\pi^4\rho^4}{8\alpha_s^2(\rho)}\langle 0|\frac{\alpha_s}{\pi}G^a_{\mu\nu}G^a_{\mu\nu}|0\rangle + O(\rho^6)\right\}. \tag{10}$$

Thus, the leading correction to the 't Hooft density is expressed in terms of the known vacuum expectation value (1). Note that we present the result in terms of the combination $\alpha_s G^2$ since this product does not depend on the normalization point. Moreover, it is evident that once the vacuum averaging is performed the answer does not depend on the orientation of the instanton considered in the color space.

It is remarkable that the correction to the instanton density is well-defined numerically and no strong dependence on poorly known parameters, such as Λ, enters (Λ is the standard scale parameter which appears in the definition of the quark-gluon coupling constant, $\alpha_s(\rho)=2\pi/b\ln(1/\Lambda\rho)$).

There is no difficulty to find next terms in the expansion (10) associated with higher powers of G. To specify them numerically one must learn the vacuum expectation values of the operators containing high powers of $G^\alpha_{\mu\nu}$. There is no independent phenomenological information on such matrix elements and we are forced to rely on some approximation. Actually we will use for this purpose the factorization hypothesis which reduces $\langle 0|(G^2)^n|0\rangle$ to $[\langle 0|G^2|0\rangle]^n$ by assuming the vacuum intermediate state dominance. Such an approximation is widely used under similar circumstances in statistical and solid state physics. Moreover, we had an opportunity [8] to check it in the case of certain four-quark operators and convinced ourselves that it works within an accuracy of several percent. Still, in the absence of a complete theory the factorization hypothesis cannot be proved, especially for the operators of high dimension.

Using the factorization and accounting for all types of contractions we find

$$\langle 0|\underbrace{\frac{2\pi^2}{g}\rho^2\bar\eta_{a_1\mu_1\nu_1}G^{a_1}_{\mu_1\nu_1}\cdots\frac{2\pi^2}{g}\rho^2\bar\eta_{a_{2k}\mu_{2k}\nu_{2k}}G^{a_{2k}}_{\mu_{2k}\nu_{2k}}}_{2k\ \text{factors}}|0\rangle$$

$$=(2k-1)!!\left[\frac{4\pi^4}{g^2}\rho^4\langle 0|\bar\eta_{a\mu\nu}\bar\eta_{b\alpha\beta}G^a_{\mu\nu}G^b_{\alpha\beta}|0\rangle\right]^k.$$

As a result, the following simple expression for the effective instanton density arises*:

$$d_{\rm eff}(\rho)=d(\rho)\exp\left[\frac{\pi^4\rho^4}{8\alpha_s^2(\rho)}\langle 0|\frac{\alpha_s}{\pi}G^a_{\mu\nu}G^a_{\mu\nu}|0\rangle\right]. \tag{11}$$

Substituting the explicit form of $d(\rho)$ gives

$$d_{\rm eff}(\rho)\simeq 1.7\cdot 10^{-3}\left(\frac{2\pi}{\alpha_s(\rho)}\right)^6\exp\left\{-\frac{2\pi}{\alpha_s(\rho)}\left[1-\frac{\pi^3\rho^4}{16\alpha_s(\rho)}\langle 0|\frac{\alpha_s}{\pi}G^2|0\rangle\right]\right\}.$$

Thus, we have derived two modified expressions for the instanton density. The modification is due to the interaction with large-scale vacuum fields and the first one is valid only as long as $\rho^4\langle 0|\alpha_s G^2|0\rangle$ is small, while the other sums up all the orders in this parameter.

To understand the range of validity of the results let us compare various factors which determine $d_{\rm eff}$. The most important factor, $\exp(-2\pi/\alpha_s)$, is due to the classical instanton action. The pre-exponential factor is associated with gaussian quantum fluctuations. Eq. (10) gives the leading correction to the standard pre-exponential factor. It assumes that the effect of the vacuum field on the instanton considered is less important than that of the quantum fluctuations. Therefore, the original one-instanton solution makes sense in the physical vacuum only as far as

$$\frac{\pi^4\rho^4}{8\alpha_s^2(\rho)}\langle 0|\frac{\alpha_s}{\pi}G^2|0\rangle<1,\qquad \rho_{\rm crit}\simeq 1/1.15\ {\rm GeV},$$

where for the numerical estimate we use eq. (1) and $\alpha_s(\rho)=2\pi/9\ln(1/\Lambda\rho)$ with $\Lambda=100$ MeV. At $\rho\gtrsim(1.1\ {\rm GeV})^{-1}$, interaction with the vacuum field becomes important.

* It is worth emphasizing that we get the exponential in eq. (11) only because of the combinatorial factors arising as a result of application of the factorization hypothesis. In particular, we do not get it if $G^a_{\mu\nu}$ is taken to be an external field and not the operator of a fluctuating field. On the other hand, in ref. [12] a very similar equation is claimed to hold for the instanton density just in the case of an external field:

$$d_{\rm ext}(\rho)=d(\rho)\exp\left[\frac{\pi^4}{2g^2}\rho^4(F^2-F\tilde F)n^2\right],\qquad F^2=F_{\mu\nu}F_{\mu\nu},\qquad F\tilde F=\tfrac{1}{2}F_{\alpha\beta}F_{\mu\nu}\epsilon_{\alpha\beta\mu\nu}.$$

Here $F_{\mu\nu}n^a\equiv G^{a\,\rm ext}_{\mu\nu}$ is an abelian external field and averaging over the instanton orientation is performed. The similarity is pure superficial, however. Moreover, we have calculated the first two terms in $d_{\rm ext}$ and have not reproduced the equation quoted above:

$$d_{\rm ext}(\rho)=d(\rho)\left[1+\frac{\pi^4\rho^4}{2g^2}(F^2-F\tilde F)n^2+\tfrac{2}{5}\left(\frac{\pi^4\rho^4}{2g^2}\right)^2(F^2-F\tilde F)^2(n^2)^2+\dots\right].$$

Note, that to derive the second term it is essential to use the identity $d^{abc}d^{dec}n^an^bn^en^d=\frac{1}{3}(n^2)^2$ for the SU(3) d^{abc} symbols. Moreover, in the next order a new non-trivial invariant $(F^2)^3(d^{abc}n^an^bn^c)^2$ emerges. Note that the actual calculation in ref. [12] does not go beyond the first term.

As a result of the summation of all the orders in the vacuum field, expression (11) for the instanton density holds for larger ρ as well. Now there is no need to require the dominance of the gaussian fluctuations near the instanton over the effect of the external vacuum fields. The relation between these two types of contribution can be arbitrary, provided that both of them are less important than the factor $e^{-2\pi/\alpha_s}$. Once the effect of the vacuum field becomes stronger than that of the classical action for the instanton solution, the quasiclassical methods cannot be used any longer. The corresponding value $\tilde{\rho}_{crit}$ can be estimated from the following equation:

$$\frac{\pi^3 \tilde{\rho}^4_{crit}}{16\alpha_s(\tilde{\rho}_{crit})} \langle 0|\frac{\alpha_s}{\pi} G^2|0\rangle = 1\,, \tag{13}$$

which differs from eq. (12) by the factor $2\pi/\alpha_s$. Numerically,

$$\tilde{\rho}_{crit} \approx 1/500\ \text{MeV}\,.$$

It is worth emphasizing that eqs. (11), (13) are based on the additional assumption of factorization of the matrix elements $\langle 0|(G^2)^n|0\rangle$. Moreover, there is some uncertainty due to possible variations in $\alpha_s(\rho)$ and $\langle 0|(\alpha_s/\pi)G^2|0\rangle$. In particular, $\langle 0|(\alpha_s/\pi)G^2|0\rangle$ seems to be known within a factor of two and the corresponding uncertainty in ρ_{crit}, $\tilde{\rho}_{crit}$ is $\sim 2^{1/4}$.

The general lesson we learn is that instanton approximation is not self-consistent. As far as we can calculate it the instanton density grows exponentially (in fact such a conclusion is not unexpected in view of the results obtained in refs. [2, 12]).

To be fully realistic, we must now switch on (nearly) massless quarks and consider the effect of the quark condensate as well. This was partly done in our previous paper [5], where we found the leading effect $\sim\langle 0|\bar{q}q|0\rangle$ which cancels the suppression of instantons in the bare vacuum due to massless quarks discovered by 't Hooft [2].

In the spirit of the present paper it would be desirable to improve the result of our previous work and perform the summation of all the terms which survive in the quasiclassical limit. As is readily seen, all such terms reduce to the matrix elements of the form $\langle 0|(\bar{q}q)^N G^k|0\rangle$, where N is the number of massless quarks, $k = 2, 3, \dots$). Higher powers of $(\bar{q}q)$ do not appear in the quasiclassical limit.

Consider first a single massless quark (and $SU(3)_{color}$). Then the first few terms in the presence of the quark condensate look like

$$d_{eff}(\rho) = d_0(\rho)\langle 0|\Big\{ -\tfrac{2}{3}\pi^2\rho^3\bar{q}q + \frac{\pi^4\rho^5}{4g}\, iG^a_{\mu\nu}\bar{q}\sigma_{\mu\nu}t^a q$$
$$-\tfrac{4}{3}\pi^2\frac{\pi^4}{2g^2}\rho^7(G^2 - G\tilde{G})\bar{q}\frac{1+\gamma_5}{2}q + O(\rho^9)\Big\}|0\rangle\,, \tag{14}$$

where t^a are the Gell-Mann SU(3) matrices and $d_0(\rho)$ is just equal to $1.3\ C_I(\rho)$. In other words, $d_0(\rho)$ coincides with the instanton density in the perturbation-theory vacuum divided by $m\rho$ (m is the mechanical quark mass tending to zero).

In the limit $\rho m \to 0$ the quark condensate, $\langle 0|\bar{q}q|0\rangle \neq 0$, determines the instanton density entirely and its effect cannot be neglected in any way. With increasing ρ, other terms enter the game. The phenomenological estimate of the second term on the right-hand side of eq. (14) is not very reliable. (A rough estimate can be obtained by considering sum rules for D-mesons [13]). Nevertheless, it shows that the second term certaintly does not dominate over the third one and, in general, is not very essential. If one neglects it and uses the factorization hypothesis for the rest, then the bound on the applicability of the one-instanton approximation turns out to be the same as above [see eqs. (10), (12)]. Moreover, neglecting operators $\sim\bar{q}\sigma_{\mu\nu}t^a qG^a_{\mu\nu}(G^2)^k$ which appear in higher orders, we are left with the operators $\sim\bar{q}q(G^2)^n$ whose matrix elements can be estimated *via* the factorization hypothesis. In this way we immediately come to

$$d_{\rm eff}(\rho) = d_0(\rho)\langle 0|-\tfrac{2}{3}\pi^2\rho^3\bar{q}q|0\rangle \exp\left[\frac{\pi^4\rho^4}{8\alpha_s^2(\rho)}\langle 0|\frac{\alpha_s}{\pi}G^2|0\rangle\right],$$

which is a direct analog of eq. (11). This equation is readily generalized to the theory with several light quarks so that the final answer for the effective instanton density contains an extra factor

$$\prod_{q=u,\,d,\,\ldots} 1,3\{m_q\rho - \tfrac{2}{3}\pi^2\langle 0|\bar{q}q|0\rangle\rho^3\}$$

as compared to the density (11).

It is clear that although we explicitly mention only the instanton configuration, all the consideration refers equally to the anti-instanton field with the obvious substitutions: $\bar{\eta}\to\eta$, $\gamma_5\to-\gamma_5$, $\tilde{G}\to-\tilde{G}$.

To summarize, vacuum fields affect the instanton density at distances which are much shorter than the characteristic size of the dominating vacuum fluctuations. Thus, to our mind, the hope that instanton-gas physics describes the vacuum structure is not realized. Note that a phenomenon of a similar type — instanton melting — has been discovered in the two-dimensional σ model [14]. The question which we succeded in answering quantitatively in the realistic case of QCD is "starting from what sizes are instantons deformed and finally destroyed?". We base our consideration on the phenomenological estimate (1) and the result is well-defined numerically. Moreover, we found that the instanton contribution to the two-point functions of the type (2) can be written out unambiguously only at very large Q^2 where it is negligibly small.

We are thankful to V.A. Fateev, B.L. Ioffe, V.A. Novikov, L.B. Okun, A.S. Schwartz and E.V. Shurjak for valuable discussions.

References

[1] A. Belavin, A. Polyakov, A. Schwartz and Yu. Tyupkin, Phys. Lett. 59B (1975) 85.
[2] C. Callan, R. Dashen and D. Gross, Phys. Rev. D17 (1978) 2717.
[3] G. 't Hooft, Phys. Rev. D14 (1976) 3432.
[4] G. 't Hooft, Phys. Rev. D18 (1978) 2199.
[5] M. Shifman, A. Vainshtein and V. Zakharov, Nucl. Phys. B163 (1980) 46.
[6] K. Wilson, Phys. Rev. 179 (1969) 1499;
A. Polyakov, ZhETF (USSR) 57 (1969) 270.
[7] A. Vainshtein, V. Zakharov and M. Shifman, ZhETF Pisma 27 (1978) 60.
[8] M. Shifman, A. Vainshtein and V. Zakharov, Nucl. Phys. B147 (1979) 385, 448.
[9] S. Eidelman, L. Kurdadze and A. Vainshtein, Phys. Lett. 82B (1979) 278.
[10] H.D. Politzer, Phys. Rev. Lett. 30 (1973) 1346;
D.J. Gross and F. Wilczek, Phys. Rev. Lett. 30 (1973) 1343.
[11] W. Caswell, Phys. Rev. Lett. 33 (1974) 244;
A. Belavin and A. Migdal, ZhETF Pisma 19 (1974) 317; Erratum *in* Elementary particles, Proc. 2nd ITEP School of Physics, Moscow, Atomizdat, (1975), vol. 3, p. 73;
D.R.T. Jones, Nucl. Phys. B75 (1974) 531.
[12] C. Callan, R. Dashen and D. Gross, Phys. Lett. 78B (1978) 307; Phys. Rev. D19 (1979) 1826.
[13] V.A. Novikov et al., Proc. Int. Conf. Neutrino-78, ed. E. Fowler, Purdue Univ., (1978), p. C278.
[14] V. Fateev, I. Frolov and A. Schwartz, Nucl. Phys. B154 (1979) 1.

Volume 168B, number 1,2 PHYSICS LETTERS 27 February 1986

COLLECTIVE-COORDINATE METHOD FOR QUASIZERO MODES

I.I. BALITSKY and A.V. YUNG

Leningrad Nuclear Physics Institute, Gatchina, Leningrad 188350, USSR

Received 19 November 1985

The collective-coordinate method for the quasizero modes is suggested. Quasizero mode means that a direction in functional space exists where the action varies slowly. As in the case of exact zero modes the corresponding integration is nongaussian and should be performed exactly. The method is illustrated by calculating the instanton–anti-instanton interaction in double-well quantum mechanics.

1. The calculation of a functional integral near the nontrivial solution of classical equations of motion is one of the central methods in modern theoretical physics. Usually the integration near this solution is performed in the gaussian approximation with the only exception of zero modes reflecting the symmetries of the system. The integration over zero modes is nongaussian but trivial and yields infinite factors such as the total volume of spacetime. Technically this integration is performed by the collective-coordinate method [1]: first one carries out the (gaussian) integrations in the directions orthogonal to zero modes and then performs the trivial integration over the collective coordinates corresponding to zero modes.

However, in many interesting cases there exists the parametrically small but nonzero mode in the functional space so that the corresponding integration is neither trivial nor gaussian and should be performed exactly. For example, the classical solution may possess one parametrically small (quasizero) mode (in addition to exact zero modes and the gaussian one). This is the case, for instance, when one considers the decay of false vacuum [2] in quantum mechanics with two non-degenerative minima potential. Then the classical solution is a wide bounce and the (negative) mode corresponding to compressing (or stretching) of the bounce is parametrically small. Another situation arises if one considers the interaction of a widely separated instanton and anti-instanton (in the symmetric double-well quantum mechanics or in a field theory). Then the mode corresponding to the change of instanton–anti-instanton separation is quasizero again. Note that in this case there is no classical solution from which the small mode starts the classical solution is reached only at infinite values of the instanton–anti-instanton separation. In general, the quasizero mode arises when there is a direction in the functional space where the action varies slowly (practically it means that changing a certain parameter in the classical solution slightly affects the action).

In this paper we suggest a general method to deal with such quasizero modes. This method will be illustrated afterwards by calculating the instanton–anti-instanton interaction in the double-well quantum mechanics. For simplicity we consider a quantum mechanical system although the treatment can easily be generalized to the case of a field theory. Suppose we calculate the matrix element of the (euclidean) evolution operator:

$$Z = \langle \exp(-HT) \rangle = N^{-1} \int D\varphi \exp[-S(\varphi)] \, , \quad (1)$$

$$S(\varphi) = \int dt \, [\dot{\varphi}^2/2 + V(\varphi)] \, , \quad (2)$$

with the boundaries to ensure the finite-action classical solution $\varphi_c(t)$ being the saddle point of the path integral:

$$\delta S/\delta\varphi|_{\varphi=\varphi_c} = -\ddot{\varphi}_c + V'(\varphi_c) = 0 \, . \quad (3)$$

Further, let the operator of the second variational derivative of the action possess one parametrically small mode [in addition to the zero one $\dot{\varphi}_c(t)$] while the other modes are gaussian:

$$\Box(\varphi) = \delta^2 S/\delta\varphi^2 = -\partial^2 + V''(\varphi)\,, \tag{4}$$

$$\Box(\varphi_c)\varphi_n = \lambda_n \varphi_n\,, \tag{5}$$

$$\lambda_0 = 0\,, \quad \lambda_1 \ll \lambda_n \quad (n > 1)\,. \tag{6}$$

Whether the quasizero mode is positive or negative is insufficient for our analysis but for definiteness we shall consider it to be positive. Then the topography of the functional space near the classical solution has the form schematically shown in fig. 1 (where only the two modes are taken into account). It resembles a canyon with steep walls corresponding to the "normal" gaussian modes and the gently sloping bottom reflecting the quasizero mode (for a moment we ignore complications due to the exact zero mode). In order to integrate over this segment of the functional space one should perform the following program.

(i) Find the line in the functional space which corresponds to the quasizero mode. One can imagine this line as the stream in the bottom of the functional canyon in fig. 1.

(ii) Perform the functional integrations in the directions orthogonal to this streamline. Since the walls of the functional canyon are steep ($\lambda_{n>1}$ are large) these integrations are almost gaussian.

(iii) Carry out a final integration along the streamline (parametrized in a certain way).

Now we shall realize this program.

2. In order to understand how to integrate over the quasizero mode it is useful to recall the usual collective-coordinate method for the exact zero mode (see e.g. ref. [3]). Suppose we calculate the one-kink contribution to the matrix element of the evolution operator (1). Expanding the action in powers of the deviation $\delta\varphi = \varphi - \varphi_c$ [note that the linear term vanishes due to the equation of motion (2)]

$$S(\varphi) = S(\varphi_c) + \tfrac{1}{2}(\delta\varphi, \Box(\varphi_c)\delta\varphi) + O(\delta\varphi^3)\,, \tag{7}$$

we obtain the gaussian functional integral over the quantum fluctuations near the classical solution φ_c:

$$N^{-1}\int \mathrm{D}\varphi \exp\{-S(\varphi_c) - \tfrac{1}{2}(\varphi, \Box(\varphi_c)\varphi)\}$$

$$= N^{-1}\int \prod_n \frac{\mathrm{d}c_n}{\sqrt{2\pi}} \exp\left(-S(\varphi_c) - \tfrac{1}{2}\sum_n \lambda_n c_n^2\right) \tag{8}$$

[we use the notation $(f,g) = \int f(t)g(t)\,\mathrm{d}t$]. Here c_n are coefficients of the expansion of the integrand function in eigenfunctions of the operator $\Box(\varphi_c)$:

$$\varphi(t) = \varphi_c(t) + \sum_{n=0}^{\infty} c_n \varphi_n(t)\,. \tag{9}$$

Since $\lambda_0 = 0$ [$\dot{\varphi}_c(t)$ is the corresponding eigenfunction] the integral over c_0 diverges. This divergent integral arises due to the translational invariance of our classical solution $\varphi_c(t-\tau)$ and should be substituted by the integral over the instanton centre which gives the factor T. The functional space in the neighbourhood of the family of stationary points $\varphi_c(t-\tau)$ is schematically shown in fig. 2. It is clear that in order to perform the integration over this segment of the functional space one should at first integrate over the direction(s) orthogonal to the tangent to the bottom $\dot{\varphi}(t-\tau)$ and then carry out the final integration over τ (which yields T due to translational invariance). In other words, for every function φ we find the "nearest" instanton $\varphi_c(t-\tau_*) = \varphi_{\tau_*}$

$$\|\varphi - \varphi_{\tau_*}\| = \min\,,$$

$$\frac{\partial}{\partial\tau}\|\varphi - \varphi_\tau\|_{\tau=\tau_*} = (\varphi - \varphi_{\tau_*}, \dot{\varphi}_{\tau_*}) = 0\,, \tag{10}$$

then integrate over all the fields satisfying the condition (10) and finally perform the trivial τ-integration. Technically it is convenient to use the Faddeev–Popov method. One writes:

$$1 = \Delta(\varphi)\int \mathrm{d}\tau\, \delta(\varphi, \dot{\varphi}_\tau)\,,$$

$$\Delta(\varphi) = S(\varphi_c) + (\varphi - \varphi_{\tau_*}, \ddot{\varphi}_{\tau_*}) \tag{11}$$

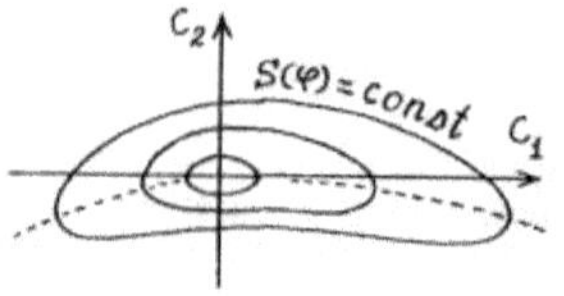

Fig. 1. The topography of the functional space near the classical solution with quasizero mode.

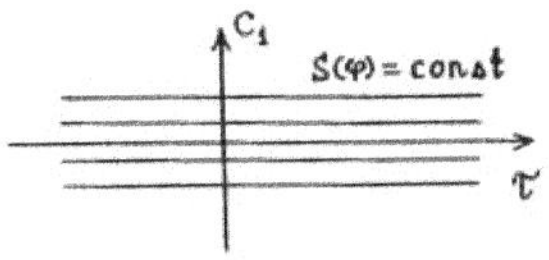

Fig. 2. The functional space near the solution with zero mode.

[here $S(\varphi_c) = \|\dot{\varphi}_c\|^2$ is the semiclassical parameter; $S(\varphi_c) \gg 1$] so

$$Z = N^{-1} \int \mathrm{d}\tau \int \mathrm{D}\varphi\, \Delta(\varphi)\delta(\varphi - \varphi_\tau, \dot{\varphi}_\tau) \exp[-S(\varphi)] \tag{12}$$

Expanding again the action $S(\varphi)$ in powers of the deviations from the nearest instanton we obtain:

$$S(\varphi) = S(\varphi_c) + \tfrac{1}{2}(\varphi - \varphi_{\tau_*}, \Box(\varphi_{\tau_*})(\varphi - \varphi_{\tau_*})) + O((\varphi - \varphi_{\tau_*})^3) \tag{13}$$

and therefore

$$Z_1 = N^{-1} S(\varphi_c) \int \mathrm{d}\tau \int \mathrm{D}\varphi\, \delta(\varphi - \varphi_\tau, \dot{\varphi}_\tau) \times \exp[-S(\varphi_c) - \tfrac{1}{2}(\varphi - \varphi_c)\Box(\varphi_c)(\varphi - \varphi_\tau)] , \tag{14}$$

where the subscript "1" means the one-kink contribution [obtained after reducing the action to the quadratic operator (13)]. This is the desired formula with a manifest integration over τ. Quite similarly one can distinguish the integral over the streamline parameter corresponding to the quasizero mode but for completeness we shall finish first the calculation of vacuum energy due to instantons.

Expanding the deviation $\varphi - \varphi_\tau$ in eigenfunctions of $\Box(\varphi_c)$

$$\varphi(t) = \varphi_c(t - \tau_*) + \sum_{n=1}^{\infty} a_n(\tau_*)\varphi_n(t - \tau_*) \tag{15}$$

[note that $a_0 = (\varphi - \varphi_{\tau_*}, \dot{\varphi}_{\tau_*}) = 0$ due to the δ-function] one has

$$Z_1 = N^{-1} \det{}^{-1/2}\Box_0\, T[S(\varphi_c)/2\pi]^{1/2} \times [\det{}'\Box(\varphi_c)/\det \Box_0]^{-1/2} , \tag{16}$$

where $\det'\Box$ denotes the product of nonzero eigenvalues. Here $\det \Box_0$ is the usual normalization determinant [e.g. $\Box_0 = -\mathrm{d}^2 + m^2$ in the case of the double-well instanton $\operatorname{th} \frac{1}{2}m(t - \tau)$]. Further, taking into account the contributions of widely separated instantons and anti-instantons one easily obtains (cf. e.g. ref. [3]):

$$Z = N^{-1} \det{}^{-1/2}\Box_0 \times \exp\{-T[S(\varphi_c)/2\pi]^{1/2} [\det{}'\Box(\varphi_c)/\det \Box_0]^{-1/2}\} , \tag{17}$$

which presents the well-known result for the vacuum energy in the dilute-gas approximation [4].

3. The integration over the quasizero mode is quite similar. We calculate the quantum fluctuations around the classical solution (3) (for simplicity we shall ignore the zero mode at first, i.e. assume that the operator $\Box(\varphi_c)$ [see eq. (4)] possesses one parametrically small eigenvalue while other modes are gaussian). The functional space near the classical solution is shown in fig. 1. In the very neighbourhood of the solution [where $S(\varphi_c + \delta\varphi) = S_0 + \frac{1}{2}(\delta\varphi, \Box\delta\varphi)$] the topography of the functional space is very simple: $S(\varphi_c + \delta\varphi) = S_0 + \frac{1}{2}\Sigma_n \lambda_n a_n^2$. Therefore "isoactas" near the origin ellypsoides stretched in the direction of the quasizero mode (see fig. 1 where $\delta\varphi = a_1\varphi_1 + a_2\varphi_2$). From inspection of fig. 1 one sees that the streamline starts from the classical solution in the direction of the quasizero mode and afterwards the tangent to the streamline is always collinear to the gradient of the action. Generalizing this conditions to the functional space we obtain the following equation for the streamline:

$$\xi(\alpha)\partial\varphi(\alpha)/\partial\alpha = \delta S/\delta\varphi|_{\varphi=\varphi(\alpha)} , \tag{18}$$

with the boundaries [‡1]

$$\varphi(\alpha)|_{\alpha=0} = \varphi_c , \quad \partial\varphi(\alpha)/\partial\alpha|_{\alpha=0} \sim \varphi_1 \tag{19, 20}$$

The coefficient function $\xi(\alpha)$ determining the parametrization of the streamline satisfies the conditions

$$\xi(0) = 0 , \quad \xi'(0) = \lambda_1 \tag{21}$$

[following from eqs. (18)–(20)] and arbitrary other-

[‡1] The two boundary conditions for the first-order differential equation may seem unfamiliar. However, the classical solution $\varphi_{\alpha=0} = \varphi_c$ is a distinguished point for eq. (18) (an infinite number of gradient line satisfying the equation pass through this point).

Volume 168B, number 1,2 PHYSICS LETTERS 27 February 1986

wise. Eq. (18) is the basic equation for our analysis and [together with eq. (24) below] the main result of the paper [‡2].

Let us integrate now over the segment of the functional space near the streamline φ_α. Similarly to the conventional method described above for every φ we find the nearest point on the streamline φ_{α_*},

$$\|\varphi - \varphi_{\alpha_*}\| = \min ,$$

$$\frac{\partial}{\partial\alpha}\|\varphi - \varphi_\alpha\|_{\alpha=\alpha_*} = \left(\varphi - \varphi_{\alpha_*}, \frac{\partial}{\partial\alpha}\varphi(\alpha_*)\right) = 0 \quad (22)$$

and expand the action in powers of the deviation $\varphi - \varphi_{\alpha_*}$

$$S(\varphi) = S(\varphi_\alpha) + (\varphi - \varphi_{\alpha_*}, \delta S/\delta\varphi_{\alpha_*})$$
$$+ \tfrac{1}{2}(\varphi - \varphi_{\alpha_*}, \Box(\varphi_{\alpha_*})(\varphi - \varphi_{\alpha_*}))$$
$$+ O((\varphi - \varphi_{\alpha_*})^3) . \quad (23)$$

Now we shall integrate first over the fields satisfying the condition (22) [so that the linear term in eq. (23) drops due to the basic equation] and then perform the final integration over α. Using the Faddeev–Popov method as described above one obtains [cf. eqs. (11)–(14)]:

$$Z_1 = N^{-1}\int d\alpha\, \|\partial\varphi_\alpha/\partial\alpha\|^2 \exp[-S(\varphi_\alpha)]$$
$$\times \int D\varphi\, \delta(\varphi - \varphi_\alpha, \partial\varphi_\alpha/\partial\alpha)$$
$$\times \exp\{-\tfrac{1}{2}(\varphi - \varphi_\alpha)\Box(\varphi_\alpha)(\varphi - \varphi_\alpha)\} \quad (24)$$

It is quite similar to the integral (14) with the only difference that the integration over α is now nontrivial. Therefore it yields not an extra volume T but a certain (parametrically large) factor. Eq. (24) is the desired formula for the integral over the segment of the functional space near the classical solution with the quasizero mode. Further, one can formally perform the functional integration in eq. (24); this results in

$$Z_1 = N^{-1}\int d\alpha\, \|\partial\varphi_\alpha/\partial\alpha\|^2 (\partial\varphi_\alpha/\partial\alpha, \Box^{-1}(\varphi_\alpha)\partial\varphi_\alpha/\partial\alpha)^-$$
$$\times \det{}^{-1/2}\Box(\varphi_\alpha)\exp[-S(\varphi_\alpha)] , \quad (25)$$

where $\Box^{-1}(\varphi_\alpha)$ is the Green function of the operator $\Box(\varphi_\alpha)$.

Now we shall take into account the exact zero mode. The functional space is then the infinite ($\sim T$) cylinder with the fig. 1 being the conic section. The streamline $\varphi_{\alpha,\tau}(t) = \varphi_\alpha(t-\tau)$ ("streamplane", strictly speaking) satisfies the same eq. (18):

$$\xi(\alpha)\partial\varphi_{\alpha,\tau}(t)/\partial\alpha = -\ddot{\varphi}_{\alpha,\tau}(t) + V'(\varphi_{\alpha,\tau}) , \quad (26)$$

$$\varphi_{\alpha,\tau}(t)|_{\alpha=0} = \varphi_c(t-\tau) ,$$
$$\partial\varphi_{\alpha,\tau}(t)/\partial\alpha|_{\alpha=0} \sim \varphi_1(t-\tau) . \quad (27)$$

Now it is easy to find the contribution of the streamline (26) to the path integral (1). Consideration is quite similar to the one above with the only difference being the two collective coordinates α and τ instead of one (α). One writes [‡3]

$$1 = \Delta(\varphi)\int d\alpha\, d\tau\, \delta(\varphi - \varphi_{\alpha,\tau}, \varphi'_{\alpha,\tau})$$
$$\times \delta(\varphi - \varphi_{\alpha,\tau}, \dot{\varphi}_{\alpha,\tau}) , \quad (28)$$

$$\Delta(\varphi) = \{\|\dot{\varphi}_{\alpha,\tau}\|^2 - (\varphi - \varphi_{\alpha,\tau}, \ddot{\varphi}_{\alpha,\tau})\}$$
$$\times \{\|\varphi'_{\alpha,\tau}\|^2 - (\varphi - \varphi_{\alpha,\tau}, \varphi''_{\alpha,\tau})\} - (\varphi - \varphi_{\alpha,\tau}, \dot{\varphi}'_{\alpha,\tau})^2 \quad (29)$$

(hereafter the prime stands for differentiation with respect to α and the dot with respect to time). Expanding then the action in powers of the deviation $\varphi - \varphi_{\alpha,\tau}$ [see eq. (23)] one obtains finally

$$Z_1 \quad \int d\alpha\, d\tau\, N^{-1}\int D\varphi\, \Delta(\varphi)\delta((\varphi - \varphi_{\alpha,\tau}, \dot{\varphi}_{\alpha,\tau}))$$
$$\times \delta((\varphi - \varphi_{\alpha,\tau}, \varphi'_{\alpha,\tau}))$$
$$\times \exp\{-S(\varphi_{\alpha,\tau}) - \tfrac{1}{2}[\varphi - \varphi_{\alpha,\tau}, \Box(\varphi - \varphi_{\alpha,\tau})]\} . \quad (30)$$

Here the linear term $[(\varphi - \varphi_{\alpha,\tau}, \delta S/\delta\varphi)]$ in the expo-

‡2 Our thanks to A. Johansen for his participation in establishing the status of eq. (18).

‡3 Writing down $\Delta(\varphi)$ we have omitted the terms proportional to $(\dot{\varphi}_{\alpha,\tau}\varphi'_{\alpha,\tau})$ since $(\dot{\varphi}_{\alpha,\tau}, \varphi'_{\alpha,\tau}) \sim (\dot{\varphi}_{\alpha,\tau}, \delta S/\delta\varphi) = \int dt\, (d/dt)[\dot{\varphi}^2_{\alpha,\tau}/2 + V(\varphi_{\alpha,\tau})] = 0$.

nential factor is absent due to the basic equation (26). Note also that the quadratic terms in $\Delta(\varphi)$ $[\sim(\varphi - \varphi_{\alpha,\tau})^2]$ are quantum corrections which should be taken into account together with cubic (and higher) terms in the expansion of the action (23). Therefore $\Delta(\varphi)$ reduces to $\|\dot{\varphi}_{\alpha,\tau}\|^2\|\varphi'_{\alpha,\tau}\|^2$ [the terms linear in $\varphi - \varphi_{\alpha,\tau}$ do not contribute to eq. (30)].

Due to the translational invariance the integral over τ is trivial and yields T. Further, the gaussian functional integral in eq. (30) can be formally calculated with the result being

$$Z_1 = N^{-1}(\det\Box_0)^{-1/2}Tk\ , \tag{31}$$

$$k = \int \frac{d\alpha}{2\pi}\ \|\varphi'_\alpha\|^2\ \|\dot{\varphi}_\alpha\|^2 \exp[-S(\varphi_\alpha)]$$
$$\times (\varphi'_\alpha, \Box_\alpha^{-1}\varphi'_\alpha)^{-1/2}(\dot{\varphi}_\alpha, \Box_\alpha^{-1}\dot{\varphi}_\alpha)^{-1/2}$$
$$\times (\det\Box_\alpha/\det\Box_0)^{-1/2}\ , \tag{32}$$

where $\varphi_\alpha = \varphi_\alpha(t)$, $\Box_\alpha = \Box(\varphi_\alpha)$ and $\det\Box_0$ is the normalization determinant. Eq. (31) is our final result for the contribution of the streamline (26) to the path integral (1). Note that $S(\varphi_\alpha)$ is minimal at $\alpha = 0$ $(\varphi_{\alpha=0} = \varphi_c)$ and $\delta^2 S/\delta\alpha^2 \sim \lambda_1$. Therefore, if we calculate the integral over α in the gaussian approximation we shall obtain the eq. (16) (but since λ_1 is parametrically small this is incorrect).

Taking into account the contributions of widely separated configurations $\varphi_{\alpha_i}(t - \tau_i)$ in the dilute-gas approximation one easily obtains for the vacuum energy

$$Z = \langle\exp(-H)\rangle = \tilde{N}\exp(-E_{\text{vac}}T)$$
$$= N^{-1}(\det\Box_0)^{-1/2}\exp(-Tk)\ . \tag{33}$$

Of course, one should verify a posteriori the validity of the dilute-gas approximation, i.e. that the mean separation of the configurations (which is $\sim K^{-1}$) is parametrically greater than the size of the configuration [determined by the characteristic α in the integral (32)].

4. Now, in order to illustrate our streamline method we shall calculate the instanton–anti-instanton contribution to the vacuum energy in quantum mechanics with the double-well anharmonic oscillator potential (this result was originally obtained in ref. [5]). The (euclidean) action of the model reads:

$$S(\varphi) = \int dt\ [\dot{\varphi}^2/2 + \lambda(\varphi^2 - V^2)]\ . \tag{34}$$

As it is shown in ref. [5] the vacuum energy in this theory can be written as

$$E(g) = \sum_{k=0}^{\infty} g^k E_k^{(0)} + \exp(-1/6g)\sum_{k=0} g^k E_k^{(1)}$$
$$+ O(\exp(-1/3g))\ , \tag{35}$$

where $g = (8\sqrt{2\lambda}v^3)^{-1} \ll 1$ is the semiclassical parameter. Note that eq. (35) is meaningful only as an analytic continuation from the Stokes line to positive values of the coupling constant g. The first term in eq. (35) is the perturbative contribution to the vacuum energy while the others are due to instantons. They represent the virial expansion of the vacuum energy: $O(\exp(-1/6g))$ is the dilute-instanton-gas contribution [4]. $O(\exp(-1/3g))$ arises when only the two-particle (instanton–anti-instanton) interactions are taken into account and higher $O(\exp(-n/6g))$ terms correspond to many-particle interactions in the (dilute-) instanton medium.

Here we shall calculate the term $\sim\exp(-1/3g)$ by the streamline method. Consider the path integral (1) with the initial and the final state being the bottom of the left well. The relevant configuration is a pair of an instanton and an anti-instanton (with arbitrary centres):

$$\varphi_c(t - \tau) = V\{\text{th}\,\tfrac{1}{2}m(t - \tau + \alpha)$$
$$- \text{th}\,\tfrac{1}{2}m(t - \tau - \alpha) - 1\}\ , \tag{36}$$

where $m = 2\sqrt{2\lambda}v$. When the separation 2α tends to infinity the configuration (36) becomes the exact solution of the equation of motion (3). [This solution has one exact zero mode $\dot{\varphi}_c(t)$ due to the translational invariance). It is clear that variation of the kink–antikink separation 2α only slightly affects the action $S(\varphi_c)$ while eigenvalues corresponding to other modes are large (provided g is small). Thus we are in a position to apply our streamline method and determine the kink–antikink interaction in a wide region of separations 2α.

The kink–antikink streamline $\varphi(t, \alpha)$ satisfying the boundary condition

Volume 168B, number 1,2 PHYSICS LETTERS 27 February 1986

$$\varphi(t,\alpha) \underset{\alpha\to\infty}{\to} \varphi_c(t) \tag{37}$$

is

$$\varphi(t,\alpha) = V\{[1 - \tfrac{3}{2}\epsilon^2 + O(\epsilon^4)] \times [\mathrm{th}\,\tfrac{1}{2}M(t+\alpha) - \mathrm{th}\,\tfrac{1}{2}M(t-\alpha)] - 1\}, \tag{38}$$

where

$$M = m[1 - \tfrac{3}{2}\epsilon^2 + O(\epsilon^4)], \quad \epsilon = 1/\mathrm{ch}\,M\alpha. \tag{39}$$

Eq. (38) gives the approximate solution of the basic equation (26):

$$3m^2\epsilon^2 \partial\varphi(t,\alpha)/\partial\alpha = -\ddot{\varphi}(t,\alpha) + V'(\varphi_\alpha). \tag{40}$$

The corresponding action is

$$S(\alpha) = S(\varphi(\alpha)) = 1/3g - (2/g)\exp(-2m\alpha) + O(\exp(-4m\alpha)/g). \tag{41}$$

Note that the accuracy of eqs. (38), (39) enables us to obtain the next term in eq. (41) [$O(\exp(-4m\alpha))$]. However, in order to calculate the $O(\exp(-1/3g))$ contribution to $E(g)$ $S(\alpha)$ in the form (41) is sufficient. Eq. (41) represents the instanton–anti-instanton interaction on the classical level (this was obtained primarily in ref. [5] by considering a certain matched ansatz).

It should be mentioned that the streamline $\varphi(t,\alpha)$ starts from the infinitely separated instanton–anti-instanton pair (at $\alpha\to\infty$) and tends to the trivial vacuum at $\alpha\to 0$, reflecting the tendency of the instanton–anti-instanton pair to collapse. Thus, $S(\alpha)\to 0$ at $\alpha\to 0$. Note that eq. (41) does not reflect this property since it becomes incorrect at small α.

Let us now take into account the pre-exponential factors in eq. (32). Since the kink–antikink separation (2α) is large we have

$$(\varphi'_\alpha, \Box_\alpha^{-1}\varphi'_\alpha)(\dot{\varphi}_\alpha, \Box_\alpha^{-1}\dot{\varphi}_\alpha)\det\Box_\alpha/\det\Box_0 \approx (\det{}''\Box_\alpha/\det\Box_0)[1 + O(\exp(-2m\alpha))], \tag{42}$$

where $\det{}''\Box_\alpha$ is the determinant with the two lowest modes excluded. Further,

$$\det{}''\Box_\alpha/\det\Box_0 = (1/12)^2[1 + O(\exp(-2m\alpha))] \tag{43}$$

(it is the standard factorization approximation, cf. ref. [3]).

Substituting now eqs. (41), (43) in eq. (32) we obtain

$$Z_2 = (m^2T^2/2\pi g)\exp(-1/3g) - \frac{m^2T}{\pi g}\exp(-1/3g)2\int_0^\infty d\alpha\,\{1 - \exp[(2/g)\exp(-2m\alpha \tag{44}$$

("2" means the two-kink contribution). The first term in eq. (44) corresponds to the dilute-gas approximation while the second is due to the kink–antikink interaction. Unfortunately, the second term is attractive reflecting the tendency of the kink–antikink pair to collapse. Therefore the main contribution to the integral (44) comes from the small α region where our analysis is incorrect. It is easy to see, however, that this (unknown) contribution is purely perturbative while we are interested in the nonperturbative part. The instanton–anti-instanton contribution to the nonperturbative part of the vacuum energy [$\exp(-1/3g)$] corresponds to the essentially singular part of the integral (44) (with respect to g). In order to extract this singular part one should [5]: (i) change the sign of the coupling g – then the attractive forces turn into repulsive, (ii) calculate the obtained integral over α along the steepest descent path (the main contribution comes now from the large α [$\sim(1/2m)\ln 1/g$] region of integration), and (iii) perform the analytic continuation of the result to positive g. One has

$$\int_0^\infty dx\,\{1 - \exp[(2/g)\exp(-x)]\}_{\text{non-pert.}} = \ln(-2/g) + C, \quad C = 0.577\ldots. \tag{45}$$

Substituting eq. (45) in (44) and using the eq. (33) we obtain the vacuum energy (35) in the form

$$E(g) = \sum_{k=0}^\infty g^k E_k^{(0)} - (m/\sqrt{\pi g})\exp(-1/6g)[1 + O(g)] + (m/\pi g)\exp(-1/3g)[\ln(-2/g) + C + O(g\ln g)] \tag{46}$$

Note that the instanton–anti-instanton contribution [the last term in eq. (46)] has an imaginary part with the sign determined by the way of the analytic continuation of the integral (45) to positive values of g. Since the ground-state energy $E(g)$ is real this imaginary part equals (up to a sign) to the perturbative one and therefore determines the (non-Borel-summable) high-order behaviour of the perturbative series [6,7].

We see that the streamline method enables us to define the notion of instanton–anti-instanton interaction in a rigorous way. In principle, the exact solution of the streamline equation (26) would yield the instanton–anti-instanton interaction at arbitrary separations α. The approximate solution (38) permits us to reobtain the instanton–anti-instanton contribution to the vacuum energy in the double-well quantum mechanics. Therefore one may hope to make use of the streamline method when studying the instanton interactions in order theories. In particular, we use this method in quantum mechanics with fermions (including SUSY case [8]) and the results will be published. Also, it can be used to introduction of instantons in models containing Yang–Mills field interacting with scalar field developing a non-vanishing vacuum expectation value. In these models instantons with finite size can be interpreted in terms of streamline solution of the equation of eq. (18) type rather than as an exact solution of equation of motion (cf., for example, the discussion of the virtual stationary points in ref. [9]).

The authors are grateful to E.B. Bogomolny, M.I. Eides, and A.A. Johansen for valuable discussions.

References

[1] J.L. Gervais and E. Sakita, Phys. Rev. D11 (1975) 2943; E. Tomboulis, Phys. Rev. D12 (1975) 1678.
[2] S. Coleman, Phys. Rev. D15 (1977) 2929; C.G. Callan and S. Coleman, Phys. Rev. D16 (1977) 1762.
[3] S. Coleman, The uses of instantons, in: The whys of subnuclear physics, Erice Lectures (1977) (Plenum, New York, 1979).
[4] E. Gildener and A. Patrascioiu, Phys. Rev. D16 (1977) 423; A.M. Polyakov, Nucl. Phys. B120 (1977) 429.
[5] E.B. Bogomolny, Phys. Lett. 91B (1980) 431; J. Zinn-Justin, Nucl. Phys. B192 (1981) 125.
[6] L.N. Lipatov, Sov. Phys. JETP 45 (1977) 216.
[7] J. Zinn-Justin, Phys. Rep. 70 (1981) 109.
[8] E. Witten, Nucl. Phys. B188 (1981) 513.
[9] W. Nahm, Phys. Lett. 96B (1980) 323.

VII. SUPERSYMMETRIC INSTANTONS

INTRODUCTION

Supersymmetric (SUSY) gauge theories are potentially one of the most important fields of application of instantons. The reason why instantons may be so important is the existence of the so-called flat directions or valleys in a large class of models with matter fields. In these models the vacuum states are degenerate along certain directions in the space of fields, and the corresponding potential energy is zero classically and to any finite order in perturbation theory. The degeneracy may or may not be lifted by nonperturbative effects. In many cases instantons result in a remarkable dynamic effect: color and/or SUSY spontaneous breaking in the *weak coupling regime* [1–7] where all approximations made are under full theoretic control and instantons are protected from the infrared disaster by induced masses of the gauge fields (for a review see Refs. [8, 9]).

Let me briefly remind the reader as to why nonperturbative symmetry breaking is considered a valuable asset in any SUSY gauge theory and is so desirable. If nature is supersymmetric at high energies, this symmetry has definitely to be broken at low energies since the observed particle spectrum is not supersymmetric. Moreover, if we want SUSY to be broken not at the Planck mass but significantly lower, in the TeV range — and, of course, this was the primary goal of model builders — we have to explain, in a natural way, a gigantic hierarchy of scales. The hierarchy might come out naturally if the breaking is only due to instantons and thus contains exponential suppression factors like $\exp(-8\pi^2/g^2)$.

The question of the dynamical SUSY breaking by instantons was raised by Witten [10] shortly after he found a necessary condition for the breaking [11], allowing one to significantly narrow down the range of models to be considered as potential candidates. (We will discuss this condition, Witten's index, in more detail below.) It was shown that in pure gauge theories and in gauge theories with nonchiral matter fields, Witten's index is nonvanishing, which automatically implies unbroken supersymmetry. Thus, if SUSY is to be dynamically broken, it can occur only in the presence of chiral matter fields. The search for a particular pattern which might realize the idea, finally crowned by success, brought a few pleasant surprises *en route*. The discovery of spontaneous gauge symmetry breaking in many instances accompanying that of SUSY, was among them.

It should be added that no phenomenologically realistic models based on instanton-induced symmetry breaking were found in the eighties. Since then the focus of theoretical investigations shifted to other ideas, leaving this not-fully-solved problem in the rear. One of the alternative strategies popular today, the introduction of *ad hoc* SUSY breaking terms by hand, is perceived by many more like a theoretical retreat compared with the elegant programs of the past, a retreat which hopefully will not last forever. Note, though, that in some instances, the breaking terms are not totally *ad hoc* but are somehow motivated by strings/supergravity.

The generation of the weak coupling regime in the valleys is not the only aspect of the supersymmetric theories where the instantons are instrumental. Historically they were first used in the strong coupling regime for calculating gluino condensates in SUSY gluodynamics [12]. The logic of this calculation is rather sophisticated and the result, although seemingly theoretically clean, harbors a mystery that defies any explanation for more than a decade.

Consider for definiteness the SUSY Yang–Mills theory with the gauge group $SU(2)$. In this case there are four fermion zero modes in the instanton field and hence, there is no direct instanton contribution to the gluino condensate $<\lambda\lambda>$ where λ denotes the gluino field. At the same time, the instanton contributes to the correlation function

$$<\lambda^a_\alpha(x)\lambda^{a\alpha}(x), \lambda^b_\beta(0)\lambda^{b\beta}(0)>, \tag{1}$$

in an apparent contradiction with supersymmetry since there is no boson analog of the correlation function (1). (Here $a, b = 1, 2, 3$ are the color indices and $\alpha, \beta = 1, 2$ are the spinor ones.) An explicit instanton calculation [12] shows that the correlation function (1) is equal to a nonvanishing constant.

Surprising as it is, supersymmetry does not forbid Eq. (1) provided that this two-point function is actually an x-independent constant. For purposes which will become clear shortly, let us now sketch the proof of the above assertion.

Three elements are of importance: (i) the supercharge $\bar{Q}^{\dot\beta}$ acting on the vacuum state annihilates it; (ii)$\bar{Q}^{\dot\beta}$ commutes with $\lambda\lambda$; (iii)the derivative $\partial_{\alpha\dot\beta}(\lambda\lambda)$ is representable as the anticommutator of $\bar{Q}^{\dot\beta}$ and $\lambda^\beta G_{\beta\alpha}$. (The spinor notations are used.) Now, we differentiate Eq. (1), substitute $\partial_{\alpha\dot\beta}(\lambda\lambda)$ by $\{\bar{Q}^{\dot\beta}, \lambda^\beta G_{\beta\alpha}\}$, and obtain zero. Thus, supersymmetry requires the derivative of (1) to vanish. This is exactly what happens if the correlator (1) is a constant.

If so, one can compute the result at short distances where it is presumably saturated by small size instantons, and then the very same constant is predicted at large distances, $x \to \infty$. On the other hand, due to the cluster decomposition property which must be valid in any reasonable theory, the correlation function (1) at $x \to \infty$ reduces to $<\lambda\lambda>^2$. Extracting the square root, we arrive at a (double-valued) prediction for the gluino condensate. [The fact that it is double-valued is in accordance with Witten's index for $SU(2)$, see below.]

Many baffling questions immediately come to mind in connection with this argument. First, if the gluino condensate is nonvanishing and shows up in a roundabout instanton calculation through (1), why is it not seen in the direct instanton calculation of $<\lambda\lambda>$? Second, the constancy of the two-point function (1) required by SUSY is ensured in the concrete calculation by the fact that the instanton size ρ turns out to be of order x. The larger the value of x, the larger ρ saturates the instanton contribution. For small x this is alright. At the same time, at $x \to \infty$, we do not expect any coherent fields with the size of order x to survive in the vacuum; such coherent fields would contradict our current ideas of infrared strong confining theories like SUSY gluodynamics. If there are no large size coherent fields in the vacuum, how can one guarantee the x independence of (1) at all distances? Third, the instanton analysis seems to give the exact prediction for $<\lambda\lambda>$. Does the fact that the exact prediction is possible in a nontrivial four-dimensional strong coupling theory imply that there is something remarkable in the spectrum of this theory?

A hypothetical answer to the first question was suggested in Ref. [9]. It was assumed that, instead of providing us with the expectation value of $\lambda\lambda$ in the given vacuum, instantons actually yield an average value of $< \lambda\lambda >$ in all possible vacuum states. Since in the case at hand, there exist two vacua and the expectation values of $\lambda\lambda$ in these vacua only differ in sign, the average expectation value of $< \lambda\lambda >$ is indeed zero.

Although, technically, this conjecture seems to go through, it produces more perplexing questions than answers. Normally, when we say "average expectation value of an operator," we do not mean averaging over different vacuum states which might exist in the theory. On the contrary, the sectors corresponding to different vacua are assumed to be totally disconnected from each other. For instance, in the scalar theory

$$\mathcal{L} = \frac{1}{2}\partial_\mu\phi\partial_\mu\phi - \lambda(\phi^2 - v^2)^2 ,$$

the vacuum expectation values and all correlation functions refer either to the vacuum with $< \phi >= v$ or $< \phi >= -v$, and in no reasonable calculation one can get an average between those two. If the instanton calculation mixes up two different vacuum states which (in the infinite volume) are not connected by tunnelings, this would be quite an unexpected property calling for an immediate explanation.

The absence of any corrections in the gluino condensate rather transparently suggests that this condensate is a nondynamical quantity and alludes to its topological nature. We assume that, instead of considering SUSY gluodynamics in the flat infinite space, we analyze this theory in a gravitational background which still preserves a part of supersymmetry. An example of this type is the theory on the (four-dimensional) Euclidean sphere. An even simpler example is provided by the theory on a torus where we, of course, lose no supersymmetry generators at all. The residual supersymmetry is needed to ensure the constancy of correlation functions of the type (1). A remarkable circumstance is that condensates obtained from instantons in the manner described above turn out to be independent of dimensionful parameters — the radius of the sphere in the first case and the periods of the torus in the second [13]. In both cases they are expressed only through ultraviolet parameters, the ultraviolet cutoff M_0 and the corresponding bare gauge coupling constant, a clear manifestation of the topological nature of the gluino condensate.

The notion was further refined and advanced in Ref. [14] (see also Ref. [15]). In ordinary SUSY theories like gluodynamics, we cannot introduce an arbitrary gravitational field without completely breaking all fermion symmetries. If they are all ruined, the instanton approach outlined above becomes completely useless. As has been suggested in Ref. [14], one can introduce "abnormal" SUSY gauge theories, the famous topological (or Chern–Simons) theories, totally deprived of any perturbative dynamics at all. The Hamiltonian is identically zero, and the only nontrivial aspect is the construction of the Hilbert space of states. In such theories, for *arbitrary* background metric, there exists a symmetry generator of the fermion type, usually referred to as the BRST charge Q. (For a review see e.g. Ref. [16].) This operator annihilates the vacuum state; moreover, $Q^2 = 0$. All "interesting" operators $\mathcal{O}_i$ in the topological field theory belong to two classes — those which anticommute with Q (call them $\mathcal{O}_1$) and those which can be written as $\mathcal{O}_2 = \{\mathcal{O}_3, Q\}$. Combining operators from these two classes, one obtains vacuum expectation values that are genuine topological invariants — for *any* background metric they do not depend on the metric at all, only on global (topological) characteristics of the manifold on which the theory is formulated.

A close parallel between the instanton calculation of $\lambda\lambda$ in SUSY gluodynamics and the treatment of condensates in the topological field theory is quite obvious. Unfortunately, the development of these ideas went deeply towards the mathematical side (e.g. the theory of knots). Many, if not all physically interesting questions, were left aside. A challenging and still open task is to try to separate a topological sector out of normal SUSY Yang–Mills theories with the nonvanishing Hamiltonian, and find out what the consequences of this sector are for the observable quantities (particle masses, etc.). At present I am aware of no attempts in this direction.

Another longstanding problem which is associated, at least in part, with instanton calculus and awaits its solution since the mid-eighties is the problem of Witten's index I_W,

$$I_W = \sum (-1)^F ,$$

where the sum runs over all states of the system and F is the fermion number of a given state. To make the sum meaningful the system is placed in a large box so that the spectrum is discrete.

As was noted in Ref. [11], if the Lagrangian of the system is supersymmetric, I_W reduces to the number of zero energy states of the boson type minus the number of zero energy states of the fermion type. Moreover, I_W in SUSY theories is a remarkable invariant — it is independent of the values of parameters of the model considered, say coupling constants, etc. In particular, I_W is independent of the volume of the box V. If so, I_W can be calculated by analyzing the theory in the limit of small volume, $V \to 0$. In this way, we reduce the problem essentially to quantum mechanics; if $I_W \neq 0$, the supersymmetry obviously cannot be spontaneously broken.

Witten's index was found by means of a quantum-mechanical reduction in SUSY gluodynamics for all gauge groups [11]; $I_W = r + 1$, where r is the rank of the group. At the same time Witten's index is calculable in a more direct way, just by counting the number of different vacuum states in the full theory. An appropriate parameter which counts the number of vacua is the gluino condensate $< \lambda\lambda >$. We have already mentioned that for $SU(2)$, it is double-valued reflecting the existence of two vacuum states. In general, the "multi-valuedness" of $< \lambda\lambda >$ yields the number of vacua and hence, I_W. Moreover, in any gauge theory the gluino condensate can be obtained using the program outlined above. To avoid possible infrared problems the corresponding calculation can be formulated in such a way that it is carried out completely in the weak coupling regime [17]. The result for Witten's index derived in this way is

$$I_W = T(G) , \tag{2}$$

where $T(G)$ is defined as

$$\mathrm{Tr}(T^a T^b) = T(G)\delta^{ab} ;$$

T^a is the generator of the gauge group G in the adjoint representation. Witten's prediction $r+1$ coincides with $T(G)$ only for unitary and simplectic groups. For orthogonal and exceptional groups, $r+1 \neq T(G)$.

Thus, the quantum-mechanical and instanton derivations of I_W cannot be simultaneously correct. The question of what went wrong and how it happened is not fully understood at present, although some work in this direction has been done [18].

To complete the discussion of the gluino condensate it is worth mentioning that in the string-inspired SUSY models, the nonvanishing $<\lambda\lambda>$ in the so-called hidden sector is considered to be a promising mechanism for SUSY breaking which eventually penetrates the observed particle spectrum [19]. This is a very interesting topic which will not be pursued in accordance with the general principle of excluding applied problems from this volume.

The gluino condensate is not the only exact result which stems from supersymmetric instantons. SUSY-based arguments fix the instanton measure exactly, to all orders in perturbation theory; the instanton analysis basically reduces to that of zero modes [12, 20].

The latter statement is a generalization of a famous theorem in Ref. [21] which reads that the sum of vacuum graphs in supersymmetric theories is zero to any finite order in perturbation theory, implying that $E_{vac} = 0$ (see also Ref. [22]). In the instanton background field a part of the supersymmetry is still preserved. This reduced symmetry is still sufficient to ensure the cancellation of boson and fermion loops. For the first loop the cancellation was observed in Ref. [23], a general proof valid in all orders is given in Ref. [24].

This remarkable situation allows one to find exactly, to all orders in the gauge coupling, the Gell–Mann–Low function of supersymmetric gluodynamics [24, 25]. The specific form of the instanton measure described above is immediately translatable into a very specific form of the Gell–Mann–Low function. Thus, the calculation of the latter turns out to be purely classic! And all coefficients in the Gell-Mann-Low function are determined by integers counting the number of zero modes (and, hence, the number of symmetries of the classical action which are nontrivially realized in the instanton). Thus, for extended supersymmetry,[12]

$$\beta(\alpha) = \frac{\alpha^2}{2\pi} \frac{(N-4)T(G)}{1 - \left(\frac{2-N}{2\pi}\right) T(G)\alpha} \,. \tag{3}$$

This can hardly be a coincidence, the phenomenon must have deep roots which are not fully revealed so far.

Zero modes are the only remnants of the one-loop calculation in the instanton background. All nonzero modes relevant to this calculation cancel each other. As has just been mentioned, higher loops in the instanton field all vanish identically. If the number of fermion and boson zero modes is balanced in a certain way, the Gell–Mann–Low function stemming from the instanton measure will be one-loop. This is, for instance, the case in $N = 2$ SUSY gauge theory. In the general case there is an imbalance in the numbers of fermion and boson zero modes, and it is precisely this imbalance that is responsible for the second and all higher order terms in the Gell–Mann–Low function (for more details see Refs. [25, 27]).

Speaking of zero modes, it is worth mentioning in passing that the existence of the 't Hooft zero modes is now explained by symmetry arguments, a desirable feature people were looking for but could not find in nonsupersymmetric models. The gluino zero modes are superpartners (and superconformal partners) to the instanton field itself, while that of the fermion matter field (the proper 't Hooft mode) is a superpartner to the solution of the equation $\mathcal{D}^2\phi = 0$ for the scalar field. Note that, to the leading order, equations on the fermion zero modes are the same in supersymmetric and nonsupersymmetric versions.

Presented below are works devoted to the superfield instanton formalism. Some of the applied topics mentioned above are discussed in more detail in the review papers [8, 9].

References

[1] I. Affleck, M. Dine and N. Seiberg, *Phys. Rev. Lett.* **51** (1983) 1026.

[2] G. Rossi and G. Veneziano, *Phys. Lett.* **138B** (1984) 195 [reprinted in *Supersymmetry, Vol. 1*, ed. S. Ferrara (North-Holland/World Scientific, 1987), p. 620].

[3] I. Affleck, M. Dine and N. Seiberg, *Nucl. Phys.* **B241** (1984) 493.

[4] Y. Meurice and G. Veneziano, *Phys. Lett.* **141B** (1984) 69.

[5] A. Vainshtein, V. Zakharov, V. Novikov and M. Shifman, *Pisma ZhETF* **39** (1984) 494 [*JETP Lett.* **39** (1984) 601].

[6] D. Amati, G. Rossi and G. Veneziano, *Nucl. Phys.* **B249** (1985) 1.

[7] I. Affleck, M. Dine and N. Seiberg, *Phys. Lett.* **137B** (1984) 187 [reprinted in *Supersymmetry, Vol. 1*, ed. S. Ferrara (North-Holland/World Scientific, 1987), p. 600]; *Phys. Rev. Lett.* **52** (1984) 1677; *Phys. Lett.* **140B** (1984) 59.

[8] A. Vainshtein, V. Zakharov and M. Shifman, *Usp. Fiz. Nauk* **146** (1985) 683 [*Sov. Phys. — Uspekhi* **28** (1985) 709].

[12] Equation (2) correctly reproduces the well-known facts that for $N = 4$ SUSY the Gell–Mann–Low function vanishes, while for $N = 2$, the Gell–Mann–Low function is one-loop. For $N = 1$, the Gell–Mann–Low function stemming from Eq. (3) coincides with the one needed to reconcile [26, 27] the chiral and scale anomalies [28] of supersymmetric gluodynamics.

[9] D. Amati, K. Konishi, Y. Meurice, G. Rossi and G. Veneziano, *Phys. Rep.* **162** (1988) 169.

[10] E. Witten, *Nucl. Phys.* **B185** (1981) 513 [reprinted in *Supersymmetry, Vol. 1*, ed. S. Ferrara (North-Holland/World Scientific, 1987), p. 443].

[11] E. Witten, *Nucl. Phys.* **B202** (1982) 253 [reprinted in *Supersymmetry, Vol. 1*, ed. S. Ferrara (North-Holland/World Scientific, 1987), p. 490].

[12] V. Novikov, M. Shifman, A. Vainshtein and V. Zakharov, *Nucl. Phys.* **B229** (1983) 407 [reprinted in *Supersymmetry, Vol. 1*, ed. S. Ferrara (North-Holland/World Scientific, 1987), p. 606].

[13] M. Shifman and A. Vainshtein, 1986, unpublished; M. Shifman, *Lecture at the Zakopane Summer School of Physics, 1986*, unpublished.

[14] E. Witten, *Comm. Math. Phys.* **117** (1988) 353.

[15] A. S. Schwarz, *Lett. Math. Phys.* **2** (1978) 247.

[16] P. Van Baal, *Acta Phys. Polon*, **B21** (1990) 73; D. Birmingham, M. Blau, M. Rakowski and G. Thompson, *Phys. Reports*, **209** (1991) 129.

[17] M. Shifman and A. Vainshtein, *Nucl. Phys.* **B296** (1988) 445; A. Morozov, M. Olshanetsky and M. Shifman, *Nucl. Phys.* **B304** (1988) 291.

[18] A. Smilga, *Nucl. Phys.* **B266** (1986) 45; **B291** (1987) 241; B. Blok and A. Smilga, *Nucl. Phys.* **B287** (1987) 589.

[19] M. Dine, R. Rohm, N. Seiberg and E. Witten, *Phys. Lett.* **156B** (1985) 55.

[20] *V. Novikov, M. Shifman, A. Vainshtein and V. Zakharov, *Nucl. Phys.* **B260** (1985) 157.

[21] B. Zumino, *Nucl. Phys.* **B89** (1975) 535 [reprinted in *Supersymmetry, Vol. 1*, ed. S. Ferrara (North-Holland/World Scientific, 1987), p. 252].

[22] P. West, *Nucl. Phys.* **B106** (1976) 219 [reprinted in *Supersymmetry, Vol. 1*, ed. S. Ferrara (North-Holland/World Scientific, 1987), p. 264].

[23] *A. D'Adda and P. Di Vecchia, *Phys. Lett.* **73B** (1978) 162.

[24] V. Novikov, M. Shifman, A. Vainshtein and V. Zakharov, *Nucl. Phys.* **B229** (1983) 381.

[25] V. Novikov, M. Shifman, A. Vainshtein and V. Zakharov, *Phys. Lett.* **166B** (1986) 329.

[26] M. Grisaru, B. Milewski and D. Zanon, *Phys. Lett.* **B157** (1985) 174; *Nucl. Phys.* **B266** (1986) 589.

[27] M. Shifman and A. Vainshtein, *Nucl. Phys.* **B277** (1986) 456.

[28] S. Ferrara and B. Zumino, *Nucl. Phys.* **B87** (1975) 207 [reprinted in *Supersymmetry, Vol. 1*, ed. S. Ferrara (North-Holland/World Scientific, 1987), p. 117]; M. Grisaru, in *Recent Developments in Gravitation*, eds. M. Levy and S. Deser (Plenum, 1979), p. 130.

Volume 69B, number 3 PHYSICS LETTERS 15 August 1977

EUCLIDEAN SUPERSYMMETRY AND THE MANY-INSTANTON PROBLEM

B. ZUMINO

CERN, Geneva, Switzerland

Received 2 June 1977

There is no Hermitean supersymmetry in Euclidean four-space. The simplest supersymmetry has complex four-component spinorial parameters. We give its algebraic structure and the automorphisms of the algebra, as well as a representation in terms of fields and an invariant Lagrangian. The results are relevant to the counting and the construction of the solutions of the many-instanton problem.

Recently there has been considerable interest [1–16] in the study of the Yang-Mills field coupled with a massless, spin one-half field in Euclidean space-time. When the Yang-Mills field has non-vanishing winding number there are non-trivial solutions for the spinor. It has been observed [14–16] that, if the spinor field is in the adjoint representation of the internal symmetry group, the equations of motion possess partial supersymmetry properties analogous to those of the corresponding equations of the supersymmetric Yang-Mills theories in Minkowski space [17–19]. We will show here that, if the theory is extended by the inclusion of a scalar and pseudoscalar field, there exists a complete supersymmetry, with transformations which form a closed algebra in the usual sense. These transformations can be used to obtain some information on the solutions of the equations of motion. In particular, some of the results of ref. [15] emerge as a special case.

In Euclidean four-space there are no real four by four gamma matrices‡ and therefore no Majorana spinors. As a consequence, the generators of the simplest supersymmetry are complex four-component spinors. Since the spinor generators are complex, the supermultiplets are larger than they would be for a Hermitean supersymmetry. We consider a supermultiplet of fields consisting of a vector V_μ, a complex spinor ψ, a real scalar A and a real pseudoscalar B. All these fields are assumed to belong to the adjoint representation of some internal gauge group of which V_μ is the gauge potential. The internal symmetry indices will not be explicitly indicated and we shall use the cross product defined by the structure constants of the group. For instance, $D_\mu A = \partial_\mu A + gV_\mu \times A$ is the covariant derivative of the field A and $F_{\mu\nu} = \partial_\mu V_\nu - \partial_\nu V_\mu + gV_\mu \times V_\nu$ is the field strength. In terms of the anticommuting (Grassmannian), complex, spinor parameter α (a singlet under the internal symmetry) the infinitesimal supersymmetry transformation is given by

$$\begin{aligned}
\delta A &= \mathrm{i}\alpha^* \psi - \mathrm{i}\psi^* \alpha \\
\delta B &= \mathrm{i}\alpha^* \gamma_5 \psi - \mathrm{i}\psi^* \gamma_5 \alpha \\
\delta V_\mu &= \mathrm{i}\alpha^* \gamma_\mu \psi - \mathrm{i}\psi^* \gamma_\mu \alpha \\
\delta \psi &= -F_{\mu\nu}\Sigma_{\mu\nu}\alpha + (D_\mu A + \gamma_5 D_\mu B)\gamma_\mu \alpha - A \times B\gamma_5 \alpha \\
\delta \psi^* &= \alpha^* F_{\mu\nu}\Sigma_{\mu\nu} + \alpha^* \gamma_\mu (D_\mu A + \gamma_5 D_\mu B) - \alpha^* \gamma_5 A \times B.
\end{aligned} \tag{1}$$

The commutator of two transformations (1) is a translation accompanied by a field dependent gauge transformation. For instance,

$$[\delta_2, \delta_1] A = \xi_\mu \partial_\mu A + g\Lambda \times A \tag{2}$$

where

$$\xi_\mu = 2\mathrm{i}\alpha_1^* \gamma_\mu \alpha_2 - 2\mathrm{i}\alpha_2^* \gamma_\mu \alpha_1 \tag{3}$$

and

$$\begin{aligned}
\Lambda = \xi_\mu V_\mu &- (2\mathrm{i}\alpha_1^* \alpha_2 - 2\mathrm{i}\alpha_2^* \alpha_1) A \\
&+ (2\mathrm{i}\alpha_1^* \gamma_5 \alpha_2 - 2\mathrm{i}\alpha_2^* \gamma_5 \alpha_1) B.
\end{aligned} \tag{4}$$

‡ Our Euclidean gamma matrices are Hermitean and satisfy $\gamma_\mu\gamma_\nu + \gamma_\nu\gamma_\mu = 2\delta_{\mu\nu}$. We take γ_1, γ_2 and γ_3 real and symmetric, γ_4 imaginary and antisymmetric. Then $\gamma_5 = \gamma_1 = \gamma_1\gamma_2\gamma_3\gamma_4$ is Hermitean, imaginary and antisymmetric. The matrix $C = \gamma_1\gamma_2\gamma_3$ is real, antisymmetric and satisfies $C\gamma_\mu = \gamma_\mu^* C = \gamma_\mu^T C$, $C^2 = -1$. We also use $\Sigma_{\mu\nu} = \frac{1}{4}[\gamma_\mu, \gamma_\nu]$, $\epsilon_{1234} = 1$.

Similarly

$$[\delta_2, \delta_1] V_\lambda = \xi_\mu \partial_\mu V_\lambda - D_\lambda \Lambda \tag{5}$$

and so on. In evaluating the commutator on the spinor field one makes use of its Grassmannian nature. Since we have not introduced auxiliary fields, the equation of motion for ψ must also be used. It is easy to verify that the Lagrangian

$$L = -\tfrac{1}{4}F_{\mu\nu}^2 + \tfrac{1}{2}(D_\mu A)^2 - \tfrac{1}{2}(D_\mu B)^2 - \frac{i}{2}(\psi^* \cdot \gamma_\mu D_\mu \psi - D_\mu \psi^* \cdot \gamma_\mu \psi) + ig\psi^* \cdot [(A - \gamma_5 B) \times \psi] + \frac{g^2}{2}(A \times B)^2 \tag{6}$$

changes by a divergence under (1). It is Hermitean. Observe, however, the sign of the kinetic term of the field A. To understand its origin we observe that the supersymmetry algebra admits as automorphisms the non-compact chiral transformations given infinitesimally by

$$\delta' A = 2B,\ \delta' B = 2A,\ \delta'\psi = \gamma_5 \psi,\ \delta'\psi^* = \psi^* \gamma_5,\ \delta' V_\mu = 0 \tag{7}$$

(we have dropped the infinitesimal parameter). This is also an invariance of the Lagrangian. There are three more automorphisms of the algebra which leave also the Lagrangian invariant. They are, on ψ,

$$\begin{aligned} &\delta_1\psi = i\psi, \qquad \delta_1\psi^* = -i\psi^* \\ &\delta_2\psi = C\psi^*, \quad \delta_2\psi^* = C\psi \\ &\delta_3\psi = -iC\psi^*, \quad \delta_3\psi^* = iC\psi \end{aligned} \tag{8}$$

while the other fields do not change, i.e., $\delta_1 A = 0$ etc. While δ_1, δ_2 and δ_3 commute with the chiral transformation δ', they form an SU(2) structure, $[\delta_1, \delta_2] = \delta_3$, etc., cyclically. All the above can be verified directly on the field supermultiplet. We are in the presence of a non-compact U(2) with a compact SU(2) subgroup. In Euclidean four-space the Lorentz transformations become compact rotations, while the chiral transformations, which are compact in Minkowski space, become non-compact. This fact has been noticed before [20, 21]. The conserved axial vector current is

$$J_\mu^{(5)} = AD_\mu B - BD_\mu A - \frac{i}{2}\psi^* \gamma_5 \gamma_\mu \psi. \tag{9}$$

The conserved spinor current of the supersymmetry (1) is

$$J_\mu = F_{\lambda\nu} \cdot \Sigma_{\lambda\nu} \gamma_\mu \psi + \gamma_\lambda D_\lambda (A + \gamma_5 B) \cdot \gamma_\mu \psi - g(A \times B) \cdot \gamma_5 \gamma_\mu \psi + \tfrac{2}{3}(\gamma_\mu \gamma_\lambda \partial_\lambda - \partial_\mu)[(A - \gamma_5 B) \cdot \psi]. \tag{10}$$

The last term, which is automatically conserved, has been added so that the spinor current satisfies also

$$\gamma_\mu J_\mu = 0. \tag{11}$$

The supersymmetry transformations δ can be split into the sum of δ_+ and δ_- by separating the parameter α into a part of positive chirality and one with negative chirality

$$\gamma_5 \alpha_+ = \alpha_+, \qquad \gamma_5 \alpha_- = -\alpha_-. \tag{12}$$

Two transformations of the type δ_+ or two of the type δ_- commute, while the commutator of a δ_+ with a δ_- gives a translation (up to field dependent gauge transformations). We mention also that, since the Lagrangian is invariant under conformal transformations one can enlarge the supersymmetry algebra by introducing conformal supersymmetry transformations. This is related to the existence of the spinor current (10) which satisfies (11).

We do not write out the obvious equations following from (6) but observe that they take a very simple form if one imposes on the solutions the restrictions, invariant under supersymmetry transformations,

$$\begin{aligned} &F_{\mu\nu} = \tilde{F}_{\mu\nu} \equiv \tfrac{1}{2}\epsilon_{\mu\nu\lambda\rho} F_{\lambda\rho}, \\ &\psi = \psi_- = \tfrac{1}{2}(1 - \gamma_5)\psi, \\ &A = -B \end{aligned} \tag{13}$$

or, alternatively,

$$F_{\mu\nu} = -\tilde{F}_{\mu\nu}, \quad \psi = \psi_+, \quad A = B. \tag{14}$$

If we take (13), for instance, the equations reduce to

$$\gamma_\mu D_\mu \psi_- = 0 \tag{15}$$

$$D_\mu^2 A + ig\psi_-^* \times \psi_- = 0, \tag{16}$$

$$F_{\mu\nu} = \tilde{F}_{\mu\nu} \tag{17}$$

and the supersymmetry transformations to

$$\begin{aligned} &\delta_+ A = 0, \quad \delta_+ \psi_- = \gamma_\mu D_\mu (2A\alpha_+), \\ &\delta_+ V_\mu = i\alpha_+^* \gamma_\mu \psi_- - i\psi_-^* \gamma_\mu \alpha_+, \end{aligned} \tag{18}$$

and

$$\delta_- A = \mathrm{i}\alpha_-^* \psi_- - \mathrm{i}\psi_-^* \alpha_-,$$

$$\delta_- \psi_- = -F_{\mu\nu}\Sigma_{\mu\nu}\alpha_-, \qquad \delta_- V_\mu = 0. \tag{19}$$

The differential eqs. (15) to (17) are covariant under (18) and (19). While our results require in general that all spinors be treated as Grassmannian variables, some partial results do not require Fierz rearrangements and are valid also if one treats the spinors as c-numbers. This applies, for instance, to the covariance of the eqs. (15) to (17) under the δ_- transformation (19). Therefore, if one starts from a solution for which $A = \psi_- = 0$, but the Yang-Mills field has non-vanishing winding number, (19) generates two independent solutions of the Dirac equation in that external field and two more are generated by the corresponding conformal supersymmetry transformation ($\alpha \to \gamma_\mu x^\mu \alpha$). For δ_+ the situation is different and the covariance of the Dirac eq. (15) is not true for c-number spinors. However, the other two equations have zero δ_+ variation if ψ_- satisfies the Dirac equation[‡]. Therefore, to each solution of the Dirac equation in a given many-instanton configuration, $\delta_+ V_\mu$ in (18) gives a fluctuation which satisfies the correct perturbation equations and which depends on α_+ (two complex components, four real parameters). In reality only two of these four solutions are independent, since the other two can be obtained by applying to ψ_- a finite automorphism δ_2 of (8) and therefore they correspond to another solution of the Dirac equation. We recover in this way the result of ref. [15], which states that the solutions with winding number n of the SU(2) Yang-Mills equation, are twice as many ($8n$) as those ($4n$) for a Dirac field belonging to the adjoint representation in a Yang-Mills field of that winding number [10–13] (more precisely $8n - 3$, subtracting the three parameters of a global gauge rotation).

The finite transformations corresponding to (18) can be used to construct certain explicit solutions of the Yang-Mills eq. (17). Since for SU(2) particular solutions of (17) with winding number n are known and the general solution (depending on $4n$ parameters) of (15) in one of those n instanton fields has also been given [16], one can exponentiate (18) to generate the finite corrections to the Yang-Mills potential. In this way one obtains an n instanton configuration depending on $8n$ parameters, which is an exact solution of (17), but only in virtue of the anticommuting Grassmannian nature of the parameters. Its form is correspondingly simpler, since the exponentiation of (18) requires only a finite number of terms. The relation of this type of solution to the pure c-number solutions and the role it may play in the quantization problem are as yet unclear.

Discussions with J. Ellis, D. Olive and W. Nahm are gratefully acknowledged.

‡ For (17), this is a consequence of the identity

$$\gamma_\nu D_\mu - \gamma_\mu D_\nu = \tfrac{1}{2}\epsilon_{\nu\mu\lambda\rho}\gamma_5(\gamma_\lambda D_\rho - \gamma_\rho D_\lambda) + \tfrac{1}{2}(\gamma_\nu\gamma_\mu - \gamma_\mu\gamma_\nu)\gamma_\lambda D_\lambda .$$

References

[1] A.A. Belavin, A.M. Polyakov, A.S. Schwarz and Yu.S. Tyupkin, Phys. Lett. 59B (1975) 85.
[2] A.M. Polyakov, Nordita preprints (1976 and 1977).
[3] G. 't Hooft, Phys. Rev. Letters 37 (1976) 8; Phys. Rev. D14 (1976) 3432.
[4] R. Jackiw and C. Rebbi, Phys. Rev. Letters 37 (1976) 172; Phys. Rev. D14 (1976) 517; Phys. Lett. 67B (1977) 189.
[5] R. Jackiw, C. Nohl and C. Rebbi, Phys. Rev. D15 (1977) 1642.
[6] E. Witten, Phys. Rev. Letters 38 (1977) 121.
[7] E. Corrigan and D.B. Fairlie, Phys. Lett. 67B (1977) 69.
[8] R. Ore, Phys. Rev. D, to be published.
[9] B. Grossmann, Phys. Lett. 61A (1977) 86.
[10] A.S. Schwarz, Phys. Lett. 67B (1977) 172; Proc. Symp. in Alushta, USSR, April 1976, p. 224.
[11] J. Kiskis, Los Alamos preprint (1976).
[12] N.K. Nielsen and B. Schroer, CERN preprint TH.2317 (1977).
[13] M.F. Atiyah, N.J. Hitchin and I.M. Singer, Oxford Mathematical Institute preprint (1977).
[14] S. Chadha, A. D'Adda, P. Di Vecchia and F. Nicodemi, Phys. Lett. 67B (1977) 103.
[15] L.S. Brown, R.D. Carlitz and C. Lee, University of Washington, Seattle, preprint (1977).
[16] R. Jackiw and C. Rebbi, MIT preprint (1977).
[17] S. Ferrara and B. Zumino, Nuclear Phys. B79 (1974) 413.
[18] A. Salam and J. Strathdee, Phys. Lett. 51B (1974) 353.
[19] B. de Wit and D.Z. Freedman, Phys. Rev. D12 (1975) 2286.
[20] J. Schwinger, Phys. Rev. 115 (1959) 721.
[21] J.D. Stack, University of Illinois preprint (1977).

Volume 73B, number 2 PHYSICS LETTERS 13 February 1978

SUPERSYMMETRY AND INSTANTONS

A. D'ADDA
The Niels Bohr Institute, University of Copenhagen, DK-2100 Copenhagen Ø, Denmark

and

P. Di VECCHIA
NORDITA, DK-2100 Copenhagen Ø, Denmark

Received 22 November 1977

We show that the eigenvalue equations for the fluctuation of scalars, fermions and gluon around any classical self-dual solution of the Yang–Mills theory have the same spectrum of non-zero eigenvalues. In the case of a supersymmetric Yang–Mills theory this implies that the one loop correction around any self dual instanton is just given by a counting of the zero modes of the gluon, fermion and ghost.

It has been proposed by Polyakov [1] that the existence of classical solutions of euclidean field theories may allow one to compute the large distance behavior of the corresponding theory. This can be done by expanding the partition function and the Green's functions around those classical solutions keeping only up to the quadratic terms in the field fluctuations. However the computation of the quantum corrections may become so complicated that it is very difficult to have an estimate of them.

One way out is to consider quantum field theories as the supersymmetric ones which have such an amount of symmetry that the calculation of the quantum corrections becomes very simple.

In this letter we consider the eigenvalue equations for the fluctuations of scalars, fermions and gluon around any classical self-dual solution of the Yang–Mills theory and we show that they have the same spectrum of eigenvalues (except for the zero modes) if the fields transform according to the same representation of the gauge group. As a consequence of this fact in the case of a supersymmetric theory, where the number of bosons is equal to the number of fermions, one gets a complete cancellation among the determinants that give the one-loop correction. Therefore the one-loop quantum correction is just given by an integral over the collective coordinates and by a logarithmic term containing the subtraction point μ whose coefficient can be determined by a counting of the zero modes of the gluon, fermions and ghost. It is given by $8N - 4NN_f$, where N is the Pontryagin number of the classical solution and N_f is the number of the fermion flavors.

The same considerations apply also to some two-dimensional theories as the supersymmetric ϕ^4 and sine-Gordon that have been recently constructed. One gets that the one-loop correction to the mass of the soliton is identically vanishing because of the cancellation between the bosonic and fermionic eigenfrequencies. These results generalize to the case of solitons and instantons the cancellation of the vacuum diagrams occurring in the vacuum sector in any supersymmetric theory [2].

If we expand the Yang–Mills action with scalars and fermions around any classical solution $\bar{A}_\mu$ of the self-duality equation:

$$\bar{F}_{\mu\nu} = \tilde{\bar{F}}_{\mu\nu} = \tfrac{1}{2}\epsilon_{\mu\nu\rho\sigma}\bar{F}_{\rho\sigma}, \tag{1}$$

one gets the following eigenvalue equations respectively for the fluctuation of the scalar, fermion and gluon:

$$\bar{\nabla}_\mu \bar{\nabla}_\mu \phi = -\lambda^2 \phi, \tag{2a}$$

$$i\gamma^\mu \bar{\nabla}_\mu \psi = \lambda \psi, \tag{2b}$$

$$V_{\nu\mu}a_\mu = \bar\nabla_\mu(\bar\nabla_\mu a_\nu - \bar\nabla_\nu a_\mu) + [\bar F_{\nu\mu}, a_\mu] = -\lambda^2 a_\nu, \quad (2c)$$

where $\bar\nabla$ is the gauge covariant derivative in the background field.

We want to show now that the fermion equation has the same spectrum of non-zero eigenvalues as the scalar provided that they transform according to the same representation of the gauge group [‡1]

If one uses the following representation for the euclidean γ-matrices:

$$\gamma^\mu = \begin{pmatrix} 0 & \alpha^\mu \\ \bar\alpha^\mu & 0 \end{pmatrix}, \quad \gamma^5 = \begin{pmatrix} 1 & 0 \\ 0 & -1 \end{pmatrix}, \quad \begin{matrix} \alpha_\mu = (-i\sigma, 1), \\ \bar\alpha_\mu = (i\sigma, 1), \end{matrix} \quad (3)$$

and if one writes ψ as:

$$\psi = \psi_+ + \psi_-, \qquad \psi_\pm = \tfrac{1}{2}(1 \pm \gamma_5)\psi, \quad (4)$$

one gets the following equations for $\psi_\pm$:

$$T\psi_- = i\alpha_\mu \nabla_\mu \psi_- = \lambda\psi_+, \quad (5a)$$

$$T^+\psi_+ = i\bar\alpha_\mu \nabla_\mu \psi_+ = \lambda\psi_-, \quad (5b)$$

where T^+ is the adjoint of T.

Using the identities

$$\bar\alpha^\mu\alpha^\nu = \delta^{\mu\nu} + 2i\sigma^{\mu\nu}, \qquad \alpha^\mu\bar\alpha^\nu = \delta^{\mu\nu} + 2i\bar\sigma^{\mu\nu}, \quad (6)$$

with

$$\sigma_{\mu\nu} = (1/4i)[\bar\alpha_\mu\alpha_\nu - \bar\alpha_\nu\alpha_\mu], \quad \sigma_{\mu\nu} = \tilde\sigma_{\mu\nu},$$
$$\bar\sigma_{\mu\nu} = (1/4i)[\alpha_\mu\bar\alpha_\nu - \alpha_\nu\bar\alpha_\mu], \quad \bar\sigma_{\mu\nu} = -\tilde{\bar\sigma}_{\mu\nu}, \quad (7)$$

and the commutator of two covariant derivatives:

$$[\bar\nabla_\mu; \bar\nabla_\nu]^{ab} = \epsilon^{acb}\bar F^c_{\mu\nu}, \quad (8)$$

it is easy to show that

$$TT^+\psi_+ = \{(\nabla_\mu\nabla^\mu)^{ab} + i\bar\sigma_{\mu\nu}\epsilon^{acb}\bar F^c_{\mu\nu}\}\psi^b_+ = -\lambda^2\psi^a_+, \quad (9a)$$

$$T^+T\psi_- = \{(\nabla_\mu\nabla^\mu)^{ab} + i\sigma_{\mu\nu}\epsilon^{acb}\bar F^c_{\mu\nu}\}\psi^b_- = -\lambda^2\psi^a_-. \quad (9b)$$

But if the background field is self-dual, then because of eq. (7) $\bar\sigma_{\mu\nu}\bar F_{\mu\nu} = 0$ and eq. (9a) becomes the same equation as the eq. (2a) for the scalar field. Therefore, starting from any solution $\phi(x)$ with eigenvalue λ^2 of the scalar, eq. (2a), one can construct an eigenfunction of the fermion eq. (2b) with the same eigenvalue λ given by

$$\psi_+(x) = \phi(x)\alpha, \quad \psi_-(x) = \lambda^{-1}T^+\psi_+(x), \quad (10)$$

where α is a constant arbitrary two-dimensional spinor. In addition the fermion eq. (9b) has also $C(T)N_fN$ zero modes (N_f is the number of fermion flavors, N is the Pontryagin number, and $C(T) = \frac{2}{3}T(T+1)(2T+1)$) whose correspondent eigenfunctions have negative chirality ($\psi_+ = 0$) [4].

Starting from the fermion eigenfunction $\psi_n(x)$:

$$\psi_n(x) = (1 + i\gamma^\mu\bar\nabla_\mu/\lambda_n)(\tfrac{1}{2}(1+\gamma^5))\phi_n(x)\alpha_n \quad (11)$$

with eigenvalues $\lambda_n \neq 0$ given in terms of the scalar eigenfunctions $\phi_n(x)$, it is possible to construct the fermion Green's function corresponding to the non-zero eigenvalues:

$$S_F(x,y) = \sum_n \frac{\psi_n(x)\psi_n^+(y)}{\lambda_n}. \quad (12)$$

Using eq. (11) into eq. (12) and summing over both positive and negative eigenfrequencies one gets:

$$S_F(x,y) = i\gamma^\mu\bar\nabla^x_\mu\Delta(x,y)(\tfrac{1}{2}(1+\gamma_5)) + (\tfrac{1}{2}(1+\gamma_5))\Delta(x,y)i\gamma_\mu\overleftarrow{\bar\nabla}^\mu_y, \quad (13)$$

where

$$\Delta(x,y) = \sum_n \frac{\phi_n(x)\phi_n^+(y)}{\lambda_n^2}. \quad (14)$$

Eq. (13) agrees with the analogous result of ref. [5].

The same procedure applies to the eigenvalue equation for the vector fluctuations. The gauge is specified by adding a gauge fixing term $\frac{1}{2}(\bar\nabla_\mu A_\mu)^2$ to the lagrangian, and the eigenvalue eq. (2c) is then modified to:

$$V_{\nu\mu}a_\mu + S_{\nu\mu}a_\mu = -\lambda^2 a_\nu, \quad (15)$$

where $V_{\nu\mu}$ is defined in eq. (2c) and $S_{\mu\nu} = \bar\nabla_\mu\bar\nabla_\nu$. Let us take $a_\mu = b_\mu + \bar\nabla_\mu\alpha$ with $\bar\nabla_\mu b_\mu = 0$, then $S_{\nu\mu}b_\mu = 0$, $V_{\nu\mu}\bar\nabla_\mu\alpha = 0$ and the eigenvalue equation becomes

$$\begin{pmatrix} V_{\nu\mu}b_\mu \\ S_{\nu\mu}\bar\nabla_\mu\alpha \end{pmatrix} = -\lambda^2\begin{pmatrix} b_\nu \\ \bar\nabla_\nu\alpha \end{pmatrix}. \quad (16)$$

Let us consider the operators:

$$T\begin{pmatrix} b \\ \bar\nabla\alpha \end{pmatrix} = \begin{pmatrix} \eta^{(-)}_{a\rho\sigma}\bar\nabla_\rho b_\sigma \\ \bar\nabla_\mu\bar\nabla_\mu\alpha \end{pmatrix}, \quad T^+\begin{pmatrix} f \\ h \end{pmatrix} = \begin{pmatrix} -\eta^{(-)}_{a\rho\sigma}\bar\nabla_\rho f_a \\ -\bar\nabla_\rho h \end{pmatrix}, \quad (17)$$

where $\eta^{(-)}_{a\rho\sigma}$ is defined in ref. [3].

‡1 The equality of the non-zero eigenvalues of eqs. (2) has also been used by 't Hooft in the calculation of the quantum fluctuations around the $N = 1$ instanton [3].

It is easy to prove that $-T^+T$ reproduces the left-hand-side of eq. (16), while TT^+ gives

$$TT^+\begin{pmatrix} f \\ h \end{pmatrix} = -\begin{pmatrix} +\bar{\nabla}_\mu \bar{\nabla}_\mu f_a \\ \bar{\nabla}_\mu \bar{\nabla}_\mu h \end{pmatrix}. \tag{18}$$

Since TT^+ and T^+T have the same spectrum of non-zero eigenvalues, it follows from eq. (18) that the eigenvalue eq. (16) has the same non-zero eigenvalues as the scalar eq. (2a). Starting from any eigenfunction $\phi(x)$ of the scalar eq. (2a), it is possible to construct the eigenfunctions of eq. (16) corresponding to the same eigenvalue by applying T^+ to

$$v(j) = \begin{pmatrix} f^a(j) = \delta_{ja}\phi \\ h = 0 \end{pmatrix}$$

and

$$v(0) = \begin{pmatrix} f_a = 0 \\ h = \phi(x) \end{pmatrix}.$$

One gets

$$a_\mu^{(i)} = -\lambda^{-1}\eta^{(-)}_{i\rho\mu}\bar{\nabla}_\rho\phi(x), \quad a_\mu^{(0)} = -\lambda^{-1}\bar{\nabla}_\mu\phi(x). \tag{19}$$

Each eigenvalue is therefore four times degenerate with respect to the scalar, but two components are killed in the functional integral by the Faddeev–Popov ghost (whose eigenvalue equation is the same as the scalar) so that only the two physical components of the gauge field give actually a contribution.

Using the eigenfunction (19) one can construct the propagator of the gluon

$$D_{\mu\nu}(x,y) = \sum_n \sum_{k=0}^{3} \frac{a_n^{(k)}(x)a_n^{(k)+}(y)}{\lambda_n^2}$$
$$= -[\delta_{\mu\rho}\delta_{\nu\sigma} + \eta^{(-)}_{i\mu\rho}\eta^{(-)}_{i\nu\sigma}]\,\vec{\bar{\nabla}}^x_\rho \Delta^2(x,y)\overleftarrow{\bar{\nabla}}^y_\sigma, \tag{20}$$

which agrees with the result of ref. [5].

In order to compute the one-loop correction around any classical solution $\bar{A}_\mu$ of the partition function of the Yang–Mills theory, one must compute the determinants of the operators (2). As in ref. [6] [‡2], we use the ζ-function regularization procedure which gives the following regularized formula for the determinant of the operator A:

$$\det((\mu\alpha)^{-1}A) = (\mu\alpha)^{-Z-\zeta_0}\,e^{-\zeta_0'}, \tag{21}$$

where μ is the subtraction point, α is containing the scales of the instantons and $\zeta_T(s)$ is defined in terms of the eigenvalues of A by:

$$\zeta_s(T) = \sum_n [\lambda_n(T)]^{-s}. \tag{22}$$

Z is the number of zero modes of A and the index T refers to the isospin of the field. Using the fact that the operators (2) have the same non-zero eigenvalues, one gets:

$$\zeta_s^{\mathrm{F}}(T) = 2\zeta_s(T), \quad \zeta_s^{\mathrm{V}}(1) = 2\zeta_s(1), \tag{23}$$

where $\zeta_s(T)$ is the ζ-function corresponding to the scalar field and the factor 2 counts the number of components of a massless spinning particle. In the ζ-function corresponding to the vector we have also included the contribution of the ghost field which behaves as the scalar.

Inserting those expressions in the partition function corresponding to the Pontryagin number N one gets:

$$Z^{(N)} = \int \mathrm{d}C\, e^{-8\pi^2 N/g^2}(\mu\alpha)^A e^B, \tag{24}$$

where $\int \mathrm{d}C$ is integral over the collective coordinates and

$$A = 8N - C(T)NN_\mathrm{f} + N_\mathrm{s}\tilde{\zeta}_0(T_\mathrm{s}) + 2\tilde{\zeta}_0(1) - 4N_\mathrm{f}\tilde{\zeta}_0(T_\mathrm{f}), \tag{25}$$

$$2B = \tilde{\zeta}_0'(T_\mathrm{s})N_\mathrm{s} + 2\tilde{\zeta}_0'(1) - 4N_\mathrm{f}\tilde{\zeta}_0'(T_\mathrm{f}). \tag{26}$$

$\tilde{\zeta}_s(T)$ is the difference between the ζ-function in the background instanton field and that in the vacuum.

In particular in a supersymmetric theory ($T_\mathrm{s} = T_\mathrm{f} = 1$) one has the same number of bosons and fermions. This implies that $N_\mathrm{s} + 2 = 4N_\mathrm{f}$ and therefore [‡3]

$$B = 0, \quad A = (8 - 4N_\mathrm{f})N. \tag{27}$$

Therefore the coefficient of the term containing the subtraction point μ is only given in terms of the zero modes of the gluon, fermions and ghost.

In the supersymmetric case the partition function becomes extremely simple

‡2 The ζ-function regularization procedure has also been used to compute the zero point energy of a dual string [7,13]. The one loop correction around the $N = 1$ instanton has been also computed in refs. [3] and [8].

‡3 The vanishing of B in the case of the $N = 1$ instanton in a supersymmetric theory has been also noticed in ref. [9].

Volume 73B, number 2 PHYSICS LETTERS 13 February 1978

$$Z^{(N)} = \int \mathrm{d}C \exp\left[-\frac{8\pi^2}{g^2} + (8 - 4N_\mathrm{f}) \log \mu\alpha\right] N. \quad (28)$$

In particular in the supersymmetric theory with an SU(4) internal symmetry containing 6 real scalars and 2 complex fermions ($N_\mathrm{f} = 2$) also the logarithmic term in eq. (28) is vanishing [‡4]. This is the consequence of the fact that this particular supersymmetric theory can be obtained as the zero slope limit of the Neveu–Schwarz–Ramond (NSR) model in ten dimensions after the compactification of the six extra dimensions [11]. It is known, in fact, that the loop corrections in the string model do not give rise to any ultraviolet divergence, and in the NSR model this property seems to be preserved also in the zero slope limit (for a general review of dual models, see, e.g., refs. [12,13]). If the same property is also valid in the closed string sector of the NSR model, one would get the vanishing of the conformal anomaly in the case of the SO(4) supergravity which contains 2 spin 0, 4 real spin 1/2, 4 real spin 3/2, 6 spin 1 and 1 spin 2 [14]. Imposing the vanishing of the trace anomaly in the SO(4) supergravity, one gets the following anomaly for a Majorana spin 3/2 field [‡5]:

$$T_\mu{}^\mu = \frac{1}{180(4\pi)^2}\{-165 R_{\mu\nu}R^{\mu\nu} + 38R^2 - \tfrac{143}{4} R_{\mu\nu\rho\sigma}R^{\mu\nu\rho\sigma}\}. \quad (29)$$

It would be interesting to study what happens to the Hawking effect [16] [‡6] in a theory without trace anomaly.

In the last part of this letter we discuss some two-dimensional sypersymmetric models with soliton solutions which have been recently constructed [18,19].

It has been shown in ref. [18] that the following action is supersymmetric:

$$S = \int \mathrm{d}x\, \mathrm{d}t \{-\tfrac{1}{2}(\partial_\mu\phi)^2 - \tfrac{1}{2}[V'(\phi)]^2 - \tfrac{1}{2}\mathrm{i}\,\bar\psi\partial\!\!\!/\psi - \tfrac{1}{2}\mathrm{i}\,\bar\psi\psi V''(\phi)\}. \quad (30)$$

In particular, if $V'(\phi) = \lambda^{-1/2}(m^2 - \phi^2)$, one gets a supersymmetric version of the ϕ^4-theory, while if $V'(\phi) = 2m^2\lambda^{-1/2}\sin(\sqrt{\lambda}\,\phi/2m)$, one gets a supersymmetric version of the sine-Gordon equation. A time independent soliton solution satisfies the following first order equation:

$$(\mathrm{d}/\mathrm{d}x)\,\phi_\mathrm{cl}(x) = V'(\phi_\mathrm{cl}). \quad (31)$$

In order to compute the lowest quantum correction to the classical mass of the soliton one must solve the following eigenvalue equations:

$$\left\{-\frac{\mathrm{d}^2}{\mathrm{d}x^2} + [V''(\phi_\mathrm{cl})]^2 + V'(\phi_\mathrm{cl})V'''(\phi_\mathrm{cl})\right\}\eta(x) = \omega_\mathrm{B}^2\eta(x), \quad (32\mathrm{a})$$

$$\left\{\gamma_1\frac{\mathrm{d}}{\mathrm{d}x} + V''(\phi_\mathrm{cl})\right\}\chi(x) = \mathrm{i}\omega_\mathrm{F}\gamma_0\chi(x). \quad (32\mathrm{b})$$

If one writes

$$\chi(x) = \chi_+(x) + \chi_-(x), \quad \chi_\pm(x) = \tfrac{1}{2}(1 \pm \gamma_1)\chi, \quad (33)$$

eq. (32b) becomes:

$$T\chi_+ = \left\{\frac{\mathrm{d}}{\mathrm{d}x} + V''(\phi_\mathrm{cl})\right\}\chi_+ = \mathrm{i}\omega_\mathrm{F}\gamma_0\chi_-, \quad (34\mathrm{a})$$

$$T^+\chi_- = \left\{-\frac{\mathrm{d}}{\mathrm{d}x} + V''(\phi_\mathrm{cl})\right\}\chi_- = \mathrm{i}\omega_\mathrm{F}\gamma_0\chi_+. \quad (34\mathrm{b})$$

It is easy to prove that

$$T^+T\chi_+ = \left\{-\frac{\mathrm{d}^2}{\mathrm{d}x^2} + [V''(\phi_\mathrm{cl})]^2 - V'(\phi_\mathrm{cl})V'''(\phi_\mathrm{cl})\right\}\chi_+ = \omega_\mathrm{F}^2\chi_+, \quad (35\mathrm{a})$$

$$TT^+\chi_- = \left\{-\frac{\mathrm{d}^2}{\mathrm{d}x^2} + [V''(\phi_\mathrm{cl})]^2 + V'(\phi_\mathrm{cl})V'''(\phi_\mathrm{cl})\right\}\chi_- = \omega_\mathrm{F}^2\chi_-, \quad (35\mathrm{b})$$

where eq. (30) has been used.

The eigenvalue equation for χ_- is identical to the one for $\eta(x)$; therefore one gets for the fermions the

[‡4] The β function of this supersymmetric theory has been recently computed and found to be vanishing up to two loops [10]. We thank S. Ferrara for communicating this result to us.

[‡5] We used the values for the conformal anomalies given in ref. [15].

[‡6] For a connection between the Hawking effect and the trace anomaly see ref. [17].

Volume 73B, number 2 PHYSICS LETTERS 13 February 1978

same spectrum of eigenvalues as in the case of the boson ($\omega_F^2 = \omega_B^2$). If $\eta(x)$ is an eigenfunction of the eq. (32a) with eigenvalue $\omega^2 \neq 0$, then one can construct the following eigenfunction of the fermion equation:

$$\chi_- = \eta(x)\alpha_-, \qquad \chi_+ = -(\mathrm{i}\omega)^{-1}\gamma^0 T^+ \chi_- . \tag{36}$$

It is also easy to see that eqs. (32a) and (35b) have a zero mode corresponding respectively to the invariance of the action under translations and supersymmetry. Eq. (35a) instead does not have any zero mode. As a consequence of the equality of the boson and fermion eigenvalues one gets that the one loop correction to the soliton mass is identically vanishing:

$$M = M_{\mathrm{cl}} + \frac{1}{2}\sum \omega_B - \frac{1}{2}\sum \omega_F = M_{\mathrm{cl}}. \tag{37}$$

We thank S. Chadha, E. Del Giudice and F. Nicodemi for discussions. We also thank N.K. Nielsen for many useful discussions on the conformal anomaly. One of us (A.D.A.) is grateful to the Danish Research Council and the Commemorative Association of the Japan World Exposition for financial support.

References

[1] A.M. Polyakov, Nucl. Phys. B120 (1977) 429.
[2] B. Zumino, Nucl. Phys. B89 (1975) 535.
[3] G. 't Hooft, Phys. Rev. D14 (1976) 3432.
[4] S. Chadha, A. D'Adda, P. Di Vecchia and F. Nicodemi, Phys. Lett. 67B (1977) 103;
R. Jackiw and C. Rebbi, Phys. Rev. D16 (1977) 1052;
N.K. Nielsen and B. Schroer, Nucl. Phys. B127 (1977) 493;
L.S. Brown, R.D. Carlitz and C. Lee, Phys. Rev. D16 (1977) 417;
J. Kiskis, Phys. Rev. D15 (1977) 2329.
[5] L.S. Brown, R.D. Carlitz, D.B. Creamer and C. Lee, Phys. Lett. 70B (1977) 180.
[6] S. Hawking, Commun. Math. Phys. 55 (1977) 133;
S. Chadha, A. D'Adda, P. Di Vecchia and F. Nicodemi, NBI-HE-77-26; Phys. Lett. 72B (1977) 103;
[7] F. Gliozzi (1976) unpublished.
[8] A.A. Belavin and A.M. Polyakov, Nucl. Phys. B123 (1977) 429;
R. Ore, MIT preprint (1977).
[9] L.F. Abbott, M.T. Grisaru and H.J. Schnitzer, Brandeis University preprint (1977).
[10] D.R.T. Jones, CERN preprint, TH 2408.
[11] F. Gliozzi, D. Olive and J. Scherk, Nucl. Phys. B122 (1977) 253;
L. Brink, J. Scherk and J. Schwarz, Nucl. Phys. B121 (1977) 77.
[12] Dual theory, ed. M. Jacob (North-Holland, Amsterdam, 1974);
J. Scherk, Rev. Mod. Phys. 47 (1975) 123.
[13] P. Di Vecchia, NORDITA preprint, to be published in Proc. of the Bielefeld Summer School (1976).
[14] A. Das, Phys. Rev. D15 (1977) 2805;
E. Cremmer, J. Scherk and S. Ferrara, Phys. Lett. 68B (1977) 234;
E. Cremmer and J. Scherk, Cambridge preprint (1977).
[15] H.S. Tsao, Phys. Lett. 68B (1977) 79;
M.J. Duff, Nucl. Phys. B125 (1977) 334.
[16] S.W. Hawking, Commun. Math. Phys. 43 (1975) 199.
[17] S.M. Christensen and S.A. Fulling, Phys. Rev. D15 (1977) 2088.
[18] P. Di Vecchia and S. Ferrara, Trieste preprint, to be published in Nucl. Phys.
[19] Hruby, Dubna preprint, to be published in Nucl. Phys.

Nuclear Physics B229 (1983) 394–406

SUPERSYMMETRY TRANSFORMATIONS OF INSTANTONS

V.A. NOVIKOV, M.A. SHIFMAN, A.I. VAINSHTEIN, V.B. VOLOSHIN
and V.I. ZAKHAROV

Institute of Theoretical and Experimental Physics, Moscow 117259, USSR

Received 8 June 1983

Instantons in the simplest supersymmetric Yang–Mills theory are considered. We introduce bosonic and fermionic collective coordinates and study how they change under the supersymmetry transformations. The instanton measure is shown to be explicitly invariant under the transformations. We discuss the relation between quantum anomalies and the functional form of the instanton measure.

1. Introduction

In this paper we consider instantons in supersymmetric gluodynamics. Needless to say that non-perturbative aspects (and the BPST instantons represent the most famous non-perturbative fluctuation) may play the key role in the theory. (For an illuminating introduction to the subject and recent development see ref. [1].) Below we investigate technical issues of instanton calculus specific for supersymmetric theories.

As is well-known, instantons of small size induce effective multifermion interaction [2]. Suspicion arises [3, 4] that this interaction breaks the symmetry between fermions and bosons expected in the supersymmetric Yang–Mills theory. For definiteness we concentrate on pure gluodynamics with SU(2) as the gauge group. (A nice review and an exhaustive list of references is given in [5].) In the Wess–Zumino (super)gauge the lagrangian takes the form

$$\mathcal{L} = -\tfrac{1}{4}(G^a_{\mu\nu})^2 + \tfrac{1}{2}i\lambda^a\gamma_\mu \mathcal{D}^{ab}_\mu \lambda^b + (\cdots)\,,$$
$$\mathcal{D}^{ab}_\mu = \partial_\mu \delta^{ab} + g\varepsilon^{acb} v^c_\mu\,, \tag{1}$$

where v^a_μ $(a=1,2,3)$ is the gluon field, $G^a_{\mu\nu}$ is the gluon field strength tensor, g is the coupling constant, λ^a is the color triplet of Majorana spinors, $\mathcal{D}^{ab}_\mu$ is the covariant derivative, and $(\cdots)$ stand for the gauge-fixing term and the ghosts.

The lagrangian (1) is invariant, up to a total derivative, under the supersymmetry transformations:

$$\delta_s v^a_\mu = i\bar{\varepsilon}\gamma_\mu\lambda^a\,, \qquad \delta_s\lambda^a = \tfrac{1}{2}\sigma_{\mu\nu}G^a_{\mu\nu}\varepsilon\,, \tag{2}$$
$$\sigma_{\mu\nu} = \tfrac{1}{2}(\gamma_\mu\gamma_\nu - \gamma_\nu\gamma_\mu)\,,$$

where ε is the transformation parameter (actually gauge fixing breaks the supersymmetry, but this is irrelevant to most of the paper; for further discussion see sect. 5).

The instanton-induced effective lagrangian is determined by fermion zero modes [2]. In our case it looks as follows [4]

$$\mathcal{L}_{\rm eff} = \tfrac{4}{3}\pi^4\left(\frac{2\pi}{\alpha_s}\right)^4 \exp\left(-\frac{2\pi}{\alpha_s}\right)\frac{d\rho}{\rho^5}\rho^8$$
$$\times\{\bar\lambda^a\lambda^a\partial_\mu\bar\lambda^b\partial_\mu\lambda^b + \bar\lambda^a\gamma_5\lambda^a\partial_\mu\bar\lambda^b\gamma_5\partial_\mu\lambda^b - \tfrac{1}{2}\bar\lambda^a\sigma_{\alpha\beta}\lambda^b$$
$$\times\partial_\mu\bar\lambda^b\sigma_{\alpha\beta}\partial_\mu\lambda^a\}\,, \tag{3}$$

where ρ is the instanton size, and it is assumed that the external momenta are much less than ρ^{-1}.

The lagrangian (3) apparently violates the symmetry under the transformations (2), and this is just the starting point of our discussion.

We will argue that it is the approximation of the effective lagrangian, or fixing the instanton size, that is inconsistent with the supersymmetry. Notice that naive supersymmetry is perfectly consistent with choosing some particular instanton size since the supersymmetry transformations do not involve dilatations. This simple argument is false, however.

The reason is that the classical lagrangian (1) possesses additional invariance: it is invariant under conformal and superconformal transformations. To begin with, consider the instanton measure μ as a function of the following collective coordinates:

$$\mu(x_0, \rho, \alpha, \bar\beta) = \text{const}\; d^4x_0\,\rho\,d\rho\,d^2\alpha\,d^2\bar\beta\,. \tag{4}$$

Here x_0 is the instanton position, ρ is its size, and α and $\bar\beta$ are the grassmannian coordinates corresponding to supersymmetric and superconformal transformations, respectively. Eq. (4) represents the result of an explicit calculation of the measure in the lowest order in the coupling constant.

We will argue that these collective coordinates naturally split into two pairs

$$(x_0, \alpha)\,, \qquad (\rho, \bar\beta)\,.$$

The first pair (x_0, α) is the instanton center coordinate, α playing the role of the supersymmetric partner of x_0. Notice that x_0 transforms under the supersymmetry transformation as $x_{\rm chiral}$, α is analogous to θ, and there is no analogue of $\bar\theta$.

The second pair $(\rho, \bar\beta)$ is, so to say, a superfield parameter labelling the instanton solution.

The central point is that supersymmetry induces simultaneous changes in ρ and $\bar\beta$. On the other hand, the effective lagrangian (3) is obtained by integration over all fermion zero modes while ρ is fixed. This is inconsistent with supersymmetry.

The supersymmetry restores itself either after integration over ρ, or, alternatively, if some particular ρ is chosen, one should necessarily fix $\bar\beta$.

2. Instanton preliminaries

Let us recall first the basic facts on instantons and supersymmetry. BPST instantons [6] are the classical solutions of the Yang–Mills equations (more exactly, duality equations). In particular,

$$(A^a_\mu)_{\rm cl}=\frac{2}{g}\,\eta_{a\mu\nu}\frac{(x-x_0)_\nu}{(x-x_0)^2+\rho^2}\,,$$

$$(G^a_{\mu\nu})_{\rm cl}=-\frac{4}{g}\,\eta_{a\mu\nu}\frac{\rho^2}{[(x-x_0)^2+\rho^2]^2}\,, \tag{5}$$

where $\eta_{a\mu\nu}$ are the 't Hooft symbols. Below we shall sometimes omit the subscript cl for brevity.

The supersymmetry transformations, when applied to the bosonic solution (5), generate nothing else but the fermion zero modes:

$$\lambda^a_{\rm ss}=\tfrac{1}{2}\sigma_{\mu\nu}G^a_{\mu\nu}\alpha\,, \tag{6}$$

where α is the spinor parameter of the transformation. Indeed, $\lambda_{\rm ss}$ satisfies the equation $\hat{\mathscr{D}}\lambda^a_{\rm ss}=0$ if $(G^a_{\mu\nu})_{\rm cl}=\pm(\tilde{G}^a_{\mu\nu})_{\rm cl}$ or $(\mathscr{D}_\mu G^a_{\mu\nu})_{\rm cl}=0$.

However, not all zero modes are constructed in this way. Two remaining modes (superconformal modes) are given by

$$\lambda^a_{\rm sc}=\tfrac{1}{2}\sigma_{\mu\nu}G^a_{\mu\nu}i(\hat{x}-\hat{x}_0)\beta\,, \tag{7}$$

with the x-dependent transformation parameter $(\hat{x}-\hat{x}_0)\beta$. One can readily check again that $\lambda^a_{\rm sc}$ satisfy the Dirac equation, $\hat{\mathscr{D}}\lambda^a_{\rm sc}=0$.

As was first explained by Zumino [7] this pattern of the zero modes is a manifestation of the superconformal symmetry of the classical equations of motion. Likewise, the existence of the parameter ρ is a reflection of the invariance under change of scale [6].

So far we have used the Majorana notation. Actually, the instanton field $G_{\mu\nu}$ is self-dual. It is more appropriate, therefore, to proceed to chiral notation. Then the equation for the instanton field is $G^{\dot\alpha\dot\beta}=0$, and only transformations with the right-handed parameters α_α generate non-trivial variations, $\lambda^a_{\rm ss}\sim G^{\alpha\beta}\alpha_\beta$. Similarly, in the case of the superconformal transformations only the left-handed parameters $\bar\beta_{\dot\alpha}$ are relevant, $\lambda^a_{\rm sc}\sim G^{\alpha\beta}x_{\beta\dot\gamma}\bar\beta^{\dot\gamma}$. Below we shall use as a rule the chiral representation. A few formulas, however, will be given in the Majorana representation. We hope that this will cause no confusion.

Two more remarks on instantons and supersymmetry are in order now. First, instantons are solutions in euclidean space–time. On the other hand, it is not straightforward to have a theory of a Majorana or Weyl spinor in euclidean space. For this reason one starts sometimes [7, 3] with the $N=2$ supersymmetric Yang–Mills theory which can be formulated directly in terms of the Dirac fields, and there are no problems with continuation to the euclidean space. All calculations become more involved, however. In particular, nobody has succeeded so far in deriving the analogue of the lagrangian (3) in this case because averaging over the instanton orientation in the color space becomes much more tedious. On the other hand, problems with continuation of the $N=1$ theory to the euclidean space do not seem to be fundamental. An explicit treatment of the issue can be found in ref. [4] where the effective lagrangian (3) has been derived.

Secondly, one might think that the supersymmetry breaking figuring in eq. (3) is a manifestation of spontaneous supersymmetry breaking induced by the instantons. This is not so; we have convinced ourselves that the instantons do not bring about spontaneous breaking of the supersymmetry in the case considered [8].

3. Supersymmetric instanton measure

Since supersymmetry is a true symmetry of the action, including quantum corrections, the instanton measure must be invariant under the supersymmetry transformations as well. In this section we find transformations of the collective coordinates that are induced by supersymmetry transformations and show that the measure (4) is indeed invariant.

As was mentioned in sect. 2 there exist both fermionic and bosonic solutions of the classical equations of motion. To keep this symmetry between fermions and bosons explicit at each step let us introduce a vector superfield $V(x_0, \rho, \alpha, \bar{\beta})$ composed of the classical solutions. We define $V(x_0, \rho, \alpha, \bar{\beta})$ in the following way

$$V(x_0, \rho, \alpha, \bar{\beta}) = e^{iPx_0} e^{-i(\alpha Q + \bar{S}\bar{\beta})} V(x_0 = 0, \rho)_{\text{inst}} e^{i(\alpha Q + \bar{S}\bar{\beta})} e^{-iPx_0}, \tag{8}$$

where Q and $\bar{S}$ are the generators of supersymmetric and superconformal transformations, respectively, P is the generator of the translations, x_0, ρ, α and $\bar{\beta}$ are the instanton parameters mentioned above, $V(x_0=0, \rho)_{\text{inst}}$ denotes an "initial" superfield containing purely bosonic solution (5) with no fermion components. Notice that $Q\alpha$ and $\bar{S}\bar{\beta}$ commute.

The spinor generators act on V in the standard way, for instance,

$$\begin{aligned}
[\alpha Q, V_{\text{inst}}] &= -i\bar{\theta}^2 \theta_\alpha \tfrac{1}{2} G_{\text{cl}}^{\alpha\beta} \alpha_\beta, \\
[\bar{S}\bar{\beta}, V_{\text{inst}}] &= -i\bar{\theta}^2 \theta_\alpha \tfrac{1}{2} G_{\text{cl}}^{\alpha\beta} x_{\beta\dot{\gamma}} \bar{\beta}^{\dot{\gamma}},
\end{aligned} \tag{9}$$

$$[\bar{Q}\bar{\alpha}, V_{\text{inst}}] = [S\beta, V_{\text{inst}}] = 0. \tag{10}$$

Now, the supersymmetry transformation of the instanton field $V(x_0, \rho, \alpha, \bar{\beta})$ is generated by $\exp(-i\varepsilon Q)$ or $\exp(-i\bar{Q}\bar{\varepsilon})$. The action of the right-handed generator is trivial and reduces to a shift of α. Indeed,

$$\begin{aligned}
&\exp(-i\varepsilon Q) \exp(iPx_0) \exp(-i\alpha Q - i\bar{S}\bar{\beta}) \\
&= \exp(iPx_0) \exp(-i(\alpha + \varepsilon) Q - i\bar{S}\bar{\beta}).
\end{aligned} \tag{11}$$

The standard instanton measure is obviously invariant under this shift

$$\mu(x_0, \rho, \alpha, \bar{\beta}) = \mu(x_0, \rho, \alpha + \varepsilon, \bar{\beta}) .$$

The effect of the left-handed generator $\bar{Q}$ is less trivial. To find the corresponding transformation of the collective coordinates we write

$$\exp(-i\bar{Q}\bar{\varepsilon}) \exp(iPx_0) \exp(-i\alpha Q - i\bar{S}\bar{\beta}) \equiv \exp(iPx_0) \exp(-i\alpha Q - i\bar{S}\bar{\beta}) \mathcal{O} \exp(-i\bar{Q}\bar{\varepsilon}) , \tag{12}$$

where $\mathcal{O}$ is an operator, to the first order in $\bar{\varepsilon}$ reducing to

$$\mathcal{O} = \exp(i(\alpha Q + \bar{S}\bar{\beta})) \exp(-i\bar{Q}\bar{\varepsilon}) \exp(-i(\alpha Q + \bar{S}\bar{\beta})) \exp(i\bar{Q}\bar{\varepsilon}) = 1 + [\alpha Q + \bar{S}\bar{\beta}, \bar{Q}\bar{\varepsilon}] + \tfrac{1}{2}i[\alpha Q + \bar{S}\bar{\beta}, [\alpha Q + \bar{S}\bar{\beta}, \bar{Q}\bar{\varepsilon}]] + \cdots . \tag{13}$$

The spinor generators Q and S, together with bosonic generators, form the superconformal algebra which is a generalization of the 15-dimensional conformal algebra of space–time (see Fayet and Ferrara, ref. [5], sect. 2.2). The algebra fixes the set of commutation relations. We reproduce here those (anti) commutators [5] which will be needed below

$$\begin{aligned}
&[Q^{\alpha}\bar{S}^{\dot{\beta}}] = 0 , \\
&[\bar{S}^{\dot{\alpha}}P^{\beta\dot{\gamma}}] = 2i\varepsilon^{\dot{\alpha}\dot{\gamma}}Q^{\beta} , \\
&\{Q^{\alpha}\bar{Q}^{\dot{\beta}}\} = 2P^{\alpha\dot{\beta}} , \\
&[\bar{S}^{\dot{\alpha}}D] = -\tfrac{1}{2}i\bar{S}^{\dot{\alpha}} , \\
&[\bar{S}^{\dot{\alpha}}\Pi] = \tfrac{3}{4}\bar{S}^{\dot{\alpha}} , \\
&\{\bar{S}^{\dot{\alpha}}\bar{Q}^{\dot{\beta}}\} = M^{\dot{\alpha}\dot{\beta}} - 2D\varepsilon^{\dot{\alpha}\dot{\beta}} - 4i\Pi\varepsilon^{\dot{\alpha}\dot{\beta}} , \\
&[\bar{S}^{\dot{\alpha}}M^{\dot{\beta}\dot{\gamma}}] = -2i(\bar{S}^{\dot{\beta}}\varepsilon^{\dot{\gamma}\dot{\alpha}} + \bar{S}^{\dot{\gamma}}\varepsilon^{\dot{\beta}\dot{\alpha}}) .
\end{aligned} \tag{14}$$

Here M, D, Π denote the following generators:

M – Lorentz rotations ,

D – dilatations ,

Π – chiral rotations of spinors .

Moreover, M acting on V_{inst} is equivalent to a color rotation and can be disregarded as far as we average over the instanton color orientation; Π does not act on V_{inst} at all.

Using eq. (14) we readily find commutators figuring in eq. (13). Notice that triple and further commutators (denoted by dots in eq. (13)) vanish identically. The result for $\mathcal{O}$ can be written as follows:

$$\mathcal{O} = 1 + 2P^{\alpha\dot\beta}\alpha_\alpha\bar\varepsilon_{\dot\beta} - 2(\bar\beta\bar\varepsilon)D + 4(\bar S\bar\beta)(\bar\beta\bar\varepsilon) + 4(\alpha Q)(\bar\beta\bar\varepsilon)$$
$$= \exp(2P^{\alpha\dot\beta}\alpha_\alpha\bar\varepsilon_{\dot\beta})\exp(4(\bar S\bar\beta)(\bar\beta\bar\varepsilon) + 4(\alpha Q)(\bar\beta\bar\varepsilon))$$
$$\times\exp(-2(\bar\beta\bar\varepsilon)D)\times(1+\mathcal{O}(\bar\varepsilon^2)).$$

Now we are finally able to answer the question how the collective coordinates are transformed. Recalling that $\exp(-i\bar Q\bar\varepsilon)$ acting on V_{inst} reduces to unity we represent eq. (12) in the following identical form

$$\exp(-i\bar Q\bar\varepsilon)\exp(iPx_0)\exp(-i\alpha Q - i\bar S\bar\beta)$$
$$= \exp(\tfrac{1}{2}iP^{\alpha\dot\beta}(x_0)_{\alpha\dot\beta} + 2P^{\alpha\dot\beta}\alpha_\alpha\bar\varepsilon_{\dot\beta})\exp(-i\alpha Q - i\bar S$$
$$\times(\bar\beta + 4i\bar\beta(\bar\beta\bar\varepsilon))\exp(-2(\bar\beta\bar\varepsilon)D). \tag{16}$$

The last exponential rescales the instanton solution $V_{\text{inst}}(x_0 = 0, \rho)$. The corresponding effect is equivalent to the following transformation of ρ:

$$\rho \to \rho(1 + 2i(\bar\beta\bar\varepsilon)). \tag{17a}$$

Moreover, occurrence of the first exponential means the redefinition of x_0, $x_{0\mu} \to x_{0\mu} - 2i\alpha\sigma_\mu\bar\varepsilon$. The measure (4) is trivially invariant under the shift of x_0. Transformations of α and $\bar\beta$ stemming from eq. (16) are

$$\alpha \to \alpha, \qquad \bar\beta \to \bar\beta(1 + 4i(\bar\beta\bar\varepsilon)). \tag{17b}$$

One can easily check that the instanton measure (4) is invariant under the changes of the collective coordinates (17). Actually, the products $d^4x_0\, d^2\alpha$ and $\rho\, d\rho\, d^2\bar\beta$ are invariant by themselves:

$$d^4x_0\, d^2\alpha \to d^4x_0\, d^2\alpha,$$
$$d\rho^2\, d^2\bar\beta \to d\rho^2(1 + 4i(\bar\beta\bar\varepsilon))\, d^2\bar\beta(1 - 4i(\bar\beta\bar\varepsilon)). \tag{18}$$

We pause here to draw the reader's attention to the fact that there is an absolute parallel between the standard chiral realization of the supergroup in the $(x, \theta, \bar\theta)$ superspace and the transformation law of (x_0, α). Indeed, supertransformations corresponding to parameters ε and $\bar\varepsilon$, respectively, are as follows

(a) $\quad \theta \to \theta + \varepsilon,\ \bar\theta \to \bar\theta, \quad x^{\text{ch}}_{\alpha\dot\beta} \to x^{\text{ch}}_{\alpha\dot\beta}, \quad x^{\text{ach}}_{\alpha\dot\beta} \to x^{\text{ach}}_{\alpha\dot\beta} - 4i\varepsilon_\alpha\bar\theta_{\dot\beta},$

(b) $\quad \theta \to \theta,\ \bar\theta \to \bar\theta + \bar\varepsilon, \quad x^{\text{ch}}_{\alpha\dot\beta} \to x^{\text{ch}}_{\alpha\dot\beta} + 4i\theta_\alpha\bar\varepsilon_{\dot\beta}, \quad x^{\text{ach}}_{\alpha\dot\beta} \to x^{\text{ach}}_{\alpha\dot\beta},$

(ch = chiral, ach = antichiral). Compare these expressions with

(a) $\quad \alpha \to \alpha + \varepsilon, \quad (-x_0) \to (-x_0),$

(b) $\quad \alpha \to \alpha, \quad (-x_0)_{\alpha\dot\beta} \to (-x_0)_{\alpha\dot\beta} + 4i\alpha_\alpha\bar\varepsilon_{\dot\beta}.$

Thus, $(-x_0)$ plays the role of x_{chiral}, while α plays the role of θ.

A similar exercise in commutation relations can be readily made with the superconformal transformations. This allows one to answer the question of how the instanton collective coordinates are transformed under the action of εS and $\bar{S}\bar{\varepsilon}$. The measure (4) is trivially invariant under the $\bar{S}\bar{\varepsilon}$ transformation. Indeed, the only change of the coordinates is a shift of α and $\bar{\beta}$, $\alpha_\alpha \to \alpha_\alpha + (x_0)_{\alpha\dot{\beta}}\bar{\varepsilon}^{\dot{\beta}}$, $\bar{\beta} \to \bar{\beta} + \bar{\varepsilon}$.

On the contrary, an infinitesimal εS transformation results in a more complicated law*:

$$\begin{aligned}
&x_0 \to x_0(1+2i(\alpha\varepsilon)),\\
&\alpha \to \alpha(1-4i(\alpha\varepsilon)),\\
&\rho \to \rho(1+2i(\alpha\varepsilon)+2i(\bar{\beta}\bar{\gamma})),\\
&\bar{\beta} \to \bar{\beta}(1-4i(\alpha\varepsilon)+4i(\bar{\beta}\bar{\gamma})),\\
&\bar{\gamma}_{\dot{\beta}} \equiv \varepsilon^\alpha (x_0)_{\alpha\dot{\beta}}.
\end{aligned} \tag{19}$$

Now the instanton measure is not invariant under these changes:

$$\mu(x_0+\delta x_0, \rho+\delta\rho, \alpha+\delta\alpha, \bar{\beta}+\delta\bar{\beta}) = \mu(x_0, \rho, \alpha, \bar{\beta})(1+24i(\alpha\varepsilon)). \tag{20}$$

This non-invariance is a manifestation of breaking of superconformal symmetry by quantum corrections.

4. Instantons and anomalies

Thus, we have shown that the standard instanton measure respects supersymmetry. The only unusual thing about instantons is that the supersymmetry is realized in a rather peculiar way, as a combination of dilatations and (super) conformal transformations none of which is a true symmetry of the quantum theory because of anomalies. In this section we discuss in more detail the relation between variations of the instanton measure and the quantum anomalies.

As mentioned in sect. 2 the super-Yang–Mills action admits dilatations and superconformal transformations. Quantum corrections do not respect these symmetries, however; taking them into account yields non-vanishing divergences of the corresponding currents. The anomalies are fixed to all orders in the coupling constant by the renormalizability and supersymmetry of the theory:

$$\begin{aligned}
&\partial_\mu(\theta_{\mu\nu}x_\nu) = \theta_{\mu\mu} = \frac{\beta(\alpha_s)}{4\alpha_s} G^a_{\mu\nu}G^a_{\mu\nu},\\
&\partial_\mu(i\hat{x}J_\mu) = i\gamma_\mu J_\mu = i\frac{\beta(\alpha_s)}{2\alpha_s} G^a_{\mu\nu}\sigma_{\mu\nu}\lambda^a,
\end{aligned} \tag{21}$$

* Apart from eq. (14) we use here the following commutation relations [5]:

$$\{S^\alpha \bar{S}^{\dot{\beta}}\} = -2K^{\alpha\dot{\beta}}, \qquad [Q^\gamma K^{\alpha\dot{\beta}}] = -2i\varepsilon^{\gamma\alpha}\bar{S}^{\dot{\beta}},$$

$$[Q^\alpha D] = \tfrac{1}{2}iQ^\alpha, \qquad [Q^\alpha \Pi] = \tfrac{3}{4}Q^\alpha.$$

where $\theta_{\mu\nu}$ is the energy-momentum tensor, J_μ is the conserved vector-spinor current (the corresponding charge was denoted above by Q), $J_\mu = \frac{1}{2}\sigma_{\alpha\beta}G^a_{\alpha\beta}\gamma_\mu\lambda^a$, and $\beta(\alpha_s)$ is the Gell-Mann–Low function,

$$\beta(\alpha_s) = -\frac{b\alpha_s^2}{2\pi} + O(\alpha_s^3)\,, \qquad b = 6 \text{ for } SU(2)_{\text{color}}\,.$$

We use in this section the Majorana notation.

The instanton measure (4) is actually a normalized pre-exponential factor in front of $\exp(-S_{\text{cl}}) = \exp(-2\pi/\alpha_{s0})$. It emerges when one performs the functional integration

$$\mu(x_0, \rho, \alpha, \bar{\beta}) = \int \mathcal{D}\lambda\, \mathcal{D}A_\mu\, \exp\{-S(A^{\text{cl}}_\mu + A_\mu, \lambda^{\text{cl}} + \lambda) + S(A^{\text{cl}}_\mu, \lambda^{\text{cl}})\}\,, \tag{22}$$

where the integration runs over quantum fluctuations around the classical solution. Recall that the solution family (8) introduces both bosonic and fermionic classical fields, see eqs. (5)–(7). Perform now a change of scale $x \to x(1+\xi)$ (and the corresponding rescaling of fields) in the original functional integral (22). Because of the conformal anomaly the action is changed under this transformation as follows

$$\Delta_D S_{\text{Mink}} = -\xi \int \frac{\beta(\alpha_s)}{4\alpha_s} G^a_{\mu\nu} G^a_{\mu\nu}\, d^4x\,. \tag{23}$$

In order to find the variation of the integral (22) we proceed to euclidean space and substitute, to leading order in α_s, the operator $G_{\mu\nu}$ by its classical part $(G_{\mu\nu})_{\text{cl}}$. Taking into account that

$$\int d^4x\, \tfrac{1}{4}(G^a_{\mu\nu} G^a_{\mu\nu})_{\text{inst}} = \frac{2\pi}{\alpha_s}\,, \tag{24}$$

we get that the right-hand side of eq. (22) is multiplied by $1+6\xi$. The instanton measure (4) must acquire this factor under the scale transformation of the collective coordinates. This, of course, happens:

$$\begin{aligned} &d^4x_0\, d\rho^2 \to (1+6\xi)\, d^4x_0\, d\rho^2\,, \\ &d^2\alpha\, d^2\bar{\beta} \to d^2\alpha\, d^2\bar{\beta}\,. \end{aligned} \tag{25}$$

Moreover, one could fix the power of ρ in the instanton measure μ in this way, without performing explicit calculation of the instanton determinant.

Let us discuss now superconformal transformation with parameter ε. Due to the superconformal anomaly (21) the action is also changed:

$$\Delta_S S_{\text{Mink}} = \int d^4x\, \bar{\lambda}^a \sigma_{\mu\nu} \varepsilon G^a_{\mu\nu} \frac{\beta(\alpha_s)}{2\alpha_s}\,. \tag{26}$$

Proceeding to euclidean space and substituting eqs. (5), (6) instead of the operators

$G^a_{\mu\nu}$, λ^a we get the variation of the integral (22) in the form

$$\exp\left(\left(\int d^4x \frac{b\alpha_s}{2\pi}(G^a_{\mu\nu}G^a_{\mu\nu})_{\rm inst}\bar{\alpha}\tfrac{1}{2}(1+\gamma_5)\varepsilon\right)\right.$$

$$=1+4b\bar{\alpha}\tfrac{1}{2}(1+\gamma_5)\varepsilon=1+24i(\alpha\varepsilon)_{\rm chiral}\,. \tag{27}$$

It is satisfying to notice that the variation (20) of the instanton measure just reproduces this anomaly.

5. Ward identities

So far we have discussed the instanton measure using only group-theoretical arguments. It is reasonable to ask whether the standard instanton calculus reproduces the results obtained. In other words, one can check the Ward identities.

The Ward identities relevant to instantons have been discussed previously [8], and an apparent violation of the Ward identities has been found. No satisfactory explanation to this observation has been given, however, and we feel it worthwhile to consider the question anew. Our conclusion is that the violation of the Ward identities disappears after integration over ρ. This is just another formulation of the result obtained in sect. 3.

First let us review briefly some points of ref. [8]. We start with the following matrix element

$$M_\mu=\langle 0|T\{J_\mu(x), j^a_\nu v^a_\nu(x_\eta)\prod_{k=2,3,4}\bar{j}^b_{(\lambda)}(x_k)\lambda^b(x_k)\}|0'\rangle\,, \tag{28}$$

and calculate its divergence

$$\frac{\partial}{\partial x_\mu}M_\mu=\langle 0|T\{\partial_\mu J_\mu(x), j^a_\nu v^a_\nu(x_1)\prod_{k=2,3,4}\bar{j}^b_{(\lambda)}\lambda^b(x_k)\}|0'\rangle$$

$$+\delta(t-t_1)\langle 0|T\{[J_0(x)v^a_\nu(x_1)]j^a_\nu\prod_{k=2,3,4}\bar{j}^b_{(\lambda)}\lambda^b(x_k)\}|0'\rangle$$

$$+\text{other commutator terms}\,. \tag{29}$$

Here $|0\rangle$ and $|0'\rangle$ are "vacuum" states with a unit difference of topological charge, J_μ is the supercurrent introduced in eq. (21), and j^a_ν, $j^b_{(\lambda)}$ are external bosonic and fermionic sources, respectively. Four fermionic sources, three $j_{(\lambda)}$ and J_μ, are needed because of the instanton zero modes. For the same reason matrix elements of the commutator terms omitted in eq. (29) actually vanish.

Naively $\partial_\mu J_\mu=0$, and substituting this equality in eq. (29) we get the desired Ward identity. Actually it should be corrected for explicit supersymmetry breaking brought in by the procedure of gauge fixing and (quite unexpectedly) for one more effect to be discussed below. Eq. (29) takes an especially simple form if we concentrate on the Ward identities for the supercharge and, to this end, integrate over

d^4x. Barring the possibility of spontaneous symmetry breaking the result of the integration should vanish (we have checked that the surface term does vanish, so that there is no spontaneous breaking [8]). Thus, one expects that the contact term in the right-hand side of eq. (29) also vanishes. This is indeed the case,

$$d^2\alpha\, d^2\bar{\beta}\, d\rho^2\, d^4x_0 \Big\langle \prod_{k=1,\dots,4} \lambda(x_k) \Big\rangle_{\text{inst}} = 0\,.$$

One can check the fact by using simple dimensional arguments and some additional results which will be published in a separate paper [9]. The latter will also give details of the corresponding scenario.

It is instructive to confront the above derivation with an explicit evaluation of M_μ in a given instanton field. Common sense says that the introduction of such a field does not affect the Ward identity since the boson field does not vary under the supersymmetry transformation if there is no external fermion field, as we assume now.

Moreover, M_μ is readily calculable (fig. 1):

$$\begin{aligned} M_\mu \sim\; & \bar{\lambda}^a_{ss1}(x-x_0)\gamma_\mu \sigma_{\alpha\beta}\mathscr{D}_\alpha D^{ab}_{\beta\gamma}(x, x_1) j^b_\gamma(x_1) \\ & \times \bar{j}_{(\lambda)}(x_2)\lambda_{ss2}(x_2)\bar{j}_\lambda(x_3)\lambda_{sc1}(x_3)\bar{j}_{(\lambda)}(x_4)\lambda_{sc2}(x_4) \\ & + \text{permutations of fermion zero modes } (\lambda_{ss1,2}, \lambda_{sc1,2})\,. \end{aligned} \tag{30}$$

Here $\lambda_{ss1,2}$, $\lambda_{sc1,2}$ denote four independent solutions of the Dirac equation in the instanton field, $D^{ab}_{\beta\gamma}$ is the gluon Green function built from non-zero gluon modes.

Differentiating (30) with respect to x_μ yields

$$\begin{aligned} \partial M_\mu/\partial x_\mu = \{ & \lambda^a_{ss1}(x-x_0)\delta^4(x-x_1)\gamma_\mu j^a_\mu(x_1) \\ & - \sum_s \lambda^a_{ss1}(x-x_0)\gamma_\mu \Phi^{a(s)}_\mu(x-x_0)\Phi^{b(s)}_\nu(x_1-x_0) j^b_\nu(x_1)\} \\ & \times \prod_{k=2,3,4} \bar{j}_{(\lambda)}\lambda(x_k-x_0)\,. \end{aligned} \tag{31}$$

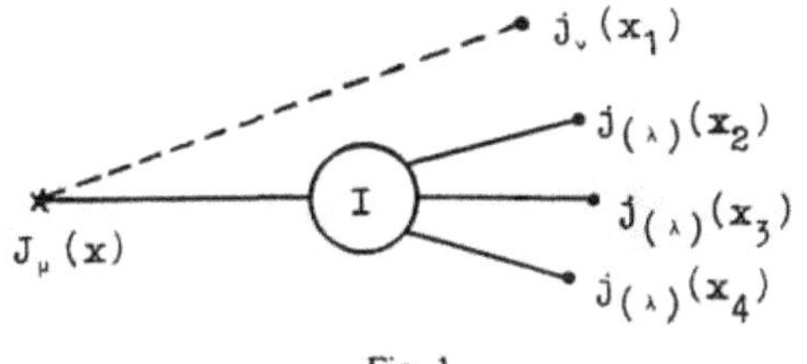

Fig. 1

where we have used the equations

$$\hat{\mathscr{D}}\lambda = 0\,, \tag{32}$$

$$-[\mathscr{D}^2 g_{\mu\lambda}\delta^{ab} + 2g\varepsilon^{abc}G^c_{\mu\lambda}]\mathscr{D}^{bd}_{\lambda\nu}(x, y)$$

$$= \delta_{\mu\nu}\delta^{ad}\delta(x-y) - \sum_s \Phi^{a(s)}_\mu(x)\Phi^{d(s)}_\nu(y)\,,$$

and the bosonic zero modes are denoted by $\Phi^{a(s)}_\mu$. The $\delta^4(x-y)$ term in eq. (32) just reproduces the contact term in the general relation (29). The sum over the bosonic zero modes produces an extra piece, however, which corresponds to nothing else but non-conservation of the supercurrent, $\partial_\mu J_\mu \neq 0$, as was first noticed in ref. [8].

To investigate further the effect of the new term let us integrate $\partial_\mu M_\mu$ over d^4x. As seen from eq. (31) the result depends on the overlap of the bosonic and fermionic zero modes. The latter look very similar, and the overlap integral turns out to be either zero or unity (in a proper normalization) depending on the modes considered.

For example, the translational modes are proportional to the gluon field strength tensor:

$$\Phi^{a(\beta)}_\mu(x-x_0) \sim G^a_{\mu\beta}(x-x_0)\,, \qquad \beta = 1, 2, 3, 4\,.$$

Since the supersymmetric zero modes are also proportional to $G^b_{\alpha\rho}$ the integration over d^4x is readily performed with the help of the following relation

$$\int G^a_{\alpha\beta}(x)G^a_{\mu\nu}(x)\,d^4x = \tfrac{1}{12}(\delta_{\alpha\mu}\delta_{\beta\nu} - \delta_{\alpha\nu}\delta_{\beta\mu}$$

$$+ \varepsilon_{\alpha\beta\mu\nu})\int G^a_{\rho\sigma}G^a_{\rho\sigma}\,d^4x\,. \tag{33}$$

On the other hand, the overlap of $\Phi^{a(\beta)}_\mu$ with the superconformal zero modes, λ_{sc}, vanishes because the product is odd in $(x-x_0)$.

In this way one can convince oneself that the extra term in the Ward identity corresponds to the substitution of the variation δv^a_μ of the classical vector field in the presence of the classical fermion field λ_{cl} in the interaction term $j^a_\mu v^a_\mu$:

$$\int d^2\alpha\, d^2\bar{\beta}\, dx\langle 0|T\Big\{\bar{\varepsilon}\partial_\mu J_\mu(x) j^a_\nu v^a_\nu(x_1) \prod_{k=2,3,4} \bar{j}^b_{(\lambda)}\lambda^b(x_k)\Big\}|0'\rangle$$

$$= \delta v^a_\mu j^a_\mu(x_1) \prod_{k=2,3,4} \bar{j}^b_{(\lambda)}\lambda^b_{cl}(x_k - x_0)\,, \qquad \delta v_\mu = -i\bar{\varepsilon}\gamma_\mu\lambda_{cl}\,. \tag{34}$$

It seems amusing that in this rather complicated way we have come back to consideration of the classical fields. Comparing the situation with the results of sect. 3 we find that we have reproduced the variations of x_0 and ρ but not of $\bar{\beta}$. However, the latter is readily recovered as well. The point is that we have arrived at the shifted bosonic field and the unshifted fermionic field

$$v_\mu(x_0+\delta x_0, \rho+\delta\rho)\,, \qquad \lambda_{ss}(x_0, \rho, \alpha)\,, \qquad \lambda_{sc}(x_0, \rho, \bar\beta)\,.$$

Redefinition of $\bar\beta$ is needed if we wish to preserve the parametrization of the instanton solution, see eq. (8).

On the other hand, we have already learned that the variation of the classical field corresponding to δx_0, $\delta\rho$, $\delta\bar\beta$, eqs. (17), is completely consistent with supersymmetry. Of course, to recover the supersymmetry it is absolutely necessary to integrate over all collective coordinates.

To summarize, the consideration of the Ward identities demonstrates once more that fixing ρ violates supersymmetry and that integrating over ρ restores it completely. Still, there is one puzzle left. How can one understand $\partial_\mu J_\mu \neq 0$ in the presence of the instanton field if $\partial_\mu J_\mu = 0$ is an operator equality? The explanation seems to be as follows.

Introduce a small mass term to the quantum vector field. This is useful since the gluon Green function in the instanton field is not defined otherwise due to zero modes. Then the divergence of the supercurrent is proportional to this mass, $\partial_\mu J_\mu \sim m^2 \bar\lambda^a \gamma_\mu v^a_\mu$. If we consider now the gluon propagation and tend $m \to 0$ the corresponding effects will be negligible everywhere except for the zero modes. In the latter case the m^2 factor just cancels out and we are left with some $\partial_\mu J_\mu \neq 0$.

Therefore, $\partial_\mu J_\mu \neq 0$ in the presence of any external field which possesses zero modes. The phenomenon is due to infrared regularization. Moreover, this supercurrent non-conservation is not manifested in the standard Ward identities. However, it is harmless in the sense that it vanishes after integration over the collective coordinates.

6. Conclusions

We have shown that the standard instanton calculus is perfectly consistent with supersymmetry provided that simultaneous integration over all collective coordinates is performed. Although we have discussed only the simplest example with the SU(2) gauge group, the assertion is of more general nature and is valid for any supersymmetric Yang–Mills theory. The notion of an instanton-induced effective lagrangian with fixed ρ is contradictory: it results in a conflict with supersymmetry*. We would like to emphasize the most general aspect: in any supersymmetric theory with conformal invariance at the classical level fixing ρ would violate supersymmetry**. There is no analogue of this phenomenon in ordinary QCD, and this might indicate that dynamics inherent to supersymmetric theories is quite specific.

* The necessity of accounting for the ρ variations was recognized first by four of us previously [8]. However, at that time the idea has not been pursued up to the logical end, and the conclusion of ref. [8] differs from what we find now.

** The opposite is also true. If the classical conformal invariance is absent, one finds complete agreement with supersymmetry for any given ρ, see ref. [8], sect. 6.

Certainly, (super)conformal invariance is not realized at the quantum level due to anomalies in $\theta_{\mu\mu}$ and $\gamma_\mu J_\mu$. Therefore, one might think that the two fermion zero modes (λ_{sc}) associated with the superconformal transformations disappear when higher-order quantum corrections are taken into account, and that the $\bar{\beta}$ integration can be peformed explicitly. This hypothesis turns out to be absolutely wrong, however. The number of fermion zero modes is fixed by topological arguments. Moreover, we have shown that the existence of the superconformal modes is necessary in order to make the instanton measure compatible with the standard anomaly relations for $\theta_{\mu\mu}$ and $\gamma_\mu J_\mu$.

The authors are grateful to M. Vysotsky and E. Witten for useful discussions.

References

[1] E. Witten, Nucl. Phys. B185 (1981) 513
[2] G. 't Hooft, Phys. Rev. D14 (1976) 3432
[3] L.F. Abbot, M.T. Grisaru and H.J. Schnitzer, Phys. Rev. D16 (1977) 2995, 3202
[4] A.I. Vainshtein and V.I. Zakharov, ZhETF Pisma 35 (1982) 258
[5] V.I. Ogievetsky and L. Mesinčesku, Usp. Fiz. Nauk 117 (1975) 637; Sov. Physics (Uspekhi) 18 (1975) 960;
P. Fayet and S. Ferrara, Phys. Reports 32 (1977) 249
[6] A. Belavin, A. Polyakov, A. Schwarz and Yu. Tyupkin, Phys. Lett. 59B (1975) 85
[7] B. Zumino, Phys. Lett. 69B (1977) 369
[8] V. Novikov, M. Shifman, A. Vainshtein and V. Zakharov, Nucl. Phys. B223 (1983) 445
[9] V. Novikov, M. Shifman, A. Vainshtein and V. Zakharov, Nucl. Phys. B229 (1983) 407

Nuclear Physics B260 (1985) 157-181

SUPERSYMMETRIC INSTANTON CALCULUS
Gauge theories with matter

V.A. NOVIKOV, M.A. SHIFMAN, A.I. VAINSHTEIN and V.I. ZAKHAROV
Institute of Theoretical and Experimental Physics, 117259 Moscow, USSR

Received 25 March 1985

Within the framework of gauge SUSY theories we discuss correlation functions of the type $\langle W^2(x), S^2(0)\rangle$ where S is the chiral matter superfield (in the one-flavor model). SUSY implies that these correlation functions do not depend on coordinates and vanish identically in perturbation theory. We develop a technique for the systematic calculation of instanton effects. It is shown that even in the limit $x \to 0$ the correlation functions at hand are not saturated by small-size instantons with radius $\rho \sim x$; a contribution of the same order of magnitude comes from the instantons of characteristic size $\rho \sim 1/v$ (v is the vacuum expectation value of the scalar field, and we concentrate on the models with $v \gg \Lambda$ where Λ is the scale parameter fixing the running gauge coupling constant). If $v \gg \Lambda$ both types of instantons can be consistently taken into account. The computational formalism proposed is explicitly supersymmetric and uses the language of instanton-associated superfields. We demonstrate, in particular, that one can proceed to a new variable, $\rho_{\rm inv}$, which can be naturally considered as a supersymmetric generalization of the instanton radius. Unlike the ordinary radius ρ, this variable is invariant under the SUSY transformations. If one uses $\rho_{\rm inv}$ instead of ρ the expressions for the instanton contribution can be rewritten in the form saturated by the domain $\rho^2_{\rm inv} = 0$. The cluster decomposition as well as x-independence of the correlation functions considered turn out to be obvious in this formalism.

1. Introduction

In this paper we discuss instantons in supersymmetric theories. We address ourselves both to some physical effects and to developing a suitable formalism. In the case of pure (SUSY) gluodynamics there exists a superfield formalism adapted for the instanton user [1]. Here we consider theories with matter fields assuming that the scalar fields have a nonvanishing vacuum expectation value. Such theories are of special interest, in our mind, since just here there was discovered a beautiful dynamic effect of SUSY and/or colour spontaneous breaking in the weak coupling regime [2-9]. This spontaneous symmetry breaking is generated by instanton effects.

One of the applications we concentrate on is the determination of various condensates in the physical vacuum. The line of reasoning is as follows. Consider for example SUSY gluodynamics with SU(2) as a colour group. Then the correlation function

$$\langle 0|T\{g^2\lambda^\alpha\lambda_\alpha(x), g^2\lambda^\beta\lambda_\beta(0)\}|0\rangle \tag{1}$$

where λ^α is the gluino field (the colour index is omitted) possesses some remarkable properties [1]. First, it vanishes in perturbation theory but receives a nonvanishing contribution in the one-instanton approximation. Second, by virtue of supersymmetry it does not depend actually on x*. In particular, if $x \to \infty$ it factorizes into the product of the bilinear condensates $\langle g^2\lambda\lambda\rangle$ because of the cluster decomposition. As for $x \to 0$, one might hope that in this limit the correlator (1) is calculable. The point is that in this limit it is contributed by the instantons of the size $\rho \sim x \to 0$ which are under theoretical control. Moreover, this contribution is perfectly supersymmetric by itself and respects all general requirements such as renormalization-group invariance.

Combining the two cases mentioned above, namely $x \to \infty$ and $x \to 0$, one might hope to determine the condensate $\langle g^2\lambda\lambda\rangle$ [3]. Such a technique was applied to a wide range of SUSY theories including models with strong coupling [3, 5, 7] - there is no apparent appeal to weak coupling in the argument above. Unfortunately, there is a loophole in the argument itself which could jeopardize the use of instantons [1]. Namely, there is no general proof that even at $x \to 0$ other fluctuations of some characteristic scale $\rho \sim \Lambda^{-1}$ do not contribute to (1).

There have been two attempts to verify that the small-size instantons do saturate (1) at $x \to 0$ within exactly solvable models. More specifically, CP($N+1$) models and gauge theories with one flavour and colour breaking were considered from this point of view in refs. [10b] and [6], respectively. The conclusion of these papers is that correlators like (1) do not receive any additional contribution from large-scale fluctuations.

On the other hand, the authors of ref. [7] have found a small-size instanton contribution which does not factorize. To find the way out the authors of ref. [7] have assumed that the instanton calculus in this particular case is unstable against introduction of a small mass term, $m \to 0$, even in the limit of the instanton size $\rho \to 0$.

We consider here theories whose running coupling constant is small at all distances. It is quite natural that in this case one can exhaustively answer what fluctuations contribute and dominate in any correlation function of interest. To this end we will consider in detail the two-point function

$$\langle 0|T\{g^2\lambda^\alpha\lambda_\alpha(x), \varphi^2(0)\}|0\rangle, \tag{2}$$

where φ^2 is $\varphi^{\alpha f}\varphi_{\alpha f}$ and $\alpha = 1, 2$ is the color index while $f = 1, 2$, is associated with SU(2) global flavour symmetry (two chiral superfields are needed for description of a single flavour). This correlator is of the same type as (1) and does not depend on x either. Since the scalar field develops a non-vanishing vacuum expectation value v there exist no instantons of the size much larger than $1/v$. Thus, if we consider the limit $x \to 0$ the two types of fluctuations which can, at least in principle, contribute to (2) are the instantons of the size $\rho \sim x$ and of the size $\rho \sim 1/v$. Both are under theoretical control. Our conclusion is as follows. The two-point function (2) is *not* saturated by the instantons with $\rho \sim x$ even for $x \to 0$. These instantons

* This is true even if the supersymmetry is realized non-linearly (spontaneous SUSY breaking), and there is a goldstino [10a].

contribute just one half of the total answer. The other half comes from the instantons of the characteristic scale $\rho \sim 1/v$. We have overlooked this factor $\frac{1}{2}$ in our earlier paper [6] where instantons of small size, $\rho \sim x$, were analysed. It is straightforward to verify then that such a situation is quite general. In particular, neglecting the contribution of the instantons of the characteristic scale results in inconsistencies in the theory with two flavors mentioned above [7].

It is amusing that correlators considered can be represented in the general case as

$$\int_0^\infty \mathrm{d}\rho(\mathrm{d}F/\mathrm{d}\rho)$$

and therefore reduce, at least formally, to the contribution of instantons of the vanishing radius, $\rho = 0$, (at $\rho = \infty$ all relevant functions fall off fast enough). The derivation relies only on supersymmetry and does not depend, say, on the instanton action. Proceeding in this way one readily realizes why the correlators of the type (1) do not depend on x. Indeed, they depend only on the ratio ρ/x which vanishes for $\rho = 0$ and any x.

The outline of the paper is as follows. In sect. 2 we introduce all necessary notations and routine instanton collective coordinates. We reiterate here also the superfield description of the instantons in pure SUSY gluodynamics. Sects. 3 and 4 are the central ones for the present paper. In sect. 3 we generalize the superfield formalism to include the theories with non-vanishing vacuum expectation values of the scalar fields. In sect. 4 we apply this technique to calculation of the two-point function (2). Sect. 5 treats the model with two flavours, while sect. 6 presents a general discussion of the hierarchy of condensates. In particular here we argue that fluctuations of the characteristic size $\rho \sim 1/\Lambda$ are important in the strong coupling regime. (The rest of the paper refers to the weak coupling regime). In appendix A the procedure of continuation to euclidean space is described. In appendix B we discuss in details the problem of mass insertions.

2. Instantons: the case $v = 0$

Let us consider SQCD with a single flavour. The lagrangian reads as

$$\mathscr{L} = \frac{1}{2g^2} \mathrm{Tr} \int \mathrm{d}^2\theta\, W^2 + \frac{1}{4} \int \mathrm{d}^2\theta\, \mathrm{d}^2\bar{\theta}\, \bar{S}_f\, \mathrm{e}^V S^f + \left(\frac{1}{4} m \int \mathrm{d}^2\theta\, S^{\alpha f} S_{\alpha f} + \text{h.c.} \right), \qquad (3)$$

where V is the general superfield containing the vector gauge field, W_α is the chiral superfield containing the gluino field λ_α and the gluon field strength tensor $G_{\alpha\beta}$:

$$\begin{aligned} W_\alpha(x_\mathrm{L}, \theta) &\equiv \tfrac{1}{8}\bar{D}^2 (\mathrm{e}^{-V} D_\alpha \mathrm{e}^V) \\ &= [i\lambda_\alpha(x_\mathrm{L}) - \theta_\alpha D(x_\mathrm{L}) - i\theta^\beta G_{\alpha\beta}(x_\mathrm{L}) + \theta^2 D_{\alpha\dot{\alpha}} \bar{\lambda}^{\dot{\alpha}}(x_\mathrm{L})]\,, \\ (x_\mathrm{L})_{\alpha\dot{\alpha}} &= x_{\alpha\dot{\alpha}} - 2i\theta_\alpha \bar{\theta}_{\dot{\alpha}}\,. \end{aligned} \qquad (4)$$

Moreover, V and W_α are matrices in the colour space, $V \equiv V^a T^a$, $W_\alpha = W_\alpha^a T^a$ and T^a are the generators of the SU(2), $T^a = \frac{1}{2}\tau^a$. The coupling constant g is included into the definition of the field V; D is an auxiliary field while $D_{\alpha\dot\alpha}$ is the covariant derivative, D_α is the spinor derivative. The chiral superfields $S^f(x_L, \theta)$, $f = 1, 2$ describe the physical particles of a single flavour. Such a theory is SU(2) globally invariant because representations of the colour SU(2) are pseudoreal. (It might be worth noting that in the literature somewhat different notations are commonly used so that $S^1 \equiv S$, $S^2 \equiv T$). In the component form,

$$S^f(x_L, \theta) = \varphi^f(x_L) + \sqrt{2}(\psi^f(x_L)\theta) + \theta^2 F^f(x_L)$$

and the colour indices $\alpha = 1, 2$ are omitted.

Let us describe now instanton collective coordinates and start with the case of the SUSY gluodynamics. The collective coordinates have a geometrical meaning and are related to the symmetries of the classical action. The full group of invariance of the classical action is the superconformal group plus global rotations in the colour space (we confine ourselves to SU(2) as the colour group).

A concise review of the superconformal group can be found in ref. [11]. Its generators are

$P_{\alpha\dot\alpha}$, translations in ordinary space,
$K_{\alpha\dot\alpha}$, special conformal transformations,
$M_{\alpha\beta}$ and $M_{\dot\alpha\dot\beta}$, Lorentz rotations,
D, dilatation,
Π, chiral rotation,
Q_α, $\bar Q_{\dot\alpha}$, supersymmetry transformations, and
S_α, $\bar S_{\dot\alpha}$, superconformal transformations.

The corresponding commutation relations are listed, for example, in refs. [11, 1]. Thus, there exist sixteen bosonic transformations (plus three colour rotations) and eight fermionic transformations.

However, the number of the collective coordinates is not necessarily this large, since there can exist a stationary group, i.e. a subset of transformations which do not affect some particular classical solution. For standard BPST instanton [12] the stationary group includes the following generators: the chiral rotations Π, the chiral supersymmetry transformations $\bar Q_{\dot\alpha}$, the chiral superconformal transformations S_α and the Lorentz rotations $M_{\dot\alpha\dot\beta}$. Moreover, the algebra of the stationary group incorporates certain combination of the Lorentz rotations $M_{\alpha\beta}$ and colour rotations $T_{\alpha\beta}$, namely,

$$M_{\alpha\beta} + T_{\alpha\beta},$$

where $T_{\alpha\beta} = T^a(\sigma^a)_\alpha^\gamma \varepsilon_{\beta\gamma}$. Finally, it includes $P_{\alpha\dot\beta} + 2\rho^2 K_{\alpha\dot\beta}$.

Thus, the introduction of collective coordinates and determination of their transformation laws reduce to a well-known problem of the action of the group in the coset space.

For simplicity below we consider only superfields which are scalars both in the Lorentz and colour spaces. Then the generators $M_{\alpha\beta}$ and $T_{\alpha\beta}$ can be disregarded though they do transform non-trivially the field of a given instanton. In other words, we do not include in the analysis explicitly collective coordinates corresponding to the global rotations in the colour space. The other collective coordinates are introduced by means of the following operator

$$\mathcal{V}(x_0, \rho, \theta_0, \bar{\beta}) = e^{iPx_0}\, e^{-i\theta_0 Q}\, e^{-i\bar{S}\bar{\beta}}\, e^{iD\ln\rho} .$$

To find out the transformation law of the collective coordinates under supersymmetry one multiplies the operator $\mathcal{V}$ by $\exp(-i\bar{Q}\bar{\varepsilon})$ on the left and then reduces it to the original form with some parameters redefined. Namely,

$$e^{-i\bar{Q}\bar{\varepsilon}}\mathcal{V}(x_0, \rho, \theta_0, \bar{\beta}) = \mathcal{V}(x_0', \rho', \theta_0', \bar{\beta}')F ,$$

with F belonging to the stationary group.

In this way there is no difficulty to establish the transformation laws of the collective coordinates [1]

$$\delta(x_0)_{\alpha\dot{\alpha}} = -4i(\theta_0)_\alpha \bar{\varepsilon}_{\dot{\alpha}}, \qquad \delta\rho^2 = -4i(\bar{\varepsilon}\bar{\beta})\rho^2 ,$$
$$\delta(\theta_0)_\alpha = \varepsilon_\alpha , \qquad \delta\bar{\beta}_{\dot{\alpha}} = -4i\bar{\beta}_{\dot{\alpha}}(\bar{\varepsilon}\bar{\beta}) , \tag{6}$$

where ε and $\bar{\varepsilon}$ are the parameters of the supersymmetry transformations. For completeness, we write down the transformation law for the ordinary coordinates as well:

$$\delta(x_L)_{\alpha\dot{\alpha}} = -4i\theta_\alpha \bar{\varepsilon}_{\dot{\alpha}} , \qquad \delta\theta_\alpha = \varepsilon_\alpha . \tag{7}$$

The supersymmetry requires that under simultaneous transformations of x_L, θ and of the collective coordinates the instanton superfield is unchanged. Starting from this requirement one arrives, for example, at the following expression for W^2 associated with the instanton:

$$(W^2)_{\rm inst} = {\rm const}\, \frac{\tilde{\theta}^2 \rho^4}{[(x_L - x_0)^2 + \rho^2]^4} , \tag{8}$$

where

$$\tilde{\theta}_\alpha = (\theta - \theta_0)_\alpha + (x_L - x_0)_{\alpha\dot{\alpha}}\bar{\beta}^{\dot{\alpha}} . \tag{9}$$

Note that the terms proportional to θ_0 and $\bar{\beta}$ in the expansion of eq. (8) correspond to the gluino zero modes*.

Once expressions for the superfield are found it is straightforward to evaluate various correlators. Upon doing so one needs also the instanton measure which in

* The gluino zero modes (as well as the matter fermion zero modes, see below) are left-handed. Traditionally, the name instantons is applied to the bosonic solutions whose fermionic modes are right-handed. Thus, here we actually discuss the anti-instanton solution, the anti-instanton contribution to correlation functions and so on. Hereafter we shall not mention this explicitly.

the case considered (pure SUSY gluodynamics) is proportional to

$$d\theta(\rho, x_0, \theta_0, \bar{\beta}) = C\Lambda^6\, d^4x_0\, d\rho^2\, d^2\theta_0\, d^2\bar{\beta} \tag{10}$$

(we have omitted here integration over the global rotations in the colour space).

3. Instantons in the models with non-vanishing vacuum expectation values of scalar fields

Let us consider now in more detail theories with matter. The simplest lagrangian is given by eq. (3). In the massless limit there exist so called valleys (flat directions) representing an infinite degeneracy of classical vacua of the theory. Indeed, the vacuum state is determined by the conditions

$$D_\mu \varphi^f = 0\,, \qquad G^a_{\mu\nu} = 0\,, \tag{11a}$$

$$D^a \equiv (\varphi^f)^+ T^a \varphi^f = 0\,, \qquad F^f = 0 \tag{11b}$$

which ensure vanishing of the vacuum energy. The solution to these equations has the form

$$\varphi^\alpha_{\dot{f}} = U^\alpha_{\dot{f}}(x)v\,,$$

$$A_\mu = iU\partial_\mu U^+\,, \tag{12}$$

where v is a complex number and $U^\alpha_{\dot{f}}$ is an arbitrary unitary unimodular matrix (generally speaking, x-dependent). In the topologically trivial case one can always impose the constraint

$$U(x) = 1\,, \qquad \varphi^\alpha_{\dot{f}}(x) = \delta^\alpha_{\dot{f}} v\,. \tag{13}$$

Such a choice corresponds to fixing the gauge in a certain way (namely, the unitary gauge). If $v \neq 0$ all fields become massive except for a single (Higgs) chiral superfield. Note that the theory can be quantized at any v. This vacuum degeneracy persists to any order in perturbation theory and is lifted only by instanton effects [13, 14, 4].

An example of a topologically non-trivial matrix U which cannot be reduced to the unit matrix is

$$U^\alpha_{\dot{f}} = x^\alpha_{\dot{f}} \frac{1}{\sqrt{x^2}}\,.$$

(We hope that the reader does not get embarassed by the fact the index f of the global symmetry has become a dotted one. Moreover, we would not dwell here on passing to euclidean space from the Minkowski one. The issue is discussed in appendix A.)

The instanton "solution" which realizes tunneling between "vacua" with different topologies is constructed in the following way (we use quotation marks on the word "solution" since instantons actually are not solutions to the classical equations of motion any longer in the presence of $v \neq 0$, see below).

The vector field is of the standard form. Written in somewhat unusual notation, it looks as

$$A^{\gamma\delta}_{\alpha\dot{\alpha}} = -i(\delta^\gamma_\alpha x^\delta_{\dot{\alpha}} + \delta^\delta_\alpha x^\gamma_{\dot{\alpha}}) \frac{1}{x^2+\rho^2}\,, \tag{14a}$$

where

$$A_{\alpha\dot{\alpha}}^{\gamma\delta} = (\sigma_\mu)_{\alpha\dot{\alpha}} A_\mu^a (T^a)_\rho^\gamma \varepsilon^{\delta\rho},$$

$$\sigma_\mu = (1, \boldsymbol{\sigma}), \qquad T^a = \tfrac{1}{2}\tau^a.$$

The scalar field satisfies the equation

$$D^2\varphi = 0$$

in the external instanton field and is given by

$$\varphi_f^\alpha = (\bar{\varphi})_f^\alpha = x_f^\alpha \frac{v}{\sqrt{x^2+\rho^2}}. \tag{14b}$$

The fields (14a) and (14b) are not exact solutions of the classical equations of motion, since there arises some source term in the equations for A_μ in the presence of $\varphi \neq 0$ and this term is actually omitted. At large distances this source term is equivalent to a mass term for the vector field, and taking it into account would result in an exponential fall-off at large distances. Note also that the field φ must also satisfy an equation with a source term proportional this time to $\psi T^a \lambda^a$ which becomes operational, however, only in the presence of zero modes of the gluino and of the matter fermion fields. The effect of this term is not omitted, however. Namely, upon introducing the fermionic collective coordinates, see below, we will come to a solution for the field φ which accounts for this source term.

The appearance of the field φ_f^α (see eq. (14b)) calls for introduction of extra collective coordinates since it is not invariant under the action of the operator $\exp(-i\bar{\varepsilon}\bar{Q})$. Indeed, the change in the field which is generated through transformation of the chiral coordinate x_L (see eq. (7)) cannot be absorbed into a redefinition of the collective coordinates introduced so far. Thus, we need to introduce a new coordinate, $\bar{\theta}_0$, which is absent in the case of pure gluodynamics.

First of all, one must convince oneself that after taking away the generators $\bar{Q}_{\dot{\alpha}}$ from the algebra of the stationary sub-group it still remains a closed algebra. Inspection of the commutation relations (see, e.g., refs. [11, 1]) shows that this is indeed the case.

As the next step, we introduce the new set of collective coordinates, x_0, ρ, θ_0, $\bar{\theta}_0$, $\bar{\beta}$. As explained in sect. 2, this is achieved through the use of the construct*

$$\exp(iPx_0)\exp(-i\theta_0 Q)\exp(-i\bar{S}\bar{\beta})\exp(-i\bar{\theta}_0\bar{Q})\exp(iD\ln\rho). \tag{15}$$

* Unlike the $v=0$ case, an explicit check demonstrates that the transformation which we would like to ascribe to the stationary group of the generalized instanton solution *in fact affects* the scalar fields φ and $\bar{\varphi}$. Thus, the reader could question the validity of the procedure described. Still, we will stick to our definition of the stationary group. The point is that for non-zero v the perturbative vacuum (i.e. the vacuum with no instanton field) is not invariant under the same transformations belonging to the "stationary group" as the fields φ, $\bar{\varphi}$ (14b). In principle, one could introduce some extra collective coordinates both for the instanton and in the no-instanton vacuum. Since the instanton amplitude is always normalized to perturbation theory, the effect of the extra coordinates would actually cancel out. An analogous case is described in detail in the two-dimensional σ model and we refer the readers to the papers [15] for further details.

Thus we proceed in the same vein as above (see the discussion preceding eq. (6)) and come to

$$\delta\bar{\theta}_0 = \bar{\varepsilon} - 4i\bar{\beta}(\bar{\theta}_0\bar{\varepsilon})\,. \tag{16}$$

Since superfields associated with the instantons are invariant under simultaneous transformations of both the collective coordinates and of (x_L, θ) it is convenient to construct first invariants of these transformations (generalizations of the combination $(x - x_0)$ pertinent to the case of space translations). As the first step in this direction, we introduce combinations

$$\begin{aligned}\tilde{\theta}_\alpha &= (\theta - \theta_0)_\alpha + (x_L - x_0)_{\alpha\dot{\alpha}}\bar{\beta}^{\dot{\alpha}}\,,\\ \tilde{x}_{\alpha\dot{\alpha}} &= (x_L - x_0)_{\alpha\dot{\alpha}} + 4i\tilde{\theta}_\alpha(\bar{\theta}_0)_{\dot{\alpha}}\,,\end{aligned} \tag{17}$$

whose supersymmetry transformations are relatively simple:

$$\begin{aligned}\delta\tilde{\theta}_\alpha &= -4i(\bar{\varepsilon}\bar{\beta})\tilde{\theta}_\alpha\,,\\ \delta\tilde{x}_{\alpha\dot{\alpha}} &= 4i\tilde{x}_{\alpha\dot{\gamma}}\bar{\beta}^{\dot{\gamma}}\bar{\varepsilon}_{\dot{\alpha}}\,,\\ \delta\tilde{x}^2 &= -4i(\bar{\varepsilon}\bar{\beta})\tilde{x}^2\,.\end{aligned} \tag{18}$$

(By x^2 we understand $x^2 = x_\mu x_\mu = \frac{1}{2}x_{\alpha\dot{\alpha}}x^{\alpha\dot{\alpha}}$.)

Moreover, we conclude from eqs. (16)–(18) that invariant superfields can depend only on the following variables:

$$\tilde{x}^2/\rho^2\,, \qquad \tilde{\theta}_\alpha/\rho^2\,, \qquad \rho^2(1 + 4i(\bar{\beta}\bar{\theta}_0))\,. \tag{19}$$

To be more precise, the above consideration refers to the case of invariant chiral fields. A similar technique can be developed for the anti-chiral and for general superfields but we will not go into detail here.

If prior to the introduction of the collective coordinates (in particular, $\rho = 1$) a given field depends on x_L^2, θ, then one has to make the substitution:

$$x_L^2 \to \tilde{x}^2/\rho^2\,, \qquad \theta_\alpha \to \tilde{\theta}_\alpha/\rho^2\,. \tag{20}$$

This recipe implies for the superfield S squared:

$$(S^2)_{\text{inst}} = (S^{\alpha\dot{f}}S_{\alpha\dot{f}})_{\text{inst}} = 2\,\frac{v^2\tilde{x}^2}{\tilde{x}^2 + \rho^2}\,, \tag{21}$$

where α is the colour index, $\dot{f}$ is the index of the global SU(2), and the ρ^2 dependence corresponds to the original bosonic solution.

Expression (8) for W^2 remains actually intact since the difference between $(x_L - x_0)$ and $\tilde{x}$ in this case is unimportant because of the factor $\tilde{\theta}^2$ equivalent to $\delta^2(\tilde{\theta})$. The superfield W^2 is obviously a function of the invariants $\tilde{x}^2/\rho^2$ and $\tilde{\theta}_\alpha/\rho^2$.

Let us emphasize once more that the new element which arises for $v \neq 0$ is the appearance of the extra collective coordinate, $\bar{\theta}_0$. To clarify its meaning let us expand eq. (21) in $\bar\theta_0$ at the point $\theta_0=\bar\beta=0$, $x_0=0$:

$$(S^2)_{\text{inst}}=2v^2\left\{\frac{x^2}{x^2+\rho^2}+4i\theta^\alpha x_{\alpha\dot\alpha}\bar\theta_0^{\dot\alpha}\frac{\rho^2}{(x^2+\rho^2)^2}-\frac{8\rho^4\theta^2\bar\theta_0^2}{(x^2+\rho^2)^3}\right\}. \tag{22}$$

Comparing eq. (22) with the general expression

$$S^2=\varphi^{\alpha f}\varphi_{\alpha f}+2\sqrt{2}(\theta\psi^{\alpha f})\varphi_{\alpha f}+(2\varphi_{\alpha f}F^{\alpha f}-(\psi^{\alpha f}\psi_{\alpha f}))\theta^2\,,$$

we find the spinor matter field to be equal to

$$\psi_\gamma^{\alpha f}=2\sqrt{2}iv(\bar\theta_0)^f\delta_\gamma^\alpha\frac{\rho^2}{(x^2+\rho^2)^{3/2}}\,, \tag{23}$$

where γ is the Lorentz index, and $F^{\alpha f}=0$.

It is amusing to observe that the spinor field (23) coincides with ordinary fermionic zero modes first found by 't Hooft [16]. The grassmannian parameter $\bar\theta_0$ which is actually a pair of parameters, $\bar\theta_0^{\dot 1}$ and $\bar\theta_0^{\dot 2}$, are proportional to the expansion coefficients in the 't Hooft modes. Thus, supersymmetry provides a *geometrical meaning* to fermionic zero modes of the matter fields.

However, such an interpretation of the fermionic zero modes provokes some natural questions. Indeed, the zero modes exist even without scalar fields (i.e. without supersymmetry). To establish the connection between our formalism which is manifestly supersymmetric and the standard approach adopted in QCD let us pass to the limit $v\to 0$. In this limit the scalar field vanishes. However, simultaneously one has to rescale the expansion coefficients in the zero modes (to normalize the zero modes to unity).

As for the instanton measure its form is fixed by supersymmetry and dimensional arguments,

$$d\mu(\rho,x_0,\theta_0,\bar\theta_0,\bar\beta)=C\frac{\Lambda^5}{v^2}\,d^4x_0\,d^2\theta_0\,d^2\bar\beta\,d^2\bar\theta_0\,\frac{d\rho^2}{\rho^2}\,, \tag{24}$$

where C is some numerical constant whose value depends on the exact definition of the parameter Λ, the scale parameter in the running coupling constant.

It is worth noting that not only the measure (24) as a whole is supersymmetric, but the products

$$(d^4x_0\,d^2\theta_0)\,,\qquad d\rho^2/\rho^2\,,\qquad (d^2\bar\beta\,d^2\bar\theta_0)$$

are invariant under transformations (6), (16) as well. For the first two combinations supersymmetry is obvious while in the latter case to prove the statement we observe that under the supersymmetry transformations

$$d^2\bar\beta\to d^2\bar\beta(1+4i\bar\varepsilon\bar\beta)\,,\qquad d^2\bar\theta_0\to d^2\bar\theta_0(1-4i\bar\varepsilon\bar\beta)\,,$$

so that the product $d^2\bar\beta\,d^2\bar\theta_0$ is indeed invariant.

Let us compare eq. (24) with the instanton measure (10) in pure gluodynamics. In the limit $v \to 0$ the fermionic coordinates $\bar{\theta}_0^{\dot{\alpha}}$ are to be replaced by η_1, η_2 where

$$\eta_1 = 4\pi\rho v \bar{\theta}_0^{\dot{1}}, \qquad \eta_2 = 4\pi\rho v \bar{\theta}_0^{\dot{2}}$$

and $\eta_{1,2}$ correspond to the fermion zero modes normalized to unity. Then $d^2\bar{\theta}_0 = 16\pi^2 v^2 \rho^2 \, d\eta_1 \, d\eta_2$ and the measure (24) goes into

$$d\mu(\rho, x_0, \theta_0, \eta_1, \eta_2, \bar{\beta}) = 16\pi^2 C \Lambda^5 \, d^4x_0 \, d^2\theta_0 \, d^2\bar{\beta} \, d\rho^2 \, d\eta_1 \, d\eta_2 ,$$

which differs from the instanton measure (10) by the differentials $d\eta_1 \, d\eta_2$. This is just what we would expect to get since these differentials correspond exactly to the appearance of two extra fermion zero modes associated with the matter fields. There is also a change in the power of the parameter Λ, $\Lambda^6 \to \Lambda^5$, reflecting the change of the first coefficient in the Gell-Mann–Low function. This change could also be anticipated on dimensional grounds since the dimension of $\eta_{1,2}$ is

$$[\eta] = m^{-1/2}, \qquad [d\eta] = m^{1/2} .$$

(For completeness we list also the dimensions of the other fermion collective coordinates:

$$[\theta_0] = m^{-1/2}, \qquad [\bar{\theta}_0] = m^{-1/2}, \qquad [\bar{\beta}] = m^{1/2}) .$$

Thus, we have succeeded in constructing invariant superfields and the instanton measure. This is still not the end of the story, however. The point is that we do not deal now with the exact solution of the classical equations of motion. Therefore, the action on the trajectory considered depends on the collective coordinates. For the bosonic fields (gluons + scalars) the action is [16]

$$S_{\text{bos}} = \frac{8\pi^2}{g_0^2} + 4\pi^2 v^2 \rho^2 . \tag{25}$$

The meaning of working with the trajectories which are "almost" solutions was clarified by 't Hooft [16] and we will not repeat the arguments here.

Our central point is that we integrate over the whole family of the field configurations which are related by the supersymmetry transformations and this guarantees that the final result is supersymmetric. We do not need to integrate over solutions to maintain supersymmetry.

It is clear that eq. (25) in a supersymmetric theory goes into

$$\frac{8\pi^2}{g_0^2} + 4\pi^2 v^2 \rho^2 \to \frac{8\pi^2}{g_0^2} + 4\pi^2 v^2 \rho_{\text{inv}}^2 ,$$

where

$$\rho_{\text{inv}}^2 = \rho^2 [1 + 4i(\bar{\theta}_0 \bar{\beta})] , \tag{26}$$

see eq. (20). The extra piece in the action $16i\pi^2 v^2 \rho^2 (\bar{\theta}_0 \bar{\beta})$ arises from the Yukawa interaction $\bar{\varphi}\lambda\psi$. One can readily verify that this Yukawa interaction does produce the extra term in the action by substituting ψ and λ by the corresponding zero modes and the field $\bar{\varphi}$ by the solution (14b).

Concluding this section we give the full expression for the instanton measure in the theory with $v \neq 0$ (after integrating over the colour rotations):

$$d\mu\, e^{-S} = C\frac{\Lambda^5}{v^2}\exp(-4\pi^2 v^2 \rho_{\rm inv}^2)\, d^4x_0\, d^2\theta_0\, d^2\bar{\theta}_0\, d^2\bar{\beta}\frac{d\rho^2}{\rho^2}. \tag{27}$$

4. Evaluation of the Two-Point Function $\langle T\{W^2(x), S^2(0)\}\rangle$

Consider the two-point function (introduced in [3])

$$T(x, \theta; y, \theta') = \langle 0|T\{W^2(x_L, \theta), S^2(y_L, \theta')\}|0\rangle \tag{28}$$

in the theory (3) in the massless limit, $m=0$. Since both W^2 and S^2 are of the same chirality T vanishes to any order in perturbation theory. On the other hand, the instanton contribution to (28) is not vanishing. In this approximation we get

$$T(x, \theta; y, \theta') = C\frac{\Lambda^5}{v^2}\int \exp(-4\pi^2 v^2\rho_{\rm inv}^2)\, d^4x_0\, d^2\theta_0\, d^2\bar{\theta}_0\, d^2\bar{\beta} \times\frac{d\rho^2}{\rho^2}\left\{\frac{\tilde{\theta}^2\rho^4}{(x_1^2+\rho^2)^4}\right\}2v^2\left\{\frac{\tilde{x}_2^2}{\tilde{x}_2^2+\rho^2}\right\}, \tag{29}$$

where $x_1 = x_L - x_0$,

$$(\tilde{x}_2)_{\alpha\dot{\alpha}} = (y_L - x_0)_{\alpha\dot{\alpha}} + 4i((\theta'-\theta_0)_\alpha + (y_L - x_0)_{\alpha\dot{\gamma}}\bar{\beta}^{\dot{\gamma}})(\bar{\theta}_0)_{\dot{\alpha}} \tag{30}$$

and $\tilde{\theta}$, $\rho_{\rm inv}$ are defined in eqs. (17) and (26), respectively. Here we substituted the instanton-generated fields W^2 and S^2 (see eqs. (8) and (21)).

Moreover, it is convenient to integrate over the collective coordinates in such a way as to keep supersymmetry explicit at each step. Thus, we integrate first over $\int d^4x_0\, d^2\theta_0$, then over $\int d^2\bar{\theta}_0^2\, d^2\bar{\beta}$ and finally over $\int d\rho^2\rho^{-2}$ (these products of the differentials are supersymmetric by themselves, see sect. 3). Upon performing the first integration we obtain

$$T(x, \theta; y, \theta') = C\tfrac{2}{3}\pi^2\Lambda^5\int\frac{d\rho^2}{\rho^2}\, d^2\bar{\theta}_0\, d^2\bar{\beta}\exp(-4\pi^2v^2\rho_{\rm inv}^2) \times\left\{1-\int_0^1\frac{2\rho^6\zeta^3\, d\zeta}{\left[\zeta(1-\zeta)\dfrac{(y_L-x_L)^2}{1+4i\bar{\theta}_0\bar{\beta}}+\rho^2\right]^3}\right\}, \tag{31}$$

where we have introduced the Feynman parameter ζ and used the fact that $\tilde{\theta}^2$ is equivalent to $\delta(\theta_0 - \theta - (x_L - x_0)\bar{\beta})$; moreover, we have accounted for the fact that the terms proportional to $(\theta-\theta')$ finally drop out anyhow and one can put $\theta-\theta'=0$ from the beginning*.

Note that if the integrand in eq. (31) were not singular at $\rho^2=0$ the result would have been identical zero. Indeed, proceeding to ρ^2_{inv} we would conclude that eq. (31) is proportional to

$$\int \frac{d\rho^2_{\text{inv}}}{\rho^2_{\text{inv}}} F(\rho^2_{\text{inv}}) \int d^2\bar{\theta}_0 \, d^2\bar{\beta} ,$$

where

$$F = \exp(-4\pi^2 v^2 \rho^2_{\text{inv}}) \left\{ 1 - \int_0^1 \frac{2\rho^6_{\text{inv}} \zeta^3 \, d\zeta}{[\zeta(1-\zeta)(y_L - x_L)^2 + \rho^2_{\text{inv}}]^3} \right\}, \tag{32}$$

and the integral over the Grassmann variables $d^2\bar{\theta}_0 \, d^2\beta$ clearly vanishes.

Actually, the result is not vanishing because of the singularity at $\rho^2=0$. True, written in this way, the result depends exclusively on $\rho^2=0$ instantons. Let us demonstrate that

$$\frac{d\rho^2}{\rho^2} d^2\bar{\theta}_0 \, d^2\bar{\beta} \, F(\rho^2_{\text{inv}}) \equiv 16 \, d\rho^2_{\text{inv}} \, \delta(\rho^2_{\text{inv}}) F(\rho^2_{\text{inv}}) \tag{33}$$

for any function F which falls off fast enough at $\rho^2 \to \infty$. To this end integrate first over $d^2\bar{\theta} \, d^2\bar{\beta}$:

$$\int d^2\bar{\theta}_0 \, d^2\bar{\beta} \, F(\rho^2_{\text{inv}}) = 16\rho^4 F''(\rho^2) .$$

Integrating over $d\rho^2/\rho^2$ by parts we reproduce eq. (33).

Assembling all the factors we find

$$T(x, \theta; y, \theta') = C\tfrac{32}{3}\pi^2 \Lambda^5 \tag{34}$$

and the function T does not depend on $(x-y)$ in accordance with the general principles, see ref. [1].

It is instructive to compare eq. (34) with the results of alternative calculations of the same amplitude in refs. [6, 7]. First of all, it is obvious that in the limit $|x-y| \to \infty$ eq. (34) does coincide with the result of ref. [6]. In this limit only the unit term survives in the braces in eq. (34).

Moreover, the technique proposed here demonstrates explicitly the factorization property. Indeed, we have shown that the result is determined in a sense by the instantons of zero size, $\rho^2=0$. These instantons are located at x, the coordinate of the operator $\lambda\lambda(x)$, and the operator $\varphi^2(0)$ reduces to its asymptotic value v^2 (since the instanton size $\rho=0$). Thus, the whole exercise can be considered as a calculation of the condensate $\lambda\lambda$ in the external field $\varphi = v$.

* In spite of x-independence of the correlator T, by no means one can put $x_L - y_L = 0$ from the very beginning. This remark is applicable to other similar computations, see below.

Let us turn now to discussion of the hypothesis [3] that only the small-size instantons, $\rho \sim |x-y|$, survive in the limit $|x-y| \to 0$. To check this hypothesis turn to the integration over ρ^2 in its original form, without use of eq. (33). Let us assume for a moment - we will demonstrate the assumption to be wrong - that at short distances it suffices to keep only instantons of the size $\rho \sim |x-y|$. Then one must put $v=0$ in eq. (31) and keep only the term proportional to $\bar{\theta}_0^2$ in the braces. This amounts to accounting for the contribution of the fermionic zero modes (and the induced scalar field; the terms proportional to v in the scalar field are neglected). The motivation is of course that the instanton field for $\rho \sim |x-y| \to 0$ is much stronger than v.

Proceeding in this way we would find

$$T(x, \theta; y, \theta')_{\text{small size inst.}}$$
$$= C\tfrac{2}{3}\pi^2 \Lambda^5 \int \frac{d\rho^2}{\rho^2} \int_0^1 d\zeta(-192) \times \frac{\rho^6 \zeta^3}{[\zeta(1-\zeta)(x-y)^2+\rho^2]^3}$$
$$\times \left\{1 - \frac{3\rho^2}{\zeta(1-\zeta)(x-y)^2+\rho^2} + 2\frac{\rho^4}{[\zeta(1-\zeta)(x-y)^2+\rho^2]^2}\right\} = C\tfrac{32}{3}\pi^2 \Lambda^5(\tfrac{1}{2}), \quad (35)$$

which is exactly $\frac{1}{2}$ of the total answer, compare to eq. (34).

Note that the calculation (35) superficially looks self-consistent since the integration over $d\rho^2$ is indeed dominated by $\rho^2 \sim |x-y|^2$. Still, instantons of the characteristic size $\rho \sim v^{-1}$ contribute exactly the same amount.

The latter statement can be checked explicitly again. Expand to this end the exponential $\exp(-4\pi^2 v^2 \rho^2(1+4i\bar{\theta}_0\bar{\beta}))$ in eq. (31) in $\bar{\theta}_0\bar{\beta}$. In the limit $|x-y| \ll v^{-1}$ the linear term in this expansion drops off while the quadratic term reduces to

$$T(x, \theta; y, \theta')_{\text{large size inst.}} \xrightarrow[|x-y| \to 0]{} C\tfrac{32}{3}\pi^2 \Lambda^5(\tfrac{1}{2})$$
$$\times \int d\rho^2\, \rho^2 (4\pi^2 v^2)^2 \exp(-4\pi^2 v^2 \rho^2) = C\tfrac{32}{3}\pi^2 \Lambda^5(\tfrac{1}{2}). \quad (36)$$

The integral over $d\rho^2$ is saturated at $\rho^2 \sim 1/4\pi^2 v^2$.

Note that the analysis of ref. [7] ignores the contribution of the instantons of the "characteristic" scale, $\rho \sim v^{-1}$ in our case. This explains, in our mind, the fact that in the case of two flavours the instanton result turned out to be inconsistent (see introduction). We will discuss this point in more detail in the next section.

5. SU(2) theory with two flavours

To build up one more flavour we add extra superfields $R^{\alpha i}$ to the lagrangian (3) (here $\alpha = 1, 2$ is the colour index while i, $i = 1, 2$, is an index of the extra global flavour symmetry). In the absence of mass terms the fields S^f, R^i actually form a representation of a global SU(4) group. The mass terms

$$\mathcal{L}_{\rm m} = \int d^2\theta \{\tfrac{1}{4} m_1 S^f S_f + \tfrac{1}{4} m_2 R^i R_i\} + \text{h.c.} \tag{37}$$

break the SU(4) to SU(2)×SU(2). We will consider the case

$$m_1 \ll m_2 \ll \Lambda .$$

As demonstrated below such a choice of the scale hierarchy results in the appearance of the classical condensate of the field S [3]:

$$\langle S \rangle \sim \Lambda (m_2/m_1)^{1/4} , \tag{38}$$

which ensures, in turn, the validity of the weak coupling approximation at all distances.

Eq. (38) can be derived by various techniques. In the language of the effective lagrangians [4] instantons generate the following effective interaction (superpotential):

$$\mathcal{L}_{\rm eff} = C\Lambda^4 m_2 \int \frac{d^2\theta}{S^2} , \tag{39}$$

where C is some calculable constant. As mentioned several times above its value depends on the definition of the parameter Λ and it is determined by the graph of fig. 1. The only difference from the case of a single flavour is that there exist now two extra matter zero modes which are annihilated by the mass insertion m_2.

Calculating furthermore the effective F-term of the superfield S and requiring this F-term to vanish in the vacuum we immediately come to the estimate (38). Since the condensate is large in the Λ scale, $\langle S \rangle \gg \Lambda$, the effective coupling constant is weak at all distances.

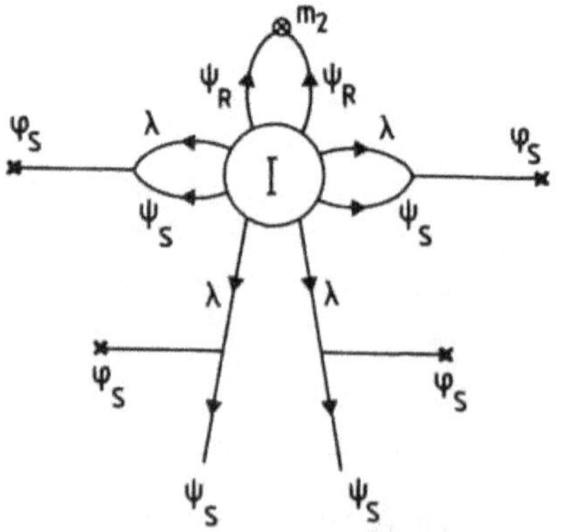

Fig. 1. The instanton-induced superpotential $Cm_2\Lambda^4/S^2$ in the SU(2) model with two flavours, S and R (m_2 is the mass parameter corresponding to R; it is assumed that $m_2 \gg m_1$).

The interesting two-point functions now are

$$K_a = \langle 0| T\{S^{\alpha f} S_{\alpha f}(x_L, \theta)\,, \qquad R^{\beta i} R_{\beta i}(y_L, \theta')\}|0\rangle\,, \tag{40a}$$

$$K_b = \langle 0| T\{S_f^{\alpha} R_{\alpha i}(x_L, \theta)\,, \qquad S^{\beta f} R_{\beta}^{i}(y_L, \theta')\}|0\rangle\,, \tag{40b}$$

they were introduced first in refs. [3, 7]. Recall that they share the virtue of being x-independent.

Let us sketch the problem raised first in ref. [7]. If $|x-y| \to 0$ then both (40a) and (40b) are contributed by the small-size instantons which give essentially coinciding expressions. On the other hand, considering the limit $|x-y| \to \infty$ one immediately arrives at the conclusion that the constant K_b can be nothing else but identical zero while a non-vanishing value of K_a violates no general principle. To explain the puzzle the authors of ref. [7] have assumed that the contribution of small-size instantons in this case is unstable against introduction of a small mass term. We will argue that mass singularities are irrelevant and the resolution of the "paradox" is given by the instantons of the characteristic sizes $\rho \sim v^{-1}$.

To substantiate our guess we need a superfield formalism suitable for the model considered. As for the superfield S^2 its description remains the same as in sects. 3 and 4. Adding the matter fields R brings in, however, extra zero modes and, therefore, extra collective coordinates. Despite the fact that the fields S and R enter the lagrangian in a perfectly symmetric way this symmetry is destroyed by the vacuum expectation values. Since the introduction of the collective coordinates, as explained in sect. 3, does depend on whether the vacuum expectation value of some field vanishes or not the description of the fields S and R in the instanton background turns out to be different as well.

In more detail, the scalar field φ_S develops a non-vanishing classical vacuum value, $\langle \varphi_S^2 \rangle \sim \Lambda^2 (m_2/m_1)^{1/2} \gg \Lambda^2$, while for the scalar field φ_R we will get $\langle \varphi_R^2 \rangle \sim \Lambda^2 (m_2/m_1)^{-1/2} \ll \Lambda^2$. Thus, the $\langle \varphi_R^2 \rangle$ condensate is a quantum effect and its small value implies the absence of the corresponding classical field in the vacuum, $\langle \varphi_R \rangle_{cl} = 0$. Thus, the expression for $(R^2)_{inst}$ can be derived starting from that for the field S^2, see eq. (21), by tending $v \to 0$. In this way we ensure vanishing $\langle \varphi_R \rangle$. Simultaneously we have to rescale the corresponding fermion collective coordinates $\bar\theta_0$. As explained in sect. 3 this rescaling is needed to render the zero modes normalized to unity. The corresponding collective coordinates are denoted by $\bar\eta_i$ and they are fully analogous to the coordinates η_1, η_2 introduced in sect. 3, see the discussion following eq. (24).

In this way we find

$$(R^2)_{inst} = -\frac{1}{\pi^2} \frac{\tilde\theta^2 \rho^2 \bar\eta^2}{[(x_L - x_0)^2 + \rho^2]^3}\,, \tag{41}$$

where we have replaced θ^2 by $\tilde\theta^2$ as it is prescribed by the general recipe (17). Note that there is no need to introduce $\tilde x^2$ instead of $(x_L - x_0)^2$ because of the factor $\tilde\theta^2 \sim \delta^2(\bar\theta)$.

Prior to plunging into calculation of the constants $K_{a,b}$ let us discuss the supersymmetry transformations on $R^2_{\rm inst}$. So far we have dealt with invariant superfields. Now the situation is a bit more subtle.

The transformation law of the new coordinates $\bar\eta$ can be readily read from eq. (16) by performing the substitution $\bar\theta_0 = (4\pi\rho v)^{-1}\bar\eta$ and tending $v \to 0$:

$$\delta\bar\eta = -2i\bar\eta(\bar\varepsilon\bar\beta) - 4i\bar\beta(\bar\eta\bar\varepsilon)\,, \tag{42a}$$

or, what is the same,

$$\delta\bar\eta = 2i\bar\eta(\bar\varepsilon\bar\beta) + 4i\bar\varepsilon(\bar\eta\bar\beta)\,. \tag{42b}$$

It is convenient to rewrite eq. (42) as

$$\bar\eta_f \to \omega^j_f\bar\eta_j\,, \tag{43}$$

where ω^j_f is the matrix of some rotation of the spinor $\bar\eta_i$ in the SU(2) space:

$$\omega^j_f(\bar\varepsilon, \bar\beta) = \delta^j_f(1 + 2i(\bar\varepsilon\bar\beta)) + 4i\bar\beta^j\bar\varepsilon_f\,. \tag{44}$$

It will be essential for the following that the matrix ω is unimodular:

$$\omega^k_f\omega^l_j\varepsilon^{fj} = \det\{\omega\}\varepsilon^{kl} = \varepsilon^{kl}. \tag{46}$$

It immediately follows that the products

$$\bar\eta^2 = \bar\eta_i\bar\eta_j\varepsilon^{ij} \qquad \text{and} \qquad d^2\bar\eta$$

are invariants of the supersymmetry transformations. Moreover, by utilizing the invariance of the combination (19) one readily checks that the superfield $R^2_{\rm inst}$ as it is given by eq. (41) is invariant as well.

As for the field RS the analysis is not so straightforward. The expression for this superfield can be obtained in the following way. Start with the known scalar field component of S, see eq. (14b), and the fermionic component of R, see eq. (23). Then we get

$$R_{\alpha i}S^\alpha_f = -\frac{i}{\pi}\, v\theta^\alpha x_{\alpha f}\bar\eta_i \frac{\rho}{(x^2+\rho^2)^2}\,.$$

Now the rest of the collective coordinates, θ_0, $\bar\beta$, $\bar\theta_0$, are introduced via supertransformations which is equivalent to the substitution (20). In this way we find

$$R_{\alpha i}S^\alpha_f = -\frac{i}{\pi}\, v\frac{\tilde\theta^\alpha}{\rho^2}\left(\frac{\tilde x_{\alpha f}}{\rho}\right)\bar\eta_i \frac{1}{(\tilde x^2/\rho^2+1)^2}\,. \tag{46}$$

The subtlety mentioned above is that the product RS – as far as the flavour indices are not contracted – *is not invariant* under the supersymmetry transformations. Indeed, since $\tilde x_{\alpha f}/\rho$ transforms as

$$\frac{\tilde x_{\alpha f}}{\rho} \to \left(\frac{\tilde x_{\alpha j}}{\rho}\right)\omega^j_f(\bar\varepsilon, \bar\beta)\,, \tag{47}$$

where the matrix ω is defined in eq. (44), we conclude that the superfield RS rotates in the flavour space under the supersymmetry transformations:

$$R_{\alpha i}S^{\alpha}_{f} \to \omega^{l}_{f}(\bar{\varepsilon},\bar{\beta})\omega^{j}_{i}(\bar{\varepsilon},\bar{\beta})R_{\alpha j}S^{\alpha}_{l}\,. \tag{48}$$

Upon observing this non-invariance the first impetus would be to introduce extra collective coordinates associated with rotations in the SU(2) × SU(2) flavour space. Further contemplation convinces us, however, that this is unnecessary, and the collective coordinates introduced so far are sufficient to perform calculations consistently. The point is that we integrate over all $\bar{\eta}$ and the fact that supersymmetry induces some rotations in the $\{\bar{\eta}\}$ space is inessential for this reason. As for the SU(2) rotations of the field S, see the footnote before eq. (15) for the explanation why these can be disregarded as well.

Let us illustrate the point by a simple example of QCD with a single flavour. For the colour group SU(2) the Dirac spinor reduces then to two left-handed Weyl spinors and there arises an extra global SU(2) symmetry associated with possible mixing of these spinors. The standard set of the collective coordinates includes, apart from the bosonic ones, two fermionic collective coordinates, η_1 and η_2. Any global SU(2) rotation is equivalent to some linear transformation in the $\{\eta_1, \eta_2\}$ space. Thus, as far as this space is linear and we integrate over the whole $\{\eta_1, \eta_2\}$ space there is no need to introduce special collective coordinates associated with these rotations.

Turning back to the problem considered we can summarize the situation as follows. Since we do not introduce collective coordinates associated with the global SU(2) × SU(2) rotations the superfield RS is not necessarily superinvariant. Rotations in the flavour SU(2) × SU(2) space are allowed under supertransformations. It is important, however, that the rotation does not depend on the concrete form of the field (and, in particular, on its coordinate x) but only on the parameters $\bar{\varepsilon}$, $\bar{\beta}$. Supersymmetry is restored of course once singlets in the flavour space are considered (see, e.g., eq. (40b)). Moreover, instantons can generate only flavour singlet amplitudes since - upon integrating over the collective coordinates - instantons carry vacuum quantum numbers.

The integration measure in the case considered is

$$\mathrm{d}\mu\, \mathrm{e}^{-S} = \mathrm{const}\,\frac{\Lambda^4}{v^2}\exp\left(-4\pi^2 v^2\rho^2_{\mathrm{inv}}\right)$$

$$\times\mathrm{d}^4x_0\,\mathrm{d}^2\theta_0\,\mathrm{d}^2\bar{\beta}\,\mathrm{d}^2\bar{\theta}_0\,\mathrm{d}^2\bar{\eta}\,\rho^{-2}\,\mathrm{d}\rho^2\,, \tag{49}$$

which differs from eq. (27) by the power of the parameter Λ and by the differential $\mathrm{d}^2\bar{\eta}$, superinvariant by itself, as explained above.

There is no need to perform the calculation of the constants (40) in great detail since the procedure is just parallel to that outlined in sect. 4. There are some minor changes but they do not modify the basic fact that the result can be expressed in terms of the zero-size instantons. After integration over $d^2\bar{\eta}$, $d^2\theta_0$ and d^4x_0 the final result depends crucially on whether the integrand is singular at $\rho^2=0$ or not. For the constant K_a the integrand is singular, and we use eq. (33) to perform the calculation. As mentioned above the factorization property is explicit within the formalism developed:

$$K_a = \langle S^2\rangle\langle R^2\rangle = (\text{a non-vanishing constant}) \times \Lambda^4 . \qquad (50)$$

Using eq. (50) and the Konishi anomaly, see appendix 2, we immediately reproduce the estimates for $\langle S^2\rangle$ and $\langle R^2\rangle$ quoted above.

As for the correlator K_b the integrand is not singular at $\rho^2=0$ if $|x_L - y_L| \neq 0$:

$$K_b \sim \Lambda^4 \int d\rho^2 \exp(-4\pi^2 v^2 \rho^2_{inv})\, d^2\bar{\beta}\, d^2\bar{\theta}_0$$
$$\times \int_0^1 d\zeta\, \zeta(1-\zeta) \frac{2\rho^2 + \zeta(1-\zeta)(x_L - y_L)^2(1+4i\bar{\theta}_0\bar{\beta})^{-1}}{[\rho^2 + \zeta(1-\zeta)(x_L-y_L)^2(1+4i\bar{\theta}_0\bar{\beta})^{-1}]^2} . \qquad (51)$$

Note that the factors $(1+4i\bar{\theta}_0\bar{\beta})$ and ρ^2 can be combined to produce ρ^2_{inv} as expected on general grounds. Due to the absence of the singularity the integration over $d^2\rho_{inv}$ is trivial, and $K_b=0$. This result is in accord with the factorization:

$$K_b = \langle R_i S_j\rangle\langle R^i S^j\rangle , \qquad \langle RS\rangle = 0 .$$

In terms of ordinary integration over the ordinary variable ρ^2 the vanishing of the correlator K_b at $|x-y| \to 0$ is due to cancellation between instantons of the size $\rho \sim |x-y|$ and $\rho \sim 1/v$ (only the former contribution is discussed in ref. [7]). As for K_a the two contributions here add up to a non-vanishing value.

6. Hierarchy of condensates

The characteristic features of all models considered in this paper is the validity of the weak coupling approximation. In the massless limit some scalar fields are not determined classically and to any order in perturbation theory. There exist vacuum valleys along which the value of the scalar fields can be changed without affecting the energy of the state. The small mass term stabilizes the theory at zero expectation values of the scalar fields. However, inclusion of instantons results in a drastic change and there arise large classical vacuum expectation values, $\langle\varphi_S^2\rangle \gg \Lambda^2$. The reason is that the small mass term is no protection against the driving force produced by instantons. Condensates of this type could be called as first class condensátes. They are not proportional directly to masses and/or to instanton density.

The large scalar field $\langle\varphi_S\rangle \neq 0$ induces some other condensates which could be called second class condensates. The examples of these are the gluino condensate $\langle\lambda\lambda\rangle$ in the on-flavour model and scalar field condensate $\langle\varphi_R^2\rangle$ in the two-flavour model (with masses satisfying the condition $m_S \ll m_R$). These condensates arise in a one-instanton approximation and are saturated by the contribution of the instantons of the characteristic sizes $\rho \sim \langle \varphi_S \rangle F^{-1}$. Their magnitudes are determined uniquely by the instanton density and by $\langle \varphi_S \rangle$. There is no explicit dependence on masses (as far as these masses are small). Thus, if $\langle \varphi_S \rangle$ is considered fixed the second class condensates arise already in the approximation of zero masses.

At the next step we find condensates whose treatment calls for a mass insertion. Thus, from general arguments it is easy to see that the correlator (1) in the one-flavour model contains one mass insertion in the one-instanton approximation. In appendix 2 we demonstrate directly that even in the limit $|x-y| \to 0$ the two-point function (1) is *not* saturated by instantons with $\rho \sim |x-y|$. To reproduce the correct result one has to account for two-instanton configurations as well. We are led to this conclusion by the following simple estimates. The one-instanton contribution to eq. (1) in the one-flavour model is of order $m\Lambda^5$. On the other hand, two-instanton configurations of the characteristic size $\rho \sim {}^1/v$ have the weight of order Λ^{10}/v^4. For the supersymmetric vacuum, on the other hand, we get $v^2 \sim \Lambda^{5/2} m^{-1/2}$ so that

$$m\Lambda^5 \sim \Lambda^{10}/v^4$$

and we see that the two contributions are comparable to each other.

This observation is quite general. Any extra mass insertion calls for inclusion of the configuration with higher topological charge for a consistent treatment.

A few remarks on pure (SUSY) gluodynamics are now in order. This theory is a strong coupling one so that there is no way to compute various correlators any longer. Still we are able to give an indirect argument demonstrating that even in the limit $|x-y| \to 0$ the correlator (1) cannot be saturated by instantons with $\rho \sim |x-y|$.

Assume that the opposite is true and the saturation does take place. Then, since the small-size instantons produce a non-vanishing contribution to (1), the factorization at $|x-y| \to \infty$ would require a non-vanishing condensate $\langle \lambda\lambda \rangle \neq 0$ as well. Then we conclude that

$$\langle 0 | T\{\lambda\lambda(x_1), \lambda\lambda(x_2), \lambda\lambda(x_3)\} | 0 \rangle \xrightarrow[|x_i - x_j| \to \infty]{} \langle \lambda\lambda \rangle^3 \neq 0 \,. \tag{52}$$

On the other hand, due to SUSY the left-hand size of eq. (52) does not depend on the coordinates. Therefore, we then may conclude

$$\lim_{|x_i - x_j| \to 0} \langle 0 | T\{\lambda\lambda(x_1), \lambda\lambda(x_2), \lambda\lambda(x_3)\} | 0 \rangle \neq 0 \,. \tag{53}$$

However, no fluctuations of small size can be responsible for this. Indeed, fluctuations with integer topological charge change the chirality modulo factor 4. Moreover, one can show that the assumption that four fermion legs are absorbed by an instanton while the other two are annihilated by some large scale fluctuation is also inconsistent. What we actually need is a genuine 6-fermion condensate produced by large-scale fluctuations.

7. Conclusions

In the present work we have considered correlation functions of gauge-invariant superfields in various models. Their common feature is the independence of coordinates stemming from SUSY as well as the fact that they emerge "for the first time" in the one-instanton approximation. A priori one can expect that having computed these correlators at short distances one can then fix the "elementary" condensates (such as $\langle\lambda\lambda\rangle$ in SUSY gluodynamics) by applying the cluster decomposition. Unfortunately, it turned out that this is not the case.

We have developed the corresponding formalism in terms of superfields and have shown that actually the complete instanton contribution reduces to instantons with $\rho^2=0$. To this end it is necessary to proceed to a new variable, $\rho^2_{\rm inv}$, whose natural name is the supersymmetric instanton size – it does not change under the supertransformations. Both the factorization property and x-independence become obvious at each step within such a procedure. From the technical point of view this seems to be the most striking phenomenon from those discussed in the paper.

As for applications of instantons to strong coupling theories, the lessons are rather unfavourable. Namely, we have shown that even at short distances, $|x-y|\to 0$, there is an essential contribution not only from small-size instantons (which was known previously) but also from instantons of characteristic size $\rho\sim v^{-1}$ where v is the vacuum expectation value of the scalar field. Since we have mostly considered the models with $v\gg\Lambda$ the fluctuations with $\rho\sim 1/v$ are under theoretical control. It is demonstrated that both types of instantons, $\rho\sim|x-y|$ and $\rho\sim 1/v$, must be taken into account to ensure theoretical selfconsistency and their weights are of the same order of magnitude.

The literal extrapolation of this result to strong coupling theories implies that even for $|x-y|\to 0$ fluctuations of characteristic size $\rho\sim\Lambda^{-1}$ are important. We were able to give also independent arguments in favour of this assertion. True, one may hope that the large-scale contributions, incalculable in the strong coupling regime at the moment, change the numbers only by order unity, which is not very essential for understanding qualitative aspects of this or that model. One cannot rule out, however, that the large-scale fluctuations result in complete cancellation of terms found at short distances. We had the chance of presenting such an example in the two-flavour model.

As a note of optimism let us add the following. It seems possible that some correlation functions are determined entirely by the zero-size instantons in the sense as elucidated in the present work in the weak coupling models. Investigation of this possibility calls for a further effort.

Appendix A

CONTINUATION TO EUCLIDEAN SPACE IN SUPERSYMMETRIC THEORIES

As is well known, the group of rotations of Minkowski space becomes SU(2) × SU(2) under euclidean continuation. However, the algebra of $N=1$ SUSY has no euclidean analogue. The reason lies in the fact that one cannot define for the chiral Fermi field the operation of involution (complex conjugation) in euclidean space. In slightly different terms one cannot introduce the notion of the Majorana spinor in euclidean space. Therefore, the problem of continuation is common to both supersymmetric and chiral theories.

Below we would like to show that the euclidean continuation is not at all equivalent to construction of euclidean supersymmetry, and this step - construction of euclidean SUSY - is simply unnecessary. The transformations which we will use are just the operations defined in a normal way in the Minkowski space. What is really needed for instanton calculations is not a euclidean field theory with SUSY but the saddle point method of computation of integrals, based on analytical continuation in t for tunneling process.

Our method of euclidean continuation of functional integrals follows the works of Berezin [17]. Let us recall the basic procedure for constructing the functional integrals. If we have a *quantum* hamiltonian given as a function of coordinate operator $\hat{q}_i$ and momentum operator $\hat{p}_i$, $\hat{H}(\hat{q}_i, \hat{p}_i)$, then we can unambiguously build a representation for the evolution operator $\exp(-i\hat{H}t)$ in terms of the functional integral

$$\langle q_i^{(+)}|e^{-i\hat{H}T}|q_i^{(-)}\rangle = \int \prod_i \mathrm{D}p_i\, \mathrm{D}q_i \exp\left\{i\int_{-T/2}^{T/2}\left[\sum p_i\dot{q}_i - H(p_i(t), q_i(t))\right]\mathrm{d}t\right\}, \qquad \text{(A.1)}$$

where the integration runs over all trajectories satisfying the following conditions

$$q_i(t=\pm\tfrac{1}{2}T) = q_i^{(\pm)}, \qquad p_i(t=\tfrac{1}{2}T) = p_i(t=-\tfrac{1}{2}T).$$

Quite an analogous representation is valid for Fermi systems with the only difference that the integration runs over anticommuting Grassmann parameters.

Now continuation to euclidean space reduces to a single step, transition to an imaginary value of the parameter T, $T=-i\tau$. In other words, instead of $\exp(-iHT)$ we consider $\exp(-H\tau)$. The functional integral representing this operator is built in the same way as above,

$$\langle q_i^{(+)}|e^{-H\tau}|q_i^{(-)}\rangle = \int \prod_i \mathrm{D}p_i\, \mathrm{D}q_i \exp\left\{-\int_{-\tau/2}^{+\tau/2}\mathrm{d}t\left[-i\sum_i p_i\dot{q}_i + H(p_i(t), q_i(t))\right]\right\}. \qquad \text{(A.2)}$$

Performing integration over $\mathrm{D}p_i(t)$ we arrive at the ordinary expression for the euclidean action. This is for Bose fields. For Fermi fields one should calculate the integral over $\mathrm{D}\bar{\psi}$, the step usually avoided.

On the other hand, arranging the expression (A.2) in the euclidean form is not at all necessary. Indeed, quantization of the theory in the Minkowski space determined the operator $\hat{H}(\hat{p}_i, \hat{q}_i)$ as well as all conserved charges (commuting operators). The action of these charges on $\hat{p}_i$ and $\hat{q}_i$ fixes their transformation law, which, in turn, generates the corresponding symmetry transformations for integration variables in the expressions like (A.2), in particular, SUSY transformations.

Algebraically, all expressions for the transformations differ from the original ones (defined in the Minkowski space) only by substitution $x_0 = -ix_4$. In the body of the paper we use the formulae referring to the Minkowski space keeping this substitution in mind; in no place do we use the assumption that x_μ is real.

To elucidate our procedure and notations let us consider the vector potential for the instanton solution (more exactly, antiinstanton)

$$A^{\alpha\gamma}_{\beta\dot\beta} = -i\frac{1}{x^2+\rho^2}(\delta^\alpha_\beta x^\gamma_{\dot\beta} + \delta^\gamma_\beta x^\alpha_{\dot\beta}) . \tag{A.3}$$

Here spinor indices are introduced running over 1, 2 both in the colour and coordinate spaces. The connection with the ordinary colour triplet index is given by the following relation

$$A^{\alpha\gamma} = A^a(\sigma^a)^\alpha_\rho \varepsilon^{\gamma\rho}, \qquad \begin{pmatrix} \alpha, \gamma, \rho = 1, 2 \\ a = 1, 2, 3 \end{pmatrix}, \tag{A.4}$$

where $(\sigma^a)^\alpha_\rho$ stand for the Pauli matrices, $\varepsilon^{\gamma\rho}$ is the antisymmetric Levi-Civita symbol, which plays the role of metric:

$$F^\alpha = \varepsilon^{\alpha\beta}F_\beta , \qquad F_\alpha = \varepsilon_{\alpha\beta}F^\beta , \qquad \varepsilon_{\alpha\beta} = -\varepsilon^{\alpha\beta} ,$$
$$\varepsilon^{12} = -\varepsilon^{21} = 1 .$$

The relation between the vector and spinor indices in the coordinate space is as follows

$$A_{\beta\dot\beta} = (\sigma_\mu)_{\beta\dot\beta}A_\nu g^{\mu\nu} , \qquad g_{\mu\nu} = \mathrm{diag}\,(1, -1, -1, -1) ,$$
$$(\sigma_\mu)_{\beta\dot\beta} = (\delta_{\beta\dot\beta}, \boldsymbol{\sigma}_{\beta\dot\beta}) . \tag{A.5}$$

In particular, this relation refers also to x_μ; explicitly

$$x_{1\dot1} = x^{2\dot2} = x_0 - x_3 ,$$
$$x_{1\dot2} = -x^{2\dot1} = -x_1 + ix_2 ,$$
$$x_{2\dot2} = x^{1\dot1} = x_0 + x_3 ,$$
$$x_{2\dot1} = -x^{1\dot2} = -x_1 - ix_2 .$$

All the above notations are obviously taken from the Minkowski space. Their euclidean nature reveals only in the fact that x_0 is imaginary, $x_0 = -ix_4$. Another compromise to the euclidean notation, which is rather unnecessary though, is in the definition of x^2,

$$x^2 = -x_\mu x_\nu g^{\mu\nu} = -\tfrac{1}{2}x^{\alpha\dot\alpha}x_{\alpha\dot\alpha} = -x_0^2 + \boldsymbol{x}^2 = x_4^2 + \boldsymbol{x}^2 .$$

One can readily convince oneself that the expression (A.3) coincides with the standard form for the BPST instanton. To this end one should take into account, apart from the expressions given above, the relation between the minkowskian and euclidean fields, A_μ and A_μ^{E}

$$A_0 = iA_\mu^{\mathrm{E}} , \qquad A_m = -A_m^{\mathrm{E}} \qquad (m = 1, 2, 3) ,$$

(the latter are used in the standard approach).

Appendix B

MASS INSERTIONS

In the one-flavour model the correlators like (1) emerge in the one-instanton approximation due to mass insertions. In the model considered the most relevant two-point function, instructive from various points of view, is

$$L = \langle 0| T\{\tfrac{1}{2}\,\mathrm{Tr}\, W^2(x, \theta), \tfrac{1}{2}\,\mathrm{Tr}\, W^2(y, \theta')\}|0\rangle . \tag{B.1}$$

From the general properties of SUSY it stems that for the lowest components L reduces to a constant, independent of x and y. One can readily convince oneself that for a single instanton this constant is proportional to m (fig. 2).

Thus, it is necessary to include in analysis the mass term $\mathcal{L}_m = mS^2|_F$. This, however, destroys the vacuum valleys, and the disappearance of the valleys simultaneously results in disappearance of supersymmetry of the measure (27) for arbitrary v. In the presence of the mass term the only supersymmetric point is $v = 0$. Unfortunately, if v vanishes, the strong coupling regime (the same as in pure gluodynamics) is restored, and theoretical calculations become uncontrollable.

For this reason, in the one-flavour model we are unable to calculate (B.1) directly for arbitrary values of $(x - y)$. Instead, we shall follow an indirect line of reasoning. In the limit $|x - y| \to \infty$ the function L factorizes,

$$L \to \langle \tfrac{1}{2}\,\mathrm{Tr}\, W^2\rangle^2 . \tag{B.2}$$

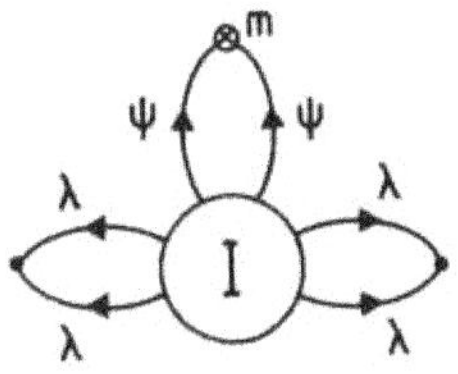

Fig. 2. The correlation function (B.1) in the one-flavour model.

The vacuum expectation value $\langle\frac{1}{2}\,\mathrm{Tr}\,W^2\rangle$ can be fixed by combining the already known result for the two-point function (28) (computed for $v \neq 0$ and $m=0$) and the anomalous Ward identity [18, 19], the so called Konishi anomaly. If now one uses the x, y-independence of L, one arrives at the value of L for arbitrary x and y.

Having thus determined the two-point function (B.1) indirectly we may again check the hypothesis of ref. [3] according to which at $|x-y| \to 0$ L is saturated by small-size instantons ($\rho \sim |x-y|$), so that the vacuum field $v \neq 0$ does not affect the answer. Within the framework of this hypothesis one should consistently put $v=0$ everywhere, which restores supersymmetry of the measure.

Let us start just from this side, from small $|x-y|$. The mass term in the action (see eq. (3)) generates the following extra factor in the instanton measure

$$S_{\mathrm{m}} = \tfrac{1}{4}m \int \mathrm{d}^2\theta\, \mathrm{d}^4x (S^2)_{\mathrm{inst}} = -4\pi^2 m v^2 \rho^2 \bar{\theta}_0^2\,, \tag{B.3}$$

where we have used eq. (21) and have performed integration over $\mathrm{d}^2\theta_0$ and d^4x_0. The $\bar{\theta}_0$ dependence which emerged in eq. (B.3) allows us to integrate over $\mathrm{d}^2\bar{\theta}_0$, i.e. to get rid of superfluous matter fermion legs (fig. 2).

For the instanton measure we get now, instead of (27)

$$\mathrm{d}\mu\, \mathrm{e}^{-S} = -8\pi^2 C m \Lambda^5 \exp(-4\pi^2 v^2 \rho^2)\, \mathrm{d}^4x_0\, \mathrm{d}^2\theta_0\, \mathrm{d}^2\bar{\beta}\, \mathrm{d}\rho^2\,. \tag{B.4}$$

This expression essentially coincides with that for the measure in pure gluodynamics [1] except for the exponent - the only place where the parameter v enters. It is obvious that just this exponent spoils the superinvariance of the measure (see eq. (6)). As was explained above, if $m \neq 0$ supersymmetry is indeed absent for arbitrary v. Putting $v=0$ we convert the exponent into unity thus arriving at the supersymmetric measure. We shall use below eq. (B.4) with $v=0$.

Moreover, it is necessary to fix the normalization of the instanton field (8). Comparison with the purely bosonic solution yields

$$\tfrac{1}{2}\,\mathrm{Tr}\, W^2_{\mathrm{inst}} = \frac{-48\tilde{\theta}^2\rho^4}{[(x_{\mathrm{L}}-x_0)^2+\rho^2]^4}\,. \tag{B.5}$$

Now there is no difficulty to find L; invoking eqs. (B.4) and (B.5) and acting in vein of the procedure presented in sect. 4 we arrive at

$$L_{\text{small size inst}} = \tfrac{16384}{10}\pi^4 C m \Lambda^5\,. \tag{B.6}$$

Let us calculate now the genuine value of L, turning to the limit $|x-y| \to \infty$. Accounting for the normalization factor (-48) in eq. (B.5) and making use of eqs. (28), (34) we learn that the product of the vacuum expectation values reduces to

$$\begin{aligned}\langle\tfrac{1}{2}\,\mathrm{Tr}\, W^2\rangle\langle S^2\rangle &= \langle 0|T\{\tfrac{1}{2}\,\mathrm{Tr}\, W^2(x,\theta),\, S^2(y,\theta')\}|0\rangle \\ &= -48 C \tfrac{32}{3}\pi^2 \Lambda^5\,. \end{aligned} \tag{B.7}$$

To extract from here the vacuum expectation values separately it is necessary to use additional information which will be supplied [3] by the Konishi anomaly. In our normalization this anomalous Ward identity can be written as

$$\bar{D}^2 \bar{S}^{\alpha f}\, e^V S^{\alpha f} = 4mS_{\alpha f}S^{\alpha f} + \frac{1}{\pi^2}\tfrac{1}{2}\,\mathrm{Tr}\, W^2\,. \tag{B.8}$$

Consequently, in the SUSY vacuum

$$\frac{1}{\pi^2}\langle \tfrac{1}{2}\,\mathrm{Tr}\, W^2\rangle = -4m\langle S^2\rangle\,. \tag{B.9}$$

Combining eqs. (B.9) and (B.7) we conclude that

$$L_{\mathrm{exact}} = \langle \tfrac{1}{2}\,\mathrm{Tr}\, W^2\rangle^2 = 2048\ Cm\Lambda^5\pi^4\,. \tag{B.10}$$

It is seen that

$$L_{\text{small size inst}} = \tfrac{4}{5}L_{\mathrm{exact}}\,. \tag{B.11}$$

The small-size instanton gives a contribution of the correct order of magnitude but still differing from the exact result. Thus, we can insist that even in the limit $|x-y|\to 0$ large-size fluctuations are essential in the correlation function (B.1). A short reflection shows that the lacking $\frac{1}{5}L_{\mathrm{exact}}$ are due to a two-instanton configuration with the characteristic size $\rho \sim v^{-1}$. Its contribution is $\sim \Lambda^{10}/v^4$, of the same order of magnitude as $m\Lambda^5$ in the supersymmetric point $v^2 \sim \Lambda^{5/2}m^{-1/2}$.

References

[1] V. Novikov et al., Nucl. Phys. B229 (1983) 394; V. Novikov et al., Nucl. Phys. B229 (1983) 407
[2] I. Affleck, M. Dine, N. Seiberg, Phys. Rev. Lett. 51 (1983) 1026
[3] G. Rossi, G. Veneziano, Phys. Lett. 138B (1984) 195
[4] I. Affleck, M. Dine, N. Seiberg, Nucl. Phys. B241 (1984) 493
[5] Y. Meurice, G. Veneziano, Phys. Lett. 141B (1984) 69
[6] A. Vainshtein et al., Pis'ma ZhETF 39 (1984) 494
[7] D. Amati, G. Rossi, G. Veneziano, preparing CERN-TH-3907, 1984
[8] I. Affleck, M. Dine, N. Seiberg, Phys. Lett. 137B (1984) p. 187; Phys. Rev. Lett. 52 (1984) 1677; Phys. Lett. 140B (1984) 59
[9] I. Affleck, M. Dine, N. Seiberg, Dynamical supersymmetry breaking and its phenomenological implications, IAS preprint, Princeton, 1984
[10a] A. Vainshtein, V. Zakharov, M. Shifman, Yad. Fiz. 42 (1985) 554
[10b] K.S. Narain, Nucl. Phys. B243 (1984) 131
[11] P. Fayet, S. Ferrara, Phys. Reports 32 (1977) 249
[12] A. Belavin et al., Phys. Lett. 59B (1975) p. 85
[13] M.J. Grisaru, W. Siegel, M. Roček, Nucl. Phys. B159 (1979) 429
[14] W. Fischler et al., Phys. Rev. Lett. 47 (1981) 757
[15] V. Novikov et al., Phys. Lett. 139B (1984) 389; Phys. Reports
[16] G.'t. Hooft, Phys. Rev. D14 (1976) 3432
[17] F.A. Berezin, The method of second quantization (Nauka, 1965); Usp. Fiz. Nauk 132 (1980) p. 497
[18] T.E. Clark, O. Piguet, K. Sibold, Nucl. Phys. B159 (1979) 1
[19] K. Konishi, Phys. Lett. 135B (1984) 439

Z. Phys. C - Particles and Fields 30, 161-174 (1986)

Zeitschrift für Physik C Particles and Fields

Instanton Induced Green Functions in the Superfield Formalism

J. Fuchs and M.G. Schmidt

Institut für Theoretische Physik der Universität, D-6900 Heidelberg, Federal Republic of Germany

Received 29 July 1985

Abstract. We discuss in detail the supersymmetric instanton calculus of NSVZ and extend it to chiral matter fields in the adjoint representation. The constant Green functions induced by the instanton of supersymmetric $SU(2)$ gauge theories are calculated systematically for the cases with and without scalar vev's bigger than the scale of the gauge theory and for nonvanishing small masses of chiral fields. One-instanton contributions to the Green functions containing four fields without large vevs would disturb clustering; but they are argued to vanish; two-instanton effects then lead to a pattern which quantitatively agrees with factorization and the anomaly relation.

1. Introduction

In supersymmetric gauge theories Green functions in the instanton field contrary to QCD can be calculated as well-defined finite expressions. In particular the Green functions of the lowest components of chiral superfields are independent of the arguments of the fields. Nonperturbative effects can give them constant values unequal to zero, and indeed this happens in instanton calculations [1-7]. Cluster decomposition ("factorization") then leads to a set of basic vacuum condensates which in the case of supersymmetric QCD (SQCD) apparently is in qualitative agreement with the results of the effective Lagrangian approach [8]. But one has to check the consistency of this picture: The condensates determined from different Green functions should agree with each other and the Konishi anomaly relation [9] should be fulfilled. Often this is a question of consistent numerical prefactors but sometimes one faces structures which seem to contradict factorization. In order to discuss such problems one needs a powerful formalism for calculations of Green functions in the instanton field. The component formalism [3-6] requires some combinatorics and easily becomes cumbersome. It is much more straightforward to calculate in the superfield formalism. This was first described for the case of scalar fields without large vevs far away from the instanton [1]. If one can reliably calculate Green functions with arguments close to each other using only small instantons at small distances, because of their constancy this result should also apply for largely different arguments where the strong interactions become effective. This is the viewpoint of [6], whereas the authors of [3] insist on staying in the weak coupling regime. The latter requires the consistent introduction of a gauge nonsinglet (Higgs) scalar field with a vev much bigger than the scale of the gauge theory Λ. This approach was recently taken up by the authors of [7] who demonstrate how to extend their superfield formalism to the case of large vev's allowing for scalar zero modes. Indeed instanton calculations without large vev's besides of the problems dealed with in [6] lead to inconsistencies in the normalization of different Green functions [7]* and require modification. The authors of [7] argue that these deficiencies are remedied in the case of a large vev. There is also the alternative approach to stay without large vev but to take into account nonzero modes. These can have tricky effects in the infrared for $m_i \to 0$, as was demonstrated in the case of nondiagonal-flavor Green functions [6].

In this paper we want to work out the supersymmetric formalism of [1, 7] in some detail and to supply some useful extensions. In particular we discuss the origin of Grassmannian parameters in the instanton calculus performing explicit SUSY-superconformal transformations instead of arguing with

* This was also noticed independently by the authors

invariants under simultaneous transformation of superfield and instanton parameters. We also write down expressions for chiral superfields in the adjoint representation. We systematically discuss the various constant Green functions always comparing results with and without large vev. Here we also remark on convenient integration tricks. Finally we consider 2-instanton contributions extending and supplementing the discussion in [7] and taking a somewhat different viewpoint, leading to a quantitatively consistent picture.

In the case of chiral fields in the adjoint representation with small vev's together with additional fields in the fundamental representation one of them with a very small mass and large vev we obtain consistency as in the case of only fundamental flavors. For an adjoint field with large vev, however, the scheme is not consistent in agreement with considerations based on the effective Lagrangian.

Appendix A explains our notation and Appendix B fixes the supersymmetric instanton integration measure. Chapter 2 contains the derivation of the classical superfields including those in the adjoint representation in the instanton background. Chapter 3 gives a systematic collection of instanton-induced Green functions. Chapter 4 checks the consistency of factorization and of the anomaly relation and discusses the two-instanton effects which are relevant to this consistency. Chapter 5 contains our conclusions.

2. Classical Superfields

In this section* we show, following [1, 7], how the solutions of the classical equations of motion can be grouped into superfields. For a pure super Yang-Mills theory the equations of motion are

$$\mathscr{D}^m v^a_{mn} = ig\varepsilon^{abc}\bar\lambda^b\bar\sigma_n\lambda^c$$
$$\bar{\mathscr{D}}\lambda^a = \mathscr{D}\bar\lambda^a = 0 \tag{2.1}$$

which possess the well-known 1-instanton solution [10] which in the so-called regular gauge reads

$$v^{a(c)}_m(x) = \frac{2}{g}\eta^a_{mn}x^n f(x-x_0)$$
$$v^{a(c)}_{mn}(x) = -\frac{4}{g}\eta^a_{mn}\rho^2 f^2(x-x_0) \tag{2.2}$$

with

$$f(x) := (x^2+\rho^2)^{-1}, \tag{2.3}$$

together with $\bar\lambda = 0$ and λ a linear combination of the normalized zero modes

$$\lambda^{a(ss)}_\alpha(x;\beta) = \frac{2}{\pi}\sigma^a{}_\alpha{}^\beta\rho^2 f^2(x-x_0),$$
$$\beta = 1,2$$
$$\lambda^{a(sc)}_\alpha(x;\dot\beta) = \frac{\sqrt{2}}{\pi}(\sigma^a\sigma^m)_{\alpha\dot\beta}(x-x_0)_m\rho f^2(x-x_0),$$
$$\dot\beta = 1,2. \tag{2.4}$$

In the presence of matter source terms are added to (2.1); it is easily seen [11] that for non-vanishing boundary conditions on the scalar fields there exist no longer solutions to the exact field equations. However, it was already argued by 't Hooft [12] that one should expand the functional integral around the solutions of the approximate field equations which are (2.1) for the gauge multiplet and

$$\mathscr{D}^2 Z = 0$$
$$(\mathscr{D}^2 H)^a = 0, \tag{2.5}$$

$$\bar{\mathscr{D}}\varphi = \mathscr{D}\bar\varphi = 0$$
$$(\bar{\mathscr{D}}\psi)^a = (\mathscr{D}\bar\psi)^a = 0 \tag{2.6}$$

for the matter fields, with $\mathscr{D}_m$ the covariant derivative in the instanton background. The source terms may then be taken into account by a systematic expansion in powers of $g^2 \sim \rho^2 v^2$ (where v is the classical expectation value of Z or H) which necessitates the introduction of constraints [11, 3].

The solutions to (2.5) are most easily found in the singular gauge which is related to the regular gauge by the singular (finite) gauge transformation

$$U_{ij}(x) = i\frac{(x-x_0)_m}{\sqrt{(x-x_0)^2}}\sigma^m_{ij}. \tag{2.7}$$

In the singular gauge the solutions to (2.5) are

$$Z^{(0)}_{iA}(x) = z\varepsilon_{iA}((x-x_0)^2 f(x-x_0))^{1/2}$$
$$H^{a(0)}(x) = h^a(x-x_0)^2 f(x-x_0) \tag{2.8}$$

and thus in the regular gauge

$$Z^{(0)}_{iA}(x) = iz\sigma^m_{iA}(x-x_0)_m f^{1/2}(x-x_0)$$
$$H^{a(0)}(x) = h^b\eta^a_{pm}\bar\eta^{bp}{}_n(x-x_0)^m(x-x_0)^n f(x-x_0). \tag{2.9}$$

The normalized solutions to (2.6) are

$$\varphi^{(0)}_{ai}(x) = \frac{1}{\pi}\rho\varepsilon_{ia}f^{3/2}(x-x_0) \tag{2.10}$$

and

* See Appendix A for our notation

$$\psi_\alpha^{a(ss)}(x;\beta)=\frac{2}{\pi}\sigma^a{}_\alpha{}^\beta\rho^2 f^2(x-x_0),$$
$$\beta=1,2$$
$$\psi_\alpha^{a(sc)}(x;\dot\beta)=\frac{\sqrt{2}}{\pi}(\sigma^a\bar\sigma^m)_{\alpha\dot\beta}(x-x_0)_m\rho f^2(x-x_0),$$
$$\dot\beta=1,2. \tag{2.11}$$

For the gauge multiplet and for matter in the fundamental representation it has been indicated by the authors of [7] how to construct the classical superfields: Start with superfields containing just the bosonic classical solutions, i.e. with

$$W_\alpha^{(0)}(y,\Theta)=-i\sigma^{mn}{}_\alpha{}^\beta\Theta_\beta v_{mn}^{(c)}(y)$$
$$\Phi^{(0)}(y,\Theta)=Z^{(0)}(y) \tag{2.12}$$

and then apply appropriate supersymmetric and superconformal transformations. We represent SUSY transformations by translations in superspace*; the authors of [7] first constructed invariants out of the various collective coordinates and then generalized the solutions (2.12) to functions of these invariants; since the resulting superfields are invariant under simultaneous transformations of both superspace coordinates and collective coordinates, these two methods are equivalent.

Since the classical action of pure super Yang-Mills is invariant under the full superconformal group [13], we have to apply to $W_\alpha^{(0)}$ both restricted (translational) SUSY transformations with generators Q_α, $\bar Q_{\dot\alpha}$ and special (conformal) SUSY transformations with generators $S_\alpha, \bar S_{\dot\alpha}$; now $W_\alpha^{(0)}$ is invariant under $\bar Q$ and S; applying Q and $\bar S$ transformations with parameters $-\Theta_0$ and $-\bar\vartheta$, respectively, one obtains the instanton superfield

$$W_\alpha^{(c)}(y,\Theta)=-i\sigma^{mn}{}_\alpha{}^\beta\tilde\Theta_\beta v_{mn}^{(c)}(y) \tag{2.13}$$

where

$$\tilde\Theta:=\Theta-\Theta_0-(y-x_0)_m\sigma^m\bar\vartheta. \tag{2.14}$$

Explicitly we have, using (A.3), in the regular gauge

$$W_\alpha^{a(c)}(y,\Theta)=\frac{8}{g}\rho^2 f^2(y-x_0)\,\sigma^a{}_\alpha{}^\beta\tilde\Theta_\beta; \tag{2.15}$$

the fermionic component of $W_\alpha^{(c)}$ thus indeed contains just the (unnormalized) gaugino zero modes (2.4)*

$$\lambda_\alpha^{a(c)}(y)=-\frac{4\pi i}{g}\lambda_\alpha^{a(ss)}(y;\beta)\Theta_{0\beta}$$
$$-\frac{4\sqrt{2}\pi i\rho}{g}\lambda_\alpha^{a(sc)}(y;\dot\beta)\bar\vartheta^{\dot\beta}. \tag{2.16}$$

As for the matter field $\Phi^{(0)}$ we must not apply S and $\bar S$ directly, since the classical expectation value of Z spoils conformal invariance; however, since $\Phi^{(0)}$ is invariant under $\bar S$ (and Q), we are allowed to apply an $\bar S$ transformation to the superfield obtained from $\Phi^{(0)}$ by a $\bar Q$ transformation (since the normalized fermionic component of this superfield does not break conformal invariance). Thus applying first a $\bar Q$ transformation with parameter $\bar\Theta_0$ and then Q and $\bar S$ transformations with the (already fixed) parameters $-\Theta_0$, $-\bar\vartheta$, one arrives at

$$\Phi^{(c)}(y,\Theta)=Z^{(0)}(\tilde y) \tag{2.17}$$

where

$$\tilde y^m:=y^m+2i\tilde\Theta\sigma^m\bar\Theta_0. \tag{2.18}$$

Explicitly, in the regular gauge,

$$\Phi_{iA}^{(c)}(y,\Theta)=iz\sigma_{iA}^m(\tilde y-x_0)_m f^{1/2}(\tilde y-x_0); \tag{2.19}$$

again the fermionic component of (2.17) is just the (unnormalized) zero mode (2.10):

$$\varphi_{\alpha iA}^{(c)}(y)=2\sqrt{2}\pi z\rho\varphi_{\alpha i}^{(0)}(y)\bar\Theta_{0A}. \tag{2.20}$$

If the above procedure is applied to a superfield containing just the bosonic solution $H^{(0)}$, only the second pair of the fermionic solutions (2.11) is reproduced; in the case of adjoint matter, we therefore have to start with a superfield that already contains the first pair of (2.11), together with $H^{(0)}$:

$$\Psi^{a(0)}(y,\Theta)=H^{a(0)}(y)+2\sqrt{2}\Theta^\alpha\psi_\alpha^{a(ss)}(y;\beta)\chi_{0\beta} \tag{2.21}$$

with some Weyl spinor parameter χ_0. The transformations with parameters $\bar\Theta_0$, $-\Theta_0$, $-\bar\vartheta$ then lead to the classical superfield

$$\Psi^{a(c)}(y,\Theta)=H^{a(0)}(\tilde y)+2\sqrt{2}\tilde\Theta^\alpha\psi_\alpha^{a(ss)}(y;\beta)\chi_{0\beta}. \tag{2.22}$$

In the regular gauge this means

$$\Psi^{a(c)}(y,\Theta)=h^b\eta^a_{pm}\bar\eta^{bp}{}_n(\tilde y-x_0)^m(\tilde y-x_0)^n f(\tilde y-x_0)$$
$$+\frac{4\sqrt{2}}{\pi}\tilde\Theta\sigma^a\chi_0\rho^2 f^2(y-x_0), \tag{2.23}$$

* Strictly speaking, these translations should be accompanied by suitable gauge transformations in order that gauge symmetry and SUSY commute. As a consequence, if one wants to get the fermionic component or the F component of a gauge variant superfield by Taylor expanding in $\Theta, \bar\Theta$, one has to compensate for having neglected this gauge transformation by using covariant instead of ordinary derivatives. In the present paper we will be only interested in gauge invariant quantities and therefore ignore these gauge transformations

* We extract the fermionic component by a formal Taylor expansion with covariant derivatives, see previous footnote

with fermionic component

$$\psi_\alpha^{a(c)}(y) = 2\psi_\alpha^{a(ss)}(y;\beta)\chi_0{}^\beta - 2\sqrt{2}h^b\bar\eta^b_{mn}(y-x_0)^n\rho^2 f^2(y-x_0)(\sigma^a\sigma^m)_{\alpha\dot\alpha}\bar\Theta_0^{\dot\alpha}. \quad (2.24)$$

In order to see explicitly that the second term in (2.24) reproduces the zero modes $\psi_\alpha^{a(sc)}$, we perform the unitary transformation

$$\bar\Theta_0^{\prime\dot\alpha} := \frac{h^a}{\sqrt{h^2}}\sigma^{a\dot\alpha}{}_\beta\Theta_0^\beta \quad (2.25)$$

to obtain

$$\psi_\alpha^{a(c)}(y) = 2\psi_\alpha^{a(ss)}(y;\beta)\chi_{0\beta} + 2i\pi\rho\sqrt{h^2}\,\psi_\alpha^{a(sc)}(y;\dot\beta)\bar\Theta_0^{\prime\dot\beta}. \quad (2.26)$$

Let us also mention that the bosonic components of $\Phi^{(c)}$ and $\Psi^{(c)}$ do not only consist of the classical solutions $Z^{(0)}$ and $H^{(0)}$, respectively; the additional terms

$$Z_{iA}^{(1)}(y) = -4z\rho^2 f^{3/2}(y-x_0)(\Theta_0 + (y-x_0)_m\sigma^m\bar\vartheta)_i\bar\Theta_{0A}$$
$$H^{a(1)}(y) = -4\rho^2 f^2(y-x_0)(\Theta_0+(y-x_0)_p\sigma^p\bar\vartheta)\sigma^a \left(\frac{\sqrt{2}}{\pi}\chi_0 + h^b\bar\eta^b_{mn}(y-x_0)^n\sigma^m\bar\Theta_0\right) \quad (2.27)$$

are solutions of the inhomogeneous field equations

$$(\mathcal{D}^2 Z)_{iA} = -ig\sqrt{2}\lambda^{\alpha a}\left(\frac{\sigma^a}{2}\right)_i^j \varphi_{\alpha jA}$$
$$(\mathcal{D}^2 H)^a = g\sqrt{2}\varepsilon^{abc}\lambda^{b\alpha}\psi^c_\alpha. \quad (2.28)$$

This can be easily checked by substituting into (2.28) the fermion zero modes (2.4), (2.10), (2.11) and the explicit form of the propagators (with zero modes omitted) [14] which in our notations are (for $x_0=0$)

$$(\mathcal{D}^{2-1})_i^j(x,x') = \frac{1}{4\pi^2(x-x')^2}f^{1/2}(x)f^{1/2}(x') \cdot [(\rho^2+x\cdot x')\delta_i^j - i\eta^a_{mn}x^m x'^n\sigma_i^{aj}] \quad (2.29)$$

for the fundamental and

$$(\mathcal{D}^{2-1})^{ab}(x,x') = \frac{1}{4\pi^2(x-x')^2}f(x)f(x') \cdot [(\rho^2+x\cdot x')^2\delta^{ab} - 2\varepsilon^{abc}\eta^c_{mn}x^m x'^n(\rho^2+x\cdot x') + (\delta^{ac}\delta^{bd}-\delta^{ab}\delta^{cd}+\delta^{ad}\delta^{bc}) \cdot \eta^c_{mn}\eta^d_{pq}x^m x^p x'^n x'^q] \quad (2.30)$$

for the adjoint representation.

Thus the bosonic components of $\Phi^{(c)}$ and $\Psi^{(c)}$ are the solutions of the full scalar field equations.

We will be always interested in the relevant case where there is at most one scalar field which acquires a large vev. The classical superfields of all other matter fields are then obtained by taking the limit $z\to 0$, $h^a\to 0$ of (2.19) and (2.22) (with the normalization of the fermionic zero modes held fixed) or, equivalently, by starting with the superfields

$$\Phi_{iA}^{(0)}(y,\Theta) = 2\Theta^\alpha\varphi_{\alpha i}^{(0)}(y)\bar\tau_A$$
$$\Psi^{a(0)}(y,\Theta) = \sqrt{2}\Theta^\alpha(2^{1/4}\rho^{-1/2}\psi_\alpha^{a(ss)}(y;\beta)\chi_\beta + 2^{3/4}\rho^{1/2}\psi_\alpha^{a(sc)}(y;\dot\beta)\bar\chi^{\dot\beta}) \quad (2.31)$$

(the normalization of the zero modes is chosen so that the constant in front of the instanton measure does not depend on the number of flavors, see Appendix B); the result is, of course,

$$\Phi_{iA}^{(c)}(y,\Theta) = 2\tilde\Theta^\alpha\varphi_{\alpha i}^{(0)}(y)\bar\tau_A = \frac{2}{\pi}\rho f^{3/2}(y-x_0)\tilde\Theta_i\bar\tau_A$$

$$\Psi^{a(c)}(y,\Theta) = \sqrt{2}\tilde\Theta^\alpha(2^{1/4}\rho^{-1/2}\psi_\alpha^{a(ss)}(y;\beta)\chi_\beta + 2^{3/4}\rho^{1/2}\psi_\alpha^{a(sc)}(y;\dot\beta)\bar\chi^{\dot\beta}) = \frac{2^{7/4}}{\pi}\rho^{3/2}f^2(y-x_0)\tilde\Theta\sigma^a\bar\chi \quad (2.32)$$

where

$$\tilde\chi := \chi + (y-x_0)_m\sigma^m\bar\chi. \quad (2.33)$$

Finally, we report the results for the gauge invariant contractions of the various superfields

$$\frac{g^2}{32\pi^2}W^{2(c)}(y,\Theta) = -\frac{6}{\pi^2}\rho^4 f^4(y-x_0)\tilde\Theta^2 \quad (2.34\,a)$$

$$\Phi^{2(c)}(y,\Theta) = \begin{cases} 2z^2(\tilde y-x_0)^2 f(\tilde y-x_0) & \text{for } z\neq 0 \quad (2.34\,b)\\ \frac{4}{\pi^2}\rho^2 f^3(y-x_0)\tilde\Theta^2\bar\tau^2 & \text{for } z=0, \quad (2.34\,c)\end{cases}$$

$$\Psi^{2(c)}(y,\Theta) = \begin{cases} h^2(\tilde y-x_0)^4 f^2(\tilde y-x_0) + \frac{48}{\pi^2}\rho^4 f^4(y-x_0)\tilde\Theta^2\chi_0^2 - \frac{16\sqrt{2}i}{\pi}h^b\bar\eta^b_{pn}(\tilde y-x_0)_m(\tilde y-x_0)^n \rho^2 f^3(y-x_0)\tilde\Theta\sigma^{pm}\chi_0 & \text{for } h\neq 0 \quad (2.34\,d)\\ \frac{3\cdot 2^{5/2}}{\pi^2}\rho^3 f^4(y-x_0)\tilde\Theta^2\bar\chi^2 & \text{for } h=0 \quad (2.34\,e)\end{cases}$$

and note that the Θ^2 components of all classical superfields constructed above vanish.

3. Instanton Green Functions

In this section we consider one instanton contributions to various Green functions for gauge invariant combinations of chiral superfields. Due to supersymmetry, these Green functions do not receive contributions from perturbation theory and their lowest components are expected to be space-time independent constants.

In an instanton background the bosonic and fermionic components of a superfield possess the same spectrum of nonvanishing eigenvalues [15]. In a naive expansion for small masses* then to lowest order in the gauge coupling only the zero modes contribute to the Green functions.

Introducing collective coordinates [16] for both bosonic and fermionic zero modes, we can therefore write

$$\langle TA(y_1,\Theta_1)B(y_2,\Theta_2)\ldots\rangle = \int d\sigma_{\text{inst}}\, A^{(c)}(y_1,\Theta_1)\,B^{(c)}(y_2,\Theta_2)\ldots \tag{3.1}$$

where $d\sigma_{\text{inst}}$ is the instanton measure derived in Appendix B, and $A^{(c)}$, $B^{(c)}, \ldots$ contain only the zero modes of $A, B\ldots$, i.e. they are classical superfields such as (2.34).

There is only a restricted number of Green functions to which instantons can contribute. In pure $SU(2)$ super Yang-Mills the only possible choice is

$$\mathscr{G}_{00}=\left\langle T\frac{g^2}{32\pi^2}W^2(y_1,\Theta_1)\frac{g^2}{32\pi^2}W^2(y_2,\Theta_2)\right\rangle \tag{3.2}$$

which according to the above remarks is given by

$$\mathscr{G}_{00}=C_0\Lambda^6\int d^4x_0\,d\rho^2\,d^2\Theta_0\,d^2\bar{\vartheta}\left(\frac{g^2}{32\pi^2}\right)^2 W^{2(c)}(y_1,\Theta_1)\,W^{2(c)}(y_2,\Theta_2) \tag{3.3}$$

where the constant C_0 is defined in (B.12) and $W^{2(c)}$ is the superfield (2.34a). $\mathscr{G}_{00}$ has been computed in [1]; in our notations the result is

$$\mathscr{G}_{00}=\frac{4}{5\pi^2}C_0\Lambda^6. \tag{3.4}$$

If a single flavour in the fundamental representation is added, the corresponding Green function is

$$\mathscr{G}_{01}=\left\langle T\frac{g^2}{32\pi^2}W^2(y_1,\Theta_1)\Phi^2(y_2,\Theta_2)\right\rangle; \tag{3.5}$$

it is given by

$$\mathscr{G}_{01}=C\frac{\Lambda^5}{z^2}\int d^4x_0\frac{d\rho^2}{\rho^2}d^2\Theta_0\,d^2\bar{\vartheta}\,d^2\bar{\Theta}_0 \cdot\exp(-4\pi^2z^2\tilde{\rho}^2)\exp(4\pi^2z^2\rho^2\bar{\Theta}_0^2m) \left(-\frac{6}{\pi^2}\right)\rho^4 f^4(\tilde{y}_1-x_0)\,\bar{\Theta}_1^2\cdot 2z^2(\tilde{y}_2-x_0)^2 f(\tilde{y}_2-x_0) \tag{3.6}$$

with $\tilde{\rho}^2$ and C defined in (B.7) and (B.11), and $\Phi^{2(c)}$ and $W^{2(c)}$ inserted according to (2.34a, b) (we substituted y_1 by $\tilde{y}_1$ in $W^{2(c)}$ which is allowed because of $\bar{\Theta}_1^2=\delta_2(\bar{\Theta}_1)$). The Θ_0-integration is most easily performed after the substitution

$$x_0^m\to x_0'^m=x_0^m-y_2^m-2i(1-2i\bar{\vartheta}\bar{\Theta}_0)^{-1}(\Theta_2-\Theta_0)\sigma^m\bar{\Theta}_0 +6\bar{\Theta}_0^2(\Theta_2-\Theta_0)\sigma^m\bar{\vartheta} \tag{3.7}$$

since then the total Θ_0-dependence is in $\bar{\Theta}$. After the Θ_0-integration the integrand depends on ρ^2, $\bar{\vartheta}$, $\bar{\Theta}_0$ only through the combination $\tilde{\rho}^2$. As shown in [7], the remaining integrations are then easily done for $z\neq 0$, since $\int d^2\bar{\vartheta}\,d^2\bar{\Theta}_0F(\tilde{\rho}^2)=4\rho^4F''(\rho^2)$ and thus for all functions F obeying $F(\rho^2)\to 0$, $\rho^2F'(\rho^2)\to 0$ at $\rho^2\to\infty$

$$\int\frac{d\rho^2}{\rho^2}d^2\bar{\Theta}_0\,d^2\bar{\vartheta}\,F(\tilde{\rho}^2)=4\cdot F(0). \tag{3.8}$$

Introducing a Feynman parameter α one easily can see that only the endpoint singularities at $\alpha=0,1$ are important for $\tilde{\rho}^2\to 0$; this one can use to integrate directly the various pole contributions in x_0'; only the f^4 singularities contribute for $\tilde{\rho}^2\to 0$ and we get

$$\mathscr{G}_{01}=-8C\Lambda^5. \tag{3.9}$$

Note that this result does not depend on whether Φ is massive or not. This technique of integrating only leading singularities can be conveniently used in all calculations where a pute $\tilde{\rho}^2$-integration appears.

Since the vev z^2 breaks chiral invariance, there is also a contribution to

$$\mathscr{G}_0=\left\langle\frac{g^2}{32\pi^2}W^2(y,\Theta)\right\rangle \tag{3.10}$$

which is easily evaluated to be

$$\mathscr{G}_0=-4C\frac{\Lambda^5}{z^2}. \tag{3.11}$$

The one-instanton contribution to $\mathscr{G}_1=\langle\Phi^2\rangle$ vanishes; thus in the one-instanton approximation $\mathscr{G}_1$ is given by its zero-instanton value $\mathscr{G}_1=2z^2$; comparing (3.9) and (3.11) we thus see that the factorization

* I.e. in contrast to the authors of [6] we do not consider contributions which are formally of higher order in the masses

$$\mathcal{G}_{01}=\mathcal{G}_0\cdot\mathcal{G}_1 \tag{3.12}$$

is trivially fulfilled.

If a second fundamental flavor is added, the relevant Green functions are (without loss of generality we choose $\Phi_{(1)}$ to have a large classical vev)

$$\mathcal{G}_{12}=\langle T\Phi^2_{(1)}(y_1,\Theta_1)\,\Phi^2_{(2)}(y_2,\Theta_2)\rangle$$
$$\mathcal{G}_2=\langle\Phi^2_{(2)}(y,\Theta)\rangle; \tag{3.13}$$

using the same methods as above, these are evaluated to be

$$\mathcal{G}_{12}=16C\Lambda^4, \tag{3.14a}$$

$$\mathcal{G}_2=8C\frac{\Lambda^4}{z^2}, \tag{3.14b}$$

again in full agreement with clustering; note that (3.14b) is consistent with our assumption of $\Phi^2_{(2)}$ having no vev classically: since by assumption $\Lambda^4/z^2\ll\Lambda^2$, the vev for $\Phi^2_{(2)}$ is a pure quantum effect.

The results (3.9) and (3.14a) do not depend on z; nonetheless they are only obtained if $z\neq 0$ is assumed. If we put $z=0$ from the beginning, the above results are replaced by

$$\mathcal{G}^{(0)}_{01}=C_0\Lambda^5\left(-\frac{6}{\pi^2}\right)\frac{4}{\pi^2}(y_2-y_1)^2$$
$$\cdot\int d^4x_0\,d\rho^2\rho^6 f^4(y_1-x_0)f^3(y_2-x_0)$$
$$\mathcal{G}^{(0)}_{12}=C_0\Lambda^5\left(\frac{4}{\pi^2}\right)^2(y_2-y_1)^2$$
$$\cdot\int d^4x_0\,d\rho^2\rho^4 f^3(y_1-x_0)f^3(y_2-x_0); \tag{3.15}$$

upon use of the formula

$$\int d^4x\,d\rho^2\rho^{2k}[(y_1-x)^2+\rho^2]^{-m}[(y_2-x)^2+\rho^2]^{-n}$$
$$=((y_2-y_1)^2)^{k-n-m+3}\pi^2$$
$$\cdot\frac{k!(k-n+2)!(k-m+2)!(m+n-k-4)!}{(n-1)!(m-1)!(2k-m-n+5)!} \tag{3.16}$$

one arrives at

$$\mathcal{G}^{(0)}_{01}=-\frac{1}{\pi^2}C_0\Lambda^5=\tfrac{1}{2}\mathcal{G}_{01}$$

$$G^{(0)}_{12}=\frac{4}{3\pi^2}C_0\Lambda^4=\tfrac{1}{3}\mathcal{G}_{12} \tag{3.17}$$

and, of course, due to chiral invariance

$$\mathcal{G}^{(0)}_0=\mathcal{G}^{(0)}_2=0. \tag{3.18}$$

The Green functions $\mathcal{G}_0$, $\mathcal{G}_{01}$, $\mathcal{G}_2$ and $\mathcal{G}_{12}$ may be also evaluated in a theory with arbitrary numbers $K\geqq 2$ of massive fundamental [5] and $L\geqq 1$ of massive adjoint flavors; the integrations over the additional collective coordinates are trivial:

$$\int d^2\bar\tau\exp(\bar\tau^2 m)=m$$

$$\int d^2\chi\,d^2\bar\chi\exp\left(\frac{\sqrt{2}}{\rho}(\chi^2+2\rho^2\bar\chi^2)M\right)=4M^2 \tag{3.19}$$

We thus get

$$\mathcal{G}_{01}(K,L)=\mathcal{G}_{01}(K=1,L=0)$$
$$\prod_{k=2}^{K}\frac{m_k}{\Lambda}\cdot\prod_{l=1}^{L}\left(\frac{2M_l}{\Lambda}\right)^2$$
$$\mathcal{G}_{12}(K,L)=\mathcal{G}_{12}(K=2,L=0)$$
$$\prod_{k=3}^{K}\frac{m_k}{\Lambda}\cdot\prod_{l=1}^{L}\left(\frac{2M_l}{\Lambda}\right)^2 \tag{3.20}$$

and analogously for $\mathcal{G}_0(K,L)$ and $\mathcal{G}_2(K,L)$.

Applying the same procedure to $\mathcal{G}_{00}$ gives

$$\mathcal{G}_{00}(K,L)=\prod_{k=2}^{K}\frac{m_k}{\Lambda}\prod_{l=1}^{L}\left(\frac{2M_l}{\Lambda}\right)^2\cdot C\frac{\Lambda^5}{z^2}$$
$$\int d^4x_0\frac{d\rho^2}{\rho^2}d^2\Theta_0\,d^2\bar\vartheta\,d^2\bar\Theta_0\exp(-4\pi^2z^2\bar\rho^2)$$
$$\cdot\exp(4\pi^2z^2\rho^2\bar\Theta_0^2m_1)\left(\frac{g^2}{32\pi^2}\right)^2$$
$$W^{2(c)}(y_1,\Theta_1)W^{2(c)}(y_2,\Theta_2)$$
$$=\frac{36}{\pi^4}C\frac{\Lambda^5}{z^2}\cdot\prod_{k=2}^{K}\frac{m_k}{\Lambda}\prod_{l=1}^{L}\left(\frac{2M_l}{\Lambda}\right)^2$$
$$\cdot\int d^4x_0\frac{d\rho^2}{\rho^2}\exp(-4\pi^2z^2\rho^2)$$
$$\cdot[4\pi^2z^2\rho^2m_1(y_2-y_1)^2+(8\pi^2z^2\rho^2)^2(\Theta_2-\Theta_1)^2]$$
$$\cdot\rho^8 f^4(y_1-x_0)f^4(y_2-x_0). \tag{3.21}$$

This is not a constant, but depends on (y_2-y_1), in contradiction to SUSY; thus only some (y_2-y_1)-independent part should be meaningful while the remaining part should be cancelled by other effects. In [7] it was argued that in order to extract this part one should set $z^2=0$; in this case the second term in (3.21) (which was not mentioned in [7]) vanishes and the first term gives

$$\mathcal{G}^{(z^2\to 0)}_{00}(K,L)=\tfrac{16}{5}C\Lambda^6\cdot\prod_{k=1}^{K}\frac{m_k}{\Lambda}\prod_{l=1}^{L}\left(\frac{2M_l}{\Lambda}\right)^2$$
$$=\mathcal{G}_{00}(0,0)\cdot\prod_{k=1}^{K}\frac{m_k}{\Lambda}\prod_{l=1}^{L}\left(\frac{2M_l}{\Lambda}\right)^2. \tag{3.22}$$

However, for small mass m_1 we expect $z^2\sim 1/\sqrt{m_1}$ to be large and $\mathcal{G}_{00}\sim m_1$ (which indeed will come out consistently at the end of this discussion). Thus

$z^2=0$ cannot be the relevant limit; the only other possibility to extract a constant part out of (3.21) is to let $z^2\to\infty$; for large z^2 the above contribution (3.21) vanishes much faster than m_1 and we conclude that in the one-instanton approximation

$$\mathscr{G}_{00}(K,L)=0 \quad \text{for } K\geq 1. \tag{3.23}$$

The same reasoning may be applied to

$$\mathscr{G}_{0k}=\left\langle T\frac{g^2}{32\pi^2}W^2(y_1,\Theta_1)\,\Phi^2_{(k)}(y_2,\Theta_2)\right\rangle,\ k\neq 1 \tag{3.24}$$

which is given by

$$\mathscr{G}_{0k}(K,L)=\prod_{\substack{k'=2\\k'\neq k}}^{K}\frac{m_{k'}}{\Lambda}\prod_{l=1}^{L}\left(\frac{2M_l}{\Lambda}\right)^2\cdot C\frac{\Lambda^4}{z^2}$$
$$\cdot\int d^4x_0\frac{d\rho^2}{\rho^2}[4\pi^2z^2\rho^2m_1(y_2-y_1)^2$$
$$+(8\pi^2z^2\rho^2)^2(\Theta_2-\Theta_1)^2]$$
$$\cdot\,\rho^6 f^4(y_1-x_0)f^3(y_2-x_0). \tag{3.25}$$

In the limit $z^2\to 0$ one gets

$$\mathscr{G}^{(z^2\to 0)}_{0k}(K,L)=-4C\Lambda^5\cdot\prod_{\substack{k'=2\\k'\neq k}}^{K}\frac{m_{k'}}{\Lambda}\prod_{l=1}^{L}\left(\frac{2M_l}{\Lambda}\right)^2$$
$$=\mathscr{G}^{(0)}_{01}(1,0)\cdot\prod_{\substack{k'=2\\k'\neq k}}^{K}\frac{m_{k'}}{\Lambda}\prod_{l=1}^{L}\left(\frac{2M_l}{\Lambda}\right)^2 \tag{3.26}$$

while in the limit $z^2\to\infty$ again

$$\mathscr{G}_{0k}(K,L)=0 \quad \text{for } k,K\geq 2. \tag{3.27}$$

Let us next consider the theory with one matter field Ψ in the adjoint representation and compute

$$\mathscr{H}_0=\left\langle\frac{g^2}{32\pi^2}W^2(y,\Theta)\right\rangle$$
$$\mathscr{H}_1=\langle\Psi^2(y,\Theta)\rangle$$
$$\mathscr{H}_{11}=\langle T\Psi^2(y_1,\Theta_1)\,\Psi^2(y_2,\Theta_2)\rangle. \tag{3.28}$$

For $\mathscr{H}_0$ and $\mathscr{H}_1$ one easily obtains

$$\mathscr{H}_0=-16CM\frac{\Lambda^4}{h^2}$$
$$\mathscr{H}_1=h^2+32C\frac{\Lambda^4}{h^2} \tag{3.29}$$

where the first term in $\mathscr{H}_1$ is the zero-instanton contribution. The one-instanton contribution to $\mathscr{H}_{11}$ is

$$\bar{\mathscr{H}}_{11}=C\frac{\Lambda^4}{h^2}\int d^4x_0\frac{d\rho^2}{\rho^2}d^2\Theta_0\,d^2\bar{\vartheta}\,d^2\chi_0\,d^2\bar{\Theta}_0$$
$$\cdot\exp(-4\pi^2h^2\tilde{\rho}^2)\exp(4M(\chi_0^2+\pi^2h^2\rho^2\bar{\Theta}_0^2))$$
$$\Psi^{2(c)}(y_1,\Theta_1)\,\Psi^{2(c)}(y_2,\Theta_2) \tag{3.30}$$

with $\Psi^{2(c)}$ given by (2.34 d); the χ_0-integration is trivial; for the Θ_0-integration we use the shift (3.7) and obtain

$$\bar{\mathscr{H}}_{11}=\frac{16h^2}{\pi^2}\cdot C\frac{\Lambda^4}{h^2}\int\frac{d\rho^2}{\rho^2}d^2\bar{\vartheta}\,d^2\bar{\Theta}_0\exp(-4\pi^2h^2\tilde{\rho}^2)$$
$$\tilde{\rho}^4\int d^4x_0'[3x_0'^4\tilde{f}^2(x_0')\tilde{f}^4(y_2-y_1-x_0')$$
$$+3(y_2-y_1-x_0')^4\tilde{f}^2(y_2-y_1-x_0')\tilde{f}^4(x_0')$$
$$-\tfrac{4}{3}\tilde{f}^3(x_0')\tilde{f}^3(y_2-y_1-x_0')$$
$$\cdot(x_0'^2(y_2-y_1-x_0')^2-4(x_0'^2-x_0'\cdot(y_2-y_1))^2)] \tag{3.31}$$

where $\tilde{f}(x):=(x^2+\tilde{\rho}^2)^{-1}$. The integrand is again a function of only $\tilde{\rho}$, thus we may use (3.8).

Only the $\tilde{f}^4$ singularities in the first two terms contribute for $\tilde{\rho}^2\to 0$ and we get

$$\bar{\mathscr{H}}_{11}=64C\Lambda^4 \tag{3.32}$$

or, including the zero-instanton contribution

$$\mathscr{H}_{11}=h^4+64C\Lambda^4. \tag{3.33}$$

As in the case of matter in the fundamental representation, the limit $h^2\to 0$ does not in general reproduce the $h^2=0$ value: if we had put $h^2=0$ from the beginning, we would have obtained

$$\mathscr{H}^{(0)}_{11}=\frac{8}{\pi^2}C_0\Lambda^4=\tfrac{1}{2}\bar{\mathscr{H}}_{11}. \tag{3.34}$$

Next we consider, still in the theory with one adjoint flavor, the Green functions

$$\mathscr{H}_{01}=\left\langle T\frac{g^2}{32\pi^2}W^2(y_1,\Theta_1)\,\Psi^2(y_2,\Theta_2)\right\rangle$$
$$\mathscr{H}_{00}=\left\langle T\frac{g^2}{32\pi^2}W^2(y_1,\Theta_1)\frac{g^2}{32\pi^2}W^2(y_2,\Theta_2)\right\rangle \tag{3.35}$$

which turn out to be

$$\mathscr{H}_{01}=-\frac{6}{\pi^2}C\frac{\Lambda^4}{h^2}\int d^4x_0\frac{d\rho^2}{\rho^2}$$
$$\Big[\exp(-4\pi^2h^2\rho^2)\cdot\frac{48}{\pi^2}\rho^8f^4(y_1-x_0)f^4(y_2-x_0)$$
$$\cdot[4M\pi^2h^2\rho^2(y_2-y_1)^2+(8\pi^2h^2\rho^2)^2(\Theta_2-\Theta_1)^2]$$
$$+\int d^2\bar{\vartheta}\,d^2\bar{\Theta}_0\exp(-4\pi^2h^2\tilde{\rho}^2)\cdot 4Mh^2$$
$$\cdot\,\tilde{\rho}^4x_0^4\tilde{f}^2(x_0)\tilde{f}^4(y_2-y_1-x_0)]$$
$$\mathscr{H}_{00}=\frac{36}{\pi^4}C\frac{\Lambda^4}{h^2}\int d^4x_0\frac{d\rho^2}{\rho^2}$$
$$\cdot\exp(-4\pi^2h^2\rho^2)\cdot\rho^8f^4(y_1-x_0)f^4(y_2-x_0)$$
$$\cdot[16M^2\pi^2h^2\rho^2(y_2-y_1)^2+4M(8\pi^2h^2\rho^2)^2$$
$$\cdot(\Theta_2-\Theta_1)^2]. \tag{3.36}$$

In the limit $h^2 \to 0$, (3.36) becomes

$$\mathscr{H}_{01}^{(h^2\to 0)} = -\frac{6}{\pi^2} C \frac{\Lambda^4}{h^2} \int d^4x_0 \frac{d\rho^2}{\rho^2} \cdot \frac{48}{\pi^2} \rho^8 f^4(y_1-x_0) f^4(y_2-x_0) \cdot 4M\pi^2 h^2 \rho^2 (y_2-y_1)^2 = -\tfrac{128}{5} CM\Lambda^4$$

$$\mathscr{H}_{00}^{(h^2\to 0)} = \frac{36}{\pi^4} C \frac{\Lambda^4}{h^2} \int d^4x_0 \frac{d\rho^2}{\rho^2} \cdot \rho^8 f^4(y_1-x_0) f^4(y_2-x_0) \cdot 16M^2\pi^2 h^2 \rho^2 (y_2-y_1)^2 = \tfrac{64}{5} CM^2\Lambda^4. \tag{3.37}$$

Had we put $h^2=0$ from the beginning, the result would have been $\mathscr{H}_{01}^{(0)} = -\frac{8\cdot 23}{5} CM\Lambda^4$, $\mathscr{H}_{00}^{(0)} = \mathscr{H}_{00}^{(h^2\to 0)}$.

For $h^2 \to \infty$ (3.36) becomes

$$\mathscr{H}_{01} = -\frac{6}{\pi^2} C \frac{\Lambda^4}{h^2} \int d^4x_0 \frac{d\rho^2}{\rho^2} d^2\bar{\vartheta} d^2\bar{\Theta}_0 \cdot \exp(-4\pi^2 h^2 \bar{\rho}^2) x_0^4 \bar{f}^2(x_0) \bar{f}^4(y_2-y_1-x_0) \cdot 4Mh^2 = -16CM\Lambda^4 = h^2 \cdot \mathscr{H}_0$$

$$\mathscr{H}_{00} = 0. \tag{3.38}$$

If a second adjoint flavor with vanishing classical vev is added, the above Green functions are multiplied by $(2M_2/\Lambda)^2$ (compare (3.19)); in addition we obtain

$$\mathscr{H}_2 \equiv \langle \Psi_{(2)}^2(y,\Theta) \rangle = 256 CM_1 M_2 \frac{\Lambda^2}{h^2} \tag{3.39}$$

and

$$\mathscr{H}_{12} = \langle T\Psi_{(1)}^2(y_1,\Theta_1)\, \Psi_{(2)}^2(y_2,\Theta_2) \rangle$$
$$\mathscr{H}_{22} = \langle T\Psi_{(2)}^2(y_1,\Theta_1)\, \Psi_{(2)}^2(y_2,\Theta_2) \rangle \tag{3.40}$$

become, for $h^2 \to 0$,

$$\mathscr{H}_{12}^{(h^2\to 0)} = \tfrac{23\cdot 64}{5} M_1 M_2 C\Lambda^2$$
$$\mathscr{H}_{22}^{(h^2\to 0)} = 128 M_1^2 C\Lambda^2 \tag{3.41}$$

while for $h^2 \to \infty$

$$\mathscr{H}_{12} = 256 CM_1 M_2 \Lambda^2 = h^2 \cdot \mathscr{H}_2$$
$$\mathscr{H}_{22} = 0. \tag{3.42}$$

Finally let us turn to a theory with one fundamental and one adjoint flavor and assume that the fundamental flavor has a classical vev (the case that the adjoint flavor possesses a classical vev is treated analogously).

We find

$$\langle \Psi^2(y,\Theta) \rangle = 64CM \frac{\Lambda^3}{z^2},$$
$$\langle T\Phi^2(y_1,\Theta_1)\, \Psi^2(y_2,\Theta_2) \rangle = 128CM\Lambda^3 = 2z^2 \langle \Psi^2 \rangle \tag{3.44}$$

while putting $z^2=0$ would yield $\langle \Psi^2 \rangle^{(0)}=0$ and $\langle T\Phi^2 \Psi^2 \rangle^{(0)} = 48CM\Lambda^3$.

4. Consistency Relations

In order to check the consistency of the results obtained in Sect. 3, one may look at the clustering properties and, in addition, use the Konishi anomaly [9].

We first assume that a fundamental flavor, $\Phi_{(1)}$, develops a large classical vev; in this case we see from (3.12), (3.14), (3.20), and (3.44) that for any numbers K, L of flavors the clustering

$$\langle TW^2 \Phi_{(1)}^2 \rangle = 2z^2 \langle W^2 \rangle = \langle W^2 \rangle \langle \Phi_{(1)}^2 \rangle$$
$$\langle T\Phi_{(1)}^2 \Phi_{(k)}^2 \rangle = 2z^2 \langle \Phi_{(k)}^2 \rangle = \langle \Phi_{(1)}^2 \rangle \langle \Phi_{(k)}^2 \rangle$$
$$\langle T\Phi_{(1)}^2 \Psi_{(l)}^2 \rangle = 2z^2 \langle \Psi_{(l)}^2 \rangle = \langle \Phi_{(1)}^2 \rangle \langle \Psi_{(l)}^2 \rangle \tag{4.1}$$

is fulfilled. In contrast, from (3.23) and (3.27) it appears that $\mathscr{G}_{00}$ and $\mathscr{G}_{0k}$, $k \neq 1$, do not cluster; however, since there are 1-instanton contributions to both $\langle W^2 \rangle$ and $\langle \Phi_{(k)}^2 \rangle$, $k \neq 1$, one expects 2-instanton contributions to $\mathscr{G}_{00}$ and $\mathscr{G}_{0k}$; clustering should hold only if these are taken into account (there are no 2-instanton contributions to $\mathscr{G}_{01}$ and $\mathscr{G}_{1k}$ since there is no 1-instanton contribution to $\mathscr{G}_1$). Note that we have to include 2-instanton effects even though for $z \neq 0$ the theory is weakly coupled.

We are thus led to perform a 2-instanton calculation for $\mathscr{G}_{00}$ (the case of $\mathscr{G}_{0k}$ is completely analogous); we work in the theory with just one fundamental flavor, since adding more matter is trivial. We then have, in the dilute gas approximation [17, 18], i.e. neglecting the interaction energy between the two instantons*

$$\mathscr{G}_{00} = \frac{1}{2}\left(C\frac{\Lambda^5}{z^2}\right)^2 \int d\sigma_{inst}^{(1)} d\sigma_{inst}^{(2)} \cdot \frac{g^2}{32\pi^2} W^{2(c)}(y_1,\Theta_1) \frac{g^2}{32\pi^2} W^{2(c)}(y_2,\Theta_2). \tag{4.2}$$

* Note that there is no contribution from the denominator of the generating functional, due to the integration over the fermionic collective coordinates. In terms of Feynman graphs the two-instanton contribution corresponds to a disconnected diagram

J. Fuchs and M.G. Schmidt: Instanton Induced Green Functions in the Superfield Formalism 16

Here $d\sigma^{(i)}_{\text{inst}}$ is the integration measure for the i-th instanton; $W^{2(c)}$ is now the classical superfield for the two-instanton case and is constructed in complete analogy to the one-instanton case, namely by starting with

$$W_\alpha^{a(0)}(y,\Theta)=-i\sigma^{mn}{}_\alpha{}^\beta\Theta_\beta v^{(c)}_{mn}(y)$$
$$=-i\sigma^{mn}{}_\alpha{}^\beta\Theta_\beta\cdot\Big[\sum_{i=1}^{2}v^{(c;i)}_{mn}(y)$$
$$+2g\varepsilon^{abc}v^{b(c;1)}_m v^{c(c;2)}_n\Big] \quad (4.3)$$

and then performing Q and $\bar S$ transformations with independent parameters $(-\Theta_{0(1)},-\bar\vartheta_{(1)})$ and $(-\Theta_{0(2)},-\bar\vartheta_{(2)})$; the result is

$$\frac{g^2}{32\pi^2}W^{2(c)}(y,\Theta)=-\frac{6}{\pi^2}\Big[\sum_{i=1}^{2}\rho^4_{(i)}f^4_{(i)}(y-x_0)\tilde\Theta^2_{(i)}$$
$$+\tfrac{4}{3}\rho^2_{(1)}\rho^2_{(2)}f^2_{(1)}(y-x_0)f^2_{(2)}(y-x_0)$$
$$\left(\frac{((y-x_{0(1)})\cdot(y-x_{0(2)}))^2}{(y-x_{0(1)})^2(y-x_{0(2)})^2}-\frac{1}{4}\right)\tilde\Theta_{(1)}\cdot\tilde\Theta_{(2)}$$
$$+2[\rho^4_{(1)}\rho^4_{(2)}f^2_{(1)}(y-x_0)f^2_{(2)}(y-x_0)$$
$$-\rho^2_{(1)}\rho^2_{(2)}f_{(1)}(y-x_0)f_{(2)}(y-x_0)$$
$$\cdot(\rho^2_{(1)}f^2_{(1)}(y-x_0)+\rho^2_{(2)}f^2_{(2)}(y-x_0))(y-x_{0(1)})$$
$$\cdot(y-x_{0(2)})]$$
$$\cdot\frac{1}{(y-x_{0(1)})^2(y-x_{0(2)})^2}\cdot\Theta^2] \quad (4.4)$$

where

$$f_{(i)}(x)=(x^2+\rho^2_{(i)})^{-1}$$
$$\tilde\Theta_{(i)}=\Theta-\Theta_{0(i)}-(y-x_{0(i)})_m\sigma^m\bar\vartheta_{(i)}. \quad (4.5)$$

Substituting (4.4) into (4.2) and performing the $\Theta_{0(i)}$-integrations, one arrives at

$$\mathscr{G}_{00}=\frac{1}{2}\left(C\frac{\Lambda^5}{z^2}\right)^2\cdot\frac{36}{\pi^4}\Big[\prod_{i=1}^{2}\int d^4x_{0(i)}\frac{d\rho^2_{(i)}}{\rho^2_{(i)}}$$
$$\cdot d^2\bar\vartheta_{(i)}d^2\tilde\Theta_{0(i)}\exp(-4\pi^2z^2\bar\rho^2_{(i)})\Big]$$
$$\cdot\rho^4_{(1)}\rho^4_{(2)}\Big[f^4_{(1)}(y_1-x_{0(1)})f^4_{(2)}(y_2-x_{0(2)})$$
$$+f^4_{(2)}(y_1-x_{0(2)})f^4_{(1)}(y_2-x_{0(1)})$$
$$-\frac{1}{2}\left(\frac{4}{3}\right)^2f^2_{(1)}(y_1-x_{0(1)})f^2_{(2)}(y_1-x_{0(2)})$$
$$\cdot f^2_{(1)}(y_2-x_{0(1)})f^2_{(2)}(y_2-x_{0(2)})$$
$$\left(\frac{((y_1-x_{0(1)})\cdot(y_1-x_{0(2)}))^2}{(y_1-x_{0(1)})^2(y_1-x_{0(2)})^2}-\frac{1}{4}\right)$$
$$\left(\frac{((y_2-x_{0(1)})\cdot(y_2-x_{0(2)}))^2}{(y_2-x_{0(1)})^2(y_2-x_{0(2)})^2}-\frac{1}{4}\right)\Big]. \quad (4.6)$$

The third term is (y_2-y_1)-dependent, so that, as argued above, one should take the limit $z^2\to\infty$ upon which its contribution vanishes; the first and second term give

$$\mathscr{G}_{00}=\frac{1}{2}\left(C\frac{\Lambda^5}{z^2}\right)^2\cdot\frac{36}{\pi^4}\cdot2\cdot\Big[\int d^4x_0\frac{d\rho^2}{\rho^2}d^2\bar\vartheta d^2\tilde\Theta_0$$
$$\cdot\exp(-4\pi^2z^2\bar\rho^2)\rho^4f^4(y-x_0)\Big]^2$$
$$=\left(4C\frac{\Lambda^5}{z^2}\right)^2; \quad (4.7)$$

in contrast, at $z=0$, the two-instanton contribution to $\mathscr{G}_{00}$ clearly vanishes. Comparing with (3.11), we thus find for $z\neq0$ the clustering

$$\langle TW^2W^2\rangle=\langle W^2\rangle\langle W^2\rangle. \quad (4.8)$$

We may also derive from (3.11), (3.14b) and (3.20) the relation

$$m_k\mathscr{G}_k(K,L)=-2\mathscr{G}_0(K,L),\quad k\neq1. \quad (4.9)$$

This is nothing but the Konishi identity inside Green functions of lowest component chiral fields:

$$m\phi^2|_{\Theta=0}=-\frac{g^2}{16\pi^2}W^2|_{\Theta=0}. \quad (4.10)$$

Since in all the examples considered above clustering is fulfilled trivially, we are not able to extract from the requirement of clustering the actual value of z^2; we need, instead, the Konishi identity (4.10) for the field $\Phi_{(1)}$; combining (3.20) and (4.10) we get consistently

$$z^2=\pm2\sqrt{C}\cdot\frac{\Lambda^3}{m_1}\cdot\Pi \quad (4.11)$$

where Π stands for

$$\Pi=\prod_{k=1}^{K}\left(\frac{m_k}{\Lambda}\right)^{1/2}\prod_{l=1}^{L}\frac{2M_l}{\Lambda}, \quad (4.12)$$

i.e.

$$\langle\Phi^2_{(k)}\rangle=\pm4\sqrt{C}\frac{\Lambda^3}{m_k}\cdot\Pi$$
$$\langle W^2\rangle=\mp\frac{64\Pi^2}{g^2}\sqrt{C}\Lambda^3\cdot\Pi. \quad (4.13)$$

The twofold degeneracy of the vev's are in agreement with Witten index calculations [19]. Also note that the assumption of one large vev $z^2\gg\Lambda^2$ that we made in Sect. 3 is fulfilled for small enough m_1 (i.e. $m_1\ll\Lambda$ for a theory with one fundamental flavor, $m_1\ll m_2$ for two fundamental flavors, etc.); in particular it is not possible to get two fields with vev's $\gg \Lambda$ as long as all masses are smaller than Λ.

We may also deduce from (3.43) the vev's for matter in the adjoint representation (for $z \neq 0$):

$$\langle \Psi^2_{(l)} \rangle = \pm 8\sqrt{C}\,\frac{\Lambda^3}{M_l} \cdot \Pi. \tag{4.14}$$

We have thus derived the form of the anomaly relation for fields in the adjoint representation:

$$M\Psi^2|_{\Theta=0} = -\frac{g^2}{8\pi^2} W^2|_{\Theta=0}. \tag{4.15}$$

It is already apparent from (4.14) that it is improbable for adjoint matter to get a large vev. However, let us make for the moment just this assumption, i.e. that $\Psi^2_{(1)}$ develops a large classical vev $h^2 \gg \Lambda^2$; then the 1-instanton contributions to $\mathscr{H}_1$ and $\mathscr{H}_{11}$ are negligible, and we find from (3.38) and (3.42)

$$\begin{aligned} &\langle T W^2 \Psi^2_{(1)} \rangle = h^2 \langle W^2 \rangle = \langle W^2 \rangle \langle \Psi^2_{(1)} \rangle \\ &\langle T \Psi^2_{(1)} \Psi^2_{(l)} \rangle = h^2 \langle \Psi^2_{(l)} \rangle = \langle \Psi^2_{(1)} \rangle \langle \Psi^2_{(l)} \rangle \end{aligned} \tag{4.16}$$

and, taking into account two-instanton contributions to $\mathscr{H}_{00}$ and $\mathscr{H}_{22}$ which can be evaluated in complete analogy to the case of $\mathscr{G}_{00}$,

$$\begin{aligned} &\langle T W^2 W^2 \rangle = \langle W^2 \rangle \langle W^2 \rangle \\ &\langle T \Psi^2_{(l)} \Psi^2_{(l)} \rangle = \langle \Psi^2_{(l)} \rangle \langle \Psi^2_{(l)} \rangle. \end{aligned} \tag{4.17}$$

However, if we now use the Konishi identity (4.15), we find from (3.29)

$$h^2 = \pm 4\sqrt{C}\,\frac{\Lambda^3}{M_1} \cdot \Pi \tag{4.18}$$

i.e. for small masses our assumption $h^2 \gg \Lambda^2$ is inconsistent.

The failure to produce a large value for h^2 can be easily understood in the effective Lagrangian approach. For a single fundamental flavor the effective superpotential due to instantons must be of the form [3, 8] $\mathscr{L}_{\text{eff}} = \int d^2\Theta F(\Phi^2)$; from the transformation law of Φ^2 under the nonanomalous combination of R symmetry and chiral symmetry, $\Phi^2 \to e^{-i \cdot 2\alpha} \Phi^2$ (and $\Theta \to e^{i\alpha}\Theta$) the function F is fixed to be $F = c \cdot \Lambda^5/\Phi^2$ with some constant c; clearly, this superpotential favours large values of Φ^2. In contrast, for a single adjoint flavor, the transformation law under the nonanomalous $U(1)$ turns out to be $\Psi^2 \to \Psi^2$; therefore no superpotential exists at all. It is easily seen that if further matter fields and interactions are added, Ψ^2 will always be either invariant or transform with a positive power of $e^{i\alpha}$, so in no case a superpotential favouring large values of Ψ^2 exists.

An additional consistency check is provided by the consideration of "non-diagonal" Green functions, i.e. Green functions which upon clustering would yield vev's for composite operators which break some global symmetry of the theory such as flavor symmetries (e.g. through a vev for $\langle T\Phi_{(1)A}\Phi_{(2)B}(y_1, \Theta_1)\Phi^A_{(1)}\Phi^B_{(2)}(y_2, \Theta_2)\rangle$) or Lorentz symmetry (e.g. through a vev for

$$\langle T W^{\alpha a} \Psi^a (y_1, \Theta_1) W^b_\alpha \Psi^b (y_2, \Theta_2) \rangle);$$

all those Green functions should turn out to be zero. Indeed, calculating the instanton contribution to such Green functions using the same methods as above, either gives identically zero or reduces to space-time dependent terms which vanish in the relevant limit $z^2 \to \infty$ (or $h^2 \to \infty$)*.

Let us finally add some remarks on the case $h = z = 0$. In this case we cannot expect that Green functions like $\langle T W^2 W^2 \rangle$ etc. cluster since due to chiral invariance instantons do not contribute to $\langle W^2 \rangle$, but one may hope that relations as

$$\frac{\mathscr{G}_{00} \cdot \mathscr{G}_{12}}{\mathscr{G}_{01} \cdot \mathscr{G}_{02}} = \frac{\langle T W^2 W^2 \rangle \langle T \Phi^2_{(1)} \Phi^2_{(2)} \rangle}{\langle T W^2 \Phi^2_{(1)} \rangle \langle T W^2 \Phi^2_{(2)} \rangle} = 1 \tag{4.19}$$

(for $K \geq 2$, L arbitrary) and

$$\frac{\mathscr{H}_{00} \cdot \mathscr{H}_{11}}{\mathscr{H}_{01} \cdot \mathscr{H}_{01}} = \frac{\langle T W^2 W^2 \rangle \langle T \Psi^2 \Psi^2 \rangle}{(\langle T W^2 \Psi^2 \rangle)^2} = 1 \tag{4.20}$$

still hold. Inspection of the results of Sect. 3 shows, however, that this is not the case:

$$\begin{aligned} &\frac{\mathscr{G}^{(0)}_{00} \cdot \mathscr{G}^{(0)}_{12}}{\mathscr{G}^{(0)}_{01} \cdot \mathscr{G}^{(0)}_{02}} = \frac{16}{15} \\ &\frac{\mathscr{H}^{(0)}_{00} \cdot \mathscr{H}^{(0)}_{11}}{\mathscr{H}^{(0)}_{01} \cdot \mathscr{H}^{(0)}_{01}} = \frac{5 \cdot 32}{23^2}. \end{aligned} \tag{4.21}$$

Of course, the results for $h = z = 0$ are then also inconsistent with the Konishi identity. These inconsistencies are not too surprising if one realizes that for $h = z = 0$ the theory is in the strong coupling regime. It has been proposed, however, [6] that for $h = z = 0$ and finite masses the non-zero modes of bosons and fermions should not cancel, signalling the existence of mass singularities associated with

* In [7] $\langle T\Phi_{(1)A}\Phi_{(2)B}(y_1, \Theta_1)\Phi^A_{(1)}\Phi^B_{(2)}(y_2, \Theta_2)\rangle$ was calculated to be identically zero, using (in our notation)

$$\Phi'_{(1)A}\Phi_{(2)\alpha B} = \frac{2i}{\pi} z\rho \bar{\Theta}^{\dot\alpha} \sigma^m_{\dot\alpha A} \bar{\tau}_B (\tilde{y} - x_0)_m f^2(\tilde{y} - x_0);$$

however, according to (2.19) and (2.32), in this expression $f^2(\tilde{y} - x_0)$ should be replaced by $f^{1/2}(\tilde{y} - x_0) \cdot f^{3/2}(y - x_0)$; this difference leads to additional terms which are precisely of the form described in the text (space-time dependent, and zero for $z^2 \to \infty$)

non-zero modes. Including such contributions*, consistency may be also obtained in the strong coupling phase.

5. Conclusions

The supersymmetric instanton calculus can be very effectively used to calculate various Green functions including also chiral matter fields in the adjoint representation. In the approach without large vev contradiction with factorization and the anomaly relation is apparent if one calculates in a naive expansion for small masses. For the case of a large vev for a field in the fundamental representation consistency can be obtained as advertised in [7]. This remains true adding matter fields in the adjoint representation without large vev. The two instanton contributions are most important for factorization in our version since we argue that the connected Green functions disturbing factorization are just zero. The Konishi anomaly relation has to be put in for the chiral fields with large scalar vev, but it comes out consistently for the rest of the fundamental and adjoint flavors.

Note added. We have not performed calculations in this paper taking into account mass singularities as proposed in [6]; indeed in the recent preprint [26] calculations at distances $\frac{1}{\Lambda} > |x-y| > \frac{1}{m}$ also lead to a consistent picture and are argued to be m-independent. However, the results in the two approaches at small m are not the same. Considering only the technical side of course superfield methods should be useful in both cases.

Acknowledgement. We would like to thank Y. Meurice for very useful discussions and correspondence.

Appendix A: Notations

Throughout the paper we work in an euclideanized version of the Wess-Bagger [20] notation. The σ matrices are

$$\sigma^m = (i, \sigma)$$
$$\bar{\sigma}^m = (i, -\sigma). \tag{A.1}$$

The combinations

$$\sigma^{mn} = \tfrac{1}{4}(\sigma^m \bar{\sigma}^n - \sigma^n \bar{\sigma}^m)$$
$$\bar{\sigma}^{mn} = \tfrac{1}{4}(\bar{\sigma}^m \sigma^n - \bar{\sigma}^n \sigma^m) \tag{A.2}$$

are the generators of Lorentz transformations for the $(\frac{1}{2}, 0)$ - and $(0, \frac{1}{2})$ - representations of the euclidean Lorentz group $SO(4) \sim SU(2) \times SU(2)$, respectively.

* We were kindly informed by Y. Meurice about work of D. Amati, Y. Meurice, G.C. Rossi, and G. Veneziano prior to the recent publication [26] where this point of view is strengthened

They are related to the generators $\frac{1}{2}\sigma^a$ of the gauge $SU(2)$ by

$$2i\sigma^a = \eta^a_{mn}\,\sigma^{mn}$$
$$2i\sigma^a = \bar{\eta}^a_{mn}\,\bar{\sigma}^{mn} \tag{A.3}$$

where η, $\bar{\eta}$ are the 't Hooft symbols [12, 21].

In the Wess-Zumino gauge the gauge multiplet is described by the chiral superfield

$$W_\alpha(y,\Theta) = -i\lambda_\alpha(y) - i\sigma^{mn}{}_\alpha{}^\beta\,\Theta_\beta v_{mn}(y) + \Theta_\alpha D(y) + \Theta^2 \sigma^m_{\alpha\dot{\alpha}}\,\mathscr{D}_m \bar{\lambda}^{\dot{\alpha}}(y) \tag{A.4}$$

where $\mathscr{D}_m$ denotes the gauge covariant derivative and y is the chiral coordinate $y^m = x^m + i\Theta\sigma^m\bar{\Theta}$. The chiral superfields for matter in the fundamental representation of the gauge $SU(2)$ are denoted by

$$\Phi(y,\Theta) = Z(y) + \sqrt{2}\,\Theta\varphi(y) + \Theta^2 F(y) \tag{A.5}$$

and those for matter in the adjoint representation by

$$\Psi(y,\Theta) = H(y) + \sqrt{2}\,\Theta\psi(y) + \Theta^2 G(y). \tag{A.6}$$

We always use the vector notation, e.g. Ψ^a, for fields in the adjoint representation rather than the matrix notation $\Psi_i^j = \left(\frac{\sigma^a}{2}\right)_i^j \Psi^a$; in particular Ψ^2 means $\Psi^a\Psi^a = 2\Psi_i^j\Psi_j^i$. Since the representations of $SU(2)$ are pseudoreal, each QCD-type flavor of matter in the fundamental representation carries, in addition to its color index i, an index A of a global $SU(2)$; by Φ^2 we will denote the field obtained from (A.5) by contraction over both color and flavor indices: $\Phi^2 = \varepsilon^{ij}\varepsilon^{AB}\Phi_{iA}\Phi_{jB}$ (in the literature frequently the notation Φ, $\tilde{\Phi}$ for the flavor components Φ_1, Φ_2 is used; note that $\Phi^2 \triangleq 2\Phi\tilde{\Phi}$).

With these notations, the euclidean Lagrangian of $SU(2)$ SUSY gauge theory with K flavors of fundamental and L flavors of adjoint matter reads

$$\mathscr{L} = \mathscr{L}_{\text{SYM}} + \mathscr{L}_{\text{matter}} + \mathscr{L}_{\text{mass}}$$

with

$$\mathscr{L}_{\text{SYM}} = -\tfrac{1}{4}\int d^2\Theta\, W^2 + \text{h.c.},$$

$$\mathscr{L}_{\text{matter}} = \int d^4\Theta \left[\sum_{k=1}^{K} \bar{\Phi}_{(k)} e^{2gV} \Phi_{(k)} + \sum_{l=1}^{L} \bar{\Psi}_{(l)} e^{2gV} \Psi_{(l)}\right],$$

$$\mathscr{L}_{\text{mass}} = -\tfrac{1}{2}\int d^2\Theta \left[\sum_{k=1}^{K} m_k \Phi^2_{(k)} + \sum_{l=1}^{L} M_l \Psi^2_{(l)}\right] + \text{h.c.} \tag{A.7}$$

(in euclidean space "h.c." means Osterwalder-Schrader conjugation [22]; in particular the fermionic components of Φ and $\bar{\Phi}$, etc., are not related by complex conjugation and thus represent independent field variables).

Our notation for the Green functions is as follows: They are denoted by $\mathscr{G}$ if a fundamental flavor has a large classical vev and by $\mathscr{H}$ if an adjoint flavor carries such a vev. The fields of which the Green functions are composed can be read off the subscripts which are 0 for $\frac{g^2}{32\pi^2} W^2$ and $k, l \geq 1$ for $\Phi^2_{(k)}$ and $\Psi^2_{(l)}$, respectively; e.g. $\mathscr{G}_0 = \left\langle \frac{g^2}{32\pi^2} W^2 \right\rangle$, $\mathscr{G}_{12} = \langle T\Phi^2_{(1)} \Phi^2_{(2)} \rangle$ etc. If it is not already clear from the context, we also specify the theory in which the Green function is evaluated by the numbers K, L of fundamental and adjoint flavors, i.e. $\mathscr{G}, \mathscr{H} \to \mathscr{G}(K,L)$, $\mathscr{H}(K,L)$.

Appendix B: The Instanton Measure

The instanton measure of pure $SU(2)$ gauge theory has been given by 't Hooft [12]:

$$d\sigma_{\text{inst}} = 2^{10} \pi^6 g^{-8} (\mu\rho)^8 \exp\left(-\frac{8\pi^2}{g^2}\right) d^4 x_0 \frac{d\rho}{\rho^5} \tag{B.1}$$

where x_0 and ρ are the collective coordinates for translations and dilatations, μ is the renormalization point and $g = g(\mu)$. If gluinos and supersymmetric matter are added, there are additional contributions from the fermion zero modes: introducing a fermionic collective coordinate γ which parametrizes the full classical solution $\Phi^{(c)}$ in terms of the fermion zero mode $f^{(0)}$,

$$f^{(0)} = \frac{\partial \Phi^{(c)}}{\partial \gamma}, \tag{B.2}$$

each fermion zero modes gives a factor of $(\mathscr{N}\mu)^{-1/2} d\gamma$ where $\mathscr{N}$ is its normalization; in the cases at hand the collective coordinates γ^1, γ^2 of two zero modes combine to a Weyl spinor coordinate so that each pair of zero modes gives a factor of

$$(\mathscr{N}\mu)^{-1} d\gamma^1 d\gamma^2 = (\mathscr{N}\mu)^{-1} \cdot 2d^2\gamma. \tag{B.3}$$

The normalizations of the various zero modes can be read off (2.16), (2.20), (2.26) and (2.31): the gluinos give

$$\left(\frac{16\pi^2}{g^2}\mu\right)^{-1} \cdot 2d^2\Theta_0 \cdot \left(\frac{32\pi^2\rho^2}{g^2}\mu\right)^{-1} \cdot 2d^2\bar{\vartheta}, \tag{B.4 a}$$

each flavor of matter in the fundamental representation gives

$$\begin{aligned} &(8\pi^2 z^2 \rho^2 \mu)^{-1} \cdot 2d^2 \bar{\Theta}_0 && \text{for } z \neq 0 \\ &(2\mu)^{-1} \cdot 2d^2 \bar{\tau} && \text{for } z = 0 \end{aligned} \tag{B.4 b}$$

and each adjoint flavor

$$(4\mu)^{-1} \cdot 2d^2 \chi_0 \cdot (4\pi^2 h^2 \rho^2 \mu)^{-1} \cdot 2d^2 \bar{\Theta}_0$$

for $h \neq 0$

$$\left(\frac{\sqrt{2}}{\rho}\mu\right)^{-1} \cdot 2d^2\chi \cdot (2\sqrt{2}\rho\mu)^{-1} \cdot 2d^2\bar{\chi}$$

for $h = 0$. (B.4 c)

In addition there are also contributions to the action, namely through the kinetic terms of the scalars, through Yukawa terms and through mass terms; the contributions of the first and second kind are evaluated to be

$$\begin{aligned} &\int d^4 x (\mathscr{D}^m Z^{(c)})^{+iA} (\mathscr{D}_m Z^{(c)})_{iA} = 2 \cdot 2\pi^2 z^2 \rho^2 \\ &\int d^4 x (\mathscr{D}^m H^{(c)})^{+a} (\mathscr{D}_m H^{(c)})^a = 4\pi^2 h^2 \rho^2, \end{aligned} \tag{B.5 a}$$

$$\begin{aligned} &\int d^4 x i\sqrt{2} g Z^{(c)+iA} \left(\frac{\sigma^a}{2}\right)_i^j \lambda^{(c)a\alpha} \varphi^{(c)}_{\alpha jA} \\ &= 16\pi^2 i z^2 \rho^2 \bar{\vartheta} \bar{\Theta}_0 \\ &\int d^4 x \sqrt{2} g \varepsilon^{abc} H^{(c)+a} \lambda^{(c)b\alpha} \psi^{(c)c}_\alpha \\ &= 16\pi^2 i h^2 \rho^2 \bar{\vartheta} \bar{\Theta}_0 \end{aligned} \tag{B.5 b}$$

while the mass terms give

$$\begin{aligned} &\int d^4 x d^2 \Theta \tfrac{1}{2} m \phi^{(c)}_{iA} \Phi^{(c)}_{jB} \varepsilon^{ij} \varepsilon^{AB} \\ &= \begin{cases} m \cdot 4\pi^2 z^2 \rho^2 \bar{\Theta}_0^2 & \text{for } z \neq 0 \\ m \cdot \bar{\tau}^2 & \text{for } z = 0, \end{cases} \end{aligned} \tag{B.5 c}$$

$$\begin{aligned} &\int d^4 x d^2 \Theta \tfrac{1}{2} M \Psi^{(c)a} \Psi^{(c)a} \\ &= \begin{cases} M \cdot (4\chi_0^2 + 4\pi^2 h^2 \rho^2 \bar{\Theta}_0^2) & \text{for } h \neq 0 \\ M \cdot \left(\frac{\sqrt{2}}{\rho}\chi^2 + 2\sqrt{2}\rho\bar{\chi}^2\right) & \text{for } h = 0. \end{cases} \end{aligned} \tag{B.5 d}$$

Combining (B.1), (B.4) and (B.5), the instanton measure for $SU(2)$ SUSY gauge theory with K massive flavors in the fundamental and L in the adjoint representation of which one fundamental flavor has a large vev z, is found to be

$$\begin{aligned} d\sigma_{\text{inst}} = {} & z^{-4} \mu^{6-K-2L} g^{-8} \exp\left(-\frac{8\pi^2}{g^2}\right) \\ & \int d^4 x_0 \frac{d\rho^2}{\rho^2} d^2\Theta_0 d^2\bar{\vartheta} d^2\bar{\Theta}_0 \exp(-4\pi^2 z^2 \tilde{\rho}^2) \\ & \cdot \exp(4\pi^2 z^2 \rho^2 \bar{\Theta}_0^2 m_1) \\ & \prod_{k=2}^{K} d^2\bar{\tau}_{(k)} \exp(\bar{\tau}^2_{(k)} m_k) \\ & \prod_{l=1}^{L} d^2\chi_{(l)} d^2\bar{\chi}_{(l)} \exp\left(\frac{\sqrt{2}}{\rho}(\chi^2_{(l)} + 2\rho^2\bar{\chi}^2_{(l)}) M_l\right) \end{aligned} \tag{B.6}$$

where [7]

$$\bar{\rho}^2 := \rho^2 \cdot (1 + 4i\bar{\vartheta}\bar{\Theta}_0). \tag{B.7}$$

If, instead, one adjoint flavor has a large vev h, we have to replace z by h, $d^2\chi_{(1)} d^2\bar{\chi}_{(1)}$ by $d^2\chi_0 d^2\bar{\tau}_{(1)}$ and

$$\exp\left(4\pi^2 z^2 \rho^2 \bar{\Theta}_0^2 m_1 + \frac{\sqrt{2}}{\rho}(\chi_{(1)}^2 + 2\rho^2 \bar{\chi}_{(1)}^2) M_1\right)$$

by

$$\exp(\bar{\tau}_{(1)}^2 m_1 + 4(\chi_0^2 + \pi^2 h^2 \rho^2 \bar{\Theta}_0^2) M_1),$$

while in the case that there is no field with large vev the measure becomes

$$d\sigma_{\text{inst}}^{(0)} = 4\pi^2 \mu^{6-K-2L} g^{-4} \exp\left(-\frac{8\pi^2}{g^2}\right) \cdot \int d^4x_0\, d\rho^2\, d^2\Theta_0\, d^2\bar{\vartheta} \prod_{k=1}^{K} d^2\bar{\tau}_{(k)} \exp(\bar{\tau}_{(k)}^2 m_k) \prod_{l=1}^{L} d^2\chi_{(l)} d^2\bar{\chi}_{(l)} \exp\left(\frac{\sqrt{2}}{\rho}(\chi_{(l)}^2 + 2\rho^2\bar{\chi}_{(l)}^2) M_l\right). \tag{B.8}$$

The instanton measure is a renormalization group invariant. This can be seen explicitly by introducing the scale parameter

$$\Lambda = \mu \cdot \exp\left(-\int_{g_0}^{g} \frac{d\bar{g}}{\beta(\bar{g})}\right) \tag{B.9}$$

where $g_0 = g(\Lambda)$. Using the one loop β-function $\beta(g) = -(6-K-2L)g^3/16\pi^2$ one has

$$\Lambda^{6-K-2L} \exp\left(-\frac{8\pi^2}{g_0^2}\right) = \mu^{6-K-2L} \exp\left(-\frac{8\pi^2}{g^2(\mu)}\right); \tag{B.10}$$

higher loop corrections to the β-function simply have the effect to convert the pre-exponential factor $g^{-4}(\mu)$ into g_0^{-4}*. The instanton measure then becomes

$$d\sigma_{\text{inst}} = C z^{-4} \Lambda^{6-K-2L} \cdot \int d^4x_0 \frac{d\rho^2}{\rho^2} \dots$$
$$C = g_0^{-4} \exp\left(-\frac{8\pi^2}{g_0^2}\right) \tag{B.11}$$

* E.g. in $SU(2)$ super Yang-Mills the "all order β-function", evaluated in a suitable [23] renormalization scheme, is [24] $\beta(g) = -\frac{3}{2}g^3(4\pi^2 - g^2)^{-1}$

for $z \neq 0$, and analogously for $h \neq 0$, and

$$d\sigma_{\text{inst}}^{(0)} = C_0 \Lambda^{6-K-2L} \cdot \int d^4x_0\, d\rho^2 \dots$$
$$C_0 = 4\pi^2 C \tag{B.12}$$

for $z = h = 0$.

Note that as a function of g_0 the constant C vanishes at $g_0 \to \infty$ (and at $g_0 \to 0$) and has a maximum $C_{\max} = (4\pi^2 e)^{-2} \cong 8.683 \cdot 10^{-5}$ at $g_0 = 2\pi$.

It is straightforward to check that the constant in front of the instanton measure is in agreement with the more conventional formalism where the influence of fermion zero modes is counted through the fermion sources rather than through the introduction of collective coordinates. E.g. one may calculate Green functions such as

$$\left\langle T \frac{g^2}{32\pi^2} \lambda\lambda(x_1) Z\hat{Z}(x_2) \right\rangle = -\tfrac{1}{2} \mathcal{G}_{01}^{(0)}\big|_{\theta_1 = \theta_2 = 0}$$

in this formalism, i.e. [6, 25]

$$\langle T\lambda\lambda(x_1) Z\hat{Z}(x_2)\rangle = C' \Lambda^5 \int d^4x_0\, \rho^2 d\rho^2 \sum_P (-1)^P \lambda^{(0)}(x_1;\alpha)\, \lambda^{(0)}(x_1;\beta)\, Z^{(1)}(x_2;\gamma)\, \hat{Z}^{(1)}(x_2;\delta) \tag{B.13}$$

where the sum runs over all permutations of $\alpha, \beta, \gamma, \delta \sim 1, 2, \dot{1}, \dot{2}$, $\lambda^{(0)}$ are the modes (2.4) and $Z^{(1)}$ the corresponding modes in (2.27) (calculated, however, with fermion modes normalized to unity) and (compare (B.1)) $C' = 2^9 \pi^6 g^{-4}(\mu) \cdot C$. The result is in complete agreement with (3.17).

References

1. V.A. Novikov, M.A. Shifman, A.I. Vainsthein, V.I. Zakharov: Nucl. Phys. **B 229**, 407 (1983)
2. V.A. Novikov et al.: Nucl. Phys. **B 229**, 394 (1983)
3. I. Affleck, M. Dine, N. Seiberg: Phys. Rev. Lett. **51**, 1026 (1983); ibid. **52**, 1677 (1984); Nucl. Phys. **B 241**, 493 (1984)
4. G.C. Rossi, G. Veneziano, Phys. Lett. **138 B**, 195 (1984); Y. Meurice, G. Veneziano: Phys. Lett. **141 B**, 69 (1984); Y. Meurice: Ph. D. thesis (1985)
5. M.G. Schmidt: Phys. Lett. **141 B**, 236 (1984)
6. D. Amati, G.C. Rossi, G. Veneziano: Nucl. Phys. **B 249**, 1 (1985)
7. V.A. Novikov, M.A. Shifman, A.I. Vainsthein, V.I. Zakharov: Preprints ITEP-31 and ITEP-45 (1985)
8. G. Veneziano, S. Yankielowicz: Phys. Lett. **113 B**, 321 (1982); T.R. Taylor, G. Veneziano, S. Yankielowicz: Nucl. Phys. **B 218**, 493 (1983); H.-P. Nilles: Phys. Lett. **129 B**, 103 (1983)
9. T.E. Clark, O. Piguet, K. Sibold: Nucl. Phys. **B 159**, 1 (1979); K. Konishi: Phys. Lett. **135 B**, 439 (1984)
10. A.A. Belavin, A.M. Polyakov, A.S. Schwartz, Yu.S. Tyupkin: Phys. Lett. **59 B**, 85 (1975)

11. I. Affleck: Nucl. Phys. **B 191**, 429 (1981)
12. G. 't Hooft: Phys. Rev. **D 14**, 3432 (1976); Phys. Rev. **D 18**, 2199 (1978)
13. S. Ferrara: Nucl. Phys. **B 77**, 73 (1974); T.E. Clark, O. Piguet, K. Sibold: Nucl. Phys. **B 143**, 445 (1978)
14. L.S. Brown, R.D. Carlitz, D.B. Creamer, C. Lee: Phys. Lett. **70 B**, 180 (1977)
15. A. d'Adda, P. di Vecchia: Phys. Lett. **73 B**, 162 (1978)
16. J.L. Gervais, B. Sakita: Phys. Rev. **D 11**, 2943 (1975); E. Tomboulis: Phys. Rev. **D 12**, 1678 (1975); C. Bernard: Phys. Rev. **D 19**, 3013 (1979)
17. C. Callan, R. Dashen, D. Gross: Phys. Lett. **66 B**, 375 (1977)
18. N. Andrei, D. Gross: Phys. Rev. **D 18**, 468 (1978)
19. E. Witten: Nucl. Phys. **B 202**, 253 (1982)
20. J. Wess, J. Bagger: Supersymmetry and supergravity. Princeton; Princeton University Press 1983
21. A.I. Vainsthein, V.I. Zakharov, V.A. Novikov, M.A. Shifman: Sov. Phys. Usp. **25**, 195 (1982)
22. K. Osterwalder, R. Schrader: Helv. Phys. Acta **46**, 277 (1973); H. Nicolai: Nucl. Phys. **B 140**, 294 (1978)
23. T.R. Morris, D.A. Ross, C.T. Sachrajda: Phys. Lett. **58 B**, 223 (1985)
24. V.A. Novikov, M.A. Shifman, A.I. Vainsthein, V.I. Zakharov: Nucl. Phys. **B 229**, 381 (1983)
25. Y. Meurice: Ph. D. thesis, Université Catolique de Louvain-La Neuve (1985)
26. D. Amati, Y. Meurice, G.C. Rossi, G. Veneziano: CERN-TH. 4201/85

Nuclear Physics B272 (1986) 677–692
North-Holland, Amsterdam

A SUPERSYMMETRIC INSTANTON CALCULUS FOR SU(N) GAUGE THEORIES

Jürgen FUCHS*

Joseph Henry Laboratories, Princeton University, Princeton, NJ 08544, USA

Received 29 January 1986

We extend a previously formulated instanton superfield method from SU(2) to SU(N) gauge theories without classical vacuum expectation values of scalar fields. We consider supersymmetric Yang-Mills, supersymmetric QCD, and a class of chiral theories with matter in the fundamental and antisymmetric tensor representation. Via cluster decomposition, our results provide non-trivial consistency checks; in particular they enable us to conclude that for most of the chiral theories the approach without classical scalar VEV's is inconsistent.

1. Introduction

Instanton effects in supersymmetric gauge theories can lead to a breakdown of perturbatively established nonrenormalization theorems and even to dynamical supersymmetry breaking, and therefore have been explored from a variety of viewpoints [1–10]. As a consequence, almost complete information on the nonperturbative behavior of both supersymmetric QCD (SQCD) and some chiral gauge theories has been obtained. On the other hand there still remain open questions; in particular there is an unresolved controversy whether a confining picture [7, 8, 10] or Higgs picture [2–4, 9] should be used to extract reliable results from instanton calculations.

On the technical side, both in the confining and in the Higgs picture considerable simplification of computations has been accomplished by using [1, 2, 9] a supersymmetric formalism which treats bosonic and fermionic zero modes on the same footing, i.e. introduces collective coordinates for both of them, and groups them together into superfields. The usefulness of this approach so far has only been discussed in the context of SU(2), first for pure supersymmetric Yang-Mills theory [1] and later for SQCD [2, 9] and for matter in the adjoint representation of SU(2) [9].

In the present paper we extend this supersymmetric formalism to SU(N) gauge theories with arbitrary N and without classical vacuum expectation values (VEV's)

* Supported by Deutsche Forschungsgemeinschaft.

0550-3213/86/$03.50

of scalar fields. We compute one-instanton contributions to Green functions in various models; the results enable us to draw some conclusions on the internal consistency of the picture without scalar VEV's. The considered models are SQCD, SU(N) with one antisymmetric tensor and $N-4$ fundamentals, and SU(N), N odd or $N=6$, with two antisymmetric tensors and $2N-8$ fundamentals; these include all models which (to our knowledge) have been discussed in the confining picture using the more conventional instanton formalism without fermionic collective coordinates (namely SQCD [7], SU(5) with one or two families of $5+\overline{10}$ [5,7], and SU(6) with matter in the $6+6+\overline{15}$ representation [10]).

In many cases we can calculate instanton-induced Green functions for several distinct combinations of a given set of elementary fields. Cluster decomposition then enables us to check the internal consistency of the results. For most theories we find that some of the required consistency conditions are violated. We discuss two complementary approaches to overcome these inconsistencies and argue that in the case of chiral theories the only way out is to consider the theory in the Higgs phase.

Our paper is organized as follows: in sect. 2 we construct the classical superfields which group bosonic and fermionic zero modes together. In sect. 3 we fix the supersymmetric integration measure and describe how the various collective coordinate integrations are performed most conveniently. We are then ready to discuss, in sect. 4, the classes of models mentioned above. Sect. 5 contains our conclusions.

In our notations we follow those of [9]. In particular we use an euclideanized version of the Wess-Bagger conventions.

2. Classical superfields

Due to the cancellation of bosonic and fermionic non-zero modes, one-instanton contributions to Green functions of gauge invariant combinations of chiral superfields can be obtained by integrating appropriate products of "classical superfields" over a set of bosonic and fermionic collective coordinates [1,2,9]. The classical gauge superfield contains the instanton together with the gaugino zero modes. For vanishing scalar VEV's, the classical matter superfields contain the matter fermion zero modes and, in addition, the scalar "zero modes" induced by the Yukawa couplings $gZ^{+}\lambda\psi$; the latter obey the inhomogeneous scalar field equation $\mathcal{D}^2 Z \sim g\lambda\psi$ with the gaugino λ and matter fermion ψ replaced by their respective zero modes. In this section we construct the explicit form of these classical superfields for the gauge multiplet and for matter in the (anti)fundamental or antisymmetric tensor representation of SU(N); matter in other representations could be treated analogously.

If the SU(2) in which the instanton resides is embedded in the "upper left-hand corner" of SU(N), the instanton and the supersymmetric and superconformal gaugino zero modes contribute to the classical gauge superfield as

$$(W_\alpha^{(c)})_i^{\ j}(z) = \frac{4}{g}\rho^2 f^2(y-x_0)\sigma^a{}_i{}^j\sigma^a{}_\alpha{}^\beta\tilde{\theta}_\beta, \qquad i,j=1,2. \tag{2.1}$$

Here x_0 and ρ denote the position and size of the instanton, $f(x) := (x^2+\rho^2)^{-1}$, and we wrote $z=(y,\theta)$ for the chiral superspace coordinates $y^m = x^m + i\theta\sigma^m\bar{\theta}$ and θ^α; σ^a are the Pauli matrices, and $\sigma^m = (i, \sigma^a)$. Finally

$$\tilde{\theta}_\alpha = \theta_\alpha - \theta_{0\alpha} - \sigma^m_{\alpha\dot{\beta}}\bar{\vartheta}^{\dot{\beta}}(y-x_0)_m, \tag{2.2}$$

where θ_0^α, $\bar{\vartheta}_{\dot{\alpha}}$ are the fermionic collective coordinates introduced for the supersymmetric and superconformal zero modes, respectively [1,9]. These four gaugino zero modes can be obtained by applying a supersymmetric and superconformal transformation to the instanton solution; in addition, the gaugino has $2N-4$ zero modes which are doublets of SU(2) and do not possess symmetry meaning. Introducing collective coordinates $\bar{\chi}_\alpha^{(i)}$, $\alpha=1,2$, $i=3\dots N$ for these additional zero modes, we get

$$(W_\alpha^{(c)})_i^{\ j}(z) = \begin{cases} 4g^{-1}\rho^2 f^2(y-x_0)\sigma^a{}_i{}^j\sigma^a{}_\alpha{}^\beta\tilde{\theta}_\beta, & i,j=1,2 \\ -4\sqrt{2}\,\pi i g^{-1}\rho^3 f^{3/2}(y-x_0)\varepsilon_{i\alpha}\bar{\chi}_1^{(j)}, & i=1,2;\ j=3\dots N \\ -4\sqrt{2}\,\pi i g^{-1}\rho^3 f^{3/2}(y-x_0)\delta_\alpha^j\bar{\chi}_2^{(i)}, & j=1,2;\ i=3\dots N \\ 0, \quad i,j=3\dots N \end{cases}. \tag{2.3}$$

For a matter superfield ϕ_i in the fundamental representation the components $i=1,2$ contain the matter fermion zero modes and the solutions of the scalar field equation induced by the supersymmetric and superconformal gaugino zero modes. The scalar solutions induced by the remaining $2N-4$ gaugino zero modes contribute to the $i=3\dots N$ components of $\phi_i^{(c)}$; they are obtained by using the explicit form of the scalar propagator $(\mathscr{D}^2)^{-1}$ in the instanton background [11,9]; the result is

$$\phi_i^{(c)}(z) = \begin{cases} \sqrt{2}\,\pi^{-1}\rho f^{3/2}(y-x_0)\tilde{\theta}_i\bar{\tau}, & i=1,2 \\ 2i\rho^2 f(y-x_0)\bar{\chi}_2^{(i)}\bar{\tau}, & i=3\dots N \end{cases}, \tag{2.4}$$

where $\bar{\tau}$ is the collective coordinate for the matter fermion zero mode. Analogously, for matter $\tilde{\phi}^i$ in the antifundamental representation one gets

$$\tilde{\phi}^{i(c)}(z) = \begin{cases} \sqrt{2}\,\pi^{-1}\rho f^{3/2}(y-x_0)\tilde{\theta}^i\bar{\bar{\tau}}, & i=1,2 \\ 2i\rho^2 f(y-x_0)\bar{\chi}_1^{(i)}\bar{\bar{\tau}}, & i=3\dots N \end{cases}. \tag{2.5}$$

(In SQCD the collective coordinates for each flavor of ϕ and $\tilde{\phi}$ may be combined to form a two-component spinor $(\bar{\tau}, \bar{\bar{\tau}}) \to (\bar{\tau}_1, \bar{\tau}_2)$; that notation was used in [9].)

The fermionic component of a matter field A^{ij} in the antisymmetric tensor representation possesses $N-2$ zero modes which we label by collective coordinates τ^i, $i=3\dots N$; the induced scalar zero modes can again be computed using the explicit form of $(\mathscr{D}^2)^{-1}$ [12]; we find

$$A^{ij(c)}=\begin{cases} -i\rho^2 f(y-x_0)\varepsilon^{ij}\sum\limits_{k=3}^{N}\bar{\chi}_2^{(k)}\tau^k, & i,j=1,2 \\ 2^{-1/2}\pi^{-1}\rho^3 f^{3/2}(y-x_0)\tilde{\theta}^i\tau^j, & i=1,2,\ j=3\dots N\,. \\ -2^{-1/2}\pi^{-1}\rho^3 f^{3/2}(y-x_0)\tilde{\theta}^j\tau^i, & j=1,2,\ i=3\dots N \\ i\rho^2 f(y-x_0)\big(\bar{\chi}_1^{(i)}\tau^j-\bar{\chi}_1^{(j)}\tau^i\big), & i,j=3\dots N \end{cases} \tag{2.6}$$

In sect. 4 we will compute the one-instanton contribution to Green functions for several gauge invariant combinations of W_α, ϕ_A, $\tilde{\phi}_A$ and X_a where $A=1\dots F$, $a=1\dots f$ are flavor indices. The relevant combinations are

$$S=\frac{g^2}{32\pi^2}W^{\alpha}{}_i{}^{j}W_{\alpha j}{}^{i},$$

$$T_{AB}=\phi_{iA}\tilde{\phi}^i{}_B,$$

$$U_{ABa}=\phi_{iA}\phi_{jB}X_a^{ij},$$

$$V_a=\varepsilon_{i_1\dots i_N}A_a^{i_1i_2}\dots A_a^{i_{N-1}i_N}\qquad (N\text{ even}),$$

$$X_{abA}=\varepsilon_{i_1\dots i_N}A_a^{i_1i_2}\dots A_a^{i_{N-2}i_{N-1}}A_b^{i_Nj}\phi_{jA}\qquad (N\text{ odd}),$$

$$Y_A=\frac{g^2}{32\pi^2}\varepsilon_{i_1\dots i_N}W^{\alpha}{}_j{}^{k}W_{\alpha k}{}^{i_1}A^{i_2i_3}\dots A^{i_{N-1}i_N}A^{jl}\phi_{lA}\qquad (f=1,\ N\text{ odd}). \tag{2.7}$$

The classical superfields for these combinations are obtained by substituting (2.3)–(2.6) into (2.7). It turns out to be useful to define

$$F_n(z)=\big(\rho^2 f(y-x_0)\big)^n\left[2\sum_{i=3}^{N}\big(\bar{\chi}^{(i)}\big)^2-\frac{n}{\pi^2\rho^2}f(y-x_0)\tilde{\theta}^2\right]. \tag{2.8}$$

In terms of F_n, we find

$$S^{(c)}(z)=F_3(z),$$

$$T_{AB}{}^{(c)}(z)=-\bar{\tau}_A\bar{\tilde{\tau}}_B F_2(z),$$

$$U_{ABa}^{(c)}(z)=4i\bar{\tau}_A\bar{\tau}_B\sum_{i=3}^{N}\bar{\chi}_2^{(i)}\tau_a^iF_3(z). \tag{2.9}$$

The expressions for $V^{(c)}$, $X^{(c)}$ and $Y^{(c)}$ are a bit more complicated:

$$V^{(c)}(z) = -\tfrac{1}{2}N\big(2i\rho^2 f(y-x_0)\big)^{N/2}\varepsilon_{i_3\dots i_N}\tau^{i_3}\dots\tau^{i_{N/2}}\bar{\chi}_1^{(i_{N/2+1})}\dots\bar{\chi}_1^{(i_{N-2})}$$

$$\times\left[\bar{\chi}_1^{(i_{N-1})}\tau^{i_N}\bar{\chi}^{2(j)}\tau^j - \frac{1}{4\pi^2\rho^2}\left(\tfrac{1}{2}N-1\right)f(y-x_0)\tilde{\theta}^2\tau^{i_{N-1}}\tau^{i_N}\right],$$

$$X^{(c)}_{abA}(z) = \tfrac{1}{8}(N-1)\big(2i\rho^2 f(y-x_0)\big)^{(N+3)/2}\bar{\tau}_A\varepsilon_{i_3\dots i_N}$$

$$\times\tau^{i_3}\dots\tau^{i_{(N+1)/2}}\bar{\chi}_1^{(i_{(N+3)/2})}\dots\bar{\chi}_1^{(i_{N-2})}$$

$$\times\Big[2\bar{\chi}_1^{(i_{N-1})}\bar{\chi}_1^{(i_N)}\bar{\chi}_2^{(j)}\tau_a^j\bar{\chi}_2^{(k)}\tau_b^k - \tau_b^{i_{N-1}}\bar{\chi}_1^{(i_N)}\bar{\chi}_2^{(j)}\tau_a^j\Sigma\left(\bar{\chi}^{(k)}\right)^2$$

$$-\frac{1}{\pi^2\rho^2}f(y-x_0)\tilde{\theta}^2\Big[\tfrac{1}{4}(N+5)\tau_a^{i_{N-1}}\bar{\chi}_1^{(i_N)}\bar{\chi}_2^{(j)}\tau_b^j$$

$$-\tfrac{1}{8}(N-3)\tau_a^{i_{N-1}}\tau_b^{i_N}\Sigma\left(\bar{\chi}^{(j)}\right)^2$$

$$-\tau_b^{i_{N-1}}\bar{\chi}_1^{(i_N)}\bar{\chi}_2^{(j)}\tau_a^j\Big]\Big],$$

$$Y^{(c)}_A = \tfrac{1}{16}i(N-1)\big(2i\rho^2 f(y-x_0)\big)^{(N+9)/2}\bar{\tau}_A\bar{\chi}_2^{(j)}\tau^j$$

$$\times\varepsilon_{i_3\dots i_N}\tau^{i_3}\dots\tau^{i_{(N+1)/2}}\bar{\chi}_1^{(i_{(N+3)/2})}\dots\bar{\chi}_1^{(i_{N-1})}$$

$$\times\Big[\bar{\chi}_1^{(i_N)}\bar{\chi}_2^{(k)}\tau^k\Sigma\left(\bar{\chi}^{(l)}\right)^2 - \frac{1}{\pi^2\rho^2}f(y-x_0)\tilde{\theta}^2$$

$$\times\Big[\tfrac{1}{8}(N+5)\tau^{i_N}\Sigma\left(\bar{\chi}^{(k)}\right)^2 + \bar{\chi}_1^{(i_N)}\bar{\chi}_2^{(k)}\tau^k\Big]\Big]. \tag{2.10}$$

(Here all repeated indices are summed from 3 to N.) However, certain products of these superfields with combinations of collective coordinates can again be expressed in terms of F_n: we get

$$V^{(c)}(z)\cdot\left(\bar{\chi}_2^{(i)}\tau^i\right)^{N/2-2} = -i^{N/2}\left(\tfrac{1}{2}N\right)!\left(\tfrac{1}{2}N-1\right)!\tau^3\dots\tau^N$$

$$\times\left(\Sigma\left(\bar{\chi}^{(i)}\right)^2\right)^{N/2-2}F_{N/2}(z),$$

$$Y^{(c)}_A(z)\cdot\left(\bar{\chi}_2^{(i)}\tau^i\right)^{(N-5)/2} = 2i^{(N+3)/2}\left(\left(\tfrac{1}{2}(N-1)\right)!\right)^2\bar{\tau}_A\tau^3\dots\tau^N$$

$$\times\left(\Sigma\left(\bar{\chi}^{(i)}\right)^2\right)^{(N-1)/2}F_{(N+9)/2}(z) \tag{2.11}$$

and, after some algebra,

$$\begin{aligned} X^{(c)}_{abA}(z_1)\,X^{(c)}_{baB}(z_2)\cdot\left(\bar{\chi}^{(i)}_2\tau^i_a\bar{\chi}^{(j)}_2\tau^j_b\right)^{(N-5)/2} \\ = (-1)^{(N+1)/2}\left(\left(\tfrac{1}{2}(N-1)\right)!\right)^4\frac{N-2}{N-1}\cdot 4\bar{\tau}_A\bar{\tau}_B \\ \times\tau^3_a\ldots\tau^N_a\tau^3_b\ldots\tau^N_b\left(\Sigma\left(\bar{\chi}^{(i)}\right)^2\right)^{N-3} \\ \times F_{(N+3)/2}(z_1)\,F_{(N+3)/2}(z_2). \end{aligned} \tag{2.12}$$

(a, b not summed.)

3. Computational methods

Here we describe how the Green functions that we will encounter in the next section can be computed conveniently. As discussed above, these Green functions take the form

$$\mathscr{G}(z_1\ldots, z_K) = \int \mathrm{d}\sigma_{\text{inst}}\, A_1(z_1)\ldots A_K(z_K), \tag{3.1}$$

where A_i are classical superfields such as (2.9), (2.10), and $\mathrm{d}\sigma_{\text{inst}}$ is the supersymmetric instanton integration measure. For pure supersymmetric SU(2) Yang-Mills theory this measure has been given in [1,9]. For SU(N) one gets an additional factor $[(N-1)!(N-2)!]^{-1}\cdot(2\pi\mu\rho/g)^{4N-8}$ obtained by integrating out the zero modes corresponding to gauge rotations of the instanton [13]; also the SU(2) doublet gaugino zero modes contribute an integration over their collective coordinates $\bar{\chi}^{(i)}$ together with a jacobian factor $(4\pi^2\rho^2/g)^{4-2N}$ (note that these zero modes are not normalized to unity) and a factor μ^{2-N} due to regularization. Here μ is the renormalization scale and $g = g(\mu)$. Using the definition of the renormalization group invariant scale parameter Λ and the two-loop β-function, the instanton measure becomes

$$\int \mathrm{d}\sigma_{\text{inst}} = \frac{4\pi^2}{(N-1)!(N-2)!}C_N\Lambda^{3N}\int \mathrm{d}^4x_0\,\mathrm{d}\rho^2\,\mathrm{d}^2\theta_0\,\mathrm{d}^2\bar{\vartheta}\prod_{i=3}^{N}\mathrm{d}^2\bar{\chi}^{(i)}, \tag{3.2}$$

with

$$C_N = g_0^{-2N}\exp\left(-\frac{8\pi^2}{g_0^2}\right) \tag{3.3}$$

and $g_0 \equiv g(\Lambda)$. Note that the numerical constant in (3.2) cannot be changed arbitrarily by a suitable redefinition of Λ since, as a function of g_0, C_N has a maximum $C_{N,\max} = (N/8\pi^2 e)^N$ at $g_0 = 2\pi\sqrt{2/N}$.

If supersymmetric matter is added, the power of Λ and the definition of C_N are appropriately changed and an integration over the collective coordinates $\bar{\tau}, \bar{\bar{\tau}}, \tau' \ldots$ for the matter fermion zero modes (which have been normalized to unity in (2.4)–(2.6)) is included in the instanton measure. If the theory contains a classical superpotential $P(\phi, \tilde{\phi}, X \ldots)$, then there is also a factor of

$$\exp\left(-\int \mathrm{d}^6 z\, P\left(\phi^{(c)}, \tilde{\phi}^{(c)}, X^{(c)} \ldots\right)\right). \tag{3.4}$$

The various collective coordinate integrations are most easily performed in the following order: first, the matter fermion collective coordinates are integrated out; for each Green function, the result is nonzero only for a limited number of terms in the series expansion of (3.4). Moreover, it turns out that the integrand then always becomes proportional to a product of factors $F_{n_i}(z_i)$ with F_n given by (2.8) and some power of $\Sigma(\bar{\chi}^{(i)})^2$. The $\bar{\chi}^{(i)}$ integrations are then immediate, and afterwards one can use

$$\int \mathrm{d}^2\theta_0\, \mathrm{d}^2\bar{\vartheta}\, \tilde{\theta}_k^2 \tilde{\theta}_l^2 = (y_k - y_l)^2. \tag{3.5}$$

After these manipulations one is left with an integral of the form

$$\begin{aligned} I(y_1, \ldots, y_K) = \int \mathrm{d}^4 x_0\, \mathrm{d}\rho^2 \rho^{-4} & \\ & \times \tfrac{1}{2} \sum_{i,j=1}^{K} n_i n_j (y_i - y_j)^2 f(y_i - x_0) f(y_j - x_0) \\ & \times \prod_{k=1}^{K} \left(\rho^2 f(y_k - x_0)\right)^{n_k}, \end{aligned} \tag{3.6}$$

with some integers n_i, $i = 1 \ldots K$. Upon use of the formula

$$\begin{aligned} \int \mathrm{d}^4 x_0\, \mathrm{d}\rho^2 \rho^{2l} \prod_{i=1}^{K} \left(f(y_i - x_0)\right)^{m_i} & \\ = \pi^2 \frac{l!(\Sigma m_i - l - 4)!}{\Pi_i((m_i - 1)!)} & \left(\prod_{j=1}^{K} \int_0^1 \mathrm{d}\alpha_j\, (\alpha_j)^{m_j - 1}\right) \\ \times \delta\left(1 - \sum_i \alpha_i\right) & \left[\tfrac{1}{2} \sum_{j,k=1}^{K} \alpha_j \alpha_k (y_j - y_k)^2\right]^{l + 3 - \Sigma m_i}, \end{aligned} \tag{3.7}$$

which is obtained by introducing Feynman parameters in the standard manner, (3.6) turns out to be a pure number:

$$I = \pi^2 \left(\left(\sum_{i=1}^{K} n_i \right) - 1 \right)^{-1} . \tag{3.8}$$

With (3.5) and (3.8), the only potentially cumbersome step in the evaluation of Green functions is the integration over the matter fermion collective coordinates; but in many cases also this step is very simple, and in the remaining cases it becomes simple once relations such as (2.11), (2.12) have been established.

4. Models

We now examine several classes of gauge theories. We compute the one-instanton contribution to Green functions of gauge invariant combinations of chiral superfields which are zero in perturbation theory and fulfill the relevant one-instanton selection rules.

4.1. SU(N) SUPER YANG-MILLS

In pure supersymmetric Yang-Mills theory with SU(N) gauge group the only Green function receiving a one-instanton contribution is

$$\mathscr{G}_0 = \langle \mathrm{T} S(z_1) \dots S(z_N) \rangle , \tag{4.1}$$

with S defined in (2.7). Using the methods described above we find

$$\mathscr{G}_0 = \frac{2^N}{(N-1)!(3N-1)} C_N \Lambda^{3N} , \tag{4.2}$$

which agrees with previous results obtained in the component formalism [7] and fixes the unspecified multiplicative constant encountered there; also, of course, for $N = 2$ the result of [9] is recovered.

4.2. SQCD

Next we consider SQCD with N colors and M flavors. If there is no superpotential, the one-instanton selection rules require $M \leqslant N$ and single out the Green function

$$\mathscr{G}_M = \left\langle \mathrm{T} S(z_1) \dots S(z_{N-M}) T_{A_1 B_1}(z_{N-M+1}) \dots T_{A_M B_M}(z_N) \right\rangle . \tag{4.3}$$

We get

$$\mathcal{G}_M = \frac{(-1)^M 2^N}{(N-1)!(3N-M-1)} C_N \Lambda^{3N-M} \varepsilon_{A_1 \dots A_M} \varepsilon_{B_1 \dots B_M}. \tag{4.4}$$

The theory without superpotential possesses a large number of flat directions along which the classical scalar potential vanishes [14]; these are all lifted by adding the most general renormalizable superpotential to the theory, i.e. mass terms for all flavors:

$$P = \sum_{A,B=1}^{M} m_{AB} \phi_{iA} \tilde{\phi}^i_B. \tag{4.5}$$

With (2.9) and (3.4), this results in an additional contribution

$$\exp\left(- \sum_{A,B=1}^{M} m_{AB} \bar{\tau}_A \bar{\bar{\tau}}_B \right) \tag{4.6}$$

to the instanton measure. Therefore one has now a one-instanton contribution to each Green function $\mathcal{G}_{M,K}$ given by (4.3) with M replaced by an arbitrary integer $K \leqslant \min(N, M)$, and there is no longer a restriction on M; we find

$$\mathcal{G}_{M,K} = \frac{(-1)^M 2^N}{(N-1)!(3N-K-1)} C_N \Lambda^{3N-M} \times \frac{1}{(M-K)!} \sum_{\substack{A_{K+1} \dots A_M, \\ B_{K+1} \dots B_M = 1}}^{M} \varepsilon_{A_1 \dots A_M} \varepsilon_{B_1 \dots B_M} \left(\prod_{i=K+1}^{M} m_{A_i B_i} \right), \tag{4.7}$$

or, after diagonalization of the mass matrix, $m_{AB} \to \delta_{AB} m_A$,

$$\mathcal{G}_{M,K} = \frac{(-1)^M 2^N}{(N-1)!(3N-K-1)} C_N \Lambda^{3N-M} \left(\prod_{\substack{A=1 \\ A \neq A_1 \dots A_K}}^{M} m_A \right) \delta^{B_1 \dots B_K}_{A_1 \dots A_K}, \tag{4.8}$$

where

$$\delta^{B_1 \dots B_K}_{A_1 \dots A_K} = \delta^{B_1}_{A_1} \dots \delta^{B_K}_{A_K} \pm \text{permutations}. \tag{4.9}$$

The results (4.4) and (4.8) are interpreted to indicate the existence of vacuum condensates $\langle S \rangle$ and $\langle T_{AB} \rangle$; problems of this interpretation have been discussed in [7] and [9]:

(i) While in the massless case the ε-tensor structure reflects the $SU(M) \times SU(M)$ flavor symmetry, in the generic massive case ($m_A \neq m_B$ for $A \neq B$) one would expect that only the first term of (4.9) (multiplied by $\prod_{i,j}(1-\delta_{A_iA_j})$) is present in (4.8), and indeed only this first term is consistent with the anomalous Ward identity known as the Konishi anomaly [15].

(ii) For any theory, if there is a one-instanton contribution to both $\langle \mathrm{T}A_1(z_1)\dots A_n(z_n)\rangle$ and $\langle \mathrm{T}B_1(z_1)\dots B_m(z_m)\rangle$ then there is a two-instanton contribution to $\langle \mathrm{T}A_1(z_1)\dots A_n(z_n)B_1(z_{n+1})\dots B_m(z_{n+m})\rangle$. Via clustering, in the massive case this provides a large number of consistency checks, the simplest one being that the ratio

$$\langle \mathrm{T}S(z_1)\dots S(z_{N-1})T_{AB}(z_N)\rangle\langle \mathrm{T}S(z_1)\dots S(z_{N-1})T_{CD}(z_N)\rangle \times\left[\langle \mathrm{T}S(z_1)\dots S(z_N)\rangle\langle \mathrm{T}S(z_1)\dots S(z_{N-2})T_{AB}(z_{N-1})T_{CD}(z_N)\rangle\right]^{-1} \tag{4.10}$$

(for $A\neq C$, $B\neq D$) should be equal to one. However, (4.8) yields (for $A\neq D$ or $B\neq C$ where the discrepancy (i) is irrelevant) the value

$$\frac{(3N-1)(3N-3)}{(3N-2)^2}\neq 1 \tag{4.11}$$

for this quantity. (4.8) is thus intrinsically inconsistent.

To remedy these inconsistencies, two distinct methods have been proposed: in [7, 8] it is argued that in the massive theory infrared singularities generated by nonzero mode contributions to propagators have to be taken into account, while in [2, 9] a classical scalar expectation value is introduced, an approach favored also by other authors [3]. Explicit calculations have been performed in the case of SU(2) [8, 9]; self-consistency and conformity with the anomaly relation are obtained with either method, but the results of the two methods are incompatible with each other. To get rid of terms which would spoil the self-consistency, both approaches require performing a particular limit at some stage of the computation, namely masses $\to\infty$ and VEV $\to\infty$, respectively. Of course it would be very interesting to investigate the case of arbitrary N along these lines, but for both methods explicit calculations become considerably more complicated.

4.3. SU(N) WITH AN ANTISYMMETRIC TENSOR AND $N-4$ FUNDAMENTALS

(i) N odd. Like SQCD, the SU(N) theory with an antisymmetric tensor A^{ij} and $N-4$ fundamentals ϕ_{iA} has flat directions if no superpotential is present. For N odd, $N\geqslant 5$, these are all lifted [14] by including the most general renormalizable superpotential, i.e.

$$P=\sum_{A,B=1}^{N-4} h_{AB}\phi_{iA}\phi_{jB}A^{ij},\qquad h=-h^{\mathrm{T}}; \tag{4.12}$$

this results in a contribution

$$\exp\left(2i\sum_{A,B=1}^{N-4} h_{AB}\bar{\tau}_A\bar{\tau}_B\sum_{i=3}^{N}\bar{\chi}_2^{(i)}\tau^i\right) \tag{4.13}$$

to the instanton measure. (For $N=5$ there is no superpotential; nevertheless there are no flat directions. Instanton effects in this model have been discussed in [4, 5, 7].) With the superpotential (4.12), the combined Konishi identities for ϕ_A and A show that the gaugino condensate $\langle S(z)\rangle|_{\theta=0}$ is an order parameter for supersymmetry breaking.

For $P=0$ only the Green function

$$\mathcal{G}_2=\left\langle \mathrm{T}S(z_1)S(z_2)U_{A_1A_2}(z_3)\dots U_{A_{N-6}A_{N-5}}(z_{(N-1)/2})Y_{A_{N-4}}(z_{(N+1)/2})\right\rangle, \tag{4.14}$$

with S, U, Y defined by (2.7) receives a one-instanton contribution. $\mathcal{G}_2$ can be computed in a straightforward way by the methods described above (in particular the second of the relations (2.11) has to be used); the result is

$$\mathcal{G}_2=\frac{(-1)^{(N-1)/2}2^{3(N-3)/2}}{(N-1)!(N+1)}\left(\frac{N-1}{2}!\right)^2 C_N\Lambda^{2N+3}\varepsilon_{A_1\dots A_{N-4}}. \tag{4.15}$$

For $P\neq 0$ also the Green functions

$$\mathcal{G}_K=\left\langle \mathrm{T}S(z_1)\dots S(z_K)U_{A_1A_2}(z_{K+1})\dots U_{A_{N-2K-2}A_{N-2K-1}}(z_{(N-1)/2})Y_{A_{N-2K}}(z_{(N+1)/2})\right\rangle \tag{4.16}$$

for $3\leqslant K\leqslant \frac{1}{2}(N-1)$ have to be considered. We find (note that with (2.9), (4.13) is just $\exp(\Sigma_{AB}h_{AB}U_{AB}^{(c)}/S^{(c)})$)

$$\mathcal{G}_K=\frac{(-1)^{(N-1)/2}2^{(3N-2K-5)/2}}{(N-1)!(N+1)}\left(\frac{N-1}{2}!\right)^2 C_N\Lambda^{2N+3} \times\sum_{A_{N-2K+1},\dots,A_{N-4}=1}^{N-4}\varepsilon_{A_1\dots A_{N-4}}\prod_{i=(N-2K+1)/2}^{(N-5)/2} h_{A_{2i}A_{2i+1}}. \tag{4.17}$$

As in SQCD, we recognize that, via clustering, the presence of a superpotential allows for a number of consistency checks. E.g., comparison of $\mathcal{G}_3\cdot\mathcal{G}_3$ and $\mathcal{G}_2\cdot\mathcal{G}_4$ shows that $h_{AB}\langle U_{AB}\rangle$ times $h_{CD}\langle U_{CD}\rangle$ should equal (for A, B, C, D all different; no sum) $(h_{AB}h_{CD}-h_{AC}h_{BD}+h_{AD}h_{BC})\langle U_{AB}\rangle\langle U_{CD}\rangle$; of course for generic choice of

h_{AB} this is not the case. Thus (4.17) is intrinsically inconsistent, just as (4.8) was (here we have to exclude the cases $N=5,7$ where no consistency check is available). Yet the situation is now somewhat different since the theory is chiral and the arguments of [7, 8] concerning infrared singularities do not apply; this seems to favor the approach of introducing classical VEV's.

Although we thus cannot really trust the result (4.17), our experience from SU(2) SQCD [8, 9] leads us to expect that (4.17) is true at least as an order of magnitude estimate and, in particular, that in a fully consistent approach $\mathscr{G}_K$ will stay nonzero for appropriate choice of the flavor indices A_i. We therefore conclude by clustering that $\langle S(z)\rangle$ is nonzero and that in this class of models supersymmetry is dynamically broken, except if the very implausible [4] possibility $\langle S\rangle = \langle U_{AB}\rangle = 0$, $\langle Y_A\rangle \to \infty$ was realized. This breakdown of supersymmetry is supported by effective lagrangian arguments [14].

(ii) N even, N ⩾ 8. For even N the superpotential (4.12) does not lift all flat directions [14]. The relevant Green functions are now

$$\mathscr{G}'_K = \left\langle \mathrm{T}S(z_1)\dots S(z_K)U_{A_1A_2}(z_{K+1})\dots U_{A_{N-2K+1}A_{N-2K+2}}(z_{N/2+1})V(z_{N/2+2})\right\rangle, \tag{4.18}$$

with $3 \leqslant K \leqslant \frac{1}{2}N+1$; we evaluate them as

$$\mathscr{G}'_K = \frac{(-1)^{N/2-1}2^{3N/2-K}}{(N-1)!(N+1)}\left(\frac{N}{2}\right)!\left(\frac{N}{2}-1\right)!C_N\Lambda^{2N+3} \times \sum_{A_{N-2K+3},\dots,A_{N-4}=1}^{N-4} \varepsilon_{A_1\dots A_{N-4}} \prod_{i=N/2-K+2}^{N/2-2} h_{A_{2i-1}A_{2i}}. \tag{4.19}$$

We thus encounter the same type of inconsistency as for odd N.

(iii) N = 6. For $N=6$ a second term can be present in the superpotential:

$$P = h\phi_{i1}\phi_{j2}X^{ij} + h'\varepsilon_{ijklmn}A^{ij}A^{kl}A^{mn}. \tag{4.20}$$

This lifts all flat directions [10]. In addition to the Green functions

$$\mathscr{G} = \left\langle \mathrm{T}S(z_1)S(z_2)S(z_3)U_{12}(z_4)V(z_5)\right\rangle,$$
$$\mathscr{G}' = \left\langle \mathrm{T}S(z_1)S(z_2)S(z_3)S(z_4)V(z_5)\right\rangle. \tag{4.21}$$

we must now also consider

$$\mathscr{G}'' = \left\langle \mathrm{T}S(z_1)S(z_2)S(z_3)S(z_4)U_{12}(z_5)\right\rangle,$$
$$\mathscr{G}''' = \left\langle \mathrm{T}S(z_1)S(z_2)S(z_3)S(z_4)S(z_5)\right\rangle. \tag{4.22}$$

We find

$$\mathcal{G} = 2\mathcal{G}'/h = 2\mathcal{G}''/h' = 4\mathcal{G}'''/hh' = \tfrac{32}{35} C_6 \Lambda^{15} . \tag{4.23}$$

Recently these Green functions have also been computed in the conventional instanton formalism [10]; our results agree with those of [10] (and fix an overall constant which there was left unspecified). If clustering is assumed, the set (4.21), (4.22) is overdetermined so that from (4.23) the values of the vacuum condensates can be deduced:

$$\begin{aligned} h\langle U_{12}(z)\rangle &= h'\langle V(z)\rangle = 2\langle S(z)\rangle \\ &= 2(hh'C_6/35)^{1/5} e^{2\pi i n/5}\Lambda^3 , \end{aligned} \tag{4.24}$$

where $1 \leqslant n \leqslant 5$. In particular, this allows even to deduce the correct form of the Konishi identities for this particular model.

Nevertheless we still have some doubt as to whether the result (4.24) is numerically reliable, since there is no consistency check of the type as for $N \geqslant 8$, and after generalizing the model to $N \geqslant 8$ the now available consistency checks fail (the fact that the $N \geqslant 8$ models possess flat directions whereas the $N = 6$ model does not, is not a valid counter argument since the inconsistencies found for N even, $N \geqslant 8$ are exactly of the same type as those for N odd, $N \geqslant 9$, where there do not exist flat directions either).

4.4. SU(N) WITH TWO ANTISYMMETRIC TENSORS AND $2N-8$ FUNDAMENTALS

(i) N odd. The most general renormalizable superpotential for this class of models is

$$P = \sum_{AB=1}^{2N-8} \sum_{a=1}^{2} h_{ABa} \phi_A \phi_B A_a , \tag{4.25}$$

with h antisymmetric in A, B. This lifts all flat directions (proof: using the methods of [14] it is easy to see that for $P = 0$ the flat directions are, up to symmetry transformations, given by $(A_1^+ A_1 + A_2^+ A_2)_i{}^j = |x_i|^2 \delta_i{}^j$ and $\phi_{iA} = v_A \delta_{iA}$ for $A = 1, \dots N$, $\phi_{iA} = 0$ otherwise, with $c := 2|x_i|^2 - |v_i|^2$ independent of i; for $P \neq 0$, the condition $\partial P/\partial A_a = 0$ gives $v_A = 0$ for all A; it then follows that $c = 2(|x_{i1}|^2 + |x_{i2}|^2)$ where x_{ia} are the eigenvalues of A_a; since these come in pairs [14], $c = 0$, and hence $x_{i1} = x_{i2} = 0$ for all i).

The relevant Green functions turn out to be

$$\begin{aligned} \mathcal{G}_{ab}^{AB} &= \langle \mathrm{T} S(z_1) X_{abA}(z_2) X_{baB}(z_3)\rangle , \\ \mathcal{G}_{abc}^{ABCD} &= \langle \mathrm{T} U_{CDc}(z_1) X_{abA}(z_2) X_{baB}(z_3)\rangle . \end{aligned} \tag{4.26}$$

(no sum on a, b). Making use in particular of (2.12), we find

$$\mathscr{G}^{ABCD}_{abc} = \frac{-(N-2)\cdot 2^{N+1}}{(N-1)!(N-1)(N+5)} \left(\frac{N-1}{2}!\right)^4 C_N \Lambda^{N+6} \varepsilon_{ab}$$

$$\times \sum_{A_5 \ldots A_{2N-8}=1}^{2N-8} \sum_{a_1 \ldots a_{N-6}=1}^{2} \varepsilon_{ABCD\,A_5 \ldots A_{2N-8}}$$

$$\times \delta^{a \ldots ab \ldots b}_{ca_1 \ldots a_{N-6}} \prod_{i=1}^{N-6} h_{A_{2i+3}A_{2i+4}a_i}$$

$$\mathscr{G}^{AB}_{ab} = \tfrac{1}{2} \sum_{CD=1}^{2N-8} \sum_{c=1}^{2} h_{CDc} \mathscr{G}^{ABCD}_{abc} \tag{4.27}$$

(here the upper indices a, b of $\delta^{\cdots}_{\cdots}$ appear $\frac{1}{2}(N-5)$ times, and $N \geqslant 7$; for $N=5$ only $\mathscr{G}^{AB}_{ab} = -\frac{8}{5}\varepsilon_{AB}\varepsilon_{ab}C_5\Lambda^{11}$ is nonzero).

As in the $N=6$ model discussed before, we do not get consistency relations such as in the previous models, but for the same reasons given there we do not trust the numerical value of (4.27). However again we interpret (4.27) as a clear evidence that for generic choice of the couplings h_{ABa} the Green functions (4.26) do not vanish, i.e. that no unexpected cancellations occur (in the particular case $N=5$ this has already been discussed in [5]). Upon clustering this means that supersymmetry is broken in this class of models, since it is easy to see [5] that the Konishi identities imply the vanishing of both $\langle S \rangle$ and $\langle X_{abA} \rangle$ and $\langle U_{ABa} \rangle$ in any supersymmetric vacuum.

(ii) $N=6$. The $N=6$ model considered above where consistency with the Konishi identities was found had the special property that the system of instanton induced Green functions was overdetermined. This property is shared by each SU(6) model with arbitrary number K of antisymmetric tensors and $2K$ fundamentals (and there are no other chiral SU(N) models with matter representations with no more than two indices which have this property). However in contrast to the $K=1$ case, the $K>1$ models also allow for consistency checks of the type found for $N \geqslant 8$. We consider the case $K=2$; the most general renormalizable superpotential is

$$P = \sum_{A,B=1}^{4} \sum_{a=1}^{2} h_{ABa}\phi_A\phi_B A_a + \sum_{a,b,c=1}^{2} h'_{abc} A_a A_b A_c, \tag{4.28}$$

with gauge indices contracted as in (4.20), h antisymmetric in A, B and h' completely symmetric.

For simplicity we consider only those Green functions which contain $A_a A_b A_c$ in the special combination $a=b=c$. Then for $P=0$ only

$$\mathscr{G} = \langle \mathrm{T} U_{ABa}(z_1) U_{CDb}(z_2) V_c(z_3) V_d(z_4) \rangle \tag{4.29}$$

receives a one-instanton contribution. We find

$$\mathcal{G} = \frac{3.2^7}{55} C_6 \Lambda^{12} \varepsilon_{ABCD} \varepsilon_{ab} \varepsilon_{cd} . \tag{4.30}$$

For $P \neq 0$ one or more of the composite matter fields U, V may be replaced by S; one gets

$$\langle \mathrm{T} S(z_1) U_{ABa}(z_2) V_c(z_3) V_d(z_4) \rangle = \tfrac{1}{2} \sum_{C,D=1}^{4} \sum_{b=1}^{2} h_{CDb} \mathcal{G} \tag{4.31}$$

and analogously for more replacements of U, V by S, up to

$$\langle \mathrm{T} S(z_1) S(z_2) S(z_3) S(z_4) \rangle$$

$$= \tfrac{1}{16} \sum_{A,B,C,D=1}^{4} \sum_{a,b,c,d=1}^{2} h_{ABa} h_{CDb} h'_{ccc} h'_{ddd} \mathcal{G} . \tag{4.32}$$

Again we find that these results are intrinsically inconsistent. We interpret this as further evidence that the results of the $K = 1$ model should not be trusted numerically although no inconsistency could be found there.

5. Summary and outlook

In this paper we have set up a supersymmetric formalism which enormously simplifies the calculation of instanton induced Green functions in supersymmetric gauge theories without classical expectation values of scalar fields. The formalism was applied to SQCD and to some chiral gauge theories with matter in the fundamental and antisymmetric tensor representation of SU(N), but could be used for models with matter in different representations as well. The only potentially cumbersome part of the computations is to set up the correct form of the "classical superfields" for the relevant gauge invariant combinations of chiral superfields (and possibly of products of these with certain combinations of fermionic collective coordinates, compare (2.11), (2.12)). Of course this part becomes more difficult for higher-dimensional representations of the gauge group.

In a particular SU(6) model we confirmed the result of [10] that the calculation without scalar VEVs is self-consistent and allows to deduce the correct form of the Konishi identities. On the other hand, in several classes of other models this approach was found to be inconsistent: we were able to identify, via clustering, a set of consistency relations among various Green functions; for generic choice of the superpotential our results violated these relations. However the outcome of two complementary approaches to solve this problem in SU(2) SQCD [8, 9] tells us that our results should be correct as order of magnitude estimates. In particular we argued that our calculations give reliable information on the vanishing or non-vanishing of various vacuum condensates; this enabled us to conclude that in $SU(N)$, N odd, with K antisymmetric tensors and $K\cdot(N-4)$ fundamentals, $K=1,2$, supersymmetry is dynamically broken.

Violation of certain consistency relations was found not only for SQCD but also for some chiral models where the approach of [7,8], i.e. inclusion of infrared singularities induced by nonzero modes, is not applicable. We therefore speculate that in order to overcome the encountered inconsistencies, classical scalar expectation values should be taken into account, even in theories which possess flat directions only if some couplings are set to zero.

It seems thus very desirable to extend the formalism presented here to the case of nonvanishing scalar VEV's. Since the solutions of scalar field equations with nonvanishing boundary condition induce contributions to the classical action which depend on the orientation of the instanton SU(2) in the gauge group, this formalism will necessarily include a nontrivial group integration (this is in contrast to the SU(2) case where the group integration is immediate). In general the calculation with scalar VEV's may therefore become considerably more complicated than those without VEV's described in the present paper.

It is a pleasure to thank I. Affleck for helpful discussions and a careful reading of the manuscript, and M.G. Schmidt for a useful correspondence.

References

[1] V.A. Novikov, M.A. Shifman, A.I. Vainshtein and V.I. Zakharov, Nucl. Phys. B229 (1983) 407; V.A. Novikov, M.A. Shifman, A.I. Vainshtein, V.B. Voloshin and V.I. Zakharov, Nucl. Phys. B229 (1983) 394
[2] V.A. Novikov, M.A. Shifman, A.I. Vainshtein and V.I. Zakharov, Nucl. Phys. B260 (1985) 157
[3] I. Affleck, M. Dine and N. Seiberg, Phys. Rev. Lett. 51 (1983) 1026; 52 (1984) 1677; Nucl. Phys. B241 (1984) 493
[4] I. Affleck, M. Dine and N. Seiberg, Phys. Lett. 137B (1984) 187
[5] Y. Meurice and G. Veneziano, Phys. Lett. 141B (1984) 69
[6] M.G. Schmidt, Phys. Lett. 141B (1984) 236
[7] D. Amati, G.C. Rossi and G. Veneziano, Nucl. Phys. B249 (1985) 1
[8] D. Amati, Y. Meurice, G.C. Rossi and G. Veneziano, Nucl. Phys. B263 (1986) 591
[9] J. Fuchs and M.G. Schmidt, Z. Phys. C30 (1986) 161
[10] A. Bicci and K. Konishi, Pisa preprint IFUP-TH-31/85
[11] L.S. Brown, R.D. Carlitz, D.B. Creamer and C. Lee, Phys. Lett. 70B (1977) 180
[12] E. Corrigan, P. Goddard and S. Templeton, Nucl. Phys. B151 (1979) 93
[13] C. Bernard, Phys. Rev. D19 (1979) 3013
[14] I. Affleck, M. Dine and N. Seiberg, Nucl. Phys. B256 (1985) 557
[15] T.E. Clark, O. Piguet and K. Sibold, Nucl. Phys. B159 (1979) 1; K. Konishi, Phys. Lett. 135B (1984) 439

Instantons in SU(N) supergluodynamics

V. A. Novikov

Institute of Theoretical and Experimental Physics, State Commission on Use of Atomic Energy

(Submitted 20 October 1986)
Yad. Fiz. **46**, 656–669 (August 1987)

A superfield formalism is developed for superinstantons in supersymmetric SU(N) gauge theories.

1. INTRODUCTION

Over the last two or three years it has become obvious that instantons[1] in supersymmetric theories result in considerably more dramatic physical effects than in the usual gauge theories, although the two cases, from the formal point of view, differ little from each other. This remarkable situation is explained by the fact that in supersymmetric theories the contribution in perturbation theory to many important quantities exactly vanishes to all orders in perturbation theory. Therefore in the absence of perturbative noise, nonperturbative fluctuations, in particular instantons, which are classical solutions in field theory, become more perceptible. The quantum field theory in this case to a remarkable degree becomes a classical theory.

Among the important achievements of superinstanton calculations[2] there should be mentioned the construction of field-theory models with weak coupling in which it is possible not only to carry out completely the calculation term by term in perturbation theory but also to calculate fundamental nonperturbative effects.[3,4] In these models the instantons, in bypassing nonrenormalization theorems, generate a nontrivial superpotential which leads to the long-expected dynamical breaking of supersymmetry[3,5] and gauge symmetry.[3,4] The appearance of four-dimensional field theories in which there is a clear structure of both the vacuum and the excitations over it is extremely important. It is not precluded that along this line it is possible to obtain still many interesting results. For example in recent work[6] it was shown that an explicit calculation of the degeneracy of the vacuum in supersymmetric models with gauge group O(N) is in contradiction with the indirect but generally accepted calculation of the Witten index.[7] This contradiction cannot be resolved without progress in our understanding of the dynamical symmetry breaking of supersymmetry.

The second interesting direction owes its beginning to superinstantons; this establishes exact relations between different anomalous dimensions, in particular, the calculation of the exact Gell-Mann–Low function in $N = 1$ supersymmetric gauge theories, which, as it turns out, depends completely nontrivially on the gauge charge.[8] Recently these relations were confirmed by a direct analysis of Feynman graphs.[9] In any case instantons in supersymmetric theories are undoubtedly interesting objects.

At the present time there are two approaches to the description of superinstantons. The first, which would seem to be the simplest and most straightforward, considers the instanton as an ordinary classical solution for a boson field (or fields) and the corresponding fermion superpartners appear as matter fields. This approach to supersymmetric theories in no way distinguishes them from the general class of gauge theories. Here the presence of supersymmetry is not at all obvious. Moreover, it turned out that instantons of fixed size actually break supersymmetry.[2a] In the second approach the gauge and fermion fields enter fully on the same basis as components of one classical superfield.[2] In such a superfield approach supersymmetry is obvious at each intermediate stage and the puzzle of the explicit breaking of supersymmetry by instantons finds an unexpected and beautiful solution. This superfield approach was developed in all detail for the SU(2) group in our work of Refs. 2a and b.

It should be noted that a substantial part of the enumerated results, which were obtained with the help of instantons, concerns theories with a larger symmetry group than SU(2). However, in these cases it was sufficient to know the very simple facts of the superinstanton formalism, say, the number of zero modes or the invariance of the integration measure, or that the instantons give a nonzero contribution. When it was necessary to make concrete calculations, an indirect way was not found and they became so cumbersome that the authors decided not to carry them out on the pages of the journals. For the SU(2) group the superfield formalism made it possible to obtain the result in one or two short lines, so that the advantages of the superfield approach are quite obvious. Attempts to develop this approach for the SU(N) group were undertaken in Ref. 10; some simple linear formulas were guessed, which directly led to significant simplifications in the calculations. However,the complete formalism for superinstantons in an arbitrary group and even more so for theories with matter (which represent the greatest interset from the physical point of view) has up to now not been developed. In this article we fill this gap by giving a detailed description of the superfield formalism for SU(N) instantons and show how to generalize the results to arbitrary groups.

In Sec. 2 it is recalled how a similar formalism was constructed for the SU(2) group, and a new formula is given for matter superfields in higher representations. In Sec. 3 a formalism is constructed for SU(N) supersymmetric gauge theories without matter. Unfortunately, in a finite size article we did not succeed in developing the corresponding apparatus for superinstantons for an arbitrary group or for gauge theories with matter. These problems will be considered in a separate article.

2. SUPERINSTANTONS IN THE SU(2) GROUP

Superinstantons of general form for the SU(2) gauge group were studied in detail in Ref. 2. We wish to construct the superfield formalism for instantons in an arbitrary group. This section is introduced as follows: here we will fix a convenient format for us; let us write the general scheme of the solution of the problem and illustrate it with the example of the known SU(2) formulas[2] (in the following, they will

 0038-5506/87/080366-08$03.80

always be used). In the course of the presentation we will correct some of the inaccuracies, which exist in Ref. 2b, and also we will show how to generalize the solution obtained to the case of the SU(2) theory with a large number of multiplets of matter in arbitrary representations.

Let us start from the classical instanton[1] in Minkowski spacetime $E^{3,1}$.[1] All instanton formulas appear in a more natural form in the Van der Waerden formalism. In this notation the tensor intensity $G^a_{\mu\nu}$ is decomposed into two symmetric spinors $G_{\alpha\beta} \sim (1,0)$ and $G_{\dot\alpha\dot\beta} \sim (0,1)$ (see for example Ref. 11) and the instanton field takes the form

$$(G_{\dot\alpha\dot\beta})_{ik}=0, \qquad (1)$$
$$(G_{\alpha\beta})_{ik}=-\frac{4i}{g}\frac{1}{\rho^2 f^2}(\varepsilon_{\alpha i}\varepsilon_{\beta k}+\varepsilon_{\alpha k}\varepsilon_{\beta i}),$$

where $f = 1 + x^2\rho^{-2}$ and $\alpha, \beta = 1, 2$ are the spatial indices; $i, k = 1, 2$ are the are the isotopic indices, and

$$\varepsilon_{\alpha i}=-\varepsilon_{i\alpha}, \quad \varepsilon_{12}=1, \quad G_{ik}=\varepsilon_{kl}G^a(\sigma^a/2)_i{}^l,$$

$$x^2=\sum_{a=1}^{3} x_a x_a - x_0 x_0.$$

The supersymmetric SU(2) theory together with the vector field $(A_{\alpha\dot\alpha})_{ik}$ contains the spinors $(\lambda_\alpha)_{ik}$ and $(\bar\lambda_{\dot\alpha})_{ik}$ and an additional scalar field $(D)_{ik}$ (in the Wess-Zumino gauge) which are different components of one superfield—the prepotential $(V)_{ik}$ or the superfield of the decomposition $(W_\alpha)_{ik}$ and $(\overline{W}_{\dot\alpha})_{ik}$:

$$(W_\alpha)_{ik}=\{\lambda_\alpha+\theta^\beta(2iG_{\alpha\beta}+D\varepsilon_{\alpha\beta})+{}^1/_2\theta^\beta\theta_\beta(2iD_{\alpha\dot\alpha}\bar\lambda^{\dot\alpha})\}_{ik},$$
$$(\overline{W}_{\dot\alpha})_{ik}=\{\bar\lambda_{\dot\alpha}+\bar\theta^{\dot\beta}(-2iG_{\dot\alpha\dot\beta}+D\varepsilon_{\dot\alpha\dot\beta})+{}^1/_2\bar\theta_{\dot\beta}\bar\theta^{\dot\beta}(2iD_{\alpha\dot\alpha}\lambda^\alpha)\}_{ik}. \qquad (2)$$

If the solution for the instanton field $G_{\alpha\beta}$ and $G_{\dot\alpha\dot\beta}$ from Eq. (1) is substituted into Eqs. (2) and $\lambda_\alpha = 0$, $\bar\lambda_{\dot\alpha} = 0$, and $D = 0$, then we obviously reproduce some part of the solution of the classical equations of motion, when only one of the components of the superfield W_α is different from zero. The problem is that, in order to construct the most general form of the supermultiplet of solutions and to clarify the number of parameters on which it depends, it is essential to introduce a convenient system of coordinates on the space of solutions and find their transformation law under the action of supersymmetry. The latter is important since it would be desirable to simplify in a maximal way the explicit integration over all instantons and make the supersymmetry of the answers, as obvious as possible at every stage.

If the total symmetry group of the equations of motion is known, then the solution of this problem is significantly simplified. With the help of the symmetry transformations it is possible to construct from a particular solution a whole family of new solutions where the transformation parameters appear as very convenient coordinates in the space of solutions. As a trivial example, with the help of the translation operator $g(x_0) = e^{iPx_0}$ it is possible to "push out" solution (1) from the origin and construct a more general solution with the center at an arbitrary point:

$$G_{\alpha\beta}(x_0)=g(x_0)G_{\alpha\beta}(0)g^{-1}(x_0).$$

The generalization of this construction is described on the first pages of any textbook on Lie groups. Let the fixed space of solutions $\{\varphi\}$ be given along with the action of the symmetry group $G = \{g\}$ on this space $\varphi_0 \to \varphi = g\varphi_0 g^{-1}$. Usually not all elements of the group move the solution φ_0 and there is a whole stability subgroup of φ_0: $H_{\varphi_0} = \{h: h\varphi_0 h^{-1} = \varphi_0\}$, i.e., the space of solutions $\{\varphi\}$ corresponds to the quotient space G/H. In order to introduce coordinates it is convenient to consider the one-parameter subgroups

$$g(\alpha_i)=e^{i\alpha_i G_i},$$

where the G_i are the generators of the group G which do not belong to H, i.e., which in fact do move the solution φ_0. Then it is possible to construct a one-parameter orbit of solutions $g_i\varphi_0 g_i^{-1}$, and by moving along different orbits to obtain the still more general solution

$$\varphi(\alpha_1,\ldots,\alpha_n)=g(\alpha_n)\ldots g(\alpha_1)\varphi_0 g^{-1}(\alpha_1)\ldots g^{-1}(\alpha_n). \qquad (3)$$

The parameters $\{\alpha_i\}$ form the whole underlying coordinate grid for all such solutions. The action of an arbitrary element g of the group on the general solution (3) can always be represented in the form

$$g\cdot g(\alpha_n)\ldots g(\alpha_1)=g(\alpha_n')\ldots g(\alpha_1')\cdot h,$$

where h is a transformation belonging to the stability group H_{φ_0}, and $g\varphi g^{-1}$ is rewritten in the same parameterization, but with new values of the coordinates $\{\alpha_i'\}$. In so far as the different generators do not commute with each other, the $\{\alpha'\}$ are related to the $\{\alpha\}$, generally speaking, nonlinearly. In supersymmetric theories one is obliged to work with graded Lie groups. In the present case this means only that some of the parameters $\{\alpha\}$ are Grassman numbers (more generally, odd elements). Correspondingly some of the commutators have to be changed to anticommutators. The simplest example of a realization of the general procedure written above is the construction of chiral superspace $(x_{\alpha\dot\alpha}, \theta_\alpha)$.

Let us now turn to our concrete problem of construction of the general solution of the classical equations of motion. The symmetry group of these equations is the SU(2) group of global isotopic rotations T_{ik} and the superconformal group, which includes the Lorentz rotations $M_{\alpha\beta}$ and $\overline{M}_{\dot\alpha\dot\beta}$, the translation $P_{\alpha\dot\alpha}$, the conformal translation $K_{\alpha\dot\alpha}$, the dilatation D, the supertransformations Q_α and $\overline{Q}_{\dot\alpha}$, the superconformal transformations S_α and $\overline{S}_{\dot\alpha}$, and the chiral rotations Π. The algebra of these generators and also the realization of the transformations on the superfields were described in Ref. 11.

The initial configuration of the superfields has the form

$$\overline{W}_{\dot\alpha}=0,$$
$$W_\alpha=\theta^\beta 2i(G_{\alpha\beta})_{ik}=-\frac{8}{g\rho^2 f^2}(\varepsilon_{\alpha i}\theta_k+\varepsilon_{\alpha k}\theta_i). \qquad (4)$$

Starting from the explicit expression (4) and the formula for the transformation of the fields one can show that $M_{\alpha\beta}$, $\overline{Q}_{\dot\alpha}$, S_α, $K_{\alpha\dot\alpha} + 2\rho^2 P_{\alpha\dot\alpha}$, and $T_{\alpha\beta} + M_{\alpha\beta}$ do not move the solution (4). Therefore as the generators of the stability group H it is possible to choose $\{M, \overline{M}, \Pi, K, S, \overline{Q}\}$, and the displacement operator, which carries the solution (4) along the whole space of solutions, can be chosen in the form

$$L=g(\omega)g(x_0)g(\theta_0)g(\bar\beta)g(\rho)$$
$$=\exp(iT_{ik}\omega^{ik})\exp[(i/2)P_{\alpha\dot\alpha}x_0^{\alpha\dot\alpha}]$$

$$\times \exp(-iQ^{\alpha}\theta_{0\alpha})\exp(-i\bar{S}_{\dot{\alpha}}\bar{\beta}^{\dot{\alpha}})\exp[iD\ln(\rho/\rho_0)]. \quad (5)$$

Simple calculations show that

$$e^{-i\bar{\epsilon}\bar{Q}}L = L'[1 - M_{\dot{\alpha}\dot{\beta}}\bar{\beta}^{\dot{\alpha}}\bar{\epsilon}^{\dot{\beta}} + \bar{\epsilon}\bar{\beta}(2D + 4i\Pi)]e^{-i\bar{\epsilon}\bar{Q}},$$

where the new coordinates θ'_0, β', and ρ' are related to the old by formulas of Ref. 2a:

$$\delta(x_0)_{\alpha\dot{\alpha}} = -4i\theta_{\alpha}\bar{\epsilon}_{\dot{\alpha}}, \quad \delta(\theta_0)_{\alpha} = \epsilon_{\alpha}, \quad (6)$$

$$\delta\rho^2 = (-4i\bar{\epsilon}\bar{\beta})\rho^2, \quad \delta(\bar{\beta})_{\dot{\alpha}} = -4i(\bar{\epsilon}\bar{\beta})\bar{\beta}_{\dot{\alpha}}.$$

The general solution for the superinstanton using the displacement L is rewritten in the form

$$V' = LV'_0L^{-1},$$

where V_0 is the supermultiplet, which corresponds to the initial superinstanton solution (4) with dimension ρ_0 and the center at $x_0 = 0$. Here one remark should be made. If for V_0 the prepotential $(V)_{ik}$ or the superdecomposition $(W_{\alpha})_{ik}$ is chosen, then the transformation $g(\theta_0)$ develops V_0 from the Wess-Zumino gauge, i.e., transforms the physical components into auxiliary fields. In order to return to the physical gauge it is necessary to supplement the supertransformation $g(\theta_0)$ with a supergauge transformation. Then the supersymmetry algebra changes and instead of the translations $P_{\alpha\dot{\alpha}} = i\partial_{\alpha\dot{\alpha}}$ the covariant derivative $iD_{\alpha\dot{\alpha}}$ appears, which somewhat complicates the solution of the problem. If V_0 is identified with the singlet superfield say $W^{\alpha}{}_{ik}W^{ik}{}_{\alpha}$, then for it $D\equiv\partial$ and the algebra reduces to the superconformal symmetry.

The explicit form of the superfields is most simply found with the help of invariants.[2b] Let us use the fact that the action of any transformation $U(\Lambda)$ can relate to either the coordinates $z = (x, \theta)$ or as we saw, to the instanton parameters $\{\alpha\}$,

$$\Phi'(z|\alpha) = U(\Lambda)\Phi(z|\alpha)U^{-1} = S(\Lambda)\Phi(u(\Lambda)z|\alpha)$$
$$= S(\Lambda)\Phi(z|v(\Lambda)\alpha),$$

where $u(\Lambda)$ and $v(\Lambda)$ are respectively the transformation matrices of coordinates and of parameters and $S(\Lambda)$ is the transformation matrix of the multiplet (for example, the transformation matrix of the spinor for W_{α}). Therefore

$$\Phi(z|\alpha) = \Phi(u(\Lambda)z|v^{-1}(\Lambda)\alpha),$$

i.e. if we make a transformation of the coordinates and the inverse transformation of the parameters, then Φ remains invariant. This indicates that the superfields must depend on the invariants. It is easy to check that

$$\delta(\theta - \theta_0) = 0, \quad \delta(\bar{\theta}_{\dot{\alpha}}/\rho^2) = 0,$$

$$\delta(x - x_0)_{\alpha\dot{\alpha}} = -4i(\theta - \theta_0)_{\alpha}\bar{\epsilon}_{\dot{\alpha}},$$

$$\delta(x - x_0)^2/\rho^2 = -4i\tilde{\theta}_{\alpha}\bar{\epsilon}_{\dot{\alpha}}(x - x_0)^{\alpha\dot{\alpha}}/\rho^2,$$

where $\tilde{\theta}_{\alpha} = (\theta - \theta_0)_{\alpha} + (x + x_0)_{\alpha\dot{\alpha}}\bar{\beta}^{\dot{\alpha}}$. Therefore there are two forms of supersymmetric fields:

$$\Phi_1^{\text{inv}} = (\theta-\theta_0)^2 f_1(x-x_0), \quad \Phi_2^{\text{inv}} = (\tilde{\theta}^2/\rho^4)f_2[(x-x_0)^2/\rho^2].$$

However, Φ_1^{inv} is not invariant relative to superconformal transformations. Thus, the form of the instanton is completely fixed. Comparing with the original solution (4), we find[2a]

$$W^{\alpha}W_{\alpha} = \frac{2^7\cdot 3}{g^2}\frac{\tilde{\theta}^2}{\rho^4}\frac{1}{f^4}. \quad (7)$$

Considering expression (7) by components, one can be convinced that it satisfies $D_{ik} = 0$, and corresponds to the four fermion zero modes $(\lambda^{\text{n}}_{\alpha})_{ik}$ and $(\lambda^{\text{sc}}_{\alpha})_{ik}$ and the former expression for $[G_{\alpha\beta}]_{ik}$. [Since zero modes transform as $(\frac{1}{2},0)$ it is impossible to construct from them a vector current $j_{\alpha\dot{\alpha}} \sim (\frac{1}{2},\frac{1}{2})$ and the Yang-Mills equation remains free.] The index theorem guarantees that the number of zero modes is exactly four. Thus, Eq. (7) gives the general solution, and of the geometric transformations of the superconformal symmetry it turns out to be sufficient to attain this general solution.

An analogous situation occurs when the chiral matter fields Q^i and $\bar{Q}_i$ are present on the same level with the gauge fields in the theory:

$$Q^i = A^i + \theta^{\alpha}\psi_{\alpha}{}^i + {}^1\!/_2\theta^2F^i, \quad \bar{Q}_i = \bar{A}_i + \theta^{\alpha}\bar{\psi}_{\alpha i} + {}^1\!/_2\theta^2\bar{F}_i$$

with the nonzero vacuum expectation values $\langle A\rangle \neq 0$. In this case there does not exist a solution of the exact equations of motion, but the equations turn out to be perfectly sensible when the influence of matter on the multiplet of gauge fields is neglected.[12] The index theorem now requires the appearance of two additional fermion matter modes. But in this case using the geometric transformations the general solutions can be obtained.

Besides the instanton field, the initial configuration contains a solution to the equations $D^2A = D^2\bar{A} = 0$ for the scalar components

$$A^i = x^i_{\dot{\alpha}}v^{\dot{\alpha}}/\rho f^{1/2}, \quad \bar{A}_i = x^{\dot{\alpha}}_i w_{\dot{\alpha}}/\rho f^{1/2}. \quad (8)$$

The vacuum expectation values $\langle A^i\rangle = v^i$ and $\bar{A}_i = w_i$ must lie in valleys in order not to break supersymmetry. Not pausing on this well-known subject let us write the solution for the valley in the present case:

$$v^i = v\delta^{i1}, \quad w_i = v\delta_{i1}. \quad (9)$$

[If an arbitrary gauge transformation is made on Eqs. (9) then again the obtained fields will also be solutions.]

The stability group H of the original configuration is changed. The supersymmetry transformation $e^{-i\bar{Q}\bar{\epsilon}}$ leads now to a variation of Q and $\bar{Q}$:

$$\delta A^i = 0, \quad \delta\psi_{\alpha}{}^i = -2i\bar{\epsilon}^{\dot{\alpha}}(D_{\alpha\dot{\alpha}}A)^i = 4i\delta_{\alpha}{}^i(v^{\dot{\alpha}}\bar{\epsilon}_{\dot{\alpha}}/\rho f^{1/2}),$$

$$\delta F^i = 0,$$

i.e., generates the desired zero fermion mode. It is different and slightly more complicated with the superconformal transformation $e^{i\bar{S}\bar{\beta}}$. The assertion, which exists in the literature, that the vacuum expectation value destroys the conformal symmetry to the extent that the transformation $g(\beta)$ does not generate solutions of the Dirac equation[10a] from a solution of $D^2A = 0$ is mistaken. A realization of the superconformal symmetry on chiral superfields is possible by determining a relation between the dilatation dimension Δ and the chiral charge r of the field Q (Ref. 11):

$$[D + 2i\Pi, Q^i(0)] = -i(\Delta - 2r)Q^i(0) = 0.$$

In this case the variation of the field equals

$$\delta\psi_\alpha{}^i = -2ix^{\dot\gamma}\beta_\gamma D_{\alpha\dot\gamma}A^i - 4i\beta_\alpha A^i$$

and, as is easily verified, satisfies the Dirac equation if $D^2A^i = 0$. However, substitution of the explicit solution (8) shows that the obtained solution is not normalizable and not suitable for our purposes. Thus, β is not a coordinate in the space of normalizable solutions and the displacement operator has the form

$$L = g(\omega)g(x_0)g(\theta_0)g(\beta)g(\bar\theta_0)g(\rho), \quad g(\bar\theta_0) = e^{-i\bar\theta_0\bar Q}.$$

The supersymmetry transformation now appears as

$$g(\bar\varepsilon)L = L'[1 - \bar\beta_{\dot\alpha}\bar\varepsilon_{\dot\beta}\bar M^{\dot\alpha\dot\beta} + \bar\varepsilon\bar\beta(2D + 4i\Pi)]g(\bar\varepsilon),$$

where $\bar\theta_0' = \bar\theta_0 + \delta\bar\theta_0$ and $\delta\bar\theta_0 = \bar\varepsilon - 4i(\bar\varepsilon\bar\theta_0)\bar\beta$ and the old parameters are transformed according to the old formulas. The new parameter $\bar\theta_0$ leads to the new invariants

$$\rho_{\rm inv}^2 = \rho^2(1 + 4i\bar\theta_0\bar\beta), \quad \bar x^2/\rho^2,$$

where $\bar x_{\alpha\dot\alpha} = (x - x_0)_{\alpha\dot\alpha} + 4i\bar\theta_\alpha\bar\theta_{0\dot\alpha}$. As a result the general solution for the singlet superfield $Q^i\bar Q_i$ is written in the form [2b]

$$Q^i\bar Q_i = \bar x^2 v^2/\rho^2 \bar f, \quad \bar f = 1 + \bar x^2/\rho^2. \qquad (10)$$

It is possible to solve this equation componentwise and be convinced that

$$A^i = \left\{\frac{x_{\dot\alpha}{}^i}{\rho f^{1/2}} - \frac{4i[\theta_0 - (x - x_0)\bar\beta]^i\bar\theta_{0\dot\alpha}}{\rho f^{3/2}}\right\}v^{\dot\alpha},$$

$$\psi_\alpha{}^i = \frac{4i}{\rho f^{3/2}}\delta_\alpha{}^i\bar\theta_{0\dot\alpha}v^{\dot\alpha}, \quad F^i = 0. \qquad (11)$$

Thus, $\bar\theta_0$ actually generates the necessary fermion mode of matter. The variation of A^i, which is quadratic in the old elements, is due to the Yukawa interaction of $(A^+)_i$ with the gauge mode $\lambda \sim \theta_0 + \chi\bar\beta$ and the matter mode $\psi \sim \bar\theta_0$.[2b] The chiral multiplet

$$Q^i_{\rm naive} = (\bar x_{\dot\alpha}{}^i/\rho)(v^{\dot\alpha}/f^{1/2}) \qquad (12)$$

does not reproduce formula (11) (for example, it gives $F^i \neq 0$), since the supersymmetric transformations evolve multiplets out of the Wess-Zumino gauge. In the present case it is easy to find an explicit expression for a supergauge transformation which returns Q^i to the physical gauge:

$$Q^i = Q^j_{\rm naive}(e^{-i\Lambda})_j{}^i, \quad \bar Q_i = (e^{+i\Lambda})_i{}^j\bar Q_{j\,\rm naive}, \qquad (13)$$

$$\Lambda_j{}^i = -2ig(V^{\rm inst}_{\alpha\dot\alpha})_j{}^i\bar\theta^\alpha\bar\theta_0{}^{\dot\alpha} = \frac{2}{f\rho^2}[\delta_j{}^i\bar\theta_\alpha\bar\theta_{0\dot\alpha}x^{\alpha\dot\alpha} - 2\bar\theta_j(\bar\theta_{0\dot\alpha}x^{\dot\alpha i})].$$

The components of the superfields Q and $\bar Q$ now reproduce the solution (11). In singlet combinations the matrix Λ, naturally, is not reproduced. It is obvious that this matrix returns to the Wess-Zumino gauge also the gauge multiplet W_α.

Let us dwell on a question, important for what follows, the treatment of which was in error in Ref. 2b. It is easy to convince oneself that[2b]

$$\left(\frac{\bar x_{ik}}{\rho}\right)' = \omega_k^i\left(\frac{\bar x_{il}}{\rho}\right), \quad \omega_k^i = \delta_k^i(1 + 2i\bar\varepsilon\bar\beta) + 4i\bar\beta^i\bar\varepsilon_k,$$

which at first glance leads to a variation of the function Q, which is invariant by construction. Let us recall, however, that

$$g(\bar\varepsilon)L = L'[1 - \bar\beta\bar M\bar\varepsilon + \bar\varepsilon\bar\beta(2D + 4i\Pi)]g(\bar\varepsilon)$$

and in addition to the coordinate transformations one should calculate the corresponding commutators of $\bar M$, D, and Π with the original configuration of fields. As already noticed, $(2D + 4i\Pi) \sim 0$ on the chiral multiplets and $\bar M \sim 0$ on scalar functions. However, there does not exist a scalar function in the usual sense as a solution for the scalar field. It is easy to convince oneself that

$$[-\bar\beta\bar M\bar\varepsilon, x_{ik}] = (-2i)[(\bar\varepsilon x)_i\bar\beta_k - (\bar\beta x)_i\bar\varepsilon_k],$$

which exactly compensates the variation of Q^i and $\bar Q_i$ from the coordinate transformations, i.e., both Q^i and $\bar Q_i$ are invariant functions. This spatial rotation is conveniently rewritten in terms of an isotopic rotation of the vacuum expectation value

$$(v')^{\dot i} = \omega_{\dot i}^{\dot k}v^{\dot i}.$$

Then the invariance of Q^i will be obvious.

It is interesting to see what happens to other matter multiplets, namely to the antichiral fields $(Q^+)_i$ and $(\bar Q^+)^i$. It is obvious that for the scalar field A_i^+ in the instanton field there is a nontrivial solution

$$A_i^+ = x_{i\dot k}v^{\dot k}/\rho f^{1/2}.$$

It is no less obvious that the fermion mode of the Dirac equation is absent.[12] However, the transformation $e^{iQ\varepsilon}$ leaves Q_i^+ in place and generates the decaying spinor field

$$\delta_\varepsilon\psi^+_{\dot\alpha i} = 2i\varepsilon^\alpha(D_{\alpha\dot\alpha}A^+)_i.$$

One is easily convinced that $\delta_\varepsilon\psi^+$ is a solution of the Dirac equation with right-hand side $D^{\alpha\dot\alpha}\delta_\varepsilon\psi^+_{\dot\alpha i} = ig[(\delta_\varepsilon\lambda_\alpha)A^+]_i$, as it must be since in the Lagrangian there appeared Yukawa terms $\mathscr{L} \sim g\psi A^+\lambda$. For us it is important that $\delta\psi^+$ is completely determined by the field $\delta_\varepsilon\lambda$. Consequently it is not necessary to introduce a new odd parameter into this fermion solution. It is evident that the stability group H was not changed from the original configuration by the inclusion of the fields Q^+ and $\bar Q^+$. Therefore all of the invariants considered remained invariant, but there arose a new one

$$z = (x + 4i\theta\bar\theta) - (x_0 + 4i\theta_0\bar\theta),$$

which corresponds to the transition to the antichiral coordinates. As a result

$$\bar Q^+Q^+ = v^2z^2/(z^2 + \rho^2_{\rm inv}). \qquad (14)$$

These examples exhaust the circle of problems in which the general solution can be obtained from a unique solution using transformations of the superconformal group. In all the following problems whole surfaces in the space of solutions must be specified so that starting from them it will be possible to reach an arbitrary solution. The motion along this surface is not described by transformations from the superconformal group; the coordinates on it must be introduced by hand. The simplest example is the matter fields Q^i and $\bar Q_i$ with vanishing vacuum expectation value. In this case $g(\bar\varepsilon) \in H$, and using supertransformations fails to excite matter modes. Therefore as the initial configuration, together with the instanton fields, we must take the zero fermion modes

$$Q^i=-\theta^i\chi/\rho^2 f^{3/2},\quad \bar Q_i=-\bar\theta_i\zeta/\rho^2 f^{3/2}.$$

The stability group H corresponds to the case of a pure gauge field. The general solution has the form

$$Q^i\bar Q_i=-(\theta^2/\rho^4)(\chi\zeta/f^3). \qquad (15)$$

From the requirement of invariance of $Q\bar Q$ it follows that the coordinates ζ and χ are not transformed under the action of supersymmetry. If we turn to the fields themselves, then

$$Q^i=-(\theta^i/\rho^2)(\chi/f^{3/2}),\quad \bar Q_i=-(\bar\theta_i/\rho^2)(\zeta/f^{3/2}).$$

For them a supersymmetry transformation is equivalent to the supergauge transformation $e^{i\Lambda(\epsilon)}$. The considered example is not representative since there is a smooth transition of the "geometrical" multiplet (10) into the one in Eq. (15) for $v\to 0$.[2b] If, however, several matter multiplets Q_p^i, $\bar Q_{iq}$, p, $q=1,\ldots,N_f$ are introduced, then it becomes clear that the parameters of the superconformal group cannot be avoided. The solution of this problem is obvious when the vacuum expectation values of the fields are zero:

$$Q_p{}^i\bar Q_{iq}=-(\theta^2/\rho^4)(\chi_p\zeta_q/f^3),\quad \delta[\chi_p\zeta_q]=0. \qquad (16)$$

The problem with nonzero vacuum expectation values is a little complicated. Admissable vacuum expectation values up to $SU(N_f)$ transformations have the form[13]

$$v_p{}^i=\langle A_p{}^i\rangle=\begin{cases} v_p\delta_p{}^i, & p=1,2\\ 0, & p\geqslant 3\end{cases},$$

$$w_{iq}=\langle\bar A_{iq}\rangle=\begin{cases}\sqrt{c^2+v_q{}^2}\,\delta_{iq}, & q=1,2\\ 0, & q\geqslant 3\end{cases}.$$

Then the general solution for the $Q_p{}^i$ can be written in the form

$$Q_1{}^i=(e^{-i\Lambda})_j{}^i(\theta^j{}_{\dot\alpha}v\bar\theta^{\dot\alpha}{}_1/\rho f^{1/2}), \qquad (17a)$$

$$Q_2{}^i=(e^{-i\Lambda})_j{}^i[\theta^j{}_{\dot\alpha}v_2\bar\theta^{\dot\alpha}/\rho f^{1/2}-\bar\theta^j\chi_2/\rho^2 f^{3/2}], \qquad (17b)$$

$$Q_p{}^i=(e^{-i\Lambda})_j{}^i[-\bar\theta^j\chi_p/\rho^2\tilde f^{3/2}],\quad p\geqslant 3. \qquad (17c)$$

Let us turn our attention to the fact that the $Q_p{}^i$ are written in Eqs. (17) in terms of $\tilde f$, and not f as in Eqs. (16). It is easy to convince oneself that

$$e^{-i\Lambda}(\bar\theta\chi/\rho^2\tilde f^{3/2})=(\theta/\rho^2)(\chi/f^{3/2}).$$

Analogous formulas can be written for $\bar Q$. The proposed solution, unfortunately, is not very symmetric; the fermion mode Q_1 is described by the parameter $\bar\theta_0$, and Q_2 by a mixture of the parameters $\bar\theta_0$ and χ_2. If the reality of the SU(2) representations is used, one can rewrite Eqs. (17) more symmetrically so that only one flavor will have a vacuum expectation value. If care is not taken to diagonalize the vacuum expectation values, then the general solution can be written as in Eq. (17a) for Q_1 and as in (17b) for all the remaining Q_p. Such a simple form of the solution is related to the fact that $\overline{M\varepsilon\beta}$ rotates all vacuum expectation values $v_p{}^i$ by the same angle, independent of the flavor, so that $\bar xv/\rho\tilde f^{3/2}$ is invariant, like $\bar\theta\chi/\rho f^{3/2}$.

Let us now consider matter in an arbitrary representation. The problem for scalar and fermion fields in principle was already solved by 't Hooft.[12] The superinstanton with matter in the adjoint representation was partially considered in Ref. 10a. We wish to direct our attention to the fact that the structure of the general solution is indeed very simple and it is already described through the known formulas for $T=\frac12$. Let us consider

$$A_{i_1\ldots i_n}=\hat P\prod_{l=1}^{n}A_{i_l,l}, \qquad (18)$$

where $A_{i_l,l}=x_{i_lk}v_l{}^k/\rho f^{1/2}$ is a solution for $T=1/2$ and $\hat P$ is the symmetrization of the indices $\{i_l\}$. Since

$$D^2A_{i_l,l}=0,$$

$$D^{\alpha\dot\alpha}A_{i_l,l}D_{\alpha\dot\alpha}A_{i_m,m}=\left(\frac{2}{\rho f^{1/2}}\right)^2(\delta^\alpha_{i_l}v_l^{\dot\alpha})(\varepsilon_{\alpha,i_m}v_{\dot\alpha,m})\sim\varepsilon_{i_l,i_m},$$

$$\hat P\varepsilon_{i_li_m}=0, \qquad (19)$$

the function (18) is a solution of the equation $D^2A=0$ and carries isospin $n/2$. In order to obtain the general solution it is sufficient to replace $\prod_{l=1}^{n}v_l^{k_l}$ by a symmetric isospinor $\eta^{k_1k_2\ldots k_n}$.

A supertransformation transforms $A_{i_l,k}$ to the zero mode $(\psi_\alpha)_{i_l,k}=(-2i\bar\varepsilon^{\dot\alpha})D_{\alpha\dot\alpha}A_{i_l,k}$ and analogously the field (18) to the mode

$$(\psi_\alpha)_{i_1\ldots i_n}=\hat P\sum_q A_{i_1,1}\ldots(\psi_\alpha)_{i_q,q}\ldots A_{i_n,n}. \qquad (20)$$

[For $n=2$ expression (20) agrees with the λ^{sc} mode of the gluino.] From formula (19) it is quite evident that Eq. (20) is a solution of the Dirac equation. [A superconformal transformation of the scalar field (18) leads to nonnormalizable solutions.] The remaining $(2n-2)$ fermion solutions can be obtained in the following way. Let us note that Eq. (20) leads to a solution since $D_{\alpha\dot\alpha}\psi^\alpha_{i_q,q}=0$ and $D^{\alpha\dot\alpha}A_{i_q,q}\sim\delta^\alpha_{i_q}v^{\dot\alpha}$. There exists still one function besides $A_{i,q}$, which by differentiation reproduces the necessary spinor structure $B_i=w_i/\rho f^{1/2}$:

$$D^{\alpha\dot\alpha}B_i=(2/\rho^2f^{1/2})\delta_i^\alpha(x_k{}^{\dot\alpha}w^k).$$

Therefore, if in Eq. (20) one of the functions $A_{i_l,l}$ is replaced by B_l, then the resulting expression will be a new solution of the Dirac equation (for $n=2$ it coincides with the λ^{sc} mode of the gluino). In order to write this fermion solution in a general form it is necessary to make the replacement

$$\hat P\sum_q\Big(\prod^{n\,\prime(r,q)}x_{i_l,k_l}v_l^{k_l}\Big)w_r(\varepsilon_{\alpha i_q}(\bar\varepsilon v_q))$$

$$\to\hat P\sum_q\Big(\prod_{l=1}^{n\,\prime(r,q)}x_{i_l,k_l}\Big)\varepsilon_{\alpha i_q}\eta_{i_r}^{(k_1\ldots k_{r-1}k_{r+1}\ldots k_{q-1}k_{q+1}\ldots k_n)},$$

where $\prod_{l=1}^{n\,\prime(r,q)}$ indicates that in the product the numbers (r and q) are excluded, and $\eta_{i_r}^{(k)}$ is an odd parameter which is symmetric in the upper isotopic indices. In order to obtain the next solutions it is necessary to make two, three, etc., replacements $A\to B$. Thus, all the fermion modes are reproduced. Now it is clear how to write the general form of the supermultiplet,

$$Q_{i_1\ldots i_n}=\Big[\prod_{l=1}^{n}\frac{\bar{x}_{i_l\dot{k}_l}}{\rho f^{1/2}}\Big]v^{\dot{k}_1\ldots\dot{k}_n}$$

$$+\hat{P}\sum_q\Big\{\frac{\bar{\theta}_{i_q}}{\rho^2 f^2}\Big[\prod_l^{n}{}'^{(r,q)}\frac{\bar{x}_{i_l\dot{k}_l}}{\rho f^{1/2}}\Big]\eta_{i_r}^{(\dot{k})}\Big\}+\cdots \quad (21)$$

Let us note that now the operator $\bar{\varepsilon}\overline{M}\bar{\beta}$ rotates not only the vacuum expectation values ν but also the parameters $\eta_i^{(k)}$, $\eta_{i_1}^{(k)},\ldots$, so that each term in Eq. (21) is invariant by itself. Thus, we have described all invariant supermultiplets in SU(2). For our work it is necessary still to construct superinvariant measures of integration. This problem was completely analyzed in Ref. 2.

3. THE SU(N) SUPERINSTANTON

Before examining superinstantons it is useful to stop and consider the classical BPST instantons in the SU(N) group.[14] An instanton implements a mapping of the subgroup of Lorentz rotations in the space of the SU(2) subgroup of the group of gauge transformations. In the present case it is obvious that by a general SU(N) rotation it is always possible to bring this subgroup (and therefore the instanton field) to the upper left corner so that

$$(G_{\alpha\beta})_i{}^k=\begin{cases}(G_{\alpha\beta})_i{}^k=\dfrac{4i}{g}\dfrac{1}{\rho^2 f^2}[\varepsilon_{\alpha i}\delta_\beta{}^k+\varepsilon_{\beta i}\delta_\alpha{}^k], & i,k=1,2,\\ (G_{\alpha\beta})_{\hat{i}}{}^{\hat{k}}=(G_{\alpha\beta})_i{}^{\hat{k}}=(G_{\alpha\beta})_{\hat{i}}{}^k=0, & \hat{i},\hat{k}=3,\ldots,N.\end{cases}$$

Then the SU($N-2$) transformations which act on the indices with the hat $\hat{i}$, $\hat{k}=3,\ldots,N$ will leave the instanton in place, and the SU(2) transformations will change the orientation of the instanton in isospace. Infinitesimal SU(2) transformations obviously produce three gauge zero modes

$$(\delta A^n_{\alpha\dot{\alpha}})_i{}^k,\qquad n=1,2,3,$$

which transform according to the adjoint representation of the SU(2) group. The remaining SU(N) transformations "deform" the instanton. As a result there arise ($N-2$) gauge modes

$$(\delta A_{\alpha\dot{\alpha}})_i{}^{\hat{k}},$$

which transform according to the fundamental representation 2 and as many modes

$$(\delta A_{\alpha\dot{\alpha}})_{\hat{i}}{}^k,$$

which transform according to the representation $\underline{2}$. Thus, altogether, SU(N) instantons have $[2(N-2)+\underline{3}]$ gauge zero modes.[14] Analogously, relative to SU(2) the classical superfield $(W_\alpha)_I{}^K$ decomposes into four types of chiral fields: 1) one superfield $(W_\alpha)_i{}^k$ in the adjoint representation, 2) (N − 2) superfields $(W_\alpha)_i^{\hat{k}}$ in the representation 2, 3) the same number of superfields $(W_\alpha)_{\hat{i}}^k$ in the representation $\underline{2}$, 4) $[(N-2)^2-1]$ fields $(W_\alpha)_{\hat{k}}^{\hat{i}}$ in the singlet representation.

The instanton field naturally lies in $(W_\alpha)_i{}^k$. The classical equations of motion for $(W_\alpha)_i^{\hat{k}}$ and $(W_\alpha)_{\hat{i}}^k$ componentwise reduce to the Dirac equation in an external instanton field for fermions and to the equation for the zero doublet gauge modes for bosons. As to $(W_\alpha)_{\hat{i}}^{\hat{k}}$, the equations of motion for the components reduce to the free equations, which have the one solution $(W_\alpha)_{\hat{i}}^{\hat{k}}=0$.

Thus, the problem of the SU(N) superinstanton without matter can always be reduced to the SU(2) instanton with matter. Another simple circumstance proves to be no less important: matter and an external field form one multiplet and transformations exist which mix one with the other. Let us begin the construction of the superinstanton by choosing an initial particular solution. In the SU(2) gauge group it was sufficient to begin with a particular boson solution. Transformations of the supersymmetry group of the equations of motion (the superconformal group) allow this particular solution to be carried over the whole space of solutions. In other words for SU(2) the space of solutions is homogeneous relative to the action of the superconformal group. This is impossible to hope for in the case of the SU(N) superinstanton. This is clear even from a calculation of the dimension of the group and of the space of solutions. In fact every solution of the Dirac equation of $(\lambda_\alpha)_i^{\hat{k}}$ or $(\lambda_\alpha)_{\hat{i}}^k$ depends on one odd parameter. Consequently, the odd dimensionality of the space of solutions equals $2N$. On the other hand, the superconformal group depends on a total of eight Grassman parameters [of which four of them then go into the construction of $(W_\alpha)_i{}^k$]. Thus, in order to reach the general solution we are forced to consider as the initial element not one boson solution, but a family of solutions which form a whole surface in the space of solutions,

$$(W_\alpha)_i^k=\theta^\beta(2iG_{\alpha\beta})_i^k,$$

$$(W_\alpha)_i{}^{\hat{k}}=(\lambda_\alpha)_i{}^{\hat{k}}=(8/g)(-\delta_\alpha{}^k)f^{-3/2}(\zeta_{\hat{i}}/\rho^2). \quad (22)$$

$$(W_\alpha)_{\hat{i}}{}^k=(\lambda_\alpha)_{\hat{i}}{}^k=(8/g)\,\varepsilon_{\alpha i}f^{-3/2}(\eta^{\hat{i}}/\rho^2).$$

Parametrization of the fermion components $(\lambda_\alpha)_i^{\hat{k}}$ and $(\lambda_\alpha)_{\hat{i}}^k$ in no way differs from the parametrization $(\psi_\alpha)_i{}^f$ of chiral matter fields for the SU(2) superinstanton considered in the preceding section. Only now the role of the flavor index is played by the group index $\hat{k}=3,\ldots,N$. In order to obtain the general solution it is necessary, as earlier, to apply the generalized displacement operator (5) to the particular solution (22). Again, in order not to complicate the problem by supergauge transformations which rotate the solution to the Wess-Zumino gauge, it is convenient to consider the singlet combination

$$(W^\alpha)_K{}^I(W_\alpha)_I{}^K=\frac{3\cdot 2^7}{g^2}\frac{1}{f^3}\frac{\theta^2}{\rho^4}-\frac{2^8}{g^2}\frac{1}{f^3}\frac{\eta^{\hat{i}}\zeta_{\hat{i}}}{\rho^4}. \quad \text{(AI)}$$

Further, following the prescription which was developed for the theory of the SU(2) superinstanton with a chiral doublet of matter, it becomes desirable to make the substitutions $\theta_\alpha\to\bar{\theta}_\alpha$ and $x_{\alpha\dot{\alpha}}\to\bar{x}_{\alpha\dot{\alpha}}$ (more precisely, $\theta/\rho^2\to\bar{\theta}/\rho^2$, $x^2/\rho^2\to\bar{x}^2/\rho^2$, and $\rho^2\to\rho^2_{\rm inv}$) and to obtain an invariant superfield. This prescription, which is certainly correct for matter fields, in the present case leads to grossly incorrect formulas. This would be clear from the fact that the Grassman parameter $\bar{\theta}_0$, which corresponds to the transformation $e^{-i\bar{Q}\bar{\theta}_0}$, is completely superfluous in the considered problem. Upon integration the superfluous Grassman parameters lead to catastrophic results.

Let us use the fact that the fields (22) form one multiplet, and calculate the stability group of the solution (22).

Let us apply the displacement operator $g(\bar{\varepsilon}) = e^{-i\bar{Q}\bar{\varepsilon}}$. The instanton field $(W_\alpha)_i{}^k$ is not changed under the action of this transformation. As concerns the "matter" fields $(W_\alpha)_i^{\hat{k}}$ and $(W_\alpha)_{\hat{i}}^{k}$, since their first components are nonzero (as also in the case of chiral matter with nonzero scalar fields), in fact there arise the variations of the solutions

$$(\delta V_{\alpha\dot{\alpha}})_i{}^{\hat{k}} = -\bar{\varepsilon}_{\dot{\alpha}}(\lambda_\alpha)_i{}^{\hat{k}} = -\frac{8}{g}\varepsilon_{\alpha i}\bar{\varepsilon}_{\dot{\alpha}}\eta^{\hat{k}}\frac{1}{\rho^2 f^{3/2}}$$

$$= -\frac{4}{g}D_{\alpha\dot{\alpha}}\left[\frac{x_{i\dot{\gamma}}\bar{\varepsilon}^{\dot{\gamma}}\eta^{\hat{k}}}{\rho^2 f^{1/2}}\right],$$

$$(\delta G_{\alpha\gamma})_i{}^{\hat{k}} = \frac{16}{g}\frac{\bar{\varepsilon}^{\dot{\alpha}}\bar{\eta}^{\hat{k}}}{\rho^4 f^{3/2}}(\varepsilon_{\alpha i}x_{\gamma\dot{\alpha}} + \varepsilon_{\gamma i}x_{\alpha\dot{\alpha}}).$$

[We introduce here also the variation of the prepotential V. The formula for the fields $(W_\alpha)_{\hat{i}}^{k}$ are obtained from the corresponding formulas for $(W_\alpha)_i^{\hat{k}}$ in the obvious way and in the following they will not be written out.] It would appear that this result clearly indicates that $e^{-i\theta_0\bar{Q}}$ does not enter the stability group of the solution and leads to a new family of solutions. In order to consider why this is not true it is necessary to recall two facts. The first is that the mode $(\lambda_\alpha)_i^{\hat{k}}$ does not differ from the mode of the matrix $(\psi_\alpha)_i$ and can be represented in the form of the derivative (see Sec. 2)

$$\delta V_{\alpha\dot{\alpha}} = -\bar{\varepsilon}_{\dot{\alpha}}(\lambda_\alpha)_i{}^{\hat{k}} = -\frac{4}{g}D_{\alpha\dot{\alpha}}\left[\frac{x_{i\dot{\gamma}}\bar{\varepsilon}^{\dot{\gamma}}\eta^{\hat{k}}}{\rho^2 f^{1/2}}\right].$$

The second is that the zero modes of the gauge field, which correspond to the global rotation $e^{ig\Lambda}$,

$$\delta A_{\alpha\dot{\alpha}} = -ig[A_{\alpha\dot{\alpha}}, \Lambda]$$

are not normalizable. It is possible to normalize them if a singular gauge transformation is admitted. As a result the zero gauge modes are represented in the form

$$(V_{\alpha\dot{\alpha}})_i{}^{\hat{k}} = (D_{\alpha\dot{\alpha}})_i{}^l\Lambda_l{}^{\hat{k}}, \qquad (D^2)_i{}^l\Lambda_l{}^{\hat{k}} = 0.$$

(This problem was considered in detail in Ref. 16.) In the present case, for us it is important that the original variation resulting from the action of $e^{i\bar{Q}\bar{\varepsilon}}$ precisely corresponds to gauge modes. From the fermion zero modes, boson gauge modes have arisen. Therefore $\delta_\varepsilon V_{\alpha\dot{\alpha}}$ can be removed by a gauge transformation with the parameter

$$\Lambda_I{}^K = \begin{cases} \Lambda_i{}^{\hat{k}} = \dfrac{4}{g\rho^2 f^{1/2}}x_{i\dot{\alpha}}\bar{\varepsilon}^{\dot{\alpha}}\eta^{\hat{k}}, \\ \Lambda_{\hat{i}}{}^{k} = \dfrac{4}{g\rho^2 f^{1/2}}x_{\dot{\alpha}}{}^{k}\bar{\varepsilon}^{\dot{\alpha}}\eta_{\hat{i}}, \end{cases} \tag{23}$$

$$g(\Lambda)g(\bar{\varepsilon})V_{\alpha\dot{\alpha}}g^{-1}(\bar{\varepsilon})g^{-1}(\Lambda) = V_{\alpha\dot{\alpha}}.$$

Let us consider the other components of the field W_α, namely, $(W_\alpha)_i{}^k$. The transformation $g(\bar{\varepsilon})$, as before, does not affect the instanton; however, the gauge transformation which compensates it leads to the variation

$$(\delta_\Lambda\lambda)_i{}^k = (-ig)(\lambda_i{}^{\hat{k}}\Lambda_{\hat{k}}{}^k - \Lambda_i{}^{\hat{k}}\lambda_{\hat{k}}{}^k)$$

$$= \frac{32i}{g\rho^4 f^2}(\varepsilon_{\alpha i}x_{\dot{\alpha}}{}^k - \delta_\alpha{}^k x_{i\dot{\alpha}})\eta^2\bar{\varepsilon}^{\dot{\alpha}}, \qquad \eta^2 = \eta^{\hat{i}}\eta_{\hat{i}}.$$

In the field $(\delta_\Lambda\lambda_\alpha)_i{}^k$ it is easy to identify the SU(2) superconformal gluinozero mode with the parameter

$$\delta_{\bar{\varepsilon}}\bar{\beta}^{\dot{\alpha}} = (4i/\rho^2)\bar{\varepsilon}^{\dot{\alpha}}\eta^2. \tag{24}$$

Consequently,

$$g(\bar{\varepsilon})V_0 = g^{-1}(\Lambda)g(\delta\beta)V_0$$

and the transformation $g(\bar{\varepsilon})$ is equivalent to the transformations $g(\Lambda)$ and $g(\bar{\beta})$. Therefore there is no need for the new parameter $\bar{\theta}_0$ in the description of the general form of the superinstanton; it is not necessary to include $g(\bar{\theta}_0)$ in the displacement operator or $d^2\bar{\theta}_0$ in the integration measure.

Thus, the general expression for the superinstanton can be written in the form

$$V = LV_0(0, \rho, \eta)L^{-1}, \quad L = g(\Omega)g(x_0)g(\theta_0)g(\beta). \tag{25}$$

The matrix Ω describes the orientation of the instanton in SU(N) space and $V_0(0,\rho,\eta)$ corresponds in the initial configuration (22) to the instanton at the origin with zero fermion modes in the spinor representation.

Let us now turn to the transformation of parameters. As before

$$g(\varepsilon)L(\Omega, x_0, \theta_0, \beta)$$

$$= L(\Omega', x_0', \theta_0', \beta_0')g(\varepsilon)[1 + \varepsilon\beta(2D + 4i\Pi) - \beta\bar{M}\bar{\varepsilon}].$$

In the case of the SU(2) instanton Π and $\bar{M}$ did not act on V_0 and $2D\bar{\varepsilon}\bar{\beta}$ reduced to the change in radius of the instanton $\rho\to\rho' = (1 - 2i\bar{\varepsilon}\bar{\beta})\rho$. Now the chiral R-transformation Π changes the parameter η:

$$(1 + 4i\bar{\varepsilon}\beta\Pi)\eta(1 - 4i\bar{\varepsilon}\beta\Pi) = \eta(1 - 3i\bar{\varepsilon}\beta),$$

which together with the dilatation $(1 + 2D\bar{\varepsilon}\bar{\beta})\eta(1 - 2D\bar{\varepsilon}\bar{\beta}) = \eta(1 - 3i\bar{\varepsilon}\bar{\beta})$ finally gives

$$\eta\to\eta' = (1 - 4i\bar{\varepsilon}\beta)\eta.$$

As a result

$$g(\bar{\varepsilon})Vg^{-1}(\bar{\varepsilon}) = L'g(\bar{\varepsilon})V_0(\rho', \eta')g^{-1}(\bar{\varepsilon})(L')^{-1},$$

where L' is described by the parameters x_0', θ_0', and $\bar{\beta}'$, which are related to x_0, θ_0, and $\bar{\beta}$ by the old formulas

$$\delta(x_0)_{\alpha\dot{\alpha}} = -4i\theta_{0\alpha}\bar{\varepsilon}_{\dot{\alpha}}, \quad \delta(\theta_0)_\alpha = \varepsilon_\alpha, \quad \delta\bar{\beta}_{\dot{\alpha}} = -4i(\bar{\varepsilon}\bar{\beta})\bar{\beta}_{\dot{\alpha}}.$$

As we have seen, $g(\bar{\varepsilon})$ reduces to a superconformal transformation and a gauge transformation,

$$g(\bar{\varepsilon})V_0g^{-1}(\bar{\varepsilon}) = g(-\Lambda)g(\delta\beta)V_0g^{-1}(\delta\beta)g(\Lambda).$$

It is easy to verify that the transformation $g(-\Lambda)$ is carried to $g(\Omega)$ without a change of other parameters. Finally

$$\delta\Omega_I{}^K = \begin{cases} -\Lambda_i{}^{\hat{k}}(x - x_0), \\ -\Lambda_{\hat{i}}{}^{k}(x - x_0), \end{cases}$$

$$(\delta x_0)_{\alpha\dot{\alpha}} = -4i(\theta_0)_\alpha\bar{\varepsilon}_{\dot{\alpha}}, \qquad \delta\eta = -4i(\bar{\varepsilon}\bar{\beta})\eta, \tag{26}$$

$$\delta\rho^2 = -4i(\varepsilon\beta)\rho^2, \quad \delta\beta = -4i(\varepsilon\beta)\beta + (4i/\rho^2)\eta^2\varepsilon.$$

This set of transformations differs from the transformations of the SU(2) superinstanton. The measure of integration over the collective coordinates can now be written in the form

$$[d\mu] = CM^{3N}[d^4x_0(d\rho\cdot\rho)d^2\theta_0\,d^2\beta]\prod_{i=3}^{N}[\rho^4 d^2\eta]$$

$$= C(M\rho)^{3N}\left[d^4x_0\frac{d\rho}{\rho^5}d^2\theta_0 d^2\beta\right]\prod_{i=3}^{N}[\rho\, d^2\eta]. \qquad (27)$$

Here M is a cutoff parameter, and the factor M^{3N} counts the number of boson and fermion modes.[8a,15] The number $3N = 2N_B - N_F$ is directly related to the first coefficient of the Gell-Mann–Low function.[8a] The degree of ρ is determined by the requirement that $[d\mu]$ be dimensionless. The coefficient C for the group SU(N) was calculated in Ref. 14. The invariant measure for the SU(2) instanton has been separated in the square brackets. It is easy to verify that a change of transformations for $\delta\bar{\beta}$ is compensated by transformations of the parameter η so that the Berezinian of the transformations (26) is equal to the identity, and the cited measure is invariant relative to the symmetry transformations. The intermediate formulas for the Berezinian are somewhat cumbersome and we will not write them out here. In this example it is interesting to go through the proof of the Adler-Bardeen theorem on the axial current anomaly. Under the action of the R-transformation $e^{i\varphi\Pi}$, $\eta\to(1+3i\varphi/4)\eta$, $\theta_0\to(1+3i\varphi/4)\theta_0$, $\bar{\beta}\to(1+3i\varphi/4)\bar{\beta}$, and $[d\mu]\to[e^{-3i\varphi/4}]^{2N}[d\mu]$. That is, the measure of integration is not invariant relative to chiral transformations. The variation of the action is equal to the divergence of the corresponding current. As a result we obtain

$$\partial_{\alpha\dot{\alpha}}(\lambda^\alpha\bar{\lambda}^{\dot{\alpha}}) = (\alpha_s/2\pi)NG\tilde{G},$$

as required by the Adler theorem.

Let us now return to the problem of the construction of the invariant supermultiplets. As in the case of SU(2), it is convenient at first to find a combination of parameters which transform relatively simply under the action of supersymmetry. It is easy to verify that

$$\delta[\eta/\rho^2] = 0, \qquad \delta z = -4i\frac{(x-x_0)_{\alpha\dot{\alpha}}}{\rho^2}\bar{\theta}^\alpha\bar{\varepsilon}^{\dot{\alpha}},$$

$$z = (x-x_0)^2/\rho^2,$$

$$\delta\left[\frac{\bar{\theta}_\alpha}{\rho^2}\right] = \frac{(x-x_0)_{\alpha\dot{\alpha}}}{\rho^2}\frac{4i}{\rho^2}\bar{\varepsilon}^{\dot{\alpha}}\eta^2,$$

$$\delta\left[\frac{\bar{\theta}^2}{\rho^4}\right] = \frac{8i}{\rho^6}\bar{\theta}^\alpha\bar{\varepsilon}^{\dot{\alpha}}(x-x_0)_{\alpha\dot{\alpha}}.$$

For $\eta = 0$ the problem of searching for the invariant singlet combination is simplified, since it exactly reduces to the case of the SU(2) instanton

$$\Phi = f(z)\bar{\theta}^2/\rho^4.$$

If $\eta \neq 0$, then an additional term arises in Φ which must have the following form:

$$\Phi = f(z)(\bar{\theta}^2/\rho^4) + \varphi(z)(\eta^2/\rho^4).$$

Now each separate term ceases to be invariant, but from the requirement $\delta\Phi = 0$ there arises a simple condition on the function $\varphi(z)$:

$$\varphi'(z) = 2f(z). \qquad (28)$$

Therefore, knowing the solution for the SU(2) instanton, we can write out the solution for the SU(N) superinstanton directly. In particular for $W^\alpha W_\alpha$ we obtain

$$W^\alpha W_\alpha = \frac{3\cdot 2^7}{g^2}\frac{1}{f^4}\frac{\theta^2}{\rho^4} - \frac{2^8}{g^2}\frac{1}{f^3}\frac{\eta^2}{\rho^4}. \qquad (29)$$

Thus, the correct answer for W^2 is obtained when to the expression for the SU(2) instanton are added "matter" modes without the substitution $x^2/\rho^2\to\tilde{x}^2/\rho^2$, which we would want to make in a naive approach. Now each term is not invariant relative to the correct supersymmetric transformations (26) and only in the sum do their variations cancel.

In conclusion it should nevertheless be noted that if one is not interested in the invariance of the multiplet W^2 and the measure of integration relative to supersymmetric transformations, then such a limited problem on the SU(N) instanton is solved directly. In fact the gluon modes do not generate a source for the gauge field. Therefore the equation of motion reduces to the Yang-Mills equation for $(V_{\alpha\dot{\alpha}})_i$ and the Dirac equation in the external instanton field for fermions in the adjoint and fundamental representations. These solutions are known,[12] and therefore componentwise W^2 is also known. If the components are written in appropriate notation, then it is possible to obtain formula (29) for W^2. This is how it was first obtained.[10b] In such an approach there remains the unexplained question of the supersymmetry of the operators. Moreover, for the somewhat more complex problems of the superinstanton with matter, the componentwise equations involve all components, and the whole approach becomes extremely awkward, not to mention the loss of explicit superinvariance.

4. CONCLUSION

In this article we have obtained a closed expression for the instantons in SU(N) supergluodynamics and a set of intermediate formulas for the matter fields in SU(2) theories. This set of formulas proves to be sufficient to develop the corresponding superfield formalism for supersymmetric theories with an arbitrary symmetry group and an arbitrary set of matter fields. This will be done in a subsequent article.

[a)] How to work with instantons in Minkowski space is explained in Ref. 2b.

[1] A. Belavin, A. Polyakov, A. Schwartz, and Y. Tyupkin, Phys. Lett. **59B**, 85 (1975).
[2] a) V. Novikov *et al.*, Nucl. Phys. **B229**, 394 (1983); b) **B260**, 157 (1985).
[3] I. Affleck, M. Dine, and N. Seiberg, Phys. Rev. Lett. **51**, 1026 (1983); **52**, 1677 (1984).
[4] A. Vainshtein *et al.*, Pis'ma v Zh. Eksp. Teor. Fiz. **39**, 494 (1984) [JETP Lett. **39**, 601 (1984)].
[5] Y. Meurice and G. Veneziano, Phys. Lett. **141B**, 493 (1984).
[6] S. F. Cordes, Nucl. Phys. **B273**, 629 (1986); S. F. Cordes and M. Dine, Nucl. Phys. **B273**, 581 (1986).
[7] E. Witten, Nucl. Phys. **B202**, 253 (1982).
[8] a) V. Novikov *et al.*, Nucl. Phys. **B229**, 381 (1983); b) V. Novikov *et al.*, Phys. Lett. **166B**, 329 (1986).
[9] M. Shifman and A. Vainshtein, Nucl. Phys. **B277**, 456 (1986).
[10] a) J. Fuchs and H. G. Schmidt, Z. Phys. **30**, 161 (1986); b) J. Fuchs, Nucl. Phys. **B272**, 677 (1986).
[11] M. F. Sohnius, Phys. Rep. **128**, 41 (1986).
[12] G. 't Hooft, Phys. Rev. D **14** 3432 (1976).
[13] F. Buccela *et al.*, Phys. Lett. **115B**, 375 (1982); I. Affleck, M. Dine and N. Seiberg, Nucl. Phys. **B241**, 493 (1984).
[14] C. W. Bernard *et al.*, Phys. Rev. D **16**, 2967 (1977).
[15] A. Vainshtein *et al.* Usp. Fiz. Nauk **136**, 553 (1982) [Sov. Phys. Usp. **25**, 195 (1982)].
[16] C. Bernard, Phys. Rev. D **19**, 3013 (1979).

Translated by L. J. Swank

Supersymmetric instanton calculus (gauge theories with matter fields in an arbitrary group)

V. A. Novikov

Institute of Theoretical and Experimental Physics, State Commission on Use of Atomic Energy
(Submitted 20 October 1986)
Yad. Fiz. **46**, 967–974 (September 1987)

A superfield formalism is developed for instantons in supersymmetric gauge theories with matter fields in an arbitrary group.

1. INTRODUCTION

Instantons,[1] the simplest nonperturbative fluctuations, lead to a number of beautiful physical effects in supersymmetric gauge theories.[2-6] The most interesting phenomena appear in supersymmetric gauge theories with a large symmetry group G and matter superfields in specially selected representations. In the present study we develop a superfield formalism for instantons in theories with an arbitrary symmetry group and matter fields with nonzero vacuum expectation values. This work is the natural extension of our studies of supersymmetric SU(2) instantons[2a,b] and the work of Ref. 7 on SU(N) instantons in supersymmetric gluon dynamics. To understand the present work, the reader should be familiar with the notation and general formalism developed in Refs. 2 and 7.

In Section 2 we show how matter fields in very simple representations can be included in supersymmetric gauge theories based on the group SU(N). In Section 3 the results are generalized to an arbitrary group. In the Conclusion we discuss our results.

2. THE INSTANTON IN SU(N) SUPERSYMMETRIC THEORIES WITH MATTER FIELDS

Let us consider a gauge theory with N_f quark chiral matter multiplets

$$Q_p{}^I\bar{Q}_{I,q},\quad I=1,\ldots,N,\quad p,q=1,\ldots,N_f,$$

in the N and $\bar{N}$ representations. We do not need to discuss the instanton solution. The situation is just the same as that in SU(2) gauge theory.

We begin with the simplest problem, where the scalar matter fields have no vacuum expectation values, and assume that the gluonic zero modes are not excited, that is, $\theta_0=\bar{\beta}=\eta_i=\eta^i=0$. The problem then reduces to the SU(2) case:

$$Q_p{}^i=\theta^\alpha(\psi_\alpha)_p{}^i=-\theta^i\chi_p/\rho^2 f^{1/2},$$

$$\bar{Q}_{iq}=\theta^\alpha[\psi_\alpha]_{iq}=-\theta_i\zeta_q/\rho^2 f^{1/2},\quad Q_p\bar{Q}_q=-\theta^2\chi_p\zeta_q/\rho^4 f^2.$$

Let us now consider the transformations $e^{-i\theta_0 Q}$ and $e^{-i\bar{S}\bar{\beta}}$. Since for $\eta=0$ the stationary group is the same as in SU(2), we have

$$Q_p{}^i\bar{Q}_{iq}=-(\tilde{\theta}^2/\rho^4)(\chi_p\zeta_q/f^2). \tag{1}$$

It is obvious from this formula that the scalar fields are no longer zero. The Yukawa interaction between λ and ψ leads to

$$A_p{}^i=(\theta_0-x\bar{\beta})^i\chi_p/\rho^2 f^{3/2}.$$

Let us now include η. Since in our approximation the matter does not affect W_α, the problem of constructing the general form of W_α is the same as the problem of constructing W_α in gluon dynamics, which has been studied earlier.[7] In particular, all the parameter transformations remain the same. The answer for W^2 is therefore the same as (29) of Ref. 7. For $Q_p\bar{Q}_q$, solving the corresponding differential equation, we find

$$Q_p\bar{Q}_q=-\tilde{\theta}^2\chi_p\zeta_q/\rho^4 f^3+(\eta^2/\rho^4)\chi_p\zeta_q(1/f^2). \tag{2}$$

We note that since χ_p and ζ_q are invariant under supersymmetry transformation in the limit $\eta=0$, they are also invariant for $\eta\neq 0$. The expression (2) solves the practical problem of the construction of the matter supermultiplet. Nevertheless, we think that it is instructive to consider some other questions. First of all, we would like to verify that the stationary groups of W_α and the matter multiplets Q and $\bar{Q}$ coincide, thereby proving that our operations are valid. Simple expressions for these multiplets follow from (2):

$$Q_p{}^I=-\frac{\tilde{\theta}^I\chi_p}{\rho^2 f^{3/2}}+\frac{\eta^I\chi_p}{\rho^2 f},\quad \bar{Q}_{Iq}=-\frac{\tilde{\theta}_I\zeta_q}{\rho^2 f^{3/2}}+\frac{\eta_I\zeta_q}{\rho^2 f}. \tag{3}$$

The initial configuration corresponds to the point $\theta_0=\bar{\beta}=0$. However, in contrast to the instanton of supersymmetric SU(2), now the scalar components of the fields are nonzero and the transformation $g(\bar{\varepsilon})$ changes Q and $\bar{Q}$:

$$\delta Q_p{}^i=4i\theta^\alpha\bar{\varepsilon}^{\dot{\alpha}}\eta^i\chi_p(x_{\alpha\dot{\alpha}}/\rho^2 f^2),$$

$$\delta\bar{Q}_{iq}=4i\theta^\alpha\bar{\varepsilon}^{\dot{\alpha}}\eta_i\zeta_q(x_{\alpha\dot{\alpha}}/\rho^2 f^2).$$

If we now make a gauge transformation with Λ_i^k and Λ_i^k, the result of the two operations reduces to the formula

$$g_1g_2Qg_2^{-1}g_1^{-1}=\frac{\eta^i\chi_p}{\rho^2 f}-\frac{4i}{\rho^2 f^{3/2}}x^i_{\dot{\alpha}}\bar{\varepsilon}^{\dot{\alpha}}\chi_p\frac{\eta^2}{\rho^2}-\frac{\theta^i\chi_p}{\rho^2 f^{3/2}}.$$

Comparing this with the general equation (3), we see that for Q and $\bar{Q}$ the transformations g_1 and g_2 reduce to a superconformal transformation:

$$g(\Lambda)g(\bar{\varepsilon})=g(\delta\bar{\beta}),\quad \delta\bar{\beta}=4i(\bar{\varepsilon}_{\dot{\alpha}}\eta^2/\rho^2).$$

Therefore, the addition of matter fields $\sim\eta\chi$ to the scalar components of the fields causes the stationary group of the matter fields to correspond to the stationary group of the vector multiplet.

Let us now discuss the integration measure. The new fermionic zero modes of the matter fields lead to a change of the power of M:

$$M^{3N}\to M^{3N-N_f}.$$

It therefore follows from dimensional considerations that

 0038-5506/87/090554-05$03.80

$$[d\mu]=C(M\rho)^{3N-N_f}\left[\frac{d^4x_0\,d\rho}{\rho^3}d^2\theta_0\,d^2\beta\right]$$
$$\times\left[\prod_i(\rho\,d^2\eta_i)\right]\left[\prod_f\rho\,d\chi\,d\zeta\right]. \quad (4)$$

Since the power of ρ is not changed, while $d\chi$ and $d\zeta$ are invariant under supersymmetric transformations by themselves, it is obvious that the full measure (4) is invariant.

Let us now consider a more complicated problem, that of instantons in supersymmetric SU(N) theories with matter fields having nonzero vacuum expectation value. It is natural to choose as the initial configuration an instanton field with "transverse" modes and the classical solution for the scalar field:

$$A^i=x_{\dot k}^i v^{\dot k}/\rho f^{1/2},\quad \bar A_i=x_i^{\dot k}w_{\dot k}/\rho f^{1/2},\quad A^{\hat i}=\bar A_{\hat i}=0,\quad v^{\dot k}w_{\dot k}=v^2.$$

Then we can use the translation operator to construct a general configuration. Before doing this, we again recall that for $\eta=0$ the problem is the same as that of the SU(2) instanton and

$$Q\bar Q=\bar x^2v^2/(\bar x^2+\rho^2),\qquad W^2=\frac{3\cdot 2^7}{g^2}\frac{\bar\theta^2}{\rho^4}\frac{1}{f^4}.$$

Since the first component of the chiral superfields is nonzero, the transformation $g(\bar\varepsilon)=e^{-i\bar Q\bar\varepsilon}$ is not contained in the stationary subgroup and we need to take the translation operator to be of the form

$$L=g(\Omega)g(x_0)g(\theta_0)g(\beta)g(\bar\theta_0)g(\rho). \quad (5)$$

This differs from the operator L in the group SU(2) only by the transformation $g(\Omega)$, which describes the orientation of the SU(2) instanton in the group SU(N). Supersymmetry transformation do not affect global rotations, so all the parameters transform as in the SU(2) case:

$$\delta x_0=-4i\theta_0\bar\varepsilon,\quad \delta\bar\theta_0=\varepsilon-4i(\bar\varepsilon\bar\theta_0)\beta,\quad \delta\bar\beta=-4i(\bar\varepsilon\beta)\beta,$$
$$\delta\rho^2=-4i(\bar\varepsilon\beta)\rho^2,\quad \delta\eta=-4i(\bar\varepsilon\beta)\eta. \quad (6)$$

Therefore, for $\eta=0$ the invariants are not changed and it is necessary only to reconstruct the terms which vanish in the limit $\eta=0$. This can be done by trial and error, starting from the requirement that the multiplets be invariant. This method works quite well, but the problem can be solved more elegantly. We recall that in the approximation in the coupling constant that we are working in, the matter fields do not affect the gauge field multiplet. The latter can therefore be treated completely independently from the matter fields. As explained in the preceding section, the action of the operator $e^{i\bar Q\bar\varepsilon}$ on a chiral multiplet is equivalent to the superconformal and gauge transformations. We now need to establish this equivalence more accurately:

$$V=g(\bar\theta_0)V_0(0,\rho,\eta)g^{-1}(\theta_0)$$
$$=V_0+[-i\bar\theta_0\bar Q,V_0]+\frac{1}{2}[-i\bar\theta_0\bar Q[-i\bar\theta_0\bar Q,V_0]]$$
$$+\cdots=g(-\Lambda)g(\Delta\beta)\left[1+\frac{2i\eta^2\bar\theta_0^2}{\rho^2}(2D+4i\Pi)\right]$$
$$\times V_0\left[1-\frac{2i\eta^2\bar\theta_0^2}{\rho^2}(2D\right.$$
$$\left.+4i\Pi)\right]g^{-1}(\Delta\beta)g(\Lambda),$$

$$\Delta\bar\beta=4i\frac{\eta^2}{\rho^2}\bar\theta_0,\quad \Lambda=\frac{4}{g}\frac{1}{\rho^2 f^{1/2}}(x_{i\dot\alpha}\bar\theta_0^{\dot\alpha}\eta^{\hat k}+x_{\dot\alpha}^{\hat k}\bar\theta_0^{\dot\alpha}\eta_{\hat i}), \quad (7)$$
$$\rho_1^2=\rho^2+8\eta^2\theta_0^2,\quad \left[1+\frac{2i\eta^2\bar\theta_0^2}{\rho^2}(2D+4i\Pi)\right]V(0,\rho)\left[1\right.$$
$$\left.-\frac{2i\eta^2\bar\theta_0^2}{\rho^2}(2D+4i\Pi)\right]=V(0,\rho_1).$$

Therefore, in terms of x_0, θ_0, $\beta_1=\beta+\Delta\beta$, and ρ_1, the problem does not differ at all from the problem of the SU(N) instanton without matter fields, which has already been solved. In particular, the rule for transformations of these parameters

$$\delta\rho_1^2=-4i(\bar\varepsilon\beta_1)\rho_1^2,\quad \delta\eta=-4i(\bar\varepsilon\beta_1)\eta,$$
$$\delta(x_0)_{\alpha\dot\alpha}=-4i\theta_{0\alpha}\bar\varepsilon_{\dot\alpha},\quad \delta\bar\beta_1=-4i(\bar\varepsilon\bar\beta_1)\bar\beta_1+4i\frac{\eta^2}{\rho^2}\bar\varepsilon \quad (8)$$

coincides with (26) of Ref. 7. [It can be verified that the transformations (8) and (6) are the same if ρ_1 and β_1 are related to ρ and β via the formulas (7).] This means that the form of W^2 in the new parameters is known:

$$W'^2=\frac{3\cdot 2^7}{g^2}\frac{1}{f_1^4}\frac{\bar\theta^2}{\rho_1^4}-\frac{2^8}{g^2}\frac{1}{f_1^3}\frac{\eta^2}{\rho_1^2}. \quad (9)$$

Here the index 1 indicates that the variables ρ_1 and β_1 are used in the definition of invariant combinations instead of ρ and β.

Let us now consider the matter multiplet. It is convenient to treat it using the original parametrization. It is easly shown that the only supersymmetry-invariant expression for $Q\bar Q$ coinciding with $x^2v^2(x^2+\rho^2)$ at the point $\eta^2=0$ has the form

$$Q\bar Q=v^2\bar x^2/(\bar x^2+\rho^2)+\frac{\eta^2}{\rho^4}\varphi(\bar x^2/\rho^2),$$

where φ is an as yet unknown function. Let us consider the point $\bar\theta_0=\beta=\theta_0=0$. Then $\bar x=x$ and the solution is obvious: the multiplet $Q\bar Q$ contains only the original scalar component (modes $\sim\eta$ cannot by themselves cause the scalar field to change), $\bar QQ=v^2x^2/(x^2+\rho^2)$, and so $\varphi\equiv 0$. This means that in the parametrization we have selected

$$Q\bar Q=v^2(\bar x^2/\rho^2)/\bar f. \quad (10)$$

It is very easy to write the matter multiplet in terms of ($\bar\beta$,ρ) and the gauge multiplet in terms of ($\bar\beta_1$, ρ_1). However, in order to work with these formulas, they must be rewritten using a common parametrization. Let us begin with the mater multiplet. One can straightforwardly make the substitution $\rho^2=\rho_1^2-8\eta^2\theta_0^2$, $\beta=\beta_1-4i(\eta^2/\rho^2)\bar\theta_0$ in (10). Then

$$Q\bar Q=v^2\left[1-\frac{1}{\bar f_1}-\frac{1}{f}\frac{8\eta^2\bar\theta_0^2}{\rho^2}+\frac{1}{f^2}\left(\frac{8\eta^2\bar\theta_0^2}{\rho^2}\right)\right], \quad (11)$$

where $\bar f_1=1+(\bar x_1^2/\rho^2)$.

In (11) the term $v^2[1-1/\bar f_1]$ corresponds to the supersymmetric SU(2) instanton and the term $8\eta^2\bar\theta_0^2/f^2\rho^2$ to the change of the scalar component due to interaction with the fermionic mode of the matter $\sim\bar\theta_0$ and the "transverse" mode of the supersymmetric SU(N) instanton. The reason for the change of the vacuum expectation value in the decreasing term $-(1/f)(8\eta^2\bar\theta_0^2/\rho^2)$ is less clear. It can only be said that the solution (11) is explicitly supersymmetric and has the correct asymptotic form at infinity $Q\bar{Q} \underset{x\to\infty}{\to} v^2$.

We could have obtained (11) bypassing the intermediate step (10). We need to construct the supersymmetry-invariant generalization of the initial configuration $Q\bar{Q} = x^2v^2/(x^2+\rho^2)$ in terms of (ρ_1, β_1). Simple algebra gives

$$\delta\left[\frac{1}{f_1}\right] = -\delta\left[\frac{8\eta^2}{\rho^4}\bar{\theta}_0^2\frac{x^2}{\rho^2}\right]$$

(in addition to the transformation (8), we have used the formulas

$$\delta\bar{\theta}_0 = \bar{\varepsilon}(1+8\bar{\theta}_0^2\eta^2/\rho^2) - 4i(\bar{\varepsilon}\bar{\theta}_0)\beta,$$

$$\delta(\bar{x}_1^2/\rho^2) = 16(\eta^2/\rho_1^2)\bar{x}_1^2\varepsilon\bar{\theta}_0).$$

The invariant generalization of $[Q\bar{Q}]$ is therefore written as

$$Q\bar{Q} = v^2\left[1 - \frac{1}{f_1} + \frac{8\eta^2\theta_0^2}{\rho^2}\frac{(\bar{x}^2/\rho^2)}{f^2}\right],$$

which, of course, coincides with (11).

Let us turn to the (β,ρ) parametrization. Here we must rewrite the multiplet W^2, the formulas for which have the seemingly horrible form

$$W^2 = \frac{3\cdot 2^7}{g^2}\left\{\frac{1}{f^4}\frac{\bar{\theta}^2}{\rho^4} + \frac{32\bar{\theta}^2}{f^3\rho^6}\bar{\theta}_0^2\eta^2 - \frac{16\bar{\theta}^2\bar{\theta}_0^2\eta^2}{f^5\rho^6} + \frac{8i\eta^2}{f^4\rho^6}(x-x_0)\bar{\theta}\bar{\theta}_0\right\} - \frac{2^8}{g^2}\frac{\eta^2}{f^3\rho^4}. \quad (12a)$$

Some of the terms in this sum contain up to six odd factors. However, if we return to the invariant variable $\tilde{x}^2/\rho^2$, (12a) takes the simple form

$$W^2 = \frac{3\cdot 2^7}{g^2}\frac{1}{f^4}\frac{\bar{\theta}^2}{\rho^4} - \frac{2^8}{g^2}\frac{\eta^2}{\rho^4}\frac{1}{f^3}. \quad (12b)$$

In these variables (just as in x, ρ_1, and $\bar{\beta}_1$) the gauge multiplet is obviously independent of the matter multiplet.

Up to now we have considered the problem where the instanton and the matter vacuum field are located in the same SU(2) group. Let us now consider the general case. It is useful to first study the configuration in which the vacuum field is an SU(2) singlet, that is,

$$Q_0{}^I = \begin{cases} -\dfrac{\theta^i\chi}{\rho^2 f^{3/2}} \\ v^{\hat{i}} - \eta^{\hat{i}}\dfrac{\chi}{\rho^2 f} \end{cases}, \quad \bar{Q}_{0I} = \begin{cases} -\dfrac{\theta_i\zeta}{\rho^2 f^{3/2}} \\ w_{\hat{i}} + \eta_{\hat{i}}\dfrac{\zeta}{\rho^2 f} \end{cases} \quad (13)$$

In addition to the plane-wave vacuum field $v^{\hat{i}}$ and $w_{\hat{i}}$, in the initial field configuration Q_0, $\bar{Q}_0$ we have included the fermionic zero modes of the matter, which are proportional to the odd parameters ζ and χ, and the variation of the scalar fields induced by the "transverse" fermionic modes of the instanton and the fermionic modes of the matter.

It is easily checked that the combination of transformations $g(-\delta\bar{\beta})g(\Lambda)g(\bar{\varepsilon})$ acting on this configuration gives a zero variation except for the constant terms $v^{\hat{i}}$ and $w_{\hat{i}}$. As a result,

$$\delta Q_0{}^I = \frac{1}{\rho f^{3/2}}x_{\dot{\alpha}}{}^i\bar{\varepsilon}^{\dot{\alpha}}(\rho\bar{\theta}_0{}^{\dot{\alpha}})\frac{\eta_{\hat{k}}v^{\hat{k}}}{\rho^2} \neq 0$$

and, consequently, $g(\bar{\varepsilon})$ no longer enters into the stationary group of the initial configuration. This means that the invariant functions must depend on $\tilde{\theta}$, $\tilde{x}$, and so on. In terms of β_1 and ρ_1, the supersymmetry-invariant generalization (13) has the form (dropping the index 1)

$$Q^I = v^{\hat{i}} - \frac{\eta^{\hat{i}}}{\rho^2}\frac{\chi}{f} - \frac{\tilde{\theta}^i\chi}{\rho^2 f^{3/2}},$$

$$\bar{Q}_I = w_{\hat{i}} + \frac{\eta_{\hat{i}}}{\rho^2}\frac{\zeta}{f} - \frac{\tilde{\theta}_i\zeta}{\rho^2 f^{3/2}}. \quad (14)$$

(As in the case of SU(2), it should be recalled that it is possible to return to the physical gauge by applying a supersymmetric gauge transformation Λ to (14).)

The resulting formula is obviously a supersymmetry invariant, although the fact that we have four odd parameters χ, ζ, and $\bar{\theta}_{0\alpha}$ associated with two fermionic modes of the matter looks suspicious. This point requires special consideration.

In the case of the group SU(2), we did not introduce special odd parameters for the fermionic zero modes of the matter if the vacuum expectation value of the field was nonzero—their role was played by the supersymmetry transformation parameter $\bar{\theta}_{0\alpha}$. For the configuration Q_0 and $\bar{Q}_0$ (13) the transformations $g(\Delta\bar{\beta})g(\Lambda)g(\bar{\theta}_0)$ also generate a fermionic mode with parameters $\delta\chi \sim \bar{\theta}_0^2(\eta w)$, $\delta\zeta \sim \bar{\theta}_0^2(\eta v)$. However, for such odd elements $\delta\chi\cdot\delta\zeta = 0$, so if fermionic modes are not included *ab initio* in the initial configurations, they will not appear:

$$Q\bar{Q} = v^2. \quad (15)$$

Although the solution obtained is supersymmetric, it obviously does not correspond to the physical problem, because configurations with singlet vacuum expectation value and fermionic modes are not distinct from the others. Fortunately, this difficulty vanishes if we allow for the appearance of an arbitrarily small condensate in the SU(2) sector:

$$Q_0{}^I = \begin{cases} x_{\dot{\alpha}}{}^i v_1{}^{\dot{\alpha}}/\rho f^{1/2} \\ v_2{}^{\hat{i}} \end{cases}, \quad (16)$$

where $v_1^{\alpha} = R_k^{\alpha}v^k$, $v_2^{\hat{i}} = R_k^{\hat{i}}v^k$, where R_k^i is an SU(N) rotation matrix. In this case the supermultiplet

$$Q\bar{Q} = v_1^2 + v_2^2 - \frac{v_1^2}{f} + \frac{1}{f^2}\frac{8\eta^2\bar{\theta}_0^2x^2v_1^2}{\rho^4}, \quad v^2 = v_1^2 + v_2^2, \quad (17)$$

already contains a fermionic mode and it is not necessary to introduce additional odd coordinates. Nevertheless, the point $V_1 = 0$ corresponding to an arbitrary rotation matrix in the group SU($N-2$) still remains singular, since it leads to the purely bosonic solution (15). The nature of this singularity is more or less clear and is related to the fact that any arbitrarily small "normal" number is still larger than any even element composed of odd elements. It is therefore necessary to first carry out all the operations on Grassmann variables (in this case to remove the integration over the fermionic modes) and only then let the normal parameters tend to zero. However, the point $v_1 = 0$ obviously has zero measure, and in the integration over all instanton orientations for fixed vacuum expectation values this singularity does not affect the result of the integration.

Finally, let us consider the problem with N_f flavors. The vacuum expectation values of Q and $\bar{Q}$ lie in a valley[8] (up to an arbitrary gauge transformation)

$$v_I{}^I = \begin{cases} v_I\delta_I{}^I, & I=1,\ldots,N_f \\ 0, & I>N_f \end{cases},$$

$$w_{If} = \begin{cases} v_I\delta_{If}, & I=1,\ldots,N_f \\ 0, & I>N_f \end{cases} \quad \text{for } N>N_f,$$

$$v_f{}^I = \begin{cases} v_f\delta_f{}^I, & f=1,\ldots,N \\ 0, & f>N \end{cases},$$

$$w_{If} = \begin{cases} \sqrt{c^2+v_f{}^2}\,\delta_{If}, & f=1,\ldots,N \\ 0, & f>N \end{cases} \quad \text{for } N<N_f. \tag{18}$$

Therefore, only two matter multiplets have vacuum expectation values in the selected SU(2) subgroup. For the present parametrization these are Q_1 and Q_2. After all this discussion the answer is more or less obvious. In the (ρ,β) parametrization it is practically the same as the SU(2) problem:

$$Q_1{}^I = \frac{\mathscr{F}_{\dot\alpha}{}^i v_1{}^{\dot\alpha}}{\rho f^{1/2}}, \quad Q_2{}^I = \frac{\mathscr{F}_{\dot\alpha}{}^i v_2{}^{\dot\alpha}}{\rho f^{1/2}} - \frac{\bar\theta^I\chi_2}{\rho^2 f^{1/2}},$$

$$Q_p{}^I = v^i - \frac{\bar\theta^i}{\rho^2}\frac{\chi_p}{f^{1/2}}, \quad p\geqslant 3. \tag{19}$$

In order to go to the (ρ_1,β_1) parametrization, it must be recalled that, in addition to the variable substitutions, it is necessary to make the supersymmetric gauge transformation (7), which distorts the vacuum expectation values relative to SU(2).

Let us conclude with a discussion of the SU(N) group. Here we have not given explicit formulas for matter fields in representations more complicated than the N and $\bar{N}$. However, it seems to us that the technique we have developed makes it easy to write down the answer for any particular case.

Finally, let us make two remarks: 1) the constant C in the integration measure (4) related to the volume of the imbedding of the group SU(2) into SU(N) has been calculated in Ref. 9 (see also Ref. 6); 2) when the instanton orientations are integrated over explicitly it is more convenient to assume that the instanton position is fixed and that the system of matter fields rotates as a whole. In this case, for all flavors a part of the vacuum expectation value v_f lies in SU(2) and a part lies in SU($N-2$). The change in the formulas (19) is obvious. The result of the integration is also obvious up to a numerical factor. However, in order to reconstruct this number, it is necessary to know the explicit parametrization of a rotation from SU($N-2$) in SU(2). This problem has been solved in the earlier studies of Refs. 6 and 9.

3. SUPERSYMMETRIC INSTANTONS IN AN ARBITRARY LIE GROUP

Let us discuss the method of constructing a supersymmetric instanton in an arbitrary group G. It must first of all be stated that from thge viewpoint of the superfield formalism, arbitrary groups do not involve anything new—the formulas are the same as those of the preceding sections. In order to construct a gauge multiplet W^2, it is necessary to imbed the SU(2) group in the group G. Then the initial configuration has the form

$$(W_\alpha)_i{}^k = \theta^\beta\cdot 2iG_{\alpha\beta}{}^a(J^a)_i{}^k,$$

where the J^a ($a=1,2,3$) are the generators of the SU(2) group. The general problem of embedding SU(2) into G has been solved by Dynkin.[10] In the present case we must select from G the SU(2) with the smallest Cartan length, that is, the smallest value of $\mathrm{Tr}\,J^aJ^a$ (Ref. 11). The solution of this problem is simply[11]

$$J_+ = E_\alpha,\ J_- = E_{-\alpha},\ J_3 = [E_\alpha, E_{-\alpha}], \tag{20}$$

where α is the root of smallest length and $\{E_\alpha, E_{-\alpha}, H_i\}$ are the generators in the Cartan-Weyl basis.[11] The tables of roots for any group G can be found from Ref. 12. Relative to (20), all the generators are decomposed into three generators in the adjoint representation $(J_\pm, J_3)$ $2[l(G)-2]$ generators in the spinor representation, and singlets. Here $l(G)$ is the Dynkin index for the adjoint representation ($l(\mathrm{SU}(N))=N$). Therefore,

$$W^2 = \frac{2^7\cdot 3}{g^2}\frac{1}{f^4}\frac{\theta^2}{\rho^4} - \frac{2^8}{g^2}\frac{1}{f^3\rho^3}\sum_{i=1}^{l(G)-2}\eta^i\eta_i.$$

The matter fields are just as easy to treat. The given multiplets in G must be expanded in SU(2) representations. Then they can be treated as independent SU(2) matter multiplets and the formulas of Section 2 can be used. If the highest weight of the representation is known, then, knowing the expansion of the generators into SU(2) singlets, spinors, and vectors, it is easy to find the expansion of this representation in SU(2) multiplets. It is also possible to first find this expansion for certain representations of G and then go to the representation itself. It is simpler to solve this problem for a particular case then to obtain a general solution. We therefore do not give it here. Difficulties arise only when the vacuum expectation values are nonzero, because the solutions of the algebraic equations for the valleys are not known in the general case. However, many solutions have been found for the simplest representations and these can be used to easily write down the answers.

4. CONCLUSION

We note that the solution we have given is somewhat unsatisfactory. In simple cases one can move through the space of solutions by means of the superconformal group, transformations from which take each solution into another. However, in the general case it is necessary to specify an entire surface and obtain a general solution starting from it. Of course, the motions on this surface form a group, but they are not described by superconformal transformations, although they must essentially be geometrical in nature. The question of whether or not it is possible to obtain a geometrical formulation of the problem in the general case remains unanswered.

The author is grateful to A. I. Vaĭnshteĭn, M. B. Voloshin, V. A. Zakharov, L. B. Okun', and M. A. Shifman for discussions.

APPENDIX 1

The spinor notation is

$$V_{\alpha\dot\alpha}=V_\mu(\sigma_\mu)_{\alpha\dot\alpha},\quad V_\mu W^\mu={}^1/_2 V_{\alpha\dot\alpha}W^{\dot\alpha\alpha},$$
$$G_{\alpha\beta}={}^1/_2\eta_{a\mu\nu}G_{\mu\nu}\sigma^a,\quad G_{\dot\alpha\dot\beta}=-{}^1/_2\bar\eta_{a\mu\nu}\sigma^a G_{\mu\nu},$$
$$\sigma_{\alpha\dot\alpha}=(I,\sigma^a)_{\alpha\dot\alpha},\quad (\bar\sigma)^{\dot\alpha\alpha}=(I,-\sigma^a)^{\dot\alpha\alpha},$$
$$(\sigma_\mu\bar\sigma_\nu)_\alpha{}^\beta=g_{\mu\nu}\delta_\alpha{}^\beta+\eta_{a\mu\nu}(\sigma^a)_\alpha{}^\beta.$$

APPENDIX 2

The $N=1$ superconformal group is

$$\{Q^\alpha\bar Q^{\dot\alpha}\}=2P^{\alpha\dot\alpha},\quad \{Q^\alpha Q^\beta\}=0,\quad [P^{\alpha\dot\alpha},Q^\beta]=0,$$
$$\{S^\alpha\bar S^{\dot\alpha}\}=-2K^{\alpha\dot\alpha},\ \{S^\alpha S^\beta\}=0,\ [K^{\alpha\dot\alpha},S^\beta]=0,$$
$$[Q^\alpha,D]=\frac{i}{2}Q^\alpha,\quad [S^\alpha,D]=-\frac{i}{2}S^\alpha,\quad [\bar Q^{\dot\alpha},D]=\frac{i}{2}\bar Q^{\dot\alpha},$$
$$[\bar S^{\dot\alpha},D]=-\frac{i}{2}\bar S^{\dot\alpha},\quad [Q^\alpha,\Pi]=\frac{3}{4}Q^\alpha,\quad [\bar S^{\dot\alpha},\Pi]=\frac{3}{4}\bar S^{\dot\alpha},$$
$$[Q^\alpha,K^{\dot\beta\beta}]=-2i\varepsilon^{\alpha\beta}\bar S^{\dot\beta},\quad [\bar S^{\dot\alpha},P^{\beta\dot\beta}]=2i\varepsilon^{\dot\alpha\dot\beta}Q^\beta,$$
$$\{\bar S^{\dot\alpha},Q^\beta\}=0,\quad \{\bar S^{\dot\alpha},\bar Q^{\dot\beta}\}=\bar M^{\dot\alpha\dot\beta}-(2D+4i\Pi)\,\varepsilon^{\dot\alpha\dot\beta}.$$

These formulas should be supplemented by the conformal algebra. Superconformal transformations of the coordinates look like

$$\delta\theta_\alpha=-2i\beta_\alpha\theta^2-x_{\alpha\dot\alpha}\bar\beta^{\dot\alpha},\quad \delta x_{\alpha\dot\alpha}=-4ix_{\dot\alpha\gamma}\beta^\gamma\theta_\alpha,$$

and those of the chiral multiplet $Q=A+\theta^\alpha\psi_\alpha+\frac{1}{2}\theta^2F$ look like

$$\delta A=0,\quad \delta\psi_\alpha=-2ix_{\gamma\dot\alpha}\beta^\gamma D_\alpha^{\dot\alpha}A-2i(\Delta+2r)\beta_\alpha A.$$
$$\delta F=2i(\Delta+2r-2)\beta\psi+2ix_{\gamma\dot\alpha}\beta^\gamma D^{\alpha\dot\alpha}\psi_\alpha,$$
$$\delta A=x_{\alpha\dot\alpha}\bar\beta^{\dot\alpha}\psi^\alpha,\quad \delta\psi_\alpha=-x_{\alpha\dot\alpha}\bar\beta^{\dot\alpha}F,\quad \delta F=0.$$

Note added in proof. After this article was sent to press, the preprint of Ref. 13 appeared. That author obtained some of the results presented here by a different method.

[i)]Other, nonminimal SU(2) groups lead to solutions corresponding to multiple aligned SU(2) instantons.[11]

[1]A. Belavin, A. Polyakov, A. Schwartz, and Y. Tyupkin, Phys. Lett. **59B**, 85 (1975).

[2]a) V. Novikov *et al.*, Nucl. Phys. **B229**, 394 (1983); b) V. Novikov *et al.*, Nucl Phys. **B260**, 157 (1985).

[3]I. Affleck, M. Dine, and N. Seiberg, Phys. Rev. Lett. **51**, 1026 (1983); **52**, 1677 (1984).

[4]A. Vaĭnshteĭn *et al.*, Pis'ma Zh. Eksp. Teor. Fiz. **39**, 494 (1984) [JETP Letters **39**, 601 (1984)].

[5]Y. Meurice and G. Veneziano, Phys. Lett. **141B**, 69 (1984).

[6]S. F. Cordes, Nucl. Phys. **B273**, 629 (1986); S. F. Cordes and M. Dine, Princeton University Preprint, 1986.

[7]V. A. Novikov, Yad. Fiz. **46**, 656 (1987) [Sov. J. Nucl. Phys. **46**, (1987)].

[8]F. Buccella *et al.*, Phys. Lett. **115B**, 375 (1982); I. Affleck, M. Dine, and N. Seiberg, Nucl. Phys. **B241**, 493 (1984).

[9]C. Bernard, Phys. Rev. D **19**, 3013 (1979).

[10]E. Dynkin, Proc. Mosk. Mat. ob-va, Moscow, Vol. 1, p. 39 (1952) [in Russian].

[11]C. W. Bernard *et al.*, Phys. Rev. D **16**, 2967 (1977).

[12]R. Slansky, Phys. Rep. **79**, 3 (1981).

[13]J. Fuchs, Princeton University Preprint, 1986.

Translated by Patricia Millard

VIII. REVIEWS

INTRODUCTION

Below the reader will find two review papers devoted to instantons in gauge theories. Many questions which are tacitly assumed or only briefly mentioned in the original publications are discussed here in detail. Along with these two reviews I would like to recommend the book "Solitons and Instantons" by R. Rajaraman (North-Holland, 1987). Although gauge theories *per se* are discussed in less detail, the book gives a broader introduction to a wide range of questions related to topology, including solitons and monopoles. It also covers several topics not included in this volume, such as instantons in two-dimensional models, instantons and high orders of perturbation theory, etc.

S. Coleman, *Aspects of Symmetry* (Cambridge University Press, 1985), pp. 265–350

7

The uses of instantons
(1977)

1 Introduction

In the last two years there have been astonishing developments in quantum field theory. We have obtained control over problems previously believed to be of insuperable difficulty and we have obtained deep and surprising (at least to me) insights into the structure of the leading candidate for the field theory of the strong interactions, quantum chromodynamics. These goodies have come from a family of computational methods that are the subject of these lectures.

These methods are all based on semiclassical approximations, and, before I can go further, I must tell you what this means in the context of quantum field theory.

To be definite, let us consider the theory of a single scalar field in four-dimensional Minkowski space, with dynamics defined by the Lagrangian density

$$\mathscr{L}=\tfrac{1}{2}\partial_\mu\phi\partial^\mu\phi-\tfrac{1}{2}m^2\phi^2-g^2\phi^4. \tag{1.1}$$

For classical physics, g is an irrelevant parameter. The easiest way to see this is to define

$$\phi'=g\phi. \tag{1.2}$$

In terms of ϕ',

$$\mathscr{L}=\frac{1}{g^2}\left(\tfrac{1}{2}\partial_\mu\phi'\partial^\mu\phi'-\tfrac{1}{2}m^2\phi'^2-\phi'^4\right). \tag{1.3}$$

Thus, g does not appear in the field equations; if one can solve the theory for any positive g, one can solve it for any other positive g; g is irrelevant. Another way of seeing the same thing is to observe that, in classical physics, g is a dimensionful parameter and can always be scaled to one.

Of course, g *is* relevant in quantum physics. The reason is that quantum physics contain a new constant, $\hbar$, and the important object (for example, in Feynman's path-integral formula) is

$$\frac{\mathcal{L}}{\hbar}=\frac{1}{g^2\hbar}\left(\tfrac{1}{2}\partial_\mu\phi'\partial^\mu\phi'+\ldots\right). \tag{1.4}$$

As we see from this expression, the relevant (dimensionless) parameter is $g^2\hbar$, and thus semiclassical approximations, small-$\hbar$ approximations, are tantamount to weak-coupling approximations, small-g approximations.

At this point you must be puzzled by the trumpets and banners of my opening paragraph. Do we not have a perfectly adequate small-coupling approximation in perturbation theory? No, we do not; there is a host of interesting phenomena which occur for small coupling constant and for which perturbation theory is inadequate.

The easiest way to see this is to descend from field theory to particle mechanics. Consider the theory of a particle of unit mass moving in a one-dimensional potential,

$$L=\tfrac{1}{2}\dot{x}^2-V(x;g), \tag{1.5}$$

where

$$V(x;g)=\frac{1}{g^2}F(gx), \tag{1.6}$$

and F is some function whose Taylor expansion begins with terms of order x^2. Everything I have said about the field theory defined by Eq. (1.1) goes through for this theory. However, let us consider the phenomenon of transmission through a potential barrier (Fig. 1). Every child knows that the amplitude for transmission obeys the WKB formula,

$$|T(E)|=\exp\left\{-\frac{1}{\hbar}\int_{x_1}^{x_2}\mathrm{d}x[2(V-E)]^{\frac{1}{2}}\right\}[1+O(\hbar)], \tag{1.7}$$

where x_1 and x_2 are the classical turning points at energy E. This is a semiclassical approximation. Nevertheless, transmission, barrier penetra-

Fig. 1

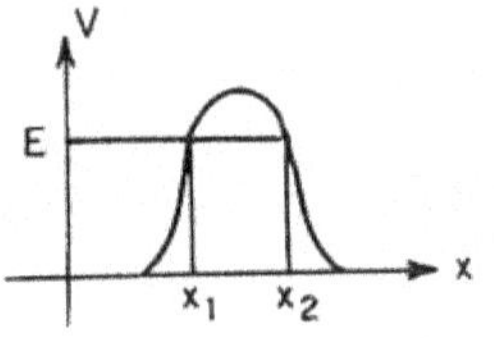

tion, is not seen in any order of perturbation theory, because Eq. (1.7) vanishes more rapidly than any power of $\hbar$, and therefore of g.

I can now make my first paragraph more explicit. There are phenomena in quantum field theory, and in particular in quantum chromodynamics, analogous to barrier penetration in quantum particle mechanics. In the last two years a method has been developed for handling these phenomena. This method is the subject of these lectures.

The organization of these lectures is as follows. In Sect. 2 I describe the new method in the context of particle mechanics, where we already know the answer by an old method (the WKB approximation). Here the instantons which play a central role in the new method and which have given these lectures their title first appear. In Sect. 3 I derive some interesting properties of gauge field theories. In Sect. 4 I discuss a two-dimensional model in which instantons lead to something like quark confinement and explain why a similar mechanism has (unfortunately) no chance of working in four dimensions. In Sect. 5 I explain 't Hooft's resolution of the U(1) problem. In Sect. 6 I apply instanton methods to vacuum decay. Only this last section reports on my own research; all the rest is the work of other hands.[1]

I thank C. Callan, R. Dashen, D. Gross, R. Jackiw, M. Peskin, C. Rebbi, G. 't Hooft, and E. Witten for patiently explaining large portions of this subject to me. Although I have never met A. M. Polyakov, his influence pervades these lectures, as it does the whole subject.[2]

A note on notation. In these lectures we will work in both Minkowski space and in four-dimensional Euclidean space. A point in Minkowski space is labeled x^μ, where $\mu=0, 1, 2, 3$, and x^0 is the time coordinate. In Minkowski space I will distinguish between covariant and contravariant vectors, $x_\mu = g_{\mu\nu}x^\nu$, where the metric tensor has signature $(+\,-\,-\,-)$. Euclidean space is obtained from Minkowski space by formal analytic continuation in the time coordinate, $x^4 = -\mathrm{i}x^0$. A point in Euclidean space is labeled x^μ, where $\mu = 1, 2, 3, 4$. The signature of the metric tensor is $(+\,+\,+\,+)$. Thus covariant and contravariant vectors are component-by-component identical, and I will not bother to distinguish between them. Note that $x\cdot y$ in Minkowski space continues to $-x\cdot y$ in Euclidean space. The Euclidean action is defined as $-\mathrm{i}$ times the continuation of the Minkowskian action. When discussing particle problems, I will use t for both Euclidean and Minkowskian time; which is meant will always be clear from the context. In Sect. 2 explicit factors of $\hbar$ are retained; elsewhere, $\hbar$ is set equal to one.

2 Instantons and bounces in particle mechanics

2.1 *Euclidean functional integrals*

In this section we will deal exclusively with the theory of a spinless particle of unit mass moving in a potential in one dimension:

$$H=\frac{p^2}{2}+V(x). \tag{2.1}$$

We will rederive some familiar properties of this much-studied system by unfamiliar methods. For the problem at hand, these methods are *much* more awkward than the standard methods of one-dimensional quantum mechanics; however, they have the great advantage of being immediately generalizable to quantum field theory.

Our fundamental tool will be the Euclidean (imaginary time) version of Feynman's[3] sum over histories:

$$\langle x_f|e^{-HT/\hbar}|x_i\rangle=N\int[dx]e^{-S/\hbar}. \tag{2.2}$$

Both sides of this equation require explanation:

On the left-hand side, $|x_i\rangle$ and $|x_f\rangle$ are position eigenstates, H is the Hamiltonian, and T is a positive number. The left-hand side of Eq. (2.2) is of interest because, if we expand in a complete set of energy eigenstates,

$$H|n\rangle=E_n|n\rangle, \tag{2.3}$$

then

$$\langle x_f|e^{-HT/\hbar}|x_i\rangle=\sum_n e^{-E_nT/\hbar}\langle x_f|n\rangle\langle n|x_i\rangle. \tag{2.4}$$

Thus, the leading term in this expression for large T tells us the energy and wave-function of the lowest-lying energy eigenstate.

On the right-hand side, N is a normalization factor, S is the Euclidean action[4]

$$S=\int_{-T/2}^{T/2}dt\left[\frac{1}{2}\left(\frac{dx}{dt}\right)^2+V\right], \tag{2.5}$$

and $[dx]$ denotes integration over all functions $x(t)$, obeying the boundary conditions, $x(-T/2)=x_i$ and $x(T/2)=x_f$. To be more specific, if $\bar{x}$ is any function obeying the boundary condition, then a general function obeying the boundary conditions can be written as

$$x(t)=\bar{x}(t)+\sum_n c_nx_n(t), \tag{2.6}$$

where the x_ns are a complete set of real orthonormal functions vanishing at the boundaries,

$$\int_{-T/2}^{T/2} dt x_n(t) x_m(t) = \delta_{nm}, \tag{2.7a}$$

$$x_n(\pm T/2) = 0. \tag{2.7b}$$

Then, the measure $[dx]$ is defined by

$$[dx] = \prod_n (2\pi\hbar)^{-\frac{1}{2}} dc_n. \tag{2.8}$$

(This measure differs in normalization from the measure defined by Feynman;[3] this is why we need the normalization constant N. However, as we shall see, we shall never need an explicit formula for N.)

The right-hand side of Eq. (2.2) is of interest because it can readily be evaluated in the semiclassical (small $\hbar$) limit. In this case the functional integral is dominated by the stationary points of S. For simplicity, let us assume for the moment that there is only one such stationary point, which we denote by $\bar{x}$,

$$\frac{\delta S}{\delta \bar{x}} = -\frac{d^2 \bar{x}}{dt^2} + V'(\bar{x}) = 0, \tag{2.9}$$

where the prime denotes differentiation with respect to x. Further, let us choose the x_ns to be eigenfunctions of the second variational derivative of S at $\bar{x}$,

$$-\frac{d^2 x_n}{dt^2} + V''(\bar{x}) x_n = \lambda_n x_n. \tag{2.10}$$

Then, in the small-$\hbar$ limit, the integral becomes a product of Gaussians, and we find

$$\begin{aligned}\langle x_f | e^{-HT/\hbar} | x_i \rangle &= N e^{-S(\bar{x})/\hbar} \prod_n \lambda_n^{-\frac{1}{2}} [1 + O(\hbar)] \\ &= N e^{-S(\bar{x})/\hbar} [\det(-\partial_t^2 + V''(\bar{x}))]^{-\frac{1}{2}} [1 + O(\hbar)].\end{aligned} \tag{2.11}$$

(Of course, we are tacitly assuming here that all the eigenvalues are positive. We shall shortly see what to do when this is not the case.) If there are several stationary points, in general one has to sum over all of them.

Equation (2.9) is the equation of motion for a particle of unit mass moving in a potential *minus* V. Thus,

$$E = \frac{1}{2}\left(\frac{d\bar{x}}{dt}\right)^2 - V(\bar{x}) \tag{2.12}$$

is a constant of the motion. This can be used to determine the qualitative features of the solutions of Eq. (2.9) by inspection.

As a simple example, consider the potential shown in Fig. 2(a). Let us choose $x_i = x_f = 0$. Figure 2(b) shows the inverted potential, $-V$. It is

Fig. 2

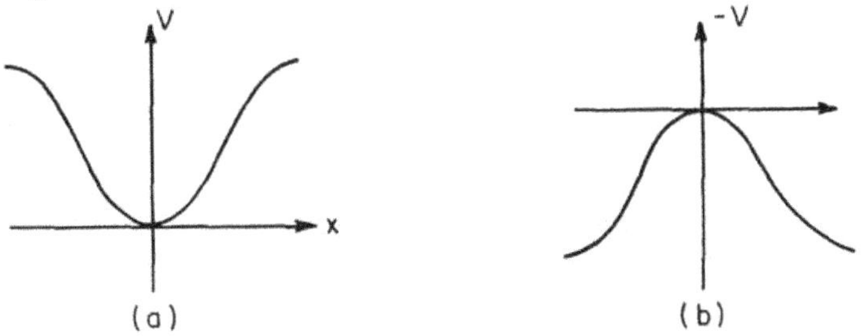

obvious from the figure that the only solution of Eq. (2.9) which obeys the boundary conditions is

$$\bar{x}=0. \tag{2.13}$$

For this solution, $S=0$. Thus, from Eq. (2.11),

$$\langle 0|e^{-HT/\hbar}|0\rangle = N[\det(-\partial_t^2+\omega^2)]^{-\frac{1}{2}}[1+O(\hbar)], \tag{2.14}$$

where

$$\omega^2 = V''(0). \tag{2.15}$$

In Appendix 1, I show that, for large T,

$$N[\det(-\partial_t^2+\omega^2)]^{-\frac{1}{2}} = \left(\frac{\omega}{\pi\hbar}\right)^{\frac{1}{2}} e^{-\omega T/2}. \tag{2.16}$$

Thus, the ground-state energy is given by

$$E_0 = \tfrac{1}{2}\omega\hbar[1+O(\hbar)]. \tag{2.17}$$

Also, the probability of the particle being at the origin when it is in its ground state is

$$|\langle x=0|n=0\rangle|^2 = (\omega/\pi\hbar)^{\frac{1}{2}}[1+O(\hbar)]. \tag{2.18}$$

These are, of course, the correct semiclassical results. In the small-$\hbar$ limit, the particle is in a harmonic-oscillator ground-state concentrated at the origin and its energy is the ground-state energy of a harmonic oscillator.

2.2 *The double well and instantons*

We now turn to a less trivial problem,[5] the double well of Fig. 3(a). I will assume the potential is even, $V(x)=V(-x)$, and will denote its minima by $\pm a$. As before, I will add a constant to V, if necessary to make V vanish at its minima, and I will denote $V''(\pm a)$ by ω^2.

We will attempt to compute both

$$\langle -a|e^{-HT/\hbar}|-a\rangle = \langle a|e^{-HT/\hbar}|a\rangle, \tag{2.19a}$$

Fig. 3

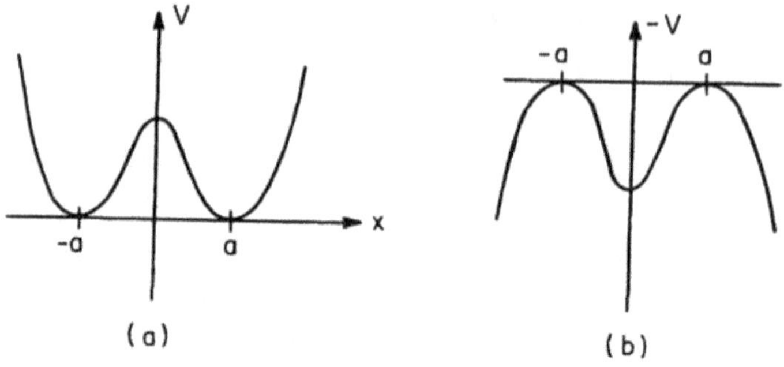

and

$$\langle a|e^{-HT/\hbar}|-a\rangle = \langle -a|e^{-HT/\hbar}|a\rangle, \tag{2.19b}$$

by approximating the functional integral by its semiclassical limit, Eq. (2.11). Just as before, the first step is to find solutions of the classical Euclidean equation of motion, (2.9), consistent with our boundary conditions.

Of course, two such solutions are those in which the particle stays fixed on top of one or the other of the two hills in Fig. 3(b). However, there is another potentially interesting solution, one where the particle begins at the top of one hill (say the left one) at time $-T/2$, and moves to the top of the right hill at time $T/2$. Since we plan eventually to take T to infinity, we will focus on the form of the solution in this limit, where the particle attains the tops of the hills at times plus and minus infinity. In this case, we are dealing with a solution of the equation of motion with vanishing E; whence

$$dx/dt = (2V)^{\frac{1}{2}}. \tag{2.20}$$

Equivalently,

$$t = t_1 + \int_0^x dx'(2V)^{-\frac{1}{2}}, \tag{2.21}$$

where t_1 is an integration constant, the time at which x vanishes.

This solution is sketched in Fig. 4; it is called 'an instanton with center at t_1'. The name 'instanton' was invented by 't Hooft. The idea is that these objects are very similar in their mathematical structure to what are called solitons or lumps,[6] particle-like solutions of classical field theories: thus the '-on'. However, unlike lumps, they are structures in time (albeit Euclidean time): thus the 'instant-'. For the same reason, Polyakov suggested the name 'pseudoparticle', also used in the literature.

Of course, we can also construct solutions that go from a to $-a$,

Fig. 4

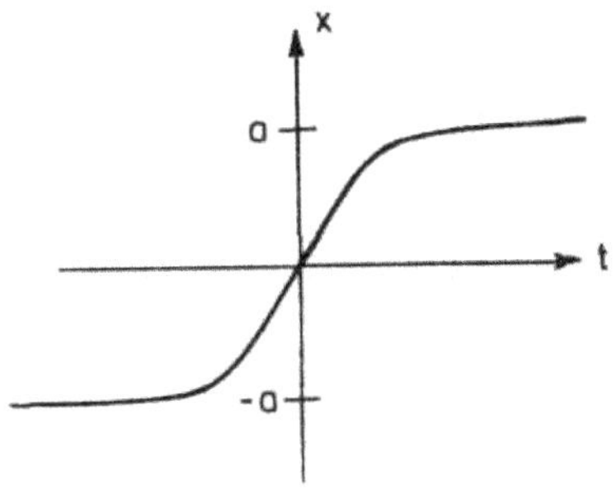

simply by replacing t by $-t$ in Eq. (2.21); these are called 'anti-instantons'.

Two properties of these solutions will be important to us:

(1) From Eq. (2.20), it is easy to derive a simple expression for S_0, the action of an instanton (or anti-instanton)

$$S_0 = \int dt[\tfrac{1}{2}(dx/dt)^2 + V] = \int dt(dx/dt)^2 = \int_{-a}^{a} dx(2V)^{\frac{1}{2}}. \tag{2.22}$$

Note that this is the same as the integral that appears in the barrier-penetration formula, Eq. (1.7). We shall see shortly that this is no coincidence.

(2) For large t, x approaches a, and Eq. (2.20) can be approximated by

$$dx/dt = \omega(a-x). \tag{2.23}$$

Thus, for large t,

$$(a-x) \propto e^{-\omega t}. \tag{2.24}$$

Thus, instantons are, roughly speaking, well-localized objects, having a size on the order of $1/\omega$.

This is of critical importance, because it means that, for large T, the instanton and the anti-instanton are not the only approximate solutions of the equation of motion; there are also approximate solutions consisting of strings of widely separated instantons and anti-instantons. (You may be troubled by the sudden appearance in the argument of approximate solutions, approximate stationary points of S. If so, bear with me; I shall give a fuller explanation of this point later.)

I shall evaluate the functional integral by summing over all such configurations, with n objects (instantons or anti-instantons) centered at $t_1 \ldots t_n$, where

$$T/2 > t_1 > t_2 \ldots > t_n > -T/2. \tag{2.25}$$

Fig. 5 shows one such configuration. T is assumed to be huge on the

Fig. 5

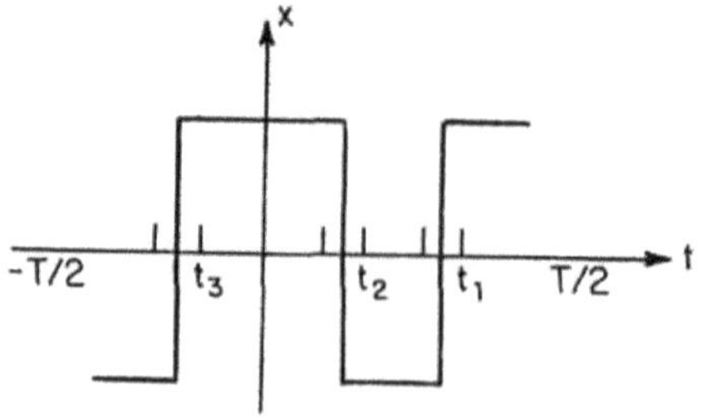

scale of the size of an instanton; thus the smooth curves of Fig. 4 appear as sharp jumps on the scale of Fig. 5. (The vertical marks on the time axis will be explained shortly.)

Now for the evaluation:

(1) For n widely separated objects, S is nS_0. This takes care of the exponential of the action.

(2) The evaluation of the determinant is a bit trickier. Let us consider the time evolution operator, e^{-HT}, as a product of operators associated with evolution between the points indicated by the vertical marks on the time axis in Fig. 5. If it were not for the small intervals containing the instantons and anti-instantons, V'' would equal ω^2 over the entire time axis, and thus we would obtain the same result we obtained for a single-well potential in Sect. 2.1,

$$\left(\frac{\omega}{\pi\hbar}\right)^{\frac{1}{2}} e^{-\omega T/2}. \tag{2.26}$$

The small intervals containing the instantons and anti-instantons correct this formula. Thus we obtain

$$\left(\frac{\omega}{\pi\hbar}\right)^{\frac{1}{2}} e^{-\omega T/2} K^n, \tag{2.27}$$

where K is defined by demanding that this formula give the right answer for one instanton. Later we shall obtain a more explicit expression for K.

(3) We must integrate over the locations of the centers:

$$\int_{-T/2}^{T/2} dt_1 \int_{-T/2}^{t_1} dt_2 \ldots \int_{-T/2}^{t_{n-1}} dt_n = T^n/n!\,. \tag{2.28}$$

(4) We are not free to distribute instantons and anti-instantons arbitrarily. For example, if we start out at $-a$, the first object we encounter must be an instanton, the next one must be an anti-instanton, etc. Furthermore, if we are to end up back at $-a$, n must be even. Likewise, if we wish to end up at a, n must be odd.

Thus,

$$\langle -a|e^{-HT/\hbar}|-a\rangle = \left(\frac{\omega}{\pi\hbar}\right)^{\frac{1}{2}} e^{-\omega T/2} \sum_{\substack{\text{even}\\ n}} \frac{(Ke^{-S_0/\hbar}T)^n}{n!}[1+O(\hbar)], \tag{2.29}$$

while $\langle a|e^{-HT/\hbar}|-a\rangle$ is given by the same expression, summed over odd ns. These sums are trivial:

$$\langle \pm a|e^{-HT/\hbar}|-a\rangle = \left(\frac{\omega}{\pi\hbar}\right)^{\frac{1}{2}} e^{-\omega T/2}\tfrac{1}{2}[\exp(Ke^{-S_0/\hbar}T) \mp \exp(-Ke^{-S_0/\hbar}T)]. \tag{2.30}$$

(From now on, to keep the page from getting cluttered, I will drop the factors of $[1+O(\hbar)]$; remember that they are omnipresent though unwritten.)

Comparing this to Eq. (2.4), we see that we have two low-lying energy eigenstates, with energies

$$E_\pm = \tfrac{1}{2}\hbar\omega \pm \hbar K e^{-S_0/\hbar}. \tag{2.31}$$

If we call these eigenstates $|+\rangle$ and $|-\rangle$, we also see that

$$|\langle +|\pm a\rangle|^2 = |\langle -|\pm a\rangle|^2 = \langle a|-\rangle\langle -|-a\rangle$$
$$= -\langle a|+\rangle\langle +|-a\rangle = \tfrac{1}{2}\left(\frac{\omega}{\pi\hbar}\right)^{\frac{1}{2}}. \tag{2.32}$$

Of course, these are the expected results: the energy eigenstates are the spatially even and odd combinations of harmonic oscillator states centered at the bottoms of the two wells; the degeneracy of the two energy eigenvalues is broken only by barrier penetration (and thus the difference of the energies is proportional to the barrier-penetration factor, $e^{-S_0/\hbar}$), and the state of lower energy, which we have denoted by $|-\rangle$, is the spatially even combination.

Our next task is to evaluate K. Before we do this, though, some comments should be made about what we have done so far:

(1) Really we have no right to retain the second term in Eq. (2.31). It is not only exponentially small compared to the first term, it is exponentially small compared to the uncomputed $O(\hbar^2)$ corrections to the first term. However, it is the leading contribution to the difference of the energies, $E_+ - E_-$; a purist would retain it only in the expression for this difference and not in the expressions for the individual energies.

(2) Our approximation has been based on the assumption that the instantons and anti-instantons are all widely separated. As a consistency check, we should verify that the major portion of our final result comes from configurations where this is indeed the case.

This check is easy to carry out. For any fixed x, the terms in the exponential series, $\sum x^n/n!$, grow with n until n is on the order of x; after this point, they begin to decrease rapidly. Applying this to the sum in Eq. (2.29), we see the important terms are those for which

$$n \lesssim KT e^{-S_0/\hbar}. \tag{2.33}$$

That is to say, for small $\hbar$, the important terms in the sum are those for which n/T, the density of instantons and anti-instantons, is exponentially small, and thus the average separation is enormous. Note that this average separation is independent of T; our approximation is indeed a small-$\hbar$ approximation; the conditions for its validity are independent of T, as long as T is sufficiently large.

This approximation of summing over widely separated instantons is called the dilute-gas approximation, because of its similarity to the approximation of that name in statistical mechanics.

(3) Finally, I want to deliver the promised fuller explanation of the idea of an approximate stationary point of S. Let us begin by studying an integral over a single variable,

$$I = \int_0^T dt\, e^{-S(t)/\hbar}, \tag{2.34}$$

where S is a function of t monotonically decreasing to some asymptotic value, $S(\infty)$. Thus the integrand has no stationary points in the region of integration. Nevertheless, it is easy to find the approximate form of the integral for small $\hbar$ and large T:

$$I \approx T e^{-S(\infty)/\hbar}. \tag{2.35}$$

Speaking loosely, the integral is dominated by the stationary point at infinity. It is straightforward to generalize this phenomenon to multidimensional integrals: we assume an integrand whose graph has a sort of trough in it; the line along the bottom of the trough flattens out as we go to infinity. Speaking less pictorially, there is a line in the multidimensional space such that the integrand is a minimum with respect to variations perpendicular to the line and approaches some limiting value as one goes to infinity along the line. Of course, the line could itself be generalized to a hyperplane, a generalized 'bottom of the trough'. This is in fact the situation for our 'approximate stationary points'; the locations of the instantons and anti-instantons are the variables along the bottom of the trough; S becomes stationary (and equal to nS_0) only when they all go to infinity.

This concludes the comments; we now turn to the evaluation of K. We must study the eigenvalue equation, Eq. (2.10), with $\bar{x}$ a single instanton. Because of time translation invariance, this equation necessarily possesses an eigenfunction of eigenvalue zero,

$$x_1 = S_0^{-\frac{1}{2}}\, d\bar{x}/dt. \tag{2.36}$$

(The normalization factor comes from Eq. (2.22).) Were we to integrate over the corresponding expansion coefficient, c_1, in Eq. (2.6), we would obtain a disastrous infinity. Fortunately, we have already done this integration, in the guise of integrating over the location of the center of the instanton in Eq. (2.28). The change of $x(t)$ induced by a small change in the location of the center, t_1, is

$$dx = (d\bar{x}/dt)\, dt_1. \tag{2.37}$$

The change induced by a small change in the expansion coefficient, c_1, is

$$dx = x_1\, dc_1. \tag{2.38}$$

Hence,

$$(2\pi\hbar)^{-\frac{1}{2}}\, dc_1 = (S_0/2\pi\hbar)^{\frac{1}{2}}\, dt_1. \tag{2.39}$$

Thus, in evaluating the determinant, we should not include the zero eigenvalue, but we should include in K a factor[7] of $(S_0/2\pi\hbar)^{\frac{1}{2}}$. Hence, the one-instanton contribution to the transition matrix element is given by

$$\begin{aligned}&\langle a|e^{-HT}|-a\rangle_{\text{one inst.}}\\ &\quad = NT(S_0/2\pi\hbar)^{\frac{1}{2}} e^{-S_0/\hbar}(\det{}'[-\partial_t^2 + V''(\bar{x})])^{-\frac{1}{2}},\end{aligned} \tag{2.40}$$

where det′ indicates that the zero eigenvalue is to be omitted when computing the determinant. Comparing this to the one-instanton term in Eq. (2.29), we find

$$K = (S_0/2\pi\hbar)^{\frac{1}{2}} \left| \frac{\det(-\partial_t^2 + \omega^2)}{\det{}'(-\partial_t^2 + V''(\bar{x}))} \right|^{\frac{1}{2}}. \tag{2.41}$$

This completes the computation.

Some remarks:

(1) To really sew things up, I should show that the formula we have obtained for the energy splitting is the same as that obtained by the traditional methods of wave mechanics. I do this in Appendix 2.

(2) I have been tacitly assuming that all the eigenvalues in Eq. (2.10) are positive, other than the zero eigenvalue associated with x_1. It is easy to prove that this is indeed the case. It is well-known that the eigenfunction of a one-dimensional Schrödinger equation (like Eq. (2.10)) of lowest eigenvalue has no nodes, the next-lowest eigenfunction has one node, etc. Because the instanton is a monotone increasing function of t, x_1, proportional to the time derivative of the instanton, has no nodes. Thus zero is the lowest eigenvalue and all the other eigenvalues are positive.

(3) K is proportional to $\hbar^{-\frac{1}{2}}$. This factor came from the zero eigenvalue

associated with time-translation invariance. Later in these lectures we will be analyzing theories that have larger invariance groups and for which the instantons have more than one zero eigenvalue associated with them. Clearly, for every zero eigenvalue there will be a factor of $\hbar^{-\frac{1}{2}}$. This rule for counting powers of $\hbar$ will be very important to us, for, as I explained in Sect. 1, counting powers of $\hbar$ is equivalent to counting powers of coupling constants.

2.3 *Periodic potentials*

Let us consider a periodic potential, like the one sketched in Fig. 6(a). (For simplicity, I have chosen the minima of V to be the integers.) If we ignore barrier penetration, the energy eigenstates are an infinitely degenerate set of states, each concentrated at the bottom of one of the wells. Barrier penetration changes this single eigenvalue into a continuous band of eigenvalues; the true energy eigenstates are the eigenstates of unit translations, the Bloch waves. Let us see how this old result can be obtained by instanton methods.

As we see from Fig. 6(b), the instantons are much the same as in the preceding problem. The only novelty is that the instantons can begin at any initial position, $x=j$, and go to the next one, $x=j+1$. Likewise, the anti-instantons can go from $x=j$ to $x=j-1$. Otherwise, everything is as before.

Thus, when doing the dilute-gas sum, we can sprinkle instantons and anti-instantons freely about the real axis; there is no constraint that instantons and anti-instantons must alternate. Of course, as we go along

Fig. 6

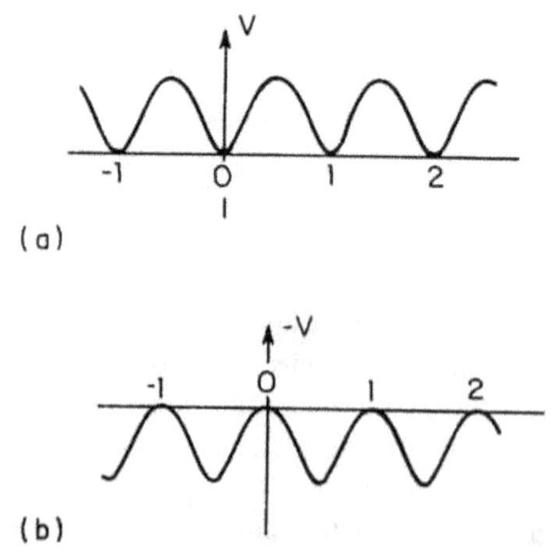

the line, each instanton or anti-instanton must begin where its predecessor ended. Furthermore, the total number of instantons minus the total number of anti-instantons must equal the change in x between the initial and final position eigenstates.

Thus we obtain

$$\langle j_+|e^{-HT/\hbar}|j_-\rangle = \left(\frac{\omega}{\pi\hbar}\right)^{\frac{1}{2}} e^{-\omega T/2} \sum_{n=0}^{\infty}\sum_{\bar{n}=0}^{\infty} \frac{1}{n!\,\bar{n}!} \times (K e^{-S_0/\hbar} T)^{n+\bar{n}} \delta_{n-\bar{n}-j_+ + j_-} \tag{2.42}$$

where n is the number of instantons and $\bar{n}$ the number of anti-instantons. If we use the identity

$$\delta_{ab} = \int_0^{2\pi} d\theta\, e^{i\theta(a-b)}/2\pi, \tag{2.43}$$

the sum becomes two independent exponential series, and we find

$$\langle j_+|e^{-HT/\hbar}|j_-\rangle = \left(\frac{\omega}{\pi\hbar}\right)^{\frac{1}{2}} e^{-\omega T/2} \int_0^{2\pi} e^{i(j_- - j_+)\theta} \frac{d\theta}{2\pi} \times \exp[2KT\cos\theta\, e^{-S_0/\hbar}] \tag{2.44}$$

Thus we find a continuum of energy eigenstates labeled by the angle θ. The energy eigenvalues are given by

$$E(\theta) = \tfrac{1}{2}h\omega + 2\hbar K \cos\theta\, e^{-S_0/\hbar}. \tag{2.45}$$

Also,

$$\langle\theta|j\rangle = \left(\frac{\omega}{\pi\hbar}\right)^{\frac{1}{4}} (2\pi)^{-\frac{1}{2}} e^{ij\theta}. \tag{2.46}$$

Hearteningly, this is just the right answer.

2.4 *Unstable states and bounces*[8]

Galilean pastiche:

SAGREDO: Let me test my understanding of these instanton methods by studying the potential of Fig. 7(a). If I neglect barrier penetration, in the

Fig. 7

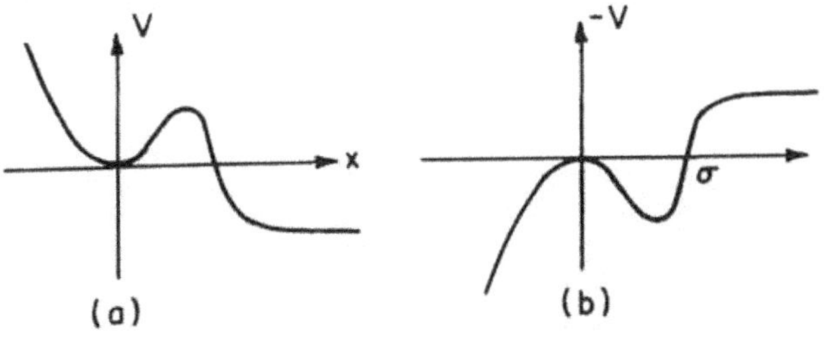

semiclassical limit, this potential has an energy eigenstate sitting in the bottom of the well. I wish to compute the corrections to the energy of this state due to barrier penetration. If I turn the potential upside down (Fig. 7(b)), I observe that the classical equation of motion has a solution in which the particle begins at the top of the hill at $x=0$, bounces off the classical turning point σ, and returns to the top of the hill (Fig. 8). I will call this motion 'the bounce'. I will compute the transition matrix element between $x=0$ and $x=0$ by summing over configurations consisting of widely separated bounces, just as one sums over instantons and anti-instantons in the study of the double well. Indeed, the sum is the same as that for the double well (with the obvious redefinitions of S_0, ω^2, etc.), save that there is no restriction to an even or odd number of bounces. Thus I obtain the complete exponential series, rather than just the odd or even terms, and I find that

$$\langle 0|e^{-HT/\hbar}|0\rangle = \left(\frac{\omega}{\pi\hbar}\right)^{\frac{1}{2}} e^{-\omega T/2} \exp[KTe^{-S_0/\hbar}], \tag{2.47}$$

and the energy eigenvalue is given by

$$E_0 = \tfrac{1}{2}\omega\hbar + \hbar K e^{-S_0/\hbar}. \tag{2.48}$$

SALVIATI: Alas, Sagredo, I fear you have erred in three ways. Firstly, the term you have computed is small compared to terms of order $\hbar^2$ which you have neglected, and thus you have no right to retain it. Secondly, I see by your sketch that the bounce has a maximum; therefore the eigenfunction x_1, which is proportional to the time derivative of the bounce, has a node. Thus it is not the eigenfunction of lowest eigenvalue, and there must be a nodeless eigenfunction, x_0, of a lower eigenvalue, that is to say, there must be a negative eigenvalue. Thus K, which is inversely proportional to the product of the square roots of the eigenvalues, is imaginary. Thirdly, the eigenvalue you attempt to compute is nowhere to be found in the spectrum of the Hamiltonian, because the state you are studying is rendered unstable by barrier penetration.

SAGREDO: Everything you say is correct, but I believe your criticisms show how to save the computation. An unstable state is one whose energy has an

Fig. 8

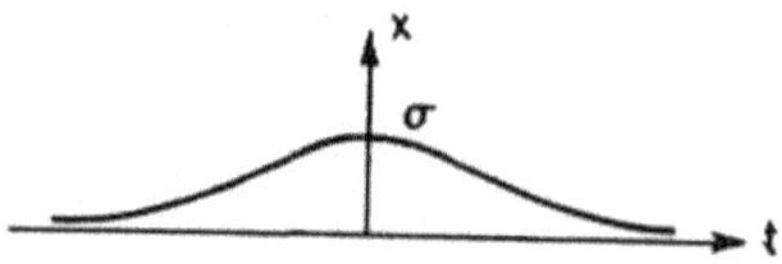

imaginary part; thus it is only to be expected that K should be imaginary. Furthermore, the term I have computed, though indeed small compared to neglected contributions to the real part of E_0, is the leading contribution to the imaginary part of E_0. Thus the correct version of Eq. (2.48) is

$$\mathrm{Im}\, E_0 = \Gamma/2 = \hbar|K|e^{-S_0/\hbar}, \tag{2.49}$$

where Γ is, as usual, the width of the unstable state.

As you can see, the Tuscan twosome are as quick-witted as ever, although (also as ever) their arguments are sometimes a bit sloppy. Sagredo has missed a factor of $\frac{1}{2}$; the correct answer is

$$\Gamma = \hbar|K|e^{-S_0/\hbar}. \tag{2.50}$$

To show that this is the case requires a more careful argument than Sagredo's. The essential point is Salviati's observation that the energy of an unstable state is not an eigenvalue of H; in fact, it is an object that can only be defined by a process of analytic continuation. I will now perform such a continuation.

To keep things as simple as possible, let us consider not an integral over all function space, but an integral over some path in function space parametrized by a real variable, z,

$$J = \int dz(2\pi\hbar)^{-\frac{1}{2}}e^{-S(z)/\hbar}, \tag{2.51}$$

where $S(z)$ is the action along the path. In particular, let us choose the path sketched in Fig. 9. This path includes two important functions that occur in the real problem: $x(t)=0$, at $z=0$, and the bounce, at $z=1$. Furthermore, the path is such that the tangent vector to the path at $z=1$ is x_0. Thus the path goes through the bounce in the 'most dangerous direction', that direction with which the negative eigenvalue is associated, and $z=1$ is a maximum of S, as shown in Fig. 10. S goes to minus infinity as z goes to infinity because the functions spend more and more time in

Fig. 9

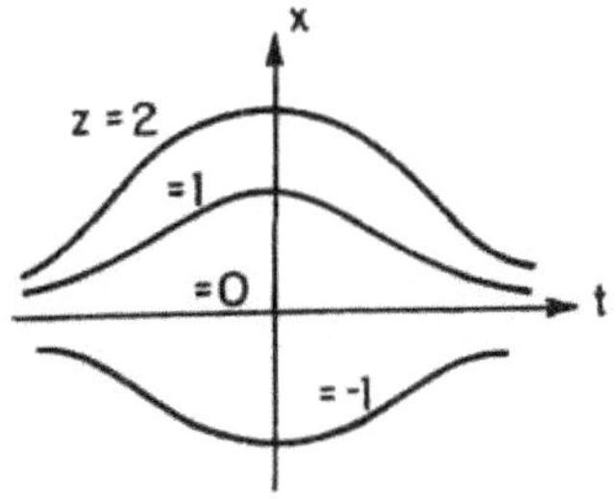

Fig. 10

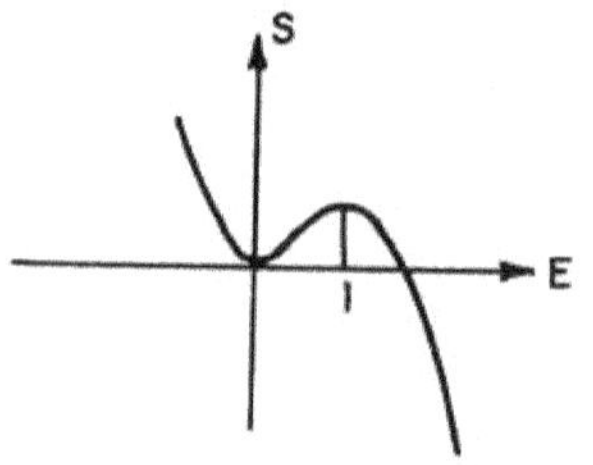

Fig. 11

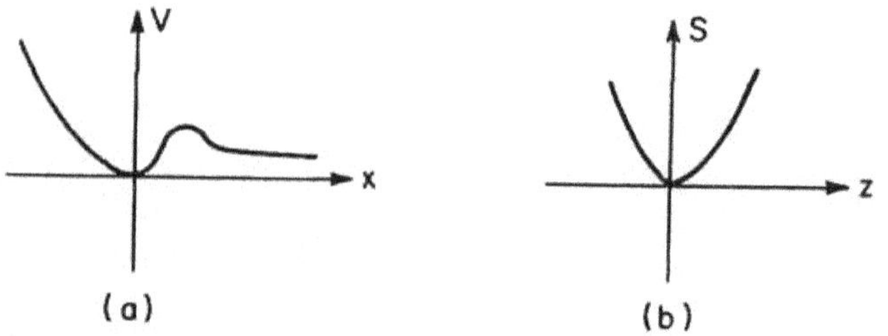

the region beyond the turning point, where V is negative; note that this implies that Eq. (2.51) is hopelessly divergent.

If $x=0$ were the absolute minimum of V, that is to say, if V were as shown in Fig. 11(a), we would have, for the same path, the situation shown in Fig. 11(b), and there would be no divergence in Eq. (2.48). Now let us suppose we analytically change V in some way such that we go from this situation back to the one of interest. To keep the integral convergent, we must distort the right-hand portion of the contour of integration into the complex plane. How we distort it depends on the details of the analytic passage from one potential to the other. In Fig. 12, I have assumed that it is distorted into the upper half plane. Following the standard procedure of the method of steepest descents, I have led the contour along the real axis to $z=1$, the saddle point, and then out along a line of constant imaginary part of S. The integral thus acquires an imaginary part; in the steepest-

Fig. 12

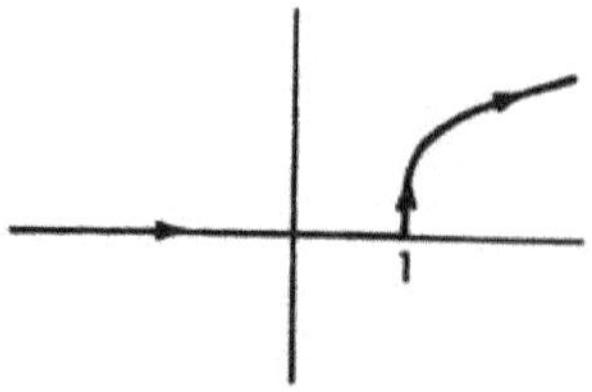

descent approximation,

$$\begin{aligned} \operatorname{im} J &= \operatorname{Im} \int_1^{1+i\infty} dz (2\pi\hbar)^{-\frac{1}{2}} e^{-S(1)/\hbar} e^{-\frac{1}{2}S''(1)(z-1)^2/\hbar} \\ &= \tfrac{1}{2} e^{-S(1)/\hbar} |S''(1)|^{-\frac{1}{2}}. \end{aligned} \tag{2.52}$$

Note the factor of $\frac{1}{2}$; this arises because the integration is over only half of the Gaussian peak.

(If we had passed from one potential to the other in the conjugate manner, the contour would have been distorted into the lower half plane, and we would have obtained the opposite sign for the imaginary part. This is just a reflection of the well-known fact that what sign you get for the imaginary part of the energy of an unstable state depends on how you do your analytic continuation.)

Now, we have studied a one-dimensional integral, but we can always reduce our functional integral to a one-dimensional integral simply by integrating (in the Gaussian approximation) over all the variables orthogonal to our path. These directions involve only positive or zero eigenvalues near the stationary point and give us no trouble. In this manner we obtain Sagredo's answer, Eq. (2.49), except that the negative eigenvalue carries a factor of $\frac{1}{2}$ with it; that is to say, we obtain Eq. (2.50).

3 The vacuum structure of gauge field theories[9]

3.1 *Old stuff*

This subsection is a telegraphic compendium of formulae from gauge field theories. Its purpose is to establish notational conventions and possibly to jog your memory. If you do not already know the fundamentals of gauge field theory, you will not learn them here.[10]

Lie algebras. A representation of Lie algebra is a set of N anti-Hermitian matrices, T^a, $a = 1 \ldots N$, obeying the equations

$$[T^a, T^b] = c^{abc} T^c, \tag{3.1}$$

where the cs are the structure constants of some compact Lie group, G. It is always possible to choose the Ts such that $\mathrm{Tr}(T^a T^b)$ is proportional to δ^{ab}, although the constant of proportionality may depend on the representation. The Cartan inner product is defined by

$$(T^a, T^b) = \delta^{ab}. \tag{3.2}$$

Thus this is proportional to the trace of the product of the matrices.

So far I have not stated a convention that gives a scale to the structure constants and thus to the Ts. For SU(2), the case I will spend most time discussing, I will choose c^{abc} to be equal to ε^{abc}. Thus, for the isospinor representation,

$$T^a = -\mathrm{i}\sigma^a/2, \tag{3.3}$$

where the σs are the Pauli spin matrices. In this case,

$$(T^a, T^b) = -2\,\mathrm{Tr}(T^a T^b). \tag{3.4}$$

Occasionally I will discuss SU(n), in particular SU(3). In this case I will choose the structure constants to agree with the preceding convention for the SU(2) subgroup composed of unitary unimodular transformations on two variables only. Thus, for SU(3), T^a is $-\mathrm{i}\lambda^a/2$, where the λs are Gell-Mann's matrices.

Gauge fields. The gauge potentials are a set of vector fields, $A^a_\mu(x)$. It is convenient to define a matrix-valued vector field, $A_\mu(x)$, by

$$A_\mu = g A^a_\mu T^a, \tag{3.5}$$

where g is a constant called the gauge coupling constant. The field-strength tensor, $F_{\mu\nu}(x)$, is defined by

$$F_{\mu\nu} = \partial_\mu A_\nu - \partial_\nu A_\mu + [A_\mu, A_\nu]. \tag{3.6}$$

Pure gauge field theory is defined by the Euclidean action,

$$S = \frac{1}{4g^2}\int d^4x (F_{\mu\nu}, F_{\mu\nu}). \tag{3.7}$$

Sometimes I will write this in a shorthand form,

$$S = \frac{1}{4g^2}\int (F^2). \tag{3.8}$$

Gauge transformations. A gauge transformation is a function, $g(x)$, from Euclidean space into the gauge group, G. In equations,

$$g(x) = \exp \lambda^a(x) T^a, \tag{3.9}$$

where the λs are arbitrary functions. (Please do not confuse $g(x)$ with the coupling constant, g.) Under such a transformation,

$$A_\mu \to g A_\mu g^{-1} + g\partial_\mu g^{-1}, \tag{3.10}$$

and

$$F_{\mu\nu} \to g F_{\mu\nu} g^{-1}. \tag{3.11}$$

Thus, S is gauge-invariant. If $F_{\mu\nu}$ vanishes, then A_μ is a gauge-transform of zero; that is to say,

$$A_\mu = g\partial_\mu g^{-1}, \tag{3.12}$$

for some $g(x)$.

Covariant derivatives. The covariant derivative of the field strength tensor is defined by

$$D_\lambda F_{\mu\nu} = \partial_\lambda F_{\mu\nu} + [A_\lambda, F_{\mu\nu}]. \tag{3.13}$$

Equation (3.7) leads to the Euclidean equations of motion

$$D_\mu F_{\mu\nu} = 0. \tag{3.14}$$

Given a field ψ that gauge-transforms according to

$$\psi \to g(x)\psi, \tag{3.15}$$

then the covariant derivative of ψ,

$$D_\mu \psi = \partial_\mu \psi + A_\mu \psi, \tag{3.16}$$

transforms in the same way.

3.2 *The winding number*

I propose to study Euclidean gauge field configurations of finite action (not necessarily solutions of the equations of motion).

Why?

The naive answer, sometimes given in the literature,[11] is that configurations of infinite action are unimportant in the functional integral, since, for such configurations, $e^{-S/\hbar}$ is zero. *This is wrong.* In fact, it is configurations of finite action that are unimportant; to be precise, they form a set of measure zero in function space. This has nothing to do with the divergences of quantum field theory; it is true even for the ordinary harmonic oscillator. (For a proof, see Appendix 3.) The only reason we are interested in configurations of finite action is that we are interested in doing semiclassical approximations, and a configuration of infinite action does indeed give zero if it is used as the center point of a Gaussian integral.

The convergence of the action integral is controlled by the behavior of A_μ for large r, where r is the radial variable in Euclidean four-space. To keep my arguments as simple as possible, I will assume that, for large r, A_μ can be expanded in an asymptotic series in inverse powers of r. (This assumption can be relaxed considerably without altering the conclusions.)[12] Thus, for the action to be finite, $F_{\mu\nu}$ must fall off faster than $1/r^2$ as r goes to infinity; that is to say, $F_{\mu\nu}$ must be $O(1/r^3)$. One's first thought is that this implies that A_μ is $O(1/r^2)$, but this is wrong: vanishing $F_{\mu\nu}$ does not imply vanishing A_μ, but merely that A_μ is a gauge transform of zero. Thus A_μ can be of the form

$$A_\mu = g\partial_\mu g^{-1} + O(1/r^2), \tag{3.17}$$

where g is a function from four-space to G of order one, that is to say, a function of angular variables only.

Thus, with every finite-action field configuration there is associated a group-element-valued function of angular variables, that is to say, a mapping of a three-dimensional hypersphere, S^3, into the gauge group, G. Of course, this assignment is not gauge-invariant. Under a gauge transformation, $h(x)$

$$A_\mu \to hA_\mu h^{-1} + h\partial_\mu h^{-1}. \tag{3.18}$$

Thus,

$$g \to hg + O(1/r^2). \tag{3.19}$$

If one could choose h to equal g^{-1} at infinity, one could transform g to one and eliminate it from Eq. (3.17). In general, though, this is not possible. The reason is that h must be a continuous function not just on the hypersphere at infinity, but throughout all four-space, that is to say, on a nested family of hyperspheres going all the way from r equals zero to r equals infinity. In particular, at the origin, h must be a constant, independent of angles. Thus, h at infinity can not be a general function on S^3, but must be one that can be obtained by continuous deformation from a constant function. Since any constant gauge transformation can trivially be obtained by continuous deformation from the identity transformation (all gauge groups are connected), we might as well say that h at infinity must be obtainable from $h=1$ by a continuous deformation.

Given two mappings of one topological space into another, such that one mapping is continuously deformable into another, mathematicians say the two functions are 'homotopic' or 'in the same homotopy class'. What we have shown is that by a gauge transformation we can transform $g(x)$ into any mapping homotopic to $g(x)$, but we can not transform it into a function in another homotopy class. Thus, the gauge-invariant quantity associated with a finite-action field configuration is not a mapping of S^3 to G but a homotopy class of such mappings. Our task is to find these homotopy classes for physically interesting Gs.

To warm up for this task, let me consider a baby version of the problem for which the geometry is somewhat easier to visualize. I will work with the simplest of all gauge groups, U(1), the group of complex numbers of unit modulus. Thus the gauge field theory is ordinary electromagnetism. (However, I will still keep to the notational conventions established in Sect. 3.1; in particular, A_μ will be an imaginary quantity, i times the usual vector potential.) Also, I will work not in Euclidean four-space but in Euclidean two-space. I will still study fields obeying Eq. (3.17), although, of course, in two-space this condition is not a consequence of finiteness of the action. Because we are working in two-space, we have, instead of a hypersphere, S^3, an ordinary circle, S^1.

Now to work:

(1) G is the unit circle in the complex plane; thus, topologically, G is also S^1, and we have to study homotopy classes of mappings of S^1 into S^1. We will label the circle in space, the domain of our functions, in the standard way, by an angle θ ranging from 0 to 2π.

(2) It will be useful to define some standard mappings from S^1 to S^1. One is the trivial mapping,

$$g^{(0)}(\theta) = 1. \tag{3.20a}$$

Another is the identity mapping,

$$g^{(1)}(\theta) = e^{i\theta}. \tag{3.20b}$$

These are both part of a family of mappings,

$$g^{(\nu)}(\theta) = [g^{(1)}(\theta)]^{\nu} = e^{i\nu\theta}, \tag{3.20c}$$

where ν is an integer (positive, negative, or zero). ν is called the 'winding number', because it is the number of times we wind around G when we go once around the circle at infinity in two-space. (By convention, winding around minus once means winding around once in the negative direction.)

(3) Every mapping from S^1 to S^1 is homotopic to one of the mappings (3.20c). We do not have the mathematical machinery to prove this rigorously, but I hope I can make it plausible. Imagine taking a rubber band and marking on it in ink a sequence of values of θ running from 0 to 2π. We then wrap the band about a circle representing G, such that each value of θ lies above the point into which it is mapped. (Fig. 13 shows such a construction.) We can continuously deform the band, first to eliminate any folds, like the one on the top of the figure, and second to stretch the band so it lies uniformly on the circle. In this way we obtain some $g^{(\nu)}(\theta)$. (In the case shown, we obtain $g^{(1)}$.) Thus we can associate a winding number with every mapping. (Note that I have not yet shown that this number is uniquely defined.)

Fig. 13

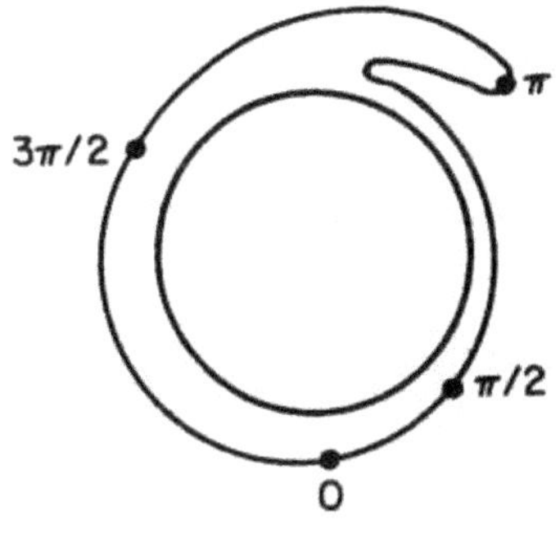

(4) I will now show that the winding number defined above is given by the integral formula

$$\nu=\frac{i}{2\pi}\int_0^{2\pi} d\theta g\, dg^{-1}/d\theta. \tag{3.21}$$

Firstly, by direct calculation, this gives the right answer for the standard mappings, Eq. (3.20c). Secondly, this quantity is invariant under continuous deformations. To prove this assertion it suffices to demonstrate invariance under infinitesimal deformations. A general infinitesimal deformation is of the form

$$\delta g = i(\delta\lambda)g, \tag{3.22}$$

where $\delta\lambda$ is some infinitesimal real function on the circle. Thus

$$\delta(g dg^{-1}/d\theta) = -i d(\delta\lambda)/d\theta, \tag{3.23}$$

and the change in ν vanishes upon integration. (We now know that all of our standard mappings are in different homotopy classes and that the winding number is uniquely defined.)

(5) If

$$g(\theta)=g_1(\theta)g_2(\theta), \tag{3.24a}$$

then

$$\nu=\nu_1+\nu_2. \tag{3.24b}$$

The proof is simple. The winding number is unchanged by continuous deformations. We can deform g_1 such that it is equal to one on the upper half of the circle ($0\leqslant\theta\leqslant\pi$) and g_2 such that it is equal to one on the lower half of the circle ($\pi\leqslant\theta\leqslant 2\pi$). The integrand in Eq. (3.21) is then the sum of a part due to g_1 (vanishing on the upper semicircle) and a part due to g_2 (vanishing on the lower semicircle).

(6) Let us define

$$G_\mu=\frac{i}{2\pi}\varepsilon_{\mu\nu}A_\nu. \tag{3.25}$$[13]

By Eqs. (3.17) and (3.21),

$$\nu=\lim_{r\to\infty}\int_0^{2\pi} r d\theta \hat{r}_\mu G_\mu, \tag{3.26}$$

where $\hat{r}_\mu$ is the radial unit vector. Thus, by Gauss's theorem,

$$\nu=\int d^2x\partial_\mu G_\mu. \tag{3.27}$$

Hence,

$$\nu=\frac{i}{4\pi}\int d^2x\varepsilon_{\mu\nu}F_{\mu\nu}. \tag{3.28}$$

I will now return to four-space, and take G to be SU(2). As we shall see, every argument will be a (mild) generalization of the arguments I have given for the baby problem.

(1) SU(2) is the group of unitary unimodular two-by-two matrices. It is well known that any such matrix can be uniquely written in the form

$$g = a + i\mathbf{b}\cdot\boldsymbol{\sigma}, \tag{3.29}$$

where $a^2 + |b|^2 = 1$. Thus, topologically, SU(2) is S^3, and we have to study homotopy classes of mappings from S^3 to S^3.

(2) It will be useful to define some standard mappings from S^3 to S^3. One is the trivial mapping,

$$g^{(0)}(x) = 1. \tag{3.30a}$$

Another is the identity mapping,

$$g^{(1)}(x) = (x_4 + i\mathbf{x}\cdot\boldsymbol{\sigma})/r. \tag{3.30b}$$

These are both part of a family of mappings,

$$g^{(\nu)}(x) = [g^{(1)}(x)]^{\nu}. \tag{3.30c}$$

where ν is an integer, called the winding number. (It is also sometimes called the Pontryagin index.) It measures the number of times the hypersphere at infinity is wrapped around G. (By convention, we say the hypersphere is wrapped around G in a negative sense if a right-handed triad of tangent vectors is mapped into a left-handed triad.)

(3) Every mapping from S^3 to S^3 is homotopic to one of our standard mappings (3.30c). We do not have the mathematical machinery to prove this assertion rigorously, but a plausibility argument can be constructed just as in the baby problem, with hyperspheres replacing circles. (If you have problems envisioning hyperspheres wrapped around hyperspheres, just accept the assertion on faith.) In this way we can associate a winding number with every mapping. (Note that I have not yet shown that this number is uniquely defined.)

(4) Let us define

$$\nu = \frac{1}{48\pi^2}\int d\theta_1\, d\theta_2\, d\theta_3\, \varepsilon^{ijk}(g\partial_i g^{-1}, g\partial_j g^{-1}\, g\partial_k g^{-1}). \tag{3.31}$$

where θ_1, θ_2 and θ_3 are three angles that parametrize S^3. How these angles are chosen is irrelevant to Eq. (3.31); the Jacobian determinant that comes from changing the angles is canceled by the Jacobian determinant from the ε-symbol. Equation (3.31) is written using the Cartan inner product, that is to say, in a representation-independent way. Of course, for any particular representation of SU(2), we can rewrite Eq. (3.31) in terms of traces; for example, for the two-dimensional representation, by

Eq. (3.4),

$$\nu = -\frac{1}{24\pi^2}\int d\theta_1\, d\theta_2\, d\theta_3\, \mathrm{Tr}\varepsilon^{ijk} g\partial_i g^{-1}\, g\partial_j g^{-1}\, g\partial_k g^{-1}. \tag{3.32}$$

I will show that this quantity is, firstly, a homotopy invariant, and secondly, agrees with the winding number as defined for our standard mappings. As before, a corollary of this proof will be that all of our standard mappings are in different homotopy classes and that the winding number is uniquely defined.

To show invariance under continuous deformations it suffices to show invariance under infinitesimal deformations. For any Lie group, a general infinitesimal transformation can be written as an infinitesimal right multiplication:

$$\delta g = g\delta\lambda^a(x)T^a \equiv g\delta T. \tag{3.33}$$

Under this transformation,

$$\delta(g\partial_k g^{-1}) = -g(\partial_k \delta T)g^{-1}. \tag{3.34}$$

The three derivatives in Eq. (3.32) make equal contributions to $\delta\nu$; thus,

$$\delta\nu \propto \int d\theta_1\, d\theta_2\, d\theta_3\, \varepsilon^{ijk}\, \mathrm{Tr} g\partial_i g^{-1}\, g\partial_j g^{-1}\, g(\partial_k \delta T)g^{-1}. \tag{3.35}$$

If we use the identity,

$$0 = \partial_i(gg^{-1}) = g\partial_i g^{-1} + (\partial_i g)g^{-1}, \tag{3.36}$$

this becomes

$$\delta\nu \propto \int d\theta_1\, d\theta_2\, d\theta_3\, \varepsilon^{ijk}\, \mathrm{Tr}\partial_i g^{-1}\, \partial_j g \partial_k \delta T, \tag{3.37}$$

which vanishes upon integration by parts, because of the antisymmetry of the ε-symbol. This completes the proof of invariance under continuous deformations.

(5) Now to evaluate Eq. (3.32) for our standard mappings. The task is easiest for $g^{(1)}$, for the integrand is here obviously a constant, and we need evaluate it only at the north pole of the unit hypersphere, $x_4 = 1$, $x_i = 0$. At this point we might as well choose θ_i to equal x_i. Thus, from Eq. (3.30b),

$$g\partial_i g^{-1} = -i\sigma_i, \tag{3.38}$$

and

$$\mathrm{Tr}\varepsilon^{ijk} g\partial_i g^{-1}\, g\partial_j g^{-1}\, g\partial_k g^{-1} = -12. \tag{3.39}$$

Since the area of a unit hypersphere is $2\pi^2$, we obtain the desired result, $\nu = 1$.

For the other standard mappings, the simplest way to proceed is to observe that if

$$g=g_1 g_2, \tag{3.40a}$$

then

$$\nu=\nu_1+\nu_2. \tag{3.40b}$$

The argument is the same as for the baby problem, with semihyperspheres replacing semicircles.

(6) Let us define

$$G_\mu=2\varepsilon_{\mu\nu\lambda\sigma}(A_\nu, \partial_\lambda A_\sigma+\tfrac{2}{3}A_\lambda A_\sigma). \tag{3.41}$$

A straightforward computation shows that

$$\partial_\mu G_\mu=\tfrac{1}{2}\varepsilon_{\mu\nu\lambda\sigma}(F_{\mu\nu}, F_{\lambda\sigma}). \tag{3.42}$$

The dual of an antisymmetric tensor (denoted by a tilde) is conventionally defined by

$$\tilde{F}_{\mu\nu}\equiv\tfrac{1}{2}\varepsilon_{\mu\nu\lambda\sigma}F_{\lambda\sigma}. \tag{3.43}$$

(The factor of $\frac{1}{2}$ is inserted in the definition so that $\tilde{\tilde{F}}=F$.) Equation (3.42) can thus be rewritten as

$$\partial_\mu G_\mu=(F_{\mu\nu}, \tilde{F}_{\mu\nu})\equiv(F, \tilde{F}). \tag{3.44}$$

From the definition of $F_{\mu\nu}$,

$$G_\mu=\varepsilon_{\mu\nu\lambda\sigma}(A_\nu, F_{\lambda\sigma}-\tfrac{2}{3}A_\lambda A_\sigma). \tag{3.45}$$

This expression is useful in evaluating

$$\int d^4x(F, \tilde{F})=\int d^3S\hat{r}_\mu G_\mu, \tag{3.46}$$

where d^3S is the element of area on a large hypersphere. The first term in Eq. (3.45) is $O(1/r^4)$ and makes no contribution to the integral; the second term simply gives (up to a multiplicative constant) the integral formula for the winding number, Eq. (3.31). Thus we obtain

$$\int d^4x(F, \tilde{F})=32\pi^2\nu. \tag{3.47}$$

Summary and generalizations. This has been a long analysis, and you may have lost track of what we were doing, so let me summarize the main results of this subsection. For a gauge field theory based on the group SU(2), every field configuration of finite action in four-dimensional Euclidean space has an integer associated with it, the Pontryagin index or winding number, ν. It is not possible to continuously deform a configuration of one winding number into one of a different winding number while maintaining the finiteness of the action. We have two integral formulae for the winding number, one in terms of a surface integral over a large sphere, Eq. (3.31), and one in terms of a volume integral over all four-space, Eq. (3.47).

How much of this depends on the gauge group being SU(2)? Firstly, if the gauge group is U(1), it is easy to see that every mapping of S^3 into U(1) is continuously deformable into the trivial mapping (all of S^3 mapped into a single point). *Thus, for an Abelian gauge field theory, there is no analog of the winding number.* Secondly, for a general simple Lie group, G, there is a remarkable theorem due to Raoul Bott[14] that states that any continuous mapping of S^3 into G can be continuously deformed into a mapping into an SU(2) subgroup of G. *Thus, everything we have discovered for* SU(2) *is true for an arbitrary simple Lie group; in particular, it is true for* SU(n). I stress that 'everything' means *everything*. In particular, not a single numerical factor in the integral formulas for the winding number needs alteration, so long as we choose the normalization of the Cartan inner product appropriately (as we have). Finally, since a general compact Lie group is locally the direct product of an Abelian group and a string of simple groups, *for a general gauge field theory, there is an independent winding number for every simple factor group.*

3.3 *Many vacua*

We have learned a lot about classical gauge field theories; now it is time to confront the quantum theory. In principle, the Euclidean functional integral tells how to go from the classical theory to the quantum theory. As I explained in Sect. 2, we can use the functional integral to study the energy eigenstates of the theory; also, by adding appropriate source terms to the Hamiltonian (equivalently, to the Euclidean action) and then differentiating with respect to the sources at the end of the computation, we can study the expectation values of strings of operators, Euclidean Green's functions. However, for gauge field theories, there is a famous complication: to make the functional integral well-defined, we must impose a gauge-fixing condition.[15]

I will choose to work in axial gauge, $A_3=0$. I have several reasons for this choice. (1) It is possible to show[16] that every non-singular gauge field configuration can be put in axial gauge by a non-singular gauge transformation. It is by no means clear whether this is true for covariant gauges, for example. (2) In axial gauge the functional integral is directly equivalent to a canonical formulation of the theory;[17] there is no need of the ghost terms that occur in covariant gauges, or of the subsidiary conditions on the space of states that are needed in such gauges as $A_0=0$. (3) Most of the treatment in the literature of the phenomena we are about to discuss is in the gauge $A_0=0$. It is nice to show explicitly that the answers do not depend on this gauge choice. (4) Although axial gauge is terribly awkward for specific computations, once we have obtained functional-integral expressions for quantities of interest, we can use the standard Faddeev–Popov methods to transform these into some more convenient gauge.

In field theory, we normally plunge directly into infinite space. However, I will here study gauge field theory in a finite box of three-volume V, with definite boundary conditions, which I shall specify shortly. Just as in Sect. 2, I will also restrict the theory to a finite range of Euclidean time, T, with appropriate boundary conditions at initial and final times. Thus we are integrating over a box in Euclidean four-space, with boundary conditions on the (three-dimensional) walls of the box. Of course, I will eventually send both V and T to infinity. I again have reasons for this choice. (1) Certainly nothing is lost by beginning in a finite box; if the transition to infinite space goes smoothly, at worst we will have wasted a little time. (2) In some theories, we can gain information about the structure of the theory by seeing how things depend on the boundary conditions imposed on the walls of the box. For example, in a scalar field theory with spontaneous symmetry breakdown, the expectation value of the scalar field in the center of the box depends on the boundary conditions on the walls, no matter how large the box; this is one of the easiest ways to see that the theory has many vacua. (3) In the canonical quantization of the theory, it is necessary to eliminate A_0 from the action. To do this, it is necessary to find A_0 from $\partial_3^2 A_0$. In infinite space, this problem has many solutions; this ambiguity is usually resolved by applying *ad hoc* conditions on the behavior of A_0 at infinity. In a box with appropriate boundary conditions, this problem always has a unique solution.

There are many possible types of boundary conditions we could impose: we could fix some components of A_μ, some components of $F_{\mu\nu}$, some combinations of these, etc. A clue to a wise choice of boundary conditions is given by the surface term in the expression for the variation of the action. For example, for a free scalar field theory,

$$\delta S = \int \mathrm{d}^3 S n^\mu \partial_\mu \phi \delta\phi + \cdots . \tag{3.48}$$

Here, $\mathrm{d}^3 S$ is the element of surface area, n^μ is the normal vector to the surface, and the triple dots denote the usual volume integral of the Euler–Lagrange equations. From this expression we see that one way to make the surface terms vanish is to fix the value of ϕ on the walls of the box. Likewise, for a gauge field theory,

$$\delta S = \frac{1}{g^2} \int \mathrm{d}^3 S n^\mu F_{\mu\nu} \delta A^\nu + \cdots . \tag{3.49}$$

From this expression we see that one way to make the surface term vanish is to fix the tangential components of A_μ on the surface. Note that there is no need to fix the normal component of A_μ; because $F_{\mu\nu}$ is antisymmetric, this makes no contribution to the surface integral.

We are not totally free to choose the tangential components of A_μ arbitrarily. Firstly, they must be chosen consistent with our gauge condition, $A_3=0$. Secondly, because we want to do semiclassical computations, we must choose our boundary conditions to be consistent with finiteness of the action, as the box goes to infinity. Equivalently, the boundary conditions must be consistent with the box being filled with a field configuration of a definite winding number. Furthermore, for fixed boundary conditions, this winding number is fixed, for only the tangential components of A_μ are needed to compute the normal component of G_μ. (See Eq. (3.41).)

Thus at least one relic of our boundary conditions remains no matter how large the box: we can not put an arbitrary finite-action field configuration in the box, but only one of a definite winding number. It turns out that the winding number is the *only* relic of the boundary conditions that survives as the box goes to infinity. The hand-waving argument for this is that the winding number is the *only* gauge-invariant quantity associated with the large-distance behavior of the fields. If you do not find this argument convincing, you will find a more careful one in Appendix 4.

Thus, for large boxes, we can forget about the boundary conditions in the functional integral and simply integrate over all configurations where the winding number, ν, has some definite value, n. I will denote the result of such an integration by $F(V, T, n)$. In equations,

$$F(V, T, n)=N\int [\mathrm{d}A]\mathrm{e}^{-S}\delta_{\nu n}. \tag{3.50}$$

where $[\mathrm{d}A]$ denotes $[\mathrm{d}A_1][\mathrm{d}A_2][\mathrm{d}A_4]$. Also, I have set $\hbar$ to one; we can always keep track of the powers of $\hbar$ by keeping track of the powers of g, as explained in Sect. 1.

$F(V, T, n)$ is a transition matrix element from some initial state to some final state (determined by our boundary conditions). What these states are will not be important to us. What is important is that for large times, T_1 and T_2,

$$F(V, T_1+T_2, n)=\sum_{n_1+n_2=n} F(V, T_1, n_1)F(V, T_2, n_2). \tag{3.51}$$

This follows from Eq. (3.47), the expression for the winding number as the integral of a local density; this tells us that the way to put total winding number n in a large box is to put winding number n_1 in one part of the box and winding number n_2 in the remainder of the box, with $n = n_1 + n_2$. (Of course, such counting misses field configurations with significant action density on the boundary between the two sub-boxes, for there is no reason for the winding-number integral for each sub-box to be an integer for such configurations. However, we expect this to be a negligible surface effect for sufficiently large boxes.)

Pretty as it is, Eq. (3.51) is not what we would expect from a transition-matrix element that has a contribution from only a single energy eigenstate. Such an object would be a simple exponential, and would obey a multiplicative composition law for large times, not the convolutive composition law of Eq. (3.51). However, it is easy enough to turn convolutions into multiplications. The technique is called Fourier transformation:

$$F(V, T, \theta) \equiv \sum_n e^{in\theta} F(V, T, n) = N \int [dA] e^{-S} e^{i\nu\theta}. \tag{3.52}$$

From Eq. (3.51),

$$F(V, T_1 + T_2, \theta) = F(V, T_1, \theta) F(V, T_2, \theta). \tag{3.53}$$

This is the correct composition law for a simple exponential. Thus we identify $F(V, T, \theta)$ as being (up to a normalization constant) the expectation value of e^{-HT} in an energy eigenstate, which we denote by $|\theta\rangle$ and call the θ vacuum.

$$F(V, T, \theta) \propto \langle\theta| e^{-HT} |\theta\rangle = N' \int [dA] e^{-S} e^{i\nu\theta}. \tag{3.54}$$

where N' is a new normalization constant.

Our analysis has been simple and straightforward (I hope), but we have been led to a very unintuitive conclusion. Our original gauge field theory seems to have split up into a family of disconnected sectors, labeled by the angle θ, each with its own vacuum. Furthermore, in each of these sectors, the computational rules are the same as those we would have naively written down if we had not gone through any of this analysis, except that an extra term, proportional to $(F, \tilde{F})$, has been added to the Lagrangian density. Probably half the people who have played with gauge field theories have thought, at one time or another, of adding such a term, and they have discarded the possibility, because the added term is a total divergence (see Eq. (3.44)) and thus has no effect on the equations of motion and therefore 'obviously' has no effect on the physics of the theory. Of course, at this stage in our investigation, it is still possible that we have been fooling ourselves, that the extra term indeed has no effect on the physics, and that all the θ vacua we think we have discovered are simply duplicates of the same state. We shall eliminate this possibility immediately.

(I should remark that what we have done here closely parallels the treatment of a periodic potential in Sect. 2.3, except the arguments are somewhat more abstract and in a different order. The winding number is something like the total change in x (the difference between the number of instantons and the number of anti-instantons) in Sect. 2.3, and the θ vacua are something like the $|\theta\rangle$ eigenstates. The two big differences are that we found the analogs of the $|\theta\rangle$ states without pausing to talk about the analogs of the $|j\rangle$ states, and that we did the Fourier transform that untangled the energy spectrum before we saturated the functional integral with instantons. The first difference is unimportant; if I had wanted to, I could have added two extra paragraphs when I was talking about $F(V, T, n)$ and discussed the analogs of the $|j\rangle$ states. (They are called n vacua.) As for the instantons, they are the subject of the next subsection.)

3.4 *Instantons: generalities*

In the next subsection I shall explicitly construct instantons, finite-action solutions of the Euclidean gauge-field equations with $\nu = 1$. Most of the qualitative consequences of these solutions are independent of their detailed structure and follow merely from the fact of their existence. Therefore, in this subsection, I will simply assume that instantons exist and draw some conclusions from this assumption.

I will denote the action of an instanton by S_0. Because S_0 is finite, the instanton can not be invariant under spatial translations. Thus there exists at least a four-parameter family of instanton solutions; I will call these parameters 'the location of the center of the instanton'. The winding number is parity-odd. Thus there must also exist at least a four-parameter family of solutions with $\nu = -1$, the parity transforms of the instanton solutions, which I will call anti-instantons. Just as in Sect. 2, we can build approximate solutions consisting of n instantons and $\bar{n}$ anti-instantons, with their centers at arbitrary widely separated locations. These approximate solutions have $\nu = n - \bar{n}$.

Again as in Sect. 2, we approximate Eq. (3.54) by summing over all these configurations. Thus we obtain

$$\langle\theta|e^{-HT}|\theta\rangle \propto \sum_{n,\bar{n}} (Ke^{-S_0})^{n+\bar{n}}(VT)^{n+\bar{n}} e^{i(n-\bar{n})\theta}/(n!\,\bar{n}!)$$
$$= \exp(2KVTe^{-S_0}\cos\theta), \tag{3.55}$$

where K is a determinantal factor, defined as in Sect. 2. Thus, the energy of a θ vacuum is given by

$$E(\theta)/V = -2K\cos\theta\, e^{-S_0}. \tag{3.56}$$

Note that, as should be the case in a field theory, the different vacua are distinguished not by different energies, but by different energy densities. (Also note the similarity with the energy spectrum of a periodic potential, Eq. (2.45).)

We can go on and compute the expectation values of various operators. A particularly easy (and particularly instructive) computation is that of the expectation value of $(F, \tilde{F})$. By translational invariance,

$$\langle\theta|(F(x), \tilde{F}(x))|\theta\rangle = \frac{1}{VT}\int d^4x\,\langle\theta|(F, \tilde{F})|\theta\rangle. \tag{3.57}$$

Thus, by Eq. (3.47),

$$\langle\theta|(F, \tilde{F})|\theta\rangle = \frac{32\pi^2\int [dA]\nu e^{-S}e^{i\nu\theta}}{VT\int [dA]e^{-S}e^{i\nu\theta}}$$

$$= -\frac{32\pi^2 i}{VT}\frac{d}{d\theta}\ln\left(\int [dA]e^{-S}e^{i\nu\theta}\right). \tag{3.58}$$

Hence there is no need to do a fresh summation over a dilute instanton–anti-instanton gas, since we have just evaluated the quantity in parentheses in Eq. (3.55). Thus in our approximation,

$$\langle\theta|(F, \tilde{F})|\theta\rangle = -64\pi^2 iKe^{-S_0}\sin\theta. \tag{3.59}$$

Some comments:

(1) The expectation value is independent of V and T, as it should be.

(2) The expectation value is an imaginary number, again as it should be. The reason is that

$$(F, \tilde{F}) = (F_{12}, F_{34}) + \text{permutations}. \tag{3.60}$$

When we continue from Euclidean space to Minkowski space, F_{12} remains F_{12}, but, just as x_4 becomes ix_0, so does F_{34} become iF_{30}. Thus, if we had obtained a real answer, we would have found that in Minkowski space (the real world) a Hermitian operator would have had an imaginary vacuum expectation value, a disaster.

(3) Both the vacuum energy density and the vacuum expectation value depend non-trivially on θ. Thus the θ-vacua are indeed all different from each other.

3.5 *Instantons: particulars*

$$\int d^4x(F, F) = \left[\int d^4x(F, F) \int d^4x(\tilde{F}, \tilde{F}) \right]^{\frac{1}{2}}$$
$$\geqslant \left| \int d^4x(F, \tilde{F}) \right|, \tag{3.61}$$

by the Schwartz inequality. Thus, for any winding number, we have an absolute lower bound on the action,

$$S \geqslant \frac{8\pi^2}{g^2} |\nu|. \tag{3.62}$$

Furthermore, equality is attained if and only if

$$F = \pm \tilde{F}, \tag{3.63}$$

where the positive (negative) sign holds for positive (negative) ν.

This inequality was first derived by Belavin, Polyakov, Schwartz, and Tyupkin,[9] who used it to search for instantons. Their idea was to look for solutions of Eq. (3.63). If such solutions exist, they are minima of the action for fixed winding number, and thus stationary points of the action under local variations, that is to say, solutions of the field equations. Furthermore, since they have lower action than any other solutions of the same winding number (if other solutions exist), they dominate the functional integral, and, for our purposes, are the only solutions we need worry about. Finally, as a bonus, Eq. (3.63) is a first-order differential equation and considerably more tractable than the second-order field equations.

Let us begin the search with $\nu = 1$. We know that any field configuration with $\nu = 1$ can be gauge-transformed such that

$$A_\mu = g^{(1)} \partial_\mu [g^{(1)}]^{-1} + O(1/r^2), \tag{3.64}$$

where

$$g^{(1)} = \frac{x_4 + i\mathbf{x} \cdot \boldsymbol{\sigma}}{r}. \tag{3.65}$$

Equation (3.64) is rotationally invariant, in the sense that the effect of any four-dimensional rotation can be undone by an appropriate gauge transformation. This is a consequence of the statement that a rotation is a continuous deformation and thus does not change the winding number. There is also a short direct proof: Under a general rotation

$$g^{(1)} \to g g^{(1)} h^{-1}, \tag{3.66}$$

where g and h are elements of SU(2) determined by the rotation. (This is a standard formula; it is the usual way of demonstrating the isomorphism between SO(4) and SU(2)⊗SU(2).) Thus,

$$A_\mu \rightarrow gA_\mu g^{-1} + O(1/r^2). \tag{3.67}$$

This, as promised, can be undone by a gauge transformation, indeed, by a gauge transformation of the first kind, a constant gauge transformation.

This suggests that we search for a solution of Eq. (3.63) that is rotationally invariant in the same sense. That is to say, we make the Ansatz,

$$A_\mu = f(r^2)g^{(1)}\partial_\mu[g^{(1)}]^{-1}, \tag{3.68}$$

where, to avoid a singularity, f must vanish at the origin. From here on it is straightforward plug-in-and-crank, which I will spare you. It turns out that we do indeed obtain a solution in this way, if

$$f = \frac{r^2}{r^2 + \rho^2}, \tag{3.69}$$

where ρ is an arbitrary constant, called 'the size of the instanton'. The existence of solutions of arbitrary sizes is a necessary consequence of the scale invariance of the classical field theory. (This fact will occasion some embarrassment shortly.)

Once we have a solution to any field theory, we can obtain new solutions by applying the invariances of the theory. In the case at hand, these are generated by (1) scale transformations, (2) rotations, (3) the four-parameter group of spatial translations, (4) the four-parameter group of special conformal transformations, and (5) gauge transformations. Scale transformations simply change the size of the instanton; thus they just shift around the members of our one-parameter family of solutions but generate no new solutions. Rotations, as I have shown, can always be undone by gauge transformations. Spatial translations generate genuinely new solutions, and give us four more parameters, the 'location of the center of the instanton'. Although I do not have time to demonstrate it here, it turns out[18] that special conformal transformations can be undone by gauge transformations and translations.

Gauge transformations, as usual, require special consideration. It is easy to see that any non-trivial gauge transformation changes (3.68). Because $g^{(1)}$ is a function of angles only, the radial component of A_μ, A_r, vanishes. Thus, under a general non-singular gauge transformation, $g(x)$,

$$A_r \rightarrow gA_r g^{-1} + g\partial_r g^{-1} = g\partial_r g^{-1}. \tag{3.70}$$

Hence, if the gauge transformation is not to change A_μ, g must be independent of r. That is to say, its value everywhere must be its value at the origin; g must be a constant gauge transformation. But the only constant gauge transformation that leaves A_μ unchanged is the identity. (Remember, the effect of a constant gauge transformation is the same as that of a rotation.)

You might think that this discussion of gauge transformations is irrelevant. After all, when we do the quantum theory, we must work in a fixed gauge, such as axial gauge, and it is commonly said that once we have fixed the gauge we have no freedom to make gauge transformations. However, although commonly said, this is not strictly true; all standard gauges still allow constant gauge transformations.[19] This is as it should be. Constant gauge transformations act like ordinary symmetries; they put particles into multiplets (if there is no spontaneous symmetry breakdown), impose selection rules on scattering processes, etc. Thus, in a sensible formulation of the theory, they should remain as manifest symmetries of the Hamiltonian. Whether you accept this philosophy or not, the fact remains that constant gauge transformation applied to an instanton solution (transformed to obey the gauge conditions) will generate a different solution still obeying the gauge conditions. Thus we have found an eight-parameter family of solutions, one parameter from scale transformations, four from translations, and three from constant gauge transformations.

Are there other solutions with unit winding number? Atiyah and Ward[20] state that there are none. I can not give their proof here because I do not understand it. Nevertheless, mathematicians I trust say that their argument is not only legitimate but brilliant, so let us assume they are right and continue.

Solutions of higher winding number (if they exist) are of no interest to us. We have used approximate solutions consisting of n widely separated objects (instantons or anti-instantons) to evaluate the functional integral. These approximate solutions depend on $8n$ parameters, 8 for each object. Now suppose there are exact solutions that can be interpreted as n objects; that is to say, they depend on $8n$ (or fewer) parameters and become our approximate solutions when some of the parameters (the separations between the objects) become large. In this case, all we learn by knowing these exact solutions exist is that the dilute-gas approximation is better than we think it is – but we already know that it is good enough for our purposes. There might also be exact solutions that can not be interpreted in this way. To have a definite example, let me suppose there were a 'binstanton', a brand-new solution of winding number two. Then in evaluating the functional integral, we would have to sum over a dilute gas of instantons, anti-instantons, binstantons, and anti-binstantons. Thus,

Eq. (3.56) would be replaced by

$$E(\theta)/V = -2K\cos\theta\, e^{-S_0} - 2K'\cos 2\theta\, e^{-S_0'}, \tag{3.71}$$

where the primed quantities are the action and determinantal factor for a binstanton. But S_0' is twice S_0, so the new term is exponentially small compared to the old one and should be neglected.[21]

3.6 *The evaluation of the determinant and an infrared embarrassment*

We now know enough to go a long way towards explicitly evaluating the right-hand side of Eq. (3.56).

(1) S_0 is $8\pi^2/g^2$.

(2) We have an eight-parameter family of solutions and thus eight eigenmodes of eigenvalue zero in the small-vibration problem. Thus K contains a factor of $(1/\hbar^{\frac{1}{2}})^8$, or, equivalently $1/g^8$. Everything else in K is independent of $\hbar$, and thus independent of g.

(3) We have already done the integral over instanton location. The integral over constant gauge transformations is an integral over a compact group and thus gives only a constant numerical factor, the volume of SU(2). The integral over instanton sizes is potentially troublesome, since ρ can be anywhere between zero and infinity, so we will, for the moment, keep it as an explicit integral.

(4) Thus we obtain

$$E(\theta)/V = -\cos\theta\, e^{-8\pi^2/g^2} g^{-8} \int_0^\infty \frac{d\rho}{\rho^5} f(\rho M), \tag{3.72}$$

where f is an unknown function and M is the arbitrary mass (more properly, arbitrary inverse wavelength) that is needed to define the renormalization prescription in a massless field theory. (I have avoided mentioning renormalization until now, but renormalization is essential in any computation that involves an infinite number of eigenmodes, as does this one. In Sect. 5 I will give a more detailed discussion of the ultraviolet divergences in determinantal factors and their removal by the usual one-loop renormalization counterterms.) The form of the integral is determined by dimensional analysis; an energy density has dimensions of $1/(\text{length})^4$.

(5) However, M and g are not independent parameters. Renormalization-group analysis[22] tells us that they must enter expressions for observable quantities only in the combination

$$\frac{1}{g^2} - \beta_1 \ln M + O(g^2), \tag{3.73}$$

where β_1 is a coefficient which can be computed from one-loop perturbation theory. In the case at hand, β_1 is $11/12\pi^2$.

(6) This fixes the form of f. Thus,

$$E(\theta)/V = -A\cos\theta\, \mathrm{e}^{-8\pi^2/g^2} g^{-8} \int_0^\infty \frac{\mathrm{d}\rho}{\rho^5} (\rho M)^{8\pi^2\beta_1} [1 + O(g^2)], \tag{3.74}$$

where A is a constant independent of g, ρ, and M.

(7) To determine A requires a lot of hard work,[23] so I shall stop the calculation here. Even though we have not been able to carry things out to the end, it is remarkable how far we have been able to go with so little effort.

No doubt you have noticed that the integral we have derived is infrared-divergent. The origin of the divergence is clear from the derivation of the integral: the effective coupling constant (in the sense of the renormalization group) becomes large for large instantons, and this makes the integrand blow up. Thus the divergence is an embarrassment but not a catastrophe. It would be a catastrophe if we obtained a divergent answer in a regime in which we trusted our approximations. This is not the situation here; the divergence arises in the regime of large effective coupling constant, where all small-coupling approximations are certainly wrong. Phrased another way, the fact that the integrand has the wrong behavior for large ρ is overshadowed by the fact that it is the wrong integrand. Thus we are free to hope that strong-coupling effects (which we can not at the moment compute) introduce some sort of effective infrared cutoff in the integrand. This hope might be wrong, but it is not ruled out by anything we have done so far.

I admit that this argument is blatant hand-waving. However, it is not some new hand-waving special to instanton calculations, but the same old hand-waving that accompanies any discussion of the large-scale behavior of non-Abelian gauge field theories. For example, there is evidence that the observed hadrons are made of weakly coupled quarks. But if the quarks are weakly coupled, why can we not knock them out of the hadron? Well, in a gauge field theory the effective coupling constant grows at large distances, etc., much hand-waving, infrared slavery and quark confinement.

Everything that we have done for SU(2) can be extended straightforwardly to SU(3). To begin with, an SU(2) instanton solution can trivially be made into an SU(3) instanton solution; all that needs to be done is to say that three of the gauge fields, those associated with an

SU(2) subgroup, are of the form given, while the other five vanish. It is believed that these exhaust the set of solutions of Eq. (3.63) with unit winding number, although, unlike the SU(2) case, there is, to my knowledge, no rigorous proof of this statement. If this is indeed the case, there are only two minor differences between the SU(3) computation and the SU(2) one. (1) Instead of three parameters associated with constant gauge transformations, we have seven. (One of the eight SU(3) generators commutes with the SU(2) subgroup and does not change the solution.) Thus the factor of g^{-8} in Eq. (3.74) is replaced by one of g^{-12}. (2) β_1 has the proper value for an SU(3) gauge theory, $11/8\pi^2$.

4 The Abelian Higgs model in 1+1 dimensions[24]

In this section I will discuss a field theory in which instanton effects drastically change the particle spectrum, the Abelian Higgs model in two-dimensional space-time.

In any number of dimensions, this is the theory of a complex scalar field with quartic self-interactions, minimally coupled to an Abelian gauge field with gauge coupling constant e, called the electric charge. In our notation, the theory is defined by the Euclidean Lagrangian density,

$$\mathscr{L} = \frac{1}{4e^2}(F, F) + \mathrm{D}_\mu \psi^* \mathrm{D}_\mu \psi + \frac{\lambda}{4}(\psi^*\psi)^2 + \frac{\mu^2}{2}\psi^*\psi, \tag{4.1}$$

where λ is a positive number and μ^2 may be either positive or negative. To this must be added renormalization counterterms; however, renormalization will play no part in our computations, and, to keep things as simple as possible, I will not distinguish between bare and renormalized parameters.

Perturbation theory tells us that for weak coupling the qualitative properties of the theory depend critically on the sign of μ^2:

(1) If μ^2 is positive, the theory is simply the electrodynamics of a charged scalar meson. The mass spectrum consists of the charged meson, its antiparticle, and a massless vector meson, the photon. The force between widely separated external charges is the ordinary Coulomb force. These statements require some modification in two dimensions. Firstly, because there are no transverse directions, there is no photon. Secondly, because the Coulomb force is independent of distance, it is impossible to separate a meson and an antimeson; in contemporary argot, the charged particles are confined. The spectrum of the theory consists of a sequence of meson–antimeson bound states, rather like the spectrum of positronium, except

that these states are all stable, since they can not decay through the emission of (nonexistent) photons.

(2) If μ^2 is negative, the Higgs phenomenon takes place. In the ground state of the theory,

$$|\langle\psi\rangle|^2 = -\mu^2/\lambda \equiv a^2. \tag{4.2}$$

The particle spectrum consists of a massive neutral scalar meson and a massive neutral vector meson. The force between widely separated external charges falls off exponentially rapidly. These statements require no modification in two dimensions.

In the remainder of this section, I will argue that the preceding sentence is a lie; contrary to the predictions of perturbation theory, the qualitative properties of the model for negative μ^2 are the same as those for positive μ^2; the two-dimensional Abelian Higgs model does not display the Higgs phenomenon. To be precise, I will show that, for negative μ^2, the theory admits instantons, and, when the effects of these instantons are taken into account, the long-range force between external charges is independent of their separations. Also, I will be able to argue, from the behavior of the long-range force, that the theory contains (confined) charged particles. There is a quantitative difference between positive and negative μ^2, though: for positive μ^2, the strength of the long-range force is independent of $\hbar$; for negative μ^2, the strength of the long-range force is exponentially small in $\hbar$, the mark of an instanton effect.

Just as in Sect. 3, we must begin the analysis by classifying classical field configurations of finite action. Of course, before doing this, we must add a constant to the Lagrangian density so the minimum of the action is zero. Thus we write

$$\mathcal{L} = \frac{1}{4e^2}(F, F) + |D_\mu\psi|^2 + \frac{\lambda}{4}(|\psi|^2 - a^2)^2. \tag{4.3}$$

This is the sum of three positive terms. In order that the third term not make a divergent contribution to the action, it is necessary that $|\psi|$ approach a as r goes to infinity. However, there is no restriction on the phase of ψ. In equations,

$$\lim_{r\to\infty} \psi(r, \theta) = g(\theta)a, \tag{4.4}$$

where g is a complex number of unit modulus, an element of U(1). In order that the second term not make a divergent contribution to the action, it is necessary that

$$A_\mu = g\partial_\mu g^{-1} + O(1/r^2). \tag{4.5}$$

(Remember, in our conventions, A_μ is an imaginary field.) The first term now automatically makes a finite contribution to the action.

The lovely thing about Eq. (4.5) is that it is identical to Eq. (3.17); that is to say, the problem of classifying finite-action configurations is the baby problem of Sect. 3.2. Thus the finite-action configurations are characterized by an integer, ν, the winding number, just as they are for four-dimensional gauge field theories. By Eq. (3.28), the integral expression for the winding number is

$$\nu = \frac{i}{4\pi} \int d^2x \varepsilon_{\mu\nu} F_{\mu\nu}. \tag{4.6}$$

Equivalently,

$$\nu = \frac{i}{2\pi} \oint A_\mu \, dx_\mu, \tag{4.7}$$

where the integral is over the circle at infinity.

Although I will not bother to explicitly display them here, it turns out that the Euclidean field equations have solutions with unit winding number, instantons, again just like four-dimensional gauge theories.[25] The only relevant difference, for our purposes, is that the Higgs model is not scale invariant; thus the instantons have a fixed size and the problems associated with integrating over scale transformations do not arise. Otherwise, though, everything is much the same as it was before, and we can copy step-by-step our earlier analysis and uncover the vacuum structure of the theory.

Thus, just as before, we have a family of θ-vacua, with energy densities given by

$$E(\theta)/L = -2Ke^{-S_0} \cos\theta. \tag{4.8}$$

Here L is the volume of (one-dimensional) space, S_0 is the action of an instanton, and K is a determinantal factor. Also, by copying the derivation of Eq. (3.59), we find that

$$\langle\theta|\varepsilon_{\mu\nu}F_{\mu\nu}|\theta\rangle = 8\pi K e^{-S_0} \sin\theta. \tag{4.9}$$

As before, this has the right reality properties; when we continue to Minkowski space, we pick up a factor of i that cancels the factor of i in our definition of A_μ. We see from this equation that the θ-vacua are characterized by a constant expectation value of the electric field F_{01}. In two dimensions, unlike four, such a constant 'background field' is not in conflict with Lorentz invariance.[26]

Now that we understand the vacuum structure, let us compute the force between widely separated external charges. To be more precise, let us introduce into the system two static charges of equal magnitude, q, and opposite sign, separated by a distance L', and let us compute (for large L') Δ, the change in the energy of a θ-vacuum caused by these charges. The standard method of computing Δ uses Wilson's loop integral,[27]

$$W = \exp\left(-\frac{q}{e}\oint A_\mu \, dx_\mu\right), \tag{4.10}$$

where the integration is over the rectangular path shown in Fig. 14. According to Wilson, the vacuum energy shift is given by

$$\Delta = -\lim_{T' \to \infty} \frac{1}{T'} \ln\langle\theta|W|\theta\rangle. \tag{4.11}$$

In our case,

$$\langle\theta|W|\theta\rangle = \frac{\int [dA][d\psi^*][d\psi] W e^{-S} e^{i\nu\theta}}{\int [dA][d\psi^*][d\psi] e^{-S} e^{i\nu\theta}}, \tag{4.12}$$

and our task is to compute these two functional integrals in our standard dilute-gas approximation, for large L' and T' (and, of course, for even larger L and T, the spatial and temporal extent of the universe). In Eq. (4.8) we have already calculated the denominator. To calculate the numerator, let us divide the sum over instantons and anti-instantons into two independent sums: one over objects lying inside the loop and one over objects lying outside the loop. By this division we neglect contributions coming from configurations in which instantons and anti-instantons overlap the loop, but, for large L, T, L', and T', this is a very small portion of the available configurations and can reasonably be neglected. (Of course, if our calculation gives zero for its answer, then these configura-

Fig. 14

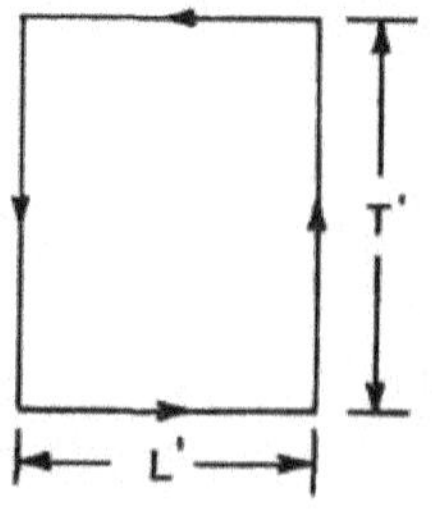

tions will be the most important ones and we will have to go back and compute them.) The functional integrand splits neatly into the product of an 'outside' term and an 'inside' term: $S = S^{\text{outside}} + S^{\text{inside}}$, $\nu = \nu^{\text{outside}} + \nu^{\text{inside}}$, while

$$W = \exp(2\pi i q \nu^{\text{inside}}/e). \tag{4.13}$$

Thus, for the outside objects, we have the same sum as for the denominator, except that the available volume of Euclidean two-space is not LT but $LT - L'T'$. For the inside objects, we also have the same sum, except that the available volume is $L'T'$, and θ is replaced by $\theta + 2\pi q/e$.

Thus,

$$\begin{aligned} \ln\langle\theta|W|\theta\rangle = 2Ke^{-S_0}[(LT - L'T')\cos\theta + L'T'\cos(\theta + 2\pi q/e) \\ - LT\cos\theta], \end{aligned} \tag{4.14}$$

where the first term comes from the outside sum, the second from the inside sum, and the third from the denominator. Hence,

$$\Delta = 2L'Ke^{-S_0}[\cos\theta - \cos(\theta + 2\pi q/e)]. \tag{4.15}$$

This is proportional to L', the separation between the external charges; thus there is a constant force between external charges at large separation. As announced, there is no quantitative difference between positive and negative μ^2. However, there is a qualitative difference. For positive μ^2, the strength of the force is proportional to q^2 for small $\hbar$; for negative μ^2, it is exponentially small in $\hbar$. (Remember, if we had not chosen our units so $\hbar$ was one, S_0 would have been $S_0/\hbar$.)

There is a simple physical interpretation of this result. For small θ and small q/e,

$$\Delta = L'Ke^{-S_0}[(\theta + q/e)^2 - \theta^2], \tag{4.16}$$

$$E(\theta) = LKe^{-S_0}\theta^2 + \text{constant}, \tag{4.17}$$

and,

$$\langle\theta|F_{12}|\theta\rangle = 4\pi Ke^{-S_0}\theta. \tag{4.18}$$

These expressions have an obvious interpretation: In a θ-vacuum, there is a background electric field, and an energy density proportional to the square of this field. Because we are in one spatial dimension, the external charges act like condenser plates in three dimensions; they induce a constant field proportional to their charge in the region between them, which is added to the pre-existing background field. Thus the energy shift is the separation multiplied by the difference of the energy density of the new field and that of the old. Equation (4.14) is just this trivial picture complicated by nonlinear terms in the expression for the energy density as a function of the field.

One aspect of these nonlinear complications is of physical import: Eq. (4.14) is periodic in q with period e. This is explicable if the theory contains charged particles of charge e. If this is the case, there is a process that can change the charge on our condenser plates by $\pm e$: a particle–antiparticle pair can materialize in the region between the plates, and the particle can fly to one plate and the antiparticle to the other. This process will occur whenever it is energetically favorable. For sufficiently large L', this is equivalent to saying that it will occur whenever it lowers the energy density, because the energetic cost of making a pair is independent of L', and the energetic gain of lowering the energy density is proportional to L'. Thus qs that are equal modulo e lead to identical physics; no matter which one you start out with, pairs are made until the charge on the plates reaches its optimum value, the one that gives minimum energy density.

What if we were to do a parallel computation in a four-dimensional gauge field theory, with non-Abelian external charges? Would we also obtain a force independent of separation? Alas, we would not. There is an L' in Eq. (4.15) because there is an $L'T'$ in Eq. (4.14), that is to say, because even an instanton deep within the loop has a non-negligible effect on the loop integral. This is precisely what does not happen in four dimensions. At large distances from an instanton, A_μ is $g\partial_\mu g^{-1}$, plus terms that fall off far too rapidly to affect the loop integral. However, the loop integral is gauge-invariant, and we can always gauge-transform g such that it is constant everywhere except within a small cone emerging from the instanton perpendicular to the plane of the loop. Whatever confines quarks, it is not instantons.

5 't Hooft's solution of the U(1) problem

5.1 *The mystery of the missing meson*

The U(1) problem is an apparent contradiction between two pieces of accepted wisdom. One is wisdom of the 1970s, that hadronic physics is quantum chromodynamics. The other is wisdom of the 1960s, that hadronic physics is approximately invariant under chiral SU(2)⊗SU(2). Let me remind you of the meaning of these two propositions.

Quantum chromodynamics is a field theory whose dynamical variables are an octet of SU(3) gauge fields and a family of SU(3) triplet Dirac bispinor fields, called quarks. In Minkowski space, the Lagrangian density is

$$\mathscr{L} = -\frac{1}{4g^2}(F_{\mu\nu}, F^{\mu\nu}) + \sum_f \bar{\psi}_f(\mathrm{i}D_\mu\gamma^\mu - m_f)\psi_f, \tag{5.1}$$

where f, called the flavour index, labels the various triplets. The usual exact and approximate symmetries of hadron physics (charge, isospin, Gell-Mann's SU(3), etc.) act only on the flavour indices; all physical hadrons are supposed to be singlets under the gauge group. (This last statement is sometimes called quark confinement; it is still far from proved, although there are some suggestive arguments.) ψ_1 and ψ_2 form an isodoublet, the non-strange quarks; ψ_3 is the strange quark; ψ_4 is the charmed quark; there may or may not be additional flavors.

Chiral SU(2)⊗SU(2) is the group generated by the strangeness-conserving weak-interaction currents and their parity transforms. Its diagonal subgroup is conventional isospin. This group is very close to being an exact symmetry of the strong interactions; it is a much better symmetry than SU(3) and roughly as good a symmetry as isospin. However, were this symmetry to be exact, only the isospin subgroup would be a manifest symmetry; the remainder of the group would be a Nambu–Goldstone symmetry, with three massless Goldstone bosons, the pions. The smallness of the pion mass (on a hadronic mass scale) is a measure of the goodness of the symmetry. This is the picture that stands in back of all the stunningly successful soft-pion computations of the mid 1960s.

Now for the apparent contradiction. In quantum chromodynamics, the limit of perfect SU(2)⊗SU(2) symmetry is the limit in which the non-strange quarks are massless. In this limit, the Lagrangian (4.1) obviously has a further chiral U(1) symmetry; it is invariant under

$$\psi_f \to e^{-i\alpha\gamma_5}\psi_f, \quad (f = 1, 2) \tag{5.2}$$

where α is a real number. The associated conserved current is

$$j^5_\mu = \sum_{f=1}^{2} \bar{\psi}_f\gamma_\mu\gamma_5\psi_f. \tag{5.3}$$

I emphasize that the appearance of this additional chiral symmetry is very special to quantum chromodynamics; for example, the σ model has no such additional symmetry in the chiral limit.

Now, either this additional symmetry is manifest or it is spontaneously broken. If it were manifest, all non-massless hadrons would occur in parity doublets. This is not the case; thus it must be spontaneously broken. But if it is spontaneously broken, Goldstone's theorem tells us there must be an associated isoscalar pseudoscalar Goldstone boson. This is the U(1) problem: *what happened to the fourth Goldstone boson*?

One's first thought is that the missing meson is the eta, but this is wrong. The chiral U(1) symmetry is broken by the same mass term that breaks chiral SU(2)⊗SU(2), and thus the fourth Goldstone boson should have roughly the same mass as the pions. The eta is far too heavy. This can be made more precise: using conventional soft-pion methods, Weinberg[28] has shown that a U(1) Goldstone boson must have a mass less than $\sqrt{3}m_\pi$. The eta grossly disobeys this inequality. Also, if we consider the approximation in which the strange quark mass also vanishes, and in which we have perfect chiral SU(3)⊗SU(3) symmetry, the eta takes its place with the pions in an octet of Goldstone bosons. But in this limit we still have an additional U(1) symmetry and we still have a missing meson.

(This should be all that I need to say about the eta. However, there is some confusion abroad on this point, and thus I emphasize that *there is no connection between the eta and the* U(1) *problem*. The eta is a red herring; it is just another hadron; it is no more a relic of a U(1) Goldstone boson than is the N**.)

It may seem that I have posed an insoluble problem; this is because I have lied to you. In fact, j_μ^5 is not a conserved current; it is afflicted with the famous Adler–Bell–Jackiw anomaly.[29] In the limit of N massless quarks,

$$\partial^\mu j_\mu^5 = \frac{N}{32\pi^2}\,\varepsilon^{\mu\nu\lambda\sigma}(F_{\mu\nu}, F_{\lambda\sigma}). \tag{5.4}$$

(Note the similarity between the right-hand side of this equation and the Pontryagin density. This will be important to us later.)

You might think that this is the end of the story; if the current is not conserved, there is no U(1) symmetry to worry about. Alas, life is not so simple. In Sect. 3, we showed that the Euclidean counterpart of the right-hand side of Eq. (5.4) could be written as the divergence of a (gauge-variant) function of A_μ and $F_{\mu\nu}$. It is easy to see that the same construction works in Minkowski space. Thus, if we define

$$J_\mu^5 = j_\mu^5 - \frac{N}{16\pi^2}\,\varepsilon_{\mu\nu\lambda\sigma}(A^\nu, F^{\lambda\sigma} - \tfrac{2}{3}A^\lambda A^\sigma), \tag{5.5}$$

this current is gauge-variant but conserved.

If we work in a covariant gauge (and why should we not?), the added term commutes with the quark fields at equal times. Thus we can derive, for Green's functions made of one J_μ^5 and a string of gauge-invariant quark multilinears, chiral U(1) Ward identities of the usual form. And since these are of the usual form, they lead to the usual conclusion: chiral U(1) is a symmetry; either Green's functions made of quark multilinears alone are U(1) symmetric, or there are Goldstone poles in Green's functions for one J_μ^5 and a string of quark multilinears.

Is there no way out? Well, there is one. The Hilbert space of a gauge field theory quantized in a covariant gauge is notoriously full of negative-norm timelike photons and similar gauge phantoms, states that never couple to gauge-invariant operators. Could it be that the Goldstone boson is such a phantom? No, this is not possible; the formulation of the question is wrong. If the Goldstone boson does not couple at all to gauge-invariant operators, it cannot produce a pole in a Green's function for one J_μ^5 and a string of gauge-invariant operators.

The proper formulation of the question was found by Kogut and Susskind,[30] who had the bright idea of looking at the Schwinger model, massless spinor electrodynamics, in $1+1$ dimensions in a covariant gauge. The Schwinger model is an exactly soluble theory that has properties very close to those we have been discussing. In particular, there is a gauge-invariant axial current with an anomalous divergence and a gauge-variant conserved axial current, and, most important, there is chiral symmetry breakdown without Goldstone poles in gauge-invariant Green's functions. What Kogut and Susskind found in the covariant-gauge Schwinger model were two free massless fields, ϕ_+ and ϕ_-. ϕ_+ creates quanta of positive norm and has the usual propagator; ϕ_- creates quanta of negative norm and has minus the usual propagator. (Remember, a covariant gauge is full of negative-norm states from the very beginning.) All gauge-invariant quantities couple to the sum of these fields, $\phi_+ + \phi_-$; this has zero propagator and produces no singularities. Thus gauge-invariant Green's functions are free of Goldstone poles. However, the gauge-variant conserved current couples to the gradient of the difference, $\partial_\mu(\phi_+ - \phi_-)$. Thus, when one considers a Green's function for one gauge-variant current and a string of gauge-invariant fields, the relative minus sign in the coupling cancels the relative minus sign in the propagators, and Goldstone poles appear where they should. This set-up is called a Goldstone dipole. (The terminology is a bit misleading, because there are only single poles in Green's functions, but I shall stick with it anyway.)

Thus according to Kogut and Susskind, the proper formulation of our question is, is the U(1) symmetry of quantum chromodynamics spontaneously broken via a Goldstone dipole? You might think that this is a question that could be asked seriously only by a field theorist driven mad by spending too many years in too few dimensions. Nevertheless, as 't Hooft[9] brilliantly showed, the answer is yes. The remainder of this section is an explanation of his computation.

5.2 *Preliminaries: Euclidean Fermi fields*

Before we can treat quantum chromodynamics by functional integration, we must know how to integrate over Euclidean Fermi fields. This section is a description of the theory of such integration, with all mathematical fine points ruthlessly suppressed.[31] I will develop the theory by defining Fermi integration as a 'natural' generalization of Bose integration. At the end, I will justify my definitions by showing that they lead to formulae equivalent to those obtained by conventional canonical quantization.

Let us begin by defining our integration variables. For Bose theories, we integrate over *c*-number Euclidean fields. These are objects that commute with each other at arbitrary separations; they can be thought of as the classical (vanishing $\hbar$) limit of quantum Bose fields. This suggests that the proper variables for a Fermi theory should be classical Fermi fields, objects which *anticommute* with each other at arbitrary separations. Thus, for example, for the theory of a single Dirac field, we would expect our integration variables to be two Euclidean bispinors, $\bar{\psi}$ and ψ, obeying

$$\{\psi(x), \psi(y)\} = \{\bar{\psi}(x), \bar{\psi}(y)\} = \{\psi(x), \bar{\psi}(y)\} = 0, \tag{5.6}$$

for all Euclidean points x and y.

The last of these relations is crucial, for it implies that $\bar{\psi}$ can not be in any sense the adjoint of ψ times some matrix. For if this were so, the last relation (multiplied by the inverse matrix) would state that the sum of two positive semi-definite objects, $\psi\psi^\dagger$ and $\psi^\dagger\psi$, was zero. This would only be possible if ψ vanished, not a happy situation for a prospective integration variable. Thus if we are to have any hope of founding a sensible integration theory, we must treat ψ and $\bar{\psi}$ as *totally independent variables.*

This independence is the main novelty of Euclidean Fermi fields; the rest of the construction is straightforward. We define the Euclidean γ-matrices to be four Hermitian matrices obeying

$$\{\gamma_\mu, \gamma_\nu\} = 2\delta_{\mu\nu}. \tag{5.7}$$

We use these to define the O(4) transformation law for ψ in the usual way, and define $\bar{\psi}$ to transform like the adjoint of ψ. We define γ_5, a Hermitian matrix, by

$$\gamma_5 = \gamma_1\gamma_2\gamma_3\gamma_4. \tag{5.8}$$

Thus, $\bar{\psi}\psi$ is a scalar, $\bar{\psi}\gamma_5\psi$ a pseudoscalar, $\bar{\psi}\gamma_\mu\psi$ a vector, etc.

The Euclidean action for a free Dirac field is

$$S = -\int d^4x\bar{\psi}(i\partial_\mu\gamma_\mu - im)\psi. \tag{5.9}$$

The minus sign is pure convention; we could always absorb it into ψ if we wanted to. (Remember, we are free to transform ψ without touching $\bar{\psi}$.) The i in front of the mass term is not conventional. It is there to insure that the Euclidean propagator is proportional to $(\not{p}+im)/(p^2+m^2)$; if it were not for the i, we would have tachyon poles. If m vanishes, Eq. (5.9) is invariant under chiral transformations,

$$\psi \to e^{-i\alpha\gamma_5}\psi, \ \bar{\psi} \to \bar{\psi} e^{-i\alpha\gamma_5}. \tag{5.10}$$

The quark part of the Euclidean action for quantum chromodynamics is obtained from Eq. (5.9) by replacing ordinary derivatives by covariant derivatives.

So much for the integrand; now for the integration. For Bose fields, we defined functional integration as iterated integration over ordinary numbers. Therefore, let us begin by defining integration for a function of a single anticommuting quantity, a. (Of course, for a single quantity, the anticommutation algebra degenerates to a single equation, $a^2=0$.)

We want to define

$$\int da\, f(a), \tag{5.11}$$

for an arbitrary function, f. We want this to have the usual linearity property: the integral of a linear combination of two functions should be the linear combination of the integrals. In addition, we would like the integral to be translation-invariant

$$\int da\, f(a+b) = \int da\, f(a), \tag{5.12}$$

where b is an arbitrary anticommuting quantity. I will now show that these conditions determine the integral, up to a normalization factor.

The reason is that there are only two linearly independent functions of a, 1 and a; all higher powers vanish. We will choose our normalization such that

$$\int da\, a = 1. \tag{5.13}$$

From this, and Eq. (5.12),

$$\int da\, 1 = 0. \tag{5.14}$$

For functions of many anticommuting variables, we define multiple integrals as iterated single integrals. Thus, for example, a complete integration table for the four linearly independent functions of two anticommuting variables, a and $\bar{a}$, is

$$\int da\, d\bar{a} \begin{Bmatrix} \bar{a}a \\ \bar{a} \\ a \\ 1 \end{Bmatrix} = \begin{Bmatrix} 1 \\ 0 \\ 0 \\ 0 \end{Bmatrix}. \tag{5.15}$$

As an application of this table, I will evaluate

$$\int da\, d\bar{a}\, e^{\lambda \bar{a} a} = \int da\, d\bar{a}(1 + \lambda \bar{a} a) = \lambda. \tag{5.16}$$

We can now define integration over Fermi fields exactly as we defined integration over Bose fields in Sect. 2. We introduce two arbitrary complete orthonormal sets of c-number functions, ψ_r and $\bar{\psi}_r$,

$$\int d^4x\, \psi_r^\dagger \psi_s = \int d^4x \bar{\psi}_r \bar{\psi}_s^\dagger = \delta_{rs}. \tag{5.17}$$

We expand the Fermi fields in terms of these functions,

$$\psi = \sum_r a_r \psi_r, \quad \bar{\psi} = \sum_r \bar{a}_r \bar{\psi}_r, \tag{5.18}$$

and define

$$[d\psi][d\bar{\psi}] = \prod_r da_r\, d\bar{a}_r. \tag{5.19}$$

As an application let me evaluate

$$\int [d\psi][d\bar{\psi}] e^{-S}, \tag{5.20a}$$

where

$$S = -\int d^4x\, \bar{\psi} A \psi, \tag{5.20b}$$

and A is some linear operator, possibly depending on external c-number fields. For simplicity, let me assume that A commutes with $A^\dagger$. (This is the case for a quark in an external gauge field.) Then we can choose the ψ_rs to be the eigenfunctions of A,

$$A\psi_r = \lambda_r \psi_r, \tag{5.21}$$

and we can choose $\bar{\psi}_r$ to be $\psi_r^\dagger$. Thus

$$S = -\sum_r \lambda_r \bar{a}_r a_r, \tag{5.22}$$

and

$$\int [d\psi][d\bar{\psi}] e^{-S} = \prod_r \lambda_r = \det A. \tag{5.23}$$

Note that this is the inverse of the answer we would have obtained had we done the identical integral with ψ and $\bar{\psi}$ complex Bose fields.

I will now show that Eq. (5.22) is the correct answer, that it is identical to the normal field-theoretic expression for the vacuum-to-vacuum transition amplitude in a theory of a quantized Dirac field interacting with external c-number fields. In this theory, this amplitude is the sum of all Feynman graphs with no external Fermi lines. This in turn is the exponential of the sum of all connected (that is to say, one-loop) graphs. Now, if ψ were a Bose field, we know that the amplitude would be the inverse determinant, because we trust functional integration for Bose fields. But the only effect of replacing bosons by fermions is to multiply the one-loop graphs by minus one. This inverts the exponential of the one-loop graphs, that is to say, it turns the inverse determinant into the determinant.

In any theory in which the Fermi fields enter the action at most bilinearly, we can always integrate over the Fermi fields, using Eq. (5.23), before we integrate over the Bose fields. In diagrammatic language, we can always sum the Fermi loops before we integrate over virtual bosons. Thus, because our definition of Fermi integration gives the right answer for a Dirac field in an external c-number field, it also gives the right answer for a Dirac field interacting with a quantum Bose field. In particular, it gives the right answer for quantum chromodynamics.

5.3 *Preliminaries: chiral Ward identities*

In this section is a discussion of the chiral Ward identities for a theory of a set of quantum Dirac fields interacting with c-number gauge fields. In the sequel, we shall use these identities in several different cases; thus it is useful to have them written down in their most general form, at hand when we need them.

Let ψ be a set of Euclidean Dirac fields, assembled into a big vector, which transforms according to some representation of SU(n), not necessarily irreducible, generated by a set of matrices, T^a. Let us define the constant C by

$$\mathrm{Tr}\, T^a T^b = -C\delta^{ab}. \tag{5.24}$$

Thus, for example, for a set of N fields each transforming according to the n-dimensional representation of SU(n),

$$C = N/2. \tag{5.25}$$

We wish to study the theory of these fields interacting with given c-number gauge fields,

$$S = -\mathrm{i}\int \mathrm{d}^4x\, \bar{\psi}(\gamma_\mu \mathrm{D}_\mu - M)\psi, \tag{5.26}$$

where D_μ is the covariant derivative defined by Eq. (3.16), and M is the mass matrix for the Dirac fields, assumed to be SU(n)-invariant. Let $\phi^{(r)}$, $r=1\ldots m$, be a set of local multilinear functions of the Dirac fields. The Euclidean Green's functions for these objects are defined by

$$\langle\phi^{(1)}(x_1)\ldots\phi^{(m)}(x_m)\rangle^A = \frac{\int [d\psi][d\bar{\psi}]e^{-S}\phi^{(1)}(x_1)\ldots\phi^{(m)}(x_m)}{\int [d\psi][d\bar{\psi}]e^{-S}} \tag{5.27}$$

where I have inserted the superscript A to remind you that we are working in an external gauge field.

Now let us perform an infinitesimal change of variables in the numerator of Eq. (5.27),

$$\delta\psi = -i\gamma_5\psi\delta\alpha, \quad \delta\bar{\psi} = -i\bar{\psi}\gamma_5\delta\alpha, \tag{5.28a}$$

where $\delta\alpha$ is an infinitesimal function of Euclidean space. Since the ϕs are functions of the Dirac fields, they will change under the change of variables; we define $\partial\phi^{(r)}/\partial\alpha$ by

$$\delta\phi^{(r)} = (\partial\phi^{(r)}/\partial\alpha)\delta\alpha. \tag{5.28b}$$

Thus, for example, $\partial\bar{\psi}\psi/\partial\alpha$ is $-2i\bar{\psi}\gamma_5\psi$. A change of variables does not change the integral; thus, taking the variational derivative with respect to $\delta\alpha$, we find

$$\begin{aligned}
&\partial^\mu\langle j^5_\mu(y)\phi^{(1)}(x_1)\ldots\phi^{(m)}(x_m)\rangle^A \\
&\quad + \langle\bar{\psi}M\gamma_5\psi(y)\phi^{(1)}(x_1)\ldots\phi^{(m)}(x_m)\rangle^A \\
&\quad + \delta^{(4)}(y-x_1)\langle\partial\phi^{(1)}(x_1)/\partial\alpha\ldots\phi^{(m)}(x_m)\rangle^A \\
&\quad + \cdots \\
&\quad + \delta^{(4)}(y-x_m)\langle\phi^{(1)}(x_1)\ldots\partial\phi^{(m)}(x_m)/\partial\alpha\rangle^A = 0,
\end{aligned} \tag{5.29}$$

where j^5_μ is $\bar{\psi}\gamma_\mu\gamma_5\psi$.

These are, of course, just the Euclidean version of the Ward identities we would have obtained in Minkowski space by studying the divergence of j^5_μ, and, of course, they are wrong, for they take no account of the Adler–Bell–Jackiw anomaly. I do not have the time here to recapitulate the theory of the anomaly, and I will simply state the correct version of Eq. (5.29): the zero on the right-hand side is replaced by

$$-\frac{iC}{8\pi^2}(F(y),\tilde{F}(y))\langle\phi^{(1)}(x_1)\ldots\phi^{(m)}(x_m)\rangle^A. \tag{5.30}$$

We can obtain a very useful equation by integrating the corrected Ward identity over y. The first term on the left vanishes by integration by parts; the theory contains no massless particles that could give a non-vanishing surface term. Also, on the right we can use

$$\int d^4y (F, \tilde{F}) = 32\pi^2 \nu. \tag{3.47}$$

Thus we obtain

$$2\left\langle \int d^4y\, \bar{\psi} M \gamma_5 \psi(y) \phi^{(1)}(x_1) \ldots \phi^{(m)}(x_m) \right\rangle^A$$

$$+ \frac{\partial}{\partial \alpha} \langle \phi^{(1)}(x_1) \ldots \phi^{(m)}(x_m) \rangle^A$$

$$= -4iC\nu \langle \phi^{(1)}(x_1) \ldots \phi^{(m)}(x_m) \rangle^A. \tag{5.31}$$

Now all our artillery is at the ready; we can begin our assault on quantum chromodynamics.

5.4 *QCD (baby version)*

I will begin by analyzing a baby version of quantum chromodynamics, in which the gauge group is SU(2), and in which there is only a single isodoublet quark, of mass zero. In equations,

$$S = \int d^4x \left[\frac{1}{4g^2} (F, F) - i\bar{\psi} D_\mu \gamma_\mu \psi \right]. \tag{5.32}$$

After we have worked out the baby theory, we will go on to the real thing.

Most of the analysis of Sect. 3 is essentially unaltered by the presence of a quark. In particular, all of our old instanton solutions are still solutions of the Euclidean equations of motion (with the quark fields set equal to zero). Thus we still have all the θ-vacua, and formulae like

$$E(\theta)/V = -2K \cos\theta\, e^{-S_0}, \tag{3.56}$$

and

$$\langle \theta | (F, \tilde{F}) | \theta \rangle = -64\pi^2 i K e^{-S_0} \sin\theta, \tag{3.59}$$

remain unaltered. The only effect of the quarks is to insert into the definition of K a term proportional to

$$\det\left[\frac{i\not{D}}{i\not{\partial}}\right] = \det\left[\frac{i(\partial_\mu + A_\mu)\gamma_\mu}{i\partial_\mu \gamma_\mu}\right], \tag{5.33}$$

where A_μ is the field of an instanton.

This is a trifling alteration, but it is a tremendous trifle, for, as we shall see, $i\not{D}$ has a vanishing eigenvalue. Thus the determinant vanishes, as does $E(\theta)/V$ and $\langle \theta | (F, \tilde{F}) | \theta \rangle$!

The vanishing eigenvalue can be demonstrated either by a short explicit computation or by a long indirect argument. I will choose the second method. Despite what you might think, this is not a perverse choice. (Well, not totally perverse.) The indirect argument will have some byproducts that will be very useful to us later.

For simplicity, I will assume (falsely) that $\mathrm{i}\not{D}$ has a purely discrete spectrum,[32]

$$\mathrm{i}\not{D}\psi_r = \lambda_r \psi_r. \tag{5.34}$$

Because $\mathrm{i}\not{D}$ is Hermitian, all the λs are real. Because γ_5 anticommutes with γ_μ,

$$\mathrm{i}\not{D}\gamma_5\psi_r = -\lambda_r \gamma_5 \psi_r. \tag{5.35}$$

Thus non-vanishing eigenvalues always occur in pairs of opposite sign. Eigenfunctions of vanishing eigenvalue, on the other hand, can always be chosen to be eigenfunctions of γ_5,

$$\gamma_5\psi_r = \chi_r \psi_r, \quad (\lambda_r = 0) \tag{5.36}$$

Because $\gamma_5^2 = 1$, $\chi_r = \pm 1$. I will denote the number of eigenfunctions of these two types by $n_\pm$.

I will now prove the remarkable sum rule,[33]

$$n_- - n_+ = \nu. \tag{5.37}$$

Thus, not only is there a zero eigenvalue in the field of an instanton, there is a zero eigenvalue in *any* gauge field of non-zero winding number, whether or not it is a solution of the Euclidean equations of motion.

The proof rests on the chiral Ward identities for the quantum theory of a *massive* quark interacting with an external gauge field.

$$S = -\mathrm{i} \int \mathrm{d}^4x\, \bar{\psi}(\not{D} - m)\psi. \tag{5.38}$$

If we take the case of *no* ϕs, Eq. (5.31) becomes

$$-2\mathrm{i}\nu = 2\left\langle \int \mathrm{d}^4y\, \bar{\psi} m \gamma_5 \psi \right\rangle^A$$

$$= \frac{2\displaystyle\int [\mathrm{d}\psi][\mathrm{d}\bar{\psi}]\mathrm{e}^{-S} \int \mathrm{d}^4y\, \bar{\psi} m \gamma_5 \psi}{\displaystyle\int [\mathrm{d}\psi][\mathrm{d}\bar{\psi}]\mathrm{e}^{-S}}. \tag{5.39}$$

(Remember, in the case at hand, $C = \frac{1}{2}$.) To evaluate the functional integrals, we need the eigenfunctions and eigenvalues of $\mathrm{i}(\not{D} - m)$. The eigenfunctions are those of $\not{D}$, and the eigenvalues are simply shifted by $-\mathrm{i}m$,

$$\mathrm{i}(\not{D} - m)\psi_r = (\lambda_r - \mathrm{i}m)\psi_r. \tag{5.40}$$

If we expand the fields in the ψ_rs, the functional integrals become trivial, and we obtain

$$-2\mathrm{i}\nu = \frac{2m\sum_r \int \mathrm{d}^4y\, \psi_r^\dagger \gamma_5 \psi_r \prod_{s\neq r} (\lambda_s - \mathrm{i}m)}{\prod_r (\lambda_r - \mathrm{i}m)}$$

$$= 2m \sum \int \mathrm{d}^4y\, \psi_r^\dagger \gamma_5 \psi_r (\lambda_r - \mathrm{i}m)^{-1}. \tag{5.41}$$

Because eigenfunctions of a Hermitian operator with different eigenvalues are orthogonal,

$$\int \mathrm{d}^4y\, \psi_r^\dagger \gamma_5 \psi_r = 0 \quad \text{if } \lambda_r \neq 0, \tag{5.42}$$

while

$$\int \mathrm{d}^4y\, \psi_r^\dagger \gamma_5 \psi_r = \chi_r \quad \text{if } \lambda_r = 0. \tag{5.43}$$

Thus,

$$-2\mathrm{i}\nu = 2\mathrm{i}(n_+ - n_-). \tag{5.44}$$

This is the desired result.

It turns out that the instanton obeys the sum rule by having one eigenfunction of vanishing eigenvalue with $\chi = -1$ and none with $\chi = +1$. (This also can be seen indirectly, without dirtying one's hands with explicit computations; see Appendix 5.) We shall never need the explicit form of the eigenfunction, but, just for completeness, I shall write it down here. For an instanton with center at X and size ρ,

$$\psi_0(x - X, \rho) = \rho[\rho^2 + (x - X)^2]^{-\frac{3}{2}} u, \tag{5.45}$$

where u is a constant spinor. Likewise, for an anti-instanton, there is one eigenfunction of vanishing eigenvalue with $\chi = +1$, the parity transform of Eq. (5.45). For n widely separated instantons and anti-instantons, there are n such eigenfunctions, one centered about each object. (More properly, I should say that there are n approximate eigenfunctions with approximately vanishing eigenvalues, but, for the dilute-gas approximation, the qualifications are irrelevant.)

What is important for our purposes is that the sum rule implies that any field configuration with non-vanishing winding number has at least one eigenfunction of vanishing eigenvalue and thus a vanishing Fermi determinant. Thus, not just in the dilute gas approximation, but to *all* orders in the semiclassical expansion, all the θ vacua have the same energy and they all have a vanishing expectation value for $(F, \tilde{F})$.

A phenomenon this general must have a deep cause. We can discover this cause if we consider the chiral Ward identities for vanishing quark mass. There is a technical obstacle to this; for vanishing quark mass, the denominator in Eq. (5.27) vanishes, at least for fields with $\nu \neq 0$. This is easily surmounted; we define denominator-free Green's functions,

$$\langle\langle \phi^{(1)}(x_1)\ldots\rangle\rangle^A \equiv \int [d\psi][d\bar{\psi}]e^{-S}\phi^{(1)}(x_1)\ldots. \tag{5.46}$$

By the same reasoning as before, these obey the Ward identities,

$$\left[\frac{\partial}{\partial\alpha}+2i\nu\right]\langle\langle \phi_1(x_1)\ldots\rangle\rangle^A = 0, \tag{5.47}$$

i.e. Eq. (5.31) without the mass term. The Green's functions of our baby version of chromodynamics are given by

$$\langle\theta|\phi^{(1)}(x_1)\ldots|\theta\rangle = \frac{\int [dA]e^{-S_g}e^{i\nu\theta}\langle\langle \phi^{(1)}(x_1)\ldots\rangle\rangle^A}{\int [dA]e^{-S_g}e^{i\nu\theta}\langle\langle 1\rangle\rangle^A}, \tag{5.48}$$

where S_g is the gauge-field part of the action. By Eq. (5.47),

$$\left[\frac{\partial}{\partial\alpha}+2\frac{\partial}{\partial\theta}\right]\langle\theta|\phi^{(1)}(x_1)\ldots|\theta\rangle = 0. \tag{5.49}$$

Thus, the effect of a chiral U(1) transformation can be undone by a change of θ. That is to say, chiral U(1) transformations turn one θ-vacuum into another; chiral U(1) symmetry is spontaneously broken, and the θ-vacua are the many vacua that appear when a symmetry suffers spontaneous breakdown. This is startling; after all, when we first met the θ-vacua in Sect. 3, they had no connection with chiral symmetry – there was no chiral symmetry for them to be connected with! Nevertheless, it is an inevitable result of our analysis, and it explains why all the θ-vacua have the same energy density and the same expectation value of $(F, \tilde{F})$; it is because these quantities are chiral U(1) invariants.

(Parenthetical remark: the factor of 2 in Eq. (5.49) is worth comment. It tells us that when we make a chiral rotation by π we return to the same θ-vacuum. This is as it should be.

$$e^{-i\pi\gamma_5} = -1. \tag{5.50}$$

Thus a chiral rotation by π has the same effect on the fields as a spatial rotation by 2π; we would be very unhappy if this symmetry suffered spontaneous breakdown.)

There is one possible loophole in the argument I have given. It remains a logical possibility that, for every Green's function, the derivative with respect to α and the derivative with respect to θ both vanish. If this happened, we would have, not spontaneous symmetry breakdown, but manifest symmetry, and the θ-vacua would be mathematical artifacts, superfluous duplicates of a single vacuum.

I will now eliminate this possibility by computing, in the dilute-gas approximation

$$\langle\theta|\sigma_\pm(x)|\theta\rangle = \frac{\int [\mathrm{d}A][\mathrm{d}\psi][\mathrm{d}\bar{\psi}]\mathrm{e}^{-S}\mathrm{e}^{\mathrm{i}\nu\theta}\sigma_\pm(x)}{\int [\mathrm{d}A][\mathrm{d}\psi][\mathrm{d}\bar{\psi}]\mathrm{e}^{-S}\mathrm{e}^{\mathrm{i}\nu\theta}}, \tag{5.51}$$

where

$$\sigma_\pm = \tfrac{1}{2}\bar{\psi}(1\pm\gamma_5)\psi. \tag{5.52}$$

These are chiral eigenfields,

$$\partial\sigma_\pm/\partial\alpha = \mp 2\mathrm{i}\sigma_\pm. \tag{5.53}$$

Thus, if we obtain a non-zero answer, we will know that spontaneous symmetry breakdown has occurred.

The computation will parallel closely that of the vacuum energy of a pure gauge field theory in Sect. 3. Indeed, as the calculation proceeds, we will accumulate all the terms that led to our earlier expression for the determinantal factor, K, as an integral over instanton size, ρ,

$$K = 2g^{-8}\int_0^\infty \frac{\mathrm{d}\rho}{\rho^5} f(\rho M), \tag{5.54}$$

where M is the renormalization mass. As these old terms come up, I shall call them to your attention, but I will not bother to write them down; I will keep explicit track only of new terms that modify the integrand in Eq. (5.54).

There is one important novelty in the dilute-gas approximation. For n widely separated instantons and anti-instantons, $\mathrm{i}\not{D}$ has n vanishing eigenvalues. Thus the integral over Fermi fields will vanish unless the integrand contains

$$\prod_{\lambda_r=0} \bar{a}_r a_r. \tag{5.55}$$

Such a term can appear only if we are computing a Green's function involving at least $2n$ Dirac fields. Hence, for any fixed Green's function, the potentially infinite sum over instantons and anti-instantons terminates.

I will first do the σ_- computation:

In the denominator of Eq. (5.51), the only configuration that does not have a surplus of vanishing eigenvalues is one of no instantons and no anti-instantons, that is to say, the classical vacuum, $A_\mu = 0$. Thus the denominator is simply the product of a Bose determinant and a Fermi determinant. The same Bose determinant appeared in the denominator in our earlier computation. The Fermi determinant, det($i\not{\partial}$), is a new factor.

In the numerator, we need a configuration with $\nu=1$, by Eq. (5.47). The only one that does not have a surplus of vanishing eigenvalues is one instanton and no anti-instantons. Let us do the Fermi integral first; this gives

$$\begin{aligned}&\tfrac{1}{2}\psi_0^\dagger(x-X,\rho)(1-\gamma_5)\psi_0(x-X,\rho)\prod_{\lambda_r\neq 0}\lambda_r\\&\quad=\psi_0^\dagger\psi_0(x-X,\rho)\,\det{}'(i\not{D}),\end{aligned} \tag{5.56}$$

where det′, as always, denotes a determinant with vanishing eigenvalues removed. The Bose integral gives a determinant and a bunch of collective-coordinate factors identical to those that go into K. Because det′($i\not{D}$) does not depend on X, the integration over the instanton location is trivial,

$$\int d^4X\,\psi_0^\dagger\psi_0(x-X)=1. \tag{5.57}$$

Finally, we have a factor of $e^{-8\pi^2/g^2}$ from the instanton action, and a factor of $e^{i\theta}$ from the $e^{i\nu\theta}$.

The σ_+ computation is almost identical to the σ_- one; the only difference is that the relevant configuration is one anti-instanton, and thus, instead of a factor of $e^{i\theta}$, we have one of $e^{-i\theta}$.

Putting all this together, we find

$$\langle\theta|\sigma_\pm(x)|\theta\rangle=e^{-8\pi^2/g^2}e^{\mp i\theta}g^{-8}\,2\int_0^\infty\frac{d\rho}{\rho^5}f(\rho M)\frac{\det{}'(i\not{D})}{\det(i\not{\partial})}. \tag{5.58}$$

(In case you have lost track of the meaning of my symbols, I remind you that $i\not{D}$ is the Dirac operator in the field of an instanton of size ρ.)

Just as before, we can use dimensional analysis to study the integrand in this formula. The eigenvalues of $i\not{D}$ have the dimensions of 1/length. One eigenvalue has been removed from the primed determinant; thus the ratio det′/det has dimensions of length, and must be of the form

$$\frac{\det{}'(i\not{D})}{\det(i\not{\partial})}=\rho h(\rho M), \tag{5.59}$$

where h is an unknown function. Note that this gives the right dimensions for the expectation values of $\sigma_\pm$, $1/(\text{length})^3$.

From here on the argument is a rerun of that of Sect. 3: we can use the renormalization group to determine the form of the integrand up to an arbitrary multiplicative constant, be embarrassed in the infrared, wave our hands about new physics giving an effective infrared cutoff, etc.

We now know spontaneous symmetry breakdown occurs. Are there

Goldstone bosons? Let us look for them in

$$\langle\theta|\sigma_+(x)\sigma_-(0)|\theta\rangle. \tag{5.60}$$

By reasoning which should now be familiar to you, only two field configurations are relevant: $A_\mu=0$, and one instanton plus one anti-instanton. The first of these just gives the usual one-loop perturbation theory expression; this has a two-quark cut, but no Goldstone pole. The second just gives the product $\langle\theta|\sigma_+|\theta\rangle\langle\theta|\sigma_-|\theta\rangle$. This also has no Goldstone pole. By similar methods one can investigate other gauge-invariant Green's functions, such as $\langle\theta|j^5_\mu\sigma_\pm|\theta\rangle$ or $\langle\theta|j^5_\mu j^5_\nu|\theta\rangle$, and again find no Goldstone poles, but really there is no need to do these computations. If Goldstone bosons appear anywhere, they should appear in (5.60), and they do not.

In the last sentence, I should have said not 'appear anywhere', but 'appear among the physical states', that is to say, as singularities in gauge-invariant Green's functions. The situation is very different if we study a gauge-variant Green's function such as

$$\langle\theta|J^5_\mu(x)\sigma_-(0)|\theta\rangle=\langle\theta|j^5_\mu(x)\sigma_-(0)|\theta\rangle$$
$$+\frac{i}{16\pi^2}\langle\theta|G_\mu(x)\sigma_-(0)|\theta\rangle, \tag{5.61}$$

where G_μ is defined in Eq. (3.41). As I have said, the first of the terms on the right has no Goldstone pole, but, as I will show, the second does. The argument is simple. In a covariant gauge, there is a Goldstone pole if and only if

$$\int d^4x\,\partial_\mu\langle\theta|G_\mu(x)\sigma_-(0)|\theta\rangle\neq 0. \tag{5.62}$$

If we use the identity,

$$\int d^4x\,\partial_\mu G_\mu=32\pi^2\nu, \tag{5.63}$$

and the fact that the only configurations that contribute to (5.62) have $\nu=1$, we find

$$\int d^4x\,\partial_\mu\langle\theta|G_\mu(x)\sigma_-(0)|\theta\rangle=32\pi^2\langle\theta|\sigma_-|\theta\rangle\neq 0. \tag{5.64}$$

On the other hand, for $\langle\theta|J_\mu J_\nu|\theta\rangle$, the contributing configurations have vanishing ν, and thus there is no Goldstone pole.

To summarize, we have found in the dilute-gas approximation: spontaneous breakdown of chiral U(1) symmetry, no Goldstone poles in gauge-invariant Green's functions, no Goldstone poles in the propagator of a gauge-variant conserved current, and a Goldstone pole in the Green's function for one gauge-variant current and one gauge-invariant operator. This is the Goldstone dipole of Kogut and Susskind.

5.5 *QCD (the real thing)*

Real quantum chromodynamics in the chiral SU(2)⊗SU(2) limit differs from our baby version in two respects. Firstly, we have triplet quarks with gauge group SU(3) rather than doublet quarks with gauge group SU(2). Secondly, we have two massless quarks, rather than one. (I will ignore the massive quarks; they are irrelevant to the U(1) problem.)

Replacing an SU(2) doublet by an SU(3) triplet makes hardly any change. If this were the only difference, we would still have instantons, and the constant C of Eq. (5.25) would still be $\frac{1}{2}$; the only thing we would need to change in Sect. 5.4 would be the integral over instanton size, where g^{-8} would become g^{-12}.

In contrast, replacing one massless triplet by two makes a profound change. C is doubled, and thus the sum rule (5.37) is changed to

$$n_- - n_+ = 2\nu. \tag{5.65}$$

Hence, $i\not{D}$ in an instanton field has two vanishing eigenvalues rather than one. (We do not really need a fancy sum rule to see this; we have two independent quark fields, so every eigenvalue occurs twice, once for ψ_1 and once for ψ_2.) Thus, two fields no longer suffice to take care of all the vanishing eigenvalues, and all quark bilinears have zero expectation values.

This is no obstacle to demonstrating the spontaneous breakdown of chiral U(1) symmetry; we just have to study quadrilinears rather than bilinears. For example, the same computation that before gave a non-vanishing expectation value for $\bar{\psi}_1(1-\gamma_5)\psi_1$ will now give a non-vanishing expectation value for $\bar{\psi}_1(1-\gamma_5)\psi_1\bar{\psi}_2(1-\gamma_5)\psi_2$.

There is a reason for this. We have found spontaneous breakdown of chiral U(1), but not of chiral SU(2)⊗SU(2); the θ-vacua are all invariant under chiral SU(2)⊗SU(2). (There are two ways to see this. (1) There are too few θ-vacua for them to be anything but invariant; for spontaneous breakdown of chiral SU(2)⊗SU(2) we need at least a three-parameter family of vacua. (2) Chiral U(1) transformations are connected to θ by the anomalous divergence of the isosinglet axial current; the isotriplet axial current is anomaly-free.) All Lorentz-invariant quark bilinears transform according to the representation $(\frac{1}{2}, \frac{1}{2})$ of SU(2)⊗SU(2), and must have vanishing expectation values. However, there are quadrilinear SU(2)⊗SU(2) singlets, such as

$$\begin{aligned}&\tfrac{1}{2}\varepsilon_{ij}\varepsilon_{kl}\bar{\psi}_i(1-\gamma_5)\psi_k\bar{\psi}_j(1-\gamma_5)\psi_l\\&\quad=\bar{\psi}_1(1-\gamma_5)\psi_1\bar{\psi}_2(1-\gamma_5)\psi_2-\bar{\psi}_1(1-\gamma_5)\psi_2\bar{\psi}_2(1-\gamma_5)\psi_1.\end{aligned} \tag{5.66}$$

These operators can have non-vanishing expectation values.

The doubling of C also changes Eq. (5.49) to

$$\left[\frac{\partial}{\partial\alpha}+4\frac{\partial}{\partial\theta}\right]\langle\theta|\phi_1(x_1)\ldots|\theta\rangle=0. \tag{5.67}$$

Thus a chiral rotation by $\pi/2$, rather than π,

$$\psi_{1,2}\rightarrow -i\gamma_5\psi_{1,2}, \tag{5.68}$$

returns us to the same θ-vacuum. Again, this is an effect of unbroken SU(2)⊗SU(2). If we multiply this by the SU(2)⊗SU(2) transformation,

$$\psi_1\rightarrow -i\gamma_5\psi_1,\quad \psi_2\rightarrow i\gamma_5\psi_2, \tag{5.69}$$

we obtain

$$\psi_1\rightarrow\psi_1,\quad \psi_2\rightarrow -\psi_2, \tag{5.70}$$

which should not be spontaneously broken.

Of course, we do not want unbroken SU(2)⊗SU(2) in quantum chromodynamics; we want spontaneous breakdown; we want pions. However, there is no reason to be disturbed that pions have not emerged from our computations. Our methods are semiclassical, valid in the limit of vanishing $\hbar$, in principle capable only of revealing those phenomena that occur for arbitrarily weak coupling. We have learned that the breakdown of chiral SU(2)⊗SU(2) is not such a phenomenon. This is no surprise. What is a surprise (and a wonderful surprise) is that the breakdown of chiral U(1) *is* such a phenomenon.

5.6 *Miscellany*[34]

There are some topics that I do not have the time to discuss in the detail they deserve but which I can not resist mentioning:

(1) For most theories with spontaneous symmetry breakdown, symmetry is restored at sufficiently high temperatures. Is this true here? This is an easy question to answer. Finite-temperature Green's functions are given by functional integrals over a Euclidean time inversely proportional to the temperature, with periodic time boundary conditions for Bose fields and antiperiodic ones for Fermi fields. Thus, as the temperature goes up, instantons of any given size eventually get squeezed out; there is no way to fit them into the available region of Euclidean space. However, no matter how high the temperature, there are always instantons so small that they barely notice the time boundary conditions. Thus, although asymmetries go to zero as a (calculable) power of the inverse temperature, symmetry is never fully restored. For extremely high temperatures, the only relevant instantons are so small that the effective coupling constant is extremely weak; thus we could make numerical computations of extreme accuracy, but only in a regime that is totally inaccessible to

experiment. I stress that this persistence of symmetry breakdown is a reflection of the scale invariance of classical chromodynamics, not of any property of instanton effects in general. For example, in the model of Sect. 4, there is a definite instanton size, and thus, at sufficiently high temperatures, all instanton effects disappear.

(2) Callan, Dashen, and Gross[1] have recently proposed a detailed picture of the dynamic structure of quantum chromodynamics. To explain their ideas, let me restrict myself to chromodynamics with two massless quarks, and let me imagine the universe cooling down from a very high temperature. Then, according to Callan, Dashen, and Gross:

(a) At very high temperatures, when the effective coupling constant is very small, chiral U(1) is spontaneously broken by instantons, but chiral SU(2)⊗SU(2) is still a good symmetry, and quarks are still unconfined. (Of course, this part is the standard picture which I have described in detail.)

(b) At somewhat lower temperatures, the effective coupling constant grows larger, and chiral SU(2)⊗SU(2) suffers spontaneous breakdown.[35] This is also an instanton effect, but an indirect one that can not be seen in the dilute gas approximation. Nevertheless, the effective coupling constant, although not tiny, is still small enough so that weak-coupling approximations are fairly reliable. (This part looks good to me.) Quarks are still unconfined.

(c) At still lower temperatures, and still larger effective couplings, new field configurations, called 'merons', become important in the functional integral. These produce a long-range force that confines the quarks.[36] (I can see nothing wrong with this idea in principle, but the details of the argument involve a stupendous amount of hand-waving. This part is just a suggestion (although a very clever suggestion) that may or may not someday become a theory of confinement.)

If you will excuse me for beating a dead horse one more time, this picture shows very sharply how misleading it is to say that 'instantons give the U(1) Goldstone boson a mass'. This implies that quarks get their masses through spontaneous symmetry breakdown, with the appearance of four Goldstone bosons, and then instantons come to the rescue. This is not what happens.

(To be fair, I should modify the last sentence and say, 'This is not what happens in the picture of Callan, Dashen, and Gross.' A skeptic might imagine replacing paragraph (c) above by, 'At still lower temperatures, and still larger effective couplings, new field configurations become important which restore chiral U(1) invariance. At a yet later stage, this suffers spontaneous breakdown and a Goldstone boson appears.' To my knowledge, there is no chromodynamic computation that offers the slightest evidence for this disgusting alternative, but it is not logically excluded.)[37]

(3) I have stressed several times that spontaneous breakdown of U(1) (without Goldstone bosons) is independent of spontaneous breakdown of $SU(2)\otimes SU(2)$ (with Goldstone bosons). In a recent paper, Crewther[38] has argued ingeniously that these phenomena are not just independent; they are inconsistent. This would be bad news if it were true, but I do not believe that it is; I think Crewther's arguments are invalid. However, since Crewther and I are at this moment entering our fourth month of correspondence on this matter, and since neither of us has yet convinced the other of the error of his ways, I will say no more about this.

(4) In all the θ-vacua, except for $\theta=0$ or π, *CP*-noninvariant operators have non-vanishing expectation values. Thus it seems that in most of the θ-vacua we have observable strong *CP* violation. Of course, this is an illusion; the θ-vacua are transformed into each other by the U(1) group, and thus all experiments must yield the same results in any vacuum. Phrased more explicitly, for every θ-vacuum there is a discrete symmetry under which the vacuum is invariant, the product of *CP* and an appropriate U(1) transformation, and we are free to redefine *CP* to be this transformation.

All this is for massless quarks. The situation changes drastically when the quarks have masses, either because we have put them in by hand, or because they have Yukawa couplings to weak-interaction Higgs mesons. Now we no longer have U(1) symmetry; there is a potential clash between the definition of *CP* selected by θ and that selected by the quark mass operator, and there is the disastrous possibility of strong *CP* violation.

(Let me dispose of a red herring. You might think that all this might be said of a theory in which U(1) breaks down in the ordinary way, with Goldstone bosons, as in the U(1) σ model. In this case, there is no problem; as soon as we add a U(1) violating interaction, no matter how weak, the order parameter, the analog of θ, automatically aligns itself with the perturbation. *This is not what happens here.* The easiest (and unfortunately also the least convincing) way of seeing this is to remember that when all the dust of Sect. 3 settled, θ emerged as effectively a coupling constant, the coefficient of a term in the action. Thus we would no more expect θ to change discontinuously in response to an external perturbation than we would expect g to.)

Several mechanisms have been suggested for avoiding this disaster.[39] At the moment I favor an up quark with vanishing bare mass, that is to say, with vanishing coupling to the Higgs fields. In this case, we still have a

U(1) symmetry, chiral U(1) acting on the up quark only, and thus we have no *CP* problem. Unfortunately, this conflicts with current-algebra estimates of the up mass; these all agree that it is somewhere between $\frac{1}{2}$ and $\frac{2}{3}$ of the down mass. However, all these estimates are based on soft-kaon and soft-eta computations, and these are notoriously less accurate than soft-pion computations. For example, only soft-pion methods are needed to compute the slope of $\eta \to 3\pi$, in good agreement with experiment; soft-eta methods are needed to compute the rate, off by a factor of three.[40] So perhaps a massless up quark is not such a silly idea. Still, I would be happier if I had a more elegant solution, and one with more predictive power.

6 The fate of the false vacuum[41]

6.1 *Unstable vacua*

In Sect. 2.4 I explained how to use instanton methods to study a particle theory with a false (that is to say, unstable) ground state. In this section I will apply these methods to a field theory with a false ground state, that is to say, a false vacuum.

For simplicity, I will restrict myself to the theory of a single scalar field in four-dimensional space-time, with dynamics defined by the Euclidean action

$$S = \int d^4x[\tfrac{1}{2}(\partial_\mu\phi)^2 + U(\phi)], \tag{6.1}$$

where U is a function of the form shown in Fig. 15. Note that U possesses two relative minima, ϕ_+ and ϕ_-, but only ϕ_- is an absolute minimum. In analogy to Sect. 2.4, I have used my freedom to add a constant to U to insure that $U(\phi_+)=0$. The state of the classical field theory for which $\phi=\phi_-$ is the unique classical state of lowest energy, and, at least for weak coupling, corresponds to the unique vacuum state of the quantum theory. The state of the classical field theory for which $\phi=\phi_+$ is also a stable classical equilibrium state. However, in the quantum theory it is rendered unstable by barrier penetration; it is a false vacuum.

Fig. 15

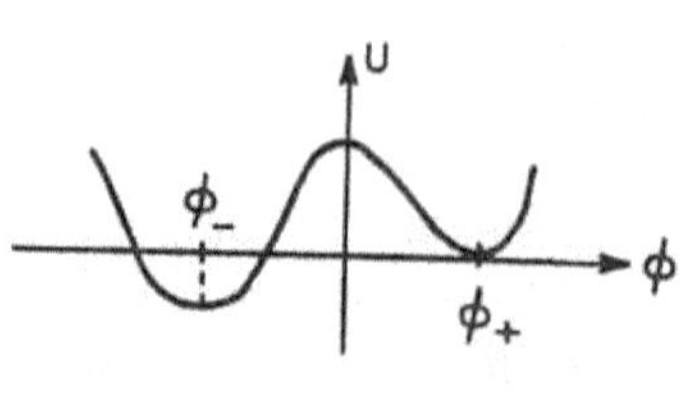

Even without any knowledge of instantons and bounces, it is easy to understand the qualitative features of the decay of the false vacuum. The decay closely parallels the nucleation processes of statistical physics, like the crystallization of a supersaturated solution or the boiling of a superheated fluid. Imagine Fig. 15 to be a plot of the free energy of a fluid as a function of density. The false vacuum corresponds to the superheated fluid phase and the true vacuum to the vapor phase. Thermodynamic fluctuations are continually causing bubbles of vapor to materialize in the fluid. If the bubble is too small, the gain in volume energy caused by the materialization of the bubble is more than compensated for by the loss in surface energy, and the bubble shrinks to nothing. However, once in a while a bubble is formed large enough so that it is energetically favorable for the bubble to grow. Once this occurs, there is no need to worry about fluctuations anymore; the bubble expands until it converts the available fluid to vapor (or coalesces with another bubble).

An identical picture describes the decay of the false vacuum, with quantum fluctuations replacing thermodynamic ones. Once in a while a bubble of true vacuum will form large enough so that it is energetically favorable for the bubble to grow. Once this happens, the bubble spreads throughout the universe, a cancer of space, converting false vacuum to true.

Thus the thing to compute is not a decay probability per unit time, Γ, but a decay probability per unit time per unit volume, Γ/V, for the probability per unit time that in a given volume a critical bubble will form is proportional to the volume (at least if the volume is much bigger than the bubble).

Of course, such a computation would be bootless were it not for cosmology. An infinitely old universe must be in a true vacuum, no matter how slowly the false vacuum decays. However, the universe is not infinitely old, and, at the time of the big bang, the universe might well have been in the false vacuum. For example, in the Weinberg–Salam model, if the mass of the Higgs meson exceeds Weinberg's lower bound, the asymmetric vacuum, in which we live, has a lower energy than the symmetric vacuum. However, if the Higgs mass is less than $\sqrt{2}$ times the lower bound, the symmetric vacuum is a local minimum of the potential, a possible false vacuum. Now we know that at high temperatures (i.e. in the early universe), symmetry breaking disappears in this model; the symmetric vacuum is the true ground state. Thus it is possible to envision a situation in which the universe gets into the false vacuum early in its history and is stuck there as it cools off; in such a situation, knowledge of Γ/V is essential if we wish to describe the future of the universe.

(I stress that I am just using the Weinberg–Salam model as an example.

I have chosen it because it is familiar and concrete, but in some ways it is a bad choice for our purposes. Firstly, the model involves, not one scalar field, but many scalar and vector fields. Secondly, the vacuum stability features I have described are not properties of the classical potential, $U(\phi)$, but require consideration of one-loop corrections. Thus the formalism I am going to develop is not applicable to this case. As long as we are talking about this model, though, you might be tempted to consider the possibility that the Higgs mass is less than Weinberg's lower bound, that we are living in the false vacuum. As Linde[41] has pointed out, this is silly; if this were the case, there would be no way for the universe to get into the false vacuum in the first place.)

The relevant parameter for cosmology is that cosmic time for which the product of Γ/V and the volume of the past light cone is of order unity. If this time is on the order of microseconds, the universe is still hot when the false vacuum decays, even on the scale of high-energy physics, and a zero-temperature computation of Γ/V is inapplicable. If this time is on the order of years, the decay of the false vacuum will lead to a sort of secondary big bang, with interesting cosmological consequences. If this time is on the order of billions of years, we have occasion for anxiety.

6.2 *The bounce*

We know from Sect. 2.4 how to compute Γ/V. We must find the bounce, $\bar{\phi}$, a solution of the Euclidean equations of motion,

$$\partial_\mu \partial_\mu \bar{\phi} = U'(\bar{\phi}), \tag{6.2}$$

that goes from the false ground state at time minus infinity to the false ground state at time plus infinity,

$$\lim_{x_4 \to \pm\infty} \bar{\phi}(\mathbf{x}, x_4) = \phi_+. \tag{6.3}$$

To these boundary conditions we can add another. It is easy to see that if the action of the bounce is to be finite,

$$\lim_{|\mathbf{x}| \to \infty} \bar{\phi}(\mathbf{x}, x_4) = \phi_+. \tag{6.4}$$

Once we have found the bounce, it is trivial to compute Γ/V. To leading order in $\hbar$,

$$\Gamma/V = K e^{-S_0}, \tag{6.5}$$

where S_0 is $S(\bar{\phi})$ and K is a determinantal factor, defined as in Sect. 2.4.

I will shortly construct the bounce. Before I do so, though, I want to make some comments:

(1) We already see the power of our method. The problem of barrier penetration in a system with an infinite number of degrees of freedom has

been reduced to a study of the properties of a single classical partial differential equation.

(2) The factor of V in the expression for Γ arises automatically in our method. No non-trivial solution of Eqs. (6.2)–(6.4) is translation invariant. Thus we must integrate over the location of the bounce. This gives us a factor of V, just as did the integration over instanton location in Sect. 3.

(3) It might be that there are many solutions to Eqs. (6.2)–(6.4). We are only interested in the solutions of minimum action, for these make the dominant contribution to the functional integral.

(4) We are not interested in the trivial solution, $\phi=\phi_+$. For this solution, $\delta^2 S/\delta\phi^2$ has no negative eigenvalues, and thus makes no contribution to the vacuum decay probability.

(5) If we imbed $\bar{\phi}$ in a one-parameter family of functions,

$$\phi_\lambda(x)=\bar{\phi}(x/\lambda), \tag{6.6}$$

then,

$$S(\phi_\lambda)=\tfrac{1}{2}\lambda^2\int d^4x(\partial_\mu\bar{\phi})^2+\lambda^4\int d^4x\, U(\bar{\phi}). \tag{6.7}$$

Because $\bar{\phi}$ is a solution of the equations of motion, this must be stationary at $\lambda=1$. Thus,

$$\int d^4x(\partial_\mu\bar{\phi})^2=-4\int d^4x\, U(\bar{\phi}), \tag{6.8}$$

and

$$S_0=\frac{1}{4}\int d^4x(\partial_\mu\bar{\phi})^2>0. \tag{6.9}$$

This is reassuring. Since U is somewhere negative, one might worry about the possibility that S_0 was negative, which would lead to a very strange dependence of the decay probability on $\hbar$. This possibility has now been eliminated. Also,

$$d^2S/d\lambda^2=-\frac{1}{2}\int d^4x(\partial_\mu\bar{\phi})^2<0. \tag{6.10}$$

Thus, at $\bar{\phi}$, $\delta^2 S/\delta\phi^2$ has at least one negative eigenvalue, and $\bar{\phi}$ does contribute to the decay probability. Of course, if there were more than one negative eigenvalue, we would have to rethink the analysis of Sect. 2.4. However, as I shall show eventually, this does not happen; there is only one negative eigenvalue.

Now for the construction of the bounce: Eqs. (6.2)–(6.4) are O(4) invariant. Thus it is not unreasonable to guess that the bounce might also be O(4) invariant, that is to say, that $\bar{\phi}$ might depend only on the distance from some point in Euclidean space. Recently, Glaser, Martin, and I were able to show that this guess is right, under mild conditions on U; there always exists an O(4)-invariant bounce and it always has strictly lower action than any O(4)-noninvariant bounce.[42] The rigor of our proof is matched only by its tedium; I would not lecture on it to my worst enemy. However, it is possible to give a sloppy argument for the first part (existence) although, unfortunately, not for the second (action minimization).

I will now give this argument.

If we choose the center of symmetry to be the origin of coordinates, then O(4) symmetry is the statement that $\bar{\phi}$ is a function only of the radial variable, r. Thus Eq. (6.2) becomes

$$\frac{d^2\bar{\phi}}{dr^2}+\frac{3}{r}\frac{d\bar{\phi}}{dr}=U'(\bar{\phi}), \tag{6.11}$$

while Eqs. (6.3) and (6.4) both become

$$\lim_{r\to\infty}\bar{\phi}(r)=\phi_+. \tag{6.12}$$

Also,

$$\left.\frac{d\bar{\phi}}{dr}\right|_{r=0}=0. \tag{6.13}$$

Otherwise, $\bar{\phi}$ would be singular at the origin.

The key to the argument is the observation that if we interpret $\bar{\phi}$ as a particle position and r as time, Eq. (3.9) is the mechanical equation for a particle moving in a potential *minus* U and subject to a somewhat peculiar viscous damping force with Stokes's law coefficient inversely proportional to the time. The particle is released at rest at time zero, Eq. (6.13); we wish to show that if the initial position is properly chosen, the particle will come to rest at time infinity at ϕ_+, that is to say, on top of the right-hand hill in Fig. 16.

I shall demonstrate this by showing that if the particle is released to the right of ϕ_-, and is sufficiently close to ϕ_-, it will overshoot and pass ϕ_+ at some finite time. On the other hand, if it is released sufficiently far

Fig. 16

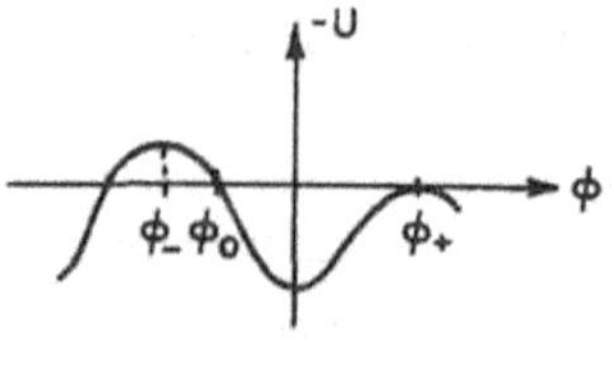

to the right of ϕ_-, it will undershoot and never reach ϕ_+. Thus (arguing in the worst tradition of nineteenth century British mathematics) by continuity there must be an intermediate initial position for which it just comes to rest at ϕ_+.

To demonstrate undershoot is trivial. If the particle is released to the right of ϕ_0, it does not have enough energy to climb the hill to ϕ_+. The damping force does not affect this argument, because viscous damping always diminishes the energy.

To demonstrate overshoot requires a little more work. For ϕ very close to ϕ_-, we may safely linearize Eq. (6.11),

$$\left(\frac{d^2}{dr^2}+\frac{3}{r}-\mu^2\right)(\bar{\phi}-\phi_-)=0, \tag{6.14}$$

where μ^2 is $U''(\phi_-)$. The solution to Eq. (6.14) is

$$\bar{\phi}-\phi_-=2[\bar{\phi}(0)-\phi_-]I_1(\mu r)/\mu r. \tag{6.15}$$

Thus, if we choose ϕ to be initially sufficiently close to ϕ_-, we can arrange for it to stay arbitrarily close to ϕ_- for arbitrarily large r. But for sufficiently large r, the viscous damping force can be neglected, since it is inversely proportional to r. But if we neglect viscous damping, the particle overshoots. Q.E.D.

We have made great progress. We have reduced the partial differential equation for the bounce to an ordinary differential equation. But we can go even farther; in the limit of small energy-density difference between the true and false vacuum, we can obtain an explicit expression for the bounce and for S_0, as I shall now show.

6.3 *The thin-wall approximation*

Let $U_+(\phi)$ be an even function of ϕ,

$$U_+(\phi)=U_+(-\phi), \tag{6.16}$$

with minima at some points $\pm a$,

$$U'_+(\pm a)=0. \tag{6.17}$$

Also, let us define

$$\mu^2=U''_+(\pm a). \tag{6.18}$$

Now let us add to U_+ a small term that breaks the symmetry,

$$U=U_+ +\varepsilon(\phi-a)/2a, \tag{6.19}$$

where ε is a positive number. This defines a theory of the sort we have been discussing. To lowest non-trivial order in ε,

$$\phi_\pm=\pm a, \tag{6.20}$$

and ε is the energy-density difference between the true and the false vacua.

from the mechanical analogy of Sect. 6.2. In order not to lose too much energy, we must choose $\phi(0)$, the initial position of the particle, very close to ϕ_-. The particle then stays close to ϕ_- until some very large time, $r=R$. Near time R, the particle moves quickly through the valley in Fig. 16, and slowly comes to rest at ϕ_+ at time infinity. Translating from the mechanical analogy back into field theory, the bounce looks like a large four-dimensional spherical bubble of radius R, with a thin wall separating the false vacuum without from the true vacuum within.

To go on, we need more information about the wall of the bubble. For r near R, we can neglect the viscous damping term and we can also neglect the ε-dependent term in U. We thus obtain

$$d^2\bar{\phi}/dr^2 = U'_+(\bar{\phi}). \tag{6.21}$$

This is the classical equation of motion for a particle in a symmetric double-welled potential, the equation we studied in Sect. 2.2, the equation that had one-dimensional instantons for its solutions. Indeed, a one-dimensional instanton centered at R is the solution we need here, for such a function goes from $-a$ to a as r increases through R, just what we want. This is our approximate description of the bounce.

The only thing missing from this description is the value of R. This is easily obtained by a variational computation:

$$S = 2\pi^2 \int_0^\infty r^3\, dr[\tfrac{1}{2}(d\bar{\phi}/dr)^2 + U]. \tag{6.22}$$

We can divide this integral into three regions: the outside of the bubble, the skin of the bubble, and the inside of the bubble. Within the accuracy of our approximation, in the outside region, $\phi=\phi_+$ and $U=0$; thus we get no contribution from this part of the integral. In the inside region, $\phi=\phi_-$ and $U=-\varepsilon$; thus from this part of the integral we get

$$-\tfrac{1}{2}\pi^2 R^4 \varepsilon. \tag{6.23}$$

Over the skin, r is approximately R, and, over this small region, the ε-dependent terms in U are negligible; thus from this part of the integral we get

$$2\pi^2 R^3 \int dr[\tfrac{1}{2}(d\bar{\phi}/dr)^2 + U_+] = 2\pi^2 R^3 S_1, \tag{6.24}$$

where S_1 is the action of a one-dimensional instanton,

$$S_1 = \int_{-a}^{a} (2U_+)^{\frac{1}{2}}\, d\phi. \tag{6.25}$$

Varying with respect to R, we find

$$dS/dR = 0 = -2\pi^2 R^3 \varepsilon + 6\pi^2 R^2 S_1. \tag{6.27}$$

Hence,

$$R = 3S_1/\varepsilon. \tag{6.28}$$

This completes the approximate description of the bounce. We also know S_0:

$$S_0 = 27\pi^2 S_1^4/2\varepsilon^3. \tag{6.29}$$

I have described what we have done as an approximation that is valid in the limit of small ε. Now that we have gone through the computation, we can phrase the condition for the validity of the approximation more precisely: the approximation is good if the radius of the bubble is much larger than the thickness of the bubble wall; R must be much larger than $1/\mu$, or, equivalently,

$$3S_1\mu \gg \varepsilon. \tag{6.30}$$

6.4 *The fate of the false vacuum*

In a particle problem like that of Sect. 2.4, we can describe the decay process in the language of the old quantum theory. The particle sits at the bottom of the potential well until, at some random time, it makes a quantum jump to the other side of the barrier, materializing at the point labeled σ in Fig. 7. At this point, the potential energy of the particle is the same as it was at the bottom of the well; thus its kinetic energy must vanish; equivalently, it has zero velocity. These conditions give the initial-value data for the subsequent motion of the particle, which is totally governed by classical mechanics. Like all descriptions of quantum-mechanical processes in the language of the old quantum theory, this one must be taken with a large grain of salt; it will certainly lead us astray if we try to use it to describe measurements made just outside the potential barrier. Nevertheless, it is very useful as an asymptotic description, for discussing what happens far from the barrier and long after the time the system decays. For example, this is the description we all use when we discuss the macroscopic detection of an alpha particle emitted by an unstable nucleus.

This description can readily be extended to a system with many degrees of freedom. The point σ becomes the point in multi-dimensional configuration space where all velocities vanish; that is to say, it is the midpoint of the bounce. Thus, for the field theory we have been studying, the description of the vacuum decay process in the language of the old quantum theory is: the classical field makes a quantum jump (say at time zero) to the state defined by

$$\phi(x_0=0, \mathbf{x}) = \bar{\phi}(\mathbf{x}, x_4=0), \tag{6.31a}$$

and

$$\partial_0 \phi(x_0=0, \mathbf{x}) = 0. \tag{6.31b}$$

Afterwards, it evolves according to the classical Minkowskian field equation,

$$(\nabla^2 - \partial_0^2)\phi = U'(\phi). \tag{6.32}$$

The first of these equations implies that the same function, $\bar{\phi}(r)$, that gives the shape of the bounce in four-dimensional Euclidean space also gives the shape of the bubble at the moment of its materialization in ordinary three-space. Indeed, it does more; because the Minkowskian field equation is simply the analytic continuation of the Euclidean field equation back to real time, the desired solution of Eqs. (6.31) and (6.32) is simply the analytic continuation of the bounce:

$$\phi(x_0, \mathbf{x}) = \bar{\phi}(r = [|\mathbf{x}|^2 - x_0^2]^{\frac{1}{2}}). \tag{6.33}$$

(As a consequence of Eq. (6.13), $\bar{\phi}$ is an even function of r, so we need not worry about which branch of the square root to take.)

We can immediately draw some very interesting consequences of Eq. (6.33):

(1) O(4) invariance of the bounce becomes O(3, 1) invariance of the solution of the classical field equations. In other words, the growth of the bubble, after its materialization, looks the same to any Lorentz observer.

(2) In the case of small ε, discussed in Sect. 6.3, there is a thin wall, localized at $r=R$, separating true vacuum from false. As the bubble expands, this wall traces out the hyperboloid

$$|\mathbf{x}|^2 - x_0^2 = R^2. \tag{6.34}$$

Typically, we would expect R to be a microphysical number, on the order of a fermi, give or take ten orders of magnitude. This means that by macrophysical standards, once the bubble materializes it begins to expand almost instantly with almost the velocity of light.

(3) As a consequence of this rapid expansion, if a bubble were expanding toward us at this moment, we would have essentially no warning of its approach until its arrival. This is shown graphically in Fig. 17. The heavy curve is the bubble wall, Eq. (6.34). A stationary observer, O, cannot tell a bubble has formed until he intercepts the future light cone, W, projected from the wall at the time of its formation. A time R later, that is to say, on the order of 10^{-10}–10^{-30} sec later, he is inside the bubble and dead. (In the true vacuum, the constants of nature, the masses and couplings of

Fig. 17

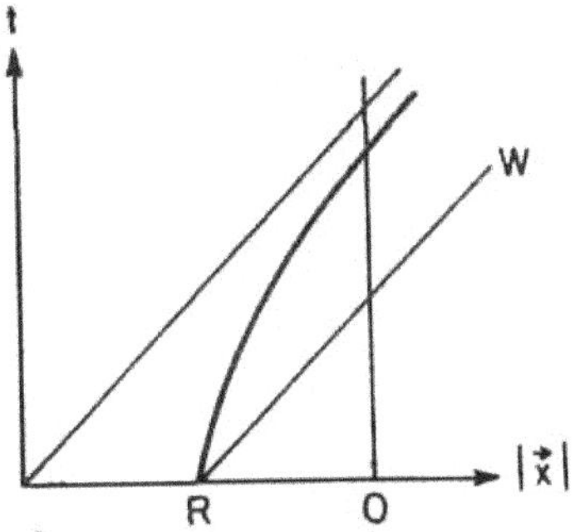

the elementary particles, are all different from what they were in the false vacuum, and thus the observer is no longer capable of functioning biologically, or even chemically.) Since even 10^{-10} sec is considerably less than the response time of a single neuron, there is literally nothing to worry about; if a bubble is coming toward us, we shall never know what hit us.

(4) The rapidly expanding bubble wall obviously carries a lot of energy. How much? A section of bubble wall at rest carries energy S_1 per unit area. Because any part of the bubble wall at any time is obtained from any other part by a Lorentz transformation, a section of wall expanding with velocity v carries energy $S_1/(1-v^2)^{\frac{1}{2}}$ per unit area. Thus, at a time when the radius of the bubble is $|\mathbf{x}|$, the energy of the wall is

$$E_{\text{wall}} = 4\pi|\mathbf{x}|^2 S_1/(1-v^2)^{\frac{1}{2}}. \qquad (6.35)$$

By Eq. (6.34),

$$v = \mathrm{d}|\mathbf{x}|/\mathrm{d}t = (1 - R^2/|\mathbf{x}|^2)^{\frac{1}{2}}. \qquad (6.36)$$

Thus,

$$E_{\text{wall}} = 4\pi|\mathbf{x}|^3 S_1/R = 4\pi\varepsilon|\mathbf{x}|^3/3. \qquad (6.37)$$

Thus, in the thin-wall approximation, all the energy released by converting false vacuum to true goes to accelerate the bubble wall. This refutes the naive expectation that the decay of the false vacuum would leave behind it a roiling sea of mesons. In fact, the expansion of the bubble leaves behind only the true vacuum.

6.5 *Determinants and renormalization*

I said earlier that the determinantal factor K in Eq. (6.5) was defined as in the particle problem of Sect. 2.4. This is basically true, but there are three technical differences. (1) In particle mechanics, we had only one infinitesimal translation, and thus one zero eigenvalue, to worry about; here we have four. (2) It was critical in the analysis of Sect. 2.4 that the second variational derivative of the action at the bounce had one and only one negative eigenvalue. Is the same true here? (3) Whenever we study a relativistic field theory, we must deal with ultraviolet divergences and renormalization. Of course, this last remark also applied to the gauge field theories of Sect. 3, where I swept renormalization problems under the rug. However, we now have a problem with a much simpler renormalization structure (only a single scalar field to worry about, no problems with gauge invariance and gauge-fixing terms, etc.), so it is worth confronting renormalization head-on.

I will deal with these three problems in the order in which I have stated them.

(*1*) *Vanishing eigenvalues.* Because we have four infinitesimal translations, we have four eigenfunctions with eigenvalue zero, proportional to $\partial_\mu \bar{\phi}$. We must determine the constant of proportionality, that is to say, the normalization of the eigenfunctions. This is easy to do. By the spherical symmetry of the bounce,

$$\begin{aligned} \int d^4x\, \partial_\mu \bar{\phi}\, \partial_\nu \bar{\phi} &= \tfrac{1}{4}\delta_{\mu\nu} \int \partial_\lambda \bar{\phi}\, \partial_\lambda \bar{\phi} \\ &= \delta_{\mu\nu} S_0 , \end{aligned} \tag{6.38}$$

by Eq. (6.9).

Thus, as far as zero eigenvalues go, the only difference between the problem at hand and the particle problem of Sect. 2.4 is that we have four factors of $(S_0/2\pi)^{\frac{1}{2}}$ rather than one. Hence,

$$K = \frac{S_0^2}{4\pi^2} \left| \frac{\det'[-\partial_\mu \partial_\mu + U''(\bar{\phi})]}{\det[-\partial_\mu \partial_\mu + U''(\phi_+)]} \right|^{-\frac{1}{2}}, \tag{6.39}$$

assuming we have no problems with negative eigenvalues or renormalization.

(2) *Negative eigenvalues.* We already know that $\delta^2 S/\delta\phi^2$ evaluated at the bounce has at least one negative eigenvalue. Can there be more than one? To answer this question I will have to steal some information from the paper by Glaser, Martin, and me that I referred to earlier.[42] There we showed that the bounce could be characterized as the absolute minimum of S for fixed

$$V = \int d^4x\, U . \tag{6.40}$$

This implies that there can not be two independent eigenvectors with negative eigenvalues; for, if there were, we could form a linear combination of the eigenvectors tangent to the surface of constant V, and the bounce would not even be a local minimum of S with fixed V, let alone an absolute minimum.

(3) *Renormalization.* Until now all of our dynamics has been expressed in terms of unrenormalized quantities. We must now recast our formulae in terms of renormalized quantities. We begin with S itself,

$$S = S_R + \sum_{n=1}^{\infty} S^{(n)}. \tag{6.41}$$

Here S_R is the renormalized action, a functional of exactly the same form as S, but with all unrenormalized quantities replaced by their renormalized counterparts, and $S^{(n)}$ is the action induced by standard renormalization counterterms computed from the sum of all n-loop graphs. To avoid excessive clutter in my equations, I will redefine ϕ to be the renormalized field, U to be the polynomial that occurs in S_R, $\bar{\phi}$ to be the bounce as computed from S_R, and S_0 to be $S_R(\bar{\phi})$.

The renormalization counterterms serve to remove all ultraviolet divergences from all one-particle irreducible Green's functions. Equivalently, they serve to remove all ultraviolet divergences from the effective action, $\gamma(\phi)$, the generating functional of these Green's functions. To one-loop order,[43]

$$\exp \gamma(\phi) = \exp[S_R(\phi) + S^{(1)}(\phi)] \det[-\partial_\mu\partial_\mu + U''(\phi)]^{\frac{1}{2}}. \tag{6.42}$$

It will be important to us shortly that (for renormalizable Us) the right-hand side of this equation is free of ultraviolet divergences for arbitrary ϕ.

Now let us imagine computing Γ/V iteratively, first treating S_R as if it were the total action, and then taking account of the renormalization counterterms perturbatively. If we had not set $\hbar$ equal to one, $S^{(n)}$ would have been proportional to $\hbar^n$. Thus, to the order in which we are working, the only counterterm we need consider is $S^{(1)}$

The first thing we must realize is that the counterterms may destroy our convention that $S(\phi_+)$ vanishes. We can take care of this trivially by replacing S_0 in Eq. (6.5) by the difference $S_0 - S(\phi_+)$.

Secondly, adding new terms to S_R will change the stationary points of S. In particular, it will change the bounce. Let us write

$$\bar{\phi} \to \bar{\phi} + \Delta\bar{\phi}. \tag{6.43}$$

Then

$$S(\bar{\phi}) \to S_0 + \int d^4x \, \frac{\delta S_R}{\delta \bar{\phi}} \Delta\bar{\phi} + S^{(1)}(\bar{\phi}) + \ldots, \tag{6.44}$$

where the triple dots indicate terms that are negligible in the order in which we are working. The second term vanishes because the bounce is a stationary point of S_R. Thus, for our purposes,

$$S(\bar{\phi}) = S_0 + S^{(1)}(\bar{\phi}). \tag{6.45}$$

By the same reasoning,

$$S(\phi_+) = S^{(1)}(\phi_+). \tag{6.46}$$

Putting all this together, we find

$$\frac{\Gamma}{V} = \frac{S_0^2}{4\pi^2} \exp[-S_0 - S^{(1)}(\bar{\phi}) + S^{(1)}(\phi_+)] \left| \frac{\det'[-\partial_\mu \partial_\mu + U''(\bar{\phi})]}{\det[-\partial_\mu \partial_\mu + U''(\phi_+)]} \right|^{-\frac{1}{2}} \tag{6.47}$$

The point of this exercise is not the simplicity of this formula. Equation (6.47) is an ugly mess, and I know no way of evaluating it for even the simplest theories without using a computer. Rather, the point is that ordinary renormalization works for instanton computations. As a good renormalized expression should be, Eq. (6.47) is free of ultraviolet divergences; each determinant is paired with an exponential of $S^{(1)}$, just as in Eq. (6.42). (That one of the factors is a primed determinant is irrelevant; omitting any finite number of eigenvalues has no effect on the ultraviolet divergence.)

6.6 *Unanswered questions*

This concludes what I know about the fate of the false vacuum. There remain many interesting unanswered questions:

(1) I have discussed the expansion of a bubble of true vacuum into false vacuum. What if the initial state of the world is not the false vacuum, but some state of non-zero particle density built on the false vacuum? What happens when a bubble wall encounters a particle?

(2) I have discussed spontaneous decay of the false vacuum. However, there is also the possibility of induced decay. In particular, in a collision of two particles of very high energy, there might be a non-negligible cross-section for the production of a bubble. How can one estimate this cross-section?

(3) If we assume that the universe starts out in a false vacuum, at some time in its expansion bubbles begin to form. Because the formation of bubbles is totally Lorentz-invariant, the average distance between bubbles at their time of formation must be of the same order of magnitude as the time at which bubbles begin to appear. Because bubble walls expand with the speed of light, after a time interval of the same order of magnitude, bubble walls begin to collide. What happens then? Can such events be accommodated in the history of the early universe?

The preceding paragraphs are taken verbatim from a paper I wrote at the end of 1976. I still do not know the answers to any of these questions; maybe you will be able to do better than I.

Appendix 1: How to compute determinants[44]

We wish to study the equation

$$(-\partial_t^2 + W)\psi = \lambda\psi, \tag{A.1.1}$$

where W is some bounded function of t. Let us define $\psi_\lambda(t)$ as the solution of this equation obeying the boundary conditions

$$\psi_\lambda(-T/2) = 0, \quad \partial_t\psi_\lambda(-T/2) = 1. \tag{A.1.2}$$

The operator $-\partial_t^2 + W$ (acting on the space of functions vanishing at $\pm T/2$) has an eigenvalue, λ_n, if and only if

$$\psi_{\lambda_n}(T/2) = 0. \tag{A.1.3}$$

As in the text, we define

$$\det(-\partial_t^2 + W) = \prod_n \lambda_n. \tag{A.1.4}$$

Now, let $W^{(1)}$ and $W^{(2)}$ be two functions of t, and let $\psi_\lambda^{(1,2)}$ be the associated solutions of Eq. (A.1.1). I will prove that

$$\det\left[\frac{-\partial_t^2 + W^{(1)} - \lambda}{-\partial_t^2 + W^{(2)} - \lambda}\right] = \frac{\psi_\lambda^{(1)}(T/2)}{\psi_\lambda^{(2)}(T/2)}. \tag{A.1.5}$$

Proof. The left-hand side of this formula is a meromorphic function of λ, with a simple zero at each $\lambda_n^{(1)}$ and a simple pole at each $\lambda_n^{(2)}$. By elementary Fredholm theory, it goes to one as λ goes to infinity in any direction except along the positive real axis. The right-hand side is a meromorphic function with exactly the same zeros and poles. By elementary differential-equation theory, it also goes to one in the same limit. Thus the ratio of the two sides is an analytic function of λ that goes to one as λ goes to infinity in any direction except along the positive real axis. That is to say, it is one. Q.E.D.

If we define a quantity N by

$$\frac{\det(-\partial_t^2 + W)}{\psi_0(T/2)} = \pi\hbar N^2, \tag{A.1.6}$$

then, by Eq. (A.1.5), N is independent of W. I will use this expression to define the normalization constant N in the functional integral. (Note that no explicit definition of this quantity was given in the text, so I am perfectly free to define it as I wish here.) Thus we have the desired formula for evaluating Gaussian functional integrals,

$$N[\det(-\partial_t^2 + W)]^{-\frac{1}{2}} = [\pi\hbar\psi_0(T/2)]^{\frac{1}{2}}. \tag{A.1.7}$$

As a specific example, for the harmonic oscillator, $W=\omega^2$,

$$\psi_0=\omega^{-1}\sinh\omega(t+T/2), \tag{A.1.8}$$

from which Eq. (2.16) immediately follows.

Appendix 2: The double well done doubly well[45]

In this appendix I shall show that the formulae derived in the text for the splitting of the ground-state energies in a double-well potential, Eqs. (2.31) and (2.41), are equivalent to the results of ordinary wave mechanics. To do this, I will have to both evaluate the determinants that appear in Eq. (2.41) (using the method of Appendix 1) and do the wave-mechanical computation. To keep my equations as simple as possible, I will choose my units such that $\omega=1$.

Evaluating determinants

We have to evaluate a primed determinant, one with the zero eigenvalue omitted. I will do this by evaluating the full determinant on a finite interval, $[-T/2, T/2]$, dividing this by its smallest eigenvalue, λ_0, and then letting T go to infinity.

Thus we must construct solutions of

$$[-\partial_t^2+U''(\bar{x})]\psi_\lambda=\lambda\psi. \tag{A.2.1}$$

We already know one solution with $\lambda=0$,

$$x_1=S_0^{-\frac{1}{2}}\,d\bar{x}/dt\to Ae^{-|t|},\quad t\to\pm\infty. \tag{A.2.2}$$

The constant A is determined by the integral expression for the instanton, Eq. (2.21),

$$t=\int_0^{\bar{x}} dx(2V)^{-\frac{1}{2}}=-\ln[S_0^{-\frac{1}{2}}A^{-1}(a-\bar{x})]+O(a-\bar{x}). \tag{A.2.3}$$

Equation (A.2.1) must have a second solution with $\lambda=0$, which I denote by y_1. It will be convenient to normalize y_1 such that its Wronskian with x_1 is given by

$$x_1\partial_t y_1-y_1\partial_t x_1=2A^2. \tag{A.2.4}$$

Thus,

$$y_1\to\pm Ae^{|t|},\quad t\to\pm\infty. \tag{A.2.5}$$

We can now construct ψ_0 of Appendix 1. For large T,

$$\psi_0(t)=(2A)^{-1}(e^{T/2}x_1+e^{-T/2}y_1). \tag{A.2.6}$$

Hence,

$$\psi_0(T/2)=1. \tag{A.2.7}$$

This takes care of the determinant. To find the lowest eigenvalue, we must find $\psi_\lambda(t)$ for small λ. This can be done by a standard method: we turn Eq. (A.2.1) into an integral equation and iterate once. This can readily be seen to yield

$$\psi_\lambda(t)=\psi_0(t)-\lambda(2A^2)^{-1}\int_{-T/2}^{t} dt'[y_1(t)x_1(t')-x_1(t)y_1(t')]\psi_0(t'), \quad \text{(A.2.8)}$$

plus terms of order λ^2, which we neglect. By Eq. (A.2.6),

$$\psi_\lambda(T/2)=1-\lambda(4A^2)^{-1}\int_{-T/2}^{T/2} dt[e^T x_1^2-e^{-T}y_1^2]. \quad \text{(A.2.9)}$$

For large T, the second term in this expression is bounded, and thus negligible compared to the first term. Thus, for large T,

$$\psi_\lambda(T/2)=1-\lambda(4A^2)^{-1}e^T, \quad \text{(A.2.10)}$$

because x_1 is properly normalized.

Thus the lowest eigenvalue is given by

$$\lambda_0=(4A^2)e^{-T}, \quad \text{(A.2.11)}$$

and, for large T,

$$\frac{\det'[-\partial_t^2+U''(\bar{x})]}{\det[-\partial_t^2+\omega^2]}=\frac{\psi_0(T/2)}{\lambda_0 e^T/2}=\frac{1}{2A^2}. \quad \text{(A.2.12)}$$

Reassuringly, this is non-zero and T-independent.

Plugging this in to Eqs. (2.31) and (2.41), we find that the lowest energy levels are given by

$$E_\pm=\hbar/2\pm A(\hbar S_0/\pi)^{\frac{1}{2}}e^{-S_0/\hbar}. \quad \text{(A.2.13)}$$

Solving the Schrödinger equation

We wish to study the solutions of

$$-\tfrac{1}{2}\hbar^2\partial_x^2\psi+V\psi=E\psi. \quad \text{(A.2.14)}$$

As long as x is not near the bottoms of the wells, we can use standard WKB solutions. Near the bottom of each well, though, there are two turning points. These are *not* separated by many wavelengths, so we can not use the standard connection formulae for a linear turning point. Fortunately, near the bottom of a well, in a region that includes both turning points, we may safety approximate V by a harmonic-oscillator potential. Thus, for example, for x near a, we may write

$$-\tfrac{1}{2}\hbar^2\partial_x^2\psi+\tfrac{1}{2}(x-a)^2\psi=E\psi. \quad \text{(A.2.15)}$$

Our strategy will be to match WKB solutions of Eq. (A.2.14) outside the wells to solutions of Eq. (A.2.15) in the bottoms of the wells. Furthermore, since we know the solutions are either even or odd, we can restrict ourselves to positive x, and only have to do this awkward matching for the right-hand well.

I will begin by constructing the even and odd WKB solutions for $0 \leqslant x < a$. If we define

$$k(x) = [2(V-E)]^{\frac{1}{2}}, \tag{A.2.16}$$

then these are

$$\psi_{\pm} = k^{-\frac{1}{2}}\left[\exp \hbar^{-1}\int_0^x k\, dx' \pm \exp -\hbar^{-1}\int_0^x k\, dx'\right]. \tag{A.2.17}$$

For the solutions we are interested in, E is itself of order $\hbar$. Thus we may ignore E in the factor of $k^{-\frac{1}{2}}$, and expand to first order in the exponential,

$$k = (2V)^{\frac{1}{2}} - E(2V)^{-\frac{1}{2}}. \tag{A.2.18}$$

To match on to the solutions of Eq. (A.2.15), we need the form of the WKB solutions as x enters the regime of validity of the quadratic approximation to V, $V = (a-x)^2/2$. In this regime $k(x)$ is just $(a-x)$, while we can compute the E-independent term in the integral by

$$\int_0^x dx(2V)^{\frac{1}{2}} = \int_0^a dx(2V)^{\frac{1}{2}} - \int_a^x dx(2V)^{\frac{1}{2}}$$
$$= \tfrac{1}{2}S_0 - \tfrac{1}{2}(a-x)^2. \tag{A.2.19}$$

For the E-dependent term in the integral, we can use Eq. (A.2.3). Thus we obtain

$$\psi_{\pm} = (a-x)^{-\frac{1}{2}}\{\exp \hbar^{-1}[\tfrac{1}{2}S_0 - \tfrac{1}{2}(a-x)^2 + E \ln S_0^{-\frac{1}{2}}A^{-1}(a-x)]$$
$$\pm \exp -\hbar^{-1}[\tfrac{1}{2}S_0 - \tfrac{1}{2}(a-x)^2 + E \ln S_0^{-\frac{1}{2}}A^{-1}(a-x)]\} \tag{A.2.20}$$

If we write

$$E = \hbar(\tfrac{1}{2} + \varepsilon), \tag{A.2.21}$$

then Eq. (A.2.20) becomes

$$\psi_{\pm} = \{e^{S_0/2\hbar}S_0^{-\frac{1}{2}}A^{-\frac{1}{2}} \exp[-(a-x)^2/2\hbar]$$
$$\pm (a-x)^{-1}e^{-S_0/2\hbar}S_0^{\frac{1}{2}}A^{\frac{1}{2}} \exp[(a-x)^2/2\hbar]\}[1 + O(\varepsilon)]. \tag{A.2.22}$$

We will hold this expression in reserve while we go on to study the solutions of Eq. (A.2.15).

We already know one solution of Eq. (A.2.15), for $\varepsilon = 0$,

$$\psi_1 = \exp[-(a-x)^2/2\hbar]. \tag{A.2.23}$$

Of course, there is another (odd, increasing) solution, ϕ_1. This does not have a simple form in terms of elementary functions, but its asymptotic form, for $|x-a| \gg \hbar$, is easily computed by the WKB approximation, or just read off from Eq. (A.2.22),

$$\phi_1 = (a-x)^{-1} \exp[(a-x)^2/2\hbar]. \tag{A.2.24}$$

It will turn out that this is all that we need. Note that I have normalized ϕ_1 such that the Wronskian of the two solutions is

$$\phi_1 \partial_x \psi_1 - \psi_1 \partial_x \phi_1 = 2/\hbar. \tag{A.2.25}$$

We wish to solve Eq. (A.2.15) for small ε. By the same argument as led to Eq. (A.2.8),

$$\psi = \psi_1 - \varepsilon \int_x^\infty \mathrm{d}x' \, \psi_1(x')[\psi_1(x')\phi_1(x) - \phi_1(x')\psi_1(x)]. \tag{A.2.26}$$

I have chosen here the solution that vanishes as x goes to plus infinity. Thus, this is the appropriate solution for matching with the decreasing WKB solution in the region $(x-a) \gg \hbar$. Thus, the only matching left to do is in the region $(a-x) \gg \hbar$.

In this region, we can use

$$\int_{-\infty}^{\infty} \mathrm{d}x \, \psi_1^2 = (\pi\hbar)^{\frac{1}{2}}. \tag{A.2.27}$$

to write

$$\psi = \exp[-(a-x)^2/2\hbar][1 + O(\varepsilon)] - \varepsilon(\pi\hbar)^{\frac{1}{2}}(a-x)^{-1} \exp[(a-x)^2/2\hbar]. \tag{A.2.28}$$

As it should be, this is proportional to Eq. (A.2.22), if we choose

$$\varepsilon = \mathrm{e}^{-S_0/\hbar} A(S_0/\pi\hbar)^{\frac{1}{2}}. \tag{A.2.29}$$

This is the desired result, and it is identical to the result of the dilute-gas approximation, Eq. (A.2.13).

Almost identical methods to these can be used to check the dilute-gas formula for the width of an unstable state, Eq. (2.50). You might find it an instructive exercise to see that things work out in this case also.

Appendix 3: Finite action is zero measure[46]

In this appendix I will show that, even for a one-dimensional harmonic oscillator, motions of finite action form a set of measure zero in function space.

If we define eigenvalues λ_n and expansion coefficients c_n as in Sect. 2.1, then, for a harmonic oscillator, the quadratic approximation to the action is exact,

$$S = \tfrac{1}{2} \sum_n \lambda_n c_n^2. \tag{A.3.1}$$

If we introduce new variables, $b_n = c_n(\lambda_n/\hbar)^{\frac{1}{2}}$, then

$$S = \frac{\hbar}{2}\sum_n b_n^2. \tag{A.3.2}$$

Let us define a slightly unconventional normalization constant, N', by

$$N'[\mathrm{d}x] = \prod_n (2\pi)^{-\frac{1}{2}}\,\mathrm{d}b_n. \tag{A.3.3}$$

This has been chosen such that

$$N'\int [\mathrm{d}x]\mathrm{e}^{-S/\hbar} = 1. \tag{A.3.4}$$

How much of this integral comes from motions of finite action? The integrand is positive, and every motion of finite action lies in a cube of side L

$$|b_n| \leqslant L \quad \text{for all } n, \tag{A.3.5}$$

for sufficiently large L. Thus, the finite-action contribution to the integral must be less than

$$\overline{\lim_{L\to\infty}} \prod_{n=1}^{\infty} (2\pi)^{-\frac{1}{2}} \int_{-L}^{L} \mathrm{d}b_n\,\mathrm{e}^{-b_n^2} = 0. \tag{A.3.6}$$

Q.E.D.

Appendix 4: Only winding number survives

This appendix is the promised (in Sect. 3.3) demonstration that for a sufficiently large box, the only relic of the boundary conditions imposed on the walls of the box is the winding number.

Consider a rectangular box in Euclidean four-space, with sides $L_1 \ldots L_4$. I will label the eight hyperplanes that bound the box by their normal vectors; thus I will refer to the upper 1-wall, the lower 1-wall, the upper 2-wall, etc. (Upper and lower here refer to greater and lesser values of the appropriate coordinate.)

On the walls of the box the tangential components of A_μ are given in a way consistent with finiteness of the action, that is to say, consistent with

$$A_\mu = g\partial_\mu g^{-1}. \tag{A.4.1}$$

Thus, giving the tangential components of A_μ on the walls is equivalent to giving g on the walls (up to an irrelevant multiplicative constant). The gauge condition $A_3 = 0$ still allows arbitrary x_3-independent gauge transformations. I will use the freedom to make such a transformation to transform g to one on the lower 3-wall. Because the vanishing of A_3 implies the vanishing of $\partial_3 g$, g is automatically one on all walls except the upper 3-wall. On this wall, g is given as a function of three variables, $g(x_1, x_2, x_4)$, equal to one on the boundary of the wall. (I stress that the only function of this gauge transformation is to simplify my subsequent arguments. Since the functional integral is gauge-invariant, anything I can prove with this gauge convention I could prove without it; it is just that the arguments would be clumsier.)

Now let us imbed our original box, with boundary conditions given by $g_1(x_1, x_2, x_4)$ in a larger box, with the same lowermost corner (chosen to be the origin of coordinates), and with the same sides L_1, L_2, and L_4, but with third side $L_3+\Delta$. Let the boundary conditions on the larger box be given by some function $g_2(x_1, x_2, x_4)$.

Theorem. If g_1 and g_2 are in the same homotopy class, then any field configuration defined inside the original box consistent with its boundary conditions can be extended to a field configuration defined inside the larger box, consistent with its boundary conditions and the gauge condition $A_3=0$, at the cost of an increase in action of order $1/\Delta$.

Before I prove this theorem I will make some comments:

(1) The theorem would certainly not be true if g_1 and g_2 were in different homotopy classes. In this case, to get from g_1 to g_2, we would have to put at least one instanton in the new volume; this would increase the action by at least $8\pi^2/g^2$, independent of the value of Δ.

(2) We are free to choose Δ to be proportional to, say, $L_3^{\frac{1}{2}}$. Thus, for a very large box, the fractional change in the volume of the box is negligible, as is the change in the action. In the language of statistical physics, changing the boundary conditions while keeping the winding number fixed is just a surface effect, not a volume effect.

(3) There is an apparent paradox that may have bothered you. For any fixed configuration of instantons and anti-instantons, $g(x_1, x_2, x_4)$ is fixed. How then can we get all configurations consistent with a fixed winding number with a single set of boundary conditions? The theorem supplies the answer. We do not get all these configurations; we get only a small portion of them. However, we do get 'close relatives' of all of them, configurations that differ only by a small distortion very close to the upper 3-wall. The difference caused by this small distortion is negligible for a sufficiently large box.

Now for the proof. By assumption, g and g' are in the same homotopy class. Thus there is a continuous function of four variables, $g(x_1, x_2, s, x_4)$, with $0\leqslant s\leqslant 1$, such that

$$g(x_1, x_2, 0, x_4)=g, \quad g(x_1, x_2, 1, x_4)=g_2. \tag{A.4.2}$$

Let $g(x)$ be a function defined in the added volume by

$$g(x)=g(x_1, x_2, (x_3-L_3)/\Delta, x_4). \tag{A.4.3}$$

If we could choose

$$A_\mu = g\partial_\mu g^{-1}, \tag{A.4.4}$$

then we could effect the desired transition at *no* cost in added action. Unfortunately, this is impossible; Eq. (A.4.4) is inconsistent with the gauge condition $A_3=0$. However,

$$\begin{aligned} A_\mu &= g\partial_\mu g^{-1}, \quad \mu\neq 3 \\ &= 0, \qquad\qquad \mu=3, \end{aligned} \tag{A.4.5}$$

is consistent with the gauge condition and will effect the transition.

We must compute the action associated with Eq. (A.4.5). If we make a gauge transformation by g^{-1}, Eq. (A.4.5) becomes

$$\begin{aligned} A_\mu &= 0, \qquad\qquad \mu\neq 3, \\ &= g^{-1}\partial_\mu g, \quad \mu=3. \end{aligned} \tag{A.4.6}$$

(A gauge transformation does not change the action.) From Eq. (A.4.6), we see that A_3 is proportional to $1/\Delta$. The only non-vanishing components of $F_{\mu\nu}$ are $F_{\mu 3}$, also proportional to $1/\Delta$. Thus the Lagrangian density is proportional to $1/\Delta^2$. However, the volume of integration is only proportional to Δ. Q.E.D.

Appendix 5: No wrong-chirality solutions[47]

In this appendix I will show that, if

$$F_{\mu\nu}=\tilde{F}_{\mu\nu}, \tag{A.5.1}$$

then the only normalizable solution of both

$$D_\mu \gamma_\mu \psi = 0, \tag{A.5.2}$$

and

$$\gamma_5 \psi = \psi, \tag{A.5.3}$$

is $\psi=0$.

From Eq. (A.5.2),

$$D_\nu\gamma_\nu D_\mu\gamma_\mu\psi = D_\mu D_\mu\psi + \tfrac{1}{2}F_{\mu\nu}\gamma_\mu\gamma_\nu\psi = 0. \tag{A.5.4}$$

Also,

$$F_{\mu\nu}\gamma_\mu\gamma_\nu\gamma_5 = -\tilde{F}_{\mu\nu}\gamma_\mu\gamma_\nu. \tag{A.5.5}$$

Thus,

$$D_\mu D_\mu \psi = 0. \tag{A.5.6}$$

Multiplying by $\psi^\dagger$ and integrating, we find

$$\int d^4x\, D_\mu\psi^\dagger D_\mu\psi = 0. \qquad \text{(A.5.7)}$$

Hence

$$D_\mu\psi = 0, \qquad \text{(A.5.8)}$$

for all μ. If we go to axial gauge, this implies, in particular, that ψ is independent of x_3. The only such normalizable function is $\psi = 0$. Q.E.D.

Notes and references

1. These topics are all drawn from the classic part of the theory. 'Classic', in this context, means work done more than six months ago. A good summary of the more recent research of one of the most active groups in this field is C. Callan, R. Dashen, and D. Gross, *Phys. Rev.* **D17**, 2717 (1978).
2. Polyakov's early work is summarized in A. M. Polyakov, *Nucl. Phys.* **B121**, 429 (1977).
3. See, for example, R. Feynman and A. Hibbs, *Quantum Mechanics and Path Integrals* (McGraw-Hill, New York, 1965).
4. See the note on notation at the end of Sec. 1.
5. It was Polyakov[2] who recognized the double well as the prototypical instanton problem.
6. For a review of lumps, see Chapter 6 in this volume.
7. This is, of course, nothing but the standard prescription for handling collective coordinates in soliton problems. See J. L. Gervais and B. Sakita, *Phys. Rev.* **D11**, 2943 (1975).
8. The treatment here follows that of C. Callan and S. Coleman, *Phys. Rev.* **D16**, 1762 (1977). The idea of handling unstable states this way goes back to Langer's analysis of the droplet model in statistical mechanics (J. S. Langer, *Ann. Phys.* (N.Y.) **41**, 108 (1967)). The factor of $\frac{1}{2}$, of which much is made below, occurs in Langer's analysis and was explained to me by Michael Peskin.
9. The order of my exposition will not be the historical order of discovery. Here is the way it happened. The topological structure of finite-action Euclidean gauge-field configurations was uncovered and the instanton solutions discovered by A. A. Belavin, A. M. Polyakov, A. S. Schwartz, and Yu. S. Tyupkin, *Phys. Lett.* **59B**, 85 (1975). The importance of the instantons was realized by G. 't Hooft (*Phys. Rev. Lett.* **37**, 8 (1976); *Phys. Rev.* **D14**, 3432 (1976)) who used them to solve the U(1) problem. (I won't get to this until Sect. 5.) 't Hooft's work was clarified and extended by R. Jackiw and C. Rebbi (*Phys. Rev. Lett.* **37**, 172 (1976)) and by C. Callan, R. Dashen, and D. Gross (*Phys. Lett.* **63B**, 334 (1976)), who discovered the properties of pure gauge field theories discussed in this section.
10. For a review of gauge field theories, see Chapter 5 in this volume.
11. And sometimes given by me. I thank Arthur Wightman for awakening me from my dogmatic slumbers.
12. It suffices to assume that the gauge field is without (gauge-invariant) singularities if we make a stereographic projection. of four-space onto a four-sphere. I would love to found the analysis on finiteness of the action, without even this assumption about the behavior of the fields at infinity, but I have not been able to do so.

13. Sign convention. In n-space, $\varepsilon_{1\ldots n}=1$. Symbols with upper indices are defined by raising with the appropriate (Euclidean or Minkowskian) metric.
14. R. Bott, *Bull. Soc. Math. France* **84**, 251 (1956).
15. See Note 10.
16. See Note 6.
17. At least in a box; see the next paragraph.
18. R. Jackiw and C. Rebbi, *Phys. Rev.* **D14**, 517 (1976).
19. Even this is true for axial gauge (in infinite space) only if we add additional gauge conditions.[6]
20. M. Atiyah and R. Ward, *Comm. Math. Phys.* **55**, 117 (1977).
21. Although I have just argued that this knowledge is irrelevant to our immediate purposes, an enormous amount *has* been learned recently about solutions to the Euclidean gauge-field equations. In fact, 'binstantons' do not exist, but $8|\nu|$-parameter families of solutions with winding number ν do. For a review (with references to the original literature) see R. Jackiw, C. Nohl, and C. Rebbi, in *Particles and Fields*, Proceedings of the 1977 Banff Summer Institute, edited D. H. Boal and A. N. Kamal, Plenum (1978).
22. For a review of the renormalization group applied to gauge theories, see Note 10.
23. Done by 't Hooft, a hard worker (second paper cited in Ref. 9). 't Hooft's computation has been somewhat simplified. See A. Belavin and A. M. Polyakov, *Nucl. Phys.* **B123**, 429 (1977); F. Ore, *Phys. Rev.* **D16**, 2577 (1977); S. Chahda, A. D'Adda, P. di Vecchia, and F. Nicodemi, *Phys. Lett.* **72B**, 103 (1977).
24. The analysis reported here is based on C. Callan, R. Dashen, and D. Gross, *Phys. Lett.* **66B**, 375 (1977). The fact that the Abelian Higgs model in two dimensions does not display the Higgs phenomenon was discovered independently by two of my graduate students, Frank De Luccia and Paul Steinhardt. They did not write up their results because I did not believe them. I take this occasion to apologize to them for my stupidity.
25. The problem is identical to that of constructing flux tubes in superconductors. See Ref. 6, and references cited therein.
26. Indeed, θ-vacua, with precisely the same interpretation (but derived in a completely different way), occur in the massive Schwinger model, quantum electrodynamics of charged fermions in $1+1$ dimensions. (See S. Coleman, R. Jackiw, and L. Susskind, *Ann. Phys.* (N.Y.) **93**, 267 (1975), and S. Coleman, *Ann. Phys.* (N.Y.) **101**, 239 (1976).) The arguments that work for the Schwinger model also work for the Higgs model when μ^2 is positive, so we also obtain θ-vacua in this case despite the absence of instantons.
27. K. Wilson, *Phys. Rev.* **D10**, 2445 (1974). The standard expression for W has a factor of $-iq$ where I have one of $-q/e$; the source of the difference is the factor of ie hidden in my definition of A_μ.
28. S. Weinberg, *Phys. Rev.* **D11**, 3583 (1975). This paper, titled 'The U(1) problem', gives a characteristically lucid description of the situation just before 't Hooft's breakthrough. (As a major unanswered question, Weinberg lists 'How does the underlying gluon-gauge invariance enforce the equal coupling of the positive- and negative-metric Goldstone bosons to gauge-invariant operators?')
29. S. L. Adler, *Phys. Rev.* **177**, 2426 (1969). J. S. Bell and R. Jackiw, *Nuovo Cimento* **60**, 47 (1969). W. Bardeen, *Phys. Rev.* **184**, 1848 (1969).
30. J. Kogut and L. Susskind, *Phys. Rev.* **D11**, 3594 (1976).
31. For more details on Fermi integration, see F. A. Berezin, *The Method of Second Quantization* (Academic Press, New York and London, 1966).

32. The easiest way to fix this up is to stereographically project Euclidean four-space onto a four-sphere; $\not{D}$ is then projected into an operator with a pure discrete spectrum. This changes the determinant, but only by a factor that is independent of the gauge field. Since, as we shall see, our final results will only depend on ratios of determinants, this change is irrelevant.
33. To my knowledge, this sum rule was first derived by A. S. Schwarz, *Phys. Lett.* **67B**, 172 (1977). The derivation in the literature closest to the one given here is that of L. Brown, R. Carlitz, and C. Lee, *Phys. Rev.* **D16**, 417 (1977).
34. This section is mainly afterthoughts; I did not know most of these things at the time these lectures were given.
35. A related picture of how instantons break SU(2)⊗SU(2) is advanced by D. Caldi, *Phys. Rev. Lett.* 39, 121 (1977).
36. An (apparently) very different picture of how merons effect confinement has been advanced by J. Glimm and A. Jaffe, *Phys. Rev.* **D18**, 463 (1978). G. 't Hooft has advocated completely different configurations (*Nucl. Phys.* **B138**, 1 (1978)).
37. The preceding paragraph is the product of conversations with Michael Peskin, who has observed that a group of two-dimensional models analyzed by C. Callan, R. Dashen, and D. Gross (*Phys. Rev.* **D16**, 2526 (1977)) display (in a certain sense) a restoration of chiral symmetry at large scales, the first half of the above scenario.
38. R. Crewther, *Phys. Lett.* **70B**, 349 (1977).
39. R. D. Peccei and H. R. Quinn, *Phys. Rev. Lett.* **38**, 1440 (1977); *Phys. Rev.* **D16**, 1791 (1977). F. Wilczek, *Phys. Rev. Lett.* **40**, 279 (1978). S. Weinberg, *Phys. Rev. Lett.* **40**, 223 (1978) and 'Instantons without axions' (unpublished).
40. These ideas are the product of discussions with S. Glashow and D. Nanopoulos.
41. The key paper on this subject is M. B. Voloshin, I. Yu. Kobzarev, and L. B. Okun, *Yad. Fiz.* **20**, 1229 (1974) (*Sov. J. Nucl. Phys.* **20**, 644 (1975)). The instanton approach to the problem was developed in S. Coleman, *Phys. Rev.* **D15**, 2929 (1977), and C. Callan and S. Coleman.[8] (Large portions of the text of this section are plagiarized from these two papers.) Similar ideas were developed independently by M. Stone, *Phys. Rev.* **D14**, 3568 (1976) and *Phys. Lett.* **67B**, 186 (1977). P. Frampton was the first to study these phenomena in the Weinberg–Salam model (*Phys. Rev. Lett.* **37**, 1378 (1976)); however, Frampton's conclusions have been criticized severely (and, I think, correctly) by A. Linde (*Phys. Lett.* **70B**, 306 (1977); **92B**, 119 (1980); *Rept. Prog. Phys.* **42**, 389 (1979).
42. S. Coleman, V. Glaser, and A. Martin, *Comm. Math. Phys.* **58**, 211 (1978).
43. See, for example, R. Jackiw, *Phys. Rev.* **D9**, 1686 (1974).
44. Formulae related to the one developed here can be found throughout the literature. Two references out of many: J. H. Van Vleck, *Proc. Nat. Acad. Sci.* **14**, 178 (1928). R. Dashen, B. Hasslacher, and A. Neveu, *Phys. Rev.* **D10**, 4114 (1974). The derivation given here was developed in conversations with Ian Affleck.
45. This appendix reports on computations done with C. Callan. A somewhat different attack on the problem (with the same conclusions) is E. Gildener and A. Patrascioiu, *Phys. Rev.* **D16**, 423 (1977).
46. I give no reference not because these results are novel but because they are a standard part of the theory of Weiner integrals.
47. This appendix is a transcription of an argument of Brown *et al.*[33]

ABC of instantons

A. I. Vaĭnshteĭn, V. I. Zakharov, V. A. Novikov, and M. A. Shifman

Institute of Theoretical and Experimental Physics, Moscow and Institute of Nuclear Physics, Siberian Branch, USSR Academy of Sciences, Novosibirsk

Usp. Fiz. Nauk 136, 553–591 (April 1982)

An attempt is made to present an instanton "calculus" in a relatively simple form. The physical meaning of instantons is explained by the example of the quantum-mechanical problem of energy levels in a two-humped potential. The nonstandard solution to this problem based on instantons is analyzed, and the reader is acquainted with the main technical elements used in this approach. Instantons in quantum chromodynamics are then considered. The Euclidean formulation of the theory is described. Classical solutions of the field equations (the Belavin-Polyakov-Shvarts-Tyupkin instantons) are obtained explicitly and their properties are studied. The calculation of the instanton density is described and the complete result is given for an arbitrary number of colors. The effects associated with fermion fields are briefly described.

PACS numbers: 03.70. + k, 11.10. – z, 12.35.Cn

CONTENTS

It appears that all fundamental interactions in nature are of the gauge type. The modern theory of hadrons—quantum chromodynamics (QCD)—is no exception. It is based on local gauge invariance with respect to the color group SU(3), which is realized by an octuplet of massless gluons. The idea of gauge invariance, however, is much older and derives from quantum electrodynamics, which was historically the first field model in which successful predictions were obtained. By the end of the forties, theoreticians had already learned how to calculate all observable quantities in electrodynamics in the form of series in $\alpha = 1/137$. The first steps in QCD at the end of the seventies were also made in the framework of perturbation theory. However, it gradually became clear that, in contrast to electrodynamics, quark-gluon physics is not exhausted by perturbation theory. The most interesting phenomena—the confinement of colored objects and the formation of the hadron spectrum—are associated with nonperturbative (i.e., not describable in the framework of perturbation theory) effects, or rather, with the complicated structure of the QCD vacuum, which is filled with fluctuations of the gluon field.

It is now clear that the construction of the complete "wave function" of the vacuum is a very difficult problem. It still remains unsolved, despite numerous attacks by theoreticians. Nevertheless, quite a lot is already known. Study of the "old," traditional hadrons gives information about the fundamental properties of the vacuum. In turn, having obtained this information, we can make a number of nontrivial predictions about gluonium and other such poorly investigated aspects of hadron phenomenology.

The corresponding approach has been developed by the authors over a number of years, but it will not be discussed here. We note only that the main element is the introduction of several vacuum expectation values. For example, the intensity of gluon fields in vacuum is obviously measured by the quantity

$$\langle 0 \mid G^a_{\mu\nu} G^a_{\mu\nu} \mid 0 \rangle,$$

where $G^a_{\mu\nu}$ is the tensor of the intensity of the gluon field ($a = 1, \ldots, 8$ is the color index). Similarly, the quark condensate expectation value $\langle 0 | \bar{q} q | 0 \rangle$ serves as a measure of the quark fields.

In a "final theory," if such is constructed, it will be possible to calculate all phenomenological matrix elements on the basis of the Lagrangian of QCD. It can

 0038-5670/82/040195-21$01.80

already be said that this will require knowledge of nonperturbative fluctuations in the physical vacuum. Here, phenomenology makes contact with the purely theoretical development, which as yet has not had great applications, though it has made it possible to reexamine a number of problems.

In 1975, there was discovered one of the most beautiful phenomena in quantum chromodynamics. We are referring to instantons—classical solutions of the field equations with nontrivial topology. The beauty of the theoretical constructions has attracted the interest of many physicists and mathematicians, and it is difficult to overestimate the popularity of instantons. The importance of instantons as the first example of fluctuations of the gluon field not encompassed by perturbation theory is in no doubt. Therefore, although one can hardly speak of any practical fruits, it appears appropriate to explain the physical essen . of the phenomenon and derive the basic formulas to enable the reader to find his (or her) way about the literature.

One of the main conclusions which we shall attempt to establish is that the original Belavin-Polyakov-Shvarts-Tyupkin solution[1] (BPST instanton) is not the fluctuation which is dominant in the vacuum wave function. However, there is no danger of its beauty being wasted. In one form or another, it will certainly play a part in the future theory of strong interactions.

We begin with a simple quantum-mechanical problem that illustrates the role of nonperturbative fluctuations. This example was analyzed in detail by Polyakov,[2] who made a major contribution to the development of the entire subject. Having studied the main technical elements, we then turn to a more complicated case—quantum chromodynamics. At the very end, we discuss the question of the importance of the BPST instanton in the real world.

1. QUANTUM MECHANICS, IMAGINARY TIME, PATH INTEGRALS

In this section, we consider the problem of the one-dimensional motion of a spinless particle in a potential $V(x)$. This problem is usually treated in all textbooks on quantum mechanics, but we shall use a somewhat unusual method to solve it. The reader may find it inconvenient, just as sum rules are "inconvenient" for finding the eigenvalues of a Schrödinger equation. But—and this is the most important property—the method can be directly generalized to field theory.

If we take the mass of the particle equal to unity, $m=1$, then the Lagrangian of the system has the simple form

$$\mathcal{L}=\frac{1}{2}\left(\frac{dx}{dt}\right)^2-V(x). \tag{1}$$

Suppose that the particle at the initial time $(-t_0/2)$ is at the point x_i and at the final time $(+t_0/2)$ at the point x_f. An elegant method of expressing the amplitude of such a process was invented by Feynman (see the book of Ref. 3). The prescription is that the amplitude is equal to the sum over *all* paths joining the world points $(-t_0/2,x_i)$ and $(t_0/2,x_f)$ taken with weight

$$e^{i(\text{action})}.$$

The action, which we shall in what follows denote by *the letter S, is related to the Lagrangian by*

$$S=\int_{-t_0/2}^{t_0/2}dt\,\mathcal{L}(x,\dot{x}). \tag{2}$$

Thus, the transition amplitude is

$$\langle x_f|e^{-iHt_0}|x_i\rangle=N\int[Dx]\,e^{iS[x(t)]}, \tag{3}$$

where H is the Hamiltonian and $\exp(-iHt_0)$ is the ordinary evolution operator of the system. The factor N on the right-hand side is a normalization factor, to the discussion of which we shall return below. $[Dx]$ denotes integration over all functions $x(t)$ with boundary conditions $x(-t_0/2)=x_i$ and $x(t_0/2)=x_f$.

Before we consider dynamical questions, we examine the left-hand side. If we go over from states with a definite coordinate to states with a definite energy,

$$H|n\rangle=E_n|n\rangle,$$

then, obviously,

$$\langle x_f|e^{-iHt_0}|x_i\rangle=\sum_n e^{-iE_nt_0}\langle x_f|n\rangle\langle n|x_i\rangle, \tag{4}$$

and we obtain a sum of oscillating exponentials. If we *are interested in the ground state (and in field theory* we are always interested in the lowest state—the vacuum), it is much more convenient to transform the oscillating exponentials into decreasing exponentials. To this end, we make the substitution $t\to-i\tau$. Then in the limit $\tau_0\to\infty$ only a single term survives in the sum (4), and this directly tells us what are the energy E_0 and the wave function $\psi_0(x)$ of the lowest level $e^{-E_0\tau_0}\psi_0(x_f)\psi_0^*(x_i)$.

In the literature, the transition to an imaginary time is frequently called the Wick rotation, and the corresponding variant of the theory of Euclidean variant. Below, we shall see that the substitution $t\to-i\tau$ is in a certain sense not only a matter of convenience, since it gives a new language for describing a very important aspect of the theory.

We now turn to the right-hand side of Eq. (3). In the Euclidean formulation, the action takes the form

$$iS[x(t)]\to\int_{-\tau_0/2}^{\tau_0/2}\left[-\frac{1}{2}\left(\frac{dx}{d\tau}\right)^2-V(x)\right]d\tau, \tag{5}$$

where we assume the boundary condition $x(-t_0/2)=x_i$, $x(\tau_0/2)=x_f$, and the origin of the energy is chosen such that $\min V(x)=0$.

We call

$$S_E=\int_{-\tau_0/2}^{\tau_0/2}\left[\frac{1}{2}\left(\frac{dx}{d\tau}\right)^2+V(x)\right]d\tau \tag{6}$$

the Euclidean action. Since $S_E\geq 0$, we have acquired an exponentially decreasing weight on the right-hand side of Eq. (3). In the present review, we shall remain in the Euclidean space and shall not return to the Minkowski space (i.e., *to a real time*); *therefore, in all* that follows we shall omit the subscript E.

The Euclidean variant of (3) is

$$\langle x_f | e^{-H\tau_0} | x_i \rangle = N \int [Dx]\, e^{-S}. \tag{7}$$

It is now time to make the next important step and explain what integration over all paths actually means. Let $X(\tau)$ be some function satisfying the boundary conditions. Then an *arbitrary* function with the same boundary conditions can be represented in the form

$$x(\tau) = X(\tau) + \sum_n c_n x_n(\tau), \tag{8}$$

where $x_n(\tau)$ is a complete set of orthonormal functions that vanish at the boundary:

$$\int_{-\tau_0/2}^{\tau_0/2} d\tau\, x_n(\tau)\, x_m(\tau) = \delta_{nm}, \quad x_n\left(\pm \frac{\tau_0}{2}\right) = 0.$$

The measure $[Dx]$ can be chosen in the form

$$[Dx] = \prod_n \frac{dc_n}{\sqrt{2\pi}}. \tag{9}$$

The coefficient of proportionality in this relation does not in general have in itself a particular meaning until the normalization factor N has been fixed.

Now suppose that in the problem under consideration the characteristic value of the action is large for certain reasons. Well known is the situation when the quasiclassical approximation, or, which is the same thing, the method of steepest descent (the latter, "mathematical" term may be more readily understood by some of the readers), "works." In other words, the entire integral in (7) is accumulated from regions near the extremum (minimum) of S. The path corresponding to the least action, which we denote by $X(\tau)$, is known in the literature as an extremal path, an extremal, or a stationary point. If there is one extremal and $S[X(\tau)] = S_0$, then

$$N \int [Dx]\, e^{-S} \sim e^{-S_0}. \tag{10}$$

Thus, to find the principal, *exponential factor* in the result, it is sufficient to put in information about a single, extremal path. (If there are several stationary points, we have in general the sum of the contributions of all the stationary points.)

There exists a standard procedure which enables us to take the next step and fix the pre-exponential factor. This operation is already somewhat more laborious. Suppose for simplicity that there is a single stationary point, $X(\tau)$. The following formula expressed in mathematical language the fact that $X(\tau)$ realizes a minimum of the action:

$$\delta S = S[X(\tau) + \delta x(\tau)] - S[X(\tau)] = \int_{-\tau_0/2}^{\tau_0/2} d\tau\, \delta x(\tau) \left[-\frac{d^2 X}{d\tau^2} + V'(X) \right] = 0,$$

where $V' = dV/dx$. The equation

$$\frac{d^2 X}{d\tau^2} = V'(X), \tag{11}$$

is of course well known to the reader from school days (we recall that "the mass multiplied by the acceleration is equal to the force"). It is the *classical* equation of motion of a particle in the potential *minus* $V(x)$.[1)]

[1)]The minus sign is due to the fact that the Euclidean formulation is considered [see (6)].

We shall shortly return to this circumstance, but first recall how the pre-exponential factor in (10) is calculated. It is determined by an entire "pencil" of paths near the extremal path, i.e., by paths with action that differs little from S_0. In other words, we take into account only the quadratic deviation:

$$S[X(\tau) + \delta x(\tau)] = S_0 + \int_{-\tau_0/2}^{\tau_0/2} d\tau\, \delta x \left[-\frac{1}{2} \frac{d^2}{d\tau^2} \delta x + \frac{1}{2} V''(X)\, \delta x \right] \tag{12}$$

(as the reader will recall, there is no term linear in the deviation).

Suppose we know a complete set of eigenfunctions and eigenvalues of the equation

$$-\frac{d^2}{d\tau^2} x_n(\tau) + V''(X)\, x_n(\tau) = \varepsilon_n x_n(\tau). \tag{13}$$

Then we can choose these functions as the orthonormalized system which occurs in (8), and the action (12) is transformed to the simple *diagonal* form

$$S = S_0 + \frac{1}{2} \sum_n \varepsilon_n c_n^2.$$

Recalling the definition (9) and the rule of Gaussian integration

$$\int_{-\infty}^{+\infty} dc \exp\left(-\frac{1}{2} \varepsilon c^2\right) = \frac{\sqrt{2\pi}}{\sqrt{\varepsilon}}$$

(it is important that after the diagonalization each such integration can be performed independently of the others), we obtain

$$\langle x_f | e^{-H\tau_0} | x_i \rangle = e^{-S_0} N \prod_n \varepsilon_n^{-1/2}. \tag{14}$$

Sometimes, instead of the product of eigenvalues one uses the notation

$$\prod_n \varepsilon_n^{-1/2} = \left[\det\left(-\frac{d^2}{d\tau^2} + V''(X(\tau)) \right) \right]^{-1/2}, \tag{15}$$

which, of course, derives from the theory of ordinary finite-dimensional matrices. In fact, the relation (15) can be regarded as the definition of the determinant of a differential operator. It is here appropriate to make three comments. First, the result (14) does not depend on the explicit form of the eigenfunctions but only on the eigenvalues. Second, we have assumed that all the ε_n are positive. In the majority of cases, this is so, but in the instanton example, which is the final aim of the present review, several eigenvalues vanish. The resulting infinity has a simple physical meaning. The problem of how it should be handled is the subject of the next section. The third and final comment is the following. The normalization factor N has still not yet been fixed. We shall not even attempt to give a general prescription but consider a simple example, which will serve us in the future too. Suppose the original particle with mass $m = 1$ is placed in the potential $V(x)$ shown in Fig. 1. We do not need the actual form of this potential, but to achieve "normalization" to the harmonic oscillator (in which the potential is usually taken to be $m\omega^2 x^2/2$), we set $V''(x=0) = \omega^2$. As the initial and final points of the motion we choose $x_i = x_f = 0$.

The rich physical intuition that we each have for potential mechanical motion enables us to find the extremal from Eq. (11) without knowing the explicit form of

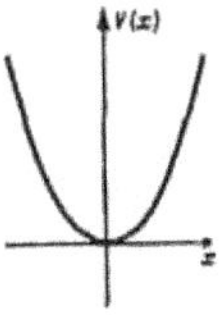

FIG. 1.

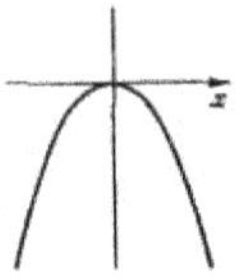

FIG. 2.

$V(x)$. Indeed, this equation describes the motion of a ball on the profile shown in Fig. 2. At the time $-\tau_0/2$, the ball is displaced from the upper point, to which it returns at the time $+\tau_0/2$. It is entirely clear that there exists only one path with such properties: $X(\tau)=0$. Any other path corresponds to an infinite motion with the ball going away to plus or minus ∞. It is also clear that the action on the path $X(\tau)=0$ vanishes.

Thus, in the given particular problem the general formula (14) becomes

$$\langle x_f=0|e^{-H\tau_0}|x_i=0\rangle = N\left[\det\left(-\frac{d^2}{d\tau^2}+\omega^2\right)\right]^{-1/2}(1+\text{following terms}),$$

and all the eigenvalues ε_n are immediately fixed by the boundary conditions $x_n(\pm\tau_0/2)=0$:

$$\varepsilon_n=\frac{\pi^2n^2}{\tau_0^2}+\omega^2,\qquad n=1,2,\ \dots\ .$$

We have now arrived at the point at which it is possible to advance further without saying what is the value of N. We split the determinant into two brackets:

$$N\left[\det\left(-\frac{d^2}{d\tau^2}+\omega^2\right)\right]^{-1/2}=\left[N\left(\prod_{n=1}^{\infty}\frac{\pi^2n^2}{\tau_0^2}\right)^{-1/2}\right]\left[\prod_{n=1}^{\infty}\left(1+\frac{\omega^2\tau_0^2}{\pi^2n^2}\right)\right]^{-1/2} \tag{16}$$

Obviously, the first square brackets corresponds to *free* motion of the particle, and therefore, it must, of course, reproduce the free result:

$$N\left(\prod_{n=1}^{\infty}\frac{\pi^2n^2}{\tau_0^2}\right)^{-1/2}=\langle x_f=0|e^{-\hat{p}^2\tau_0/2}|x_i=0\rangle=\sum_n|\langle p_n|x=0\rangle|^2e^{-p_n^2\tau_0/2}$$
$$=\int_{-\infty}^{+\infty}\frac{dp}{2\pi}e^{-p^2\tau_0/2}=\frac{1}{\sqrt{2\pi\tau_0}}. \tag{17}$$

Of course, Eq. (17) is somewhat symbolic, but it can be regarded as the definition of the normalization factor N. We now consider the second, less trivial brackets. For the infinite product which occurs in it we have the well-known formula [see, for example, formula (1.431.2) in Ref. 4][2)]

$$\pi y\prod_{n=1}^{\infty}\left(1+\frac{y^2}{n^2}\right)=\mathrm{sh}\,\pi y,$$

where in our case $y=\omega\tau_0/\pi$.

We now collect together all the factors, take into account (16) and (17), and write down the final result:

$$\langle x_f=0|e^{-H\tau_0}|x_i=0\rangle=N\left[\det\left(-\frac{d^2}{d\tau^2}+\omega^2\right)\right]^{-1/2}$$
$$=\frac{1}{\sqrt{2\pi\tau_0}}\left(\frac{\mathrm{sh}\,\omega\tau_0}{\omega\tau_0}\right)^{-1/2}=\left(\frac{\omega}{\pi}\right)^{1/2}(2\,\mathrm{sh}\,\omega\tau_0)^{-1/2}. \tag{18}$$

[2)] *Translator's Note*. The Russian notation for the trigonometric, inverse trigonometric, hyperbolic trigonometric functions, etc., is retained here and throughout the article in the displayed equations.

Going to the limit $\tau_0\to\infty$, we find

$$\langle x_f=0|e^{-H\tau_0}|x_i=0\rangle\xrightarrow[\tau_0\to\infty]{}\left(\frac{\omega}{\pi}\right)^{1/2}e^{-\omega\tau_0/2}\left(1+\frac{1}{2}e^{-2\omega\tau_0}+\dots\right),$$

from which it follows that for the lowest state $E_0=\omega/2$, $[\psi_0(0)]^2=(\omega/\pi)^{1/2}$. The next term in the expansion corresponds to the level of an oscillator with $n=2$ [the odd n do not contribute, since for them $\psi_n(0)=0$]. The results are exact for the harmonic oscillator and serve as a zeroth approximation for a potential with small anharmonicity, say $(\omega^2/2)x^2+\lambda x^4$.

2. TWO-HUMPED POTENTIAL. TUNNELING

In the previous section, we reformulated in the language of Euclidean space and path integrals one of the most fundamental problems—an oscillator system near the equilibrium position. This problem provides the basis of all field theory. In fact, we have taken into account small vibrations—small deviations from the equilibrium position—and have made the first step to ordinary perturbation theory. For more than 20 years, right up to the middle of the seventies, all field-theoretical models (apart from the small exception of exactly solvable two-dimensional models) were developed in this and only this direction. The field variables were regarded as a system of an infinitely large number of oscillators coupled to each other and each possessing zero-point vibrations; one then considered small deviations, with respect to which perturbation theory was constructed successively. In this sense, the "infant" period of quantum chromodynamics, when quark–gluon perturbation theory was created, did not introduce anything fundamentally new. It was only the discovery of instantons which showed that it contains effects which cannot be described if one does not go beyond the framework of small deviations from the equilibrium position. It is in principle impossible to describe these effects by expansions in the coupling constant. Here, we again turn to a simple quantum-mechanical analogy, in which, however, all the main features are already present.

Thus, we again consider the one-dimensional potential motion of a spinless particle with unit mass. The potential

$$V(x)=\lambda(x^2-\eta^2)^2 \tag{19}$$

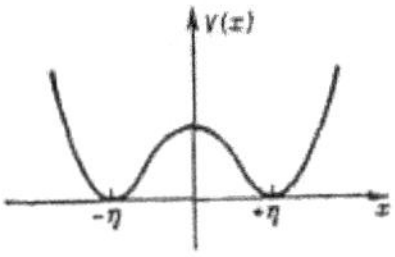

FIG. 3.

is shown in Fig. 3. We fix the parameters λ and η in such a way that $8\lambda\eta^2=\omega^2$, where ω is the frequency introduced in the previous section. Then near each of the minima, which are indicated by the symbols $\pm\eta$, the curve is identical to the potential of the previous section. If $\lambda \ll \omega^3$, then the wall separating the two minima is high. Its height is $\omega^4/64\lambda$. Suppose for a moment that it is actually equal to infinity. Then the lower state of the system has a twofold degeneracy—the particle may be in the right-hand well or in the identical left-hand well, i.e., it executes small vibrations near the point $+\eta$ or $-\eta$. At first glance, the solution to our problem should be constructed in exactly the same way. The expectation value of the coordinate in the ground state should be

$$\langle x\rangle_0=+\eta\,(1+\text{corrections})\quad\text{or}\quad\langle x\rangle_0=-\eta\,(1+\text{corrections}),$$

the original symmetry of the system with respect to the substitution $x\to-x$ is broken, $E_0=(\omega/2)(1+\text{corrections})$ in both cases, and at small λ the corrections are small. In fact, it is known from courses of quantum mechanics that this picture is *qualitatively* incorrect. The symmetry is *not* broken, the expectation value of x for the ground level is *exactly* zero, and there is *no* degeneracy:

$$\begin{aligned} E_0&=\frac{\omega}{2}-\sqrt{\frac{2\omega^3}{\pi\lambda}}\,e^{-\omega^3/12\lambda}\,\frac{\omega}{2},\\ E_1&=\frac{\omega}{2}+\sqrt{\frac{2\omega^3}{\pi\lambda}}\,e^{-\omega^3/12\lambda}\,\frac{\omega}{2}. \end{aligned}\tag{20}$$

We note the fact that $E_1-E_0\sim\exp(-\omega^3/12\lambda)$ and this quantity cannot be expanded in a series in λ. [It is assumed that $\omega^3/\lambda\gg1$. In reality, Eqs. (20) begin to "work" when $\omega^3/12\lambda\gtrsim6$.]

Thus, we have gone wrong and failed to take into account an important element that leads to qualitative changes. What is this element? Everyone knows the standard answer given in courses of quantum mechanics. If at the initial time the particle is concentrated in, say, the left-hand minimum, it nevertheless feels the existence of the right-hand well despite the fact that the latter is inaccessible according to the classical laws of motion. Quantum-mechanical tunneling transfers the wave function from one well to the other and, in Polyakov's terminology, "mixes" the ground states. The correct wave function of the ground state is an even superposition of the wave functions in each of the wells.

We now consider how this phenomenon appears in the imaginary time and how the technique presented in the previous section is changed. It turns out—and this is a great good fortune—that all the fundamental technical elements remain unchanged. It is only necessary to take into account the fact that the classical equations of motion in the imaginary time have not only the trivial solutions $X(\tau)=\text{const}$ considered earlier but also additional nontrivial topological solutions which extend far from both the minima. These solutions connect the points $\pm\eta$, and they are entirely responsible for the phenomenon under discussion. We emphasize that in real time there are no additional classical solutions, since the transition from the one minimum to the other occurs below the barrier and is classically forbidden.

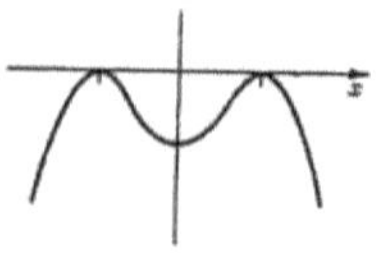

FIG. 4.

The solutions arise only after the Euclidean rotation.

We consider the calculation of the amplitudes

$$\langle\eta\,|\,e^{-H\tau_0}\,|-\eta\rangle\quad\text{and}\quad\langle\eta\,|\,e^{-H\tau_0}\,|\,\eta\rangle.$$

The first step, as the reader may still recall, consists of solving Eq. (11). The "mechanical profile" for this equation is shown in Fig. 4. We are interested in solutions of Eq. (11) that have finite action in the limit $\tau_0\to\infty$, since it is such solutions that are important in the quasiclassical approximation we are discussing. Most of the paths correspond to either vibrational motion or to $x\to\infty$ as $\tau\to\infty$, and they have infinite action.

A finite action in the limit $\tau_0\to\infty$ is obviously obtained when the particle stays at the top of a hump, i.e., $X(\tau)=\eta$ and $X(\tau)=-\eta$. The contribution of these trajectories was considered above. Another interesting motion leading to a finite action as $\tau_0\to\infty$ corresponds to the particle sliding from one hump and stopping on the other. Thus, we are interested in a path which begins at $-\tau_0/2$ at the point $-\eta$ and ends at the point η at the time $\tau_0/2$.[3)] Physical intuition suggests that such trajectories exist, though their explicit form for finite τ_0 is complicated. We are always interested in only the lowest state, and therefore we can directly assume that $\tau_0\to\infty$. In this limit, the solution is very simple:

$$X(\tau)=\eta\,\mathrm{th}\,\frac{\omega(\tau-\tau_c)}{2}\tag{21}$$

[it corresponds to mechanical motion with zero energy, $E=(1/2)\dot{x}^2-V(x)=0$, so that the equations can be readily integrated].

Such a solution is called an instanton (Polyakov proposed the name "pseudoparticle," which can also be found in the literature); the arbitrary parameter τ_c indicates its center. Of course, there also exist anti-instantons, which begin at $+\eta$ and end at $-\eta$. They are obtained from (21) by the substitution $\tau\to-\tau$.

Since all the integrals can be calculated, it is easy to obtain a closed expression for the action of the instanton [we recall that for the instanton $\frac{1}{2}\dot{x}^2=V(x)$]:

$$S_0=S\,[X(\tau)]_{\text{inst}}=\int_{-\infty}^{+\infty}d\tau\,\dot{X}^2=\frac{\omega^3}{12\lambda}.\tag{22}$$

We recall that the principal exponential factor in the amplitude is $e^{-\text{action}}$ [see Eq. (10)]. The exponential which occurs in (20) has emerged. Of course, we still have a long way to go before we can reproduce the complete answer.

[3)]Here we have allowed a slight inaccuracy. If τ_0 is large but not infinite, the path begins just to the right of $-\eta$ and ends just to the left of $+\eta$. It is only in the limit $\tau_0\to\infty$ that the end points coincide with $\pm\eta$.

We draw attention to one further property of an instanton, which has far reaching consequences. The center of the solution may be at any point, and the action of the instanton does not depend on the position of the center. This circumstance obviously reflects the symmetry of the original problem. Namely, the Lagrangian of the system is invariant with respect to shifts in time, and the time origin can be chosen arbitrarily. Each concrete solution (21) has a definite position with respect to the origin, and thus there exists an infinite family of solutions distributed arbitrarily with respect to the origin. Intuitively, it is clear that the instanton must occur in any physical quantity in the form of an integral over the position of its center. How does this integral arise formally and what weight is then obtained? Answers to these questions are given in the following section.

3. DETERMINANT AND ZERO-FREQUENCY MODES

In this section, we find the one-instanton contribution to $\langle -\eta | e^{-H\tau_0} | \eta \rangle$. We shall not, of course, be concerned with the exponential factor, which has actually already been found, but rather the pre-exponential factor, whose calculation presents a more laborious problem. It is true that in the case under consideration one can employ various devices that significantly simplify the problem and are sometimes discussed in the literature.[5] However, we shall proceed in a "head on" manner, which is closest to the method used by 't Hooft[6] to calculate the instanton determinant in QCD. We hope that this will subsequently enable the reader to reproduce for himself all details of 't Hooft's work, which is central for the entire instanton problem.

The original formula (14) is conveniently rewritten as

$$\langle -\eta | e^{-H\tau_0} | \eta \rangle = N\left[\det\left(-\frac{d^2}{d\tau^2}+\omega^2\right)\right]^{-1/2}\left\{\frac{\det[-(d^2/d\tau^2)+V''(X)]}{\det[-(d^2/d\tau^2)+\omega^2]}\right\}^{-1/2} e^{-S_0}(1+\text{corrections}).$$

We have multiplied and divided by a known number—the determinant for the harmonic oscillator [see (18)]. The harmonic oscillator will serve as a "point of reference" for manipulations with the more complicated determinant in the numerator. Substituting the explicit expression $X(\tau)=\eta\tanh(\omega\tau/2)$ in $V''(X)$, we arrive at the eigenvalue equation

$$-\frac{d^2}{d\tau^2}x_n(\tau)+\left(\omega^2-\frac{3}{2}\omega^2\frac{1}{\operatorname{ch}^2(\omega\tau/2)}\right)x_n(\tau)=\varepsilon_n x_n(\tau). \tag{23}$$

It can be regarded as a certain Schrödinger equation, which, *fortunately*, is very well studied. Indeed, Eq. (23) is described in detail in, for example, the textbook of Landau and Lifshitz (Ref. 7, pp. 97 and 105), and we shall use this source. We recall that the boundary conditions are $x_n(\pm\tau_0/2)=0$ and $\tau_0\to\infty$. These conditions are automatically satisfied with exponential accuracy for bound levels, i.e., for the truly discrete spectrum.[4)]

[4)] Without boundary conditions, the complete spectrum is in fact discrete. The genuine discrete levels can however be readily distinguished from the quasidiscrete levels formed from the continuum after the system has been enclosed in the "box" $x(\pm\tau_0/2)=0$. The former are separated by intervals of order ω^2, while the latter are at a distance of order $1/\tau_0^2$ from their neighbors.

There are two such levels in Eq. (23). One of them corresponds to the eigenvalue $\varepsilon_1=(3/4)\omega^2$, and the other to

$$\varepsilon_0=0.$$

The wave function of the latter, normalized to unity, is

$$x_0(\tau)=\sqrt{\frac{3\omega}{8}}\frac{1}{\operatorname{ch}^2(\omega\tau/2)}. \tag{24}$$

The vanishing of the eigenvalue may discourage the reader, since the answer contains $\varepsilon_n^{-1/2}$! However, this result, $\varepsilon_0=0$, cannot be regarded as a surprise. Indeed, Eq. (23) actually describes the response of the dynamical system under consideration to small perturbations imposed on $X(\tau)$. Since $X(\tau)$ is a solution which realizes a "local" minimum of the action, a perturbation of $X(\tau)$ increases the action. Accordingly, the ε_n are positive. However, we already know that there is one direction in the function space along which the solution can be perturbed without changing the action. We have in mind a shift of the center. By virtue of the translational invariance,

$$S\{X(\tau,\tau_c)\}-S\{X(\tau,\tau_c+\delta\tau_c)\}=0.$$

The so-called zero-frequency mode (i.e., the mode with $\varepsilon=0$) is obviously proportional to $X(\tau,\tau_c)=X(\tau,\tau_c+\delta\tau_c)$. The correctly normalized zero-frequency mode has the form

$$x_0(\tau)=S_0^{-1/2}\left(-\frac{d}{d\tau_c}\right)X(\tau,\tau_c),$$

or, which is the same thing,

$$x_0(\tau)=S_0^{-1/2}\frac{d}{d\tau}X(\tau). \tag{25}$$

The correctness of the normalization follows from the expression (22). It is readily seen that (25) is identical to (24), and we now see that this agreement is not fortuitous but a consequence of the translational invariance.

Thus, integration with respect to the coefficient c_0 corresponding to the zero-frequency mode [see (8) and (9)] is non-Gaussian, and the integral between infinite limits does not exist at all. The way out of the dilemma is simple. We shall not calculate this integral explicitly. It is clear that the integration over dc_0 is the same as integration over $d\tau_c$ apart from a coefficient of proportionality. We have here the same integral over the position of the center of the instanton whose appearance our intuition required. In the literature, this trick is sometimes called the introduction of a collective coordinate.

We determine the coefficient of proportionality. If c_0 changes by Δc_0, then $x(\tau)$ changes by

$$\Delta x(\tau)=x_0(\tau)\Delta c_0$$

[see (8)]. On the other hand, the change $\Delta x(\tau)$ on a shift $\Delta\tau_c$ of the center is

$$\Delta x(\tau)=\Delta X(\tau)=\frac{dx}{d\tau_c}\Delta\tau_c=-\sqrt{S_0}\,x_0(\tau)\Delta\tau_c.$$

Equating the two increments, we obtain

$$dc_0=\sqrt{S_0}\,d\tau_c. \tag{26}$$

[In Eq. (26), we have not inserted the minus sign to ensure that as c_0 varies from $-\infty$ to $+\infty$ the parameter τ_c changes in the same interval.] This is not yet everything, since we agreed to normalize the result to the ordinary oscillator (we recall that we are interested in the ratio of determinants). In the oscillator problem, the minimal eigenvalue is $\omega^2+\pi^2/\tau_0^2 \to \omega^2$ in the limit $\tau_0 \to \infty$. Finally,

$$\left\{\frac{\det[-(d^2/d\tau^2)+V''(X)]}{\det[-(d^2/d\tau^2)+\omega^2]}\right\}^{-1/2}=\sqrt{\frac{S_0}{2\pi}}\,\omega\, d\tau_c\left\{\frac{\det'[-(d^2/d\tau^2)+V''(X)]}{\omega^{-2}\det[-(d^2/d\tau^2)+\omega^2]}\right\}^{-1/2}, \tag{27}$$

where det′ denotes the reduced determinant with the zero-frequency mode removed.

We emphasize that although we have analyzed only a single specific example with the simplest instanton $\eta\tanh(\omega\tau/2)$, the method of dealing with zero-frequency modes is in fact general. Thus, in the BPST instanton any invariance will generate a zero-frequency mode, and the integration with respect to the corresponding coefficient must be replaced by integration with respect to some collective variable. We have already learned how to find the Jacobian of the transformation.

We now consider positive-frequency modes. It is easiest to deal with the second discrete level, whose eigenvalue is $(3/4)\omega^2$. If we denote by Φ the ratio

$$\Phi=\frac{\det'[-(d^2/d\tau^2)+V''(X)]}{\omega^{-2}\det[-(d^2/d\tau^2)+\omega^2]}, \tag{28}$$

then the contribution of this level to Φ as $\tau_0\to\infty$ is obviously

$$\frac{3}{4}. \tag{29}$$

We now turn to other modes with $\varepsilon>\omega^2$. If we did not have the boundary condition $x(\pm\tau_0/2)=0$, Eq. (23) in this region would have a continuous spectrum. Let us forget the boundary conditions for a moment. The general solution of (23) is given in the book of Landau and Lifshitz; however, we do not require its explicit form. It is sufficient to know the following. First, the solutions with $\varepsilon>\omega^2$ are labeled by a continuous index p. This index is related to the eigenvalue ε by $p=\sqrt{\varepsilon_p-\omega^2}$ and ranges over the entire inverval $(0,\infty)$. Second, for the values of the parameters that occur in (23) there is no reflection. In other words, choosing one of the linearly independent solutions in such a way that

$$x_p(\tau)\to e^{ip\tau}\quad \text{as}\quad \tau\to+\infty,$$

we have in the other asymptotic region the same exponential:

$$x_p(\tau)\to e^{ip\tau+i\delta_p}\quad \text{as}\quad \tau\to-\infty.$$

The second exponential, $e^{-ip\tau}$, which should in principle arise, is absent, and the entire dynamical effect has been reduced to the phase

$$e^{i\delta_p}=\frac{1+(ip/\omega)}{1-(ip/\omega)}\,\frac{1+(2ip/\omega)}{1-(2ip/\omega)} \tag{30}$$

(we have used here the formula from the textbook of Ref. 7 on p. 106). The second linearly independent solution can be chosen in the form $x_p(-\tau)$. The general solution is $Ax_p(\tau)+Bx_p(-\tau)$, where A and B are arbitrary constants.

This information is already sufficient to find the spectrum if we recall the boundary condition $x(\pm\tau/2)=0$. The equations for A and B,

$$Ax_p\left(\frac{\tau_0}{2}\right)+Bx_p\left(-\frac{\tau_0}{2}\right)=0,\quad Ax_p\left(-\frac{\tau_0}{2}\right)+Bx_p\left(\frac{\tau_0}{2}\right)=0,$$

have nontrivial solutions if and only if

$$\frac{x_p(\tau_0/2)}{x_p(-\tau_0/2)}=\pm 1.$$

This gives an equation for p:

$$e^{ip\tau_0-i\delta_p}=\pm 1,$$

or, which is the same thing,

$$p\tau_0-\delta_p=\pi n,\qquad n=0,\,1,\,\ldots\,. \tag{31}$$

We denote the nth solution by $\bar{p}_n$. In the case of $\det[-(d^2/d\tau^2)+\omega^2]$, by which we normalize, the equation is $p\tau_0=\pi n$ and the nth solution $p_n=\pi n/\tau_0$. We need to calculate the product[5)]

$$\prod_{n=1}^{\infty}\frac{\omega^2+\bar{p}_n^2}{\omega^2+p_n^2}.$$

For any preassigned n, the ratio $(\omega^2+\bar{p}_n^2)/(\omega^2+p_n^2)$ is arbitrarily close to unity as $\tau_0\to\infty$. Only the multiplication of a very large number of factors with $n\sim\omega\tau_0$, each of them differing from 1 by an amount of order $1/\omega\tau_0$, gives an effect. (For $n\gg\omega\tau_0$, the difference between $\omega^2+\bar{p}_n^2$ and $\omega^2+p_n^2$ again becomes unimportant, in complete agreement with our physical intuition.) Under these conditions, we can write

$$\prod\frac{\omega^2+\bar{p}_n^2}{\omega^2+p_n^2}=\exp\left(\sum_n\ln\frac{\omega^2+\bar{p}_n^2}{\omega^2+p_n^2}\right)\approx\exp\left[\sum_n\frac{2p_n(\bar{p}_n-p_n)}{\omega^2+p_n^2}\right],$$

where we have made an expansion with respect to the small difference $\bar{p}_n-p_n$. Going over from summation over n to integration over p_n and using (31) for $\bar{p}_n-p_n$, we obtain on the right-hand side

$$\exp\left[+\frac{1}{\pi}\int_0^\infty\frac{\delta_p\cdot 2p\,dp}{p^2+\omega^2}\right]=\exp\left[-\frac{1}{\pi}\int_0^\infty\frac{d\delta_p}{dp}\ln\left(1+\frac{p^2}{\omega^2}\right)dp\right].$$

Differentiating the phase by means of (30) and introducing the dimensionless variable $y=p/\omega$, we transform this expression identically to

$$\exp\left[-\frac{2}{\pi}\int_0^\infty dy\left(\frac{1}{1+y^2}+\frac{2}{1+4y^2}\right)\ln(1+y^2)\right]=\frac{1}{9}. \tag{32}$$

Finally, combining (32) and (29), we find that

$$\Phi=\frac{1}{12}. \tag{33}$$

We have now made all the necessary preparations, namely, we have derived formulas (33), (28), (27), (22), and (18), and we write down the result for the one-instanton contribution:

$$\langle-\eta|e^{-H\tau_0}|\eta\rangle_{\text{one-inst}}=\left(\sqrt{\frac{\omega}{\pi}}e^{-\omega\tau_0/2}\right)\left(\sqrt{\frac{6}{\pi}}\sqrt{S_0}\,e^{-S_0}\right)\omega\, d\tau_c. \tag{34}$$

[5)] The reader may recall that we have already "taken up" in the denominator two eigenvalues, $\omega^2+\pi^2/\tau_0^2$ and $\omega^2+4\pi^2/\tau_0^2$, in calculating the contribution of the discrete modes with $\varepsilon=0$ and $\varepsilon=3\omega^2/4$. Therefore, it would be more correct in the denominator to write $\omega^2+p_{n+2}^2$. However, as we shall see very shortly, it is the region of very large n, of order $\omega\tau_0$, that is important, so that the difference between p_{n+2} and p_n is immaterial.

This result can be trusted as long as

$$\sqrt{S_0}\, e^{-S_0} \omega\tau_0 \ll 1.$$

At large τ_0, when this condition is violated, it is necessary to take into account paths constructed from many instantons and anti-instantons, and this will be done in the following section.

It is here appropriate to make some comments. The factor in the first square brackets corresponds to a simple harmonic oscillator. By separating it, we have been able to normalize, or regularize, the instanton calculations. A similar device for regularization is used in quantum chromodynamics. The factor in the second square brackets can naturally be called the instanton density. Besides the exponential factor e^{-S_0}, the density contains the pre-exponential $\sqrt{S_0}$, which is associated with the existence of the zero-frequency mode. This circumstance is also of a general nature. In quantum chromodynamics too, each zero-frequency mode is associated with $\sqrt{S_0}$. Finally, the existence of the zero-frequency mode leads to the appearance of a regularization frequency and of integration over the collective coordinate $\omega d\tau_c$.

We wish to emphasize that it is worth remembering the lessons we have learned, since they can be directly transferred to the BPST instanton. The only thing specific in the present case is the number $\sqrt{6/\pi}$. If this number is not particularly important (and in QCD, as we shall see below, this is indeed the case), all the remaining result can be reconstructed almost at once, without calculations. We have taken so much bother with the relatively simple determinant for a pedagogical reason—to avoid greater boredom in the case of the BPST *instanton*.

4. INSTANTON GAS

It remains for us to make the final, small step to reproduce formula (20). The energy of the lower state is determined by the transition to the limit $\tau_0 \to \infty$. We cannot go to this limit directly in Eq. (34). At very large τ_0, paths constructed of many instantons and anti-instantons are important. If the distance between their centers is large, such a path is also a classical solution.

Suppose we have n instantons or anti-instantons with centers $\tau_1, \tau_2, \ldots, \tau_n$ (Fig. 5). The points τ_i satisfy the condition

$$-\frac{\tau_0}{2} < \tau_1 < \tau_2 < \ldots < \tau_n < \frac{\tau_0}{2},$$

and otherwise can be distributed arbitrarily. If the characteristic intervals satisfy $|\tau_i - \tau_j| \gg \omega^{-1}$ (we shall verify the condition *a posteriori*), then the action corresponding to such a configuration is nS_0, where S_0 is the action of one instanton. With regard to the determinant, it is obvious that if we did not have the n narrow transition regions (near $\tau_1, \tau_2, \ldots, \tau_n$) we should obtain the same result as in the case of the harmonic oscillator, $\sqrt{\omega/\pi}\, e^{-\omega\tau_0/2}$. The transition regions lead to a correction, and we now know in what way:

$$\sqrt{\frac{\omega}{\pi}}\, e^{-\omega\tau_0/2} \to \sqrt{\frac{\omega}{\pi}}\, e^{-\omega\tau_0/2} \left(\sqrt{\frac{6}{\pi}}\sqrt{S_0}\, e^{-S_0}\right)^n \prod_1^n (\omega\, d\tau_i).$$

Finally, the contribution of the n-instanton configuration can be written in the form

$$\sqrt{\frac{\omega}{\pi}}\, e^{-\omega\tau_0/2} d^n \int_{-\tau_0/2}^{\tau_0/2} \omega\, d\tau_n \int_{-\tau_0/2}^{\tau_n} \omega\, d\tau_{n-1} \cdots \int_{-\tau_0/2}^{\tau_2} \omega\, d\tau_1 = \sqrt{\frac{\omega}{\pi}}\, e^{-\omega\tau_0/2} d^n \frac{(\omega\tau_0)^n}{n!},$$

where we have denoted by d the instanton density,

$$d = \sqrt{\frac{6}{\pi}}\sqrt{S_0}\, e^{-S_0}. \qquad (35)$$

The amplitudes $\langle -\eta | e^{-H\tau_0} | \eta\rangle$ and $\langle \eta | e^{-H\tau_0} | \eta\rangle$ are obtained by summation over n. In the first case, we start from $-\eta$ and arrive at $+\eta$ and therefore the number of pseudoparticles is odd. In the second case, conversely, only an even number of pseudoparticles works:

$$\langle -\eta | e^{-H\tau_0} | \eta\rangle = \sum_{n=1,3,\ldots} \sqrt{\frac{\omega}{\pi}}\, e^{-\omega\tau_0/2} \frac{(\omega\tau_0 d)^n}{n!} = \sqrt{\frac{\omega}{\pi}}\, e^{-\omega\tau_0/2}\, \mathrm{sh}\,(\omega\tau_0 d), \qquad (36)$$

$$\langle \eta | e^{-H\tau_0} | \eta\rangle = \sum_{n=0,2,\ldots} \sqrt{\frac{\omega}{\pi}}\, e^{-\omega\tau_0/2} \frac{(\omega\tau_0 d)^n}{n!} = \sqrt{\frac{\omega}{\pi}}\, e^{-\omega\tau_0/2}\, \mathrm{ch}\,(\omega\tau_0 d).$$

Going to the limit $\tau_0 \to \infty$, we immediately reproduce formula (20) for the energy of the lowest state. Denoting the ground state of the system by $|0\rangle$, we see that $\langle \eta | 0\rangle = \langle -\eta | 0\rangle = (\omega/4\pi)^{1/4}$, i.e., the symmetry between the right- and left-hand well is indeed not broken.

We now return to the assumption that the characteristic distances between the centers of the instantons are large,

$$|\tau_i - \tau_j| \gg \omega^{-1},$$

and consider how well it works. It is clear that the sums in (36) converge well, and all terms with number $n \gg d\omega\tau_0$ are unimportant. Thus, $n_{\mathrm{char}} \sim d\omega\tau_0$ and $|\tau_i - \tau_j|_{\mathrm{char}} \sim d^{-1}\omega^{-1}$. Having at our disposal the free parameter λ, we can achieve an arbitrary smallness of d, since $d \to 0$ as $e^{-\omega^3/12\lambda}$ in the limit $\lambda \to 0$.

Thus, for $\lambda \ll 1$ we are fully justified in "stringing" instantons and anti-instantons on one another, forming thereby a chain of noninteracting pseudoparticles. Noninteracting in the sense that they are all far from one another, know nothing about the remaining partners, and the total weight function is obtained by multiplying the individual weight functions [d^n in formulas (36)].

Such an approximation is called a dilute instanton gas. In quantum chromodynamics, it has been exploited particularly by Callan, Dashen, and Gross.[8] Unfortunately, in QCD we do not dispose of free parameters like λ that can be kept small. Therefore, a dilute

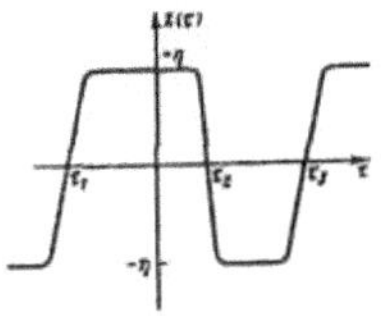

FIG. 5.

instanton gas is not suitable from the quantitative point of view in QCD, and the most we can extract from it are are heuristic indications.

To conclude the section, we note that a somewhat more extensive exposition of the instanton approach to the two-humped potential is contained in Coleman's lecture.[5] The reader interested in special questions, for example, situations not covered by the gas approximation, must consult Ref. 9.

5. EUCLIDEAN FORMULATION OF QCD

Thus, in the simple example of the two-humped potential we have seen that if there exist nontrivial solutions of the classical equations qualitatively new effects occur in the theory. Tunneling from one well to another makes the vacuum wave function quite different from the one obtained in perturbation theory. Our aim in this review is, of course, chromodynamics and not quantum mechanics. However, in chromodynamics too there is a similar phenomenon, which we shall discuss in this and all the following sections of the review.

As we said above, we are concerned with the solution of classical equations in *Euclidean* space. Therefore, we first formulate the Euclidean version of QCD. We give the formulas for the transition from Minkowski to Euclidean space. The spatial coordinates x_1, x_2, x_3 are not changed. For the time coordinate x_0, we make the substitution

$$x_0 = -ix_4. \tag{37}$$

Clearly, when x_0 is continued to imaginary values the zeroth component of the vector potential A_μ also becomes imaginary.

We define the Euclidean vector potential $\hat{A}_\mu$ as follows:

$$A_m = -\hat{A}_m \quad (m = 1, 2, 3), \quad A_0 = i\hat{A}_4 \tag{38}$$

(in this section, we shall use the caret to denote all quantities defined in the Euclidean space). With this definition, the quantities $\hat{A}_\mu$ ($\mu = 1, \ldots, 4$) form a Euclidean vector. The difference between formulas (38) and the corresponding relations for the vector x_μ [the difference is in the common sign of $\hat{A}_\mu$ ($\mu = 1, \ldots 4$)] is introduced for convenience in the expression of the following formulas.[6]

Thus, for the operator of covariant differentiation

$$D_\mu = \partial_\mu - igA_\mu^a T^a, \tag{39}$$

where T^a are the matrices of the generators in the representation being considered, we obtain

$$D_m = -\hat{D}_m, \quad D_0 = i\hat{D}_4,$$
$$\hat{D}_\mu = \frac{\partial}{\partial \hat{x}_\mu} - ig\hat{A}_\mu^a T^a. \tag{40}$$

We recall that the operator ∂_μ in Minkowski space has the form $\partial_\mu = (\partial/\partial x_0, -\partial/\partial x_m)$.

[6] If we use the definition $\hat{A}_m = A_m$ ($m = 1, 2, 3$), then in all the following connection formulas it is necessary to make the substitution $g \to -g$.

For the intensities $G_{\mu\nu}$ we obtain the formulas

$$G_{mn}^a = \hat{G}_{mn}^a \quad (m, n = 1, 2, 3), \quad G_{0n}^a = -i\hat{G}_{4n}^a, \tag{41}$$

where the Euclidean intensities $\hat{G}_{\mu\nu}^a$,

$$\hat{G}_{\mu\nu}^a = \frac{\partial}{\partial \hat{x}_\mu}\hat{A}_\nu^a - \frac{\partial}{\partial \hat{x}_\nu}\hat{A}_\mu^a + gf^{abc}\hat{A}_\mu^b\hat{A}_\nu^c \qquad (\mu, \nu = 1, \ldots, 4). \tag{42}$$

can be expressed in terms of $\hat{A}_\mu$ and $\partial/\partial\hat{x}_\mu$ in the same way as the Minkowskian $G_{\mu\nu}^a$.

To complete the transition to the Euclidean space, it remains to give the formulas for the Fermi fields. We begin with the definition of four Hermitian γ matrices $\hat{\gamma}_\mu$:

$$\hat{\gamma}_4 = \gamma_0, \quad \hat{\gamma}_m = -i\gamma_m \qquad (m = 1, 2, 3),$$
$$(\hat{\gamma}_\mu, \hat{\gamma}_\nu) = 2\delta_{\mu\nu} \qquad (\mu, \nu = 1, \ldots, 4), \tag{43}$$

where γ_0 and γ_m are the ordinary Dirac matrices.

The fields ψ and $\bar{\psi}$ are regarded as independent anticommuting variables, with respect to which integration is performed in the functional integral. On the transition to the Euclidean space, it is convenient to define the variables $\hat{\psi}$ and $\hat{\bar{\psi}}$ by

$$\psi = \hat{\psi}, \quad \bar{\psi} = -i\hat{\bar{\psi}}. \tag{44}$$

Note that under rotations of the pseudo-Euclidean space, $\bar{\psi}$ transforms as $\psi^+\gamma_0$. In the Euclidean space, $\hat{\bar{\psi}}$ transforms as $\hat{\psi}^+$. Indeed, under infinitesimal rotations of the pseudo-Euclidean space characterized by the parameters $\omega_{\mu\nu}$ ($\mu, \nu = 0, 1, \ldots, 3$) the spinor ψ acquires the addition

$$\delta\psi = -\frac{1}{4}(\gamma_\mu\gamma_\nu - \gamma_\nu\gamma_\mu)\,\omega_{\mu\nu}\psi.$$

For the change in $\bar{\psi} = \psi^+\gamma_0$ we deduce from this

$$\delta(\psi^+\gamma_0) = -\frac{1}{4}\psi^+\gamma_0\gamma_0(\gamma_\nu^+\gamma_\mu^+ - \gamma_\mu^+\gamma_\nu^+)\,\gamma_0\omega_{\mu\nu} = \frac{1}{4}(\psi^+\gamma_0)(\gamma_\mu\gamma_\nu - \gamma_\nu\gamma_\mu)\,\omega_{\mu\nu},$$

so that $\psi_1^+\gamma_0\psi_2$ is a scalar and $\psi_1^+\gamma_0\gamma_\mu\psi_2$ a vector.

On the transition to the Euclidean space, the parameters ω_{mn} ($m, n = 1, 2, 3$) do not change, and $\omega_{0n} = i\omega_{4n}$ (because of the substitution $x_0 = -ix_4$). For the variations of $\hat{\psi}$ and $\hat{\psi}^+$ under rotations, we obtain

$$\delta\hat{\psi} = \frac{1}{4}(\hat{\gamma}_\mu\hat{\gamma}_\nu - \hat{\gamma}_\nu\hat{\gamma}_\mu)\,\hat{\omega}_{\mu\nu}\hat{\psi}, \quad \delta\hat{\psi}^+ = -\frac{1}{4}\psi^+(\hat{\gamma}_\mu\hat{\gamma}_\nu - \hat{\gamma}_\nu\hat{\gamma}_\mu)\,\hat{\omega}_{\mu\nu},$$

so that $\psi_1^+\psi_2$ and $\psi_1^+\gamma_\mu\psi_2$ are a scalar and vector, respectively.

Finally, we can write down an expression for the Euclidean action:

$$iS = -\hat{S},$$
$$S = \int d^4x\left[-\frac{1}{4}G_{\mu\nu}^a G_{\mu\nu}^a + \bar{\psi}(i\gamma_\mu D_\mu - M)\psi\right],$$
$$\hat{S} = \int d^4\hat{x}\left[\frac{1}{4}\hat{G}_{\mu\nu}^a\hat{G}_{\mu\nu}^a + \hat{\bar{\psi}}(-i\hat{\gamma}_\mu\hat{D}_\mu - iM)\hat{\psi}\right], \tag{45}$$

where it is assumed that $\hat{\psi}$ is a column in the space of flavors (with color index), and M is a matrix in this space.

Below, we shall use the Euclidean space and omit the caret. The formulas given below make it possible to relate the quantities in the pseudo-Euclidean and Euclidean spaces.

To conclude the section, we note that if we are considering quantities such as the vacuum expectation values of the time-ordered products of currents for spacelike external momenta, i.e., when the sources do not produce real hadrons from the vacuum, the Euclidean formulation is not only merely possible but in fact is more adequate than the pseudo-Euclidean. The region of timelike momenta, where there are singularities, can be reached by means of analytic continuation. Such an approach is particularly necessary for quantum chromodynamics, for which the fundamental objects of the theory—the quarks and gluons—have meaning only in the Euclidean domain, and the real singularities corresponding to hadrons have to be obtained.

6. BPST INSTANTONS. GENERAL PROPERTIES

a) Finiteness of the action and the topological charge

It was already clear in the quantum-mechanical example discussed above what an important part is played by solutions that give a minimum of the Euclidean action in the limit $\tau_0 \to \infty$. In general, the action increases unboundedly in the limit $\tau_0 \to \infty$, and the condition that it be finite imposes strong restrictions on the paths.

Thus, in the one-dimensional example we have analyzed, the finite-action condition means that the function $x(\tau)$ as $\tau \to \pm\infty$ must have the limits $\pm\eta$. In this way there arises naturally a topological classification of functions giving a finite action on the basis of their limiting values. Formally, a topological charge can be introduced as follows:

$$Q = \frac{1}{2\eta} \int_{-\infty}^{+\infty} dt\, \dot{x}(t) = \frac{x(+\infty) - x(-\infty)}{2\eta}.$$

It is obvious that Q can take on the values 0, +1, −1. Functions with different Q cannot be carried into one another by a continuous deformation that leaves the action finite. Therefore, in each of the classes $Q = 0$, +1, −1 there exists a corresponding minimum of the action and corresponding functions that realize it. The instanton and anti-instanton realize minima for $Q = \pm 1$.

We now turn to "gluodynamics"—the theory of a non-Abelian vector field—and consider first the case of the group SU(2). We pose the same question: What must be the behavior of the vector fields A_μ^a as $x \to \infty$ if the action is to be finite? (We have in mind the Euclidean action $\tilde{S}$; see (45).] It is clear that the intensities $G_{\mu\nu}^a$ must decrease more rapidly than $1/x^2$. But this by no means implies that the fields A_μ^a must decrease faster than $1/x$. Indeed, suppose A_μ^a in the limit $x \to \infty$ has the form

$$A_\mu = \frac{\tau^a}{2} A_\mu^a \xrightarrow[x\to\infty]{} iS\, \partial_\mu S^+, \tag{46}$$

where we have introduced matrix notation: S is a unitary unimodular matrix that depends on the angles in the Euclidean space. Although the angular components of A_μ are proportional to $1/x$, it is clear that in the region in which the expression (46) holds the intensities $G_{\mu\nu}^a$ vanish, since A_μ^a has a purely gauge form.

Thus, the behavior of A_μ^a at large x is determined by the matrix S, which depends on the angles. Under a gauge transformation of A_μ defined by the matrix $U(x)$:

$$A_\mu \to U^+ A_\mu U + iU^+ \partial_\mu U,$$

the matrix S is replaced by $U^+(x\to\infty)S$. It would appear that one can always choose $U(x)$ such that $U(x\to\infty) = S$ and thus remove the terms $1/x$ from A_μ. However, this argument is correct only if the matrix $U(x)$ does not have singularities at any value of x. Otherwise, the problem of the behavior of $A_\mu(x)$ is transferred from the point at infinity to the position of the singularity of $U(x)$.

As a result, the problem of classifying the fields A_μ^a which give finite action reduces to the topological classification of the matrices S. We shall not present this classification, which was obtained in the pioneering paper of Ref. 1, but rather give examples of nontrivial (not reducible to the unit matrix) matrices S. For example, we have the matrix

$$S_1 = \frac{x_4 + i\mathbf{x}\boldsymbol{\tau}}{\sqrt{x^2}}. \tag{47}$$

It corresponds to unit topological charge (there is a one-to-one correspondence between the space of unitary unimodular matrices and the points of the hypersphere in Euclidean space). To topological charge n there corresponds a matrix of the form

$$S_n = (S_1)^n, \qquad n = 0, \pm 1, \pm 2, \ldots \tag{48}$$

Of course, one could choose a different form of the matrix S corresponding to the charge n, but the difference between it and S_n reduces to a gauge transformation.

For n, there exists the gauge-invariant integral representation

$$n = \frac{g^2}{32\pi^2} \int d^4x\, G_{\mu\nu}^a \tilde{G}_{\mu\nu}^a, \tag{49}$$

where

$$\tilde{G}_{\mu\nu}^a = \frac{1}{2} \varepsilon_{\mu\nu\gamma\delta} G_{\gamma\delta}^a, \quad \varepsilon_{1234} = 1. \tag{50}$$

The validity of Eq. (49) can be verified by using the fact that $G_{\mu\nu}^a \tilde{G}_{\mu\nu}^a$ can be represented in the form of a total derivative,

$$G_{\mu\nu} \tilde{G}_{\mu\nu} = \partial_\mu K_\mu,$$
$$K_\mu = 2\varepsilon_{\mu\nu\gamma\delta} \left(A_\nu^a \partial_\gamma A_\delta^a + \frac{1}{3} g \varepsilon^{abc} A_\nu^a A_\gamma^b A_\delta^c \right),$$

so that the volume integral (9) can be transformed into an integral over a distant surface, where A_μ^a has the form (46).

b) The distinguished role of the group SU(2)

Hitherto, we have discussed the group SU(2). For groups different from SU(2), the construction of instanton solutions with $n = 1$ reduces to the case of SU(2) by means of separation of SU(2) subgroups. Why is the group SU(2) distinguished? We shall attempt to explain this without using topological terminology.

The possibility of deformation of the matrices S is determined by the gauge invariance discussed above. We attempt to fix the gauge, for which we represent an arbitrary field A_μ in the form

$$A_\mu(x) = S(x) \bar{A}_\mu(x) S^+(x) + iS(x)\, \partial_\mu S^+(x), \tag{51}$$

where the field $\bar{A}_\mu$ satisfies definite gauge conditions [for example, $\tilde{A}_0=0$ or $\partial_m\tilde{A}_m=0$ $(m=1,2,3)$]. This fixing does not completely determine the transition to the new fields $\tilde{A}_\mu(x)$ and $S(x)$, since A_μ is invariant under global transformations of the form

$$S(x)\to S(x)U_2^+,\quad \tilde{A}_\mu(x)\to U_2\tilde{A}_\mu U_2^+ \tag{52}$$

with matrix U_2 that does not depend on x.

In addition, even after the fixing of the gauge the theory is still invariant with respect to global isotopic rotations for A_μ, which in terms of the new fields $\tilde{A}_\mu(x)$ and $S(x)$ is equivalent to the transformations

$$S(x)\to U_1S,\quad \tilde{A}_\mu(x)\to\tilde{A}_\mu(x). \tag{53}$$

Thus, the isotopic SU(2) invariance of the theory together with the gauge invariance reduce to the set of *global* transformations (52) and (53), which obviously form the group SU(2)×SU(2). The field $S(x)$ transforms in accordance with the representation (1/2,1/2), and $\tilde{A}_\mu(x)$ in accordance with the representation (1,0).

On the other hand, the group of rotations of four-dimensional Euclidean space is again, as is well known, SU(2)×SU(2), and the generators of the SU(2) subgroups have the form

$$\begin{aligned} I_1^a&=\tfrac{1}{4}\eta_{a\mu\nu}M_{\mu\nu},\\ I_2^a&=\tfrac{1}{4}\bar{\eta}_{a\mu\nu}M_{\mu\nu}\end{aligned}\qquad\begin{pmatrix}a=1,\,2,\,3\\ \mu,\,\nu=1,\,\ldots,\,4\end{pmatrix}, \tag{54}$$

where $M_{\mu\nu}=-ix_\mu\partial/\partial x_\nu+ix_\nu\partial/\partial x_\mu$ + spin part are the operators of infinitesimal rotations in the (μ,ν) plane, and $\eta_{a\mu\nu}$ are the numerical symbols

$$\eta_{a\mu\nu}=\begin{cases}\epsilon_{a\mu\nu}, & \mu,\,\nu=1,\,2,\,3,\\ -\delta_{a\nu}, & \mu=4,\\ \delta_{a\mu}, & \nu=4,\\ 0, & \mu=\nu=4.\end{cases} \tag{55}$$

(The symbols $\bar{\eta}_{a\mu\nu}$ differ from η by a change in the sign of δ.) The coordinate vector x_μ transforms in accordance with the representation (1/2,1/2). This is conveniently seen by considering transformations of the matrix

$$x_4+i\mathbf{x}\boldsymbol{\tau}=i\tau_\mu^+x_\mu, \tag{56}$$

where we have introduced the notation

$$\tau_\mu^\pm=(\boldsymbol{\tau},\,\mp i). \tag{57}$$

For $\tau_\mu^\pm$, we have

$$\tau_\mu^+\tau_\nu^-=\delta_{\mu\nu}+i\eta_{a\mu\nu}\tau^a,\quad \tau_\mu^-\tau_\nu^+=\delta_{\mu\nu}+i\bar{\eta}_{a\mu\nu}\tau^a. \tag{57'}$$

It is not difficult to find the law of transformation of the matrix (56),

$$e^{i\varphi_1^aI_1^a+i\varphi_2^aI_2^a}i\tau_\mu^+x_\mu=e^{-i\varphi_1^a(\tau/2)^a}(i\tau_\mu^+x_\mu)\,e^{i\varphi_2^a(\tau^a/2)},$$

where φ_1^a and φ_2^a are parameters of the rotations, i.e., there is multiplication by unitary unimodular matrices from the left and the right.

The choice of S in the form $S_1=ix_\mu\tau_\mu^+/\sqrt{x^2}$ distinguishes certain directions in the isotopic and coordinate spaces. However, under rotation through the same angles in the spatial SU(2)×SU(2) group and in the SU(2)×SU(2) group given by the transformations (52) and (53), the matrix S_1 obviously does not change. In other words, if instead of I_1^a and I_2^a we call $I_1^a+T_1^a$ and $I_2^a+T_2^a$, where $T_{1,2}^a$ are the operators of the infinitesimal transformations (52) and (53), the angular momentum operators, the introduced object has spin zero.

Thus, we see that the group SU(2) is distinguished on account of the dimension of the coordinate space.

c) Value of the action for instanton solutions

Although we do not yet have the explicit form of the instanton solution, we can nevertheless calculate the value of the action for it. Indeed, for positive values of the topological charge n, the Euclidean action can be rewritten in the form

$$\begin{aligned}S=\int d^4x\,\tfrac{1}{4}G_{\mu\nu}^aG_{\mu\nu}^a&=\int d^4x\left[\tfrac{1}{4}G_{\mu\nu}^a\tilde{G}_{\mu\nu}^a+\tfrac{1}{8}(G_{\mu\nu}^a-\tilde{G}_{\mu\nu}^a)^2\right]\\ &=n\frac{8\pi^2}{g^2}+\frac{1}{8}\int d^4x\,(G_{\mu\nu}^a-\tilde{G}_{\mu\nu}^a)^2.\end{aligned} \tag{58}$$

It is clear from this formula that in the class of functions with given positive n the minimum of S is attained for $G_{\mu\nu}^a=\tilde{G}_{\mu\nu}^a$ and is equal to $(8\pi^2/g^2)$. We recall that specification of n does not signify that we seek a conditional extremum, since functions with different n cannot be related by a continuous deformation if the action is to remain finite.

The case of negative n is obtained from (58) by the reflection $x_{1,2,3}\to -x_{1,2,3}$, under which $G_{\mu\nu}\tilde{G}_{\mu\nu}\to -G_{\mu\nu}\tilde{G}_{\mu\nu}$ and accordingly $n\to -n$. Thus, the minimum of the action for negative n is $(8\pi^2/g^2)|n|$, and it is attained when $G_{\mu\nu}^a=-\tilde{G}_{\mu\nu}^a$.

As can be seen from this discussion, fulfillment of the self-duality and antiself-duality conditions $G_{\mu\nu}^a=\pm\tilde{G}_{\mu\nu}^a$ automatically leads to satisfaction of the equations of motion $D_\mu G_{\mu\nu}=0$. This can also be seen directly; indeed, for a self-dual field, say, we have

$$D_\mu G_{\mu\nu}^a=D_\mu\tilde{G}_{\mu\nu}^a=\tfrac{1}{2}\epsilon_{\mu\nu\gamma\delta}D_\mu G_{\gamma\delta}^a=\tfrac{1}{6}\epsilon_{\mu\nu\gamma\delta}(D_\mu G_{\gamma\delta}^a+D_\gamma G_{\delta\mu}^a+D_\delta G_{\mu\gamma}^a)=0,$$

where we have used the Bianchi identity:

$$D_\mu G_{\gamma\delta}+D_\delta G_{\mu\gamma}+D_\gamma G_{\delta\mu}=0.$$

7. EXPLICIT FORM OF THE BPST INSTANTON

a) Solution with $n=1$

As discussed in the previous section, the asymptotic behavior of A_μ^a for this solution is

$$\begin{aligned}&g\frac{\tau^a}{2}A_\mu^a\xrightarrow[x\to\infty]{}iS_1\partial_\mu S_1^+,\\ &S_1=\frac{i\tau_\mu^+x_\mu}{\sqrt{x^2}},\end{aligned} \tag{59}$$

where the matrices $\tau_\mu^\pm$ are defined in (57). We shall also use the symbols $\eta_{a\mu\nu}$ and $\bar{\eta}_{a\mu\nu}$ defined by Eqs. (55). These numerical coefficients are frequently called the 't Hooft symbols, and some useful relations for $\eta_{a\mu\nu}$ are given in subsection c) of this section.

The expression for the asymptotic behavior of A_μ^a can be rewritten in terms of the 't Hooft symbols as follows:

$$A_\mu^a\xrightarrow[x\to\infty]{}\frac{2}{g}\eta_{a\mu\nu}\frac{x_\nu}{x^2}.$$

For an instanton with center at the point $x=0$, it is natural to assume the same angular dependence of the field for all x, i.e., to seek the solution in the form

$$A^a_\mu(x) = \frac{2}{g}\eta_{a\mu\nu}x_\nu\frac{f(x^2)}{x^2}, \tag{60}$$

where $f(x^2) \xrightarrow[x^2\to\infty]{} 1$, $f(x^2) \xrightarrow[x^2\to 0]{} \mathrm{const}\cdot x^2$. The last condition corresponds to the absence of a singularity at the origin. A justification for the assumption (60) will be the construction of a self-dual expression for $G^a_{\mu\nu}$. From (60), we obtain for $G^a_{\mu\nu}$

$$G^a_{\mu\nu} = -\frac{4}{g}\left\{\eta_{a\mu\nu}\frac{f(1-f)}{x^2} + \frac{x_\mu\eta_{a\nu\gamma}x_\gamma - x_\nu\eta_{a\mu\gamma}x_\gamma}{x^4}\,[f(1-f) - x^2f']\right\}. \tag{61}$$

In deriving (61), we have used the relation for $\varepsilon^{abc}\times\eta_{b\mu\gamma}\eta_{c\nu\delta}$ from the list of formulas in subsection c) at the end of this section. Using the formula for $\varepsilon_{\mu\nu\gamma\delta}\eta_{a\delta\rho}$ from the same list, we obtain for $\tilde G^a_{\mu\nu}$ the expression

$$\tilde G^a_{\mu\nu} = -\frac{4}{g}\left\{\eta_{a\mu\nu}f' - \frac{1}{x^4}(x_\mu\eta_{a\nu\gamma}x_\gamma - x_\nu\eta_{a\mu\gamma}x_\gamma)\,[f(1-f) - x^2f']\right\}.$$

The condition of self-duality, $G^a_{\mu\nu} = \tilde G^a_{\mu\nu}$, requires fulfillment of the equation $f(1-f) - x^2f' = 0$, which determines the function f:

$$f(x^2) = \frac{x^2}{x^2+\rho^2}, \tag{62}$$

where ρ^2 is a constant of integration; ρ is called the scale of the instanton. The translational invariance guarantees the obtaining of a solution with center at an arbitrary point x_0, for which it is necessary to replace x by $x - x_0$.

Thus, the final expression for the instanton with center at the point x_0 and scale ρ has the form

$$\begin{aligned} A^a_\mu &= \frac{2}{g}\eta_{a\mu\nu}\frac{(x-x_0)_\nu}{(x-x_0)^2+\rho^2}, \\ G^a_{\mu\nu} &= -\frac{4}{g}\eta_{a\mu\nu}\frac{\rho^2}{[(x-x_0)^2+\rho^2]^2}. \end{aligned} \tag{63}$$

It can now be verified that the action for the instanton is $8\pi^2/g^2$, as was shown in general form. The anti-instanton is obtained by the substitution $\eta_{a\mu\nu} \to \bar\eta_{a\mu\nu}$.

b) Singular gauge. The 't Hooft ansatz

It is frequently convenient to use the expression for A^a_μ in the so-called singular gauge, when the "bad" behavior *of A^a_μ is transferred from the point at infinity* to the center of the instanton. As was discussed in the previous section, such a transfer can be realized by a gauge transformation with a matrix $U(x)$ which becomes identical with $S(x)$ as $x \to \infty$.[7)] We write down the formulas of the gauge transformation,

$$\begin{aligned} g\frac{\tau^a}{2}\bar A^a_\mu &= U^+ g\frac{\tau^a}{2}A^a_\mu U + iU^+\partial_\mu U, \\ g\frac{\tau^a}{2}\overline{G^a_{\mu\nu}} &= U^+ g\frac{\tau^a}{2}G^a_{\mu\nu}U, \end{aligned} \tag{64}$$

and for an instanton with center at x_0 take a matrix of the form

$$U = \frac{i\tau^+_\mu (x-x_0)_\mu}{\sqrt{(x-x_0)^2}}. \tag{64'}$$

Then for the potential $\bar A^a_\mu$ and the intensities $\bar G^a_{\mu\nu}$ in the singular gauge we obtain

$$\begin{aligned} \bar A^a_\mu &= \frac{2}{g}\bar\eta_{a\mu\nu}(x-x_0)_\nu\frac{\rho^2}{(x-x_0)^2[(x-x_0)^2+\rho^2]}, \\ \bar G^a_{\mu\nu} &= -\frac{8}{g}\left[\frac{(x-x_0)_\mu(x-x_0)_\rho}{(x-x_0)^2} - \frac{1}{4}\delta_{\mu\rho}\right]\bar\eta_{a\nu\rho}\frac{\rho^2}{[(x-x_0)^2+\rho^2]^2} - (\mu \leftrightarrow \nu). \end{aligned} \tag{65}$$

It is obvious that the quantities $G^a_{\mu\nu}G^a_{\gamma\delta}$ are invariants of the gauge transformation (see, however, the last footnote). Note also the circumstance that (65) contains the symbols $\bar\eta_{a\mu\nu}$ but not $\eta_{a\mu\nu}$. This difference is due to the fact that in the singular gauge the topological charge (49) is accumulated in the neighborhood of $x = x_0$ and not at infinity.

The expression (65) for $\bar A^a_\mu$ can be rewritten in the form

$$\bar A^a_\mu = -\frac{1}{g}\bar\eta_{a\mu\nu}\partial_\nu \ln\left[1 + \frac{\rho^2}{(x-x_0)^2}\right]. \tag{66}$$

As was noted by 't Hooft, this expression can be generalized to a topological charge n greater than unity. Indeed, if

$$A^a_\mu = -\frac{1}{g}\bar\eta_{a\mu\nu}\partial_\nu \ln W(x), \tag{67}$$

then for $G^a_{\mu\nu} - \tilde G^a_{\mu\nu}$ we obtain [see the properties of the η symbols in subsection c)]

$$G^a_{\mu\nu} - \tilde G^a_{\mu\nu} = \frac{1}{g}\eta_{a\mu\nu}\frac{\partial_\gamma\partial_\gamma W}{W}.$$

The self-duality of $G^a_{\mu\nu}$ requires fulfillment of the equation $\partial_\gamma\partial_\gamma W/W = 0$. The solution with topological charge n has the form

$$W = 1 + \sum_{i=1}^{n}\frac{\rho_i^2}{(x-x_i)^2}, \tag{68}$$

i.e., it describes instantons with centers at the points x_i. The effective scale of an instanton with center at the point x_i is obviously

$$\rho_i^{eff} = \rho_i\left[1 + \sum_{k\neq i}\frac{\rho_k^2}{(x_k-x_i)^2}\right]^{-1/2}$$

It should be noted that the choice of A^a_μ in the form (67) did not give the most general solution with charge n, since all n instantons have the same orientation in the isotopic space (for the construction of the general solution, see Ref. 10).

c) Relations for the η symbols

We give a list of relations for the symbols $\eta_{a\mu\nu}$ and $\bar\eta_{a\mu\nu}$ defined by Eqs. (55):

$$\eta_{a\mu\nu} = \frac{1}{2}\varepsilon_{\mu\nu\alpha\beta}\eta_{a\alpha\beta},$$

$$\eta_{a\mu\nu} = -\eta_{a\nu\mu}, \qquad \eta_{a\mu\nu}\eta_{b\mu\nu} = 4\delta_{ab},$$

$$\eta_{a\mu\nu}\eta_{a\mu\lambda} = 3\delta_{\nu\lambda}, \qquad \eta_{a\mu\nu}\eta_{a\mu\nu} = 12,$$

$$\eta_{a\mu\nu}\eta_{a\gamma\lambda} = \delta_{\mu\gamma}\delta_{\nu\lambda} - \delta_{\mu\lambda}\delta_{\nu\gamma} + \varepsilon_{\mu\nu\gamma\lambda},$$

$$\varepsilon_{\mu\nu\lambda\sigma}\eta_{a\gamma\sigma} = \delta_{\gamma\mu}\eta_{a\nu\lambda} - \delta_{\gamma\nu}\eta_{a\mu\lambda} + \delta_{\gamma\lambda}\eta_{a\mu\nu},$$

$$\eta_{a\mu\nu}\eta_{b\mu\lambda} = \delta_{ab}\delta_{\nu\lambda} + \varepsilon_{abc}\eta_{c\nu\lambda},$$

$$\varepsilon_{abc}\eta_{b\mu\nu}\eta_{c\gamma\lambda} = \delta_{\mu\gamma}\eta_{a\nu\lambda} - \delta_{\mu\lambda}\eta_{a\nu\gamma} - \delta_{\nu\gamma}\eta_{a\mu\lambda} + \delta_{\nu\lambda}\eta_{a\mu\gamma},$$

$$\eta_{a\mu\nu}\bar\eta_{b\mu\nu} = 0, \quad \eta_{a\gamma\mu}\bar\eta_{b\gamma\lambda} = \eta_{a\gamma\lambda}\bar\eta_{b\gamma\mu}.$$

[7)] More precisely, this transformation should be called a quasigauge transformation, since at the point where $U(x)$ has a singularity (and there must be such a singularity) this transformation changes the gauge-invariant quantities, for example, $G^a_{\beta\nu}G^a_{\mu\nu}$. To use such transformations, it is necessary to consider a space with the neighborhoods of the singular points deleted. This we shall do, remembering that the physical quantities are nonsingular at the singular points.

To go over from the relations for $\eta_{a\mu\nu}$ to those for $\bar{\eta}_{a\mu\nu}$ it is necessary to make the substitution

$$\eta_{a\mu\nu}\to\bar{\eta}_{a\mu\nu},\qquad \varepsilon_{\mu\nu\gamma\delta}\to-\varepsilon_{\mu\nu\gamma\delta}.$$

8. CALCULATION OF THE PRE-EXPONENTIAL FACTOR FOR THE BPST INSTANTON

a) Expansion near a saddle point. Choice of the gauge and regularization

As in the quantum-mechanical example, to calculate the pre-exponential factor in the instanton contribution to the vacuum-vacuum transition, it is necessary to represent the field A^a_μ in the form

$$A^a_\mu=A^{a(\mathrm{ins})}_\mu+a^a_\mu \tag{69}$$

and expand the action $S(A)$ with respect to the deviation a^a_μ from the instanton field $A^{a(\mathrm{ins})}_\mu$:

$$\begin{aligned}S(A)&=S_0+\frac{1}{2}\int d^4x\,a^a_\mu L^{ab}_{\mu\nu}(A^{\mathrm{ins}})\,a^b_\nu\\&=\frac{8\pi^2}{g^2}+\frac{1}{2}\int d^4x\,a^a_\mu\,[D^2a^a_\mu-D_\mu D_\nu a^a_\nu-2g\varepsilon^{abc}G^b_{\mu\nu}a^c_\nu],\end{aligned} \tag{70}$$

where the instanton field is substituted in D_μ and $G_{\mu\nu}$. As in the one-dimensional case, the integration with respect to the deviations a_μ reduces to calculation of the determinant of the operator $L^{ab}_{\mu\nu}$. There are however two important differences from the one-dimensional case:

The operator L is degenerate due to the gauge invariance. Indeed, fields a^a_μ of the form $a^a_\mu=(D_\mu\lambda)^a$ with arbitrary function $\lambda^a(x)$ make the quadratic form (70) vanish. In order to have the possibility of working with a degenerate form of this kind, it is necessary to fix the gauge. This can be done conveniently by adding to the action the term

$$\Delta S=\frac{1}{2}\int d^4x\,(D_\mu a^a_\mu)^2=\frac{1}{2}\int d^4x\,a^a_\mu(\Delta L)^{ab}_{\mu\nu}a^b_\nu, \tag{71}$$

which lifts the degeneracy. To avoid changing the content of the theory, we must, as is well known, simultaneously add Faddeev–Popov ghosts:

$$\Delta S_{gh}=\int d^4x\,\bar{\Phi}^aD^2\Phi^a=\int d^4x\,\bar{\Phi}^aL^{ab}_{gh}\Phi^b, \tag{72}$$

where Φ^a is a complex anticommuting field. As a result, the instanton contribution can be written in the form

$$\langle 0\,|\,0_T\rangle_{\mathrm{ins}}=[\det(L+\Delta L)]^{-1/2}(\det L_{gh})\,e^{-S_0}, \tag{73}$$

where $|0_T\rangle$ is the vacuum after time T, $|0_T\rangle=e^{-HT}|0\rangle$, H is the Hamiltonian, $S_0=8\pi^2/g^2$, $(L+\Delta L)^{ab}_{\mu\nu}$ is the operator in the quadratic form of the fields a^a_μ, and L_{gh} acts on the ghost fields. The determinant of L_{gh} occurs in a positive power, since $\Phi^a,\bar{\Phi}^a$ are anticommuting fields.

A second difference from the one-dimensional case is the presence in the theory of ultraviolet divergences. By virtue of the renormalizability, all the divergences must be eliminated by a renormalization of the coupling constant, but it is first necessary to regularize the expressions under consideration. The regularization can be done as follows. Instead of the determinant of the operator $L+\Delta L$ we consider the ratio $\det(L+\Delta L)/\det(L+\Delta L+M^2)$, where the introduction of the cutoff parameter M can be interpreted as the addition to the theory of a Pauli–Villars vector field with mass M. The determinant of L_{gh} is regularized similarly. Thus, it is necessary to calculate

$$\langle 0\,|\,0_T\rangle^{\mathrm{Reg}}_{\mathrm{ins}}=\left[\frac{\det(L+\Delta L)}{\det(L+\Delta+M^2)}\right]^{-1/2}\frac{\det L_{gh}}{\det(L_{gh}+M^2)}\,e^{-S_0}, \tag{74}$$

or, more precisely, the ratio of $\langle 0|0_T\rangle^{\mathrm{Reg}}_{\mathrm{ins}}$ to the corresponding perturbation-theoretical quantity $\langle 0|0_T\rangle_{\mathrm{p.th}}$, which differs in having $A^a_\mu\equiv0$ substituted instead of the instanton field. For $A^a_\mu=0$, it is obvious that $S_0=0$, while for the instanton $S_0=8\pi^2/g_0^2$, where the subscript in the coupling constant g_0 emphasizes that this is the unrenormalized coupling constant normalized by the cutoff parameter M, $g_0=g(M)$.

We shall not go into a detailed exposition of 't Hooft's calculations for $\langle 0|0_T\rangle/\langle 0|0_T\rangle_{\mathrm{p.th}}$ but obtain the result up to a numerical factor. Study of the zero-frequency modes plays the main part in obtaining the result.

b) Zero-frequency modes

As was shown in the one-dimensional example, each zero-frequency modes leads in $[\det(L+\Delta L)]^{-1/2}$ to a factor proportional to $\sqrt{S_0}$ and an integral with respect to a corresponding collective coordinate. What are the collective coordinates in the case of the BPST instanton in the group SU(2)?

First, there are the four coordinates of the center x_0, then the scale ρ, and, finally, the three Eulerian angles θ,φ,ψ, which specify the orientation of the instanton in the isospace. The spatial rotations need not be counted, since they are equivalent to isorotations (see Sec. 6b).

As a result of the regularization, $[\det(L+\Delta L)]^{-1/2}$ is multiplied by $[\det(L+\Delta L+M^2)]^{1/2}$, i.e., each zero-frequency mode gives rise to a factor M. Thus, from all (since we have listed *all* collective coordinates) zero-frequency modes there arises in $\langle 0|0_T\rangle^{\mathrm{Reg}}_{\mathrm{ins}}$ the factor

$$\int d^4x_0\,d\rho\,\sin\theta\,d\theta\,d\varphi\,d\psi\,M^8(\sqrt{S_0})^8\rho^3. \tag{75}$$

The factor ρ^3 arises from the Jacobian of the transition to integration over θ,φ,ψ and is recovered on the basis of dimensional considerations.

Using (75) we rewrite $\langle 0|0_T\rangle^{\mathrm{Reg}}_{\mathrm{ins}}/\langle 0|0_T\rangle_{\mathrm{p.th}}$ in the form

$$\frac{\langle 0\,|\,0_T\rangle^{\mathrm{Reg}}_{\mathrm{ins}}}{\langle 0\,|\,0_T\rangle_{\mathrm{p.th}}}=\mathrm{const}\int\frac{d^4x\,d\rho}{\rho^5}\left(\frac{8\pi^2}{g_0^2}\right)^4\exp\left(-\frac{8\pi^2}{g_0^2}+8\ln M\rho+\Phi_1\right), \tag{76}$$

where $\exp\Phi_1$ denotes the contribution of the positive-frequency modes.

c) Positive-frequency modes. Effective charge

The quantity Φ_1 depends on the dimensionless parameter $M\rho$ and in the limit $M\rho\gg1$ can be readily found by means of ordinary perturbation theory. Indeed, calculation of the pre-exponential factory by retaining the terms quadratic in the deviation from the external field corresponds to calculation of the single-loop corrections in perturbation theory. We are here referring to diagrams of the form

 (77)

where the cross denotes vertices of the interaction with the external field, and the broken lines correspond to the propagators of the fields a^a_μ (plus similar loops with the ghosts $\Phi^a, \bar{\Phi}^a$); the external field has the form $A^{a(\mathrm{ins})}_\mu$.

It is clear that complete calculation of the contribution of the zero-frequency modes requires summation of a complete chain of diagrams—the zero-frequency modes do not appear in any finite order. A manifestation of this nonanalyticity is the presence of the term $\ln(8\pi^2/g_0^2)$ in $\ln\langle 0|0_T\rangle_{\mathrm{ins}}$. It is also clear that there is no nonanalyticity of this kind for the positive-frequency modes.

In the limit in which we are interested, $M\rho \gg 1$, only the first of the diagrams (77) is important in the calculation, since all the following diagrams are convergent and do not give a dependence on the cutoff parameter M [they change the constant in (76)]. Moreover, in the second order in the external field it can be seen that the contribution of the positive-frequency modes is given by an unsubtracted dispersion relation for the polarization operator $\Pi^{ab}_{\mu\nu}$.

The imaginary part of $\Pi^{ab}_{\mu\nu}$ is obtained by cutting the first diagram (77) and is well defined. In its calculation, it is necessary to take into account only quanta with three-dimensionally transverse polarization states; the unphysical polarizations and ghosts are not necessary. Omitting the details of this simple calculation, we give the result for $\mathrm{Im}\,\Pi_{\mu\nu}$:

$\mathrm{Im}\,\Pi^{ab}_{\mu\nu} =$  $= \delta^{ab}(g_{\mu\nu}k^2 - k_\mu k_\nu)\frac{g^2}{16\pi}\cdot\frac{2}{3}.$

Writing down the unsubtracted dispersion representation for $\Pi^{(1)}_{\mu\nu}$ (the part of the polarization operator associated with the positive-frequency modes), we obtain

$$\Pi^{ab(1)}_{\mu\nu} = \delta^{ab}(g_{\mu\nu}k^2 - k_\mu k_\nu)\frac{1}{\pi}\int\frac{ds}{s-k^2}\cdot\frac{2}{3}\frac{g^2}{16\pi}$$
$$= \delta^{ab}(g_{\mu\nu}k^2 - k_\mu k_\nu)\frac{2}{3}\frac{g^2}{16\pi^2}\ln\frac{M^2}{-k^2}, \quad (78)$$

where we have terminated the integration over s at M^2, since the regularization involves a subtraction of an analogous contribution with Pauli–Villars particles of mass M.

The result (78) for the contribution of the positive-frequency modes means that the action for the external field acquires from these quantum corrections the effective addition

$$\Delta S^{\mathrm{Mink}} = \frac{2}{3}\frac{g^2}{16\pi^2}\ln M^2\rho^2\int d^4x\left[-\frac{1}{4}(G^a_{\mu\nu})^2\right]. \quad (79)$$

where we use the notation of pseudo-Euclidean space and have replaced $1/(-k^2)$ by the square ρ^2 of the characteristic scale of the field (strictly speaking, we ought to write a differential operator, but for the calculation of the coefficient of $\ln M\rho$ this is not important). Going over to the Euclidean action and substituting the instanton $G^a_{\mu\nu}$, we obtain the result for Φ_1:

$$\Phi_1 = \frac{2}{3}\ln M\rho. \quad (80)$$

Thus, allowance for the zero-frequency and positive-frequency modes has the consequence that $8\pi^2/g_0^2$ in the argument of the exponential (76) is replaced by the effective charge $8\pi^2/g^2(\rho)$:

$$\frac{8\pi^2}{g^2(\rho)} = \frac{8\pi^2}{g_0^2} - 8\ln M\rho + \frac{22}{3}\ln M\rho = \frac{8\pi^2}{g_0^2} - \frac{22}{3}\ln M\rho. \quad (81)$$

Of course, this result is a direct consequence of the renormalizability, and we have wasted time on its derivation only to emphasize the very beautiful explanation of the antiscreening of the charge in a non-Abelian theory which arises when the zero-frequency modes are considered.

Indeed, both the sign and the magnitude of the coefficient of the "antiscreening" logarithm (76) are obvious consequences of the above—the coefficient is simply the number of zero-frequency modes.

In the framework of the perturbation-theoretical calculations, the "antiscreening" result can be most clearly explained in the framework of the ghostless Coulomb gauge, which was used in calculations by Khriplovich[11] as early as 1969. Besides the "dispersion" part, the calculation of which we have discussed above, the polarization operator in this gauge contains a contribution that does not have an imaginary part and arises when one of the virtual quanta has a three-dimensionally transverse polarization and the second is a Coulomb quantum. The opposite signs of the "nondispersion" and "dispersion" parts of $\Pi_{\mu\nu}$ correspond to the opposite signs of interactions due to the exchange of a Coulomb quantum and a transverse quantum (electric forces repel charges of the same sign, while magnetic forces attract currents of the same type).

The calculation of the "nondispersion" part in the Coulomb gauge *requires care, since it is necessary* to use the noncovariant Hamiltonian formalism, and the coefficient of the logarithm is not, of course, known *a priori*. As we have seen, none of these problems arise in the determination of the contribution of the zero-frequency modes. With this we conclude our panegyric to the zero-frequency modes.

d) Two-loop approximation

The above calculations led to replacement of the unrenormalized coupling constant g_0 in the classical action by the effective constant $g(\rho)$. However, the *unrenormalized constant still remains in the factor* $(8\pi^2/g_0^2)^4$ [see (76)], though it is clear that, because of the renormalizability, it should not occur in the result. The reason for this is that the accuracy obtained by using the single-loop approximation is inadequate to distinguish the factor $(8\pi^2/g_0^2)^4$ from $[8\pi^2/g^2(\rho)]^4$, and we require a two-loop calculation.

We show that from the two-loop calculation we actually require only the expression for the effective charge; such an expression is known from perturbation theory,[12]

$$\frac{8\pi^2}{g^2(\rho)} = \frac{8\pi^2}{g^2(\rho_0)} + N\left[\frac{11}{3}\ln\frac{\rho_0}{\rho} + \frac{17}{11}\ln\left(1 + \frac{11}{3}N\frac{g^2(\rho_0)}{8\pi^2}\ln\frac{\rho_0}{\rho}\right)\right], \quad (82)$$

where we have given the result for the group SU(N) (without the contribution of fermions). The unrenormalized constant is $g_0 = g(\rho_0 = 1/M)$. The instanton contribution to the vacuum–vacuum transition for the group SU(2) has the form

$$\frac{\langle 0 \mid 0_T\rangle^{\rm Reg}_{\rm ins}}{\langle 0 \mid 0_T\rangle_{\rm p.\,th}} = {\rm const}\cdot\left[\frac{8\pi^2}{g^2(\rho)}\right]^4 e^{-8\pi^2/g^2(\rho)}\,(1+O(g^2(\rho))), \tag{83}$$

where $g^2(\rho)$ is given by the expression (82) with $N=2$. For the factor $[8\pi^2/g^2(\rho)]^4$, we can restrict ourselves to the single-loop expression for $g^2(\rho)$, the difference being of the order of the ignored terms which give relative corrections of order $g^2(\rho)$. Note that the complete two-loop calculation of the instanton contribution would determine these corrections.

The proof of the correctness of (83) is based on the renormalizability of the theory and the method of effective Lagrangians. In the functional integral, we integrate in the spirit of Wilson over fields of small scale (less than ρ_0), i.e., over configurations corresponding to instantons with small $\rho < \rho_0$. As a result, we obtain an effective Lagrangian of the fields with scales greater than ρ_0. In this Lagrangian, the small-scale fluctuations are taken into account in the coefficients of the expansion with respect to the operators.

The calculation of the contribution of the instantons to the vacuum–vacuum transition is equivalent to determination of their contribution to the coefficient of the identity operator. The calculation of the coefficients of the other operators will be considered in Sec. 10. A specific feature of the identity operator is the fact that its matrix elements are independent of the normalization point; ρ_c is the zero-frequency anomalous dimension. Therefore, the coefficient of it, expressed in terms of $g(\rho)$, cannot contain ρ_c (for operators with positive-frequency anomalous dimension the factor $[g^2/(\rho_c)/g^2(\rho)]^b$ arises).

It now only remains to express $g^2(\rho)$ in terms of $g^2(\rho_0)$ by means of the renormalization-group equations, and the retention of the two-loop correction in (82) is fully valid.

e) Density of instantons in the group SU(*N*)

How does the number of zero-frequency modes change on the transition to the group SU(N)? We have already said that the instanton field uses only a SU(2) subgroup of the complete group. Suppose this subgroup occupies the top left-hand corner in the $N\times N$ matrix of generators. It is clear that the five zero-frequency modes associated with shifts and dilatations remain the same as in the group SU(2), and only the modes associated with group rotations are changed. In SU(2) there were three, and in SU(N) they correspond to three generators in a 2×2 matrix at the top left (Fig. 6). Those of the remaining generators that occur in the $(N-2)\times(N-2)$ matrix in the bottom right obviously do not rotate the instanton field. Thus, to the three SU(2) rotations there are added a further $4(N-2)$ unitary rotations. The total number of zero-frequency modes is $5+3+4(N-2)=4N$. Of course, this number $4N$ exactly corresponds to the coefficient of the "antiscreening" logarithm in the formula for $8\pi^2/g^2(\rho)$. Finally, we write down an expression for the reduced instanton density $d(\rho)$, which is defined as follows:

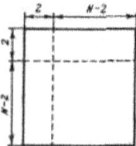

FIG. 6.

$$\frac{\langle 0 \mid 0_T\rangle^{\rm Reg}_{\rm ins}}{\langle 0 \mid 0_T\rangle_{\rm p.\,th}} = \int \frac{d^4x\,d\rho}{\rho^5}\,d(\rho). \tag{84}$$

The function $d(\rho)$ is equal to

$$d(\rho) = \frac{C_1}{(N-1)!\,(N-2)!}\left[\frac{8\pi^2}{g^2(\rho)}\right]^{2N} e^{-[8\pi^2/g^2(\rho)]-C_2N}, \tag{85}$$

where $g^2(\rho)$ is expressed in terms of $g_0^2=g^2(\rho_0=1/M)$ by formula (82), and the constants C_1 and C_2 can be found by a certain modification of 't Hooft's calculations.[13] Concretely, it is necessary to take into account a further $4(N-2)$ vector fields with the above quantum numbers in both the zero-frequency and the positive-frequency modes. In addition, we require the embedding volume of SU(2) in SU(N); the factor $[(N-1)!\,(N-2)!]^{-1}$ is associated with it. This part of the modification proved to be the most complicated (see Ref. 13). The result for C_1 and C_2 has the form

$$C_1 = \frac{2e^{5/6}}{\pi^2} = 0.466,$$

$$C_2 = \frac{5}{3}\ln 2 - \frac{17}{36} + \frac{1}{3}(\ln 2\pi + \gamma) + \frac{2}{\pi^2}\sum_{s=1}^{\infty}\frac{\ln s}{s^2} = +1.679. \tag{86}$$

Note that the constant C_2 depends on the method of regularization, which actually provides the definition of the unrenormalized constant. Instead of Pauli–Villars regularization (PV scheme), so-called dimensional regularization is frequently used. Instead of logarithms of the cutoff parameter, poles with respect to the dimension of space arise in this method, $\ln M - 1/(4-D)$. Use of the minimal scheme[14] (MS) for determining the coupling constant leads to an expression of the form (85) with the substitution

$$g(\rho) \to g_{\rm MS}(\rho) \qquad C_2 \to C_{2\rm MS},$$

$$C_{2\rm MS} = C_2 - \frac{5}{36} - \frac{11}{6}(\ln 4\pi - \gamma) = C_2 - 3.721. \tag{87}$$

The numerical coefficient in $d(\rho)$ for the MS scheme is $e^{3.72N}$ times greater than in the PV scheme, which for SU(3) gives the factor $\sim 7\cdot 10^4$.

Of course, the relations between the observable amplitudes do not depend on the definition of g^2—the same conversion constants associated with the change of regularization occur, for example, in the corrections in g^2 to the cross section of e^+e^- annihilation into hadrons (though there, it is true, the dependence on them is not exponential). We note in this connection that in perturbation theory the $\overline{\rm MS}$ scheme has proved helpful, since in it too large coefficients of the expansion in g^2 do not arise.[15] The difference between the $\overline{\rm MS}$ scheme and the MS scheme reduces to the substitution

$$\frac{8\pi^2}{g^2_{\overline{MS}}} = \frac{8\pi^2}{g^2_{MS}} - \frac{11}{6}N(\ln 4\pi - \gamma),$$
$$C_{2\overline{MS}} = C_2 - \frac{5}{36} \approx 1.54. \tag{88}$$

We give finally the explicit form of the dependence on ρ for the function $d(\rho)$:

$$d(\rho)=\frac{0.466}{(N-1)!\,(N-2)!}\left(\frac{\rho}{\rho_0}\right)^{11N/3}\left[1+\frac{11}{3}N\frac{g^2(\rho_0)}{8\pi^2}\ln\frac{\rho}{\rho_0}\right]^{5N/11}\times\left[\frac{8\pi^2}{g^2(\rho_0)}\right]^{2N}e^{-(8\pi^2/g^2(\rho_0))-1.679N}. \quad (89)$$

9. INSTANTON GAS AND GENERAL THEOREMS

The calculated instanton contribution is proportional to $\int d^4x_0=V_4$, the volume of the considered region of the Euclidean space. As long as $V_4\rho^{-4}d(\rho)$, the probability of finding an instanton of scale ρ in the considered volume, is a small quantity, one can ignore fluctuations for which there are two or more instantons of scale ρ in this volume. But with increasing V_4, we naturally arrive at the need to consider an instanton gas.

As in the one-dimensional case, the vacuum-vacuum transition has the form

$$\langle 0\,|\,0_T\rangle=\langle 0\,|\exp\left(-\int d^4x\mathscr{H}\right)|\,0\rangle=e^{-\varepsilon V_4}, \quad (90)$$

where $\mathscr{H}$ is the Hamiltonian density, and ε can be called the vacuum energy density. Clearly, for the summation it is convenient to consider the logarithm of (90), i.e., the quantity ε:

$$\begin{aligned}\varepsilon&=-\frac{1}{V_4}\ln\langle 0\,|\,0_T\rangle=-\frac{1}{V_4}\ln\left[\langle 0\,|\,0_T\rangle_{\mathrm{p.th}}+\langle 0\,|\,0_T\rangle_{\mathrm{ins}}\right]\\&\approx-\frac{1}{V_4}\left[\ln\langle 0\,|\,0_T\rangle_{\mathrm{p.th}}+\frac{\langle 0\,|\,0_T\rangle_{\mathrm{ins}}}{\langle 0\,|\,0_T\rangle_{\mathrm{p.th}}}\right]\approx\varepsilon_{\mathrm{p.th}}-\int\frac{d\rho}{\rho^5}d(\rho).\end{aligned} \quad (91)$$

Thus, the correction to the vacuum energy density in *the gas approximation is negative and given by the integral* $\int d\rho\,\rho^{-5}d(\rho)$.

Due to the power-law growth $d(\rho)\sim\rho^{11N/3}$, this integral is determined by large ρ, and the formal expression diverges as a power.

Unfortunately, $d(\rho)$ is known only in the region of fairly small ρ, which must be such as to guarantee that the ignored quantum corrections $\sim g^2(\rho)$ are small. In addition, ε contains a contribution of fluctuations with topological charge $|n|>1$, which, roughly speaking, is proportional to $[d(\rho)]^n$. Both these effects have the consequence that formula (91) does not hold at large ρ. Of particular interest is the possibility that fluctuations with large topological charge become important in the region of scales for which the corrections $\sim g^2(\rho)$ are still small. Such a situation appears all the more plausible because $d(\rho)$ increases with ρ much more rapidly than $g^2(\rho)$. The two-dimensional models analyzed in the interesting papers of Ref. 16 provide an example in which dense fluctuations with large topological charge are dominant in the vacuum wave function. In Ref. 16, this antigas situation was called melting of instantons.

The approximation of a dilute instanton gas was developed in Ref. 8. The approximation is based on the hypothesis that the phenomena associated with large ρ reduce effectively to the appearance of an upper limit ρ_m in the integral over ρ, and for all $\rho<\rho_m$ one can use the one-instanton formula (91) for $d(\rho)$.

In this subsection we shall demonstrate that the dilute gas hypothesis is not self-consistent by giving an example which violates a general relation. In the following section, we shall explicitly find the region of ρ in which the one-instanton expressions are valid on the basis of phenomenological information about the fields in the QCD vacuum. We shall see that the admissible ρ *are too small to make a claim to a description of the vacuum structure in the region of the main scales even* in order of magnitude.

The exact relation whose verification we have in mind is the connection between the vacuum energy density and the mean square intensity of the gluon field in the vacuum. For the derivation, we consider the vacuum expectation value of the energy-momentum tensor $\theta_{\mu\nu}(x)$. By virtue of relativistic invariance,

$$\langle 0\,|\,\theta_{\mu\nu}\,|\,0\rangle=g_{\mu\nu}\varepsilon, \quad (92)$$

from which, after summation, we deduce an expression for ε in terms of the vacuum expectation value of the trace of the energy-momentum tensor:

$$\varepsilon=\frac{1}{4}\langle 0\,|\,\theta_{\mu\mu}\,|\,0\rangle. \quad (93)$$

For $\theta_{\mu\mu}$ the gluodynamics with group SU(N) the following operator expression holds[17]:

$$\theta_{\mu\mu}=\frac{\beta(\alpha_s)}{4\alpha_s}G^a_{\mu\nu}G^a_{\mu\nu}\qquad(a=1,\ldots,N^2-1), \quad (94)$$

where $\alpha_s=g^2/4\pi$, and $\beta(\alpha_s)$ is the Gell-Mann-Low function,

$$\beta(\alpha_s)=\mu\frac{d\alpha_s(\mu)}{d\mu}=-\frac{11}{3}N\frac{\alpha_s^2}{2\pi}+O(\alpha_s^3). \quad (95)$$

The expression (94) *for* $\theta_{\mu\mu}$ *is called the trace anomaly* of the energy-momentum tensor. The point is that for a classical massless vector field $\theta_{\mu\mu}=0$. The difference from zero appears at the single-loop level and is associated with the need to introduce a guage-invariant cutoff.

The appearance in $\theta_{\mu\mu}$ of the function $\beta(\alpha_s)$, which controls the charge renormalization, can be explained as follows. The stretching $x\to\lambda x$ of all scales is determined in the infinitesimal form of the transformation by the dilatation operator D:

$$D=\int d^3x D_0(x),\qquad D_\mu(x)=\theta_{\mu\nu}x_\nu.$$

It is readily seen that there is invariance with respect to dilatations only when the divergence of the dilatation current D_μ vanishes. This divergence is

$$\partial_\mu D_\mu=\theta_{\mu\mu},$$

i.e., the operator $\theta_{\mu\mu}$ determines the noninvariance under dilatations. The noninvariance indicates the existence of a certain distance scale. In a massless theory, the only possibility for a scale to appear is associated with the need to introduce the cutoff parameter M when considering the quantum effects. Under the simultaneous transformations $x\to\lambda x, M\to M/\lambda$ the theory is invariant, i.e., dilatations are equivalent to a change in M. It is for this reason that $\theta_{\mu\mu}$ is proportional to $\beta(\alpha_s)=M\,d\alpha_s/dM$.

Considering the action of dilatation transformations

on transition amplitudes expressed in the form of path integrals, we can readily deduce the relation (94). We shall not give this derivation but restrict ourselves to two comments about it:

a) energy–momentum conservation, $\partial_\mu \theta_{\mu\mu} = 0$, has the consequence that the right-hand side of the relation is independent of the normalization point μ of the operator $G^a_{\mu\nu} G^a_{\mu\nu}$, and the effective charge α_s must be taken at the same point, $\alpha_s = \alpha_s(\mu)$;

b) the quantum corrections also lead to a cutoff-dependent c-number part in $\theta_{\mu\mu}$. Therefore, the more accurate expression is

$$\theta_{\mu\mu} = \langle 0 | \theta_{\mu\mu} | 0 \rangle_{\text{p. th}} + \frac{\beta(\alpha_s)}{4\alpha_s} G^a_{\mu\nu} G^a_{\mu\nu}, \tag{96}$$

where the c-number part is separated by averaging over the perturbation theory vacuum $|0\rangle$ (which differs from the exact physical vacuum $|0\rangle$).

Substituting the expression (96) for $\theta_{\mu\mu}$ in (93), we arrive at the desired connection between ε and the mean square of the field intensity in the vacuum:

$$\varepsilon = \varepsilon_{\text{p. th}} + \frac{\beta(\alpha_s)}{16\alpha_s} \langle 0 | G^a_{\mu\nu} G^a_{\mu\nu} | 0 \rangle. \tag{97}$$

It is clear from the derivation that in $\langle 0 | G^a_{\mu\nu} G^a_{\mu\nu} | 0 \rangle$ it is necessary to take into account only those fluctuations not given by perturbation theory.

Instantons are an example of such fluctuations. The one-instanton contribution to $\langle 0 | G^a_{\mu\nu} G^a_{\mu\nu} | 0 \rangle$ can be readily found, for which it is necessary to go over to the Euclidean space (above, we have used the notation of Minkowski space), replace the field $G^a_{\mu\nu}$ by the instanton field, and add the factor $d^4x_0 d\rho\, \rho^{-5} d(\rho)$, the probability of finding an instanton with scale ρ with center at x_0. The result for the one-instanton contribution to $\varepsilon - \varepsilon_{\text{p. th}}$, integrated over x_0 [the integral is $\int d^4x_0 G^a_{\mu\nu}(x - x_0) G^a_{\mu\nu}(x - x_0) = \int d^4x G^a_{\mu\nu}(x) G^a_{\mu\nu}(x) = 4 \cdot 8\pi^2/g^2$], is

$$\varepsilon - \varepsilon_{\text{p. th}} = -\frac{11}{12} N \int \frac{d\rho}{\rho^5} d(\rho) \tag{98}$$

[for $\beta(\alpha_s)$ we have used the single-loop approximation]. On the other hand, the one-instanton contribution to $\varepsilon - \varepsilon_{\text{p. th}}$ is given by the expression (91) obtained earlier and differs from (98) by the absence of the factor $11N/12$.

What does this mean? Since the integral over ρ is determined by large ρ, the paradox is resolved by noting that the one-instanton approximation does not give the possibility of finding $\varepsilon - \varepsilon_{\text{p. th}}$ and $\langle 0 | G^a_{\mu\nu} G^a_{\mu\nu} | 0 \rangle$. Moreover, the attempt to take into account the effects other than the one-instanton effects by introducing a cutoff in ρ is inconsistent in that this cannot be done in a unified manner even for quantities associated with general relations—the cutoff in them is effectively different.

To conclude the section, we note that the inadequacy of the one-instanton approximation for quantities such as ε or $\langle 0 | [\beta(\alpha_s)/\alpha_s] G^a_{\mu\nu} G^a_{\mu\nu} | 0 \rangle$ can also be proved by a somewhat different argument. Physical quantities, of course, are independent of the normalization point. For such quantities having dimension m^4, the normalization point μ can occur only in the (μ-independent) combination

$$\left\{ \mu \left[\frac{-2\pi}{\alpha_s(\mu)} \right]^{b_1/b} e^{-(1/b)2\pi/\alpha_s} \right\}^4 [1 + O(\alpha_s(\mu))], \tag{99}$$

where b and b_1 are the first and second coefficients in the expansion of the Gell-Mann–Low function:

$$\beta(\alpha_s) = -b\frac{\alpha_s^2}{2\pi} - b_1 \frac{\alpha_s^3}{(2\pi)^2} + O(\alpha_s^4). \tag{100}$$

In SU(N) gluodynamics [see (82)],

$$b = \frac{11}{3} N, \quad \frac{b_1}{b} = \frac{17}{11} N. \tag{100a}$$

On the other hand, the one-instanton approximation with a cutoff of the integral over ρ at the upper limit at ρ_m gives for the same parameters a result proportional to

$$\frac{1}{\rho_m^4} d(\rho_m) \sim \frac{1}{\rho_m^4} \left[\frac{2\pi}{\alpha_s(\rho_m)} \right]^{2N} e^{-2\pi/\alpha_s(\rho_m)}, \tag{101}$$

where we have used the fact that the main contribution is made by the region of ρ near ρ_m. Comparing (101) and (99) for $\mu = 1/\rho$, we see that the dependence on ρ_m does not agree with that required by renormalization invariance. The power of ρ_m is greater by the same $b/4 = 11N/12$ times, and the power of $\ln \rho_m$ also does not agree.

10. INSTANTONS IN THE QCD VACUUM

As we have already said, the main fluctuations in the QCD vacuum are those of large scales of the order of the confinement radius or, which is the same thing, the radius of hadrons. Unfortunately, we are not yet able to treat such fluctuations quantitatively.

The quasiclassical methods that have been developed apply to the study of nonperturbation-theoretical fluctuations of small scale, among which the instantons are dominant.

In this subsection we take into account the influence on the small-scale instantons of the fields due to the characteristic long-wavelength fluctuations in the vacuum.[18]

Since we distinguish fields of two types, namely, the fields of small-scale instantons and the fields of the characteristic vacuum fluctuations, it is convenient to introduce an effective Lagrangian. In it, as usual, the contribution of the rapidly varying fields is included in the coefficients of the various operators that act on the space of the slowly varying fields.

Thus, the effect of a distinguished instanton with scale ρ and center at x_0 reduces to the following correction to the effective Lagrangian of the long-wavelength fluctuations:

$$\Delta L(x_0) = \frac{d\rho}{\rho^5} \sum_n C_n(\rho) O_n(x_0),$$

where $C_n(\rho)$ are numerical coefficients and $O_n(x_0)$ are local operators constructed from the gluon fields (we consider pure gluodynamics; the changes introduced by fermions are discussed in the following section).

The probability of finding the instanton under consideration in the physical vacuum is given by averaging ΔL over this state. On the other hand, to find the coefficients C_n, it is necessary to consider the matrix elements of ΔL between perturbation-theory states (with different number of free gluons with momenta $q \ll 1/\rho$). These matrix elements can be calculated by quasiclassical methods.

Concretely, we consider the instanton contribution to the vacuum $\to n$ gluons transition and apply to it the reduction formula

$$\langle n\ \text{gluons} \,|\, \Delta L \,|\, 0\rangle = \langle 0 \,|\, T \prod_{k=1}^{n} \int dx_k e^{iq_k x_k} \varepsilon_{\mu_k}^{a_k} q_k^2 A_{\mu_k}^{a_k}(x_k) \,|\, 0\rangle, \quad (102)$$

where q_k and $\varepsilon_{\mu_k}^{a_k}$ are the 4-momentum and the polarization of the kth gluon, and $A_\mu^a(x)$ is the operator of the gluon field. For $n=0$, i.e., for the vacuum-vacuum transition, the right-hand side of (102) was already calculated in Sec. 8 and is equal to $d\rho\rho^{-5}d(\rho)$; the left-hand side is obviously equal to the coefficient of the unit operator: $C_I d\rho/\rho^5$.

For $n \neq 0$, the prescription of the quasiclassical calculation of the expression (102) reduces to

a) the transition to the Euclidean space (see the equations of Sec. 5);

b) replacement of the Euclidean $\hat{A}_\mu^a(x)$ by the instanton field $\overline{A}_\mu^a(x-x_0)$ given by formula (65). The singular gauge is used because the reduction formula (102) is valid only for rapidly decreasing fields $A_\mu^a(x)$. For a nonsingular gauge, the reciprocal propagator q^2 is replaced by a more complicated expression;

c) multiplication by the $\langle 0|0_T\rangle_{\text{1.s.}}$ transition amplitude, which is equal to $d\rho\rho^{-5}d(\rho)$. Thus, for the matrix element (102) we obtain

$$\langle n\ \text{gluons} \,|\, \Delta L(x) \,|\, 0\rangle = \frac{d\rho}{\rho^5} d(\rho) e^{-ix\Sigma q_k} \prod_{k=1}^{n} \left[\int dx_k e^{-iq_k x_k} (-q_k^2) \varepsilon_{\mu_k}^{a_k} \overline{A_{\mu_k}^{a_k}}(x_k) \right], \quad (103)$$

where all the quantities on the right-hand side are Euclidean.

The Fourier transform of the instanton solution, which we want in the limit $q\rho \to 0$, is readily found:

$$\int dx e^{-iqx} (-q^2) A_\mu^a(x) = \frac{4\pi i}{g} \bar{\eta}_{a\mu\nu} q_\nu \rho^2. \quad (104)$$

After this, it is easy to recover the complete operator form of ΔL:

$$\Delta L(x) = \frac{d\rho}{\rho^5} d(\rho) \exp\left[-\frac{2\pi^2}{g} \rho^2 \bar{\eta}_{a\mu\nu}^{M} G_{\mu\nu}^a(x) \right],$$

$$\bar{\eta}_{a\mu\nu}^{M} = \begin{cases} \bar{\eta}_{amn}, & \mu=m,\ \nu=n;\quad m, n=1, 2, 3, \\ i\bar{\eta}_{a4n}, & \mu=0,\ \nu=n;\quad n=1, 2, 3, \end{cases} \quad (105)$$

where $G_{\mu\nu}^a(x)$ is the operator of the large-scale gluon field. The factorials which occur in the expansion of the exponential cancel against the combinatorial coefficients when the matrix element (103) is taken.

The expression (105) for the interaction of an instanton with an external field was obtained for the first time by Callen, Dashen, and Gross[8] by a different and more complicated method. An important point is that we, in contrast to them, have not fixed the external $G_{\mu\nu}^a(x)$ "by hand" but have related it to the field of the large-scale fluctuations.

This is achieved by averaging the Lagrangian (105) over the physical vacuum. The term linear in $G_{\mu\nu}^a$ obviously vanishes as a result of such averaging, and the first nonvanishing correction to the effective density of the instantons is proportional to G^2:

$$\langle 0 \,|\, \Delta L \,|\, 0\rangle = \frac{d\rho}{\rho^5} d_{\text{eff}}(\rho) = \frac{d\rho}{\rho^5} d(\rho) \left[1 + \frac{\pi^3\rho^4}{(N^2-1)\alpha_s} \langle 0 \,|\, G_{\mu\nu}^a G_{\mu\nu}^a \,|\, 0\rangle + O(\rho^8) \right], \quad (106)$$

where in the averaging we have used the relation

$$\langle 0 \,|\, G_{\mu\nu}^a G_{\mu'\nu'}^{a'} \,|\, 0\rangle = \frac{\delta^{aa'}}{N^2-1} \cdot \frac{1}{12} (g_{\mu\mu'} g_{\nu\nu'} - g_{\mu\nu'} g_{\nu\mu'}) \langle 0 \,|\, G_{\alpha\beta}^b G_{\alpha\beta}^b \,|\, 0\rangle. \quad (107)$$

Note that the constant α_s and the operator $(G_{\mu\nu}^a)^2$ which occur here are normalized at the point ρ. A quantity that does not depend on the renormalization point [to accuracy $\alpha_s(\rho)$] is the product $\alpha_s G_{\mu\nu}^a G_{\mu\nu}^a$ (see the previous section).

To obtain a quantitative estimate of the correction, it is necessary to know the mean square of the intensity of the gluon field in the physical vacuum. This was found in Refs. 19 by analyzing the influence of the vacuum fields on the charmonium states, and it was found to be

$$\langle 0 \,|\, \frac{\alpha_s}{\pi} G_{\mu\nu}^a G_{\mu\nu}^a \,|\, 0\rangle \approx 0.012\ \text{GeV}^4. \quad (107')$$

For the group SU(3), the relative correction to $d(\rho)$ can be written in the form

$$\frac{\pi^4\rho^4}{8\alpha_s^2(\rho)} \langle 0 \,|\, \frac{\alpha_s}{\pi} G_{\mu\nu}^a G_{\mu\nu}^a \,|\, 0\rangle. \quad (108)$$

It reaches unity at a value of ρ equal to

$$\rho_{\text{crit}} \approx \frac{1}{1.15\ \text{GeV}}, \quad (109)$$

if for α_s we take $\alpha_s(\rho) = 2\pi/9 \ln(1/\Lambda\rho)$ with $\Lambda = 100$ MeV. For $\rho = \rho_{\text{crit}}$, the interaction of the instanton with the vacuum fields of the other fluctuations becomes 100% important. This ρ_{crit} is very small compared with the characteristic hadron dimensions 1/(200–300) MeV. The word "very" can indeed be used if one bears in mind the fact that $d(\rho)$ is proportional to a high power of ρ; the $\rho \leq \rho_{\text{crit}}$ contribution to, say, the vacuum energy density is extremely small.

The smallness of ρ_{crit} given by the estimate (109) can also be seen in a different way by calculating the contribution of the instantons to the correlation function

$$i \int dx e^{iqx} \langle 0 \,|\, TA(x) B(0) \,|\, 0\rangle, \quad (110)$$

where A and B are certain local operators. At large Euclidean q, the instantons make contributions of two types to (110). First, there is the contribution of the fluctuations of a fixed (q-independent) scale to the coefficients of the regular expansion in powers of $1/q^2$. Second, there is the contribution from instantons with scales $\rho \sim \rho_{\text{eff}} = C/q$, which is proportional to $d(\rho_{\text{eff}})$, i.e., to a high (and not necessarily integral) power of $1/q^2$. The constant of proportionality C can be determined by the method of steepest descent and because of the high power of ρ in $d(\rho)$ is approximately equal to 5.

Thus, we can calculate the one-instanton contribution to (110) in terms of q using the ordinary expressions only when $q^2 > (5.5\text{ GeV})^2$. It is clear that such q^2 are considerably greater than the characteristic hadron masses.

We conclude this subsection by giving a formula that takes into account the higher powers of $G^a_{\mu\nu}$ in the effective instanton density. This formula is based on the hypothesis of dominance of the vacuum intermediate state, which makes it possible to reduce $\langle 0|(G^2)^n|0\rangle$ to $(\langle 0|G^2|0\rangle)^n$. This approximation is analogous to one used in many-body theory and for some 4-quark operators for which it can be verified has an accuracy of the order of a few percent.

The factorization leads to the relation

$$\langle 0|\left(\frac{2\pi^2}{g}\rho^2\bar{\eta}^M_{a\mu\nu}G^a_{\mu\nu}\right)^{2k}|0\rangle=(2k-1)!!\left[\frac{4\pi^4}{g^2}\rho^4\langle 0|(\bar{\eta}^M_{a\mu\nu}G^a_{\mu\nu})^2|0\rangle\right]^k,$$

by means of which we obtain for the effective instanton density the result

$$d_{\text{eff}}(\rho)=d(\rho)\exp\left[\frac{\pi^4\rho^4}{(N^2-1)\alpha_s^2(\rho)}\langle 0|\frac{\alpha_s}{\pi}G^a_{\mu\nu}G^a_{\mu\nu}|0\rangle\right], \qquad (111)$$

which can be represented as the replacement in the expression for $d(\rho)$ of $2\pi/\alpha_s(\rho)$ by

$$\frac{2\pi}{\alpha_s(\rho)}\to\frac{2\pi}{\alpha_s(\rho)}\left[1-\frac{\pi^3\rho^4}{2(N^2-1)\alpha_s^2(\rho)}\langle 0|\frac{\alpha_s}{\pi}(G^a_{\mu\nu})^2|0\rangle\right]. \qquad (112)$$

Using for $d(\rho)$ the expression (111), we can advance in ρ to $\rho > \rho_{\text{crit}}$. However, when the interaction with the vacuum fields changes the classical action strongly, i.e., when (112) vanishes, the quasiclassical methods cannot be used. This limit under the same assumptions about α_s and $\langle 0|G^2|0\rangle$ is $\rho < 1/500$ MeV.

Despite the numerical uncertainties in the value of $\langle 0|G^2|0\rangle$ (which are of the order of a factor 2) and in α_s (the uncertainty in Λ is also of the order of a factor 2), it can be said that the vacuum fields deform the instantons at scales much smaller than the characteristic scales of the fluctuations that are dominant in the vacuum.

11. FERMIONS IN AN INSTANTON FIELD

In this section, we shall relatively briefly discuss how the instanton contribution to the vacuum–vacuum transition amplitude changes when fermions are included in the theory.

It is immediately clear that for a fluctuation with a given scale ρ the influence of "heavy" quarks with mass $m \gg \rho^{-1}$ is small; for in this case the quarks appear at times and distances $\sim 1/m \ll \rho$, at which perturbation theory can be used to calculate the quark loops of the form shown in Fig. 7. We give the first few terms of the effective Lagrangian that takes into account the fermion loops:

$$\Delta L_F=\frac{1}{2}\operatorname{Tr}\left\{-\frac{1}{4}G^2_{\mu\nu}\times\frac{g^2}{24\pi^2}\ln\frac{M^2}{m^2}+\frac{1}{16\pi^2}\left(\frac{ig^3}{180m^2}G_{\mu\nu}G_{\nu\gamma}G_{\gamma\mu}\right.\right.$$
$$\left.\left.+\frac{g^4}{288m^4}\left[-(G_{\mu\nu}G_{\mu\nu})^2+\frac{7}{10}\{G_{\mu\gamma},G_{\gamma\nu}\}^2_+ +\frac{29}{70}[G_{\mu\gamma},G_{\gamma\nu}]^2_- -\frac{8}{35}[G_{\mu\nu},G_{\gamma\delta}]^2_-\right]\right)\right\}, \qquad (113)$$
$$G_{\mu\nu}=G^a_{\mu\nu}t^a,\quad \operatorname{Tr}t^at^b=2\delta^{ab}.$$

The first term in this expression contains the cutoff parameter M and, obviously, describes the contribution of the quark under consideration to the change in

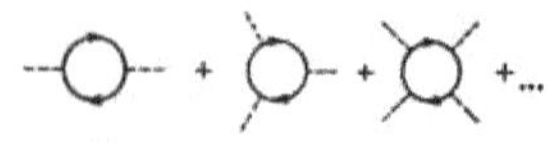

FIG. 7.

the charge g. Therefore, it is automatically taken into account when the result is expressed in terms of the charge at distances greater than $1/m$.

The following terms in (113) give a series in powers of $1/m^2\rho^2$ on the transition to the Euclidean space and substitution of the instanton field.

We now turn to the limiting case of "light" quarks, $m\rho \ll 1$. We note that for sufficiently small instantons all quarks are light. We calculate the integral over the Fermi fields in the path integral that determines the vacuum–vacuum transition: $\langle 0|0_T\rangle$. In the Euclidean action, a fermion with mass m adds a term of the form [see (45)]

$$S_F^{(E)}=\int d^4x\bar{\psi}(-i\gamma_\mu D_\mu-im)\psi,$$

and integration of this with respect to the anticommuting fields leads to

$$\operatorname{Det}(-i\gamma_\mu D_\mu-im).$$

The determinant can be understood as a product of the eigenvalues of the corresponding operator,

$$\operatorname{Det}(-i\gamma_\mu D_\mu-im)=\prod_n(\lambda_n-im),$$

where the real numbers λ_n are the eigenvalues of the Hermitian operator $-i\gamma_\mu D_\mu$:

$$-i\gamma_\mu D_\mu u_n(x)=\lambda_n u_n(x). \qquad (114)$$

Of fundamental importance in the study of the limit $m=0$ is the question of whether certain λ_n vanish, i.e., the question of zero-frequency modes of the fermion field. We shall show that the interaction with the instanton field leads to the appearance of one such mode u_0,

$$-i\gamma_\mu D_\mu u_0=0. \qquad (115)$$

We go over to two-component spinors $\chi_{L,R}$ (we use the standard representation for the γ matrices):

$$u_0=\begin{pmatrix}1\\-1\end{pmatrix}\chi_L+\begin{pmatrix}1\\1\end{pmatrix}\chi_R,\quad \sigma^+_\mu D_\mu\chi_L=0,\quad \sigma^-_\mu D_\mu\chi_R=0, \qquad (116)$$

where $\sigma^{\pm}_\mu=(\sigma,\mp i)$. To the equations for χ_L,χ_R we apply the operators $\sigma^-_\mu D_\mu,\sigma^+_\mu D_\mu$, respectively. Using the relations (57), the commutator $[D_\mu D_\nu]=-(ig/2)\tau^aG^a_{\mu\nu}$, and the explicit form of $G^a_{\mu\nu}$ [see (63)], we obtain

$$-D^2_\mu\chi_L=0,\qquad -D^2_\mu\chi_R=-4\sigma\tau\frac{\rho^2}{[(x-x_0)^2+\rho^2]^2}\chi_R.$$

The operator $-D^2_\mu$ is a sum of the squares of Hermitian operators: $-D^2=(-iD_\mu)^2$, i.e., it is positive definite. Therefore, it does not have vanishing eigenvalues (the boundary conditions are imposed at a large but finite distance R) and, therefore $\chi_L=0$.

In the equation for χ_R, we use a basis in the space of spinor and color indices that diagonalizes the matrix $\sigma\tau$. We recall that σ acts on the spinor indices, and τ on the color indices. This basis corresponds to addition of the ordinary spin and the color spin to a total angular momentum equal to zero (when $\sigma\tau=-3$) or unity ($\sigma\tau=+1$). It again follows from the positive definiteness of $-D_\mu^2$ that the only suitable case for us is when the total spin is equal to zero, which completely determines the dependence of χ_R on the indices:

$$(\sigma+\tau)\chi_R=0, \quad \chi_R^{\alpha m}\sim\varepsilon^{\alpha m}, \tag{117}$$

where $\alpha=1,2$ and $m=1,2$ are the spin and color indices, respectively.

The dependence on the coordinates can be readily found from the explicit form of D_μ^2, and the final result for the zero-frequency mode $u_0(x-x_0)$ (normalized by the condition $\int u^* u\, dx=1$) has the form

$$u_0(x)=\frac{1}{\pi}\frac{\rho}{(x^2+\rho^2)^{3/2}}\begin{pmatrix}1\\1\end{pmatrix}\varphi, \quad \varphi^{\alpha m}=\frac{1}{\sqrt{2}}\varepsilon^{\alpha m}. \tag{118}$$

We also write down the expression for the zero-frequency mode in the singular gauge, $u_0^{\text{sing}}(x-x_0)$ (which we shall require),

$$u_0^{\text{sing}}(x)=\frac{1}{\pi}\frac{\rho}{(x^2+\rho^2)^{3/2}}\frac{x_\mu\gamma_\mu}{\sqrt{x^2}}\begin{pmatrix}1\\-1\end{pmatrix}\varphi, \tag{119}$$

it being obtained by multiplication of (118) by the gauge transformation matrix (64a).

We now turn to the instanton part of the vacuum–vacuum transition amplitude. In it, we have the factor

$$F=\frac{m}{M}\frac{\text{Det}'(-i\gamma_\mu D_\mu)}{\text{Det}'(-i\gamma_\mu D_\mu-iM)}\frac{\text{Det}(-i\gamma_\mu\partial_\mu-iM)}{\text{Det}(-i\gamma_\mu\partial_\mu)},$$

where Det′ denotes the determinant without the zero-frequency mode and we have taken into account the regularization and also the normalization by perturbation theory. In all the positive-frequency modes, m is taken equal to zero, so that after the separation in F of the factor m/M the remaining part depends only on the dimensionless parameter $M\rho$. As in pure gluodynamics (see Sec. 8), this dependence must be such that the cutoff parameter M is removed by a renormalization of the coupling constant, i.e., the dependence of F on $M\rho$ must give the renormalization of the coupling constant due to the fermions in the factor $e^{-8\pi^2/g_0^2}$,

$$\Delta_F\frac{8\pi^2}{g^2}=-\ln\frac{F}{m\rho\cdot\text{const}}=\ln M\rho-\frac{1}{3}\ln M\rho. \tag{120}$$

The first logarithm derives from the zero-frequency mode, the second from the positive-frequency modes. Comparing the result with formula (81) for gluons, we see that the situation has been changed because of the anticommutativity: The zero-frequency modes of the light quarks lead to screening of the charge, and the positive-frequency modes to antiscreening.

In ordinary perturbation theory, the splitting in (120) can be associated with the spin-dependent part of the interaction (the first logarithm) and the "charge" part, which is not associated with the spin (the second logarithm). Indeed, the imaginary part of the gluon polarization operator, which derives from the intermediate $q\bar{q}$ state, can be represented in the form

$$\begin{aligned}\operatorname{Im}\Pi_{\mu\nu}^{F\,ab}&=\delta^{ab}\frac{g^2}{2}\int\frac{d\sigma}{32\pi^3}[q_\mu q_\nu-g_{\mu\nu}q^2-(p_1-p_2)_\mu(p_1-p_2)_\nu]\\&=\delta^{ab}\frac{g^2}{16\pi}(q_\mu q_\nu-g_{\mu\nu}q^2)\left(1-\frac{1}{3}\right).\end{aligned} \tag{121}$$

In this formula, p_1 and p_2 are the particle and antiparticle momenta, $q=p_1+p_2$, and the integration is over the directions of $\mathbf{p}_1=-\mathbf{p}_2$ in the center-of-mass system. The second term in (121) differs by only the factor -2 from the contribution of a spinless color doublet. The factor 2 corresponds to the two polarization states, and the minus to the anticommutativity.

We note that for the vacuum polarization there is also an analogous relation between the spin part of the polarization and the zero-frequency modes. This is readily seen in the "background" gauge obtained by adding the term (71) to the action. In perturbation theory, one can take as the "external" field, for example, a potential that has only a third color component, and in the loop only "charged" components will propagate. The three-gluon vertex in this gauge has the form of a sum of a color part and a magnetic part, which do not interfere in the polarization operator. The spin part gives the "antiscreening" logarithm, and the charge part (together with the Higgs particles) the "screening" part.

What has been obtained from the inclusion in the theory of a light quark? In the limit $m\to 0$, the vacuum–vacuum transition amplitude tends to zero. Does this mean that for $m=0$ there are no tunnel transitions? By no means. The point is that now the instanton fluctuation couples the vacuum to states of a quark–antiquark pair.

To see this, we consider the crossing process—the transition from a single-quark state to a single-quark state; we shall assume that the quark momenta p and p' are small compared with $1/\rho$. Proceeding as in Sec. 10, we use the reduction formula

$$\langle p'|p_T\rangle=-\int dx\,dx'\,e^{ip'x'-ipx}\bar{v}_\alpha^m(\hat{p}')_{\alpha\gamma}\langle 0|T\{q_\gamma^m(x')\bar{q}_\beta^k(x)\}|0\rangle_{\text{inst}}(\hat{p})_{\beta\delta}v_\delta^k, \tag{122}$$

where $\bar{v}_\alpha^m$ and v_δ^k are the spinors that describe the final and the initial quark (the superscript is the color index, the subscript the spinor index).

We find the instanton contribution to the fermion Green's function by using the relation

$$\langle 0|T\{q_\gamma^m(x')\bar{q}_\beta^k(x)\}|0\rangle_{\text{inst}}\xrightarrow[x_4\to -ix_4]{}\sum_n\frac{u_{(n)\gamma}^m(x')u_{(n)\beta}^{+k}(x)}{m+i\lambda_n}\langle 0|0_T\rangle_{\text{inst}}. \tag{123}$$

In the limit $m\to 0$, the zero-frequency mode makes the main contribution, and (123) is finite at $m=0$.

Using the explicit form (119) of the zero-frequency mode in the singular gauge, we can now readily obtain the result. We formulate it in the form of the expression for the effective Lagrangian that describes all transitions which arise from an instanton fluctuation with scale ρ:

$$\begin{aligned}\Delta L(x)=\prod_q\Big[m_q\rho-2\pi^2\rho^3\bar{q}_R\Big(1+\frac{i}{4}\tau^a\bar{\eta}_{a\mu\nu}\sigma_{\mu\nu}\Big)q_L\Big]\\\times\exp\Big(-\frac{2\pi^2}{g}\rho^2\bar{\eta}_{b\gamma\delta}G_{\gamma\delta}^b\Big)d_0(\rho)\frac{d\rho}{\rho^5}d\bar{o}.\end{aligned} \tag{124}$$

This contains a product over all species of light ($m_q\rho \ll 1$) quarks, and $\sigma_{\mu\nu}=(\gamma_\mu\gamma_\nu-\gamma_\nu\gamma_\mu)/2$. In Minkowski space, the symbols $\eta_{a\mu\nu}$ differ from the Euclidean symbols only when μ or $\nu=0$, and then by a factor i. By $d\bar{o}$ we denote the differential corresponding to the color orientation of the instanton, and it is normalized to unity, $\int d\theta = 1$. A dependence on the orientation enters through the substitution $\bar{\eta}_{a\mu\nu} \to h_{aa'}\bar{\eta}_{a'\mu\nu}$ (h is the matrix of rotations in the color space), which must be made in (124). The quantity $d_0(\rho)$ differs from $d(\rho)$ (85) in pure gluodynamics by multiplication by the factor

$$\exp F\left[-\frac{1}{3}\ln 2 - \frac{17}{36} + \frac{1}{3}(\ln 2\pi + \gamma) + \frac{2}{\pi^2}\sum_{s=1}^{\infty}\frac{\ln s}{s^2}\right] = e^{0.292F},$$

where F is the number of light fermions. This is for Pauli–Villars regularization; for the MS scheme, the 0.292 is replaced by −0.495 and by 0.153 for the $\overline{\text{MS}}$ scheme. In addition, in the expression (82) for $8\pi^2/g^2(\rho)$ it is necessary to include the fermion contribution.

For the anti-instanton, ΔL is obtained from (124) by the substitution $\bar{\eta}_{a\mu\nu} \to \eta_{a\mu\nu}$, $q_{L,R} \to q_{R,L}$. Note also that all the operators, the constant g, and the masses m_q in ΔL are normalized at the point ρ, so that besides the dependence given explicitly there is a logarithmic dependence on ρ, which is determined by the anomalous dimension of the operator term in ΔL under consideration.

Of particular interest are the instanton-generated fermion vertices; this interaction is frequently called the 't Hooft determinant interaction. The point is that it explicitly demonstrates the breaking of the $U(1)$ symmetry associated with transformations of the form $q' = e^{i\alpha\gamma_5}q$. Naively, such a symmetry holds in a theory with massless quarks. The nontrivial nature of the breaking of this symmetry can be seen from the fact that, for example, in a theory with one quark ΔL describes the transition of a "left-handed" quark into a "right-handed" one, which is impossible in any finite order of perturbation theory for $m = 0$.

[1]A. A. Belavin, A. M. Polyakov, A. S. Schwartz (Shvarts), and Yu. S. Tyupkin, Phys. Lett. **B59**, 85 (1975).
[2]A. M. Polyakov, Nucl. Phys. **B120**, 429 (1977).
[3]R. P. Feynman and A. R. Hibbs, Quantum Mechanics and Path Integrals, New York (1965) [Russian translation published by Mir, Moscow (1968)].
[4]I. Gradshtein and I. A. Ryzhik, Tablitsy integralov, summ, ryadov i proizvedenii, Fizmatgiz, Moscow (1962); English translation: Tables of Integrals, Series, and Products, Academic Press, New York and London (1965).
[5]S. Coleman, "The uses of instantons," Preprint HUTP-78/004 (1977).
[6]G. 't Hooft, Phys. Rev. D **14**, 3432 (1976).
[7]L. D. Landau and E. M. Lifshitz, Kvantovaya mechanika, Fizmatgiz, Moscow (1963); English translation: Quantum Mechanics, 2nd ed., Pergamon Press, Oxford (1965).
[8]C. G. Callan, R. Dashen, and D. J. Gross, Phys. Rev. D **17**, 2717 (1978); **19**, 1826 (1979).
[9]J. F. Willemsen, Phys. Rev. D **20**, 3292 (1979).
[10]M. F. Atiah, N. J. Hitchin, V. G. Drinfeld, and Yu. I. Manin, Phys. Lett. **A65**, 185 (1978); V. G. Drinfel'd and Yu. I. Manin, Yad. Fiz. **29**, 1646 (1979) [Sov. J. Nucl. Phys. **29**, 845 (1979)].
[11]I. B. Khriplovich, Yad. Fiz. **10**, 409 (1969) [Sov. J. Nucl. Phys. **10**, 235 (1970)].
[12]D. R. T. Jones, Nucl. Phys. **B75**, 531 (1974).
[13]C. Bernard, Phys. Rev. D **19**, 3013 (1979).
[14]G. 't Hooft and M. Veltman, Nucl. Phys. **B144**, 189 (1972); G. 't Hooft, Nucl. Phys. **B162**, 444 (1973).
[15]W. A. Bardeen, A. J. Buras, D. W. Duke, and T. Muta, Phys. Rev. D **18**, 3998 (1978).
[16]V. A. Fateev, I. V. Frolov, and A. S. Schwartz (Shvarts), Nucl. Phys. **B154**, 121 (1979); Yad. Fiz. **30**, 1134 (1979) [Sov. J. Nucl. Phys. **30**, 590 (1979)].
[17]R. Crewther, Phys. Rev. Lett. **28**, 1421 (1972); M. Chanowitz and J. Ellis, Phys. Lett. **B40**, 397 (1972); J. Collins, L. Dunkan, and S. Joglekar, Phys. Rev. D **16**, 438 (1977).
[18]M. A. Shifman, A. I. Vainshtein, and V. I. Zakharov, Nucl. Phys. **B163**, 46 (1980); **165**, 45 (1980).
[19]M. A. Shifman, A. I. Vainshtein, and V. I. Zakharov, Nucl. Phys. **B147**, 385 448 (1979).

Translated by Julian B. Barbour

www.ingramcontent.com/pod-product-compliance
Lightning Source LLC
Chambersburg PA
CBHW070741220326
41598CB00026B/3722
* 9 7 8 9 8 1 0 2 1 8 2 6 3 *